Communications
in Computer and Information Science 1652

More information about this series at https://link.springer.com/bookseries/7899

Silvia Chiusano · Tania Cerquitelli ·
Robert Wrembel · Kjetil Nørvåg ·
Barbara Catania · Genoveva Vargas-Solar ·
Ester Zumpano (Eds.)

New Trends in Database and Information Systems

ADBIS 2022 Short Papers, Doctoral Consortium and Workshops:
DOING, K-GALS, MADEISD, MegaData, SWODCH, Turin, Italy
September 5–8, 2022, Proceedings

 Springer

Editors
Silvia Chiusano (iD)
Politecnico di Torino
Turin, Italy

Tania Cerquitelli (iD)
Politecnico di Torino
Turin, Italy

Robert Wrembel (iD)
Poznań University of Technology
Poznań, Poland

Kjetil Nørvåg (iD)
Norwegian University of Science
and Technology
Trondheim, Norway

Barbara Catania (iD)
University of Genoa
Genoa, Italy

Genoveva Vargas-Solar (iD)
CNRS
Villeurbanne Cedex, France

Ester Zumpano (iD)
University of Calabria
Rende, Italy

ISSN 1865-0929   ISSN 1865-0937 (electronic)
Communications in Computer and Information Science
ISBN 978-3-031-15742-4   ISBN 978-3-031-15743-1 (eBook)
https://doi.org/10.1007/978-3-031-15743-1

This Springer imprint is published by the registered company Springer Nature Switzerland AG
The registered company address is: Gewerbestrasse 11, 6330 Cham, Switzerland

Preface

This year ADBIS – the Conference on Advances in Databases and Information Systems – celebrated its 26th anniversary.

The first ADBIS conference was held in Saint Petersburg, Russia (1997). Since then, ADBIS has taken place annually, with previous editions were held in: Poznan, Poland (1998); Maribor, Slovenia (1999); Prague, Czech Republic (2000); Vilnius, Lithuania (2001); Bratislava, Slovakia (2002); Dresden, Germany (2003); Budapest, Hungary (2004); Tallinn, Estonia (2005); Thessaloniki, Greece (2006); Varna, Bulgaria (2007); Pori, Finland (2008); Riga, Latvia (2009); Novi Sad, Serbia (2010); Vienna, Austria (2011); Poznan, Poland (2012); Genoa, Italy (2013); Ohrid, North Macedonia (2014); Poitiers, France (2015); Prague, Czech Republic (2016); Nicosia, Cyprus (2017); Budapest, Hungary (2018); Bled, Slovenia (2019); Lyon, France (2020); and Tartu, Estonia (2021).

The official ADBIS portal – http://adbis.eu – provides up to date information on all ADBIS conferences, committees, publications, and issues related to the ADBIS community.

The 26th ADBIS was held in Turin, Italy, during September 5–8, 2022, as a hybrid event. It received significant attention from both the research and industrial communities, as 90 papers were submitted to the conference. The papers were reviewed by an international Program Committee consisting of 85 members. The Program Committee selected 28 short papers from the main ADBIS conference, an acceptance rate of 42%, for inclusion in this volume. These papers cover the following topics: data modeling and visualization, data processing (fairness, cleaning, pre-processing), information and process retrieval, data access optimization, graph data, data science, and machine learning.

Additionally, this volume includes papers accepted at the five workshops and Doctoral Consortium, which accompanied ADBIS 2022. Each workshop had its own international Program Committee, whose members served as the reviewers of papers included in this volume. The maximum paper acceptance rate at each of these events did not exceed 50%. In total, 53 papers were submitted to these workshops, out of which 26 were selected for presentation and publication in this volume, giving an overall acceptance rate of 49%.

The following five workshops were run at ADBIS 2022:

DOING: 3rd Workshop on Intelligent Data – From Data to Knowledge, chaired by Mirian Halfeld-Ferrari (Université d'Orléans, France) and Carmem S. Hara (Universidade Federal do Parana, Brazil). The idea underlying DOING is to gather researchers in natural language processing, databases, and artificial intelligence to discuss two main problems: (1) how to extract information from textual data and represent it in knowledge bases, and (2) how to propose intelligent methods for handling and maintaining these databases with new forms of requests, including efficient, flexible, and secure analysis mechanisms, adapted to the user, and with quality and privacy preservation guarantees.

K-GALS: 1st Workshop on Knowledge Graphs Analysis on a Large Scale, chaired by Mariella Bonomo (University of Palermo, Italy) and Simona E. Rombo (University of Palermo, Italy). The goal of K-GALS is to provide participants with the opportunity to introduce and discuss new methods, theoretical approaches, algorithms, and software tools that are relevant to the knowledge graphs based research, especially when it is focused on a large scale. Relevant issues include how knowledge graphs can be used to represent knowledge, how systems managing knowledge graphs work, and which applications can be provided on top of a knowledge graphs, in a distributed environment.

MADEISD: 4th Workshop on Modern Approaches in Data Engineering and Information System Design, chaired by Ivan Luković (University of Belgrade, Serbia), Slavica Kordić (University of Novi Sad, Serbia), and Sonja Ristić (University of Novi Sad, Serbia). The main goal of MADEISD is to address open questions and real potentials for various applications of modern approaches and technologies in data engineering and information system design so as to develop and implement effective software services in support of information management in various organization systems.

MegaData: 2nd Workshop on Advanced Data Systems Management, Engineering, and Analytics, chaired by Feras M. Awaysheh (University of Tartu, Estonia), Fahed Alkhabbas (Malmö University, Sweden), and Said Alawadi (Uppsala University, Sweden). MegaData aims to report on the advances and trends in database deployment models and environments from both the infrastructure and application levels. This year, questions related to how to develop data as the foundation for advanced databases and information systems and what the next generation of data systems will look like led the discussion on the latest trends in modern data systems.

SWODCH: 2nd Workshop on Semantic Web and Ontology Design for Cultural Heritage, chaired by Antonis Bikakis (University College London, UK), Roberta Ferrario (ISTC-CNR, Italy), Stéphane Jean (University of Poitiers and ISAE-ENSMA, France), Béatrice Markhoff (University of Tours, France), Alessandro Mosca (Free University of Bozen-Bolzano, Italy), and Marianna Nicolosi Asmundo (University of Catania, Italy). The foundational purpose of SWODCH is to gather original research work about both application and foundational issues emerging from the design of conceptual models, ontologies, and Semantic Web technologies for the digital humanities, here understood according to its broader definition including cultural heritage, digital history, archaeology, and related fields. The application-oriented focus of SWODCH, on the other hand, aims at bringing together stakeholders from various scientific fields involved in the development or deployment of Semantic Web solutions, including computer scientists, data scientists, and digital humanists.

The ADBIS 2022 Doctoral Consortium (DC) provided a forum for PhD students to present their research projects to the scientific community. DC papers describe the status of PhD students' research, a comparison with relevant related work, their results, and plans on how to experiment, validate and consolidate their contribution. Through the Doctoral Consortium, PhD students receive feedback from a selected group of mentors about potential opportunities and perspectives and suggestions to pursue their work. The PhD Consortium allows PhD students to establish international collaborations with members and participants of the ADBIS community.

The 2022 edition of the ADBIS DC included seven papers accepted after a very selective peer-review process, which are published in this volume. Additionally, the ADBIS 2022 DC program included a panel about strategies for preparing for post-PhD life with professionals with accomplished careers in academia, industry, and entrepreneurship.

The ADBIS 2022 Organizing Committee supported diversity and inclusion by offering some grants, supporting a few researchers to participate in the conference and become part of the ADBIS community. All the grants were assigned based on the underrepresented community, gender, and role/position. The grants include:

- two free regular registrations, assigned to researchers from Argentina and Brazil,
- three regular registration fee discounts of 200 Euros, assigned to researchers from Estonia, Lebanon, and Italy, and
- four regular registration fee discounts of 150 Euros, assigned to researchers from Croatia, France, and Italy.

We would like to wholeheartedly thank all participants, authors, PC members, workshop organizers, session chairs, doctoral consortium chairs, workshop chairs, volunteers, and co-organizers for their contributions in making ADBIS 2022 a great success. We would also like to thank the ADBIS Steering Committee and all sponsors.

July 2022

Silvia Chiusano
Tania Cerquitelli
Robert Wrembel
Kjetil Nørvåg
Barbara Catania
Genoveva Vargas-Solar
Ester Zumpano

Organization

General Chair

Silvia Chiusano   Polytechnic University of Turin, Italy

Program Committee Chairs

Tania Cerquitelli   Polytechnic University of Turin, Italy
Robert Wrembel   Poznan University of Technology, Poland

Workshop Chairs

Kjetil Nørvåg   Norwegian University of Science and Technology, Norway
Barbara Catania   University of Genoa, Italy

Doctoral Consortium Chairs

Genoveva Vargas-Solar   CNRS, LIRIS, France
Ester Zumpano   University of Calabria, Italy

Local Organization Committee Chairs

Luca Cagliero   Polytechnic University of Turin, Italy
Paolo Garza   Polytechnic University of Turin, Italy
Bartolomeo Vacchetti   Polytechnic University of Turin, Italy
Giovanni Malnati   Polytechnic University of Turin, Italy

Publicity Chair

Oscar Romero   Polytechnic University of Catalonia - BarcelonaTech, Spain

Proceedings Chair

Khalid Belhajjame   Paris Dauphine University - PSL, France

Special Issue Chair

Ladjel Bellatreche   ISAE-ENSMA, France

Diversity and Inclusion Chair

Jérôme Darmont   University of Lyon 2, France

Steering Committee

Andreas Behrend	TH Köln, Germany
Ladjel Bellatreche	ISAE-ENSMA, France
Maria Bielikova	Kempelen Institute of Intelligent Technologies, Slovakia
Barbara Catania	University of Genoa, Italy
Jérôme Darmont	University of Lyon 2, France
Johann Eder	Universität Klagenfurt, Austria
Johann Gamper	Free University of Bozen-Bolzano, Italy
Tomáš Horváth	Eötvös Loránd University, Hungary
Mirjana Ivanović	University of Novi Sad, Serbia
Marite Kirikova	Riga Technical University, Latvia
Manuk Manukyan	Yerevan State University, Armenia
Raimundas Matulevicius	University of Tartu, Estonia
Tadeusz Morzy	Poznan University of Technology, Poland
Kjetil Nørvåg	Norwegian University of Science and Technology, Norway
Boris Novikov	National Research University, Higher School of Economics, Saint Petersburg, Russia
George Papadopoulos	University of Cyprus, Cyprus
Jaroslav Pokorný	Charles University in Prague, Czech Republic
Oscar Romero	Polytechnic University of Catalonia - BarcelonaTech, Spain
Sergey Stupnikov	Russian Academy of Sciences, Russia
Bernhard Thalheim	University of Kiel, Germany
Goce Trajcevski	Iowa State University, USA
Valentino Vranić	Slovak University of Technology in Bratislava, Slovakia
Tatjana Welzer	University of Maribor, Slovenia
Robert Wrembel	Poznan Unviersity of Technology, Poland
Ester Zumpano	University of Calabria, Italy

Program Committee

Alberto Abelló	Polytechnic University of Catalonia - BarcelonaTech, Catalonia
Cristina D. Aguiar	University of São Paulo, Brasil
Syed Muhammad Fawad Ali	IBM, Germany
Bernd Amann	Sorbonne University, France
Witold Andrzejewski	Poznan University of Technology, Poland
Daniele Apiletti	Polytechnic University of Turin, Italy
Costin Badica	University of Craiova, Romania
Sylvio Barbon Junior	University of Trieste, Italy
Andreas Behrend	University of Bonn, Germany
Khalid Belhajjame	Paris Dauphine University - PSL, France
Ladjel Bellatreche	ISAE-ENSMA, France
András Benczúr	Eötvös Loránd University, Hungary
Fadila Bentayeb	University of Lyon 2, France
Maria Bielikova	Kempelen Institute of Intelligent Technologies, Slovakia
Sandro Bimonte	INRAE, France
Pawel Boiński	Poznan University of Technology, Poland
Zoran Bosnić	University of Ljubljana, Slovenia
Omar Boussaid	University of Lyon 2, France
Drazen Brdjanin	University of Banja Luka, Bosnia and Herzegovina
Damiano Carra	University of Verona, Italy
Jérôme Darmont	University of Lyon 2, France
Claudia Diamantini	Marche Polytechnic University, Italy
Christos Doulkeridis	University of Piraeus, Greece
Johann Eder	University of Klagenfurt, Austria
Markus Endres	University of Passau, Germany
Javier A. Espinosa-Oviedo	University of Lyon, France
Georgios Evangelidis	University of Macedonia, Greece
Flavio Ferrarotti	Software Competence Centre Hagenberg, Austria
Alessandro Fiori	Polytechnic University of Turin, Italy
Flavius Frasincar	Erasmus University Rotterdam, The Netherlands
Johann Gamper	Free University of Bozen-Bolzano, Italy
Matteo Golfarelli	University of Bologna, Italy
Marcin Gorawski	Silesian University of Technology, Poland
Jānis Grabis	Riga Technical University, Latvia
Francesco Guerra	University of Modena and Reggio Emilia, Italy
Giancarlo Guizzardi	Federal University of Espirito Santo, Brazil
Tomas Horvath	Eötvös Loránd University, Hungary

Mirjana Ivanović	University of Novi Sad, Serbia
Stefan Jablonski	University of Bayreuth, Germany
Aida Kamišalić	University of Maribor, Slovenia
Zoubida Kedad	University of Versailles, France
Attila Kiss	Eötvös Loránd University, Hungary
Julius Köpke	Alpen-Adria University of Klagenfurt, Austria
Dejan Lavbič	University of Ljubljana, Slovenia
Yurii Litvinov	St. Petersburg State University, Russia
Audrone Lupeikiene	Vilnius University, Lithuania
Federica Mandreoli	University of Modena and Reggio Emilia, Italy
Yannis Manolopoulos	Open University of Cyprus, Cyprus
Patrick Marcel	University of Tours, France
Sara Migliorini	University of Verona, Italy
Angelo Montanari	University of Udine, Italy
Lia Morra	Polytechnic University of Turin, Italy
Tadeusz Morzy	Poznan University of Technology, Poland
Boris Novikov	National Research University, Higher School of Economics, Russia
Kjetil Nørvåg	Norwegian University of Science and Technology, Norway
Andreas Oberweis	Karlsruhe Institute of Technology, Germany
George Papadopoulos	University of Cyprus, Cyprus
András Pataricza	Budapest University of Technology and Economics, Hungary
Jan Platos	Technical University of Ostrava, Czech Republic
Jaroslav Pokorný	Charles University in Prague, Czech Republic
Giuseppe Polese	University of Salerno, Italy
Alvaro E. Prieto	University of Extremadura, Spain
Elisa Quintarelli	University of Verona, Italy
Miloš Radovanović	University of Novi Sad, Serbia
Franck Ravat	University of Toulouse, France
Stefano Rizzi	University of Bologna, Italy
Oscar Romero	Polytechnic University of Catalonia, Spain
Gunter Saake	University of Magdeburg, Germany
Kai-Uwe Sattler	TU Ilmenau, Germany
Milos Savic	University of Novi Sad, Serbia
Claudia Steinberger	University of Klagenfurt, Austria
Sergey Stupnikov	Russian Academy of Sciences, Russia
Bernhard Thalheim	University of Kiel, Germany
Goce Trajcevski	Iowa State University, USA
Raquel Trillo-Lado	University of Zaragoza, Spain
Genoveva Vargas Solar	CNRS, LIRIS, France

Goran Velinov	Ss. Cyril and Methodius University, North Macedonia
Peter Vojtas	Charles University in Prague, Czech Republic
Isabelle Wattiau	ESSEC and Cnam, France
Marek Wojciechowski	Poznan University of Technology, Poland
Vladimir Zadorozhny	University of Pittsburgh, USA
Ester Zumpano	University of Calabria, Italy

Additional Reviewers

Stylianos Argyrou
Paul Blockhaus
Miroslav Blšták
Andrea Brunello
Gabriel Campero Durand
Andrea Chiorrini
Francesco Del Buono
Chiara Forresi
Marco Franceschetti
Matteo Francia
Verena Geist

Joseph Giovanelli
Nicolas Labroche
Christos Mettouris
Simone Monaco
Federico Motta
Uchechukwu Njoku
Thomas Photiadis
Matúš Pikuliak
Nicola Saccomanno
Vladimir A. Shekhovtsov
Emanuele Storti

Contents

ADBIS Short Papers: Data Access Optimization

ADBIS Short Papers: Data Pre-processing and Cleaning

ADBIS Short Papers: Modeling and Querying of Graph Databases

ADBIS Short Papers: Data Science and Machine Learning

ADBIS Short Papers: Data Understanding, Modeling and Visualization

On an Extensible Conceptual Data Model by a Non-formal Example

Manuk G. Manukyan[✉][iD]

Yerevan State University, 0025 Yerevan, Armenia
mgm@ysu.am

Abstract. The investigation subject of this paper is an extensible conceptual data model. Formal semantics of an algebra of hierarchical relations is proposed. Extensibility concept of conceptual data model is discussed. A non-formal example of extending the conceptual data model is considered. Conceptual schema is defined as a set of OPENMath attribution objects.

Keywords: Conceptual data model · Hierarchical relations algebra · Data integration · Reasoning rules · OPENMath

1 Introduction

Conceptual modeling has been the subject of intense research since the late 1970s. Prerequisites for such research is that database systems usually have limited knowledge about the *meaning* of the data stored in them [1]. In fact, they allow to manipulate data of certain simple types. Any more complex interpretation is left to the user. In this context, the issues of formal knowledge representation in the form of a set of concepts of some subject domain and relations between them are topical. Such representations are used for reasoning about entities of the subject domains, as well as for the domains description. Thus, conceptualization of subject domain assumes accessing and managing the data in terms of the conceptual entities. The emergence of new data management paradigms, as well as the many directions in which modern database systems are evolving, leads to the idea of developing an extensible conceptual data model.

In particular, the data integration concept is an example of a very important direction in which modern database systems are evolving. Analysis of existing approaches to data integration can be found in [2]. Basically, these works are devoted to the problems of integrating homogeneous data sources. Typically, an extended relational or object data model was used as the target data model. In this connection, using the XML data model as the target one is preferred, as this model is some compromise between conventional and semistructured data models.

This work was supported by the RA MES State Committee of Science, in the frames of the research project No. 21T-1B326.

S. Chiusano et al. (Eds.): ADBIS 2022, CCIS 1652, pp. 3–13, 2022.
https://doi.org/10.1007/978-3-031-15743-1_1

The emergence of the polyglot persistence approach is another argument to develop an extensible conceptual data model. This is explained by the fact that different databases are designed to solve different problems. Using a single database engine for all of the requirements usually leads to non-performant solutions (for more details see [3]).

This work is based on the paper [4] in which we proposed an extensible conceptual data model. The considered data model uses a single concept for data modeling, namely hierarchical relation. An algebra of hierarchical relations has been developed. In the frame of this paper, formal semantics of hierarchical relations algebra is proposed. In addition, a non-formal example of extension of the considered conceptual data model is proposed. The result of such extension is a conceptual data model with the ability to model data integration concept. An example of a data warehouse is constructed which is an instance of the proposed conceptual data model.

The paper is organized as follows: An extensible conceptual data model and its formal bases are considered in Sects. 2 and 3. Formal semantics of a hierarchical relations algebra is proposed in Sect. 4. Extensibility concept of conceptual data model is discussed in Sect. 5. A non-formal example of extending the conceptual data model is offered in Subsect. 5.1. Related work is presented in Sect. 6. The conclusion is provided in Sect. 7.

2   Formal Bases

OPENMath Objects. OPENMath is an extensible formalism for representing mathematical objects so that software packages can exchange these objects without losing semantic content [5,6]. The possibility of the extensibility of the considered formalism is explained by the fact that the mathematical notation is constantly evolving. Moreover, mathematics and its applications are a growing field of knowledge, new ideas and notations appear constantly. Formally, an OpenMath object is a labeled tree whose leaves are basic OpenMath objects. Examples of basic OPENMath objects are: Integer, Symbol and Variable. The compound objects are defined in terms of *binding* and *application* of the λ-calculus [7]. The following recursive rules for constructing compound OPENMath objects are proposed:

- Basic OPENMath objects are OPENMath objects.
- If $A_1, A_2, ..., A_n$ $(n \geq 1)$ are OPENMath objects, then $application(A_1, A_2, ..., A_n)$ is an OPENMath *application object*.
- If $S_1, S_2, ..., S_n$ are OPENMath symbols, and $A, A_1, A_2, ..., A_n$ $(n \geq 1)$ are OPENMath objects, then $attribution(A, S_1\ A_1, S_2\ A_2, ..., S_n\ A_n)$ is an OPENMath *attribution object* and A is the object *stripped* of *attributions*.
- If B and C are OPENMath objects, and $v_1, v_2, ..., v_n$ $(n \geq 0)$ are OPENMath variables or attributed variables, then $binding(B, v_1, v_2, ..., v_n, C)$ is an OPENMath *binding object*.

Types in OPENMath. A type system is built from basic types which are predefined as typed OPENMath objects (for example, *integer*, *string*, *boolean*, etc.) and the following rules, by means of which typed OPENMath objects are constructed:

Attribution Rule. If v is an OPENMath variable and t is a typed OPENMath object, then *attribution(v, type t)* is a typed OPENMath object. It denotes a variable with type t.

Abstraction Rule. If v is an OPENMath variable and t, A are typed OPEN-Math objects, then *binding(lambda, attribution(v, type t), A)* is a typed OPEN-Math object and denotes the function that assigns to the variable v of type t the object A.

Application Rule. If F and A are typed OPENMath objects, then *application(F, A)* is a typed OPENMath object.

3 An Extensible Conceptual Data Model

In the frame of the considered conceptual data model (as in the relational data model), a single concept is used to model subject domains, namely, hierarchical relation. Below we introduce the definitions of the hierarchical relation schema and the hierarchical relation. These definitions can be considered as strengthening the definitions of the relation schema and relation of the relational databases.

3.1 Formalization of Hierarchical Relations

Definition 1. *A hierarchical relation schema X is an attribution object and is interpreted by a finite set of attribution objects $\{A_1, A_2, \ldots, A_n\}$. Corresponding to each attribution object A_i is a set D_i (a finite, non-empty set), $1 \leq i \leq n$, called the domain of A_i.*

In the frame of the proposed extensible conceptual data model, the following OPENMath representation of the hierarchical relation schema X is accepted:

$$attribution(X,\ type\ A,\ S_1\ A_1,\ S_2\ A_2, ...,\ S_k\ A_k),\ k \geq 0$$

Here $S_1, S_2, ..., S_k$ are OPENMath symbols, X is an OPENMath variable, and $A, A_1, A_2, ..., A_k$ are OPENMath objects. In this representation, X is the name of the attribution object, A represents the type of the attribution object (basic or composite), and by means of the $\langle S_i\ A_i \rangle$ pair, it defines one property of the modeled object ($1 \leq i \leq k$). To construct composite types, we introduced the following type constructors: *sequence* and *choice*. The conceptual entities that are created using these type constructors have analogous semantics as the *sequence* and *choice* elements of the XML Schema language. The arguments of these functions are typed attribution objects. We distinguish two types of typed attribution objects: basic and composite. A composite attribution object is defined by a type constructor.

Definition 2. *Let* $D = D_1 \cup D_2 \cup ... \cup D_n$. *A hierarchical relation* x *on hierarchical relation schema* X *is a finite set of mappings* $\{t_1, t_2, ..., t_k\}$ *from* X *to* D *with the restriction that for each mapping* $t \in x$, $t[A_i]$ *must be in* D_i, $1 \leq i \leq n$. *The mappings are called hierarchical tuples or simply tuples.*

Definition 3. *A key of a hierarchical relation* x *on hierarchical relation schema* X *is a minimal subset* K *of* X *such that for any distinct tuples* $t_1, t_2 \in x$, $t_1[K] \neq t_2[K]$.

A database D on database schema S is a collection of hierarchical relations $\{x_1, x_2, ..., x_n\}$ such that for each schema of the hierarchical relation schema $s \in S$ there is a hierarchical relation $x \in D$ such that x is a hierarchical relation with schema s that satisfies every constraint defined in s.

3.2 Conceptual Schema

The conceptual schema is an instance of conceptual data model and is intended for formal knowledge representation in the form of a set of concepts of some subject domain and relations between them. Such representations are used for reasoning about entities of the subject domains, as well as for the domains description. A conceptual schema is defined as a set of OPENMath attribution objects. A distinguishing feature of the conceptual level is its stratification of the local and global levels to model the conceptual entities. On the local level, data sources are represented with hierarchical relations. The global level is intended to support derived hierarchical relations. The hierarchical relations of the global conceptual level are defined by algebraic programs. The conceptual level data definition language is based on the OPENMath formalism. We use the mechanism of the OPENMath Content Dictionaries (CD) to support the proposed conceptual data model concept. They are used to assign formal and informal semantics to all symbols (concepts) used in the OPENMath objects. A CD (titled as *cdm*) was developed. This CD represents the kernel of the extensible conceptual data model.

4   Formal Semantics of Hierarchical Relations Algebra

Preliminaries. Firstly, we introduce several concepts and notations that will be used in the definitions of the algebraic operations.
Let R and S be hierarchical relations schemas. We say that R and S are similar[1], if

1. $|R| = |S|$
2. The i-th attribution object in R and the i-th attribution object in S are defined on identical domains. Similarity does not assume using symbols of identical cardinality in the corresponding attribution objects.

---

[1] This is an analogous concept for the union compatibility of the relational algebra.

Let R and S be hierarchical relations schemas. We say that $R \subseteq S$, if

1. $|R| \leq |S|$
2. $\forall e_r \in R$, $\exists e_s \in S$ such that e_r and e_s are similar.

Let R and S be hierarchical relations schemas, then
$R \cap S = \{A | \exists B (A \in R \wedge B \in S \wedge (name(A) = name(B)))\}^2$
Let $u = <u_1, u_2, ..., u_k>$ and $v = <v_1, v_2, ..., v_m>$ be tuples. The concatenation of u and v is a tuple defined as follows:

$$\widehat{uv} = <u_1, u_2, ..., u_k, v_1, v_2, ..., v_m>$$

Let us define the following operations $\oplus$, $\otimes$ and $\ominus$ over cardinality symbols $(?, *, +, \perp)^3$, which we will use in determining the resulting schemas of the algebraic operations:

$$? \oplus ? = ? \quad ? \oplus * = * \quad ? \oplus + = * \quad ? \oplus \perp = ? \quad \perp \oplus * = * \quad \perp \oplus + = +$$
$$? \otimes ? = ? \quad ? \otimes * = ? \quad ? \otimes + = ? \quad ? \otimes \perp = ? \quad \perp \otimes * = ? \quad \perp \otimes + = ?$$
$$? \ominus ? = ? \quad ? \ominus * = ? \quad * \ominus ? = * \quad ? \ominus + = ? \quad + \ominus ? = * \quad ? \ominus \perp = ?$$
$$+ \ominus \perp = * \quad \perp \ominus \perp = ? \quad * \ominus + = * \quad + \ominus * = * \quad * \ominus * = * \quad + \ominus + = *$$
$$* \ominus \perp = * \quad + \oplus + = + \quad + \otimes + = * \quad \perp \ominus + = ? \quad \perp \oplus \perp = \perp \quad * \oplus + = *$$
$$\perp \otimes \perp = ? \quad * \otimes + = * \quad \perp \ominus ? = ? \quad \perp \ominus * = ? \quad * \oplus * = * \quad * \otimes * = *$$

Operations. The following operations over hierarchical relations are proposed below:

Set-Theoretic Operations. In definitions of set-theoretic operations union, intersection and difference, it is assumed that the schemas of operands are similar. Let r and s be hierarchical relations with R and S similar schemas correspondingly. The union, intersection and difference of r and s are the hierarchical relations defined as follows:

$$r \cup s = \{t | t \in r \vee t \in s\}$$
$$r \cap s = \{t | t \in r \wedge t \in s\}$$
$$r - s = \{t | t \in r \wedge t \notin s\}$$

Notice that the union and intersection are associative and commutative operations. Let q be a hierarchical relation, then $sch(q)$ denotes the schema of that relation q, and $card(q)$ - cardinality number of $sch(q)$. Below the $sch(r \cup s)$, $sch(r \cap s)$, $sch(r - s)$ and their cardinalities are defined as follows (Cardinalities for all the nested attribution objects are computed similarly. The value of $dom(Z)$ is the domain of the attribution object Z. Finally, the value of $card(Z)$ is the cardinality symbol of the attribution object Z.):

---

^{2} The value of $name(Z)$ is the name of the attribution object Z.
^{3} Using a cardinality symbol $\perp$ in the attribution object means that this object may occur exactly one time.

$sch(r \cup s) = \{A_R | \exists A, B (A \in R \wedge B \in S \wedge (name(A) = name(B) =$
$name(A_R)) \wedge (dom(A_R) = dom(A)) \wedge card(A_R) = card(A) \oplus card(B))\}$
$sch(r \cap s) = \{A_R | \exists A, B (A \in R \wedge B \in S \wedge (name(A) = name(B) =$
$name(A_R)) \wedge (dom(A_R) = dom(A)) \wedge card(A_R) = card(A) \otimes card(B))\}$
$sch(r - s) = \{A_R | \exists A, B (A \in R \wedge B \in S \wedge (name(A) = name(B) =$
$name(A_R)) \wedge (dom(A_R) = dom(A)) \wedge card(A_R) = card(A) \ominus card(B))\}$
$card(r \cup s) = card(r) \oplus card(s)$
$card(r \cap s) = card(r) \otimes card(s)$
$card(r - s) = card(r) \ominus card(s)$

Cartesian Product. Let r and s be hierarchical relations with R and S schemas correspondingly. The Cartesian product of r and s is a hierarchical relation defined as follows:

$$r \times q = \{\widehat{uv} | u \in r \wedge v \in q\}.$$

Let $R_1 = \{R.A | A \in R\}$ and $S_1 = \{S.A | A \in S\}$. Below the $sch(r \times s)$ and its cardinality are defined as follows:

$$sch(r \times s) = R_1 \cup S_1$$
$$card(r \times s) = card(r) \oplus card(s)$$

Selection. Let r be a hierarchical relation with R schema, and F be a predicate. The result of the operation of selection from r by F is a hierarchical relation defined as follows:

$$\sigma_F(r) = \{t | t \in r \wedge F(t)\}$$

Below the $sch(\sigma_F(r))$ and its cardinality are defined as follows:

$$sch(\sigma_F(r)) = R$$
$$card(\sigma_F(r)) = card(r) \otimes \text{``}*\text{''}$$

Projection. Let r be a hierarchical relation with R schema, and $L \subseteq R$. The result of the projection operation is a hierarchical relation defined as follows:

$$\pi_L(r) = \{t[A] | t \in r\}$$

Below the $sch(\pi_L(r))$ and its cardinality are defined as follows:

$$sch(\pi_L(r)) = L$$
$$card(\pi_L(r)) = card(r) \otimes \text{``}+\text{''}$$

Natural Joins. Let r and s be hierarchical relations with R and S schemas correspondingly, such that $R \not\subseteq S$ and $S \not\subseteq R$ and $R \cap S \neq \emptyset$. The natural join of r and s is a hierarchical relation defined as follows:

$$r \bowtie s = \{\widehat{uv[\bar{L}]} | u \in r \wedge v \in s \wedge u[L] == v[L]\}, \text{where } L = R \cap S, \bar{L} = S - L$$

Below the $sch(r \bowtie s)$ and its cardinality are defined as follows:

$$sch(r \bowtie s) = R \cup S$$
$$card(r \bowtie s) = (card(r) \oplus card(s)) \otimes \text{ "}*\text{ "}$$

Grouping. Let r be a hierarchical relation with R schema and $\gamma_L(r)$ be a grouping operation, where L is a list of elements. For simplicity let us assume that $L = [A, f(B) \to C]$, where $A, B \in R$, $f \in \{min, max, sum, count, average\}$, and C is the name of the aggregation result (a typed attribution object). The result of the grouping operation is a hierarchical relation defined as follows:

$$\gamma_L(r) = \{u[\widehat{A]v[C]}|u \in r \wedge v[C] = f(\pi_B(\sigma_{A=u[A]}(r)))\}$$

Below the $sch(\gamma_L(r))$ and its cardinality are defined as follows:

$$sch(\gamma_L(r)) = \{A\} \cup \{C\}$$
$$card(\gamma_L(r)) = \text{ "}*\text{ "}$$

In the schema of the resulting relation the cardinality symbol of the attribution object C is defined as "$\perp$".

5 The Conceptual Data Model Extensibility Concept

The extensibility concept of the considered conceptual data model coincides with the analogous concept of OPENMath. In other words, extension of the conceptual data model is reduced to defining new symbols and formalizing these symbols using the CD mechanism. For applying a *symbol* on the conceptual level, the following rule is proposed: Concept $\leftarrow$ *symbol* OPENMath object. To support hierarchical relations concept on the conceptual level, the following *local, global* and *source* symbols are introduced to the CD *cdm*. The following construction to define hierarchical relation schemas on the local conceptual level is proposed: *attribution(local, source application(set, S_1, S_2, ..., S_n))*. It is assumed that there are n source data ($n \geq 1$). The value of *source* symbol is a set of source data schemas and each element of this set is defined as a set of the following application objects: *application*(typeConstructor, $A_1, A_2, ..., A_k$), where A_i is a typed attribution object ($1 \leq i \leq k$). To support derived hierarchical relations on the global conceptual level, reasoning rules are used. The following construction to define derived hierarchical relation schemas is proposed: *attribution(global, C V_C, R V_R)*. It is assumed that C is a new modeling concept and R - a reasoning rule for supporting this concept. Finally, V_C is an attribution object and V_R - an algebraic program.

5.1 Non-formal Example of Extending the Conceptual Data Model

In this Section an informal example of extending the proposed conceptual data model is considered. Namely, the conceptual data model is extended to model the data integration concept. Our approach to data integration assumes formalizing the mediator, data warehouse and data cube concepts. Formalization of these concepts will be the mathematical basis for constructing the reasoning rules (for more details, see [4]). The following reasoning rules are considered: *mediator rule, warehouse rule* and *cube rule*. To support the data warehouse concept on the global conceptual level, the conceptual data model is extended by means of the following *whse* and *rule* symbols. The following construction to define a derived hierarchical relation schema is proposed: $attribution(global,$ *whse* $attrObj,$ *rule* $appObj)$. The value of the *whse* symbol is an attribution object and the value of the *rule* symbol is an application object (an algebraic program) by means of which a mapping from data sources into data warehouse is defined. In a similar way the reasoning rules of the mediator and data cube are defined. To support the reasoning rules, we developed a new CD (titled as *dic*) to assign informal and formal semantics to data integration concepts. The result of such extension is a new conceptual data model with orientation to data integration. Below, an example of a warehouse for an automobile company database is adduced [8] which is an instance of the conceptual data model with orientation to data integration. Suppose for simplicity that there are only two dealers in the Aardvark system and they respectively use the schemas

Cars = {serialNo, model, color, autoTrans}, and

Autos = {serial, model, color}, Options = {serial, option}

It is assumed that a warehouse is created with the schema

AutosWhse = {serialNo, model, color, dealer}

Following is the definition of the local conceptual schema:

$$attribution(local, \, source \, application(set, S_1, S_2))$$

$S_1 \leftarrow application(set, attribution(\text{Cars}, type \, T_1))$

$T_1 \leftarrow application(sequence, attribution(\text{serialNo}, type \, int),$
    $attribution(\text{model}, type \, int), attribution(\text{color}, type \, string),$
    $attribution(\text{AutoTrans}, type \, string))$

$S_2 \leftarrow application(set, attribution(\text{Autos}, type \, T_2),$
    $attribution(\text{Options}, type \, T_3))$

$T_2 \leftarrow application(sequence, attribution(\text{serialNo}, type \, int, rename \, \text{serial}),$
    $attribution(\text{model}, type \, string, retype \, int),$
    $attribution(\text{color}, type \, string))$

$T_3 \leftarrow application(sequence, attribution(\text{serialNo}, type \, int, rename \, \text{serial}),$
    $attribution(\text{option}, type \, string))$

The value of the *rename* symbol is the name of the attribution object in the terms of the source schema. Analogously, the value of the *retype* symbol is the name of the type of the attribution object in the terms of the source schema. These symbols are defined in the *dic* content dictionary.

Following is the definition of the global conceptual schema:

$$attribution(global, whse\, attribution(\text{AutosWhse}, type\, T, rule\, R))$$
$$T \leftarrow application(sequence, attribution(\text{serialNo}, type\, int),$$
$$attribution(\text{model}, type\, int), attribution(\text{color}, type\, string),$$
$$attribution(\text{dealer}, type\, string))$$
$$R \leftarrow application(\cup, application(\times, P_1, D1), application(\times, P_2, D2))$$
$$P_1 \leftarrow application(\pi, application(list, \text{SerialNo}, \text{model}, \text{color}), F_1)$$
$$F_1 \leftarrow application(\sigma, \text{Cars}, application(=, \text{autoTrans}, \text{'yes'}))$$
$$P_2 \leftarrow application(\pi, application(list, \text{SerialNo}, \text{model}, \text{color}), J_1)$$
$$J_1 \leftarrow application(\bowtie, F_2, \text{Autos})$$
$$F_2 \leftarrow application(\sigma, \text{Options}, application(=, \text{option}, \text{'autoTrans'}))$$

Here $D1$ and $D2$ are constant hierarchical relations and are defined as follows:
$D1 : \{\langle \text{'dealer1'}: dealer\rangle\}$, $D2 : \{\langle \text{'dealer2'}: dealer\rangle\}$

6   Related Work

In this Section we will consider popular notations for describing database designs, namely the E/R, UML and XML models [8]. Furthermore, a brief discussion of the data integration approach based on the conceptual data model is provided, along with known data integration approaches. The most common conceptual data model is the E/R model in which two concepts (*entity set* and *relationship*) are used to model the subject domain. By means of this model, it is impossible to define the behavior of the conceptual entities. In addition, this model does not support extension possibilities. UML offers much the same capabilities as the E/R model, with the exception of multiway relationships. Basic construction for modeling the subject domain is *class*. In contrast to the E/R model, it provides the ability to define the behavior of conceptual entities. UML also does not support extension means. In the XML data model, basic construction to model the subject domain is *element*. The XML data model, like the E/R model, does not provide means to define the behavior of the conceptual entities. In contrast to the above considered data models, the XML data model supports extension means. Extension is reduced to the creation of a new DTD. Due to this property, the use of the XML data model as the conceptual data model is preferred. In our case, a single concept (*attribution*) is used to model the conceptual entities. This concept is associated with the concept of *class* in the object data model. Finally, like the XML data model, the considered data model is extensible. Extension is reduced to the creation of new CDs. To support the conceptual data model and the data integration concept, the following *cdm* and *dic* CDs are developed. The extensibility property is an argument to use this conceptual data model as a canonical data model for integrating arbitrary data sources. One of the first works in the area of justifiable data models mapping for heterogeneous databases integration is [9]. In this paper, the concept of reversible mapping of an arbitrary source data model into a target one (canonical data model) is proposed. The canonical data model is expanded axiomatically [10]. In works [11–15], relational data sources are considered in the frame of the traditional approach of the data integration as well as in the frame of the paradigm of ontology-based data access and integration. In the context of these works a query language on the ontology level SPARQL is considered in the papers [16,17]. In [18] an analysis to use machine learning techniques to solve different problems to integrate unstructured and semistructured data is provided. In [4] we used the considered conceptual data model as a canonical data model for supporting the data integration concept. It is important that the proposed approach allows to generate a mapping from arbitrary data sources into conceptual schema. Mapping from data sources into conceptual schema is defined as an algebraic program. Modelling means of the conceptual level are insensitive to the extension of the canonical data model. The conceptual level means are sufficient to model the data integration concepts proposed in the above considered works.

7   Conclusions

The investigation subject of this paper is an extensible conceptual data model. A distinctive trait of the considered conceptual data model is its stratification into local and global levels for modeling both basic and derived conceptual entities. On the local level, data sources are represented with basic conceptual entities (hierarchical relations). The global level is intended to define derived conceptual entities. To support such conceptual entities, an algebra of hierarchical relations was developed. Formal semantics of the developed algebra is proposed. An important feature of the considered conceptual data model is its extensibility. The extension is achieved by introducing new concepts into the conceptual data model. Support for these concepts leads to the creation of new content dictionaries. Thus, the result of extension of the conceptual data model is a new conceptual data model, which is defined as the kernel of the considered conceptual data model extended with new concepts. A non-formal example of extending the conceptual data model is considered. Outcome of such extension is a conceptual data model with orientation to data integration. It is important that the proposed concept of the conceptual data model allows to generate a mapping from arbitrary data sources into conceptual schema.

References

1. Date, C.J.: An Introduction to Database Systems, 8th edn. Addison-Wesley, USA (2004)
2. Golshan, B., Halevy, A., Mihaila, G., Tan, W.: Data integration: after the teenage years. In: Proceedings of the 36th ACM SIGMOD-SIGACT-SIGAI Symposium on Principles of Database Systems, pp. 101–106 (2017)
3. Sadalage, P.J., Fowler, M.: NoSQL Distilled, 2nd edn. Addison-Wesley, USA (2013)

4. Manukyan, M.G.: On a conceptual data model with orientation to data integration. In: Pozanenko, A., et al. (eds.) Analytics and Management in Data Intensive Domains 2021, CEUR-WS, Russia, vol. 3036, pp. 39–53 (2021)
5. Dewar, M.: OpenMath: an overview. ACM SIGSAM Bull. **34**(2), 2–5 (2000)
6. Davenport, J.H.: A small OpenMath type system. ACM SIGSAM Bull. **34**(2), 16–21 (2000)
7. Hindley, J.R., Seldin, J.P.: Introduction to Combinators and λ-Calculus. Cambridge University Press, Great Britain (1986)
8. Garcia-Molina, H., Ullman, J., Widom, J.: Database Systems: The Complete Book, 2nd edn. Prentice Hall, USA (2009)
9. Kalinichenko, L.A.: Data model transformation method based on axiomatic data model extension. In: Bing Yao, S. (ed.) Fourth International Conference on Very Large Data Bases, September 13-15, West Berlin, Germany, VLDB 1978, pp. 549–555. Springer, Germany (1978)
10. Kalinichenko, L.A.: Methods and tools for equivalent data model mapping construction. In: Bancilhon, F., Thanos, C., Tsichritzis, D. (eds.) EDBT 1990. LNCS, vol. 416, pp. 92–119. Springer, Heidelberg (1990). https://doi.org/10.1007/BFb0022166
11. Calvanese, D., Giacomo, G.D., Lembo, D., Lenzerini, M., Rosati, R.: Tractable reasoning and efficient query answering in description logics: the $DL - Lite$ family. JAR **39**(3), 385–429 (2007)
12. Calvanese, D., De Giacomo, G., Lembo, D., Lenzerini, M., Rosati, R.: Ontology-based data access and integration. In: Liu, L., Özsu, M.T. (eds.) Encyclopedia of Database Systems, pp. 1–7. Springer, New York (2017). https://doi.org/10.1007/978-1-4614-8265-9_80667
13. Calvanese, D., De Giacomo, G., Lenzerini, M., Vardi, M.Y.: Query processing under GLAV mappings for relational and graph databases. Proc. VLDB Endow. **6**(2), 61–72 (2012)
14. Calvanese, D., et al.: The MASTRO system for ontology-based data access. Semant. Web **2**(1), 43–53 (2011)
15. Console, M., Lenzerini, M.: Data quality in ontology-based data access: the case of consistency. In: Proceedings of the 28th AAAI Conference on Artificial Intelligence, AAAI 2014, vol. 28, no. 1, pp. 1020–1026 (2014)
16. Xiao, G., Hovland, D., Bilidas, D., Rezk, M., Giese, M., Calvanese, D.: Efficient ontology-based data integration with canonical IRIs. In: Gangemi, A., et al. (eds.) ESWC 2018. LNCS, vol. 10843, pp. 697–713. Springer, Cham (2018). https://doi.org/10.1007/978-3-319-93417-4_45
17. Xiao, G., Kontchakov, R., Cogrel, B., Calvanese, D., Botoeva, E.: Efficient handling of SPARQL OPTIONAL for OBDA. In: Vrandečić, D., et al. (eds.) ISWC 2018. LNCS, vol. 11136, pp. 354–373. Springer, Cham (2018). https://doi.org/10.1007/978-3-030-00671-6_21
18. Dong, X.L., Rekatsinas, T.: Data integration and machine learning: a natural synergy. Proc. VLDB Endow. **11**(12), 2094–2097 (2018). https://doi.org/10.14778/3229863.3229876

Understanding Misinformation About COVID-19 in WhatsApp Messages

Antônio Diogo Forte Martins$^{(\boxtimes)}$, José Maria Monteiro, and Javam C. Machado

Computer Science Department, Federal University of Ceará, Fortaleza, Brazil
{diogo.martins,jose.monteiro,javam.machado}@lsbd.ufc.br

Abstract. During the COVID-19 pandemic, the misinformation problem arose again through social networks, like a harmful health advice and false solutions epidemic. In Brazil, one of the primary sources of misinformation is the messaging application WhatsApp. Thus, the automatic misinformation detection (MID) about COVID-19 in Brazilian Portuguese WhatsApp messages becomes a crucial challenge. Recently, some works presented different MID approaches for this purpose. Despite this success, most explored MID models remain complex black boxes. So, their internal logic and inner workings are hidden from users, which cannot fully understand why a MID model assessed a particular WhatsApp message as misinformation or not. Thus, in this article, we explore a post-hoc interpretability method called LIME to explain the predictions of MID approaches. Besides, we apply a textual analysis tool called LIWC to analyze WhatsApp messages' linguistic characteristics and identify psychological aspects present in misinformation and non-misinformation messages. The results indicate that it is feasible to understand relevant aspects of the MID model's predictions and find patterns on WhatsApp messages about COVID19. So, we hope that these findings help to understand the misinformation phenomena about COVID-19 in WhatsApp messages.

Keywords: Explainable machine learning · Misinformation detection · COVID19 · WhatsApp

1   Introduction

Misinformation is a major issue in our society, and unfortunately, during the coronavirus pandemics, it arose intensely through social networks. The misinformation concept can be defined as a process of intentional production of a communicational environment based on false, misleading, or decontextualized information to cause a communicational disorder [15]. The United Nations (UN) stated in April 2020 that there is a "dangerous misinformation epidemic" responsible for disseminating misleading advice and solutions about the coronavirus.[1]

---

[1] UN. "Hatred going viral in 'dangerous epidemic of misinformation' during COVID-19 pandemic". 14 April, 2020. Available in: https://news.un.org/en/story/2020/04/1061682. Accessed on: April 25, 2020.

© Springer Nature Switzerland AG 2022
S. Chiusano et al. (Eds.): ADBIS 2022, CCIS 1652, pp. 14–23, 2022.
https://doi.org/10.1007/978-3-031-15743-1_2

In February 2020, the Brazilian Health Ministry reported that among 6,500 messages analyzed by it, between January 22 and February 27, 90% were related to the new virus. From the messages about coronavirus, 85% were false[2].

WhatsApp instant messaging application is very popular in Brazil, with more than 120 million users in a population of about 210 million people. Besides, it is currently the main misinformation spread channel [10]. A very relevant WhatsApp feature is the public groups. These public groups are accessible through invitation links published on popular websites and social networks, such as Facebook and Twitter. Usually, they have specific topics for discussion, such as health, politics, and sports. Each group can put together a maximum of 256 members. So, WhatsApp public groups are very similar to social networks. Thus, they have been used to spread misinformation.

In this context, the automatic misinformation detection (MID) about COVID-19 in Brazilian Portuguese WhatsApp messages becomes a crucial challenge. MID is the task of assessing the appropriateness (truthfulness, credibility, veracity, or authenticity) of claims in a piece of information [15]. Early detection of misinformation could prevent its spread, thus reducing its damage. Recently, some works presented and evaluated different MID approaches to detect misinformation in Brazilian Portuguese WhatsApp messages [7,8]. Experimental results evidenced the suitability of these MID approaches. However, the MID approaches explored in [7,8] have a significant drawback: the lack of transparency behind their behaviors. Thus, users have little understanding of how these models make particular decisions. For example, users have no idea why a MID model assessed a particular WhatsApp message as misinformation. In this scenario, providing some explanation can help users better understand the misinformation problem, the data set and why a MID model might fail.

On the other hand, there are several methods, called interpretable machine learning techniques, to help users understand the intricate working of machine learning techniques, such as Local Interpretable Model-Agnostic Explanations (LIME) [11]. Another interesting approach is to analyze texts in its psychological and linguistic structure. Linguistic Inquiry and Word Count (LIWC) [9] is a text analysis framework that calculates the usage for different words categories based on their linguistic dimensions. LIWC's output is useful to analyze the psychological and linguistic features of the speech of individuals from different groups. In this paper, we first apply LIME to explain the predictions of the best MID approaches evaluated in [7,8]. Next, we explore LIWC to analyze WhatsApp messages' linguistic characteristics and identify psychological aspects present in misinformation and non-misinformation messages.

The remainder of this paper is organized as follows. Section 2 presents the main related work. Section 3 describes the data set used in the experiments, called COVID19.BR. Section 4 details the experimental evaluation performed to understand the misinformation about COVID-19 and its results. Conclusions and future work are presented in Sect. 5.

---

[2] Available in: https://www.saude.gov.br/fakenews. Accessed in: April 25, 2020.

2 Related Work

The study presented in [4] proposes a misleading-information detection model that relies on several contents about COVID-19 collected from the World Health Organization, UNICEF, and the United Nations, as well as epidemiological material obtained from a range of fact-checking websites. Ten machine learning algorithms, with seven feature extraction techniques, were used to construct a voting ensemble machine learning classifier. The research presented in [1] proposed a set of machine learning techniques to classify information and misinformation. They achieved a classification accuracy of 86.7% with the Decision Tree classifier and 86.67% with the Convolutional Neural Network model. In [6], the authors introduced CoVerifi, a web application that combines the power of machine learning and human feedback to assess the credibility of news about COVID-19. By allowing users to "vote" on news content, the CoVerifi platform will allow the data labeling in an open and fast way.

In [7], the authors presented the COVID-19.BR, a large-scale, labelled, anonymized, and public data set formed by WhatsApp messages in Brazilian Portuguese (PT-BR) about coronavirus pandemic, collected from public WhatsApp groups using the platform proposed in [13]. In that work, they conduct a series of classification experiments using nine different machine learning methods to build an efficient MID for WhatsApp messages. The best result reached by [7] had an F1 score of 0.778, considering the full corpus of COVID-19.BR data set.

In [8], the authors presented a new approach, called MIDeepBR, based on BiLSTM neural networks, pooling operations and attention mechanisms. MIDeepBR can automatically detect misinformation in PT-BR WhatsApp messages. MIDeepBR automatically detects misinformation at the Digital Lighthouse [13] platform. Experimental results evidenced the suitability of the proposed approach to automatic misinformation detection. Their best results achieved an F1 score of 0.834, while in work presented in [7], the best results achieved an F1 score of 0.778. Thus, MIDeepBR outperformed the approaches evaluated in [7]. MIDeepBR makes use of BiLSTM, BERT Embeddings, Poolings, and Linear layers. MIDeepBR combines these different tools and algorithms to improve the automatic misinformation detection performance.

In [16] the authors present a novel XAI approach for BERT-based fake news detectors. They state that this new XAI method can be used as an extension to existing fake news detectors. Their XAI method is based on LIME and Anchors focusing on local explanations. In [3] the authors propose a framework to analyze how political polarization affects opposed groups behavior. They used as case study the Brazilian COVID-19 pandemic polarized scenario being the two opposed groups the pro and against social isolation on Twitter. Their proposed framework provides techniques to infer political orientation, topic modeling to discover common concerns of a group of users, network analysis and community detection, and analysis linguistic characteristics to identify psychological aspects. They performed their linguistic characteristics analysis using LIWC.

3   COVID-19.BR Data Set

In this section, we describe the data set used in the experiments, called COVID-19.BR [7], and the adaptations that we performed on it. An important aspect to consider while applying methods to understand the misinformation phenomenon about COVID-19 in WhatsApp messages is the need for a large-scale labelled data set. The COVID-19.BR [7] is a large-scale, labelled, anonymized, and public data set formed by WhatsApp messages in PT-BR about coronavirus pandemic, collected from public WhatsApp groups. COVID19-BR is inspired by [14] and the authors built it following the methodological guideline for building corpora of deceptive content of [12].

COVID-19.BR contains messages from 236 open WhatsApp groups with at least 100 members. The data set has messages collected between April and June 2020. Alongside the messages, the data set other columns are date, hour, phone number, international phone code, if the user is Brazilian its state, word count, character count, and if the message contained media (audio, image, or video). Another feature of this data set is the definition of "viral messages" that are messages with more than five words that appear more than once in it. COVID-19.BR tackles users' privacy issues by anonymizing their names and cell phone numbers. Using a hash function to create a unique and anonymous identifier for each user using the cell phone number as input. It sets an alias for each group to achieve their anonymization. Since these groups are publicly available, this approach does not violate WhatsApp's privacy policy.

We revised the labels, removed the messages that have less than five words, messages not related to the coronavirus pandemics, messages containing only daily news summaries, and messages with only *url* as text content. The final corpus contains 2043 messages being 865 labelled as misinformation (label 1) and 1178 labelled as non-misinformation (label 0).

Table 1. Data set basic statistics.

Statistics	Non-misinformation	Misinformation
Count of unique messages	1178	865
Mean and std. dev. of number of tokens in messages	82.80 ± 57.59	169.72 ± 243.76
Minimum number of tokens	5	5
Median number of tokens	22	52
Maximum number of tokens	2210	1666
Mean and std. dev. of number of types in messages	57.59 ± 96.59	109.03 ± 133.15
Average size of words (in characters)	5.82	5.12
Type-token ratio	0.696	0.642
Mean and std. dev. of shares	2.02 ± 4.17	1.89 ± 2.76

Table 1 presents basic statistics about the corpus, including some traditional NLP features based on the number of tokens, types, characters, as well as the average number of shares, i.e., the frequency of the message in the original data set. We have a class ratio of 1.36, meaning that the data set is slightly imbalanced. Messages containing misinformation, on average, have more words, and their length varies more than the messages without misinformation, indicating that this type of message is disseminated in different writing styles. Number of shares has similar values in messages of both classes in COVID-19.BR data set.

4 Understanding Misinformation About COVID-19 Using LIME and LIWC

As mentioned before, MID models are complex black boxes, and users have great difficulty understanding the rationale behind their predictions. Besides, finding psychological patterns present in WhatsApp messages can help users comprehend the misinformation phenomenon about COVID-19.

In this section, we explore LIME to explain the predictions of the best MID approaches evaluated in [7,8]. Next, we apply LIWC to analyze WhatsApp messages' linguistic characteristics and identify psychological aspects present in misinformation and non-misinformation messages. All the experiments and the COVID-19.BR data set are available at our public repository[3].

4.1 Applying LIME

In order to better understand the predictions of MID approaches evaluated in [7,8], we used *ELI5*[4] which implements the LIME [11] algorithm. LIME gives machine learning models the ability to explain or present their behaviors in understandable terms to human beings. Using *ELI5* we can analyze the impact of each word in the prediction and get a list of words that most contribute to a message being classified as misinformation. Table 2 and 3 show the top 10 words and their weights for predictions from the SVM with linear kernel model trained using TF-IDF as feature extractor, trigram, removing stop words, and performing lemmatization. Positive weight values means that this word contributes to the MID model classifying the message as misinformation, in the other hand, negative weight values means that this word contributes to classifying the message as non-misinformation.

Analyzing the top words contributing for misinformation from Table 2, we can observe that *Chinese* and *Communist* are the words that affirm that a given message is a misinformation. The majority of the misinformation messages from COVID-19.BR data set states that China created the COVID-19 virus and it is a communist threat created to dominate the world. Another topic of misinformation on COVID-19.BR data set is about the treatment using *hydroxychloroquine*,

---

[3] https://gitlab.com/jmmonteiro/misinformation_covid19_xai.
[4] https://github.com/eli5-org/eli5.

Table 2. Top 10 words contributing for misinformation from the TFIDF-TRIGRAM-LEMMA-LSVM model.

Word	English translation	Weight
chinês	chinese	+1.747
!	!	+1.663
comunista	communist	+1.275
governador	governor	+1.179
hidroxicloroquina	hydroxychloroquine	+1.166
:angry-face:	:angry-face:	+1.151
* *	* *	+1.079
vírus	virus	+1.076
"	"	+0.980
médico	doctor	+0.974

Table 3. Top 10 words contributing for non-misinformation from the TFIDF-TRIGRAM-LEMMA-LSVM model.

Word	English translation	Weight
corona	corona	−1.470
coronavírus	coronavirus	−1.257
. br	. br	−1.146
br	br	−1.128
. . br	. . br	−0.980
casar	marry	−0.738
com	com	−0.704
- 19	- 19	−0.698
. com	. com	−0.690
kkk	lol	−0.678

that is why it appears in the top 10 alongside *doctor*. The *!* sign, ** * and "* markers, and the *:angry-face:* emoji also figure in the top 10. Misinformation messages tend to have an alarmist nature, so the messages are written in an aggressive style with many emojis, bold letters (** ** markers on WhatsApp) and citations, when possible, to give it some credibility.

Analyzing the top words contributing for non-misinformation shown in Table 3, we can see that even though we are developing MID models in the context of the COVID-19 pandemics, the two words *coronavirus* and *corona* highly contributes for a message to be predicted as non-misinformation. This fact stems from the large number containing veracious news on the COVID-19.BR data set. Besides, it is essential to highlight that some tokens such as *".br"* and *".com"* also contribute for a message to be predicted as non-misinformation since they are related to veracious sites.

4.2   Applying LIWC

– Cohesion and Union: this group contains the *we* and *assent* LIWC categories. From this group, we can analyze if the message has collective and agreement features. Messages with a higher rate of *we* words indicate a group cohesion while *assent* words indicate a group consensus [17]. This group also analyses the idea of engaging [2].
– Personal Concerns: this group contains the *work, money, leisure, home, health, death*, and *relig* (religion) categories. With this group, we can analyze which topics are being are treating in the messages [3].
– Emotions: this group contains the *anger, sad, anx* (anxiety), *negemo* (negative emotion), and *posemo* (positive emotion) LIWC categories. From this group, we can analyze if the message contains emotion related features. They are important because the expression of words from emotional features represent a semantic level relevant to polarization [2].
– Cognitive Refinement: this group contains the *excl* (exclusive), *conj* (conjunction), *preps* (prepositions), and *cogmech* (cognitive mechanisms) categories. Messages containing a high rate of words in these categories are generally sophisticated written. It is also associated to how refined is an individual thought [17].

In this experiment, we used the LIWC version available for PT-BR [5]. First, we divided the WhatsApp messages from COVID-19.BR data set into two groups: misinformation and non-misinformation. Next, we created a counter for each LIWC category for each message group. Then, we go through each message in the misinformation group and for each word, we check its LIWC categories (since a word can belong to many categories) and increase the respective counters. Then, we apply the same process to the non-misinformation group. We can now evaluate if the same LIWC category presents different values for both message groups (misinformation and non-misinformation) to understand better how misinformation is crafted. In this sense, we applied the Chi-squared statistical test with 95% of significance level to verify if there is a significant statistical difference between the values obtained for a specific LIWC category in both groups of WhatsApp messages. Considering all the 64 LIWC categories, 44 have a significant difference between misinformation and non-misinformation groups of messages. Considering only the 18 LIWC categories used in the four psychological groups, 10 have significant statistical differences between misinformation and non-misinformation groups. Table 4 shows the results for the 18 LIWC categories used in the four psychological groups. Next, we discuss the results for each psychological group.

– Cohesion and Union: The *we* category occurs just 5 and 6 times in the misinformation and non-misinformation groups, respectively. The *assent* category occurs 275 and 216 times in the misinformation and non-misinformation groups, respectively. So, we can not state that a message group has a sense of union and cohesion greater than the other. The diversity of topics discussed on the WhatsApp groups can explain these results.

- Personal Concerns: In this psychological group we can observe significant difference between misinformation and non-misinformation groups in the *money*, *home*, and *death* categories. About *money*, there are misinformation messages related to politicians receiving bribe payments and concerns about losing jobs during the pandemic. The *home* category, in turn, occurs more in the non-misinformation group due to the stay-at-home movement and people warning others to avoid the virus spreading. The *death* category occurs more in the misinformation group because some messages disbelieve the veracity of COVID-19 deaths, and other messages state that people will die from starvation if they stop working to stay home.
- Emotions: In this psychological group we can observe significant difference between misinformation and non-misinformation groups in the *anger*, *sad*, *anx*, and *negemo* categories. In general, misinformation messages are more aggressive and full of negative emotions. It is explained by the alarmist nature of this type of message.
- Cognitive Refinement: In this psychological group we can observe significant difference between misinformation and non-misinformation groups in the *conj*, *preps*, and *cogmech* categories. We can observe that misinformation messages are well written and use a refined and complex language. They are more capable to deceive the general population because they prefer to believe in sophisticated messages.

Table 4. LIWC results.

Category	Mis.	Non-Mis.	Diff.	Mis. Percent	Non-Mis. Percent	Diff. Percent	Sig. Diff.
we	5	6	1	0.006	0.010	0.004	No
assent	275	216	59	0.322	0.347	0.025	No
work	5726	4159	1567	6.704	6.682	0.022	No
money	**2201**	1398	803	**2.577**	2.246	0.331	Yes
leisure	1412	1035	377	1.653	1.663	0.010	No
home	315	**387**	72	0.369	**0.622**	0.253	Yes
health	1685	1237	448	1.973	1.987	0.015	No
death	**587**	335	252	**0.687**	0.538	0.149	Yes
relig	915	709	206	1.071	1.139	0.068	No
anger	**2293**	1246	1047	**2.685**	2.002	0.683	Yes
sad	**1490**	814	676	**1.744**	1.308	0.437	Yes
anx	**966**	622	344	**1.131**	0.999	0.132	Yes
negemo	**5295**	3022	2273	**6.199**	4.855	1.344	Yes
posemo	6042	4286	1756	7.074	6.886	0.188	No
excl	693	500	193	0.811	0.803	0.008	No
conj	**613**	260	353	**0.718**	0.418	0.300	Yes
preps	757	**801**	44	0.886	**1.287**	0.401	Yes
cogmech	**13593**	9600	3993	**15.914**	15.424	0.490	Yes

5   Conclusion

In these days of pandemics, the automatic misinformation detection (MID) about COVID-19 in PT-BR WhatsApp messages is a crucial challenge. The early detection of misinformation can prevent its spread, thus reducing its damage. Recently, some works presented and evaluated different MID approaches to detect misinformation in Brazilian Portuguese WhatsApp messages [7,8]. However, these MID approaches have a significant drawback: the lack of transparency behind their behaviors. So, their internal logic and inner workings are hidden to users, which cannot fully understand why a MID model assessed a particular WhatsApp message as misinformation or not.

Thus, in this article, we explored a post-hoc interpretability method called LIME to explain the predictions of the MID approaches. Besides, we applied a textual analysis tool called LIWC to analyze WhatsApp messages' linguistic characteristics and identify psychological aspects present in misinformation and non-misinformation messages. The results indicate that it is feasible to understand relevant aspects of the MID model's predictions and find patterns on WhatsApp messages about COVID19. So, we hope that these findings help to understand the misinformation phenomena about COVID-19 in WhatsApp messages. As future works, we want to apply XAI tools to our MIDeepBR model to analyze how this more complex model behave predicting misinformation. We also want to analyze how misinformation spread among the groups using complex networks.

Acknowledgements. This research was funded by CAPES/UFC and conducted at LSBD/ARIDA.

References

1. Choudrie, J., Banerjee, S., Kotecha, K., Walambe, R., Karende, H., Ameta, J.: Machine learning techniques and older adults processing of online information and misinformation: a COVID 19 study. Comput. Hum. Behav. **119**, 106716 (2021). https://doi.org/10.1016/j.chb.2021.106716
2. Demszky, D., et al.: Analyzing polarization in social media: method and application to tweets on 21 mass shootings. In: Proceedings of the 2019 Conference of the North American Chapter of the Association for Computational Linguistics: Human Language Technologies, Volume 1 (Long and Short Papers), Minneapolis, Minnesota, pp. 2970–3005. Association for Computational Linguistics, June 2019. https://doi.org/10.18653/v1/N19-1304, https://aclanthology.org/N19-1304
3. Ebeling, R., Sáenz, C., Nobre, J., Becker, K.: Quarenteners vs. cloroquiners: a framework to analyze the effect of political polarization on social distance stances. In: Anais do VIII Symposium on Knowledge Discovery, Mining and Learning, pp. 89–96. SBC, Porto Alegre (2020). https://doi.org/10.5753/kdmile.2020.11963, https://sol.sbc.org.br/index.php/kdmile/article/view/11963
4. Elhadad, M.K., Li, K.F., Gebali, F.: Detecting misleading information on COVID-19. IEEE Access **8**, 165201–165215 (2020). https://doi.org/10.1109/ACCESS.2020.3022867

5. Filho, P.P.B., Pardo, T.A.S., Aluísio, S.M.: An evaluation of the Brazilian Portuguese LIWC dictionary for sentiment analysis. In: STIL (2013)
6. Kolluri, N.L., Murthy, D.: CoVerifi: a COVID-19 news verification system. Online Soc. Netw. Media **22**, 100123 (2021). https://doi.org/10.1016/j.osnem.2021.100123
7. Forte Martins, A.D., Cabral, L., Chaves Mourão, P.J., Monteiro, J.M., Machado, J.: Detection of misinformation about COVID-19 in Brazilian Portuguese WhatsApp messages. In: Métais, E., Meziane, F., Horacek, H., Kapetanios, E. (eds.) NLDB 2021. LNCS, vol. 12801, pp. 199–206. Springer, Cham (2021). https://doi.org/10. 1007/978-3-030-80599-9_18
8. Martins, A., Cabral, L., Mourão, P.J., Monteiro, J., Machado, J.: Detection of misinformation about COVID-19 in Brazilian Portuguese WhatsApp messages using deep learning. In: Anais do XXXVI Simpósio Brasileiro de Bancos de Dados, pp. 85–96. SBC, Porto Alegre (2021). https://doi.org/10.5753/sbbd.2021.17868, https://sol.sbc.org.br/index.php/sbbd/article/view/17868
9. Pennebaker, J., Francis, M., Booth, R.: Linguistic inquiry and word count (LIWC) (1999)
10. Resende, G., et al.: (Mis)information dissemination in WhatsApp: gathering, analyzing and countermeasures (2019). https://doi.org/10.1145/3308558.3313688
11. Ribeiro, M.T., Singh, S., Guestrin, C.: "Why should I trust you?": explaining the predictions of any classifier (2016)
12. Rubin, V.L., Chen, Y., Conroy, N.K.: Deception detection for news: three types of fakes. Proc. Assoc. Inf. Sci. Technol. **52**(1), 1–4 (2015)
13. de Sá, I.C., Monteiro, J.M., da Silva, J.W.F., Medeiros, L.M., Mourão, P.J.C., da Cunha, L.C.C.: Digital lighthouse: a platform for monitoring public groups in WhatsApp. In: Filipe, J., Smialek, M., Brodsky, A., Hammoudi, S. (eds.) Proceedings of the 23rd International Conference on Enterprise Information Systems, ICEIS 2021, Online Streaming, 26–28 April 2021, vol. 1, pp. 297–304. SCITEPRESS (2021). https://doi.org/10.5220/0010480102970304
14. Silva, R.M., Santos, R.L., Almeida, T.A., Pardo, T.A.: Towards automatically filtering fake news in Portuguese. Expert Syst. Appl. **146**, 113199 (2020)
15. Su, Q., Wan, M., Liu, X., Huang, C.R.: Motivations, methods and metrics of misinformation detection: an NLP perspective. Nat. Lang. Process. Res. **1**, 1–13 (2020). https://doi.org/10.2991/nlpr.d.200522.001
16. Szczepański, M., Pawlicki, M., Kozik, R., Choraś, M.: New explainability method for BERT-based model in fake news detection. Sci. Rep. **11** (2021). https://doi. org/10.1038/s41598-021-03100-6
17. Tauszik, Y.R., Pennebaker, J.W.: The psychological meaning of words: LIWC and computerized text analysis methods. J. Lang. Soc. Psychol. **29**(1), 24–54 (2010). https://doi.org/10.1177/0261927X09351676

Automatic Inference of Taxonomy Relationships Among Legal Documents

Irene Benedetto[1,2]([✉]) [ID], Luca Cagliero[1]([✉]) [ID], and Francesco Tarasconi[2]

[1] Politecnico di Torino, 10129 Turin, Italy
{irene.benedetto,luca.cagliero}@polito.it
[2] H-Farm Innovation, 10121 Turin, Italy
irene.benedetto@h-farm.it

Abstract. Exploring legal documents such as laws, judgments, and contracts is known to be a time-consuming task. To support domain experts in efficiently browsing their contents, legal documents in electronic form are commonly enriched with semantic annotations. They consist of a list of headwords indicating the main topics. Annotations are commonly organized in taxonomies, which comprise both a set of is-a hierarchies, expressing parent/child-sibling relationships, and more arbitrary related-to semantic links. This paper addresses the use of Deep Learning-based Natural Language Processing techniques to automatically extract unknown taxonomy relationships between pairs of legal documents. Exploring the document content is particularly useful for automatically classifying legal document pairs when topic-level relationships are partly out-of-date or missing, which is quite common for related-to links. The experimental results, collected on a real heterogeneous collection of Italian legal documents, show that word-level vector representations of text are particularly effective in leveraging the presence of domain-specific terms for classification and overcome the limitations of contextualized embeddings when there is a lack of annotated data.

Keywords: Legal judgments annotation · Legal text modeling · Natural language processing · Deep learning

1 Introduction

Legal databases contain a vast and heterogeneous collection of documents of different types among which legislative documents, regulations, judgments and maxims [7]. To support legal experts, such as lawyers and judges, in the navigation of these electronic data sources it is necessary to design efficient and effective retrieval systems. However, the inherent complexity of the legal terminology, the increasing number of resources available in electronic form, and the variability of the data sources across different countries make the problem of efficiently retrieving and exploring legal documents particularly challenging [27].

Content browsing and retrieval in legal databases is commonly driven by human-generated annotations organized in domain-specific taxonomies

© Springer Nature Switzerland AG 2022
S. Chiusano et al. (Eds.): ADBIS 2022, CCIS 1652, pp. 24–33, 2022.
https://doi.org/10.1007/978-3-031-15743-1_3

(e.g., [7]). A legal taxonomy consists of a set of is-a hierarchies describing the parent/child-sibling relationships between topical headwords. Each legal document is potentially annotated with many topical headwords. For example, according to the taxonomy depicted in Fig. 1 documents ranging over *Rights of exploitation* also belong to the parent category *Collective exploitation of property*, which, in turn, includes the documents ranging over the *Public domain* as well. Based on their domain knowledge, legal experts also exploit arbitrary *Related-to* taxonomy relationships such as those linking *Public domain* to *Immovable property* categories.

Domain experts can leverage taxonomy relationships to browse legal documents, jumping from one to another according to their specific needs. For example, exploring two complementary documents linked by a related-to relationship can be deemed as useful for covering different aspects of the same topic. However, annotation-based legal content retrieval is hindered by the following issues:

- Taxonomies evolve with legal systems [18] according to a *temporal concept drift*, which is typical of legal topic classification [7]. Therefore, existing taxonomy-based document relationships may become unreliable.
- Taxonomy relationships are often incomplete. Specifically, the *Related-to* relationships are unlikely to be available in several legal domains.
- Most electronic documents and the related annotations are written in English. Hence, there is a lack of solutions tailored to languages other than English.

To overcome the above issues, this paper proposes a classification system, based on Deep Natural Language Processing techniques, to automatically infer the taxonomy relationships holding between pairs of legal documents. Starting from a proprietary dataset collecting Italian legal judgments, mainly in the domain of private property, it learns supervised machine learning techniques to automatically predict the type of relationship holding between the document pair, i.e., *Parent-of/Child-of*, *Sibling-of*, or *Related-to*. The automatic annotation of document pairs is instrumental for improving the efficiency and effectiveness of legal content retrieval, especially when the taxonomy content is incomplete or partly out-of-date.

The preliminary results acquired in a real use case show that 1) traditional word-level vector representations of text, such as Doc2Vec [15], are particularly effective in the considered domain as capture syntactic properties between domain terms, such as "Patents" and "Trademarks", that are useful for classification purposes. 2) Contextualized embeddings [8] achieve performance than Doc2Vec on Italian documents due to the lack of domain-specific training data.

The rest of the paper is organized as follows: Sect. 2 overviews the prior works. Section 3 describes the classification system. Section 4 summarizes the main experimental results, whereas Sect. 5 draws the conclusions of the present work.

2 Related Work

In recent years the interest in legal AI has constantly grown. It encompasses legal judgment prediction [5,29], entity recognition [1,22], legal document classification [2,6,7], legal question answering [11,13], and legal summarization [12,28]. To analyze legal documents' similarities prior works adopt rule-based approaches [9,24], document-level text similarities [17], graph-based methods [25] and machine learning-based solutions [14,21]. All of the aforesaid works focus their analyses on English-written documents. Conversely, this work applies machine learning-based strategies to analyze Italian document relationships. To the best of our knowledge, the only attempts to perform a similarity analysis between multilingual legal documents have been presented in [19,20], where the authors analyze a multilingual corpus of 43 directives. Unlike [19,20], this work focuses on inferring taxonomy relationships among topical headwords.

3 The Classification System

3.1 Preliminaries

Let d_i be an arbitrary legal document, $\mathcal{L}$ be the set of labels present in a legal taxonomy $\mathcal{T}$, and $L_i \subseteq \mathcal{L}$ be the subset of labels used to annotate d_i.

The taxonomy $\mathcal{T}$ is a set of is-a hierarchies built over labels in $\mathcal{L}$. Each label consists of a set of headwords describing a document topic. A *Is-a* taxonomy relationship indicates the specular *Child-of* and *Parent-of* relationships holding between pairs of annotations $l_i, l_j \in \mathcal{L}$. Instead, the *Sibling-of* relationship holding between a pair of annotations indicates that l_i and l_j have a parent in common. Furthermore, an arbitrary *Related-to* relationship links a pair of label l_i and l_j that are semantically related one to another.

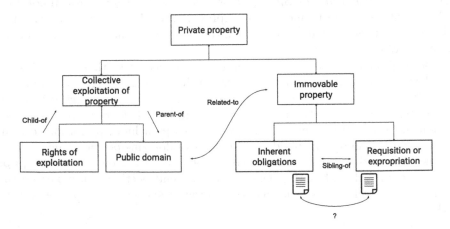

Fig. 1. Example of the taxonomy: in red we highlighted the relationship types we aim at identifying with our classification system. (Color figure online)

3.2   Problem Statement

Given an arbitrary pair of annotated documents (d_i, d_j), we aim at inferring the pairwise taxonomy relationships between labels in L_i and L_j (and vice versa).

More specifically, given (d_i, d_j), the purpose is to define a classification function f that predicts the type of the relationships holding between labels in L_i and labels in L_j. Thus, the target of the prediction is the relationship type, which takes values *Sibling-of*, *Parent-of/Child-of*, *Related-to* or *Other* (if a relationship either does not hold or is not relevant to the domain under analysis). For the sake of simplicity, hereafter we will cast the problem under analysis to a single-label task, i.e., the relationship cannot belong to multiple types at the same time.

f is computed as the composition of two functions $(h \circ g)(d_i, d_j) = h(g(d_i, d_j))$. Specifically, $g(d_i, d_j)$ produces a high-dimensional vector representation $e_i, e_j \in \mathbb{R}^N$ of the given documents. On top of the generated text representations, we train a classification model to estimate the function $h(e_i, e_j)$ and detect the corresponding relationship type r_{l_i, l_j}. At inference time, we predict the relationship type based on the document content, without any prior knowledge on the existing document annotations l_i and l_j.

3.3   Text Representations

We consider the following established text representations of the input legal documents:

- Term Frequency-Inverse Document Frequency (TF-IDF, in short) [26]: a occurrence-based, word-level text representation.
- Doc2Vec [15]: a sentence-level embedding model based on a Word2Vec extension.
- Multilingual BERT [8]: a contextualized embedding model.[1]

3.4   Classifiers

We integrate and test the following established classification approaches available in the scikit-learn library [23][2]: Support Vector Machines (SVMs), K-Nearest Neighbor (k-NN), Random Forest classifier (RF), and Logistic Regression (LR) [10].

---

[1] Due to the lack of a sufficient amount of domain-specific data in the Italian language, BERT cannot be retrained from scratch on the input data collection.

[2] Neural Network models are not suited to the prediction task under analysis due to the lack of a sufficient amount of training data.

4   Experimental Results

4.1   Experimental Design

Dataset. The proprietary dataset used in the empirical validation contains Italian legal judgments and maxims, mainly in the context of property law. Each legal document is annotated with one or more labels, corresponding to the legal principle of reference. Pairs of those legal principles are partitioned into groups according to the type of relationship between them (i.e., *Sibling-of*, *Parent-of/Child-of*, *Related-to* or *Other*).

The training dataset consists of a set of triples (d_i, d_j, r_{l_i, l_j}). Each triple represents a given pair of documents (d_i, d_j) annotated with the corresponding legal topics' relationship.

The complete dataset consists of more than 1300 examples for each relationship type and the relationship types are roughly equally distributed. For testing purposes, we apply a holdout strategy (80% train, 10% validation, 20% test).

Metrics. To evaluate classifiers' performance we compute the Precision, Recall and F1-score scores [10]. The goal is to evaluate the ability of the classification system to correctly identify positive examples separately for each class. Their definitions follow.

– *Precision (Pr) of class c*: it indicates the fraction of document pairs correctly classified as c among all the pairs labeled as c.
– *Recall (Rc) of class c*: it indicates the fraction of document pairs classified as c that have been retrieved over the total number of pairs labeled as c.
– *F1-score (F1) of class c*: it is the harmonic mean of precision and recall of class c.

To compute the similarity between a pair of legal documents in the vector space we use the established cosine similarity [10]. Specifically, let $e_i = f(d_i)$ and $e_j = f(d_j)$ be the encodings of d_i and d_j, respectively. The document similarity is defined by

$$sim_{d_i, d_j} = sim(e_i, e_j) = \frac{e_i \cdot e_j}{\|e_i\| \|e_j\|} \tag{1}$$

4.2   Results Discussion

Analysis of the Text Representations. We compute the pairwise similarity between each document pair, group the results by relationship type, and test the difference in mean between the per-type similarity values' distributions. The main goal is to understand whether the text encoding phase is able to clearly separate the groups related to different relationship types.

To verify the initial hypothesis of having a clear separation among relationship types for each pair we compute the two-sided t-test [16] for the difference between their means (with a type I error of 0.05). The outcomes of the significance tests are summarized in Table 1. As expected, the differences in mean between *Parent-of/Child-of* and *Sibling-of* are never significant. Among all the candidate representations, Doc2Vec achieves the best performance. Unlike Doc2Vec, contextualized embedding models do not achieve performance superior to the others mainly due to the lack of in-domain training data.

Analysis of Classifiers-Performance. In Table 2 we report the values of the classifier performance metrics achieved on the test set. The joint use of the Doc2Vec text representation and of the Logistic Regression classifier has shown to be highly beneficial, probably due to the inherent characteristics of the input data.

Model Explainability. To gain insights into the classification problem we leverage the characteristics of the decision tree models, which allow us to evaluate the influence of the input features on the output prediction.

Figure 2 shows that the headwords related to the legal domain, such as "Patents", "Trademarks", are actually very discriminating. This property is particularly helpful for modelling the input data, as less pertinent features can be early pruned.

Table 1. Significance test for the difference of two means computed on documents' similarities groups $\alpha = 0.05$.

Method	Type 1	Type 2	P-value
BERT	Sibling-of	Parent-of/Child-of	0.073
BERT	Related-to	Parent-of/Child-of	<0.001
BERT	Related-to	Sibling-of	<0.001
BERT	Other	Parent-of/Child-of	0.848
BERT	Other	Sibling-of	0.047
BERT	Other	Related-to	<0.001
Doc2Vec	Sibling-of	Parent-of/Child-of	0.308
Doc2Vec	Related-to	Parent-of/Child-of	<0.001
Doc2Vec	Related-to	Sibling-of	<0.001
Doc2Vec	Other	Parent-of/Child-of	<0.001
Doc2Vec	Other	Sibling-of	<0.001
Doc2Vec	Other	Related-to	<0.001
TFIDF	Sibling-of	Parent-of/Child-of	0.839
TFIDF	Related-to	Parent-of/Child-of	0.889
TFIDF	Related-to	Sibling-of	0.746
TFIDF	Other	Parent-of/Child-of	<0.001
TFIDF	Other	Sibling-of	<0.001
TFIDF	Other	Related-to	<0.001

Table 2. Classification results

Representation	Classifier	Precision	Recall	F1-score
Doc2vec	Logistic Regression	0.740	0.744	0.740
Doc2vec	SVM	0.731	0.735	0.733
Doc2vec	Random Forest	0.721	0.724	0.722
TF IDF	Random Forest	0.702	0.706	0.703
TF-IDF	SVM	0.690	0.695	0.005
TF-IDF	Logistic Regression	0.666	0.669	0.667
Doc2vec	KNN	0.660	0.664	0.662
BERT	Random Forest	0.650	0.660	0.648
BERT	SVM	0.639	0.647	0.642
BERT	Logistic Regression	0.610	0.620	0.614
TF-IDF	KNN	0.583	0.591	0.584
BERT	KNN	0.429	0.460	0.433

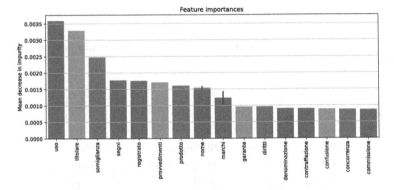

Fig. 2. Feature importance of Random forest classifier. This barplot shows that the most relevant terms for the classifier are mainly in-domain terms, such as "trademark" ("marchio"), "commission" ("commissioni").

Predictability of Different Relationship Types. Table 3 reports the per-class precision, recall and F1-score results achieved on the test set. The relationships of type *Related-ot* appear to be simpler to predict, whereas *Parent-of/Child-of* and *Sibling-of* turn out to be particularly challenging (see also the confusion matrix plot in Table 3.

Table 3. Predictability of different relationship types

	Parent-of/Child-of	Sibling-of	Related-to	Other
Precision	0.658	0.675	0.905	0.663
Recall	0.546	0.738	0.937	0.690
F1-score	0.60	0.706	0.920	0.672

5 Takeaways and Future Directions

The present work focused on overcoming the main issues of annotation-based legal content retrieval systems by proposing a classification-based to document pair annotation. The main takeaways can be summarized below.

- *Concept drift*: The updates of the original document collection and the presence of a relevant drift in the covered topics triggers the periodic retraining of the entire classification model. Such an activity can be labour-intensive and time-consuming. To overcome the aforesaid issue, we recommend to first explore the graphical distributions of the pairwise vector similarities and set up the classification pipeline accordingly.
- *Missing Related-to relationships*: *Related-to* is the most unconventional relationship in legal taxonomies. Since it is not easy to map it to known concepts or to existing data structures its correct prediction is particularly challenging. The proposed classification system achieved 90% F1-score on class *Related-to* thus confirming its effectiveness and usability in real application contexts (see Table 3).
- *Portability to languages other then English*: The increasing availability of state-of-the-art multilingual pretrained models (e.g., MultiLingual BERT [8]) and the recent advances in cross-lingual approaches [3,4] allow the direct processing of the raw data without the need to perform automatic machine translation. The preliminary results achieved on Italian legal documents confirm the feasibility of the multilingual extension.

As future works, we envision (1) The application of the proposed method to generate annotations in complex scenarios, e.g., zero-shot classification and active learning. (2) The application of eXplainable AI methods to increase model transparency. (3) The organization of a crowdsourcing validation process, based on domain experts, to study the applicability of the system in real scenarios (e.g., law firms, courthouse rooms). (4) The application of the proposed classification system to legal documents written in low-resource languages (e.g., Hindi, Vietnamese, Zulu).

Acknowledgements. The research leading to these results has been partly supported by the SmartData@PoliTO center for Big Data and Machine Learning technologies. The dataset has been provided by Giuffrè Francis Lefebvre S.p.A.

References

1. Angelidis, I., Chalkidis, I., Koubarakis, M.: Named entity recognition, linking and generation for Greek legislation. In: JURIX (2018)
2. Luz de Araujo, P.H., de Campos, T.E., Ataides Braz, F., Correia da Silva, N.: VICTOR: a dataset for Brazilian legal documents classification. In: Proceedings of the 12th Language Resources and Evaluation Conference, Marseille, France, pp. 1449–1458. European Language Resources Association, May 2020. https:// aclanthology.org/2020.lrec-1.181
3. Cagliero, L., Quatra, M.L.: Inferring multilingual domain-specific word embeddings from large document corpora. IEEE Access **9**, 137309–137321 (2021). https://doi. org/10.1109/ACCESS.2021.3118093
4. Cagliero, L., Quatra, M.L., Garza, P., Baralis, E.: Cross-lingual timeline summarization. In: Fourth IEEE International Conference on Artificial Intelligence and Knowledge Engineering, AIKE 2021, Laguna Hills, CA, USA, 1–3 December 2021, pp. 45–53. IEEE (2021). https://doi.org/10.1109/AIKE52691.2021.00014
5. Chalkidis, I., Androutsopoulos, I., Aletras, N.: Neural legal judgment prediction in English. In: Proceedings of the 57th Annual Meeting of the Association for Computational Linguistics, Florence, Italy, pp. 4317–4323. Association for Computational Linguistics, July 2019. https://doi.org/10.18653/v1/P19-1424, https:// aclanthology.org/P19-1424
6. Chalkidis, I., Fergadiotis, E., Malakasiotis, P., Androutsopoulos, I.: Large-scale multi-label text classification on EU legislation. In: Proceedings of the 57th Annual Meeting of the Association for Computational Linguistics, Florence, Italy, pp. 6314–6322. Association for Computational Linguistics, July 2019. https://doi.org/ 10.18653/v1/P19-1636, https://aclanthology.org/P19-1636
7. Chalkidis, I., Fergadiotis, M., Androutsopoulos, I.: MultiEURLEX - a multi-lingual and multi-label legal document classification dataset for zero-shot cross-lingual transfer. In: EMNLP (2021)
8. Devlin, J., Chang, M.W., Lee, K., Toutanova, K.: BERT: pre-training of deep bidirectional transformers for language understanding. ArXiv arXiv:1810.04805 (2019)
9. Geist, A.: Using citation analysis techniques for computer-assisted legal research in continental jurisdictions. SSRN Electron. J. (2009). https://doi.org/10.2139/ssrn. 1397674
10. Han, J., Kamber, M.: Data Mining. Concepts and Techniques, 2nd edn. Morgan Kaufmann (2006)
11. Hendrycks, D., Burns, C., Chen, A., Ball, S.: CUAD: an expert-annotated NLP dataset for legal contract review. CoRR arXiv:2103.06268 (2021)
12. Kanapala, A., Pal, S., Pamula, R.: Text summarization from legal documents: a survey. Artif. Intell. Rev. **51**(3), 371–402 (2019). https://doi.org/10.1007/s10462-017-9566-2
13. Kim, M.-Y., Xu, Y., Goebel, R., Satoh, K.: Answering yes/no questions in legal bar exams. In: Nakano, Y., Satoh, K., Bekki, D. (eds.) JSAI-isAI 2013. LNCS (LNAI), vol. 8417, pp. 199–213. Springer, Cham (2014). https://doi.org/10.1007/ 978-3-319-10061-6_14
14. Landthaler, J., Waltl, B., Holl, P., Matthes, F.: Extending full text search for legal document collections using word embeddings. In: JURIX (2016)
15. Le, Q.V., Mikolov, T.: Distributed representations of sentences and documents. CoRR arXiv:1405.4053 (2014)

16. Limentani, G.B., Ringo, M.C., Ye, F., Bergquist, M.L., McSorley, E.O.: Beyond the t-test: statistical equivalence testing (2005)
17. Mandal, A., Chaki, R., Saha, S., Ghosh, K., Pal, A., Ghosh, S.: Measuring similarity among legal court case documents. In: Proceedings of the 10th Annual ACM India Compute Conference, Compute 2017, pp. 1–9. Association for Computing Machinery, New York (2017). https://doi.org/10.1145/3140107.3140119
18. Mattei, U.: Three patterns of law: taxonomy and change in the world's legal systems. Am. J. Comp. Law **45**(1), 5–44 (1997). https://doi.org/10.2307/840958
19. Nanda, R., Caro, L.D., Boella, G.: A text similarity approach for automated transposition detection of European union directives. In: JURIX (2016)
20. Nanda, R., et al.: Unsupervised and supervised text similarity systems for automated identification of national implementing measures of European directives. Artif. Intell. Law **27**(2), 199–225 (2018). https://doi.org/10.1007/s10506-018-9236-y
21. Ostendorff, M., Ash, E., Ruas, T., Gipp, B., Moreno-Schneider, J., Rehm, G.: Evaluating document representations for content-based legal literature recommendations, pp. 109–118. Association for Computing Machinery, New York (2021). https://doi.org/10.1145/3462757.3466073
22. Papaloukas, C., Chalkidis, I., Athinaios, K., Pantazi, D., Koubarakis, M.: Multi-granular legal topic classification on Greek legislation. CoRR arXiv:2109.15298 (2021)
23. Pedregosa, F., et al.: Scikit-learn: machine learning in Python. J. Mach. Learn. Res. **12**, 2825–2830 (2011)
24. Raghav, K., Reddy, K., Reddy, V.B.: Analyzing the extraction of relevant legal judgments using paragraph-level and citation information (2016)
25. Wagh, R.S., Anand, D.: Legal document similarity: a multi-criteria decision-making perspective. PeerJ Comput. Sci. **6**, e262 (2020). https://doi.org/10.7717/peerj-cs.262
26. Sammut, C., Webb, G.I. (eds.): TF-IDF, pp. 986–987. Springer, Boston (2010). https://doi.org/10.1007/978-0-387-30164-8_832
27. Van Opijnen, M., Santos, C.: On the concept of relevance in legal information retrieval. Artif. Intell. Law **25**(1), 65–87 (2017). https://doi.org/10.1007/s10506-017-9195-8
28. Wu, Y., et al.: De-biased court's view generation with causality. In: Proceedings of the 2020 Conference on Empirical Methods in Natural Language Processing (EMNLP), pp. 763–780. Association for Computational Linguistics, November 2020. https://doi.org/10.18653/v1/2020.emnlp-main.56, https://aclanthology.org/2020.emnlp-main.56
29. Xu, N., Wang, P., Chen, L., Pan, L., Wang, X., Zhao, J.: Distinguish confusing law articles for legal judgment prediction. In: Proceedings of the 58th Annual Meeting of the Association for Computational Linguistics, pp. 3086–3095. Association for Computational Linguistics, July 2020. https://doi.org/10.18653/v1/2020.acl-main.280, https://aclanthology.org/2020.acl-main.280

Visualisation of Numerical Query Results on Industrial Data Streams

Miran Ismaiel Nadir$^{(\boxtimes)}$ and Kjell Orsborn

Department of Information Technology, Uppsala University, Uppsala, Sweden
miran.nadir@it.uu.se

Abstract. The capability to efficiently handling and analysing data streams in industrial processes and industrial cyber-physical systems (ICPS) is critical for digitalisation and renewal of current manufacturing industry. A key problem within this context is to provide scalable capability to collect, process, analyse, and visualise data streams to support these ICPSs. The status of these systems continuously changes, and analysts must understand and act upon such changes often in real time. Visualisation tools are increasingly used by analysts to get insights from these changes, but inconveniently nowadays, analysts have to store the data from the Industrial Data Stream to a data storage system and then use some visualisation tools to be able to visualise the data to help them understand and act promptly. In this paper, we propose to integrate visualisation and analysis primitives transparently into the query language of the Data Stream Management System (DSMS) and we show a proof of concept by a successful integration of an operator that can execute query-based visualisation methods that support processing of continuous numerical queries over streaming data in industrial analytics applications. We will also show how we are benefiting from the meta data in DSMSs to perform dimensionality reduction in real time to identify a two, three, four, or five-dimensional representation of the numerical query results on the data stream which preserve the salient relationships in the results and how the operator can suggest the most appropriate visualisation of the data to the analyst.

Keywords: Visualisation · Query · Data stream · Data analytic · Data management · Edge analytic · Industrial · Cyber-physical systems

1 Introduction

Several major international research and development initiatives such as Industrial Internet [1], Industry 4.0 [2,3], and Made in China 2025 [4], are focusing on transforming the current manufacturing industry through digitalisation, with the overall goal of improving productivity and quality of industrial processes and products. They are promoting the idea of intelligent manufacturing, which simply means analysing the data from the Internet of Things in time and feed the

© Springer Nature Switzerland AG 2022
S. Chiusano et al. (Eds.): ADBIS 2022, CCIS 1652, pp. 34–42, 2022.
https://doi.org/10.1007/978-3-031-15743-1_4

data back into the industry loop which then will be used to efficiently engage in mass production and product customisation.

Current cyber-physical systems have processing power, which makes it possible to deploy data processing both as edge analytics in these systems [5] and at an aggregated level in the cloud. In addition to processing performance, high level access to data and data analytics (i.e. query-based analytical operators) capabilities are vital and more efficient for the overall process performance [6]. In this paper we argue that to fully exploit this possibility a change in the process and in the underlying technologies is required.

The left part of Fig. 1 illustrates the initial idea of Industrial Internet where (aggregated) data analytics is supplied in a cloud environment, the proprietary data is extracted from the industrial machines then transferred to data systems were algorithms and analysis are applied and the results visualised and shared with the right people to create insights. These new findings will be fed back into the industrial machines to improve or correct a behaviour.

The right part of Fig. 1 provides a complete edge-based perspective with a tighter analytic loop that can react to sudden changes in industrial processes: through online analytics capabilities connected to equipment, one can provide a much tighter monitoring and feedback loop in comparison to conventional Extract-Transfer-Load methods [9,10]. The Analyse-Visualise-Feedback loop is illustrated to the right in Fig. 1.

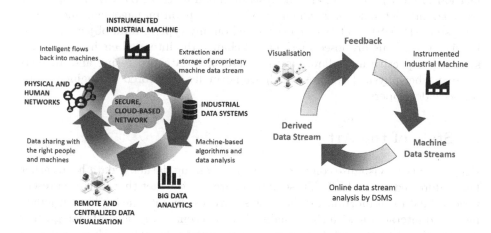

Fig. 1. The Industrial Internet conventional data analysis cycle, adapted from [20], versus an edge analytics architecture for online data stream analytics.

A key problem within this context is to provide and handle scalable capability to collect, process, analyse, and visualise data streams to support cyber-physical systems [5]. Within this context, there is an increasing demand for data analysis tools that support query-based access for management of data [5–8,21].

Query-based access to continuous streaming data and data analysis can efficiently enable high-level data management for engineering analysts compared to more conventional programming-oriented access. Additionally, for analytics, an important capability is the effective visualisation of numerical data and models, including spatial and/or temporally distributed quantities in the form of numerical arrays, matrices, and streams.

The general goal of our research is to study and develop methods for scalable processing of numerical queries involving real-time continuous streaming data and visualisation and analysis primitives for advanced numerical applications. In particular, the objective of this paper is to introduce query based visualisation and analysis primitives transparently into the query-language level of the data stream management and analysis system such that query processing can exploit meta-data of the visualisation and analysis operations as well as conventional meta-data in the database to guide the optimisation of the data processing. This means engineering analysts can have immediate access to visual analytic capabilities avoiding unnecessary data shipping between systems, by for example analysing and displaying results of continuous queries over streaming data as dynamic graphs or in some more complex model-based view displaying dynamic contour plots or trajectories over a geometric model.

The current work provides a preliminary answer to the following research question: How can visualisation and analysis primitives be transparently integrated into a query language making visualisation and analysis of streaming numerical data intuitive and easy to use. In particular, we are proposing a new method of data processing and analysis by integrating a "DISPLAY" operator transparently into the query language of a data stream management and analysis system (DSMS), that can be overloaded on any type of data object that can be visualised. This proposed method will change the intended environment and system perspective and allows us to deploy such data processing and analysis in edge cyber-physical systems which are normally monitored by engineers, not software developers.

2   State of the Art

Functionality to visualise conventional non-streaming database data have earlier been addressed in [7,11,13], These papers agree on the fact that visual representation of data is important for the analysis process by integrating human judgement and interaction with the visualisation. A recent study suggests integrating visualisation into the database management system [8], this work is related to our proposal for streaming data. Other studies focused on ways to handle scaling issues when trying to visualise query results from conventional DBMSs and they have proposed ways to dynamically perform resolution reductions when trying to visualise large data sets [16,17]. They suggest using data reduction methods like binned aggregation and sampling and also interactive querying, but again most of these methods are not usable for streaming data which have different temporal and scalar values. Some studies also focused on the laborious and time consuming process of finding the right visualisation to apply on large volumes of data and on how to build a system that can automate this task [18].

For real-time streaming data there are more recent works which integrate a Visual data flow programming environment with a DSMS using a Visual Data Stream Monitor [9,10], but functionality is limited to simple forms of numerical data. In [19] they presented an interactive development environment to coordinate running clusters to be able to process and transfer the data points which must be visualised, and they have presented an algorithm for real-time visualisation of a time series, however we are considering more than data points in our proposed environment, for example complex continuous scalar, vector, tensor of time-series data that normally is related to some basic analysis model.

Conventional data stream visualisation options are normally "hard-wired" into some programmed procedure [23,24] or through some external visualisation tools like Microsoft Power BI, Tableau Desktop, etc. [25], and cannot easily be adapted to changing needs and quick acting which is essential for industrial instrumental live streams. By using Ad-hoc queries for retrieving information to visualise and analyse, one adds a powerful mechanism for selecting what information to show since all data can be accessed and composed through the query language.

To our knowledge, there is no system that tightly integrates visualisation and analytical operators together with numerical arrays and matrices transparently into a query language and the corresponding query processing. Even if data can be accessed through a query language, there is normally a need for additional coding e.g. algorithms in some programming or scripting language. The aim of this work is to minimise the need for coding experts and put more high-level interfaces in the hands of scientists and engineers through a simple data manipulation query language and promote the "low-code" trend in industrial data analysis [21].

3   The DISPLAY Operator

Through our work we show a proof of concept of query-based visualisation methods that support processing of continuous numerical queries over streaming data in engineering analytics applications. More specifically, we show how a DISPLAY operator transparently has been introduced into the query language of the data stream management and analysis system and can be overloaded on any type of data object that can be visualised.

The Operator has been designed as follows: first it receives the conventional meta-data from the query processor regarding the resulted field types, number of fields, and their order in the Continuous Query (QC), then it processes the data results of the QC running on the stream, finds out the amount of records in the specified window size and calculates some aggregations like minimum, maximum, and mean values of each resulted field. These values will represent metadata for the visualisation and analysis method when selected. To find and suggest the best visualisation method to use, the Operator will (for the first batch of results) remove all non-numeric fields and order the remaining resulted fields according to a relevant feature selection using histogram analysis algorithms

[26,27]. The visualisation parameters generated will have to take into account the previous parameters from the previous windows in the time-series so that the observer will not lose context, then the Operator updates only a part of the generated real-time visualisation output according to the current query results. The Operator is acting like a compiler that generates visualisation code from the data it receives. The generated visualisation is vector graphics and it will be viewed on any browser without the need of any external libraries in most cases. The analyst will be able to interact with parameters of the QC in real-time changing the visualisation accordingly as needed.

The mechanisms presented provide self-service simplicity technique, a "low-code" method which is trending across industry currently, to facilitate innovation for engineering analysts to access advanced and ad-hoc visualisation and analytical capabilities without reliance on software developers [21]. Visualisation and analytical primitives can be implemented by Implementation Experts and Engineering Analysts can apply these and even compose new functionality by combining these primitives in ad-hoc analytical queries.

4   Experimental Validation

The following results show how our "DISPLAY" visualisation operator could be introduced as a polymorphic overloaded function into the AmosQL query language of the Amos II data stream management and analysis system, and can be overloaded on any type (or combination of types) of data objects that can be visualised. Initially, some basic types of geometric elements are supported such as points and lines. Spatial and temporal scalar and vector-valued quantities can be modelled as metadata in the database and can then be queried and visualised through built-in visualisation primitives using various types of charting elements. Currently there are 16 methods of two- and three-dimensional visualisation types. Furthermore, queries can incorporate more or less advanced analytical operations.

The following examples show how easily spatial data can be extracted and visualised through queries using this type of mechanisms in the query language:

```
display(select temperature(p), p from point p);
display(select temperature(p), p from point p
        where (temperature(p) > 2500) or (temperature(p) < 300));
display(select temperature(l), l from line l);
display(select temperature(l), l from line l
        where (x(l)>0.1) and (y(l)>0.2));
```

The first two queries display temperatures for points (point-based data), whereas the subsequent two queries display temperatures along a set of line entities of a grid-based model illustrating the convenience and need for vector-based graphics in these applications. Figure 2 shows the result of executing such queries at a given point in time.

Fig. 2. A query visualising the temperature for a complete set of points and for points with temperatures over 2500 and below 300 °C (left), and temperatures for a complete set of lines and for a set of lines with x coordinates bigger than 0.1 and y coordinates bigger than 0.2 (right).

Two additional examples show how time-series data from a pressure sensor from a hydraulic power system is visualised in a point chart that exemplifies how numerical operations (greater than) and metadata (e.g. specification of legend) can be controlled from within the query expression (Fig. 3). Notice that the legend is dynamically determined by the current result of the query.

```
display(select time(p), value(p) from Pressure p
        where (charttype(p)='PlotChart'));
display(select time(p), value(p) from Pressure p
        where (value(p)>180) and (charttype(p)='PlotChart'));
```

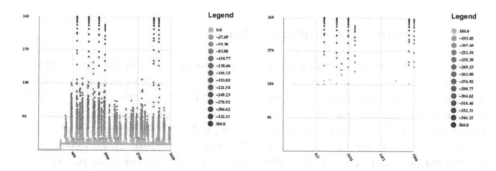

Fig. 3. Two queries visualising a point chart (left) and point chart with threshold (right) over a time series of pressure values (unit bar) from sensor readings (vertical axis) from a hydraulic power system over time (horizontal axis, unit seconds).

One additional example in Fig. 4 shows how the Display Operator decides and selects a relevant feature set from earthquake data in Italy using the below query results and suggests two different visualisation methods to the analyst. We can observe that a map background can be added interactively to the 2D visualisation method on the left.

```
display(select source(e), magnitude(e), longitude(e),
        latitude(e), -depth(e), gap(e), distance(e) from Earthquake e
        where (magnitude(e)>3) and (charttype(e)='suggestvm=2'));
```

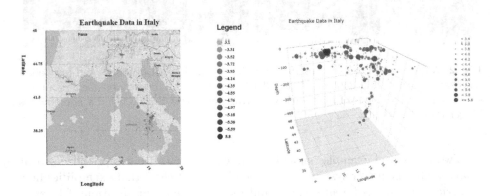

Fig. 4. The Operator automatically removes non-numeric fields from the results and reorders the remaining fields according to histogram analysis to select relevant features for the visualisation, a minus operator has been added manually for the depth field

Additional forms of data and features are being added, such as faces, bodies, contour plots, streaming syntax and additional analytical capabilities. Requirements for query processing of these types of numerical queries will also be addressed in continued work.

5   Conclusions and Contributions

This work proposes an Operator and presents a proof of concept for query-based visualisation methods to support scalable processing of continuous numerical queries over streaming data in engineering analytics applications. Query-based visualisation and analysis primitives are transparently enabled at the query-language level of a data stream management and analysis system. Query processing can exploit metadata of the visualisation and analysis operations as well as conventional meta-data in the database to guide the optimisation of the data processing. This means that engineering analysts can have immediate access to efficient visual analytics capabilities avoiding unnecessary data shipping between systems. Ad-hoc queries can be expressed and visualised in a graphic format suitable for the application such as dynamic contour plots, graphs or trajectories on analytics models such as spatial geometric models.

This work also raises new research questions. Firstly: Exploiting metadata of the visualisation and analysis operations and the conventional meta-data in the database to guide the optimisation of the data processing must be investigated further. Secondly: What methods and algorithms are available for data approximation and reduction (without loosing the relational context in the data stream) to cope with visualisation when the volume of the arriving data is big. The work will also continue with developing new functionality and syntax for visualisation of online streams and more detailed investigation of optimisation will follow for these types of visual and analytical queries.

Acknowledgements. This project is supported by eSSENCE through the Swedish Foundation for Strategic Research, grant RIT08-0041.

Thanks and appreciation to Matteo Magnani who reviewed this paper and provided valuable feedback and insights.

References

1. Evans, P.C., Annunziata, M.: Industrial internet: pushing the boundaries of minds and machines. In: General Electric (2012)
2. Lee, J., Bagheri, B., Kao, H.-A.: A cyber-physical systems architecture for Industry 4.0-based manufacturing systems. Manuf. Lett. **3**, 18–23 (2015)
3. Brettel, M., Friederichsen, N., Keller, M., Rosenberg, M.: How virtualization, decentralization and network building change the manufacturing landscape: an Industry 4.0 perspective. Int. J. Mech. Aerosp. Indust. Mechatron. Manuf. Eng. **8**(1), 37–44 (2014)
4. Kennedy, S.: Made in China 2025, Center for Strategic & International Studies, 1 June 2015. [Online]. http://csis.org/publication/made-china-2025. Accessed 21 Oct 2015
5. Abadi, D., et al.: The Beckman report on database research. ACM SIGMOD Record **43**(3), 61–70 (2014). Also in CACM **59**(2) (2016)
6. Markl, V.: Breaking the chains: on declarative data analysis and data independence in the Big Data era. In: Proceedings of the VLDB Endowment from the 40th International Conference on Very Large Data Bases, Hangzhou, China, vol. 7, no. 13 (2014)
7. Keim, D.A., Mansmann, F., Schneidewind, J., Thomas, J., Ziegler, H.: Visual analytics: scope and challenges. In: Simoff, S.J., Böhlen, M.H., Mazeika, A. (eds.) Visual Data Mining. LNCS, vol. 4404, pp. 76–90. Springer, Heidelberg (2008). https://doi.org/10.1007/978-3-540-71080-6_6
8. Wu, E., Battle, L., Madden, S.R.: The case for data visualization management systems: vision paper. Proc. VLDB Endowm. **7**(10), 903–906 (2014)
9. Melander, L., Orsborn, K., Risch, T.: Visualization of Continuous Queries Using a Visual Data Flow Programming Language. (submitted to DASFAA 2017)
10. Melander, L.: Integrating visual data flow programming with data stream management, PhD thesis, Department of Information Technology, Uppsala University, Uppsala, Sweden (2016). ISBN:978-91-506-2583-7
11. Egenhofer, M.J.: Spatial SQL: a query and presentation language. IEEE Trans. Knowl. Data Eng. **6**(1) (1994)
12. Chan, E.P.F., Wong, J.M.T.: Querying and visualization of geometric data. In: Proceedings of the Fourth ACM Workshop on Advances in Geographic Information Systems, Rockville, pp. 129–138 (1996)
13. Kurc, T., Chang, C., Ferreira, R., Sussman, A., Saltz, J.: Querying very large multi-dimensional datasets in ADR, conference on high performance networking and computing. In: Proceedings of the 1999 ACM/IEEE conference on Supercomputing, Portland, 13–19 November 1999

14. Pourabbas, E., Rafanelli M.: PQL: an extended pictorial query language for querying geographical databases using positional and OLAP operators. In: Proceedings of the 7th International Symposium on Advances in Geographic Information Systems (ACM-GIS 1999), Kansas City, 2–6 November 1999
15. Keahey, T.A., McCormick, P.S., Ahrens, J.P., Keahey, K.: Qviz: a framework for querying and visualizing data. In: Erbacher, R.F., Chen, P.C., Roberts, J.C., Wittenbrink, C.M., Groehn, M. (eds.) Proceedings of SPIE: Visual Data Exploration and Analysis VIII, vol. 4302, pp. 259–267 (2001)
16. Battle, L., Stonebraker, M., Chang, R.: Dynamic reduction of query result sets for interactive visualization. IEEE BigData (2013)
17. Liu, Z., Jiang, B., Heer, J.: imMens: real-time visual querying of Big Data. In: Eurographics Conference on Visualization (EuroVis) 2013, vol. 32(203), no. 3 (2013)
18. Vartak, M., Madden, S., Parameswaran, A., Polyzotis, N.: SEEDB: automatically generating query visualizations. In: Proceedings of the VLDB Endowment from the 40th International Conference on Very Large Data Bases, Hangzhou, vol. 7, no. 13 (2014)
19. Traub, J., Steenbergen, N., Drulich, P.M., Rabl, T., Markl, V.: I^2: interactive real-time visualization for streaming data. In: Proceedings of 20th International Conference on Extending Database Technology, Venice (2017)
20. Industrial Digitalization in China. http://www.gcis.com.cn/china-insights-en/industry-articles-en/238-industrial-digitalization-in-china. Accessed Mar 2022
21. Thunman, M.: How edge analytics can help manufacturers overcome obstacles associated with more equipment data (2021). https://www.automation.com/en-us/articles/may-2021/how-edge-analytics-help-manufacturers-data
22. Dasgupta, A., Arendt, D.L., Franklin, L.R., Wong, P.C., Cook, K.A.: Human factors in streaming data analysis: challenges and opportunities for information visualization. Comput. Graph. Forum **37**, 254–272 (2018). https://doi.org/10.1111/cgf.13264
23. Kaur, J.: Real-time streaming data visualizations (2021). https://www.xenonstack.com/blog/streaming-data-visualizations
24. What is Streaming Analytics: Stream Processing, Data Streaming, and Real-time Analytics (2020). https://www.altexsoft.com/blog/real-time-analytics/
25. Baker, P.: The best data visualization tools for 2020 (2019). https://uk.pcmag.com/cloud-services/83744/the-best-data-visualization-tools-for-2020
26. Lutu, P.E.N.: The use of histogram analysis to support fast selection of predictive features for data stream mining. In: 2018 International Conference on Advances in Big Data, Computing and Data Communication Systems (icABCD), pp. 1–6 (2018). https://doi.org/10.1109/ICABCD.2018.8465411
27. Casillas, J., Wang, S., Yao, X.: Concept drift detection in histogram-based straightforward data stream prediction. In: 2018 IEEE International Conference on Data Mining Workshops (ICDMW), pp. 878–885 (2018). https://doi.org/10.1109/ICDMW.2018.00129

ADBIS Short Papers: Fairness in Data processing

Clustering-Based Subgroup Detection for Automated Fairness Analysis

Jero Schäfer[✉][iD] and Lena Wiese[iD]

Institute of Computer Science, Goethe University, Frankfurt am Main, Germany
{jeschaef,lwiese}@cs.uni-frankfurt.de

Abstract. Fairness in Artificial Intelligence is a major requirement for trust in ML-supported decision making. Up to now fairness analysis depends on human interaction – for example the specification of relevant attributes to consider. In this paper we propose a subgroup detection method based on clustering to automate this process. We analyse 10 (sub-)clustering approaches with three fairness metrics on three datasets and identify SLINK as an optimal candidate for subgroup detection.

Keywords: Artificial intelligence · Fairness · Clustering

1 Introduction

Nowadays a great variety of AI systems are spread over the digital world and affect the lives of millions of people every day. Over the past years there has been a strive for optimizing performance of such systems by better and faster technologies – but modern ML requires for other inalienable objectives, too. The societal impact of decisions of AI systems has to be considered along the objective of maximizing the prediction accuracy. As a consequence, ML models should also be checked carefully for providing equal treatment of individuals from different ethnics, races, or sexes. Especially, the intersections of sensitive characteristics – or those characteristics not obviously involved in discrimination – make judging the model behavior challenging when facing complex data. Additionally, the huge number of possible subgroups growing exponentially with the number of features inside data makes it infeasible to test the model's behavior towards each subgroup. Thus, automation of fairness testing is required to solve this issue.

We propose two subgroup detection methods based on an unsupervised clustering. The computed clusters serve as subgroups for the fairness evaluation and prototypes for the generation of patterns defining subgroups. Furthermore, we compare different clustering algorithms on their performance to identify subgroups for a fairness assessment of a binary classifier under three common fairness criteria and three fairness-related datasets. In Sect. 2 we give an overview over related work on automated subgroup fairness and Sect. 3 introduces the theory for our methods of subgroup detection, that are explained in Sect. 4. Section 5 describes our experimental setting and discusses the results. Finally, we the summarize key findings and give an outlook into future work in Sect. 6.

© Springer Nature Switzerland AG 2022
S. Chiusano et al. (Eds.): ADBIS 2022, CCIS 1652, pp. 45–55, 2022.
https://doi.org/10.1007/978-3-031-15743-1_5

2   Related Work

There has been an uprising number of tools being developed to aid data scientists and developers with the investigation and improvement of their developed ML models. They usually support the model selection or optimization phase by diverse visualizations of data and the model performance. Recently, there evolved an advance towards the assessment of model fairness to meet the demands of society for equality in AI. However, many tools require export knowledge as they partly rely on user interaction via controls or parameters.

The tools Boxer [5] and Fairkit [7] let the user interactively explore and compare the behavior of multiple models. They opt for the identification of intersectional bias in the model but require the selection of subgroups to investigate for discrimination. The What-If tool [13] yields insights into local and global modal behavior in various scenarios, performs an intersectional analysis for a chosen fairness objective and automatically adapts the model's classification threshold. The framework of Morina et al. [9] comprises a suite of metrics for evaluating and estimating intersectional fairness but also does not discover subgroups automatically. The FairVis tool [2] was designed to identify the intersectional bias of ML models by visualizations. Despite possible user interaction, automatically generated subgroups are suggested and the subgroup performance and fairness of the model are presented. These subgroups are found by a k-means clustering and extracting patterns that describe the makeup of the cluster members. The dominant features in a cluster are ranked by the feature entropy quantifying the feature's uniformity. Our approach uses a similar technique that directly uses the clustering results and the feature entropy with a threshold instead of a ranking.

In contrast, the Divexplorer project [10] provides automatic subgroup detection by frequent-pattern mining. An exhaustive search through possible itemsets (i.e. patterns to match data instances to) is carried out and only itemsets above a support threshold are considered while pruning others from the search tree. The divergence of the model behavior between a subgroup of instances complying to a mined pattern and the full dataset is assessed by an outcome function for classification or ranking tasks that evaluates fairness by the difference between the FPR or FNR of a subgroup and the global rate. The DENOUNCER [8] system discovers subgroups with a low prediction accuracy by pattern graph traversal, applies a support threshold to the attribute-value patterns filtering out insignificant patterns and prunes for the most general patterns.

3   AI Fairness

Generally, one can distinguish *individual* (similarity-based) and *group* (statistical) fairness criteria. Individual fairness refers to the discrimination by the model on an individual level (per instance) and is expressed as the different behavior of the model wrt. similar individuals although they should be treated similarly. This work is focused on the assessment of group fairness of a given classification model that tests whether the model systematically discriminates against a certain subgroup of instances [6]. The subgroups of interest are therefore usually defined for a set of protected attributes (such as sex or nationality) but generally involve the intersection of multiple protected attributes. Hence we formalize a dataset as $\mathcal{D} = \{x_1, \ldots, x_n\}$ with the set of attributes $\mathcal{A} = \{A_1, \ldots, A_p\}$ that comprises n instances $x_i, \forall i \in \{1, \ldots, n\}$. The active domain of an attribute $A_j \in \mathcal{A}$ is then denoted as $Dom(A_j)$ and describes all the possible values for the feature A_j. The active domain $Dom(\mathcal{D})$ is then the cartesian product of all its attributes' active domains. Thus, each instance $x \in \mathcal{D}$ is from the active domain $Dom(\mathcal{D})$ and we write the value of x for attribute $A_j \in \mathcal{A}$ as $x(A_j)$. To define metrics for measuring group fairness, we first introduce protected attributes and patterns to match instances to certain groups similar to the definitions in [8, 10].

Definition 1. *Pattern.* *Let $\mathcal{D}$ a dataset and $A = \{A_1, \ldots, A_q\} \subseteq \mathcal{A}$ a nonempty subset of the dataset attributes. Then, a tuple of attribute values $P = (a_1, \ldots, a_q) \in Dom(A)$ is a pattern over dataset $\mathcal{D}$. An instance $x \in Dom(\mathcal{D})$ satisfies such a pattern P if the respective attribute values of x match the attribute values of P: If $\forall A_j \in A : x(A_j) = a_j$, then $x \vDash P$*

A classification model $\hat{M}$ can be trained on a dataset $\mathcal{D}$ labeled by the true classes $\mathcal{Y} = \{y_1, \ldots, y_z\}$ by $\mathcal{M} : \mathcal{D} \mapsto \mathcal{Y}$ to learn predicting the class $\hat{y} = \hat{M}(x)$ of any new input instance x. The model $\hat{M}$ serves as an approximation of the real mapping $M : Dom(\mathcal{D}) \mapsto \mathcal{Y}$ on $Dom(D)$. We also call $\hat{M}$ the predictor and in the following assume a binary classification model, i.e. $\mathcal{Y} = \{0, 1\}$, where $y = 1$ corresponds to the positive or favorable class label and $y = 0$ to the negative or unfavorable class label. In case of a score $\hat{M}(x) \in [0, 1]$ estimating the probability of an instance to belong to the favorable class a t-threshold rule [3] rule can be used to discretize the prediction as $\hat{y} = 1$ if $\hat{M}(x) \geq t$ or $\hat{y} = 0$ otherwise.

A pattern $P = (a_1, \ldots, a_q)$ partitions dataset $\mathcal{D}$ into two disjoint subgroups based on the protected attribute values. This partitions are the *protected* (P satisfied) and *unprotected* (P not satisfied) subgroups $\mathcal{D}_P = \{x \in \mathcal{D} \mid x \vDash P\}$ and $\mathcal{D}_{\bar{P}} = \{x \in \mathcal{D} \mid x \nvDash P\} = \mathcal{D} \setminus \mathcal{D}_P$, respectively. The different behavior regarding the prediction of the favorable or unfavorable class label on the subgroups by a classification model $\hat{M}$ is tested for fairness violations. The probabilities of a model $\hat{M}$ to predict the positive or negative class label are denoted as $\mathbb{P}(\hat{y} = 1)$ and $\mathbb{P}(\hat{y} = 0)$, respectively. Given a pattern P over dataset $\mathcal{D}$ the probability for instances from the protected subgroup to be predicted the class label $c \in \{0, 1\}$ is written as $\mathbb{P}(\hat{y} = c \mid x \in \mathcal{D}_P)$. Furthermore, the probability for a correct or wrong prediction of class c given the groundtruth class label g and one of the subgroups is expressed by $\mathbb{P}(\hat{y} = c \mid y = g, x \in \mathcal{D}_P)$. Notations for the probabilities of true classes (according to mapping M) and the unprotected group are analogous.

3.1 Subgroup Fairness Metrics

There exist various subgroup fairness metrics that mostly rely on the rates computed from confusion matrices [6,12] such as the positive predictive value (PPV) or TPR to estimate the chances for $\mathcal{D}_P$ and $\mathcal{D}_{\bar{P}}$. We do not focus on any specific group fairness metric but consider multiple of them as there are various, sometimes opposing opinions on the justification of equal treatment. Instead, we define in the following three common fairness criteria that we will use in our evaluation to ensure a broad comparison.

Statistical parity (Definition 2) is a fairness definition based on the predicted outcome $\hat{y} = \hat{M}(x)$ and is satisfied if $\mathcal{D}_P$ has the same probability of getting a positive prediction ($\hat{y} = 1$) from the model as $\mathcal{D}_{\bar{P}}$ [12]. A fair classifier predicts the favorable label with a probability independent from the protected attribute values but the bias against instances belonging to multiple protected groups might be magnified [11]. Subgroup fairness based on equal opportunity (Definition 3) is achieved if the TPR of $\mathcal{D}_P$ is equal to the TPR of $\mathcal{D}_{\bar{P}}$. Regardless of their subgroup membership, the chance for each individual $x \in \mathcal{D}$ to get a positive prediction if they actually belong to the favorable class should be the same. As a consequence of Eq. 2, every individual from the subgroup should have the same probability of being assigned the unfavorable class label if they actually belong to the favorable class. The equalized odds subgroup fairness (Definition 4) is a generalization of equal opportunity as it requires the equality of the TPRs and FPRs of the both subgroups. In addition to the equal chance of getting a correct positive prediction also the chance of being incorrectly assigned the favorable class label has to be equal between the protected and unprotected subgroups.

Definition 2. *Statistical parity.* *Let $\mathcal{D}$ a dataset. A classifier $\hat{M}$ satisfies statistical parity wrt. a pattern P over $\mathcal{D}$ if:*

$$\mathbb{P}(\hat{y} = 1 \mid x \in \mathcal{D}_P) = \mathbb{P}(\hat{y} = 1 \mid x \in \mathcal{D}_{\bar{P}}) \tag{1}$$

Definition 3. *Equal opportunity.* *Let $\mathcal{D}$ a dataset and M the groundtruth mapping. A classifier $\hat{M}$ satisfies equal opportunity wrt. a pattern P over $\mathcal{D}$ if*

$$\mathbb{P}(\hat{y} = 1 \mid y = 1,\ x \in \mathcal{D}_P) = \mathbb{P}(\hat{y} = 1 \mid y = 1,\ x \in \mathcal{D}_{\bar{P}}). \tag{2}$$

Definition 4. *Equalized odds.* *Let $\mathcal{D}$ a dataset, M the groundtruth mapping and $g \in \{0, 1\}$. A classifier $\hat{M}$ satisfies equalized odds wrt. pattern P over $\mathcal{D}$ if*

$$\mathbb{P}(\hat{y} = 1 \mid y = g,\ x \in \mathcal{D}_P) = \mathbb{P}(\hat{y} = 1 \mid y = g,\ x \in \mathcal{D}_{\bar{P}}) \tag{3}$$

Commonly, the strict equality of fairness definitions is relaxed to accept also similar chances for predictions by $\hat{M}$ as fair, e.g., by ϵ-differential fairness definitions [4,9]. We prefer a simpler relaxation as provided by the "AI Fairness 360" toolkit [1] that relies on the difference between the probabilities for $\mathcal{D}_P$ and $\mathcal{D}_{\bar{P}}$ (Table 1). The fairness of $\hat{M}$ wrt. stat. parity and eq. opportunity is calculated as the difference between the probabilities given $x \in \mathcal{D}_{\bar{P}}$ or $x \in \mathcal{D}_P$. As equalized odds (Eq. 3) requires the equality of two probabilities, the average of the probability differences denoted as F_{aod} in Table 1 is calculated. Each metric $F \in \{F_{spd}, F_{eod}, F_{aod}\}$ yields a value in $[-1, 1]$. If $F = 0$, the evaluated classifier $\hat{M}$ is considered perfectly fair wrt. P and the fairness definition as the confusion matrix rates to estimate the probabilities coincide for $\mathcal{D}_P$ and $\mathcal{D}_{\bar{P}}$ (i.e., under the same conditions the prediction is independent of the membership in $\mathcal{D}_P$ or $\mathcal{D}_{\bar{P}}$). A value $F > 0$ corresponds to discrimination against individuals in $\mathcal{D}_P$ or favoritism of individuals in $\mathcal{D}_{\bar{P}}$ by $\hat{M}$ and $F < 0$ indicates the opposite.

Table 1. Subgroup fairness metrics

Definition	Fairness metric
Statistical parity	$F_{spd} = \mathbb{P}(\hat{y} = 1 \mid x \in \mathcal{D}_{\bar{P}}) - \mathbb{P}(\hat{y} = 1 \mid x \in \mathcal{D}_P)$
Eq. opportunity	$F_{eod} = \mathbb{P}(\hat{y} = 1 \mid y = 0, x \in \mathcal{D}_{\bar{P}}) - \mathbb{P}(\hat{y} = 1 \mid y = 0, x \in \mathcal{D}_P)$
Equalized odds	$F_{aod} = \frac{1}{2}\big[\mathbb{P}(\hat{y} = 1 \mid y = 0, x \in \mathcal{D}_{\bar{P}}) - \mathbb{P}(\hat{y} = 1 \mid y = 0, x \in \mathcal{D}_P)$
	$\qquad + \mathbb{P}(\hat{y} = 1 \mid y = 1, x \in \mathcal{D}_{\bar{P}}) - \mathbb{P}(\hat{y} = 1 \mid y = 1, x \in \mathcal{D}_P)\big]$

4   Automatic Subgroup Detection

Assigning instances of a dataset to meaningful groups that mirror high similarities between the instances is challenging. Clustering algorithms provide unsupervised techniques to compute such a grouping $\mathcal{C}$, called *clustering*, that assigns the instances $x \in \mathcal{D}$ to clusters $C_1, \ldots, C_k$. Each pair $x, y \in C_i$ for $i \in \{1, \ldots, k\}$ shares some similarity as defined by the type of clustering, the algorithm parameters and the similarity/distance measure. For example, a centroid-based clustering expresses similarity by cluster membership wrt. the proximity to the computed centroids representing the clusters and a density-based clustering distinguishes dense regions of instances, that are considered the clusters, from the sparse regions, which are marked as containing outliers. With the notion of a clustering, we automatically evaluate the fairness of a classification model with our previous fairness metrics in two ways. A clustering $\mathcal{C}$ specifies a set of clustering-based patterns $P^{\mathcal{C}}$ (Definition 5) defining $\mathcal{D}_P$ and $\mathcal{D}_{\bar{P}}$ according to the established clusters. The fairness of a classifier $\hat{M}$ can then be evaluated for the subgroups of instances that comply to $\mathcal{C}$. To this end, the cluster labels of the instances $x \in \mathcal{D}$ are added as an artificial attribute $A_{\mathcal{C}}$ to $\mathcal{D}$. For each pattern $r = P_i^{\mathcal{C}} \in P^{\mathcal{C}}$ we can assess the fairness of $\hat{M}$ with any of the mentioned fairness metrics by comparing the treatment of the clustering-based protected and unprotected subgroup $\mathcal{D}_r = \{x \in \mathcal{D} \mid x \vDash P_i^{\mathcal{C}}\}$ and $\mathcal{D}_{\bar{r}} = \{x \in \mathcal{D} \mid x \nvDash P_i^{\mathcal{C}}\}$, respectively. The metric values are aggregated over all pairs of subgroups $\mathcal{D}_r$ and $\mathcal{D}_{\bar{r}}$ as defined by the patterns (i.e. over all clusters).

Definition 5. *Clustering-based pattern.* Let $\mathcal{C} = \{C_1, \ldots, C_k\}$ *a clustering of dataset* $\mathcal{D}$ *with an attribute set* $\mathcal{A} \cup \{A_{\mathcal{C}}\}$ *that was extended by the attribute* $A_{\mathcal{C}}$ *of the clustering labels of* $\mathcal{C}$. *We call* $P_i^{\mathcal{C}} = (i)$ *a clustering-based pattern over* $\mathcal{D}$ *and denote the set of all clustering-based patterns over* $\mathcal{D}$ *as* $P^{\mathcal{C}} = \{P_i^{\mathcal{C}}\}_{i=1}^k$. *An instance* $x \in \mathcal{D}$ *satisfies* $P_i^{\mathcal{C}}$ *if* x *belongs to cluster* $C_i \in \mathcal{C}$.

Furthermore, our system extracts more general patterns from $\mathcal{C}$ to perform the subgroup fairness analysis. We use the cluster feature entropy [2] to identify dominant features in $C_1, \ldots, C_k$ from which patterns are extracted. The cluster feature entropy $H_{i,j}$ quantifies the distribution of values for attribute A_j in cluster C_i and is calculated for each cluster and feature separately. An entropy value $H_{i,j}$ close to zero indicates a single dominant value at attribute A_j in C_i

whereas high values indicate a frequent occurrence of multiple values. A uniform distribution of all values across the cluster has maximal entropy.

Definition 6. Normalized feature entropy. Let $C = \{C_1, \ldots, C_k\}$ a clustering of dataset $\mathcal{D}$ with attributes $\mathcal{A} = \{A_1, \ldots, A_p\}$. We define the normalized cluster feature entropy for cluster $C_i \in C$ and feature $A_j \in \mathcal{A}$ where $N_i = |C_i|$ and $N_{i,j,v} = |\{x \in C_i \mid x(A_j) = v\}|$ as

$$H_{i,j} = -\frac{1}{\log_2 |Dom(A_j)|} \cdot \sum_{v \in Dom(A_j)} \frac{N_{i,j,v}}{N_i} \cdot \log_2 \left(\frac{N_{i,j,v}}{N_i} \right) \tag{4}$$

However, it is impossible to define an appropriate global entropy threshold t when using the definition of Cabrera et al. [2] as it does not account for the varying sizes of the active domains $\mathcal{A}$. As a consequence, it often fails to classify non-dominant features with larger active domains as such in clusters without a clear dominant value but potentially multiple of them when t was tuned for smaller active domains. To improve their definition, we also normalize the entropy by the logarithm $\log_2(|Dom(A_j)|)$ of the number of possible values for feature A_j in Definition 6. This ensures entropy values between 0 and 1 such that t can be picked independently of the size of the active domain of a feature.

For example, consider the three value distributions (frequencies) $a = [0.025, 0.025, 0.025, 0.025, 0.025, 0.025, 0.025, 0.025, 0.8]$, $b = [0.1, 0.1, 0.1, 0.1, 0.6]$ and $c = [0.25, 0.25, 0.5]$, respectively. The first distribution a clearly shows a dominant feature in the cluster (80%), b also expresses a single value making up most of the feature but less significantly (60%), and c represents a scenario where still one value occurs more often than others but the feature is not really dominant. The feature entropies are $H_a \approx 1.32$, $H_b \approx 1.77$ and $H_c = 1.5$, respectively. Other than expected, the H_b does not reflect a dominant feature and, in contrast, H_c is lower although the distribution c does not dominate the feature as much as b. Instead of this misleading values, one can normalize them to obtain $\frac{H_a}{\log_2 9} \approx 0.42$, $\frac{H_b}{\log_2 5} \approx 0.76$ and $\frac{H_c}{\log_2 3} \approx 0.95$ accounting for the feature diversity.

Definition 7. Entropy-based pattern. Let $C = \{C_1, \ldots, C_k\}$ a clustering of dataset $\mathcal{D}$ with attribute set $\mathcal{A} = \{A_1, \ldots, A_p\}$ and threshold $t \geq 0$. We then define an entropy-based pattern over $\mathcal{D}$ as $P_i^t = (a_1, \ldots, a_q) \in Dom(A^i)$ where $A^i = \{A_j \in \mathcal{A} \mid H_{i,j} < t\}$ contains the dominant features of cluster $C_i \in C$ and $a_j \in Dom(A_j)$ is the most frequent or dominant value of $A_j \in A^i$ in C_i. The set of entropy-based patterns for all clusters of C, $\{P_i^t\}_{i=1}^k$, is denoted as P^t.

From all dominant features of a cluster, one pattern is created for each cluster and used for the fairness metric calculation: all features A_j with $H_{i,j} \leq t$ are collected in a subset of attributes $A^i = \{A_j \in \mathcal{A} \mid H_{i,j} < t\}$ for each $C_i \in C$. An entropy-based pattern P_i^t (Definition 7) is then derived from each C_i comprising the most frequent value of each dominant attribute $A_j \in A^i$. In contrast to a clustering-based pattern P_i^C, that is satisfied by an instance $x \in \mathcal{D}$ if x belongs to C_i, an entropy-based pattern P_i^t is satisfied by x according to Definition 1. Hence, an entropy-based pattern is evaluated wrt. specific attribute values as determined by $\mathcal{C}$ and t whereas clustering-based patterns are evaluated value-agnostic and specific attribute values are considered only implicitly via the computation of $\mathcal{C}$. For a hard partitional clustering, where each instance belongs to exactly one of the pairwise disjoint clusters, it holds that $\mathcal{D}_{P_i} \cap \mathcal{D}_{P_j} = \emptyset$ for each pair of clustering-based patterns $P_i, P_j \in P^{\mathcal{C}}$. However, two entropy-based patterns P_i^t and P_j^t extracted from different clusters $C_i, C_j \in \mathcal{C}$ might coincide as the clusters might have the same dominant features and values, i.e., $A^i = A^j$ and $P_i^t = P_j^t$. For our evaluation, we remove the duplicate entropy-based patterns before the subgroup fairness assessment and report the duplication rate.

5   Results

We evaluated our system on three fairness-related datasets: ProPublica's COM-PAS dataset[1], the South German Credit dataset[2] and the Medical Expenditure Panel Survey[3] dataset of the year 2015 (panel 19). We refer to them as COM-PAS, Credit and MEPS19, respectively. The COMPAS data ($n = 6172$, $p = 7$) was taken as provided in the FairVis [2] repository incl. predictions. We trained LightGBM classifiers (gradient boosting decision tree) for the Credit ($n = 1000$, $p = 20$) and MEPS dataset ($n = 15830$, $p = 40$). Our system first preprocessed the data by encoding categorical features and min-max normalizing before clustering. To analyze the suitability of clustering models for our subgroup detection task, we compared the proposed method for the following (subspace) clustering techniques of different types: k-Means, DBSCAN, OPTICS, Spectral Clustering, SLINK, Ward, BIRCH, SSC-BP, SSC-OMP, and EnSC. We tested a small set of parameter values individually on each algorithm and dataset. The fairness metrics producing the best combination of fairness violation indication and clustering performance was then reported. We selected the run that maximizes the product of silhouette score S_C and mean absolute error of the clustering-induced subgroup prediction accuracy $Acc_{\hat{M}}(\mathcal{D}_{P_i^c})$, $i = 1, \ldots, k$, as compared to the global accuracy $Acc_{\hat{M}}(\mathcal{D})$: $\arg\max\limits_{C_i \in \mathcal{C}} \left(S_C \cdot \frac{1}{|\mathcal{C}|} \cdot \sum_{P_i^c \in P^C} \left| Acc_{\hat{M}}(\mathcal{D}_{P_i^c}) - Acc_{\hat{M}}(\mathcal{D}) \right| \right)$

The experimental results are shown in Table 2, 3 and 4. Each table represents a dataset and both subgroup detection methods with one row for the best run. The columns display the mean (Avg), standard deviation (Std) and absolute mean (Abs) values across all clustering- or entropy-induced subgroups as measured by the fairness metrics. The clustering model with the highest absolute mean is highlighted for each fairness criterion by bold numbers for the three reported quantities. On the COMPAS dataset (Table 2) the best results were obtained for SLINK throughout all fairness metrics and both subgroup detection methods. Especially the entropy-based subgroups detected by SLINK clearly outperformed in the absolute mean of the metric values when compared to the

---

[1] https://www.propublica.org/article/machine-bias-risk-assessments-in-criminal-sentencing.

[2] https://archive.ics.uci.edu/ml/datasets/South+German+Credit+%28UPDATE%29.

[3] https://meps.ahrq.gov/mepsweb/data_stats/download_data_files_detail.jsp?cboPufNumber=HC-183.

Table 2. Clustering- (top) & Entropy-based (bottom) subgroup fairness COMPAS

Algorithm	Statistical parity			Equal opportunity			Equalized odds		
	Avg	Std	Abs	Avg	Std	Abs	Avg	Std	Abs
k-Means	−0.0074	0.3017	0.2513	0.0592	0.3100	0.2454	−0.0071	0.2877	0.2315
DBSCAN	−0.1446	0.1914	0.1992	−0.0880	0.1516	0.1491	−0.1242	0.1763	0.1733
OPTICS	−0.0592	0.1778	0.1580	0.0102	0.1790	0.1439	−0.0466	0.1651	0.1415
Spectral	0.0328	0.3558	0.3081	0.1388	0.3755	0.3198	0.0254	0.3305	0.2749
SLINK	0.0343	**0.4057**	**0.3580**	**0.2474**	**0.4373**	**0.4217**	**0.0767**	**0.3457**	**0.2821**
Ward	−0.0376	0.2711	0.2233	0.0163	0.2487	0.1859	−0.0316	0.2460	0.1984
BIRCH	−0.1032	0.1744	0.1710	−0.0506	0.1376	0.1274	−0.0810	0.1449	0.1411
SSC-OMP	−0.0160	0.2115	0.1725	0.0205	0.1679	0.1386	−0.0167	0.1802	0.1460
SSC-BP	−0.0993	0.2078	0.1949	−0.0490	0.1699	0.1467	−0.0851	0.1863	0.1712
EnSC	−0.1356	0.1978	0.2018	−0.0798	0.1538	0.1470	−0.1172	0.1808	0.1765
k-Means	−0.0985	0.2323	0.2022	−0.0459	0.2076	0.1700	−0.0918	0.2269	0.1891
DBSCAN	−0.2115	0.1860	0.2265	−0.1332	0.1417	0.1630	−0.1985	0.1875	0.2143
OPTICS	−0.1517	0.1830	0.1850	−0.0842	0.1509	0.1498	−0.1263	0.1737	0.1580
Spectral	−0.0601	0.2997	0.2496	0.0290	0.3288	0.2477	−0.0581	0.2728	0.2188
SLINK	**−0.0158**	**0.4009**	**0.3487**	**0.1725**	**0.4312**	**0.3761**	**0.0138**	**0.3524**	**0.2783**
Ward	−0.1211	0.2274	0.2081	−0.0748	0.1620	0.1517	−0.1190	0.2148	0.1920
BIRCH	−0.1860	0.1854	0.2069	−0.1010	0.1417	0.1423	−0.1665	0.1794	0.1885
SSC-OMP	−0.0781	0.1920	0.1664	−0.0305	0.1525	0.1295	−0.0724	0.1677	0.1456
SSC-BP	−0.0993	0.2128	0.1861	−0.0566	0.1502	0.1249	−0.0914	0.1968	0.1629
EnSC	−0.1839	0.1842	0.2117	−0.1167	0.1358	0.1500	−0.1713	0.1795	0.1959

Table 3. Clustering- (top) & Entropy-based (bottom) subgroup fairness credit

Algorithm	Statistical Parity			Equal Opportunity			Equalized Odds		
	Avg	Std	Abs	Avg	Std	Abs	Avg	Std	Abs
k-Means	−0.0096	0.2210	0.2062	0.0120	0.0738	0.0628	−0.0665	0.1838	0.1556
DBSCAN	−0.0059	0.2911	0.2097	0.0578	0.1562	0.1000	−0.0876	0.2957	0.1928
OPTICS	−0.0051	0.2616	0.2029	0.0253	0.0524	0.0317	0.0035	0.1379	0.0988
Spectral	−0.0323	0.3207	0.2480	0.0761	0.3091	0.1340	−0.1124	0.3216	0.2560
SLINK	**−0.0120**	**0.4338**	**0.3123**	**0.3691**	**0.5306**	**0.4103**	**0.0006**	**0.4249**	**0.3332**
Ward	−0.0006	0.1744	0.1442	0.0055	0.0980	0.0655	−0.0320	0.1569	0.1232
BIRCH	−0.0002	0.1787	0.1509	0.0051	0.0967	0.0632	−0.0361	0.1580	0.1291
SSC-OMP	0.0032	0.0461	0.0375	0.0033	0.0277	0.0232	−0.0011	0.0279	0.0229
SSC-BP	−0.0117	0.1286	0.0898	−0.0003	0.0428	0.0299	−0.0097	0.0514	0.0359
EnSC	−0.0217	0.1377	0.1070	−0.0034	0.0298	0.0227	−0.0239	0.0824	0.0677
k-Means	−0.0558	0.2130	0.1940	−0.0094	0.0591	0.0469	−0.0969	0.2065	0.1694
DBSCAN	−0.0171	0.2516	0.1717	0.0907	0.2258	0.1434	−0.0967	0.2829	0.1814
OPTICS	−0.0878	0.3209	0.2763	−0.0122	0.0694	0.0575	**−0.2370**	**0.3103**	**0.3179**
Spectral	−0.0504	0.3193	0.2469	0.0631	0.3119	0.1282	−0.1268	0.3190	0.2617
SLINK	**−0.0108**	**0.4339**	**0.3135**	**0.3605**	**0.5374**	**0.4017**	−0.0293	0.4155	0.3033
Ward	−0.0164	0.1915	0.1682	0.0080	0.1180	0.0786	−0.0845	0.2022	0.1661
BIRCH	−0.0354	0.2032	0.1775	0.0023	0.1177	0.0754	−0.0989	0.2122	0.1783
SSC-OMP	0.0407	0.0612	0.0499	0.0048	0.0173	0.0151	0.0223	0.0324	0.0240
SSC-BP	−0.0307	0.1802	0.1326	−0.0010	0.0352	0.0253	−0.0824	0.2285	0.1469
EnSC	−0.0407	0.1384	0.0857	−0.0001	0.0333	0.0232	−0.0684	0.1997	0.1152

other clustering results with absolute mean values from $\approx$ 0.28–0.42. The eq. opportunity mean values revealed a skew towards the detection of mainly discriminated subgroups or subgroups with a higher degree of discrimination than the degree of favorization for the other subgroups by SLINK in both detection methods. The other fairness metrics were balanced between discriminated and favored subgroups as indicated by the mean metric values close to zero. The subspace clustering algorithms showed no improvement over the conventional clustering algorithms and the spectral clustering also performed quite good on the COMPAS dataset. The duplication rate of the entropy-induced subgroups was between 0 and 0.5.

Table 3 displays again outstanding results (Abs $\approx$ 0.30–0.41) of SLINK across both detection methods and all fairness criteria. Only OPTICS achieved a higher absolute mean value of 0.3179 under eq. odds and entropy-induced subgroups on the Credit dataset. For the SLINK clustering, again we observed an even more significant skew towards the detection of discriminated subgroups over favored ones for the eq. opportunity criterion. The OPTICS clustering, in contrast, showed a skew towards the favored subgroups with an average eq. odds value of –0.2370 across the clustering-induced subgroups. The spectral clustering also performed well as measured by stat. parity and eq. odds for both detection methods. For the Credit dataset the observed performance of the subspace clustering algorithms was weaker than the other algorithms. We observed a duplication rate of 0 for all reported trials except for the SCC-OMP model (single duplication).

The MEPS19 dataset (Table 4) yielded more variety regarding the best performance. We observed for both detection methods the best performance in stat. parity for SLINK ($\approx$ 0.80), in eq. opportunity for OPTICS ($\approx$ 0.58) and in eq. odds for spectral clustering ($\approx$ 0.60 and 0.58). The spectral clustering produced similarly good results as the SLINK clustering for each of the detection methods. In contrast to the other two datasets, our experiments showed more shift towards favored or discriminated subgroups as detected by any of the computed clusterings for the MEPS19 dataset. Furthermore, we observed (nearly) equality of the absolute value of the mean and the absolute mean metric value for selected models. This indicates that few or no favored subgroups were detected by the clustering algorithm if the skew occurred towards discriminated subgroups and vice versa. This behavior was often observed for the subspace clustering algorithms, that again did not perform differently than the conventional algorithms. Only for SSC-BP, we observed a duplication rate of 0.4 whereas the other algorithms had maximally one collision on the entropy-induced subgroups.

Table 4. Clustering- (top) & Entropy-based (bottom) subgroup fairness MEPS19

Algorithm	Statistical parity			Equal opportunity			Equalized odds		
	Avg	Std	Abs	Avg	Std	Abs	Avg	Std	Abs
k-Means	−0.0279	0.3180	0.2050	0.2343	0.3675	0.4002	0.0862	0.3058	0.2630
DBSCAN	−0.0566	0.4154	0.3009	0.5054	0.0395	0.5054	0.1176	0.2836	0.2573
OPTICS	0.1323	0.0002	0.1323	**0.5806**	**0.0118**	**0.5806**	0.3097	0.0058	0.3097
Spectral	−0.6962	0.4504	0.7958	−0.2681	0.4133	0.4830	**−0.5098**	**0.3938**	**0.6090**
SLINK	**−0.7121**	**0.4276**	**0.8031**	−0.1229	0.4810	0.4834	−0.4814	0.3629	0.5461
Ward	0.0029	0.2636	0.1810	0.2988	0.3222	0.4111	0.1311	0.2565	0.2549
BIRCH	−0.0336	0.2998	0.2068	0.2503	0.3432	0.4003	0.0930	0.2835	0.2625
SSC-OMP	0.1175	0.0585	0.1306	0.4805	0.2388	0.5339	0.2613	0.1299	0.2904
SSC-BP	0.1174	0.0623	0.1321	0.4805	0.2415	0.5350	0.2613	0.1321	0.2913
EnSC	0.1127	0.0662	0.1293	0.4429	0.2386	0.4927	0.2419	0.1311	0.2700
k-Means	0.0082	0.3082	0.2031	0.3665	0.3484	0.4679	0.1574	0.3043	0.2981
DBSCAN	−0.0866	0.3894	0.2709	0.5370	0.0027	0.5370	0.1264	0.2910	0.2661
OPTICS	0.1322	0.0001	0.1322	**0.5826**	**0.0002**	**0.5826**	0.3106	0.0001	0.3106
Spectral	−0.7765	0.2673	0.7902	−0.2653	0.4100	0.4788	**−0.5528**	**0.2827**	**0.5818**
SLINK	**−0.7906**	**0.2507**	**0.8043**	−0.1163	0.4895	0.4900	−0.5222	0.2675	0.5262
Ward	0.0306	0.2562	0.1889	0.3955	0.3076	0.4796	0.1832	0.2535	0.2930
BIRCH	−0.0344	0.3370	0.2373	0.3123	0.3866	0.4807	0.1148	0.3271	0.3180
SSC-OMP	0.1310	0.0016	0.1310	0.5341	0.0010	0.5341	0.2656	0.1119	0.2866
SSC-BP	0.1408	0.0102	0.1408	0.5400	0.0085	0.5400	0.2571	0.1404	0.2893
EnSC	0.1405	0.0092	0.1405	0.5411	0.0083	0.5411	0.2962	0.0061	0.2962

6   Conclusion and Future Work

In our research we have proposed two techniques to identify subgroups in data
to perform a subgroup fairness analysis on. The experimental results proved
the ability of our clustering- and entropy-based approach to detect subgroups
in datasets on which a given classifier violates common fairness criteria, namely
statistical parity, equal opportunity and equalized odds. We found a strong overall performance when employing the SLINK clustering algorithm in our subgroup detection methods as it identifies subgroups with a high deviation from
what would be considered fair. Future research could investigate on relationships between classification models and the subgroup detection performance or
extend the proposed subgroup detection. We currently work on the integration
of our method into a graphical, web-based tool allowing users to perform an
automatic subgroup fairness analysis on their dataset and classifier in a user-
friendly manner. With the help of the fairness analysis tool, we want to provide
deeper insights into the composition of the detected subgroups straightforward
to users.

References

1. Bellamy, R.K.E., et al.: AI Fairness 360: an extensible toolkit for detecting, understanding, and mitigating unwanted algorithmic bias, Oct 2018. arxiv.org/abs/1810.01943
2. Cabrera, A.A., Epperson, W., Hohman, F., Kahng, M., Morgenstern, J., Chau, D.H.: FAIRVIS: Visual analytics for discovering intersectional bias in machine learning. In: 2019 IEEE Conference on Visual Analytics Science and Technology (VAST), Oct 2019
3. Castelnovo, A., Crupi, R., Greco, G., Regoli, D., Penco, I.G., Cosentini, A.C.: A clarification of the nuances in the fairness metrics landscape. Sci. Rep. **12**(1), 1–21 (2022)
4. Foulds, J.R., Islam, R., Keya, K.N., Pan, S.: An Intersectional Definition of Fairness. In: 2020 IEEE 36th International Conference on Data Engineering (ICDE), pp. 1918–1921. IEEE (2020)
5. Gleicher, M., Barve, A., Yu, X., Heimerl, F.: Boxer: Interactive comparison of classifier results. In: Computer Graphics Forum, vol. 39, pp. 181–193. Wiley Online Library (2020)
6. Hertweck, C., Heitz, C.: A systematic approach to group fairness in automated decision making. In: 2021 8th Swiss Conference on Data Science (SDS), pp. 1–6. IEEE (2021)
7. Johnson, B., Brun, Y.: Fairkit-learn: a fairness evaluation and comparison toolkit. In: 44th International Conference on Software Engineering Companion (ICSE 2022 Companion) (2022)
8. Li, J., Moskovitch, Y., Jagadish, H.: DENOUNCER: detection of unfairness in classifiers. Proc. VLDB Endowm. **14**(12), 2719–2722 (2021)
9. Morina, G., Oliinyk, V., Waton, J., Marusic, I., Georgatzis, K.: Auditing and Achieving Intersectional Fairness in Classification Problems. arXiv preprint arXiv:1911.01468 (2019)
10. Pastor, E., de Alfaro, L., Baralis, E.: Looking for trouble: analyzing classifier behavior via pattern divergence. In: Proceedings of the 2021 International Conference on Management of Data, pp. 1400–1412 (2021)
11. Teodorescu, M.H., Morse, L., Awwad, Y., Kane, G.C.: Failures of fairness in automation require a deeper understanding of human-ML augmentation. MIS Q. **45**(3) (2021)
12. Verma, S., Rubin, J.: Fairness definitions explained. In: 2018 IEEE/ACM International Workshop on Software Fairness (FairWare), pp. 1–7. IEEE (2018)
13. Wexler, J., Pushkarna, M., Bolukbasi, T., Wattenberg, M., Viégas, F., Wilson, J.: The what-if tool: Interactive probing of machine learning models. IEEE Trans. Visual Comput. Graphics **26**(1), 56–65 (2019)

Open Science in the Cloud:
The CloudFAIR Architecture
for FAIR-compliant Repositories

João Pedro C. Castro[1,2(✉)] , Lucas M. F. Romero[1] ,
Anderson Chaves Carniel[3] , and Cristina D. Aguiar[1]

[1] Department of Computer Science, University of São Paulo, São Paulo, Brazil
lucasromero@usp.br, cdac@icmc.usp.br
[2] Information Technology Board, Federal University of Minas Gerais,
Belo Horizonte, Brazil
jpcarvalhocastro@ufmg.br
[3] Department of Computer Science, Federal University of São Carlos,
São Carlos, Brazil
accarniel@ufscar.br

Abstract. Fulfilling the FAIR Principles is a challenge that requires the management of research data and metadata considering the inherent big data complexity of volume, variety, and velocity. A suitable solution to deal with this problem is to combine a software reference architecture with a cloud computing environment. In this paper, we propose Cloud-FAIR, a novel Open Science architecture with an infrastructure located entirely in the cloud, which unburdens scientists in data and metadata management and improves performance. CloudFAIR also addresses security issues related to data encryption. We conducted performance tests with a real-world dataset to assess the efficiency of CloudFAIR. Compared to BigFAIR, the proposed architecture provided performance gains that ranged from 41.03% up to 75.95% when the issued queries required the retrieval of both data and metadata.

Keywords: Open science · FAIR principles · Cloud computing

1 Introduction

Collaboration plays a core role in the development of valuable scientific research. This fact became evident during the COVID-19 pandemic: vaccine development was unprecedentedly accelerated due to the sharing of scientific data between research institutions across the globe. This kind of collaboration is the foundation behind the concept of *Open Science*, which requires that every digital output of research objects is made available and usable free of charge [9].

The *FAIR Principles* provide a standardized protocol that complies with Open Science. These principles encompass the Findability, Accessibility, Interoperability, and Reusability of scientific data objects and their respective metadata [10]. However, scientific data can easily reach the concept of *big data* [4],

© Springer Nature Switzerland AG 2022
S. Chiusano et al. (Eds.): ADBIS 2022, CCIS 1652, pp. 56–66, 2022.
https://doi.org/10.1007/978-3-031-15743-1_6

according to their volume, variety, and velocity. Hence, it is necessary to guarantee the FAIR principles while being concerned with the traits of big data.

The combination of a *software reference architecture* (SRA) [1] with a *cloud computing* environment [4] can assist scientists in implementing FAIR-compliant big data repositories. An SRA can serve as a baseline to develop workflows adapted to specific requirements. A cloud environment transfers the clustering complexity and resource management to the cloud provider. As a result, scientists can focus on the data under analysis rather than on infrastructure details.

In Example 1, we introduce an application aimed at developing an SRA for implementing a FAIR-compliant repository in the cloud. We describe a repository whose goal is to store several scientific data objects from different sources, along with the integration of their metadata. We employ this application as a running example throughout the paper.

Example 1. Consider that a research foundation responsible for supporting several clinics and hospitals is building an open science repository, which should guarantee the following requirements. The repository should store, share, and analyze data and metadata generated by the funded institutions. It also should be compliant with the FAIR Principles. The data and metadata refer to those of COVID-19 patients, including their individual (e.g., age and gender) and exam (e.g., COVID-19 test results) data. Since each institution treats many patients daily and stores their data heterogeneously, a huge volume and variety of data are generated at a high velocity. Thus, the repository should provide good performance results when dealing with big data. Furthermore, the repository should unburden the funded institutions in maintaining their infrastructure.   □

A FAIR-compliant cloud SRA is a solution for guiding the implementation of the requirements described in Example 1. To this end, there are three groups of state-of-the-art solutions in the literature: general-purpose big data architectures [5]; FAIR implementations [6,8]; and FAIR-compliant big data architectures [3]. Despite the benefits of these solutions, they face several limitations. Solutions in the first group are not designed to conform the FAIR principles, while those in the second group are domain-specific and do not meet the concept of a generic SRA. To the best of our knowledge, there is only one solution in the third group, referred to as BigFAIR. This architecture takes advantage of existing repositories and supports data ownership. As a result, its repository infrastructure manages only metadata, leaving data management to the local infrastructure of the data providers. BigFAIR also does not tackle data security issues because the owners should be responsible for their data. Furthermore, there is no performance evaluation regarding BigFAIR. The characteristics of this architecture call for improvements to support applications that impose the requirements described in Example 1.

In this paper, we introduce a novel SRA to implement a FAIR-compliant repository in the cloud that extends BigFAIR, called CloudFAIR. The distinctive features of the proposed architecture are described as follows. First, data and metadata are managed in the same infrastructure, unburdening data providers about its maintenance. Second, security is carefully addressed through an encryption component since sensitive data objects are manipulated. Third, performance is of paramount importance in defining the components and their functionalities.

We also provide a performance evaluation comparing the CloudFAIR and the BigFAIR architectures. The workload retrieves real data and metadata from multiple sources, such as hospitals and clinics that compose the application described in Example 1. We analyze the impact of accessing different storage components and obtaining data from distinct providers in both architectures. The results demonstrated no difference between the architectures when queries required only metadata. However, CloudFAIR outperforms BigFAIR considerably when queries returned both data and metadata.

The remainder of this paper is organized as follows. Section 2 summarizes BigFAIR. Section 3 introduces the CloudFAIR architecture. Section 4 validates CloudFAIR through performance tests. Finally, Sect. 5 concludes the paper.

2   The BigFAIR Architecture

BigFAIR is a FAIR-compliant SRA that deals with big data. Figure 1a depicts this architecture, which includes the Local and the Repository infrastructures. Regarding the Local Infrastructure, the User Layer encompasses data providers and consumers, who respectively provide and retrieve information from the repository. Data providers store their scientific data and metadata in the Personal Storage Layer, comprised of personal repositories located within their institutions. Thus, their management falls under the owners' responsibility. Metadata from this layer is extracted and loaded into the Metadata Storage Layer located in the Repository Infrastructure. This layer contains: (i) a Metadata Lake, storing raw metadata; (ii) a Metadata Warehouse, containing transformed metadata; and (iii) a Metadata Governance Repository, maintaining provenance information and other types of metadata from the extracted metadata.

The Repository Infrastructure also encompasses the following layers. The Data Retrieval Layer is responsible for querying and processing data and metadata. The Knowledge Mapping Layer contains associations between generic data models and the data models implemented in the other layers. The Data Publishing Layer functions as the single access point for data consumers to obtain data and metadata from the repository, handling user authentication and data anonymization specifications. The Data Insights Layer generates descriptive, predictive, and prescriptive analyses of data and metadata.

Figure 1b depicts the Metadata Warehouse Generic Model introduced by Big-FAIR, modeled in a star schema that contains one fact table and eight dimension tables. The event represented by the fact table is the extraction of the metadata of a *data cell* (i.e., the intersection of an attribute and a tuple) at a given *date* and *time*, considering its *data repository*, *data provider*, and the associated *status*, *permissions*, and *license*. The fact table contains, besides the keys of each dimension, a numeric measure that represents the size of the data cell, which can be expressed in characters, bytes, or similar measurement units.

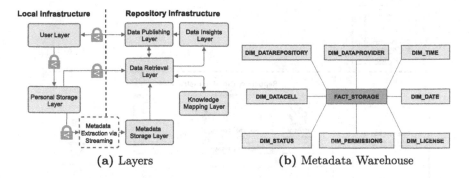

(a) Layers (b) Metadata Warehouse

Fig. 1. The BigFAIR Architecture.

3   The CloudFAIR Architecture

In this section, we present CloudFAIR, a novel SRA to implement a FAIR-compliant repository in the cloud. Figure 2 depicts the *layers* of the architecture, where boxes with a solid border represent *components* and boxes with a dashed border indicate *processes*. Further, directional arrows describe the *data flow* between layers, components, and processes. A padlock in an arrow specifies that the data flow must be *end-to-end encrypted*. This is necessary to guarantee that sensitive data is not leaked since data and metadata are uploaded into the repository infrastructure without prior anonymization. CloudFAIR is an extension of BigFAIR; thus, it fulfills the requirements imposed by the FAIR Principles.

Except for the User Layer, the CloudFAIR components are located in the Repository Cloud Infrastructure. The User Layer represents the repository data consumers and data providers, i.e., scientists that provide and retrieve information from the repository. Example 2 illustrates these roles.

Example 2. In the context of Example 1, data providers are hospitals and clinics that provide scientific data and metadata to the repository. Data consumers can be any user who consumes these data, such as the general public or the research foundation manager responsible for the repository.   □

Data providers and data consumers interact with the Repository Cloud Infrastructure through the User Interaction Layer. This layer contains two interfaces: one for data providers to upload their scientific data and associated metadata into the repository (Data/Metadata Upload Interface), and another for data consumers to retrieve this data (Data/Metadata Retrieval Interface). These interfaces can be implemented either through a graphical user interface or an application programming interface. The User Interaction Layer also inherits some components from the BigFAIR Data Publishing Layer, as follows. A Search Engine Registering exposes the repository in the network. Moreover, an Authentication Software and a User Permissions Database allow the content of the repository to be accessed differently based on the user profile.

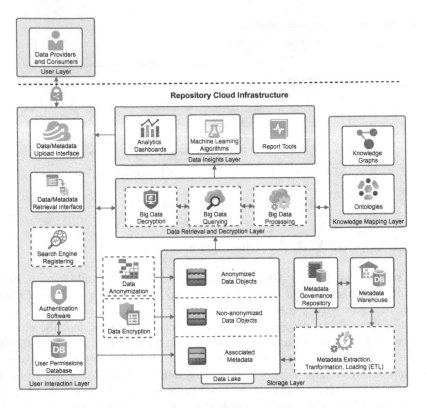

Fig. 2. The CloudFAIR Architecture.

Once the scientific data and their associated metadata are uploaded to the repository through the User Interaction Layer, they are loaded into a Data Lake in the Storage Layer. The Data Lake is specially organized to enable different loading processes for data and metadata. The data loading process encompasses the duplication of every data object, with one copy undergoing data anonymization and the other undergoing data encryption. Data anonymization refers to the irreversible removal of information that could lead to an individual being identified, while data encryption consists of the translation of data into another form so that only data consumers with access to a secret key can access their content. The goal is to create a hierarchy of anonymized data in the Data Lake, from non-anonymized data to completely anonymized data, enabling different levels of data access depending on the permissions of data consumers.

Metadata, on the other hand, is loaded directly into the Data Lake without undergoing anonymization or encryption. This is due to the fact that the objective of a FAIR-compliant repository is to expose the metadata to the general public. Hence, humans and machines can effortlessly find scientific data.

The Storage Layer also contains a Metadata Warehouse and a Metadata Governance Repository, whose features follow the same components described in Sect. 2. Furthermore, the warehouse design adopts the Metadata Warehouse Generic Model. The metadata is extracted from the Associated Metadata in the Data Lake and transformed before being loaded into the Metadata Warehouse. The metadata provenance from this process is stored in the Metadata Governance Repository, including other meta-metadata, such as the metadata that describes the content of the Associated Metadata in the Data Lake. Example 3 illustrates instances of these components, according to Fig. 1b.

Example 3. A typical numeric measure stored in the fact table can be the size of a patient's name (i.e. the data cell) in bytes. Furthermore: DIM_DATAPROVIDER stores data from the hospital that provided the data cell; DIM_DATAREPOSI-TORY contains connection data encompassing the path to the source data objects; DIM_DATACELL keeps the metadata related to the source data objects (e.g., a patients' file), its attributes (e.g., the column containing the patients' names), and its instances (e.g., the row of a specific patient); DIM_STATUS specifies if the data cell still exists in the data lake; DIM_PERMISSIONS maintains the role required to access a data cell (e.g., hospital manager) and a Boolean attribute to inform if data anonymization is needed; DIM_LICENSE contains the data cell licensing information; and DIM_DATE and DIM_TIME represent the date and time when the data cell metadata was extracted from the Data Lake. Examples of information stored in the Metadata Governance Repository are the previous value of the metadata before the transformation, the current value, and the date and time when the transformation occurred.   □

With the loading process complete, the Data Retrieval and Decryption Layer can perform its functionality. CloudFAIR implements this functionality by using the big data infrastructure in the repository, along with a parallel and distributed framework, to process and query the data and metadata stored in the Storage Layer. The Big Data Processing and Querying components interact with the Big Data Decryption to decrypt the non-anonymized data objects whenever these are necessary to answer an authorized data consumer request.

In addition, the Data Retrieval and Decryption Layer leverages the ontologies and knowledge graphs from the Knowledge Mapping Layer to translate consumers' requests to the data model adopted by the Storage Layer. Further, it accesses the content of every other layer to generate different types of intelligence for the Data Insights Layer. The Knowledge Mapping and the Data Insights layers are the same as those presented in Sect. 2.

4   Performance Evaluation

In this section, we evaluate the efficiency of CloudFAIR by comparing it with BigFAIR. To this end, we instantiate a subset of components in the architectures to support the execution of analytical queries. For both architectures, we used Hadoop Distributed File System (HDFS) to maintain the Metadata Warehouse and Apache Spark to implement the Data Retrieval Layer. We also used HDFS to store the BigFAIR Metadata Lake and the CloudFAIR Data Lake.

Table 1. Data and metadata size of each data provider in the real-world dataset of COVID-19 Brazilian patients, expressed in number of tuples.

Data provider	Scientific data objects		Associated metadata	
	# Patients	# Exams	# Tuples metadata warehouse	# Tuples metadata lake and data lake
USP	3,751	2,498,650	389,445	–
AE	79,863	3,415,155	2,357,416	–
SL	14,673	2,952,999	2,442,606	–
BPSP	39,000	6,329,103	6,046,764	–
FG	725,284	39,567,768	–	361,186,900

The personal repositories of BigFAIR were implemented using PostgreSQL. The remaining components were not instantiated since they are out of the scope of the present evaluation. We implemented two separate infrastructures. The Repository Infrastructure was simulated by an Apache YARN cluster of five machines (one master and four workers), whereas the Local Infrastructure was composed of a single machine. Each node in the cluster and the local infrastructure machine had a 6-core processor and approximately 8GB of RAM running on an Ubuntu Server. This configuration leads to a cluster default of 4 executors, 4 cores per executor, and 7GB of main memory per executor.

We used a real-world dataset of COVID-19 Brazilian patients. This dataset can be obtained in the COVID-19 DataSharing/BR repository [7], which is a FAIR-compliant Open Science repository developed by the State of São Paulo Research Foundation (FAPESP). The data objects available in the repository are associated with their metadata, including a data dictionary that describes the data type and the meaning of every attribute. The dataset consisted of five different data providers (BPSP, USP, FG, AE, and SL), each contributing with different data objects, such as patients' data, medical exams, and outcomes. In our performance evaluation, we considered that FG has recently been added to the environment. Hence, its associated metadata has not yet been loaded into the Metadata Warehouse, being present only in the BigFAIR Metadata Lake and in the CloudFAIR Data Lake. The remaining data providers had their respective metadata stored in the Metadata Warehouse. Table 1 depicts the data and metadata size for each data provider, which is the same for both architectures.

The workload consists of three different queries, each involving a different number of storage components. In CloudFAIR, we issued the queries against the anonymized version of the dataset, thus not requiring the application of data encryption and decryption to the result. BigFAIR, on the other hand, does not store an anonymized version of its data objects. Instead, it performs data anonymization during query processing. The workload is described as follows:

- **Q1:** "Retrieve the average data cell size of data that has the type DATE (DD/MM/YYYY), contains the Alive status, and requires anonymization. Group the size by data provider, data provider storage type, date, time, and license type". This query requires data from one storage component: the Metadata Warehouse. Thus, Q1 processing is the same in CloudFAIR and BigFAIR, i.e., the query must access the Metadata Warehouse and join the dimension tables with the fact table.

- **Q2:** "Retrieve the average numeric results of basophils exams in patients from the BPSP provider, grouped by the patient birth year". This query requires data from two storage components since it manages data and metadata. The first component is the Metadata Warehouse, which contains connection information to the source data objects. In CloudFAIR, the second component is the Data Lake, whereas in BigFAIR, the second component is the BPSP Personal Repository.

- **Q3:** "Retrieve the average numeric results of basophils exams in patients from the FG and AE providers, grouped by the patient birth year". This query also uses more than one storage component and requires data and metadata. Q3 processing starts by accessing the Metadata Warehouse to retrieve the connection information to the source data objects of the AE provider. This occurs in CloudFAIR and BigFAIR. However, the same procedure cannot be applied to the FG provider since its metadata is not stored in the Metadata Warehouse. CloudFAIR obtains the connection information to FG and all required source data objects from the Data Lake. Thus, Q3 processing accesses two components in CloudFAIR: Metadata Warehouse and Data Lake. Regarding BigFAIR, the Metadata Lake only stores the connection information to FG. The source data objects are managed by the FG and AE personal repositories located in each local infrastructure of the provider. Hence, Q3 processing accesses three components in BigFAIR: Metadata Warehouse, Metadata Lake, and Personal Storage Layer.

We collected the elapsed time in seconds, which was recorded by issuing each query 5 times, removing outliers, and calculating the average time and the standard deviation. All cache and buffers were flushed after finishing each query.

4.1   Queries Involving Different Storage Components

Figure 3a shows the performance results when processing each type of query previously introduced. Hence, the goal is to analyze how the architectures behave when different numbers of storage components are accessed to answer a query. For Q1, there is no significant difference in query performance between Cloud-FAIR and BigFAIR. Both architectures access the same component (Metadata Warehouse), process the query similarly, and return the required metadata.

On the other hand, CloudFAIR distinguished itself in efficiently processing Q2 and Q3, which require data and metadata. It provided performance gains of 41.03% for Q2 and 72.22% for Q3 when compared to BigFAIR. A performance gain refers to the reduction of the elapsed time provided by an architecture when compared with another.

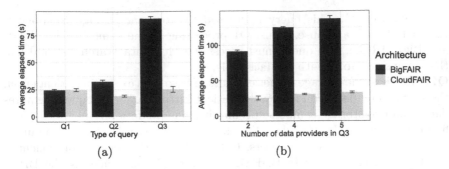

Fig. 3. Performance results of the architectures by processing each type of query (a) and by varying the number of data providers when processing Q3 (b).

The performance gains obtained by CloudFAIR are due to the following factors. First, both Q2 and Q3 access source data objects. In BigFAIR, the source data objects are stored in the Personal Storage Layer, which is located in a Local Infrastructure apart from the Repository Infrastructure. On the other hand, CloudFAIR stores the source data objects in the Data Lake located in the Repository Cloud Infrastructure, which unburdens the architecture in regards to accessing an external infrastructure to retrieve these objects.

Second, CloudFAIR stores the source data objects using HDFS in the same cluster that Spark is employed. Thus, it allows Spark to read and write the data in the same nodes where the processing is performed, significantly improving performance. Meanwhile, BigFAIR stores the source data in a different environment than Spark, not enabling this type of advantage.

4.2   Effect of the Number of Data Providers

In this section, we analyze how the number of data providers affects the performance of the architectures. For this, we pick Q3 since it is the most costly query to be processed according to the experiments reported in Sect. 4.1. Further, we generated three configurations to process Q3 using: (i) 2 data providers (FG and AE); (ii) 4 data providers (FG, AE, USP, and SL); and (iii) 5 data providers (FG, AE, USP, SL, and BPSP). The data volume increase from one configuration to the next is approximately 11%.

Figure 3b depicts the obtained results. In all configurations, CloudFAIR outperformed BigFAIR by showing performance gains of 72.22%, 75.72%, and 75.95% when accessing 2, 4, and 5 data providers, respectively. In addition, there is a growth in the difference between BigFAIR and CloudFAIR as the number of data providers increases. The reason behind this growth resides mostly in the fact that increasing the number of data providers accessed by a query does not only imply an increase in the volume of data. The architectures also need to access the metadata of a new data provider to retrieve its connection information and then use this metadata to connect to the location in which the source data object is stored. For CloudFAIR, reading a data object from a new data provider implies reading new files from the Data Lake only. On the other hand, our implementation of BigFAIR requires that Spark starts a new connection with PostgreSQL to read each new table, which can consume a significant amount of time.

5 Conclusions and Future Work

We have proposed CloudFAIR, a novel software reference architecture to implement FAIR-compliant repositories in the cloud. CloudFAIR manages data and metadata in the same environment, unburdening data providers from maintaining their local infrastructure. It also enables the storage of sensitive data objects through encryption. Further, CloudFAIR prioritizes performance.

We have conducted a performance evaluation that compared CloudFAIR with BigFAIR when accessing different storage components and obtaining data and metadata from a different number of providers. CloudFAIR was the most efficient when queries required retrieving both data and metadata, providing performance gains ranging from 41.03% up to 75.95%. Regarding queries accessing only metadata, CloudFAIR and BigFAIR delivered similar results.

Future work will include the design of guidelines to assist the implementation of CloudFAIR. Further, we plan to extend our experiments by using more layers in the architectures. We also aim to employ our architecture in other application contexts, such as sharing performance results collected by the tool detailed in [2].

Acknowledgments. This work was supported by the São Paulo Research Foundation (FAPESP), the Brazilian Federal Research Agency CNPq, and the Coordenação de Aperfeiçoamento de Pessoal de Nível Superior, Brasil (CAPES), Finance Code 001. J. P. C. Castro was supported by UFMG (PRODIS) and C. D. Aguiar by FAPESP grant #2018/22277-8. A. C. Carniel was supported by Google as a recipient of the 2022 Google Research Scholar.

References

1. Angelov, S., Grefen, P., Greefhorst, D.: A framework for analysis and design of software reference architectures. Inf. and Soft. Technology **54**(4), 417–431 (2012)
2. Carniel, A.C., Ciferri, R.R., Ciferri, C.D.A.: FESTIval: a versatile framework for conducting experimental evaluations of spatial indices. MethodsX **7**, 100695 (2020)
3. Castro, J.P.C., et al.: FAIR Principles and Big Data: A Software Reference Architecture for Open Science. In: International Conference on Ent. Information Systems, pp. 27–38 (2022)
4. Chen, M., Mao, S., Liu, Y.: Big data: a survey. Mobile Netw. Appli. **19**(2), 171–209 (2014)
5. Davoudian, A., Liu, M.: Big data systems: a software engineering perspective. ACM Comput. Surv. **53**(5), 1–39 (2020)
6. Devarakonda, R., et al.: Big federal data centers implementing FAIR data principles: ARM data center example. In: Proceedings of IEEE Big Data, pp. 6033–6036 (2019)

7. FAPESP: COVID-19   Data   Sharing/BR   (2020).   https://
   repositoriodatasharingfapesp.uspdigital.usp.br zch6lannom2020fair
8. Lannom, L., Koureas, D., Hardisty, A.R.: FAIR data and services in biodiversity
   science and geoscience. Data Intell. **2**(1–2), 122–130 (2020)
9. Medeiros, C.B., et al.: IAP input into the UNESCO Open Science Recommendation
   (2020). www.interacademies.org/sites/default/files/2020-07/Open_Science_0.pdf
10. Wilkinson, M.D., et al.: The FAIR Guiding Principles for scientific data management and stewardship. Sci. Data **3**(1), 1–9 (2016)

Detecting Simpson's Paradox: A Step Towards Fairness in Machine Learning

Rahul Sharma[1](✉)[iD], Minakshi Kaushik[1][iD], Sijo Arakkal Peious[1][iD],
Markus Bertl[2][iD], Ankit Vidyarthi[3][iD], Ashwani Kumar[4][iD],
and Dirk Draheim[1][iD]

[1] Information Systems Group, Tallinn University of Technology,
Akadeemia tee 15a, 12618 Tallinn, Estonia
{rahul.sharma,minakshi.kaushik,sijo.arakkal,dirk.draheim}@taltech.ee
[2] Department of Health Technologies, Tallinn University of Technology,
Akadeemia tee 15a, 12618 Tallinn, Estonia
mbertl@taltech.ee
[3] Jaypee Institute of Information Technology, Noida, India
[4] Sreyas Institute of Engineering and Technology, Hyderabad, India

Abstract. In the last two decades, artificial intelligence (AI) and machine learning (ML) have grown tremendously. However, understanding and assessing the impacts of causality and statistical paradoxes are still some of the critical challenges in their domains. Currently, these terms are widely discussed within the context of explainable AI (XAI) and algorithmic fairness. However, they are still not in the mainstream AI and ML application development scenarios. In this paper, first, we discuss the impact of Simpson's paradox on linear trends, i.e., on continuous values, and then we demonstrate its effects via three benchmark training datasets used in ML. Next, we provide an algorithm for detecting Simpson's paradox. The algorithm has experimented with the three datasets and appears beneficial in detecting the cases of Simpson's paradox in continuous values. In future, the algorithm can be utilized in designing a certain next-generation platform for fairness in ML.

Keywords: Big data · Artificial intelligence · Machine learning · Data science · Simpson's paradox · Explainable AI

1 Introduction

The outcomes of artificial intelligence (AI) and machine learning (ML) applications are explicitly dependent on the correctness of algorithms and training datasets. However, like statistics and mathematics, handling statistical paradoxes, cause and effect together in datasets is still not in the mainstream AI and ML application development. There are many cases where the outcome of

This work has been partially conducted in the project "ICT programme" which was supported by the European Union through the European Social Fund.

© Springer Nature Switzerland AG 2022
S. Chiusano et al. (Eds.): ADBIS 2022, CCIS 1652, pp. 67–76, 2022.
https://doi.org/10.1007/978-3-031-15743-1_7

AI applications is observed to be biased [13,18]. Like fossil fuels, data is considered a new fuel in the 21st century, but it needs to be properly cleaned for fair results. Nowadays, increased usages of AI and ML in healthcare, social media, digital advertising, search engines, etc., directly or indirectly impact human life and their decisions. Therefore, understanding causal relationships and evaluating the existence of statistical paradoxes should be an essential part of AI application development scenarios for better, fair and unbiased AI applications.

Statistical paradoxes, causality, selection bias, confounding and information bias have been debated in statistics and mathematics for a long time; expert statisticians and mathematicians have effectively handled their severe consequences. Several statistical paradoxes include Simpson's Paradox, Tea Leaf Paradox, Berkson's Paradox, Latent Variables, Law of Unintended Consequences, etc. The term "paradox" denotes a fundamental link between several statistical issues and mathematical reasoning, e.g., causal inference [19,20], Lord's paradox [28], propensity score matching [24], suppressor variables [4], the ecological fallacy [16,23], conditional independence [5], p-technique [3] and partial correlations [9], mediator variables [17], etc.

Handling statistical paradoxes and causality will not only build trust in artificial applications but also serve as the foundation for fairness in AI. In this paper, we explicitly focus on Simpson's paradox, which has also been discussed in various data mining techniques [10], e.g., association rule mining [1,6,7] and numerical association rule mining [14,15,27]. The main aim of this article is to develop an algorithm for detecting Simpson's paradox in continuous values. The algorithm is tested with the three benchmark datasets and appears beneficial in detecting the cases of Simpson's paradox in linear trends. In the future, the algorithm may be used to create a specific next-generation platform for trustworthy AI and fairness in ML.

The paper is organized as follows. In Sect. 2, we discuss the background of Simpson's Paradox. In Sect. 3, we discuss ways to detect Simpson's paradox and propose an algorithm for detecting the paradox in linear trends. In Sect. 4, three benchmark datasets are used to experiment with the algorithm. Finally, a discussion on future work and conclusion is given in Sect. 5 and Sect. 6, respectively.

2   Yule-Simpson's Paradox

In the year 1899, Karl Pearson et al. [21] demonstrated a statistical paradox in marginal and partial associations between continuous variables. Further, in 1903, Udny Yule [29] presented "the theory of association of attributes in statistics" and revealed the existence of an association paradox with categorical variables. Later in 1951, Edward H. Simpson [26] presented the concept of reversing results and in 1972, Colin R. Blyth coined the term "Simpsons Paradox" [2]. Therefore, this paradox is known by different names and is well-known as the Yule-Simpson effect, amalgamation paradox, or reversal paradox [22].

We have used a real-world dataset from Simpson's article [26] to discuss the paradox. In this example, analysis for medical treatment is described. Table 1 provides the number that shows the effect of the medical treatment for the entire population ($N = 52$) as well as for men and women separately in subgroups. The treatment is suitable for both male and female subgroups; however, the treatment appears unsuitable for the entire population.

Table 1. A real life case of Simpson's Paradox: The numbers in the table are taken from Simpson's original article [26].

	Full population N = 52			Women (F) = 20			Men (M), N = 32		
	Success (S)	Failure (¬S)	Succ.%	Success (S)	Failure (¬S)	Succ.%	Success (S)	Failure (¬S)	Succ.%
T	20	20	50%	8	5	61%	12	15	44%
¬T	6	6	50%	4	3	57%	2	3	40%

The Simpson's paradox scenario can also be described via probability theory and conditional probabilities. Let $T = treatment$, $S = successful$, $M = Male$, and $F = Female$ then the $P(S \mid T)$ can be described as:

$$P(S \mid T) = P(S \mid \neg T) \tag{1}$$

$$P(S \mid T, M) > P(S \mid \neg T, M) \tag{2}$$

$$P(S \mid T, F) > P(S \mid \neg T, F) \tag{3}$$

This reversal of results between the male, female population and the entire population has been referred to as Simpson's Paradox. In statistics, these concepts have been discussed widely and named differently by several authors [21, 29].

3 Detecting Simpson's Paradox

The Simpson's paradox instances are investigated for both categorical and continuous values. However, we investigate the paradox in linear trends. The Pearson correlation index is used in the algorithm to determine the correlations between two variables and further define the function of the confounding variable. A confounder can be defined as a factor that affects both the dependent and independent variables, resulting in an incorrect association. The Pearson correlation index allows us to measure the strength of the linear association between two variables. In Eq. 4, Pearson correlation coefficient is represented by r and x, y are input vectors, $\bar{x}$ and $\bar{y}$ are the means of the variables, respectively. The output value always lies between -1 and 1. Values greater than 0 imply a positive correlation, while the values 1 and 0 indicate the exact positive association and no correlation, respectively. Values less than 0 suggest a negative association, and -1 indicates a clear negative association.

$$r = \frac{\sum_{i=1}^{n}(x_i - \overline{x})(y_i - \overline{y})}{\sqrt{\sum_{i=1}^{n}(x_i - \overline{x})^2(y_i - \overline{y})^2}} \tag{4}$$

3.1 Algorithm

To identify an instance of the Simpson's paradox in a continuous dataset with n continuous variable and m discrete variables, we compute a correlation matrix $(n \times n)$ for all the data. Then for m discrete variable with k_m levels, an additional $(n \times n)$ matrix needs to be calculated for each level of variables. Therefore, we need to calculate the $1 + \sum_i^m = k_i$ correlation matrices of size $(n \times n)$ and compare it with the lower half of $\sum_i^m = k_i$ for subgroup levels.

The algorithm's initial step is to determine the correlation between x and y variables with the values of the corresponding columns in the dataset. In this way, we learn the direction of the relationship between the variables. We next walk through the list of remaining variables, compute the Pearson index conditional on each subgroup (category), count the percentage of subgroups where the correlation index is reversed with respect to the correlation index in the aggregate data, and store the value key pairs in an array. We further get the array element where the value (ratio) is the highest. A value greater than 0 implies the existence of Simpson's paradox and the maximal value of 1 indicates a full reversal effect.

4 Experiments

Python programming language is used to implement the algorithm on a personal computer with the Windows 10×64 operating system and an Intel(R) Core(TM) i5-8265U CPU running at $1.60\,\text{GHz}$, $1800\,\text{MHz}$, 4 Cores, and 8 Logical Processors. We evaluate the algorithm with three benchmark datasets that are widely used to train various ML models. The performance of the algorithm strongly correlates with the size of the datasets. The programming code, datasets, and other necessary instructions for the algorithm are available in the GitHub repository [25].

4.1 Datasets

Iris dataset, Miles per gallon (MPG) dataset and Penguin dataset are used to demonstrate the presence of Simpson's paradox in continuous data.

Algorithm 1: Identification of the confounding variable in continuous values to identify Simpson's paradox

---

Input: dataset D, variable x, variable y
Output: a pair consisting of confounding variable and ratio of reversed
          association signs
aggreg_index = Pearson(d[x], d[y]) // *calculate corr. index between columns*
indexes = [] // *initialize index array to store key value pairs where the key is*
 column and value is the number of reversed subgroups
cols = columns(D) // *initialize array of all columns of D*
foreach *column* $\in$ *cols* **do**
    if *Column Is Not Categorical(column)* **then**
     | Continue
    end
    else
        subgroups = Categories(column) // *get the categories of a column*
        coefficients = [] // *initialize empty array to store the correlation*
        indexes **foreach** *subgroup* $\in$ *subgroups* **do**
            disaggreg_index = Pearson(D[x]: where D[column] = subgroup,
            D[y]: where D[column] = subgroup) *calculate corr. index between*
            columns for current subgroup
            Add index of disaggregated to correlation indexes array
        end
    end
    reversed_subgroups = RatioReversedSubgroups(aggreg_index, coefficients)
    // *calculate ratio of the correlation indexes reversed with respect to the*
    correlation index for the aggregated data
    Add $\{$*column, reversed_subgroups*$\}$ values into *indexes*
end
Store the max values of *indexes* pairs into *result*
Return *result*

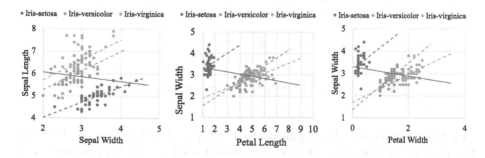

Fig. 1. Scatter plot with trend lines for Iris dataset.

Iris Dataset: In 1936 Ronald Fisher introduced the iris dataset in one of his research papers [8]. In this dataset, there are 50 data samples for the three different iris species, i.e., 'Setosa', 'Versicolor', and 'Virginicare'. In the dataset, species names are categorical, while length and breadth are continuous values.

We visualize the possible associations between the length and breadth of each pair of candidate attributes to identify the instances of Simpson's paradox. Table 2 demonstrates the Pearson correlation index returned by the algorithm 1 between two continuous variables ('sepal length' and 'sepal width').

We identify the existence of Simpson's paradox for three pairs of measurements. 1. sepal length and width, 2. sepal width and petal length, and 3. sepal width and petal width. In Fig. 1, the correlation between sepal width and sepal length is positive (dashed line) for each species. However, the correlation between sepal width and sepal length for the entire population is negative (solid red trend line). Similarly, the pair of petal length and width and the pair of petal width and sepal width have positive trends for each species. However, the overall trend for the length and width for the entire population is negative in both cases.

Table 2. The Pearson correlation index for Iris dataset by the Algorithm 1.

Agg. correlation	Variable 1	Variable 2	Sub group	Group correlation
−0.1093	Sepal length	Sepal width	Iris-setosa	0.7467
−0.1093	Sepal length	Sepal width	Iris-versicolor	0.5259
−0.1093	Sepal length	Sepal width	Iris-virginica	0.4572

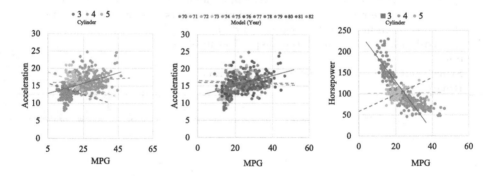

Fig. 2. Scatter plot with trend lines for MPG dataset.

The MPG Dataset: Ross Quinlan used the Auto MPG dataset in 1993 [22]. The dataset contains 398 automobile records from 1970 to 1982, including the vehicle's name, MPG, number of cylinders, horsepower, and weight. The dataset includes three multi-valued discrete attributes and five continuous attributes. We visualize the relationship between MPG, acceleration and horsepower for two categorical attributes (number of cylinders and model year). The goal of analyzing the dataset is to know the factors that influence each car's overall fuel consumption. The dataset consists of fuel consumption in mpg, horsepower, number of cylinders, displacement, weight, and acceleration.

Table 3. The Pearson correlation index for MPG dataset by the proposed algorithm.

Agg. correlation	Variable 1	Variable 2	Sub group	Group correlation
0.4230	mpg	Acceleration	Cylinders	−0.8190
0.4230	mpg	Acceleration	Cylinders	−0.3410
0.4230	mpg	Acceleration	Model year	−0.0510

In the MPG dataset, the existence of Simpson's paradox is discovered in three pairs of measurements. 1. MPG with acceleration according to the engine cylinders, 2. MPG with acceleration with respect to their model year, and 3. MPG with horsepower according to the engine cylinders. In Fig. 2, it is visualized that there is a negative correlation between MPG and acceleration for three-cylinder engines and six-cylinder engines; however, the overall trend between MPG and acceleration is positive (solid red line). Similarly, the overall trend is the opposite for MPG with acceleration with respect to the model year and MPG with horsepower according to the engine cylinders. Table 3 demonstrates the Pearson correlation index returned by the Algorithm 1 between two continuous variables ('mpg' and 'acceleration').

Penguin Dataset. Palmer penguins dataset [11, 12] is also a well-known dataset used as an alternative to the Iris dataset. The dataset contains the descriptions of three species of penguins (Adelie, Chinstrap, Gentoo) in the islands of Palmer, Antarctica. The dataset contains 344 data rows with columns: 'species', 'island' , 'culmen_length_mm', 'culmen_depth_mm', 'flipper_length_mm' , 'body_mass_g' and 'sex'. To investigate the instances of Simpson's paradox in the dataset, we set x as 'culmen_length_mm' and y as 'culmen_depth_mm'. As per the results from the algorithm, there is an instance of Simpson's paradox in data as the association between the culmen_length and culmen_depth reverses when data is disaggregated by the species. Figure 3 demonstrates a positive correlation between the culmen_length_mm and culmen_depth_mm of each species. However, it is negative for the aggregate data.

5   Discussion and Future Work

The presence of Simpson's paradoxes in benchmark datasets provides a direction to understand the causality in decision making. We noticed that most ML and deep learning algorithms focus only on identifying correlations rather than identifying the real or causal relationships between data items. Therefore, understanding and evaluating causality is an important term to be discussed in big data, data science, AI and ML. In future, we plan to develop a framework to simplify the impacts of Simpson's paradox and address various other statistical paradoxes (e.g., Berkson's paradox) that have severe implications for big data, data science, AI and ML.

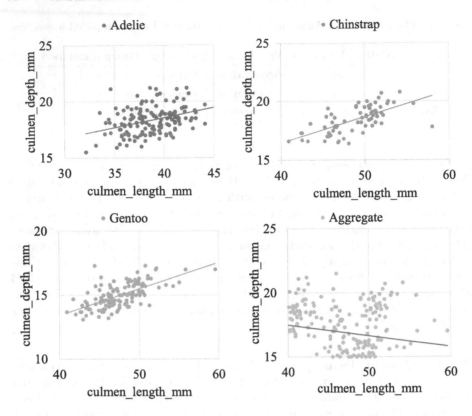

Fig. 3. Scatter plot with trend lines for Penguin dataset.

6   Conclusion

Handling statistical paradoxes is a complex challenge in automatic data mining, specifically in AI and ML techniques. In this paper, we discussed a strong need for statistical evaluations of datasets and demonstrated the impacts of Simpson's paradox on AI and ML via some benchmark training datasets. We argue that if confounding effects are not properly addressed in automatic data mining, the outcomes of data analysis can be completely opposite. However, with the right tools and data analysis, a good analyst or data scientist can handle it in a better way. Further, we provided an algorithm to detect Simpson's paradox in linear trends (continuous values). The algorithm is evaluated on three benchmark datasets and performed well in each experiment. This algorithm can be a part of developing a platform to detect Simpson's paradox in different data (continuous, categorical) and enable data scientists to explore the impacts of confounding variables.

References

1. Agrawal, R., Srikant, R.: Fast algorithms for mining association rules in large databases. In: Proceedings of VLDB 1994 - the 20th International Conference on Very Large Data Bases, pp. 487–499. Morgan Kaufmann (1994)
2. Blyth, C.R.: On Simpson's paradox and the sure-thing principle. J. Am. Stat. Assoc. **67**(338), 364–366 (1972)
3. Cattell, R.B.: P-technique factorization and the determination of individual dynamic structure. J. Clin. Psychol. **8**, 5–10 (1952)
4. Conger, A.J.: A revised definition for suppressor variables: a guide to their identification and interpretation. Educ. Psychol. Meas. **34**(1), 35–46 (1974)
5. Dawid, A.P.: Conditional independence in statistical theory. J. Roy. Stat. Soc. Ser. B (Methodol.) **41**(1), 1–15 (1979). https://doi.org/10.1111/j.2517-6161.1979.tb01052.x
6. Draheim, D.: DEXA 2019 keynote presentation: future perspectives of association rule mining based on partial conditionalization, Linz, Austria, 28th August 2019. https://doi.org/10.13140/RG.2.2.17763.48163
7. Draheim, D.: Future perspectives of association rule mining based on partial conditionalization. In: Hartmann, S., Küng, J., Chakravarthy, S., Anderst-Kotsis, G., Tjoa, A.M., Khalil, I. (eds.) Proceedings of DEXA'2019 - the 30th International Conference on Database and Expert Systems Applications. LNCS, vol. 11706, p. xvi. Springer, Heidelberg (2019)
8. Fisher, R.A.: The use of multiple measurement in taxonomic problems. Ann. Eugenics **7**(2), 179–188 (1936). https://doi.org/10.1111/j.1469-1809.1936.tb02137.x
9. Fisher, R.A.: Iii. the influence of rainfall on the yield of wheat at rothamsted. Phil. Trans. Roy. Soc. Lond. Ser. B Containing Papers Biol. Charact. **213**(402–410), 89–142 (1925)
10. Freitas, A.A., McGarry, K.J., Correa, E.S.: Integrating bayesian networks and simpson's paradox in data mining. In: Texts in Philosophy. College Publications (2007)
11. Gorman, K.B., Williams, T.D., Fraser, W.R.: Ecological sexual dimorphism and environmental variability within a community of antarctic penguins (genus pygoscelis). PLOS ONE **9**(3), 1–14 (2014). https://doi.org/10.1371/journal.pone.0090081
12. Horst, A.M., Hill, A.P., Gorman, K.B.: palmerpenguins: Palmer Archipelago (Antarctica) penguin data (2020). https://doi.org/10.5281/zenodo.3960218, https://allisonhorst.github.io/palmerpenguins/, r package version 0.1.0
13. Julia, A., Jeff, L., Surya, M., Lauren, K.: Machine Bias, www.propublica.org/article/machine-bias-risk-assessments-in-criminal-sentencing?token=TiqCeZIj4uLbXl91e3wM2PnmnWbCVOvS
14. Kaushik, M., Sharma, R., Peious, S.A., Shahin, M., Ben Yahia, S., Draheim, D.: On the potential of numerical association rule mining. In: Dang, T.K., Küng, J., Takizawa, M., Chung, T.M. (eds.) FDSE 2020. CCIS, vol. 1306, pp. 3–20. Springer, Singapore (2020). https://doi.org/10.1007/978-981-33-4370-2_1
15. Kaushik, M., Sharma, R., Peious, S.A., Shahin, M., Yahia, S.B., Draheim, D.: A systematic assessment of numerical association rule mining methods. SN Comput. Sci. **2**(5), 1–13 (2021). https://doi.org/10.1007/s42979-021-00725-2
16. King, G., Roberts, M.: Ei: a (n r) program for ecological inference. Harvard University (2012)

17. MacKinnon, D.P., Fairchild, A.J., Fritz, M.S.: Mediation analysis. Ann. Rev. Psychol. **58**(1), 593–614 (2007). https://doi.org/10.1146/annurev.psych.58.110405.085542

18. O'Neil, C.: Weapons of Math Destruction: How Big Data Increases Inequality and Threatens Democracy. Crown Publishing Group, New York (2016)

19. Pearl, J.: Causal inference without counterfactuals: comment. J. Am. Stat. Assoc. **95**(450), 428–431 (2000)

20. Pearl, J.: Understanding Simpson's paradox. SSRN Electron. J. **68** (2013). https://doi.org/10.2139/ssrn.2343788

21. Pearson Karl, L.A., Leslie, B.M.: Genetic (reproductive) selection: inheritance of fertility in man, and of fecundity in thoroughbred racehorses. Phil. Trans. Roy. Soc. Lond. Ser. A **192**, 257–330 (1899)

22. Quinlan, J.: Combining instance-based and model-based learning. In: Machine Learning Proceedings 1993, pp. 236–243. Elsevier (1993). https://doi.org/10.1016/B978-1-55860-307-3.50037-X

23. Robinson, W.S.: Ecological correlations and the behavior of individuals. Am. Sociol. Rev. **15**(3), 351–357 (1950)

24. Rosenbaum, P.R., Rubin, D.B.: The central role of the propensity score in observational studies for causal effects. Biometrika **70**(1), 41–55 (1983)

25. Sharma, R., Peious, S.A.: Towards unification of decision support technologies: Statistical reasoning, OLAP and association rule mining. https://github.com/rahulgla/unification

26. Simpson, E.H.: The interpretation of interaction in contingency tables. J. Roy. Stat. Soc. Ser. B (Methodol.) **13**(2), 238–241 (1951)

27. Srikant, R., Agrawal, R.: Mining quantitative association rules in large relational tables. In: Proceedings of the 1996 ACM SIGMOD International Conference on Management of Data, pp. 1–12 (1996)

28. Tu, Y.K., Gunnell, D., Gilthorpe, M.S.: Simpson's paradox, lord's paradox, and suppression effects are the same phenomenon-the reversal paradox. Emerg. Themes Epidemiol. **5**(1), 1–9 (2008)

29. Yule, G.U.: Notes on the theory of association of attributes in statistics. Biometrika **2**(2), 121–134 (1903)

ADBIS Short Papers: Data Management Pipeline, Information and Process Retrieval

FusionFlow: An Integrated System Workflow for Gene Fusion Detection in Genomic Samples

Federica Citarrella[1], Gianpaolo Bontempo[2,3] ⓘ, Marta Lovino[3(✉)] ⓘ, and Elisa Ficarra[3] ⓘ

[1] Politecnico di Torino, Corso Duca degli Abruzzi 24, Torino, Italy
`federica.citarrella@studenti.polito.it`
[2] University of Pisa, via Lungarno Pacinotti 43, Pisa, Italy
`gianpaolo.bontempo@phd.unipi.it`
[3] Enzo Ferrari Engineering Department, University of Modena and Reggio Emilia, Via Vivarelli 10/1, Modena, Italy
`{marta.lovino,elisa.ficarra}@unimore.it`

Abstract. Gene fusion is a genomic alteration where two genes after a break event are juxtaposed to form a new hybrid gene, leading to possible cancer development and progression. However, identifying gene fusions is not a trivial process as it requires the management and processing countless amounts of data. Genomic data (particularly DNA and RNA) can reach up to 300 GB per sample. Furthermore, specific software and hardware architectures are required to correctly process this type of data. Although many tools are available for detecting gene fusions, to date, systematic workflows that are free and easily usable even by non-specialists are hardly available.

This paper presents an integrated system for identifying gene fusions in RNA and DNA genomic samples, focusing on hardware and software architectural aspects. The proposed workflow is easy-to-use, scalable, and highly reproducible. It includes five gene fusion detection tools, three mainly intended for RNA samples (EricScript, Arriba, FusionCatcher) and two for DNA samples (INTEGRATE and GeneFuse). The workflow runs on servers exploiting Nextflow (a DSL for data-driven computational pipelines), Docker containers, and Conda virtual environments.

Keywords: Gene fusions · Gene fusion detection · Genomic samples

1 Introduction

Gene fusion is a phenomenon that occurs when two or more genes become juxtaposed, forming a single hybrid gene or transcript. Gene fusions remarkably contribute to the evolutionary process by providing a continuous source of new genes. However, at the same time, they often lead to genomic disorders or cancer. Numerous gene fusions have been recognized as essential drivers for various

ⓒ Springer Nature Switzerland AG 2022
S. Chiusano et al. (Eds.): ADBIS 2022, CCIS 1652, pp. 79–88, 2022.
https://doi.org/10.1007/978-3-031-15743-1_8

cancer types. Thus, the discovery of novel gene fusions can better comprehend tumour development and progression [29]. For these reasons, gene fusion identification employing gene fusion detection tools has become crucial in bioinformatics research [30]. Recent advances in deep learning and convolutional networks [3,4,6,8–10,21–23] have also progressively spread to tools for gene fusion detection [17,19].

Although many gene fusions detection tools have been developed over the past years, it is still challenging to use them. In addition, the RNA seq artefacts, introduced by library preparation and sequence alignment, make gene fusions predictions hardly reliable [15].

The typical practice is executing multiple tools and using the union or intersection of their results. Unfortunately, this approach is computationally demanding. There are several limitations in traditional tools usage:

- each tool has specific installation requirements and version dependencies that must be precisely adhered to;
- downloading files and databases and executing tools is time-consuming;
- distinct tools can require different input data formats;
- multiple complementary fusion detection tools are needed to improve sensitivity.

During the last years, bioinformatics workflows (which consist of a wide array of algorithms executed in a predefined sequence) were developed to deal with multiple bioinformatic issues (e.g., RNA data processing and CNA detection) [24]. However, only a limited number of gene fusion detection workflows is available, and no one of them can simultaneously handle both RNA and DNA sequencing data [28].

This paper presents FusionFlow, an easily reproducible and scalable bioinformatics workflow for detecting gene fusions from RNA and DNA data. It processes numerous sequence data and their associated metadata through multiple transformations using a series of software components, databases, and operation environments (hardware and operating system). It includes five gene fusion detection tools executed through multiple processes. The processes are built using Nextflow Groovy/JVM-based framework exploiting Docker and Conda technologies. Indeed, Nextflow allows running tools downloads, installation, and execution concurrently in the interest of time constraints. At the same time, Docker and Conda engines are used to create virtual environments precisely configured for each tool. Finally, the pipeline inputs standard data formats and eventually converts them directly inside specific converter processes.

2   The Workflow

FusionFlow includes five fusion detection tools: EricScript [5], Arriba [26], FusionCatcher [20], GeneFuse [7] and Integrate [32]. Three of them, EricScript, Arriba, and FusionCatcher, accept as input just RNA-seq data. Concerning DNA tools, GeneFuse takes just DNA data, while Integrate has two input options: 1) just RNA data or 2) both RNA and DNA data. All gene fusion detection tools are made up of three steps: 1) preliminary alignment of the reads (a row in the genomic input files) to the transcriptome to build specific gene fusion references; 2) alignment of previously unmapped reads to gene fusion references to support the gene fusion detection; 3) cleaning filters to discard false positives. The main differences between the tools consist of the alignment type (e.g., BLAST vs BWA) and the properties of the cleaning filters. Although a proper gold standard procedure for gene fusion detection has not been established, the most widely used approach involves applying multiple gene fusion detection tools, unifying the results obtained. Ericscript, FusionCatcher, and Arriba have been selected for this workflow due to their spread and unique characteristics. Ericscript and FusionCatcher have been selected due to the differences in the cleaning filters. The former exploits, among the others, heuristic filters to remove analysis artefacts, while the latter removes false positives using known and novel criteria, which make biological sense. In the end, Arriba has been chosen since it can find aberrations that the competitors hardly find (e.g. intragenic and intergene duplications/inversions/translocations). Since the DNA sequencing method has only recently spread on a large scale [16], the panorama in DNA gene fusion detection tools includes a few software available. GeneFuse and INTEGRATE deserve to be mentioned for their user experience. GeneFuse can detect gene fusions from DNA samples alone, while INTEGRATE requires both RNA and DNA data from the same sample to provide the gene fusion list. At the same time, it can reconstruct gene fusion junctions and genomic breakpoints by split-read mapping in a complete way.

In order to make the pipeline usage as simple as possible, the only mandatory inputs are the RNA or DNA files to be analyzed. In this case, the workflow looks for tools' required files in default paths. The gene fusion detection tools start processing data if the files are present. Otherwise, the pipeline downloads and installs all the necessary tools and files before the tools' execution. FusionFlow receives input RNA only, DNA only, or both RNA and DNA data.

The FusionFlow pipeline produces several files divided into two categories: tools' required files and gene fusions' output files. The first category includes all the files needed to execute the tools. These files can be directly provided to the workflow, skipping their downloads processes, or can be downloaded while running the workflow for the first time. Then, the files will be saved in a specific path to be available to the pipeline for the subsequent runs.

The second category of output includes the files produced as output from the gene fusion tools. Each tool gives as output one or more files in specific formats. The most diffused formats are Tab Separated Value (TSV), Variant Call Format (VCF), and standard text format [2].

In the following, the general workflow architecture is described.

2.1 Architecture

The general workflow structure is based on Nextflow, a dataflow programming model that simplifies writing complex distributed pipelines.

Nextflow Groovy/JVM-based framework is selected among a series of workflow management systems (e.g., Galaxy [11], Toil [27], Snakemake [14], Bpipe) due to its peculiar features. In particular it allows:

- the existence of several processes written in different languages. Nextflow recognizes the script's language automatically, and it generates a launch file per process dynamically;
- to process data as stream atop by step. Indeed, each process can communicate through the input/output channel definition. These channels can also be used for synchronization mechanisms in order to make the pipeline sequential;
- integration with sharing platforms such as GitHub. Nextflow can notice if the repository is not installed and, in that case, it downloads all the requirements, environments included;
- integration with the most famous containers as Docker and Singularity. This feature is crucial for gene fusion tools since they often require conflicting packages. The current pipeline has considered each process in a separate environment;
- integration with several schedulers as SLURM. Due to the substantial memory boundaries requested by the gene fusion tools, the pipeline can be executed basically on large systems servers. Rarely are they used without a scheduler.

The workflow is composed of fifteen processes. These processes can be divided into three main categories:

- **downloaders**: they are responsible for the tools installation and download input files. The downloaders processes are: *referenceGenome_downloader, arriba_downloader, ericscript_downloader, fusioncatcher_downloader, integrate_downloader* and *genefuse_downloader*;
- **converters**: they are responsible for the file preparation and format conversion if needed. The converters processes are: *integrate_converter* and *genefuse_converter*;
- **runners**: they allow the code and tools execution. The runner processes are: *arriba, ericscript, fusioncatcher, integrate, genefuse, referenceGenom_index, integrate_builder*.

The fifteen processes are structured into six main parallel lines shown in green in Fig. 1.

Executing the script with Nextflow, the algorithm will look for the required files in the paths specified in *nextflow.config* configuration file or the paths specified in the command line. The associated downloader is skipped if the files exist, and the following processes can start processing.

Nextflow processes usually are executed concurrently. Nextflow queue channels are used to execute downloaders, converters, and runners sequentially and provide inter-communication between processes. A queue channel creates an asynchronous unidirectional FIFO queue and allows to connect processes or operators. Using a combination of queue channels permits the creation of predefined sequences of processes. The processes expect to receive input data from the channels specified in the input block. When the inputs are emitted, the processes run.

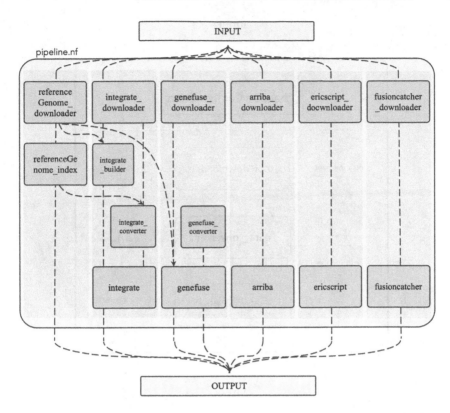

Fig. 1. Pipeline architecture parallelization: each tool is composed of multiple sub-units (shown in blue) executed concurrently through six main parallel lines (shown in green) to optimize the workflow performances. (Color figure online)

The five fusion detection tools included in FusionFlow are managed through Nextflow queue channels that provide inter-communication between the workflow processes. All processes have the same structure since they are triggered by input and, after the script block execution, provide output to trigger the subsequent processes. As illustrated in Fig. 2, for each channel, the first step consists of describing the channel configuration (e.g., DNA files, RNA files, the tool installation path, and further databases necessary for gene fusion tools). If the user does not define tool databases, a separate channel is used to download it. A data channel passes the database to the next process triggered at the moment. Then, the tool/database is installed if not present yet, and the data is converted in the correct format if the user requests it. Finally, the data is passed to the gene fusion tool for the tool execution.

In the end, the workflow provides as output the list of candidate gene fusions for each tool to be investigated by the user.

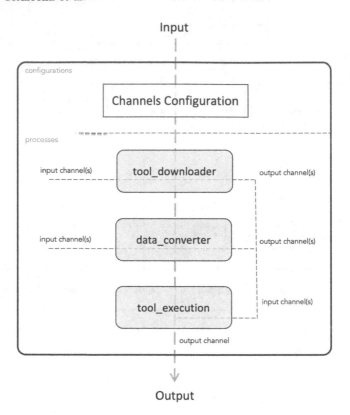

Fig. 2. General flow for each tool. Given the data as input, 1) the required configuration is set; 2) the tool/database is installed if not present; 3) the data is converted if with the wrong format; 4) the data is passed to the gene fusion tool.

3   Workflow Test and Discussion

The files used to test the pipeline are the same proposed in the FusionCatcher tool and publicly available both at https://github.com/ndaniel/fusioncatcher/ tree/master/test and in the FusionFlow GitHub repository.

They are fastq compressed paired-end files (a standard file data format used to store genetic information) where the reads were manually selected to cover 17 already known fusion genes: FGFR3 - TACC3, FIP1L1 - PDGFRA, GOPC - ROS1, IGH - CRLF2, HOOK3 - RET, AKAP9 - BRAF, EWSR1 - ATF1, TMPRSS2 - ETV1, EWSR1 - FLI1, ETV6 - NTRK3, ETV6 - NTRK3, ETV6 - NTRK3, BRD4 - NUTM1, CD74 - ROS1, CIC - DUX4, DUX4 - IGH, DUX4 - IGH, EML4 - ALK, MALT1 - IGH, NPM1 - ALK.

Initially, each tool was tested separately on a linux operating system and inside a docker environment checking the setup (e.g. paths, files, profiles, libraries). Then, after making sure that all the tools worked, the entire Nextflow workflow was tested inside a single docker environment. In the scenario without docker, Conda virtual environments were manually created. Otherwise, with docker the setup is prepared automatically through the use of the dockerfile. The test files and the local profile were specified in the command line to execute these tests.

Each gene fusion detection tool gives output one or more files in specific formats. Generally, a summary file is produced in output to allow a quick predictions overview.

The outputs obtained from the tools are concordant with the gene fusions previously specified. All the tools in the pipeline recognize at least ten predictions out of seventeen fusions, except for GeneFuse, which recognizes just three of them. Although GeneFuse performances should be investigated on additional data, the poor result could be explained by the specific DNA filters implemented in GeneFuse.

In order to select the final gene fusions prediction drivers, different approaches can be used. The typical practice is to use the union or intersection of tools predictions. The union of the results gives numerous sets of predictions. This approach increases the probability of including the real drivers of cancer processes. However, it enhances the possibility of incorporating false positives or passenger mutations. Using the intersection approach, conversely, decreases the number of predictions radically. This approach allows discarding false positives and passenger mutations. However, this selection could also cause the discarding of the cancer drivers.

In this test case, the union of the results contains nineteen gene fusions predictions, while the intersection includes just two of them (ETV6-NTRK3 and GOPC-ROS1).

4   Conclusions

FusionFlow is an easy-to-use, flexible, highly reproducible, and integrated workflow. The workflow includes five gene fusion discovery tools that input both RNA and DNA data. Docker and Conda technologies allow performing tools installations, avoiding version conflicts. In addition, the Nextflow framework allows the execution of the five tools in parallel, optimizing time and resources usage and managing the tool's installations and the file allocation. The workflow was tested using publicly available test files. The tests were performed using a local profile in two conditions: on a private server and the private server inside a docker container. In both cases, the outputs were satisfactory. Thus, the Fusion-Flow pipeline is available for further validation over additional DNA and RNA genomic data.

This work represents a foundation on which improvements and future works can be built. Indeed, one of the main problems related to gene fusion detection is determining which gene fusions are drivers of cancer processes and not just passenger mutations. The fusion detection tools already provide a first step for solving this problem. Indeed, fusion detection tools filter the candidate gene fusions based on the sample's reads, trying to decrease as much as possible the number of false positives. However, generally, this step is insufficient to determine the cancer drivers, and an additional step can be required. It consists of post-processing tools (called prioritization tools) that can predict a gene fusion's oncogenic potential. There is a high number of prioritization tools such as Oncofuse, Pegasus, DEEPrior, and ChimerDriver [1,17–19,25]. These tools are based on machine learning (ML) algorithms trained with the protein domains of the fusion proteins and allow the selection of the most probable cancer drivers. The post-processing step could also be completed by adding a different algorithm. This algorithm performs comparisons between the outputs of the tool and selects the more probable driver of cancer processes by analyzing the union and the intersection and taking into account the different characteristics of the gene fusion detection tools.

Another crucial question is related to visualization tools. Humans can efficiently distinguish true positives from false positives if the evidence is provided in an easily interpretable form. These tools also better interpret the potential consequence of gene fusion events. Several visualization tools were released in the last years, such as INTEGRATE-vis [31], FGviewer [13], and FuSpot [12].

Funding Information. This study was funded by the European Union's Horizon 2020 research and innovation programme DECIDER under Grant Agreement 965193.

References

1. Abate, F., et al.: Pegasus: a comprehensive annotation and prediction tool for detection of driver gene fusions in cancer. BMC Syst. Biol. **8**, 97 (2014). https://doi.org/10.1186/s12918-014-0097-z
2. Ahmed, S., Ali, M.U., Ferzund, J., Sarwar, M.A., Rehman, A., Mehmood, A.: Modern data formats for big bioinformatics data analytics (2017). https://www.ijacsa.thesai.org
3. Allegretti, S., Bolelli, F., Cancilla, M., Pollastri, F., Canalini, L., Grana, C.: How does connected components labeling with decision trees perform on GPUs? In: International Conference on Computer Analysis of Images and Patterns, pp. 39–51. Springer (2019). https://doi.org/10.1007/978-3-030-29888-3_
4. Allegretti, S., Bolelli, F., Pollastri, F., Longhitano, S., Pellacani, G., Grana, C.: Supporting skin lesion diagnosis with content-based image retrieval. In: 2020 25th International Conference on Pattern Recognition (ICPR), pp. 8053–8060. IEEE (2021)
5. Benelli, M., Pescucci, C., Marseglia, G., Severgnini, M., Torricelli, F., Magi, A.: Discovering chimeric transcripts in paired-end rna-seq data by using ericscript. Bioinformatics **28**, 3232–3239 (2012). https://doi.org/10.1093/bioinformatics/bts617
6. Bolelli, F., Baraldi, L., Pollastri, F., Grana, C.: A hierarchical quasi-recurrent approach to video captioning. In: 2018 IEEE International Conference on Image Processing, Applications and Systems (IPAS), pp. 162–167. IEEE (2018)
7. Chen, S., Liu, M., Huang, T., Liao, W., Xu, M., Gu, J.: Genefuse: detection and visualization of target gene fusions from dna sequencing data. Int. J. Biol. Sci. **14**, 843–848 (2018). https://doi.org/10.7150/ijbs.24626

8. Cirrincione, G., Randazzo, V., Kumar, R.R., Cirrincione, M., Pasero, E.: Growing curvilinear component analysis (GCCA) for stator fault detection in induction machines. In: Esposito, A., Faundez-Zanuy, M., Morabito, F.C., Pasero, E. (eds.) Neural Approaches to Dynamics of Signal Exchanges. SIST, vol. 151, pp. 235–244. Springer, Singapore (2020). https://doi.org/10.1007/978-981-13-8950-4_22

9. Cirrincione, G., Randazzo, V., Pasero, E.: Growing curvilinear component analysis (GCCA) for dimensionality reduction of nonstationary data. In: Esposito, A., Faudez-Zanuy, M., Morabito, F.C., Pasero, E. (eds.) Multidisciplinary Approaches to Neural Computing. SIST, vol. 69, pp. 151–160. Springer, Cham (2018). https://doi.org/10.1007/978-3-319-56904-8_15

10. Cirrincione, G., Randazzo, V., Pasero, E.: A neural based comparative analysis for feature extraction from ECG signals. In: Esposito, A., Faundez-Zanuy, M., Morabito, F.C., Pasero, E. (eds.) Neural Approaches to Dynamics of Signal Exchanges. SIST, vol. 151, pp. 247–256. Springer, Singapore (2020). https://doi.org/10.1007/978-981-13-8950-4_23

11. Goecks, J., Nekrutenko, A., Taylor, J.: Galaxy: a comprehensive approach for supporting accessible, reproducible, and transparent computational research in the life sciences. Genome Biol. 11(8), 1–13 (2010)

12. Killian, J.A., Topiwala, T.M., Pelletier, A.R., Frankhouser, D.E., Yan, P.S., Bundschuh, R.: Fuspot: a web-based tool for visual evaluation of fusion candidates. BMC Genom. 19, 139 (2018). https://doi.org/10.1186/s12864-018-4486-3

13. Kim, P., Yiya, K., Zhou, X.: Fgviewer: an online visualization tool for functional features of human fusion genes. Nucleic Acids Res. 48, W313–W320 (2021). https://doi.org/10.1093/NAR/GKAA364

14. Köster, J., Rahmann, S.: Snakemake-a scalable bioinformatics workflow engine. Bioinformatics 28(19), 2520–2522 (2012). https://doi.org/10.1093/bioinformatics/bts480, https://doi.org/10.1093/bioinformatics/bts480

15. Latysheva, N.S., Babu, M.M.: Discovering and understanding oncogenic gene fusions through data intensive computational approaches. Nucleic Acids Res. 44, 4487–4503 (2016). https://doi.org/10.1093/nar/gkw282

16. Lovino, M., Bontempo, G., Cirrincione, G., Ficarra, E.: Multi-omics classification on kidney samples exploiting uncertainty-aware models. In: Huang, D.-S., Jo, K.-H. (eds.) ICIC 2020. LNCS, vol. 12464, pp. 32–42. Springer, Cham (2020). https://doi.org/10.1007/978-3-030-60802-6_4

17. Lovino, M., Ciaburri, M.S., Urgese, G., Di Cataldo, S., Ficarra, E.: Deeprior: a deep learning tool for the prioritization of gene fusions. Bioinformatics 36(10), 3248–3250 (2020)

18. Lovino, M., Montemurro, M., Barrese, V.S., Ficarra, E.: Identifying the oncogenic potential of gene fusions exploiting mirnas. J. Biomed. Inform. 129, 104057 (2022)

19. Lovino, M., Urgese, G., Macii, E., Di Cataldo, S., Ficarra, E.: A deep learning approach to the screening of oncogenic gene fusions in humans. Int. J. Mol. Sci. 20(7), 1645 (2019)

20. Nicorici, D., et al.: Fusioncatcher - a tool for finding somatic fusion genes in paired-end rna-sequencing data. bioRxiv, p. 011650 (2014). https://doi.org/10.1101/011650

21. Paviglianiti, A., Randazzo, V., Pasero, E., Vallan, A.: Noninvasive arterial blood pressure estimation using abpnet and vital-ecg. In: 2020 IEEE International Instrumentation and Measurement Technology Conference (I2MTC), pp. 1–5. IEEE (2020)

22. Ponzio, F., Deodato, G., Macii, E., Di Cataldo, S., Ficarra, E.: Exploiting "uncertain" deep networks for data cleaning in digital pathology. In: 2020 IEEE 17th International Symposium on Biomedical Imaging (ISBI), pp. 1139–1143. IEEE (2020)
23. Ponzio, F., Villalobos, A.E.L., Mesin, L., de'Sperati, C., Roatta, S.: A human-computer interface based on the "voluntary" pupil accommodative response. Int. J. Hum. Comput. Stud. 126, 53–63 (2019)
24. Roy, S., et al.: Standards and guidelines for validating next-generation sequencing bioinformatics pipelines: a joint recommendation of the association for molecular pathology and the college of american pathologists, Jan 2018. https://doi.org/10.1016/j.jmoldx.2017.11.003
25. Shugay, M., Mendíbil, I.O.D., Vizmanos, J.L., Novo, F.J.: Oncofuse: a computational framework for the prediction of the oncogenic potential of gene fusions. Bioinformatics 29, 2539–2546 (2013). https://doi.org/10.1093/bioinformatics/btt445
26. Uhrig, S., et al.: Accurate and efficient detection of gene fusions from rna sequencing data
27. Vivian, J., et al.: Toil enables reproducible, open source, big biomedical data analyses. Nat. Biotechnol. 35(4), 314–316 (2017)
28. Wang, Q., Xia, J., Jia, P., Pao, W., Zhao, Z.: Application of next generation sequencing to human gene fusion detection: Computational tools, features and perspectives. Briefings Bioinf. 14, 506–519 (2013). https://doi.org/10.1093/bib/bbs044
29. Wang, Y., Shi, T., Song, X., Liu, B., Wei, J.: Gene fusion neoantigens: Emerging targets for cancer immunotherapy May 2021. https://doi.org/10.1016/j.canlet.2021.02.023
30. Williford, A., Betrán, E.: Gene fusion, May 2013. https://doi.org/10.1002/9780470015902.a0005099.pub3, https://onlinelibrary.wiley.com/doi/10.1002/9780470015902.a0005099.pub3
31. Zhang, J., Gao, T., Maher, C.A.: Integrate-vis: A tool for comprehensive gene fusion visualization. Scientific Reports 7, 17808 (2017). https://doi.org/10.1038/s41598-017-18257-2
32. Zhang, J., et al.: Integrate: gene fusion discovery using whole genome and transcriptome data. Genome Res. 26, 108–118 (2016). https://doi.org/10.1101/gr.186114.114

ProSA Pipeline: Provenance Conquers the CHASE

Tanja Auge$^{(\boxtimes)}$, Moritz Hanzig, and Andreas Heuer

University of Rostock, Rostock, Germany
{tanja.auge,andreas.heuer}@uni-rostock.de

Abstract. One of the main problems in data minimization is the determination of the relevant data set. Combining the CHASE—a universal tool for transforming databases—and data provenance, a (anonymized) minimal sub-database of an original data set can be calculated. To ensure reproducibility, the evaluations performed on the original data set must be feasible on the sub-database, too. For this, we extend the CHASE&BACKCHASE with additional *why*-provenance to handle lost attribute values, null tuples, and duplicates occurring during the query evaluation and its inversion. In this article, we present the ProSA pipeline, which describes the method of data minimization using the CHASE&BACKCHASE extended with additional provenance.

Keywords: Research data management · Data minimization · Data provenance · CHASE&BACKCHASE

1 Introduction

Research institutions around the world produce research data in large quantities. Collecting, evaluating, analyzing, archiving and publishing these data are major tasks of research data management. The effort required to manage and store these data volumes is often underestimated, especially when certain criteria, such as the traceability or reproducibility of a published result, are to be guaranteed. This can be supported substantially by adding additional provenance.

Let's imagine the following situation: We are writing an article for a journal or conference in which we have evaluated a larger data set. Since the conference or journal promotes reproducibility and traceability of new research results to be published, we are asked to publish our underlying data as well. Open data, as demanded in our example, is desirable in the interest of good science, but sometimes not feasible. Some reasons for not publishing data in detail are of privacy-related, economic, financial or military nature. We aim to limit the amount of data to the part that is relevant for publication. To determine this data set ProSA (**Pro**venance Management using **S**chema Mappings with **A**nnotations) uses a version of the CHASE&BACKCHASE. Additional provenance information such as *why*-provenance and additional side tables allows to reconstruct as much information as possible. However, we try to keep the amount of data to be stored as small as possible.

© Springer Nature Switzerland AG 2022
S. Chiusano et al. (Eds.): ADBIS 2022, CCIS 1652, pp. 89–98, 2022.
https://doi.org/10.1007/978-3-031-15743-1_9

Systems for determining *why*-provenance are already existing. A list of these can be found in [3], where we tested the best-known data provenance systems for their functionality and effectiveness. Most of them determine witness basis [7] based on a given SQL query. However, we provide a better suited approach. In ProSA we use the determined provenance in a subsequent step to classify not only the IDs of the original tuples. Instead, we determine the entire instance and then anonymize it as necessary. Some early ideas can be found in [1].

The foundation for our approach is a variant of the CHASE, a family of algorithms for reasoning with constraints [5]. It modifies a database instance I by incorporating a set of dependencies Σ represented as *(source-to-target) tuple generating dependencies* ((s-t) tgds) or *equality generating dependencies* (egds). While chasing (s-t) tgds creates new tuples, chasing egds "cleans up" the database by substituting null values. We regard in the following the application of the CHASE as chasing. We take advantage of this and use the CHASE to evaluate queries on a given database instance.

For evaluating queries, we define a query Q as a set of (s-t) tgds Σ. Then $\mathcal{M} = (S, T, \Sigma)$ with S source schema, T target schema, and Σ set of database dependencies specifying the relationship between S and T formalizes a *schema mapping* which can be processed by the CHASE. Thus, the CHASE can be used for evaluating conjunctive queries or simple SQL queries, such as SELECT n,m FROM R1, R2 WHERE R1.m = R2.m. Even aggregate functions such as MAX and evolution operators such as PARTITION Table, which can be formulated as sets of (s-t) tgds, can also be processed with our approach.

Applying the CHASE twice we call CHASE&BACKCHASE. This technique is used among others by [8] for query optimization or query reformulation. In these two cases, provenance supports also the optimization process of queries. We, on the other hand, do not use the CHASE&BACKCHASE for optimization but for evaluating queries and generating the minimal sub-database. In addition to evaluating queries, the CHASE can be used to evolve databases. A corresponding algorithm can be found in [2].

Consider again our research data management use case from the very beginning. Given a query Q and a database instance I, ProSA uses the CHASE&BACKCHASE to determine the minimal sub-database I^* relevant for publication. However, an exact inverse cannot always be computed without additional information. Thus, we extend the CHASE&BACKCHASE with *why*-provenance and additional side tables. But first, let's consider this with a concrete example. For this, let I be a database instance containing of two relations employee(id,name) and salary(id,sal), defined in Table 1a and 1b. Let further Q_1 and Q_2 be two SQL queries defined as:

Query Q_1	Query Q_2

```
SELECT name, sal
FROM employee NATURAL JOIN salary
WHERE name = "Stefan";
```

```
SELECT MAX(sal)
FROM salary
```

Then ProSA calculates in a first step, the CHASE *phase*, the query result $Q(I)$ and in a second step, the BACKCHASE *phase*, the reconstructed instance I^*. We call I^* *sub-database* of I if there is a homomorphism from I^* to I that maps all null values to corresponding constants in I. Furthermore, we call the sub-database I^* *minimal* if it is free of duplicates and solves the optimization problem $min \ |I^*|$ under the constraint $\text{CHASE}_{\mathcal{M}}(I^*) = \text{CHASE}_{\mathcal{M}}(I)$, i.e. the number of tuples of I^* is minimal respect to $Q(I) = Q(I^*)$.

After generating the minimal sub-database ProSA allows to perform a final anonymization, if necessary. For this, we generalize the sub-database based on user-defined domain generalization hierarchies. In doing so, we ensure k-anonymity and l-diversity, the most common anonymization measurements. The generated anonymized minimal sub-database we identify as I^*_{anon}.

Our Contribution. The ProSA pipeline describes a way to minimize data without loosing reproducibility of its corresponding evaluations. Our approach is particularly suitable for research data management, which more and more often requires the publication of data. For this, we combine the CHASE&BACKCHASE with additional provenance information such as ***why***-provenance and side tables. We not only determine the relevant source IDs, as in some known systems, but also determine and anonymize the resulting minimal sub-database.

In Sect. 2, we introduce the necessary definitions regarding the CHASE and data provenance. In Sect. 3.2, we present our ProSA pipeline. The theory behind we introduce in Section and close with a short description of our implementation in Sect. 3.3.

2 Basics and Related Work

Chase. The CHASE is a technique used in a number of data management tasks such as data exchange, data cleaning or answering queries using views [5,6]. It takes a set of dependencies Σ and an instance I as input and returns an instance J that satisfies all dependencies of Σ represented as (s-t) tgds and egds.

A *source-to-target tuple generating dependency* (s-t tgd) is a formula of the form $\forall \mathbf{x} : (\phi(\mathbf{x}) \rightarrow \exists \mathbf{y} : \psi(\mathbf{x}, \mathbf{y}))$, with $\mathbf{x}, \mathbf{y}$ tuples of variables, and $\phi(\mathbf{x})$, $\psi(\mathbf{x}, \mathbf{y})$ conjunctions of atoms, called *body* and *head*. A s-t tgd can be seen as inter-database dependency, representing a constraint within a database. An intra-database dependency is called *tuple generating dependency* (tgd) and a dependency with equality atoms in the head *equality generating dependency* (egd). Specifically, an egd is a formula of the form $\forall \mathbf{x}(\phi(\mathbf{x}) \rightarrow (x_1 = x_2))$ with $\mathbf{x}$ tuple of variables and $\phi(\mathbf{x})$ conjunction of atoms.

This means chasing (s-t) tgds create new tuples and chasing egds clean the database by replacing null values η_i (by other null values η_j or constants c_j). The CHASE can be used to compute a target instance J from a source instance I via a schema mapping $\mathcal{M}$ [9]. A *schema mapping* is a triple $\mathcal{M} = (S, T, \Sigma)$ with S source schema, T target schema and Σ set of constraints describing the relationship between S and T. It is semantically identified with the binary

Table 1. Example relations with associated provenance: source instance I with relations employee (a) and salary (b), query result $Q_1(I)$ resp. $Q_2(I)$ with relation result (c) resp. (d), minimal sub-database I^* with relations employee (e) and salary (f) resp. (h) as well as side table interim (g).

(a) employee

id	name	
1	Stefan	e_{id_1}
2	Stefan	e_{id_2}
3	Mia	e_{id_3}
4	Anna	e_{id_4}

(b) salary

id	sal	
2	1500	s_{id_1}
3	1500	s_{id_2}
4	1250	s_{id_3}
1	1470	s_{id_4}
2	1470	s_{id_5}

(c) result of S_1

name	sal	
Stefan	1500	$\{\{e_{id_2}, s_{id_1}\}\}$
Stefan	1470	$\{\{e_{id_1}, s_{id_4}\}, \{e_{id_2}, s_{id_5}\}\}$

(d) result of S_2

sal	
1470	$\{\{s_{id_1}\}, \{s_{id_2}\}\}$

(e) employee

id	name	
η_1	Stefan	e_{id_1}
η_2	Stefan	e_{id_2}

(f) salary

id	sal	
η_2	1500	s_{id_1}
η_1	1470	s_{id_4}
η_2	1470	s_{id_5}

(g) interim

name	
1250	e_{id_3}
1470	e_{id_4}
1470	e_{id_5}

(h) salary

id	sal	
η_1	1500	s_{id_1}
η_2	1500	s_{id_2}

relation: $\text{Inst}(\mathcal{M}) = \{(I, J) \mid (I, J) \models \Sigma\}$, whereby I is S-instance, J is T-instance and Σ defined as set of (s-t) tgds and egds.

The CHASE produces a target instance J, denoted as $\text{CHASE}_{\mathcal{M}}(I)$, as follows: For every s-t tgd $\forall \mathbf{x} : (\phi(\mathbf{x}) \rightarrow \exists \mathbf{y} : \psi(\mathbf{x}, \mathbf{y}))$ in Σ and for every tuple t_1 of constants of I we add to $\text{CHASE}_{\mathcal{M}}(I)$ all facts in $\psi(t_1, \eta)$ where η is a tuple of new, distinct labeled nulls interpreting the existential quantified variables $\mathbf{y}$. However, if there is a tuple t_2 of constants c_i or labeled null values η_i such that $\psi(t_1, t_2)$ already exists in J, no new tuple is created.

Let's take a look at a concrete example. Let S be source schema with two relations employee(id,name) and salary(id,sal) and T target schema with one relation result(name,sal). Let further I be the source instance defined in Table 1a and 1b. Let $\Sigma = \{$employee(id, "Stefan") $\wedge$ salary(id, sal) $\rightarrow$ result("Stefan", sal)$\}$ be formalization of Q_1. Then the corresponding schema mapping is defined as triple $\mathcal{M} = (S, T, \Sigma)$. Let employee(1, Stefan) and salary(1, 1470) be two tuples of I. The CHASE of I respectively $\mathcal{M}$ will then produces a target instance $J = \text{CHASE}_{\mathcal{M}}(I)$ that consists the fact result(1, 1470).

Provenance. As described in [11], "provenance generally refers to any information that describes the production process of an end product, which can be anything from a piece of data to a physical object." For this, let I be a database instance and Q conjunctive query. It describes (1) ***where*** a result tuple $r \in Q(I)$ comes from, (2) ***why*** and (3) ***how*** r exists in the result $Q(I)$. In this paper, we focus on ***why***-provenance [7], which specifies a witness basis identifying the tuples involved in the calculation of r. This tuple-based basis contains all IDs necessary for reconstructing r.

For example, let employee(1, Stefan, e_{id_1}), employee(2, Stefan, e_{id_2}), salary(1, 1470, s_{id_1}) and salary(2, 1470, s_{id_2}) be four tuples of I defined in Table 1a and 1b. The corresponding witness basis $\{\{e_{id_1}, s_{id_4}\}, \{e_{id_2}, s_{id_5}\}\}$ consists of two witnesses $\{e_{id_1}, s_{id_4}\}$ and $\{e_{id_2}, s_{id_5}\}$. The first duplicate uses the tuples e_{id_1} and s_{id_e}, while the second uses the tuples e_{id_2} and s_{id_5}.

3   The ProSA-Pipeline

As seen in Fig. 1, the ProSA pipeline consists of ten steps. The CHASE&
BACKCHASE is executed in step ④ (evaluating query Q) and step ⑦ (evaluating inverse query Q^{-1}), outsourced in ChaTEAU [4]. The specification of the
inverse query Q^{-1} is done by the Inverter in step ⑤. However, an *exact inverse*—
a function f with $f \circ f = $ id—cannot always be computed without additional
information. This is especially the case for highly selective and summarizing
queries. But provenance can solve this problem.

Preparing provenance takes place in the Provenancer in step ③. Here provenance IDs and additional side tables are introduced. The calculation of **why**-
provenance is part of the CHASE phase and its evaluation is considered in the
BACKCHASE phase. Subsequently, the Controller in step ⑧ verifies the correctness of the calculated minimal sub-database before it is anonymized in the
Anonymizer in step ⑨.

3.1   The Theory Behind ProSA

Chase & Backchase. In this article, we use the CHASE to evaluate queries on
instances. Let Q be a query formalized as a set of (s-t) tgds Σ and I over
S a given database instance. Let further be $Q(I)$ query result over T. Then
the corresponding schema mapping is defined as $\mathcal{M} = (S, T, \Sigma)$ and applying
the CHASE yields $\text{CHASE}_{\mathcal{M}}(I) = I(Q)$. On the other hand, let Q^{-1} be the
query inverse to Q, which is also formalized as a set of (s-t) tgds Σ^{-1}. Let further be $\mathcal{M}^* = (T, S, \Sigma^{-1})$ the corresponding schema mapping with I^* instance
over S. Then applying the CHASE yields $I^* = \text{CHASE}_{\mathcal{M}^*}(Q(I))$. Altogether the
CHASE&BACKCHASE yields: $\text{CHASE}_{\mathcal{M}^*}(\text{CHASE}_{\mathcal{M}}(I)) = \text{CHASE}_{\mathcal{M}^*}(Q(I)) = I^*$.
For simplicity, we denote Σ instead of $\mathcal{M}$ and Σ^{-1} instead of $\mathcal{M}^*$ in the following. Additionally, we use Σ_P and Σ_P^{-1} to refer to the dependency set extended by
provenance annotations p. For the CHASE& BACKCHASE with additional prove-
nance this means $\text{CHASE}_{\Sigma_P^{-1}}(\text{CHASE}_{\Sigma_P}(I)) = \text{CHASE}_{\Sigma_P^{-1}}(Q(I)) = I^*$.

If one extends both the database instance I and the query Q with additional
provenance annotations (step ③), these are automatically integrated into the
query result $Q(I)$ in the CHASE phase. In the BACKCHASE phase, this information is used to reconstruct the minimal sub-database. Again, this is done
automatically, provided that the inverse query has been extended by provenance
annotations before (step ⑤).

Combining **Chase** *and* **Provenance.** To include additional provenance information in the CHASE, we require globally unique provenance IDs. These we
construct using the relation name (or its first letter) with the suffix id and a
continuous integer number. Specifically, the i-th tuple $R(x_{i_1}, ..., x_{i_n})$ of relation
R receives the provenance ID R_{id_i}, which is stored as an additional, new attribute
in R. So, each tuple of I has the form $R(x_{i_1}, ..., x_{i_n}, R_{\text{id}_i})$.

Additionally, we modify the tgds and s-t tgd by adding provenance IDs to
the dependencies body and head. Analogous to the tuples of I, the body's relations receive additional IDs R_{id_i}. These new attributes are also appended to

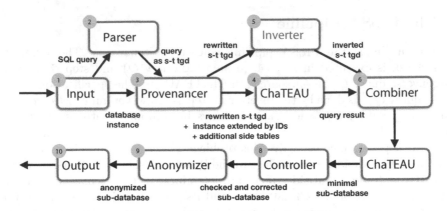

Fig. 1. ProSA pipeline

the result tuple in the head. In summary, this results in dependencies such as $R(x_{i_1}, ..., x_{i_n}, R_{\mathrm{id}_i}) \wedge R'(x_{j_1}, ..., x_{j_m}, R'_{\mathrm{id}_j}) \rightarrow \mathtt{result}(r_1, ..., r_k, R_{\mathrm{id}_i}, R'_{\mathrm{id}_j})$. So, the ID attributes correspond to the witness set $\{R_{\mathrm{id}_i}, R'_{\mathrm{id}_j}\}$ [7], which is created by joining the two relations R and R'. Due to the attribute-wise storage of the provenance IDs in the result tuple, duplicates need not be considered.

By extending the source tuples in I and the dependencies with additional provenance IDs, the calculation of witness sets is outsourced to the CHASE phase. The evaluation of the **why**-provenance, i.e. the integration of additional information based on witness sets, can also be outsourced to the CHASE. This information is used to reconstruct lost attribute values, null tuples and duplicates.

When the query is inverted, redundant null values and tuples may be created in the BACKCHASE phase. These are subsequently eliminated by global variable substitution. For this, we make use of the key properties of the Provenance IDs internally defined by equality generating dependencies (egds) such as $R(x_{i_1}, ..., x_{i_n}, R_{\mathrm{id}}) \wedge R(x_{j_1}, ..., x_{j_n}, R_{\mathrm{id}}) \rightarrow x_{i_1} = x_{j_1}, ..., x_{i_n} = x_{j_n}$. No further input is required from the user. The egds perform global data cleaning internally. At the end, ProSA returns a minimal sub-database I^* of I, which is computed using the CHASE&BACKCHASE extended by additional provenance information.

For queries with highly condensed information, we need to store concrete attribute values. This values are stored in a separate side table, which consists of the attribute value itself as well as the corresponding provenance ID: $R(x_i, R_{\mathrm{id}_i})$. The tables are calculated during the CHASE phase and used in the BACKCHASE phase to reconstruct the lost attribute values. The extension is done by the Provenancer by $R(x_{i_1}, ..., x_{i_n}, R_{\mathrm{id}_i}) \rightarrow \mathtt{result}(r_1, ..., r_k, R_{r_{\mathrm{id}_1}}) \wedge \mathtt{side}(x_{i_j}, R_{\mathrm{id}_i})$. The combination of side tables and **why**-provenance is sufficient to check the results of SPJU queries extended by aggregation, grouping and more.

Chase *Execution by ChaTEAU*. The CHASE is a technique used in a number of data management tasks such as data exchange, data cleaning or answering queries using views [5]. The versatile applicability of the CHASE is due to the fact that one can pass different types of objects and parameters as input. As described in [4], this object can be an instance as needed here, or alternatively a query. ChaTEAU (**Cha**se for **T**ransforming, **E**volving, and **A**dapting databases and queries, **U**niversal Approach) abstracts instances and queries to a general CHASE object and parameter. It thus provides a multi-purpose implementation of the CHASE. In ProSA, we use ChaTEAU for processing the CHASE as well as the BACKCHASE phase.

Anonymizing the Minimal Sub-database. In the ProSA pipe-line, it is feasible to anonymize at different points in time: It can be done in the input, i.e. the source instance I or the query Q can be anonymized. Anonymizing the source database is theoretically possible, but the effort involved is disproportionate. However, anonymizing the query violates reconstructability and can lead to problems in the ProSA pipeline. Anonymizing the intermediate ProSA steps, i.e., the provenance annotations Σ_P as well as the query result $I(Q)$, is not very useful, too. These would make using the *why*-provenance obsolete. We therefore choose to anonymize the minimal sub-database I^*. As anonymization measure, we use k-anonymity [13] as well as l-diversity, and as anonymization method, generalization using domain generalization hierarchies based on [12].

Limitations of ProSA. The key component of ProSA is the CHASE. Hence we are limited to processing (s-t) tgds and/or egds. Thus, queries that cannot be formulated as a set of (s-t) tgds cannot be evaluated with ProSA.

3.2   The Steps of the ProSA Pipeline

1 + **10** *Input and Output.* The input—a database instance I and a SQL query Q—and output—the minimal sub-database I^*_{anon}—is organized via a *GUI*. It is the central interface between the user and ProSA. It is composed of five tabs: First of all, we need a connection to the database to be minimized. This can be established via the *DB Configuration* tab. The calculation of the minimal sub-database is then done in the next three tabs (CHASE, BACKCHASE and *Anonymizer*). Finally, the *Log* offers additional information about CHASE execution such as the results of the single CHASE steps after applying an (s-t) tgd or egd. The single ProSA steps can be tracked here, too.

2 *Parsing SQL Queries.* ProSA uses the CHASE for evaluating queries. However, the CHASE deals with dependencies such as egds, tgds or s-t tgds and not with SQL queries, which we want to use as input for a better user experience. Hence, the *Parser's* main task is to transform these SQL queries into a set of (s-t) tgds. It transforms conjunctive SPJU queries, aggregate functions such as `Min`/`Max`, `Count`, `Sum` and `Avg`, nested queries as well as grouping.

For rendering simple conjunctive queries, an s-t tgd is usually required. So, a SQL query `SELECT` $\psi(\mathbf{x})$ `FROM` $\phi(\mathbf{x})$ corresponds to an s-t tgd $\phi(\mathbf{x}) \rightarrow \psi(\mathbf{x})$ with $\psi(\mathbf{x})$ query result. However, there are queries formalized as two s-t tgds. This includes, for example, union, nested queries or aggregate functions. For example, the query (`SELECT` $\psi(\mathbf{x})$ `FROM` $\phi_1(\mathbf{x})$) `UNION` (`SELECT` $\psi(\mathbf{x})$ `FROM` $\phi_2(\mathbf{x})$) corresponds to $\phi_1(\mathbf{x}) \rightarrow \psi(\mathbf{x}), \phi_2(\mathbf{x}) \rightarrow \psi(\mathbf{x})$. In the case of Q_1 we get:

$$\texttt{employee}(\text{id}, \text{"Stefan"}) \wedge \texttt{salary}(\text{id}, \text{sal}) \rightarrow \texttt{result}(\text{"Stefan"}, \text{sal}).$$

3 *Including **Provenance**.* The *Provenancer* prepares the input of the CHASE phase in such a way that provenance information can be collected. For this, he adds global unique provenance IDs to the (s-t) tgds of Σ as well as to the source tuples of I. We thus obtain Σ_P and I_P as provenance extensions of Σ and I. If necessary, additional side tables are created here, too. Afterwards, the Provenances rewrites the (s-t) tgds in Σ_P as described in Sect. 3.1.

In the case of Q_2, the SQL query is transformed into two dependencies. The tgd creates an additional side table `interim`, which stores all intermediate results (see Table 1g). The s-t tgd provides the maximum salaries using the side table as a filter. After that, the Provenancer adds global unique provenance IDS (highlighted in purple) to the s-t tgd of Σ as well as to the source tuples of I. In summary, the associated rewritten query $Q_{2,P}$ is:

$$\texttt{salary}(\text{id}_1, \text{sal}_1, s_{\text{id}_1}) \wedge \texttt{salary}(\text{id}_2, \text{sal}_2, s_{\text{id}_2}) \wedge \text{sal}_1 < \text{sal}_2 \rightarrow \texttt{interim}(\text{sal}_1, s_{\text{id}_1}), \quad \text{(tgd)}$$

$$\texttt{salary}(\text{id}, \text{sal}, s_{\text{id}}) \wedge \neg \texttt{interim}(\text{sal}, s_{\text{id}}) \rightarrow \texttt{result}(\text{sal}, s_{\text{id}}). \quad \text{(s-t tgd)}$$

And I is extended to $I_P = \{\texttt{salary}(2, 1500), s_{\text{id}_1}), ..., \texttt{salary}(2, 1470), s_{\text{id}_5})\}$.

4 *ChaTEAU (**Chase Phase**).* In the CHASE phase, *ChaTEAU* determines the query result $Q(I)$ by incorporating the dependencies from Σ_P into the database instance I, i.e. $Q(I) = \text{CHASE}_{\Sigma_P}(I)$. In addition to the query result extended by **why**-provenance, the CHASE phase provides the results of **where**- and **how**-provenance. The provenance types are calculated for the sake of completeness, but are no longer used in the following.

The query results can be found in Table 1c and Table 1d. Note that a result tuple such as $\texttt{result}(\text{"Stefan"}, 1470) \in Q_{1,P}(I_P)$ is stored internally as $\texttt{result}(\text{Stefan}, 1470, e_{\text{id}_1}, s_{\text{id}_4})$. In this way, duplicates are retained, but not displayed to the user. The same is done for the tuples of $Q_{2,P}(I_P)$.

5 *Inverting s-t tgds.* In this module, a parsed query Q is inverted. The inverse query Q^{-1} always corresponds to one s-t tgd, which then be processed in the BACKCHASE phase. For inverting the query, we extend the *maximum-extended-recovery-algorithm* of [10] by grouping the inverted dependencies into a minimal number of s-t tgds. In the case of $Q_{1,P}^{-1}$, for example, the Inverter yields:

$$\texttt{result}(\text{"Stefan"}, \text{sal}, e_{\text{id}}, s_{\text{id}}) \rightarrow \exists \text{ Id} : \texttt{employee}(\text{Id}, \text{"Stefan"}, e_{\text{id}}) \wedge \texttt{salary}(\text{Id}, \text{sal}, s_{\text{id}}).$$

6 *Preparing the* **Backchase**. The *Combiner* combines the inverse Q_P^{-1} with the query result $Q_P(I_P)$ to obtain a valid input for the BACKCHASE phase.

7 *ChaTEAU (**Backchase Phase**).* The technical design of the BACKCHASE phase does not differ from the CHASE phase. Only the input differs, i.e. Σ_P^{-1} instead of Σ_P and $Q(I)$ instead of I. Here, *ChaTEAU* determines the minimal sub-database I^* by incorporating the dependencies from Σ_P^{-1} into the query result $Q(I)$, i.e. $I^* = \text{CHASE}_{\Sigma_P^{-1}}(Q(I))$ with Σ_P^{-1} formalization of Q^{-1}. In doing so, the provenance information P is automatically integrated into I^*.

By applying the CHASE, lost attribute values are filled with null values and duplicates are reconstructed. Thus I^* contains regarding to Q_1 the tuples employee(η_1,Stefan, e_{id_1}), employee(η_2,e_{id_2}), and employee(η_3,Stefan, e_{id_2}).

8 *Proving the Calculated Sub-database and Correcting Duplicates.* The controller checks whether the calculated minimal sub-database I^* is correct concerning reproducibility of the query result $Q(I_P)$. This is important, because only then the calculated sub-database satisfies the requirements of reproducibility at all and can be co-published along with the research result $Q(I_P)$. For this, the controller calculates $Q(I^*)$ and compares it with $Q(I_P)$ calculated in the CHASE phase. If the check fails, the calculated minimal sub-database has been incorrectly determined and an error message is returned.

In addition, the controller eliminates redundant tuples, which can be created in the BACKCHASE phase. For this, the key properties of the provenance IDs are exploited as described in Sect. 3.1 by a directly implemented global variable substitution. So, replacing null values in I^* resp. Q_1 leads to the deletion of the superfluous and wrong tuple employee(η_3,Stefan,e_{id_2}) as shown in Table 1e by

$$\texttt{employee}\big(\mathrm{id}_1,\,\mathrm{name}_1,\,e_{\mathrm{id}}\big) \wedge \texttt{employee}\big(\mathrm{id}_2,\,\mathrm{name}_2,\,e_{\mathrm{id}}\big) \rightarrow \mathrm{id}_1 = \mathrm{id}_2 \wedge \mathrm{name}_1 = \mathrm{name}_2.$$

9 *Anonymizing the Sub-database.* ProSA guarantees an k-anonym and l-diverse anonymized minimal sub-database I^*_{anon}. For this, we anonymize I^* by generalization using previously defined domain generalization hierarchies. The calculation of possible quasi-identifiers is done by ProSA itself.

3.3   Implementing ProSA

ProSA is implemented as a Maven project in Java 11. It provides a user interface that allows connecting a database and answering simple conjunctive queries as well as any query represented as a set of (s-t) tgds. For processing the CHASE, ProSA calls ChaTEAU, an implementation of the CHASE [4]. ProSA itself performs query parsing, adding provenance information, preparing the CHASE execution and interpreting the CHASE result. In addition, the minimal sub-database is checked for correctness and then be anonymized by generalization in compliance with k-anonymity [13] and l-diversity. The software, examples, and further information about ChaTEAU and ProSA are available at our Git repositories[1],[2].

---

[1] ChaTEAU repository: https://git.informatik.uni-rostock.de/ta093/chateau-demo.
[2] ProSA repository: https://git.informatik.uni-rostock.de/ta093/prosa-demo.

4   Conclusion

In this article, we introduced the ProSA pipeline for minimizing research data before publishing. Therefore, we combine a variant of the CHASE&BACKCHASE with additional provenance such as *why*-provenance and side tables to reconstruct lost attribute values, null tuples and duplicates. Subsequently we anonymize the calculated sub-database to satisfy possibly occurring privacy aspects.

Acknowledgments. We thank all students involved in implementing ProSA: Leonie Förster, Melinda Heuser, Ivo Kavisanczki, Judith-Henrike Overath, Tobias Rudolph, Nic Scharlau, Tom Siegl, Dennis Spolwind, Anne-Sophie Waterstradt, Anja Wolpers, and Marian Zuska. Also, thanks to Bertram Ludäscher for comments and suggestions.

References

1. Auge, T., Heuer, A.: ProSA—using the CHASE for provenance management. In: Welzer, T., Eder, J., Podgorelec, V., Kamišalić Latifić, A. (eds.) ADBIS 2019. LNCS, vol. 11695, pp. 357–372. Springer, Cham (2019). https://doi.org/10.1007/978-3-030-28730-6_22
2. Auge, T., Heuer, A.: Tracing the history of the Baltic sea oxygen level. In: BTW, LNI, vol. P-311, pp. 337–348. Gesellschaft für Informatik, Bonn (2021)
3. Auge, T., Heuer, A.: Testing provenance systems. Technical report CS 01-22, Computer Science Division, University of Rostock (2022)
4. Auge, T., Scharlau, N., Görres, A., Zimmer, J., Heuer, A.: ChaTEAU: a universal toolkit for applying the CHASE. https://arxiv.org/abs/2206.01643
5. Benedikt, M., et al.: Benchmarking the chase. In: PODS, pp. 37–52. ACM (2017)
6. Benczúr, A., Kiss, A., Márkus, T.: On a general class of data dependencies in the relational model and its implication problems. Comput. Math. Appl. **21**(1), 1–11 (1991)
7. Cheney, J., Chiticariu, L., Tan, W.C.: Provenance in databases: why, how, and where. Found. Trends Databases **1**(4), 379–474 (2009)
8. Deutsch, A., Hull, R.: Provenance-directed chase&backchase. In: Tannen, V., Wong, L., Libkin, L., Fan, W., Tan, W.-C., Fourman, M. (eds.) In Search of Elegance in the Theory and Practice of Computation. LNCS, vol. 8000, pp. 227–236. Springer, Heidelberg (2013). https://doi.org/10.1007/978-3-642-41660-6_11
9. Fagin, R., Kolaitis, P.G., Popa, L., Tan, W.C.: Schema mapping evolution through composition and inversion. In: Bellahsene, Z., Bonifati, A., Rahm, E. (eds.) Schema Matching and Mapping. Data-Centric Systems and Applications. Springer, Heidelberg (2011). https://doi.org/10.1007/978-3-642-16518-4_7
10. Fagin, R., Kolaitis, P.G., Popa, L., Tan, W.C.: Reverse data exchange: coping with nulls. ACM Trans. Database Syst. **36**(2), 11:1–11:42 (2011)
11. Herschel, M., Diestelkämper, R., Ben Lahmar, H.: A survey on provenance - what for? what form? what from? VLDB J. **26**(6), 881–906 (2017)
12. Han, J., Cai, Y., Cercone, N.: Data-driven discovery of quantitative rules in relational databases. IEEE Trans. Knowl. Data Eng. **5**(1), 29–40 (1993)
13. Samarati, P.: Protecting respondents' identities in microdata release. IEEE Trans. Knowl. Data Eng. **13**(6), 1010–1027 (2001)

A Tabular Open Data Search Engine Based on Word Embeddings for Data Integration

Alberto Berenguer(ID), Jose-Norberto Mazón(✉)(ID), and David Tomás(ID)

Department of Software and Computing Systems,
University of Alicante, Alicante, Spain
{aberenguer,jnmazon,dtomas}@dlsi.ua.es

Abstract. Nowadays, open data has become a prominent information source for creating value-added product and services. Actually, open data portal initiatives are adopted by most of the governments to supply their public sector information, usually in the form of tabular data such as spreadsheets or CSV files. Most open data portals allow reusers to retrieve tabular open data by means of a keyword-based search engine over metadata. However, these search engines rely on the (not so often good enough) metadata quality, which must be complete, descriptive, and representative of the tabular open data content. Moreover, keyword-based search is not always an adequate solution for retrieving open data, since it does not consider the tabular nature of (most) open data and search results can be useless for reusers (e.g., when they attempt to find open data to be integrated with a given tabular dataset). To overcome these problems, this paper presents *Search!*, a search engine that enables users to pose an input query table to retrieve adequate tabular open data to be integrated with. To do so, semantic searches are performed by leveraging word embeddings to compute the similarity between column names and cell contents of tabular data. The relevance criteria established in the search engine aims to retrieve a ranking of tabular open datasets suitable for completion and augmentation, and thus, enabling integration with the input query table.

Keywords: Information retrieval · Search engine · Open data ·
Tabular data · Data integration · User interfaces

1 Introduction

Currently the amount of open data available on the Web is increasing due to the strong interest of governments and institutions around the world in adopting open data initiatives [1]. Within these initiatives, open data portals are developed to publish public sector information under the appropriate licence to encourage its re-use (e.g., by the development of value-added products and services).

Importantly, the ultimate goal of open data portals is to provide Linked Open Data (LOD), allowing consumers to use semantic web technologies to identify

© Springer Nature Switzerland AG 2022
S. Chiusano et al. (Eds.): ADBIS 2022, CCIS 1652, pp. 99–108, 2022.
https://doi.org/10.1007/978-3-031-15743-1_10

relationships among data [2] and unleash the full potential of data to be reused. Unfortunately, LOD is not always available in open data portals, since the most used formats are tabular (e.g., CSV), accounting for 46.5% in government data portals as stated by [12]. Furthermore, prevalence of tabular formats in open data government portals has been highlighted in recent studies, such as the Open Government Report from Organisation for Economic Co-operation and Development (OECD).[1]

This scenario hampers users of open data government portals, since additional efforts must be done to find datasets that can be successfully integrated with their data [6]. For example, imagine the following motivating example: a data journalist that wishes to expand an initial dataset containing the number of refugees arriving in France last year with information about refugees arriving in Spain. This open data can be found in the World Bank data portal,[2] which includes a keyword-based search engine. Data journalist poses the query "refugees in Spain" in the search engine, but there are no relevant results. Modifying the query to "refugees by country" also returns no results. Therefore, the data journalist does not retrieve the required open data (although World Bank data portal indeed contains it in the "World Development Report" dataset), as metadata does not represent the content of the dataset, so it does not match the keywords that were intuitive a priori.

From this motivating example, two main problems are illustrated:

1. Tabular open data search engines suffer from high dependence on metadata quality (e.g., title or description of datasets must be complete and accurate) in order to be able to provide reliable results. Keeping updated and accurate metadata is a major challenge for data publishers (as stated by the European Comission in its 2020 Open Data Maturity Report[3]) that can seriously affect the performance of search engines.
2. Furthermore, it is not easy for open data reusers to express their intentions with a keyword-based search. If reusers want to expand an initial dataset with additional tabular data (e.g., using union and join operations), representing the input query as a table instead of as keywords is more appropriate.

To overcome these problems, this paper presents *Search!*,[4] a tabular open data search engine optimised for dataset integration. The system provides a table-based interface and uses word embeddings [10] to calculate the semantic similarity between tables, improving the system recall and avoiding the need for exact matches between queries and metadata to retrieve relevant datasets.

The rest of this paper is structured as follows: Sect. 2 presents related work; Sect. 3 describes our approach for tabular open data search and integration; an example of the use of the search engine is described in Sect. 4; finally, in Sect. 5, conclusions and future work are presented.

---

[1] https://www.oecd.org/gov/open-government-data-report-9789264305847-en.htm.
[2] https://databank.worldbank.org.
[3] https://data.europa.eu/en/dashboard/2020.
[4] https://wake.dlsi.ua.es/datasearch.

2   Related Work

Data search is a task studied for decades. However, there are still open problems like disconnected datasets, meeting user needs, and the availability and reliability of data [3]. Another problem is that of how to rank the results, including approaches that use keyword-based search [13] and others employing more sophisticated techniques based on supervised learning [9].

Regarding how the data is processed, organisations responsible for publishing datasets generally provide metadata like title, description, column names, author, creation date or language. To carry out this task, organisations can use standards like DCAT,[5] a vocabulary that proposes metadata to describe datasets and data services in an open data catalog. Open data portals frequently utilise data publishing platforms aligned with DCAT, such as CKAN,[6] Socrata[7] or OpenDataSoft.[8] These platforms include search engines that make use of (previously unified) metadata from datasets. They are called centralised search engines [3]) and they present limited searching capabilities as they focus on a unique data catalogue. Also, they entirely depends on metadata availability and quality (e.g., ability of metadata to correctly describe the data content).

Decentralised search engines, on the other hand, go beyond the boundaries of a single open data portal and provide ways to discover the datasets across multiple catalogues [3], e.g., searching on LOD. There are also decentralized search engines on tabular open data, such as the general-purpose Google Dataset Search[9] that uses crawlers to search and index datasets that follow schema.org or DCAT standards. Similarly, there are search tools focused on specific domains, such as DataMed,[10] which implements a crawler that identifies and indexes sets of scientific biomedical data. Decentralised search engines aims to adequately manage and enrich of metadata for each dataset in order to improve retrieval results. However, they are keyword-based search engines, which is not always an adequate solution for retrieving open data, since it does not consider the tabular nature of (most) open data and search results can be useless for reusers (e.g., when they attempt to find open data to be integrated with a given tabular dataset).

Unlike existing tabular open data search engines, our system (i) makes use of word embeddings to provide semantic similarity capabilities that improve the recall of relevant datasets beyond metadata-based search, and (ii) provides a suitable tabular-based interface for the task of tabular open data retrieval and integration for a given query in tabular form beyond keyword-based search.

---

[5] https://www.w3.org/TR/vocab-dcat-2/.
[6] https://ckan.org.
[7] https://dev.socrata.com.
[8] https://www.opendatasoft.com.
[9] https://datasetsearch.research.google.com/.
[10] https://datamed.org.

3   Tabular Open Data Search and Integration with Word Embeddings

Our search engine consist of a microservice, a crawler, an API, and a Web application (see Fig. 1). Its architecture can be divided into two main parts. The first one is responsible for information search and indexation through a crawler that accesses short-listed open data portals, obtains tabular datasets (currently only in CSV format) and metadata, and calculates the corresponding word embeddings for each table. The stored information is available through an API.

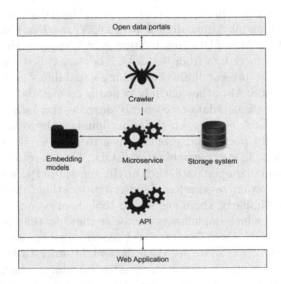

Fig. 1. System conceptual architecture

The second part corresponds to the Web application that uses the afore-mentioned API to, given an input query table, retrieve relevant stored tabular datasets for further integration. The front-end was built using HTML, CSS, and JavaScript. The back-end API was created using Node.js, providing a way to communicate the core functionalities of the search engine and the Web application.

3.1   Crawler

This component consist of a Python script which is responsible for collecting suitable datasets in short-listed open data portals. This crawler aims to inspect and extract metadata information from the datasets, as well as preparing metadata for further processing and storage. It uses DCAT to recognise and analyse the metadata.

3.2 Storage

The storage system is composed of two elements: Solr,[11] a powerful information retrieval system used to index metadata and contents of the tabular datasets crawled from the Web; and Faiss,[12] an indexation tool optimised for vector operations with capacity for billions of vectors and compatible with GPU architectures.

Faiss is used to store the word embedding vectors of the indexed tabular datasets and to calculate their similarity with the vectors representing the input query table, providing the information required to build a ranked list of relevant results (see Sect. 3.3). It is worth noting that, a previous solution for vector comparison was implemented using a Solr plugin, but the calculations were excessively time-consuming, making the system unsuitable for its deployment in a production environment. To overcome this issue, Faiss was used, thus speeding up the comparison between word embedding vectors using clustering techniques.

3.3 Microservice

This component is the core of the search engine, where developed algorithms run. It carries out three main tasks: obtaining the word embedding vectors that represent each tabular dataset using fastText,[13] comparing tabular datasets by means of these vectors, and searching for related datasets given an input query table.

In the first task, for each tabular dataset indexed in the system, two word embedding vectors are retrieved for each column following the procedure described in [6]: one represents the names of the columns (headers) and the other the content of the cells. First, the CSV files representing the tabular datasets are preprocessed to normalised the content (e.g. lowercase and split camelCase words). Then, the fastText model is employed to obtain the vector representation of each token in the tabular dataset. In order to obtain a single vector for multiple tokens (e.g., the content of all the cells in the column) the average of the individual vectors is computed [8]. All this process is done offline, when new tabular dataset are crawled, and does not affect during search time.

The second task carried out is tabular dataset comparison based on word embedding vectors. When the user introduces an input query table, the corresponding word embedding vectors are obtained following the procedure mentioned above. To calculate the similarity between the query table and the stored tabular datasets, the similarity between each column is calculated in Faiss using the following equation:

$$sim(c1, c2) = \alpha \cdot sim(cn1, cn2) + (1 - \alpha) \cdot sim(cc1, cc2), \qquad (1)$$

where $sim(cn1, cn2)$ represents the similarity between column names (headers) and $sim(cc1, cc2)$ between cell contents. The parameter α is a value in the range

---

[11] https://solr.apache.org.
[12] https://faiss.ai.
[13] https://fasttext.cc.

[0, 1] that indicates the importance of each part (column names and cell contents) when the final similarity value is calculated [6].

The similarity between word embedding vectors is calculated using the cosine similarity [11], obtaining a value between 0 and 1, where 0 means no similarity and 1 means maximum likeness. In those situations where there are out-of-vocabulary tokens, with no vector representation in fastText, the Levenshtein distance is used as a backup strategy. The final output of this task is a JSON structure containing the similarity of each column of the query table with each column of the stored tabular datasets, sorted by column similarity.

Finally, the third task takes as an input the JSON structure computed before to carry out the search functionality. Once the similarity between column vectors is calculated, the final similarity between the input query table and each indexed tabular dataset is computed as follows:

$$sim(t1, t2) = \frac{\sum_{i=1,j=i}^{i \leq n, j \leq m} sim(c_{1i}, c_{2j})}{|C1||C2|}, \tag{2}$$

where $C1 = \{c_{11}, c_{12}...c_1 n\}$ and $C2 = \{C_{21}, C_{22}...C_{2m}\}$ are the set of columns from $t1$ and $t2$ respectively. That is, the similarity between two tables (i.e., the input query table and the tabular dataset) is computed as the average similarity of their columns. Once these values are computed, tables are sorted by descending order to provide the user with a ranked list.

3.4   Web Application

The Web application offers the user a novel way to search tabular data (see Fig. 2). The landing page shows an empty table that can be manually filled with values, or it can be automatically filled by uploading a CSV file from the local computer. Once the table is filled with values, it can be used as an input query table to search for related tabular datasets stored in the system.

The results page shows a ranking of the most similar tabular datasets stored in the system for the given input query table. Each result includes metadata information like title, description, license type, and publication day. This page also shows the similarity score for each table retrieved. Three options are provided for each result: preview the content, download and go to the integration interface.

In the integration interface (see Fig. 3) the users can easily integrate the input query table and the selected retrieved tabular dataset. The query table appears at the top of the page, whereas the retrieved tabular dataset is below this. When the interface is loaded, a quick arrangement in the column order is done by taking into account the semantic similarity of the columns, suggesting which columns of the input query table can be integrated with those on the retrieved tabular dataset. If necessary, this arrangement can be changed manually by users, selecting the columns that they want to be involved in the integration process. Once the columns are selected, the users choose which operation to carry out (according to our previous work on data integration [6]): union, if

Fig. 2. Screenshot of the tabular open data search interface.

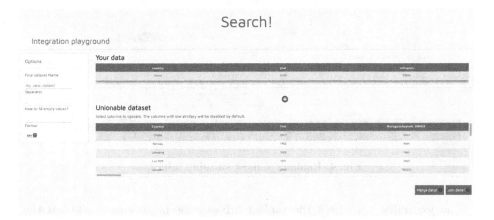

Fig. 3. Screenshot of the data integration interface.

they want to perform a row extension of the query table, or join, if they look for column augmentation. This interface also provides integration parameters for the merged tabular dataset than can be downloaded, such as name, output format (e.g., CSV), etc.

4 Using *Search!*

The data journalist case mentioned in Sect. 1 is used here to explain how to use *Search!*. The journalist wanted to expand their dataset containing the number of refugees arriving in France last year with information about refugees arriving in Spain, but unfortunately she did not make it using open data portals. For this reason, she decides to try *Search!*.

To use *Search!*, she only needs to provide a CSV file with initial data; alternatively, she can manually fill the table in the landing page. She chooses the first option and uploads a CSV with the country name, year, and the number of refugees from France (see Fig. 2). Two search options are then offered depending on the goal: one focused on table completion and the other on table augmentation. In the case concerning the journalist, the selected option is the first one, since she needs to complete her dataset with information from another country. Once the search option is selected, a ranking with the most suitable tables to integrate is shown (see Fig. 4), including metadata and similarity score.

Fig. 4. Ranking of most suitable tables.

The journalist can check the ranked tables content to ensure she has the content she wants to integrate and decide which result is more appropriate. The first result, called "World Development Report 2011", contains precisely the information she is looking for. She can download this dataset directly or, if she decides to integrate her data directly using *Search!*, she can use the integration interface with quick similar columns pairing. In this interface the journalist can select or edit the columns she wants to integrate and choose the operation to perform (see Fig. 3). Finally, she can download a CSV file with the integrated data.

5   Conclusions and Future Work

Within an open data scenario, development of tools that support retrieval and integration of tabular datasets beyond keyword-based search on metadata are highly required for boosting open data reuse [6]. This work has described *Search!*,[14] a tabular open data retrieval and integration system. It uses word

---

[14] https://wake.dlsi.ua.es/datasearch.

embeddings to calculate the semantic similarity between tabular information, providing a ranking of candidate datasets to be integrated with an input query table. The system includes a novel interface to facilitate table-based search and integration. *Search!* provides open data reusers with a new way to search and integrate compatible information with their data, offering a more flexible approach to these tasks than classic keyword-based systems do.

Future work includes improving the current crawling system, making it compatible with more open data portals to index a wider range of datasets. Also, it is planned to admit different tabular formats (besides CSV) in the system in order to increase its coverage. As word embedding models are language-dependent, the version of fastText described in *Search!* only works for English. As a future work, models in other languages can be used to make the system suitable for a wider audience. Another possibility is to use multilingual word embeddings [7] that can be applied seamlessly to different languages. The use of contextual word embeddings like BERT [5] instead of static models like fastText is another issue that is worth exploring in the future. Contextual models have demonstrated to obtain state-of-the-art results in many tasks involving language processing and have been recently used also in the task of table retrieval [4]. Finally, our plan is to conduct an evaluation of *Search!* with potential users.

Acknowledgements. This research has been funded by project "Desarrollo de un ecosistema de datos abiertos para transformar el sector turístico" (GVA-COVID19/2021/103) funded by "Conselleria de Innovación, Universidades, Ciencia y Sociedad Digital de la Generalitat Valenciana".

References

1. Altayar, M.S.: Motivations for open data adoption: an institutional theory perspective. Government Inf. Q. **35**(4), 633–643 (2018)
2. Bizer, C., Heath, T., Berners-Lee, T.: Linked data: The story so far. In: Semantic Services, Interoperability and Web Applications: Emerging Concepts, pp. 205–227. IGI global (2011)
3. Chapman, A., Missier, P., Simonelli, G., Torlone, R.: Capturing and querying fine-grained provenance of preprocessing pipelines in data science. Proc. VLDB Endow. **14**(4), 507–520 (2020)
4. Chen, Z., Trabelsi, M., Heflin, J., Xu, Y., Davison, B.D.: Table search using a deep contextualized language model. In: Proceedings of 43rd International ACM SIGIR Conference on Research and Development in Information Retrieval, pp. 589–598. Association for Computing Machinery (2020)
5. Devlin, J., Chang, M., Lee, K., Toutanova, K.: BERT: pre-training of deep bidirectional transformers for language understanding. In: Proceedings of Conference of the North American Chapter of the ACL: Human Language Technologies, NAACL-HLT 2019, Minneapolis, USA, 2–7 June 2019, pp. 4171–4186. Association for Computational Linguistics (2019)
6. González-Mora, C., Tomás, D., Garrigós, I., Zubcoff, J.J., Mazón, J.N.: Model-driven development of web apis to access integrated tabular open data. IEEE Access **8**, 202669–202686 (2020)

7. Grave, E., Bojanowski, P., Gupta, P., Joulin, A., Mikolov, T.: Learning word vectors for 157 languages. In: Proceedings of the International Conference on Language Resources and Evaluation (LREC 2018) (2018)
8. Gupta, S., Kanchinadam, T., Conathan, D., Fung, G.: Task-optimized word embeddings for text classification representations. Front. Appli. Math. Stat. **5**, 1–67 (2020)
9. Gysel, C., Rijke, M., Kanoulas, E.: Neural vector spaces for unsupervised information retrieval. ACM Trans. Inf. Syst. **36** (2017)
10. Mikolov, T., Chen, K., Corrado, G., Dean, J.: Efficient estimation of word representations in vector space. In: 1st International Conference on Learning Representations, ICLR 2013 pp. 1–12 (2013)
11. Mikolov, T., Sutskever, I., Chen, K., Corrado, G., Dean, J.: Distributed representations of words and phrases and their compositionality. In: Proceedings of 26th International Conference on Neural Information Processing Systems, NIPS 2013, Red Hook, NY, USA, vol. 2, pp. 3111–3119 (2013)
12. Neumaier, S., Umbrich, J., Polleres, A.: Automated quality assessment of metadata across open data portals. J. Data Inf. Qual. (JDIQ) **8**(1), 1–29 (2016)
13. Zhang, S., Balog, K.: Ad hoc table retrieval using semantic similarity. In: Proceedings of the 2018 World Wide Web Conference, WWW 2018, International World Wide Web Conferences Steering Committee, Republic and Canton of Geneva, CHE, pp. 1553–1562 (2018)

The Use of M2P in Business Process Improvement and Optimization

Meriem Kherbouche[✉][iD], Yossra Zghal[iD], Bálint Molnár[iD],
and András Benczúr[iD]

Information Systems Department, Faculty of Informatics Pázmány Péter 1/C,
Eötvös Loránd University (ELTE), Budapest 1117, Hungary
{meriemkherbouche,p599ba,molnarba,abenczur}@inf.elte.hu

Abstract. The goals of business process automation and improvement
are to increase productivity, satisfy clients and workers, and reduce costs.
In this paper, we improved and redesigned processes to use electronic documents, improve the velocity of activities, and enhance transparency of
processes using the DMAIC method (Define, Measure, Analyze, Improve,
and Control). First, we used Papyrus to represent the activity diagram
for the insurance claim process. Then, using Acceleo, an algorithm was
constructed using the retrieved XML to change the non-improved XML
file into an improved one. Finally, using the updated XML, we produced
the improved UML activity diagram.

Keywords: M2P (Model to Program) · BPM · Business process
improvement · Model transformation

1 Introduction

Due to the fast-growing market, organizations are undergoing rapid, driven by
such pressures as customer expectations, product quality, new technology, agility,
and growing global competition. As a result, business processes within an organization are constantly changing. To survive in such environments, it's important
for organizations to continually measure, analyze, improve, control, and redesign
their business process to quickly respond to changes.

There are some techniques and tools that aim to help businesses improve
their processes. However, Business Process Modelling plays a significant role to
identify weaknesses within the process.

The unified modeling language (UML) is a business process modeling language that helps to describe the elements that make up a specific software system
and describe how elements interact with each other. Moreover, UML is known
as the most transformed language defined as a meta-model. This transformation

Supported by Development and Innovation Fund of Hungary, financed under the Thematic Excellence Programme TKP2020-NKA-06, TKP2021-NVA-29, COST Action
'CA19130.

is a mechanism for deriving the source model (Input) into another model (as output) while executing a set of rules and algorithms.

In this paper, we will present a novel transformation from a non-improved UML activity diagram model to an improved UML activity diagram by constructing a set of transformation algorithms.

2   Related Works

Various research studies focus on process improvement [12], several solutions were investigated for this purpose [15]. The authors in [16] focuses on business process improvement techniques. There are several methodologies available, which causes organizations to become confused while deciding on the best approach. All methodologies have common features. They only differ in their efficacy. The purpose of their article is to raise awareness and perspectives of organizations and assist them in making the correct decision in selecting the appropriate technique based on organizational goals.

The Book of [10] carries on the topic of improvement. They define lines to improve the process including setting the stage for process improvement, organizing for process improvements, using flowcharting techniques to draw the process, understanding the process characteristic, streamlining the process, and measurement and Benchmarking Process. The paper of [19] outlines a case study of business process design that has been optimized using the state-of-the-art multi-objective optimization method NSGA2.

The paper [11] introduce as a first step a roadmap that supports the goal-oriented execution of BPI projects as well as the systematic development of suggestions to overcome process weaknesses. Based on the BPI roadmap, a domain-specific modeling method (DSMM) [7–9] is developed. The third contribution is the development of the DSMM as a prototype modeling tool that facilitates the application of the BPI roadmap, analyzes information defined in the form of conceptual models, and creates reports automatically based on the findings generated.

[18] investigate the impact of continuous improvement on the design of activity based cost systems. The main contribution of [19] is that it identifies business process models and optimization. Another paper [5] aims at carrying out business process modelling and business process improvement using TAD methodology. [14] uses the Unified Modelling Language (UML) as a process-modelling technique for clinical-research process improvement.

Recent researches are made in the domain of BPI. The authors in [15] represented a literature review about different pattern based approaches for business process improvement. [4] propose an Object-Process Methodology for business process improvement and they applied it to optimize an aviation manufacturing company study case. The authors in [2] aim to identify which activities are essential in improvement projects depending on culture, organizational size, and resources. Using a multiple-case study approach discovering how improvement is to be addressed in organizations of different contexts.

3 Preliminaries

3.1 Business Process Management (BPM)

Business Process Management (BPM) is the discipline that combines knowledge from information technology and knowledge from management sciences and applies this to operational business processes [1]. BPM aims to improve a business process. This approach creates a more efficient organization better able to deliver products and services and adapt to changing needs. And as far as the complexity of the business process is increasing, organizations are obliged to deal with the complexity and focus on BMP.

The BPM lifecycle standardizes the process of implementing and managing business processes within the company as a series of stages. The BPM lifecycle has five stages: design, model, execute, monitor, and optimize [17].

- Process Design: The goal of process Design is to analyze and understand the process. In this situation, a mock-up can be useful it helps in collecting and display of data.
- Business process modeling: Modelling means identifying, defining, and representing the new process to support the current business rules for various stack holders.
- Process Execution: The goal of Process Execution is to see the process in action. It refers to the actual run of a process by a process engine.
- Process monitoring: The goal of Process Monitoring is to perform business operations by prepared and implementing process descriptions.
- Process Optimization: Business process Optimization is redesigning the process to improve efficiencies and strengthen individual business processes.

3.2 Business Process Improvement

Business process improvement (BPI) is an operational practice used to identify and evaluate inefficiencies within the companies. This approach redesigns existing business tasks, implements, and optimizes their existing processes, and improves their effectiveness. The goal of business process improvement and business process optimization is customer and employee satisfaction, productivity, reduced risk, compliance, agility, and technology integration are identifying wasted resources, improving the quality of products, achieving regulatory compliance, reducing friction in the process, and reducing process completion time.

3.3 Business Process Modelling

Business process modeling is a fundamental step of business process management. With Process Modeling, we make a comprehensive analysis of all processes and it helps to identify weaknesses within the process. Implementing Business process modeling help to optimize the business process and identify the challenges in the project. There are several techniques used in Business process modeling to establish the organizational goal.

3.4 Unified Modelling Language

Unified Modeling Language (UML) is a business process modeling technique that helps to describe the elements that make up a specific software system and describe how elements interact with each other.

Activity Diagram. An activity diagram provides a view of the behavior of a system. It describes the sequence of actions in a process. An activity diagram is composed of nodes and edges. Activity nodes consist of invocation nodes, object nodes, and control nodes. Invocation nodes are basic activity nodes in which some piece of behavior is invoked, Object nodes denote the availability of objects, and control nodes are used to specify choice and parallelism [6]. There are several kinds of control nodes such as Decision Nodes that choose one of the outgoing flows, Merge Nodes that merge different incoming flows, Fork nodes that split a flow into several parallel flows, the initial node that starts the global flow, and Final nodes that stop the current local flow. There are two kinds of ActivityEdge to link the nodes: ObjectFlow and ControlFlow. ObjectFlow edges connect ObjectNodes and can have data passing along it. ControlFlow edges constrain the desired order of execution of the ActivityNodes [13].

3.5 Model Transformation

Model transformation is an automated way to generate and modify models. The goal of using model transformation is to reduce errors by automating the building and modifying the models. The model transformation takes models as input and generates an output based on the modification.

There are different types of model transformation:

– Model To Text Transformation (M2T): M2T refers to a large class of transformations that translate models to text. The text can be generated code, other models in textual syntax, or other textual artifacts such as reports or documentation [3].
– Model To Model Transformation (M2M): Transformations are executed by transformation engines that are plugged into the Eclipse Modelling infrastructure

4 Contribution

The aim of this research is the use of business process modeling for the improvement and optimization of the business process. First, we modeled the insurance claim process activity diagram using Papyrus. It is a graphical editing tool for UML as defined by OMG that provides editors for all UML diagrams. It also provides complete support for SysML to enable a model-based system. Then, extracted the corresponding XML. After that, we created an algorithm that transform the non-improved XML file into an improved one using Acceleo. It is a template-based technology that includes authoring tools for custom code generators. It produces any kind of source code from any data source available in EMF format.

The improvement algorithm extracts the keyword from the XML format. We analyze the non-improved diagram and apply the transformation based on the keyword with the use of Acceleo and JAVA programming languages. Finally, we generated the improved UML activity diagram from the improved XML as shown in Fig. 1.

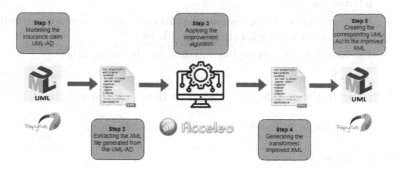

Fig. 1. Summary of the insurance claim improvement process.

By applying a transformation algorithm on the non-improved UML Activity Diagram, we get an improved UML Activity Diagram.

4.1   Transformation Table

The UML Activity Diagram contains the following Components: INITIAL NODE, EDGES, OPAQUE ACTION, JOIN NODE, FORK NODE, DECISION NODE, and FINAL NODE.

The Table 1 represents the used components from UML activity diagram in the algorithm.

Table 1. UML AD components

Components	Name in the algorithm
INITIAL NODE	IN
EDGE	ED
OPAQUE ACTION	OA
JOIN NODE	JN
FORK NODE	FK
DECISION NODE	DN
FINAL NODE	FN
PACKAGE	PK

We denote the Input UML Activity Diagram NonImproved as (UM) and the Output UML Activity Diagram Improved as (Y).

Table 2 represents the variables used to clearly understand the algorithm.

Table 2. Variables used in the algorithm

Variable	Meaning
i	i refers to the current Line in the file (xml)
z,w	Intermediaries file
xf	Xml line generated after calling the function with the specific attributes Xf contains the line with the changed keyword
i'	The Xml line after fixing the outgoing and incoming Edge

Table 3 refers to the specific words used in the improved algorithm.

Table 3. Specific words used in the algorithm

Specific word	Meaning
read	We read from the XML File
write	We write in the Final XML File
call	Used to call function or procedure
Change	Change used to change the keyWord
Fix	Fix the outgoing and Incoming Edge. Since we delete some Opaque Action so the order of edges will change
generate	Each Line has outgoing and Incoming so we need to generate Edge based on the Incoming and the Outgoing Directions of the Line
nextLine	We read the file line by Line
specificWord	The algorithm of improvement based on the specific Word the opaque action. Once we find specific Word in the Opaque Action we change it
type	The type of the Components
"XXXX"	XXXX is a keyword used in case the Line need to be removed and no need for this action

4.2   The Algorithm of Transformation

The final algorithm for transformation from a non-improved UML activity diagram to an improved UML activity diagram is demonstrated in Algorithm 1.

4.3   Insurance Claim Case Study

The insurance industry is always evolving. There has been an increase in client expectations in the area of new insurance products. Customers have become

Algorithm 1. Algorithm :UML Activity Diagram (XML) not improved to UML Activity improved

---

Input A UML Activity Diagram UM=(IN,ED,OA,JN,FK,DN,FN,PK)
Output Improved Activity Diagram Y=(IN, ED,OA,JN,FK,DN,FN)
Initiation
F:=read file(UM) ; generate_Components(UM)

    while nextLine(F) =true **do**
      read i in UM
      if $i \in$PK **then**
        write *ii*in Y
      else {$i \in$OA and *ii*specificWord }
        change OA:keyword ;generate xf ; write xf in Z
      end if
    end while
    while nextLine(Z) =true **do**
      read i in Z
      if type i =IN **then**
        fix IN:OutgoingEdge ; write i' in W ; call proc_fixEdgeForInitialNode
      else {type i =OA and name i "XXXX" }
        fix OA:IncommingEdge && OA:OutgoingEdge ; write i' in W ; call proc_fixNextLineEdge
      else {type i =JN or type i = DN }
        fix JN_DN:IncommingEdge && JN_DN:OutgoingEdge ; write i' in W ; call proc_checkNextOutgoingEdge
      else {type i=FK }
        fix FK:OutgoingEdge ; write i' in W
      else {type i =FN }
        write i in W
      end if
    end while
    while nextLine(W) =true **do**
      read i in W
      if type I =OA or type I = FK or type I=DN or type I=FN **then**
        generate Edge ; write Edge in K
      end if
    end while
    while NextLine K =true **do**
      read i in k ; Write i in Y
    end while
    while NextLine W =true **do**
      read i in W ; Write i in Y
    end while

---

more demanding and have learned to expect the greatest level of service. In this way, insurance companies are facing considerable changes in terms of providing speedier responses, handling insurance claims, lowering operating costs, pleasing customers to retain clients, and selling insurance products. To gain customer satisfaction, it's important for insurance to improve their process.

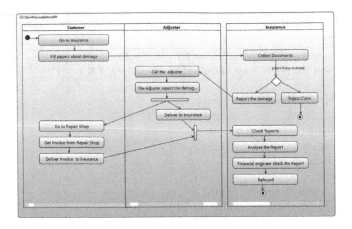

Fig. 2. Insurance Company Activity Diagram before the improvement

Improving processes demands identifying internal weaknesses, and business process modeling is the best way to do it. Figure 2 represent the UML Activity Diagram representation of the earlier Insurance Company. The Claim process within the insurance company has a lot of issues that result:

1. Customer needs to wait for a long time. Sometimes the process takes more than one month which is not acceptable for part of customers.
2. Insurance Employee wastes time since he spends a long time searching and regrouping documents.
3. Waste of resources since the Insurance company needs a third party (adjuster) to check the car status.
4. There is no online system for the enterprise. Coordination between employees is difficult.

The mechanism of the Insurance Company must be emphasized. As a result, we extracted the corresponding XML, and we made an algorithm that transform the non-improved XML file into an improved one using Acceleo and Java Programming language and as output, we get the XML corresponding to the improved UML activity diagram. The improved UML activity diagram for the claim process insurance company is represented in Fig. 3. With the use of modeling, we replace the weaknesses within an insurance company's processes to increase customer satisfaction, technology, and competitiveness. The improved UML activity diagram is toward the use of technology and digitalization with insurance companies.

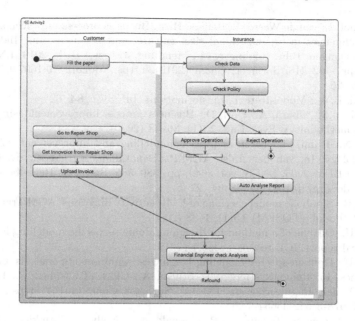

Fig. 3. Insurance Company Activity Diagram after the improvement

5   Conclusion and Future Works

In this paper, we use the activity diagram Unified Modelling Language to improve and optimize business processes by demonstrating a transformation from a non-improved UML-AD to an improved UML-AD. We performed a sequence of model-to-text transformation stages, implementing a set of transformation rules and some algorithms.

For future work, two directions are planned, firstly the transformation will be expanded to incorporate other UML diagrams. Secondly, the transformation from XML to UML Activity Diagram must be performed using automated tools.

Acknowledgements. The project was supported by the "Application Domain-Specific Highly Reliable IT Solutions" project that has been implemented with the support provided by the National Research, Development and Innovation Fund of Hungary, financed under the Thematic Excellence Programme TKP2020-NKA-06, TKP2021-NVA-29 (National Challenges Subprogramme) funding scheme. Furthermore, this article is partly based upon work from COST Action 'CA19130—Fintech and Artificial Intelligence in Finance—Towards a transparent financial industry', supported by COST (European Cooperation in Science and Technology), https://www.cost.eu/actions/CA19130/, accessed on 2021-12-20.

References

1. Van der Aalst, W.M.: Business process management: a comprehensive survey. International Scholarly Research Notices 2013 (2013)

2. Beerepoot, I., van de Weerd, I., Reijers, H.A.: Business process improvement activities: differences in organizational size, culture, and resources. In: Hildebrandt, T., van Dongen, B.F., Röglinger, M., Mendling, J. (eds.) BPM 2019. LNCS, vol. 11675, pp. 402–418. Springer, Cham (2019). https://doi.org/10.1007/978-3-030-26619-6_26
3. Asowski, B.W.: Model-to-text transformations. Inf. Syst. **84**, 62
4. Casebolt, J.M., Jbara, A., Dori, D.: Business process improvement using object-process methodology. Syst. Eng. **23**(1), 36–48 (2020)
5. Damij, N., Damij, T.: Business process modelling and improvement using TAD methodology. In: van der Aalst, W.M.P., Benatallah, B., Casati, F., Curbera, F. (eds.) BPM 2005. LNCS, vol. 3649, pp. 380–385. Springer, Heidelberg (2005). https://doi.org/10.1007/11538394_27
6. Eshuis, R.: Symbolic model checking of UML activity diagrams. ACM Trans. Softw. Eng. Methodol. (TOSEM) **15**(1), 1–38 (2006)
7. Frank, U.: Outline of a method for designing domain-specific modelling languages. Technical report, ICB-research report (2010)
8. Frank, U.: Domain-specific modeling languages: requirements analysis and design guidelines. In: Reinhartz-Berger, I., Sturm, A., Clark, T., Cohen, S., Bettin, J. (eds.) Domain Engineering, pp. 133–157. Springer, Cham (2013). https://doi.org/10.1007/978-3-642-36654-3_6
9. Gray, J., Neema, S., Tolvanen, J.P., Gokhale, A.S., Kelly, S., Sprinkle, J.: Domain-specific modeling. In: Handbook of Dynamic System Modeling, vol. 7, p. 7-1 (2007)
10. Harrington, H.J., et al.: Business process improvement. Association for Quality and Participation (1994)
11. Johannsen, F., Fill, H.-G.: Meta modeling for business process improvement. Bus. Inf. Syst. Eng. **59**(4), 251–275 (2017). https://doi.org/10.1007/s12599-017-0477-1
12. Kashfi, H., Aliee, F.S.: Business process improvement challenges: a systematic literature review. In: 2020 11th International Conference on Information and Knowledge Technology (IKT), pp. 122–126. IEEE (2020)
13. Kherbouche, M., Mukashaty, A.A., Molnár, B.: An operationalized transformation for activity diagram into YAWL. In: Developments in Computer Science, pp. 17–19 (2021)
14. Kumarapeli, P., De Lusignan, S., Ellis, T., Jones, B.: Using unified modelling language (UML) as a process-modelling technique for clinical-research process improvement. Med. Inform. Internet Med. **32**(1), 51–64 (2007)
15. Missaoui, N., Ayachi Ghannouchi, S.: Pattern-based approaches for business process improvement: a literature review. In: Park, J.H., Shen, H., Sung, Y., Tian, H. (eds.) PDCAT 2018. CCIS, vol. 931, pp. 390–400. Springer, Singapore (2019). https://doi.org/10.1007/978-981-13-5907-1_42
16. Rashid, O.A., Ahmad, M.N.: Business process improvement methodologies: an overview. J. Inf. Syst. Res. Innov. **5**, 45–53 (2013)
17. Szelągowski, M.: Evolution of the BPM lifecycle. In: Federated Conference on Computer Science and Information Systems, pp. 205–211 (2018)
18. Tunney, P., Reeve, J.M.: The impact of continuous improvement on the design of activity based cost systems. J. Cost Manag., 43–50 (1992)
19. Vergidis, K., Tiwari, A., Majeed, B.: Business process improvement using multi-objective optimisation. BT Technol. J. **24**(2), 229–235 (2006). https://doi.org/10.1007/s10550-006-0065-2

Embedding Process Structure in Activities for Process Mapping and Comparison

Andrea Chiorrini[1]($\boxtimes$), Claudia Diamantini[1], Laura Genga[2], Martina Pioli[1], and Domenico Potena[1]

[1] Dipartimento di Ingegneria dell'Informazione, Universitá Politecnica delle Marche, Ancona, Italy
a.chiorrini@pm.univpm.it, {c.diamantini,d.potena}@univpm.it
[2] Department of Industrial Engineering and Innovation Sciences, Eindhoven University of Technology, Eindhoven, The Netherlands
l.genga@tue.nl

Abstract. Today's organizations often have to manage hundreds of process models. This requires organizations to be able to efficiently manage process models as a kind of organizational data. Most of previous approaches for process model representation exploit graph data structures to represent (part of) the process control-flow. While these representations work well to analyze the overall process behavior and its KPIs, they are complex data structures which pose some challenges for other kind of analyses requiring, e.g., model comparison. In this paper we explore an alternative approach to representing process structure. We introduce a set of features that describe the context of an activity inside a process, thus embedding the information about the process structure in the activity representation. This representation enables the use of standard vector techniques to capture certain structural similarities among activities and is then suited to support tasks like similarity-based comparison and mapping of processes without resorting to graph-based approaches.

Keywords: Business Process Management · Process similarity · Process model clustering · Activity similarity

1 Introduction

Business Process Management has rightfully entered among the best business management practices. It helps analyzing the work flow, recognizing interdependencies among activities or bottlenecks, monitoring and improving performance. It mainly relies on a high-level description of the structure of business processes (in the following business process models) which provide an invaluable picture of the (operational) status of an organization and, as such, represent a kind of organizational data (like organizational charts or database schemes) to be managed. The management of (possibly hundreds) business process models is also relevant

© Springer Nature Switzerland AG 2022
S. Chiusano et al. (Eds.): ADBIS 2022, CCIS 1652, pp. 119–129, 2022.
https://doi.org/10.1007/978-3-031-15743-1_12

in the context of information sharing in supply chains and open science [5,7]. It requires operations like retrieving a process in a process repository, comparing different versions of a process, either generated as the result of the evolution of the same process over time or representing alternative operational practices in different organizations, finding common sub-processes for integration purposes, and so forth. Most Business Process Modeling notations (e.g. BPMN, Petri Nets) focus on the representation of the relations between activities with a control-flow perspective, and are thus based on graph data structures. While these representations work well to analyze the overall process behavior and its KPIs, these are complex data structures which pose some challenges for operations like those listed before. In this paper we introduce a set of features that describe the context of an activity inside a process, thus embedding the process structure in the activity representation. We introduce the features, explain the rationale under them, and discuss how they can be used to derive a measure of similarity among activities. Processes using standard vector processing. As case study, we apply the approach to address the problem of activity identification and comparison within and between process models. Results demonstrates that these features are able to capture certain structural similarities among activities and are then suited to support tasks like similarity-based comparison and mapping of processes without resorting to graph-based approaches. In the following Sections, Sect. 2 discusses related work about business process representation and similarity assessment approaches, in Sect. 3 the features are described. Section 4 is devoted to assess the effectiveness of the proposed features in identifying similar activities either in the same or in a different process. Finally, Sect. 5 concludes the paper and draws future work.

2   Related Work

A number of process modeling notations are commonly used to represent organization processes. Well-known examples are Petri nets or BPMN, which represent a process in the form of a graph. When event logs tracking process executions are available, process mining techniques can be used to derive alternative representations of process executions. This can be done either by combining information from the event log and the process model, as done by, e.g., alignments or trace replay techniques, or inferring directly follows mappings directly from the event log [1,2]. Over time, the need to browse and query processes has induced the scientific community to investigate further possible representations. In [8] the authors present the usage and comparison of various possible techniques, in particular graph matching techniques, alignment techniques and causal footprint, i.e. an abstract representations of the behavior captured by a process model to define similarity metrics at a process model level of granularity. In contrast with these techniques, the present work focuses on a finer granularity, i.e. at the level of single activities. Lately, due to the attention that has been devoted to predictive tasks in the process mining field, how to encode the information regarding a business process is becoming crucial. Several works [2,3,13] have tackled the problem of encoding traces, i.e. the sequence of activities or events that represent a particular process execution. Other works have also tried to encode the information regarding only a partial trace, in order to leverage such information to perform predictions [4,14,17]. Another recent proposal [6] consists in the usage of neural networks to learn a representation of activities, logs, and process models in order to derive an informative and low-dimensional embedding. In [11] word2vec has been used in conjunction with an LSTM network to label nodes in business process models. It is worth noting that approaches based on neural network representations don't have a clear semantic. Differently from those approaches we purposely design each feature to measure a well-defined structural characteristic and to be more suited to human comprehension.

3   Methods

This section describes the proposed features. They have been designed for Petri nets, which is the best investigated formalism for process modelling [16]. Each set of such features represent a single process activity describing its execution context, intended as the control-flow constructs in which the activity is involved (e.g., parallelism, choice, loop), so that activities with similar structural properties will be close to each other in the feature space.

Figure 1 shows an example of Petri net. Squared boxes represent process activities, while circles (aka, places) are used to represent possible *states* of the process. Edges model the activity ordering relations. Black boxes represent *hidden* activities, i.e. activities that are not observed by the information systems and are mainly used for routing purposes.

Path Length. This feature represents the position of the activity w.r.t. the beginning of the process. Given an activity a_i in the set of all process activities $\mathcal{A}$, we define the longest path $LP(a_i)$ as the maximum number of activities in a path from the beginning of the process to a_i, excluding hidden activities and loops. The Path Length for a_i is then computed as follows:

$$PL(a_i) = \frac{LP(a_i)}{\max_{a_j \in \mathcal{A}} LP(a_j)} \tag{1}$$

The Path Length assumes values in $]0, 1]$, where small values indicate that the activity is located near the beginning of the process and values close to 1 are of activity near the end of the process. For the first activity in the process the value is $\dfrac{1}{\max\limits_{a_j \in \mathcal{A}} LP(a_j)}$, while for the farthest one it is 1. It should be noted that the farthest activity is not necessarily the last activity in the process, although the two concepts often coincide. The feature is computed using the longest path so to ensure coherent order among all process activities representation, w.r.t. the process model. To better understand this, consider activities c and e in

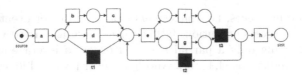

Fig. 1. Example Petri net 1

Fig. 1, we can notice that if we used the shortest instead of the longest we would obtain a value of $\frac{3}{4}$ for activity c and of $\frac{2}{4}$ for activity e, but activity e is always executed after activity c, according to the model. Hence, wanting to maintain in the feature the relative order of the activities in the process the shortest path would be inadequate.

Example 1. In Fig. 1, the maximum length of the longest path is 6, namely $< a, b, c, e, f, h >$ (or $< a, b, c, e, g, h >$). Hence, for the first and last activities we have $PL(a) = \frac{1}{6} = 0.17$ and $PL(h) = 1$, which correspond to the lowest and highest value of the feature respectively. The activities b and d have the same value because they both have to wait for one activity to be executed before they can be activated ($PL(b) = PL(d) = 0.33$), whereas c has to wait for both a and b to be executed ($PL(c) = 0.5$). Regarding the activity e, it should be noted that it can be started only if the parallel block preceding e is terminated, that is both branches $< b, c >$ and $< d >$ are executed. So the delay of e depends on the longest branch and is $PL(e) = \frac{4}{6} = 0.67$.

Optionality. This feature captures the degree of optionality of a specific activity. Optionality is computed as the reciprocal of the number of the maximum alternative branches in which an activity is involved, as follows:

$$Opt(a_i) = 1 - \frac{1}{n_{\#}(a_i)} \tag{2}$$

where $n_{\#}(a_i)$ represents the number of alternative branches for the activity a_i.

The feature assumes values in $[0, 1[$, where 0 represents a sequential activity and values close to 1 are related to activities in choice block with many possible branches. The feature is not computed for hidden activities, and backward branches (i.e., defining loops) are non taken into account in the formula.

Example 2. let's consider again the example in Fig. 1. Here we have two choice blocks: after the activity a and before the activity h. For the former, we have $Opt(b) = Opt(c) = Opt(d) = 0.67$ because there are three possible alternatives: execute the activities b and c, perform d or neither (passing through the hidden activity $t1$ in the lowest branch). For the latter, we do not consider the branch with $t2$ because it represents a loop, so $Opt(h) = 0$. The optionality for all other activities in the process model assumes the value 0.

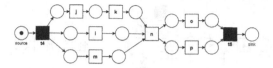

Fig. 2. Example Petri net 2

Parallelism. This feature is used to calculate the degree of parallelism of an activity, i.e. with how many activities it is in parallel. It is calculated evaluating the number of parallel branches in the parallel block containing the activity. The Parallelism (Par) for the activity a_i is computed as follows:

$$Par(a_i) = 1 - \frac{1}{n_{\parallel}(a_i)} \qquad (3)$$

where $n_{\parallel}(a_i)$ represents the number of parallel branches for the activity a_i. In the case of nested parallel blocks, the maximum number of possible activities in parallel is considered. The feature assumes values in $[0, 1[$, where 0 represents a sequential activity and the closer the value is to 1, the higher the number of parallel branches. The feature is not computed for hidden activities.

Example 3. In Fig. 2, we have two parallel blocks: one from hidden activity $t4$ to activity n, and the other from n to $t5$. As for the former, we have $Par(j) = Par(k) = Par(l) = Par(m) = 1 - \frac{1}{3} = 0.67$, whereas in the latter block $Par(o) = Par(p) = 0.5$. The feature for the activity n assumes the value 0, since it can not be executed in parallel with any other activity.

Parallelism Path Length. This feature is based on the PL feature; namely, it captures the position of an activity within the parallel block in which the activity is located. The formula is the same, except that the starting point from which to compute the longest path is the beginning of the parallel block. The feature is computed as follows:

$$PPL(a_i) = \frac{LP_{\parallel}(a_i)}{\max_{a_j \in \mathcal{A}} LP_{\parallel}(a_j)} \qquad (4)$$

where, $LP_{\parallel}(a_j)$ is the length of the longest path for the activity a_j. The path takes into account activities in any parallel branch plus the last activity of the parallel block (i.e., the synchronization point); so the activity from which the parallelism stems is not considered. For activities that aren't in a parallelism's branch the value of PPL is equal to 0. The same considerations made for PL take to this feature, but limited to the sub-process related to the parallel block.

Example 4. Let us consider the two blocks in Fig. 2. For the former (from $t4$ to n), the longest path is formed by 3 activities (i.e., $< j, k, n >$), hence $PPL(j) =$

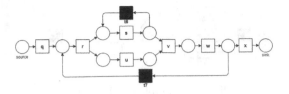

Fig. 3. Example Petri net 3

$PPL(l) = PPL(m) = 0.33$ and $PPL(k) = 0.67$. In the latter parallel block (from n to $t5$), the length of the longest path is 1 because the path starts after the first activity of the block and the synchronization point is and hidden activity which is not considered. Hence, $PPL(o) = PPL(p) = 1$.

Self Loopable. This feature evaluates if an activity can be consecutively repeated. In particular, the Self Loopable (SL) feature for the activity a_i is equal to 1 if the activity a_i is in a self-loop, 0 otherwise.

Long Loopable. This feature is concerned with representing if an activity belongs to a sub-process that can be repeated. In this case the Long Loopable (LL) feature assumes the value 1, otherwise 0. The same considerations made for a Straightly Loopable activity also hold here.

Example 5. Let us consider Fig. 3, which involves two loops. In the former the activity s can be repeated consecutively, whereas in the latter the sub-process from r to w can be repeated. Since only s is in a self-loop, we have $SL(s) = 1$ and for all others activities SL takes the value 0. The LL, instead, takes the value 1 for all activities in the loop block from r to w (i.e., $LL(r) = LL(s) = LL(u) = LL(v) = LL(w) = 1$). For all others activities, LL is equal to 0.

4   Case Study

This section discusses a case study which serves as a proof of concept for the features introduced in the present work. We focus on two versions of a loan application process of a Dutch financial institute, the first one from 2012 [9] and the second one from 2017 [10] (hereafter referred to as bpi2012 and bpi2017, respectively). The latter model has some substantial differences from the former, since a new workflow system has been implemented in the company. We argue that this context represents a good case study for our approach. First, the two models under analysis represent the same process, possibly with some variation. Therefore, it is reasonable to assume that the semantics of (part of) the activities overlaps, at least to some degree, which simplifies the analysis of the detected structural similarities and differences. Second, there are several studies available in the literature on these processes, which provide us with useful insights to elaborate upon the observations derived by the structural comparison.

4.1   Settings

The embedding strategy has been implemented as a python library.[1] Note that the implementation of the features discussed in Sect. 3 requires to investigate the state space of the process. To this end, here we exploit so-called *process trees* [15], which provide a block-structured representation of a process from which the control-flow relations for each activity can be inferred. We use the pm4py library[2] to derive a process tree from a given Petri net. The construction and analysis of process trees is a more efficient alternative to the exploitation of reachability graphs. As a drawback, it requires a block-structured process model. We argue that this is not a strong limitation, since block-structures are the default representation of well-known state-of-the-art process discovery methods (e.g., Inductive Miner [12]). Figure 4a and 4b show the Petri Nets for bpi2012 and bpi2017, respectively. The models have been extracted by applying the Inductive Miner algorithm with the default settings on the event logs provided at [9,10].

The two Petri nets look quite different from each other, thus making it challenging for a human to identify meaningful mappings of similar portions. This highlights the needs for a set of features able to guide the analyst. We first embedded each activity in both processes in a feature vector composed of the features previously introduced. Then, we applied k-means to determine clusters that group activities of both processes. We leveraged a grid search in [4; 20] to determine the best k, that resulted in $k = 14$.

4.2   Results

Table 1 shows the detected activity clusters, grouping on columns activities belonging to different processes. In this way, we define a mapping from the two sets of activities. Note that if a cluster contains only activities of the same process, such a mapping is not defined. Hence, for the sake of space, we limit our analysis to inter-process clusters, that is clusters containing activities from both processes These clusters are highlighted in Figures 4a and 4b with a labeled dotted red rectangle. Hereafter we briefly discuss these clusters.

c1: Process Initialization. The first cluster involves the first sequence of activities in both processes, interrupted with the occurrence of a XOR operator.

c2 and c4: Offer Management. The second and fourth clusters show two activity mappings related to the management of the offer. In both cases, one activity from bpi2012 activities has been mapped to a set of activities in bpi2017, which may indicate, for instance, that a more fine-grained modelling has occurred for the second model. Furthermore, the order between the activities of the first and of the second mapping reflects the order of the original activities $OSELECTED$ and $OSENT$. The activities $A\_validation$ and $W\_Call\ incomplete\ files$ are not included in the mapping to $OSENT$ because of the presence of a self-loop.

c3: Decision on Application. These clusters maps together activities related to decisions taken over an application. It is worth noting that the cluster of

---

[1] https://github.com/KDMG/Embedding-Structure-in-Activities.
[2] https://pm4py.fit.fraunhofer.de/.

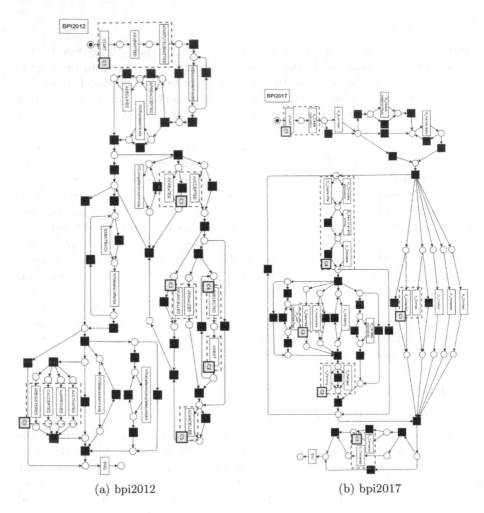

(a) bpi2012   (b) bpi2017

Fig. 4. Process model for a loan application in 2012 (a) and 2017 (b)

activities in bpi2012 it's not straightforward to derive. Since *AACCEPTED*
and *OCREATED* occur before in the process and are in parallel with a different activity, analyzing the overall process structural it would seem natural to
either separate these activities in different clusters or merge them with the other
occurring nearby. However, by focusing on structural features from the activity
perspective, their similarities emerge; indeed, all of these activities are optional,
occur in a similar position in the process, have a similar degree of parallelism and
do not occur in a self-loop. These characteristics also justify the inter-process
mapping with the corresponding activities in bpi2017.

c5: Application Finalization. Here we have multiple finalization activities
from bpi2012 mapped to two finalization activities in bpi2017. This is in line with
the fact that in the 2017 model the closure of the process has been significantly

simplified w.r.t. the 2012 model. The inclusion of the activity $ACANCELED$ in this mapping set shows consistency between the mappings; indeed, in the original process this activity occurs immediately after the $OSENT$ activity, and in the new process it occurs immediately after the activities mapped to $OSENT$. Such relation would be quite challenging to identify from a visual inspection.

Table 1. Cluster table

Cluster	2012	2017
c1	START; ASUBMITTED; APARTLYSUBMITTED	START; A_Create Application
c2	OSENT	O_Refused; A_Pending; O_Returned; A_Incomplete
c3	AFINALIZED; OCREATED; OCANCELLED; AACCEPTED	A_Denied; O_Accepted
c4	OSELECTED	O_Create Offer; W_Complete application; O_Sent (mail and online); A_Complete
c5	AACTIVATED; OACCEPTED; AAPROVED; ACANCELLED; AREGISTERED	A_Cancelled; O_Cancelled

While this case study is mostly intended as a preliminary proof of concept for the introduced features, the clusters discussed above show that our local notion of similarity allows to generate non-contiguous partitions of the process model, thus uncovering relations among activities occurring in different section of the model that would be easily missed by a human analyst or by adopting graph-based, global similarity metric.

5 Conclusions

This paper introduced a set of features characterizing process activities on the basis of structural properties of the control-flow constructs they are involved in. We discussed how these features can be used to address the problem of activities identification and comparison between process models. In turns, this enables more sophisticated analyses, ranging from retrieving process models from a process repository starting from a given reference process, to the evaluation of the changes that have occurred in a model over time.

We applied our embedding strategy on a case study involving two process models describing the same real-world loan management application process at a different times. In particular, we used such an embedding to realize an activity clustering between the two process models. The results are promising and show the capabilities of the adopted features to highlight similarities among activities which would be hard to grasp adopting name-based or global similarity features.

Nevertheless, these experiments represent a first preliminary study to investigate the potential of the approach. For future work, we plan first to extend the experimental set. Furthermore, we intend to test the feasibility of the approach for different application contexts, e.g., process querying in process repositories. Finally, we plan to investigate extensions to the set of features to include behaviors stored in the event log.

References

1. van der Aalst, W., et al.: Workflow mining: discovering process models from event logs. IEEE Trans. Knowl. Data Eng. **16**(9), 1128–1142 (2004)
2. Barbon Junior, S., Ceravolo, P., Damiani, E., Marques Tavares, G.: Evaluating trace encoding methods in process mining. In: Bowles, J., Broccia, G., Nanni, M. (eds.) DataMod 2020. LNCS, vol. 12611, pp. 174–189. Springer, Cham (2021). https://doi.org/10.1007/978-3-030-70650-0_11
3. Bose, R.P.J.C., et al.: Context aware trace clustering: towards improving process mining results, pp. 401–412 (2009)
4. Chiorrini, A., et al.: Exploiting instance graphs and graph neural networks for next activity prediction. In: Munoz-Gama, J., et al. (eds.) Process Mining Workshops, pp. 115–126. Springer, Cham (2022). https://doi.org/10.1007/978-3-030-98581-3_9
5. Corradini, F., et al.: RePROSitory: a repository platform for sharing business PROcess modelS. In: Proceedings of the Dissertation Award, Doctoral Consortium, and Demonstration Track at BPM 2019 Co-located with 17th International Conference on Business Process Management (BPM 2019), Vienna, Austria, vol. 2420, pp. 149–153. CEUR-WS (2019)
6. De Koninck, P., vanden Broucke, S., De Weerdt, J.: act2vec, trace2vec, log2vec, and model2vec: representation learning for business processes. In: Weske, M., Montali, M., Weber, I., vom Brocke, J. (eds.) BPM 2018. LNCS, vol. 11080, pp. 305–321. Springer, Cham (2018). https://doi.org/10.1007/978-3-319-98648-7_18
7. Diaz, J., et al.: Workflow hunt: combining keyword and semantic search in scientific workflow repositories, pp. 138–147 (2017)
8. Dijkman, R., et al.: Similarity of business process models: metrics and evaluation. Inf. Syst. **36**(2), 498–516 (2011). Special Issue: Semantic Integration of Data, Multimedia, and Services
9. van Dongen, B.: BPI Challenge 2012 (2012). https://data.4tu.nl/articles/dataset/BPI_Challenge_2012/12689204
10. van Dongen, B.: BPI Challenge 2017 (2017). https://data.4tu.nl/articles/dataset/BPI_Challenge_2017/12696884
11. Hake, P., Zapp, M., Fettke, P., Loos, P.: Supporting business process modeling using RNNs for label classification. In: Frasincar, F., Ittoo, A., Nguyen, L.M., Métais, E. (eds.) NLDB 2017. LNCS, vol. 10260, pp. 283–286. Springer, Cham (2017). https://doi.org/10.1007/978-3-319-59569-6_35
12. Leemans, S.J.J., Fahland, D., van der Aalst, W.M.P.: Discovering block-structured process models from event logs - a constructive approach. In: Colom, J.-M., Desel, J. (eds.) PETRI NETS 2013. LNCS, vol. 7927, pp. 311–329. Springer, Heidelberg (2013). https://doi.org/10.1007/978-3-642-38697-8_17

13. Leontjeva, A., Conforti, R., Di Francescomarino, C., Dumas, M., Maggi, F.M.: Complex symbolic sequence encodings for predictive monitoring of business processes. In: Motahari-Nezhad, H.R., Recker, J., Weidlich, M. (eds.) BPM 2015. LNCS, vol. 9253, pp. 297–313. Springer, Cham (2015). https://doi.org/10.1007/978-3-319-23063-4_21
14. Pasquadibisceglie, V., et al.: A multi-view deep learning approach for predictive business process monitoring. IEEE Trans. Serv. Comput. (2021)
15. Polyvyanyy, A., Vanhatalo, J., Völzer, H.: Simplified computation and generalization of the refined process structure tree. In: Bravetti, M., Bultan, T. (eds.) WS-FM 2010. LNCS, vol. 6551, pp. 25–41. Springer, Heidelberg (2011). https://doi.org/10.1007/978-3-642-19589-1_2
16. Van Der Aalst, W.: Process Mining: Data Science in Action, vol. 2. Springer, Heidelberg (2016). https://doi.org/10.1007/978-3-662-49851-4
17. Venugopal, I., et al.: A comparison of deep learning methods for analysing and predicting business processes. In: Proceedings of International Joint Conference on Neural Networks, IJCNN (2021)

ADBIS Short Papers: Data Access Optimization

Bulk Loading of the Secondary Index in LSM-Based Stores for Flash Memory

Wojciech Macyna(✉) and Michal Kukowski

Department of Computer Science, Faculty of Information and Communication Technology,
Wrocław University of Science and Technology, Wrocław, Poland
{wojciech.macyna,michal.kukowski}@pwr.edu.pl

Abstract. NoSQL databases have gained much attention recently. Due to utilizing Log Structured Merge (LSM) tree, they support fast write throughput and fast lookups on primary keys. Nevertheless, the implementation of the secondary index for these databases is still a challenging task. Modern data storage technologies like: flash memory or phase change memory make the situation even more difficult. The most important problems of such memory types are: limited write endurance and asymmetry between write and read latency. These limitations affect both the index structure and index modification methods.

In this paper, we propose a new bulk loading of the secondary index in LSM-based stores for flash memory. The bulk loading happens when many insert operations are performed in one batch. The method works on a new LSM tree variant optimized for flash memory called Flash Aware LSM (FA-LSM) tree. To reach the optimal performance, the method can be adapted to the changing workload. We conduct several experiments which confirm that our method outperforms the traditional LSM insert strategy by about 30% while preserving high search efficiency.

Keywords: LSM tree · Flash memory · NoSQL database

1 Introduction

NoSQL databases have gained great popularity recently. Due to their scalability and flexibility, they are treated as an alternative to relational databases. Such systems are particularly useful in social network applications where high volume and variety of data must be processed. The most popular NoSQL database systems are: HBase, Cassandra, Voldemort, MongoDB, BigTable, AsterixDB, Spanner and RocksDB. In most cases, NoSQL databases hold the data in the Log-Structured Merge-Tree (LSM) [1]. Typically, LSM tree consists of several levels. Each level contains key-value entries sorted by the key. A new entry is added to the top level. When the number of entries in the level exceeds the predefined limit, this level is merged with the level below.

In general, LSM-based stores support fast write throughput and fast lookups on primary keys. However, this is not enough to support effective query processing on non-key attributes. Let us consider a tweeter database. A tweet is of the form:

© Springer Nature Switzerland AG 2022
S. Chiusano et al. (Eds.): ADBIS 2022, CCIS 1652, pp. 133–143, 2022.
https://doi.org/10.1007/978-3-031-15743-1_13

$e = <id, \{user, text\} >$. It means that $user$ sent a tweet with the message $text$ to the database. In this case, a tweet identifier (id) is the key of entry e. Unfortunately, to find all tweets sent by the particular user, the whole LSM tree must be scanned. Clearly, the secondary index on $user$ should accelerate this query significantly.

In this paper, we consider flash memory as the main storage type. A distinguishing feature of flash memory is the asymmetry of read and write latency. The read operation is significantly faster than the write one. The characteristics of flash memory impacts the data storage strategy and index maintenance. Several types of flash-aware indexes have been proposed. Many of them, like BFTL [2] are write-optimized, but suffer poor search performance. Other index structures like: FD-tree [3] and LA-tree [4] are read and write optimized. All flash optimized index types have several common features. They reduce the number of writes and erases as much as possible. Moreover, they prefer to facilitate the sequential write over the random one, because the former is significantly faster.

NoSQL databases adopt two different strategies for secondary indexing: stand-alone index table and embedded secondary indexing (see [5]). In the first strategy, the secondary index is separated from the data. It uses B+-tree (MongoDB) or LSM index table (BigTable, Cassandra, AsterixDB, Spanner). The second strategy relies on storing the attribute information inside the original data blocks.

The bulk loading happens when many entries are inserted into the index in one batch. In most cases, typical index structures are not suited to perform the bulk loading efficiently. For example, inserting many entries into the traditional B+tree would drastically increase the number of node split operations. This has a very negative effect on system performance in the case of flash memory where the write operation is much slower than the read one. Some interesting bulk loading algorithms for spatial data are proposed in [6] and [7]. In [8], the authors consider the bulk loading in LSM tree data stores. Nevertheless, they mainly focus on efficient item distribution between data nodes. Apart from that, the flash memory limitations are not considered.

Several efforts of improvements of the original LSM tree have been made. In [9], the authors introduce Lazy Leveling as a new design that removes merge operations from all levels of the LSM-tree but the largest. Lazy Leveling improves the worst-case complexity of update cost while maintaining the efficient search performance. A persistent LSM-tree-based key-value store that separates keys from values is presented in [10]. This approach drastically minimizes I/O amplification. The optimization for Solid Stare Disks has been proposed in [11] and [12]. A nice comparison of the LSM-based storage techniques can be found in [13]. However, all the research papers concentrate on improving the storage in the LSM tree. Our work, by contrast, focuses on storing of the secondary index in the LSM-based auxiliary structure. Additionally, we consider optimization of LSM tree for bulk loading on flash memory.

The aim of this work is to create an efficient bulk loading of the secondary index in the LSM-tree optimized for flash memory. For the secondary index, we use the Flash Aware LSM (FA-LSM) tree. Each level of the FA-LSM can contain two sections: sorted and unsorted. In the sorted section, the entries are sorted by the key. In the unsorted section entries are stored in an unsorted way. So, the new entries can be inserted to the unsorted section leaving the sorted section intact. In this way, we reduce the number of merging on the LSM tree and avoid to sort all entries within the level. Clearly, the penalty is the need of scanning the whole unsorted section during the search. The idea is not new: many LSM trees utilize partially sorting of entries inside the level (see [14], [15]). However, we introduce a novel level merge policy in the FA-LSM. We claim that the FA-LSM tree is optimized for flash memory and well suited for the bulk loading method.

The main **contribution** of the paper is a new bulk loading method of the secondary index for flash memory. It enables loading many entries into the FA-LSM in one batch. Unlike in the traditional LSM tree, the insertion is not started from the top of the index. This would be very unfavorable in the situations where the number of entries in the batch is large. In our approach, the entries are inserted directly to the level that has enough space to hold them all. The method is equipped with the threshold parameter that can be changed depending on the workload specification. We experimentally prove the efficiency of the proposed method in terms of flash memory limitations.

2   Bulk Loading of the Secondary Index

In this section, we outline a system architecture and point out the critical issues of the problem.

2.1   LSM Tree

The LSM tree consists of one in-memory level (L_0) called MemTable and a few on-disk levels (L_i) referred as SSTables. Each level consists of several data files and each data file contains many data blocks. The entries are ordered by the key within each level. The insert operation works as follows. First, the entries are written into L_0. If L_0 exceeds the size limit, the data are flushed to the first on-disk level (L_1). When L_1 is over the size limit, the data from L_1 are merged with the data from L_2. Then, all the entries are sorted by the key and written to L_2. Clearly, when the limit of L_2 is exceeded, L_2 is merged with L_3 and so on. As a result, the on-disk levels increase i.e. the size of L_{i+1} is much larger than the size of L_i. The LSM tree is optimized for writes as the writes update only in-memory level L_0.

2.2   Bulk Loading Outline

We consider a system architecture presented in Fig. 1. The data are held in the main LSM storage as a set of key-value entries $e = < k, v >$, where k denotes a primary key and v is a value. An entry of the secondary index is a tuple: $ind(e) = < key, ptr >$. In this case, key denotes a key of the secondary index and ptr is a pointer to e. The bulk loading happens when many entries are inserted into the LSM secondary index in one batch. Obviously, inserting is not the only operation in the secondary index. If an entry e is deleted from the main storage and has previously been indexed in the secondary index, then the corresponding entry $ind(e)$ in the secondary index must be deleted as well. Similarly, the update in the main storage have to be reflected in the secondary index. All the cases are described below. It may also happen that the tuple $ind(e)$ is

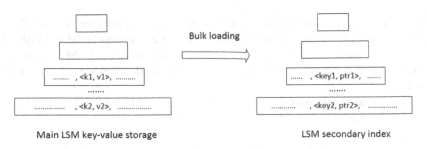

Fig. 1. Bulk loading architecture

inserted into the secondary index several times (for example in a few batches). In this situation, the duplicates are removed during the level merge.

Deletion in the main LSM key-value storage is performed as follows. When an entry $e = < k, v >$ is deleted, the deleted version of this entry $e_d = < k, null >$ is inserted into MemTable. Then, e_d is propagated to the level that contains the original entry. As a result of the merge, both entries: deleted and original are removed from LSM tree. The deletion of e from the main storage can invoke a deletion from the LSM secondary index. First, we check if the corresponding secondary index entry $ind(e) = < key, ptr >$ exists in the LSM secondary index (ptr is a pointer to e). If so, a deleted entry $ind_d(e) = < key, ptr, del >$ is inserted into MemTable of the LSM secondary index. It means that the entry $ind(e)$ is marked as deleted and are ignored during query processing. Please note that the deleted entry $ind_d(e)$ must contain ptr, since key is not unique in the secondary index.

Update is performed in the similar way. When the entry $e = < k, v >$ is updated in the main LSM key-value storage, a new entry $e' = < k, v' >$ is inserted into MemTable. As the query processing starts from the top of the main LSM tree, it finds the most actual version of the entry with the required key. The update in the main storage can trigger the update in the LSM secondary index. If the secondary index entry $ind(e) = < key, ptr >$ for e exists, a deleted entry $ind_d(e) = < key, ptr, del >$ and a new entry $ind(e') = < key, ptr' >$ are inserted into MemTable of the LSM secondary index. In this case, ptr and ptr' point to e and e', respectively. When a query on key is processed, both entries: $ind(e)$ and $ind_d(e)$ are ignored and only the new entry $ind(e')$ can be fetched.

2.3 Bulk Loading for Flash Memory

Unfortunately, intensive bulk loading demands a lot of modifications and merge operations in the LSM tree. Such operations have negative impact on flash memory, since they increase a number of writes and erases in memory cells and slow down the overall system performance. On the other hand, inserting entries in an unsorted manner spoils search efficiency, because the binary search cannot be applied.

The goal of the proposed method is to reach the optimal balance between search and modification so that the system works efficiently for a specific workload. To achieve this, the traditional insert method described in the previous section cannot be applied, as it may invoke many merge operations between LSM tree levels. Let us suppose that levels L_0 and L_1 can hold 10 and 40 entries, respectively. If we load 20 entries, the traditional insert works as follows. The batch containing 20 entries must be split into two equal 10 entries packages. Then, the first package is loaded into L_0. Clearly, the number of entries in L_0 reaches the limit, and the merge of levels L_0 and L_1 is invoked. After that, the second package is loaded into L_0 and, consequently, the same steps are repeated. As we can see, bulk loading of 20 entries triggers at least two merge operations between L_0 and L_1. Apart from this, one merge of L_1 and L_2 may happen.

To avoid this overhead, we propose a method that loads the batch directly into the level with sufficient space. In this case, the batch containing 20 entries could be loaded directly into L_1. As a result, at most one merge of L_1 and L_2 is performed.

3 Proposed Solution

In this section, we describe the FA-LSM tree in more detail and propose a new bulk loading approach of the secondary index for flash memory.

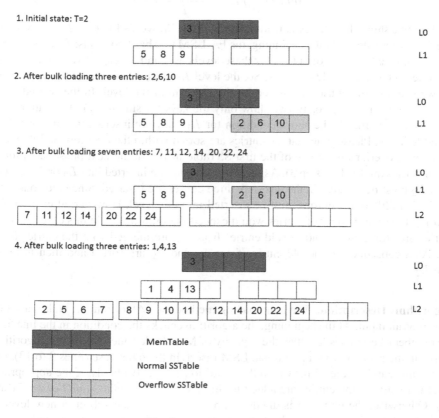

Fig. 2. Bulk loading example

In the FA-LSM tree, we distinguish two kinds of SSTables: normal and overflow. Each FA-LSM tree level can contain many normal and many overflow SSTables. The entries in a normal SSTable are sorted by the key, but their key range cannot overlap with any entry range of the other normal SSTable in the same level. The entries in an overflow SSTable are sorted by the key as well. However, their key range may overlap with the key ranges of the other SSTables. Thus, inserting many entries into the FA-LSM tree level does not invoke the sorting procedure of the entire level: the new entries are simply inserted into the overflow SSTable. In this way, a number of writes on flash memory is drastically reduced. Additionally, the FA-LSM tree is equipped with the sparse index stored in RAM. The entry of that index holds a key range of each SSTable. It allows an effective search because only SSTables that can contain the required key are selected.

The key point of the method is the parameter T. It determines the level where a batch of entries should be loaded. Let $numEntry$ and $maxEntry(l)$ denote the number of entries to insert in the batch and the maximal number of entries in the level l, respectively. The batch is loaded into l when the following condition holds:

$$numEntry < \frac{maxEntry(l)}{T} \tag{1}$$

Figure 2 shows how the bulk loading in the FA-LSM works for $T = 2$. The example consists of four steps. At the beginning, the FA-LSM tree has two levels: L_0 and L_1. We fix the maximal number of entries for these levels as 4 and 8, respectively. We assume that one SSTable can hold 4 entries. So, the level L_1 consists of at most two SSTables. As we can see, the entries are sorted by the key within the level. In the second step, we load three entries: 2, 6, 10 (we show only the keys for simplicity). As a number of the new entries fulfills the above condition for L_1, they are inserted into the overflow SSTable in L_1. Please note that the entries are sorted within the overflow SSTable, but their range overlaps the range of the first SSTable in L_1. In the next phase, additional seven entries are loaded (step 3). As the entries cannot be inserted into L_0 or L_1, a new level L_2 must be created. In the last step, three entries are inserted. Since the condition for L_1 is fulfilled, the entries must be inserted into this level. However, after inserting into L_1, the total number of entries would exceed the limit of L_1. In this case, the new entries are stored in L_1 and the old entries from L_1 are merged with the entries from L_2. As a consequence, the old entries from L_1 and L_2 are sorted and then inserted into L_2.

Algorithms Description. The bulk loading method (see Algorithm 1.1) takes a set of entries as an input. At the beginning, the algorithm checks the condition in the line 5. If the number of entries is less than the capacity of MemTable (lines 5 to 7), the algorithm works in the same way as a traditional LSM insert. In the other case (lines 9 to 13), the algorithm searches a level for which the condition 1 (line 10) holds. If the appropriate level is found, all the entries are added to this level by the method $addToLevel$ (line 11). Otherwise, the entry set is divided into subsets. For each subset, a new level is created and the entries of this subset are inserted into this level. This is reflected in lines 15 to 22. Please note that the number of new levels depends on the batch size and the

FA-LSM topology. We assume that the fixed ratio between the number of entries of each two consecutive levels must be held.

Algorithm 1.1. bulkLoading(input: Entry $entriesToInsert[\ ]$)

---

1 Let $height$ be the height of the FA-LSM tree.
2 Let $numEntries$ be the size of $entriesToInsert$ set
3 Let $maxEntry[k]$ denote the maximal number of entries in level k
4 Let $maxEntry[0]$ denote the maximal number of entries in MemTable in RAM
5 **if** $numEntries < maxEntry[0]$ **then**
6   insertMemTable($entriesToInsert$)
7   **return**

8 $lvl := 1$
9 **while** $lvl <= height$ **do**
10   **if** $numEntries < \left\lceil \frac{maxEntry[lvl]}{T} \right\rceil$ **then**
11   addToLevel($entriesToInsert, lvl$)
12   **return**
13     $lvl := lvl + 1$

14 Int pos := 0
15 **while** $numEntries > 0$ **do**
16   createLevel(lvl)
17     $n := min(maxEntry[lvl], numEntries)$
18   Entry $entriesToLoad[\ ] := truncateEntrySet(entriesToInsert, pos, n)$
19   addToLevel($entriesToLoad, lvl$);
20     $numEntries := numEntries - n$
21     $pos := pos + n$
22     $lvl := lvl + 1$

The algorithm $addToLevel$ takes an entry set and a level number as arguments. If the maximal number of entries of the level lvl is exceeded (line 4), the merge is invoked. In that case, all the existing entries from lvl are added to $entriesToMerge$. If the next level ($lvl + 1$) does not exists, it is created (line 11), $entriesToMerge$ are sorted (line 12) and the whole entry set is inserted into $lvl + 1$ (line 13). If the level $lvl + 1$ exists, the entries from $lvl + 1$ are also fetched and added to $entriesToMerge$ (lines 7 and 8). In lines 9 and 10, the entries from $lvl + 1$ are removed and the level is destroyed. After that, the lines from 11 to 13 are executed in the same way as described above. At the end of the algorithm (lines 14 and 15), $entriesToInsert$ are packed to the new created SSTables which are added to lvl.

Algorithm 1.2. addToLevel(input: Entry $entriesToInsert[\ ]$, int lvl)

---

1 Let $numEntriesToInsert$ be the size of $entriesToInsert$ set
2 Let $maxEntry[k]$ denote the maximal number of entries in level k
3 Let $numEntry[k]$ denote the current number of inserted entries in level k
4 **if** $numEntry[lvl] + numEntriesToInsert > maxEntry[lvl]$ **then**
5   Entry $entriesToMerge[]$:= getEntries(lvl)
6   **if** $exists\ lvl + 1$ **then**
7   Entry $entries[]$:= getEntries($lvl + 1$)
8         $entriesToMerge$:= $entriesToMerge \cup entries$
9   removeEntries($lvl + 1$)
10   removeLevel($lvl + 1$)
11   createLevel($lvl + 1$)
12   Entry $sortedEntries[]$:= sortEntries($entriesToMerge$)
13   addToLevel($sortedEntries, lvl + 1$)
14 $SSTable\ tab[] := createSSTables(lvl)$
15 addToSSTables($tab, entriesToInsert$)

---

4 Experiments

The experiments were carried out on Intel Core i7-9700F (cache L1 512 KB, L2 2 MB, L3 12 MB) equipped with 8CPUs (4700 MHz per core) and 32 GB RAM. The implementation is written in C compiled with gcc 9.3.0. For the experiments, we chose the disk SSD Samsung 840 with the page size and block size equals to 8192 bytes and 64 pages, respectively. The disk exhibits random read latency, random write latency, sequential read time and sequential write time estimated as 95 KIOPS, 44 KIOPS, 540 MB/s and 520 MB/s, respectively. Each experiment was conducted on table $Warehouse$ from TPC-C [16] that contains one key column (8 bytes) and eight non-key columns of the fixed width. So, the total record size is 113 bytes. We assume that the size of both MemTable and SSTable is 2MB. We fix the ratio between two subsequent levels as 5. It means that the level $i + 1$ contains five times more entries than the level i.

In the experiments (see Fig. 3, 4, 5 and 6), we load secondary index entries in batches into the FA-LSM. Each batch contains between 50000 and 100000 entries. After each batch, a set of range queries with selectivity 1% is executed. The number of the range queries depends on the specific workload. All the experiments run until 10 millions entries are inserted into the FA-LSM. We observe the elapsed time for the different workloads and different T value. We measure the bulk loading time as the summarized time of all batches and search time as the time of all range queries of the workload. The total time is obtained as a sum of the bulk loading and search time. On the charts, we compare the total time of our approach with the total time of the traditional LSM tree for the specific workload. Obviously, parameter T affects the FA-LSM tree topology. When T grows, the number of overflow SSTables in the level is increased and, consequently, less merge operations must be accomplished. As a result, the bulk loading executes faster, but the search gets slower. We can observe that the optimal T

for each workload may be different. It is worth to see that if T is very large (for example 50), the proposed method can be less efficient than the traditional one. However, when T is fixed properly, the total time of using the FA-LSM with bulk loading is better than the LSM total time by about 20% to 40% depending on the workload.

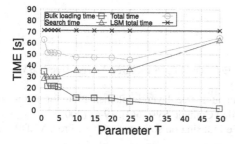

Fig. 3. After each batch, 20 range queries

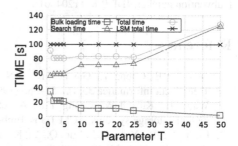

Fig. 4. After each batch, 40 range queries

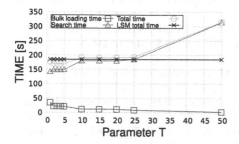

Fig. 5. After each batch, 100 range queries

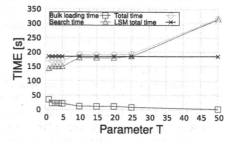

Fig. 6. After each batch, 250 range queries

We conducted several experiments which are out of the scope of this paper. We measure bulk loading time for a different batch size and different T value. W notice that loading the large batches decreases the bulk loading time. It is due to the fact, that the bulk loading of large batches implies creating new FA-LSM tree levels. Apart from that, we check an impact of the SSTable and MemTable size on the system performance. We observe that the increasing SSTable and MemTable size slightly slows down search performance, but does not affect the bulk loading time.

5 Conclusions

In this work, we invent a new bulk loading method of the secondary index in LSM-based stores for flash memory. As the secondary index structure, we utilize the new LSM tree variant (FA-LSM) optimized for flash memory. Our bulk loading approach inserts new index entries not into the top of the FA-LSM, but directly to the level with sufficient space. The bulk loading method can be adapted to the different workload specification by changing a threshold parameter T. The experiments confirm that our approach drastically outperforms the traditional LSM tree in terms of flash memory.

Acknowledgment. The paper is supported by Wroclaw University of Science and Technology (subvention number: IDUB/8211204601).

References

1. O'Neil, P.E., Cheng, E., Gawlick, D., O'Neil, E.J.: The Log-Structured Merge-Tree (LSM-Tree). Acta Informatica **33**(4), 351–385 (1996)
2. Wu, C.H., Kuo, T.W., Chang, L.P.: An efficient B-tree layer implementation for flash-memory storage systems. ACM Trans. Embedded Comput. Syst. **6**(3), 19-es (2007)
3. Li, Y., He, B., Yang, R.J., Luo, Q., Yi, K.: Tree indexing on solid state drives. Proc. VLDB Endow. **3**(1–2), 1195–1206 (2010)
4. Agrawal, D., Ganesan, D., Sitaraman, R., Diao, Y., Singh, S.: Lazy-adaptive tree: an optimized index structure for flash devices. Proc. VLDB Endow. **2**(1), 361–372 (2009)
5. Qader, M.A., Cheng, S., Hristidis, V.: A comparative study of secondary indexing techniques in LSM-based NoSQL databases. In: Proceedings of the 2018 International Conference on Management of Data, SIGMOD Conference 2018, Houston, TX, USA, 10–15 June 2018, pp. 551–566. ACM (2018)
6. Roumelis, G., Fevgas, A., Vassilakopoulos, M., Corral, A., Bozanis, P., Manolopoulos, Y.: Bulk-loading and bulk-insertion algorithms for xBR$^+$-trees in Solid State Drives. Computing **101**(10), 1539–1563 (2019)
7. Roumelis, G., Vassilakopoulos, M., Corral, A., Manolopoulos, Y.: An efficient algorithm for bulk-loading xBR$^+$-trees. Comput. Stand. Interfaces **57**, 83–100 (2018)
8. Zhu, Y., Zhang, Z., Cai, P., Qian, W., Zhou, A.: An efficient bulk loading approach of secondary index in distributed log-structured data stores. In: Candan, S., Chen, L., Pedersen, T.B., Chang, L., Hua, W. (eds.) DASFAA 2017. LNCS, vol. 10177, pp. 87–102. Springer, Cham (2017). https://doi.org/10.1007/978-3-319-55753-3_6
9. Dayan, N., Idreos, S.: Dostoevsky: better space-time trade-offs for LSM-tree based key-value stores via adaptive removal of superfluous merging. In: Proceedings of the 2018 International Conference on Management of Data, SIGMOD Conference 2018, Houston, TX, USA, 10–15 June 2018, pp. 505–520. ACM (2018)
10. Lu, L., Pillai, T.S., Gopalakrishnan, H., Arpaci-Dusseau, A.C., Arpaci-Dusseau, R.H.: WiscKey: separating keys from values in SSD-conscious storage. ACM Trans. Storage **13**(1), 5:1–5:28 (2017)
11. Wu, S., Lin, K., Chang, L.: KVSSD: close integration of LSM trees and flash translation layer for write-efficient KV store. In: 2018 Design, Automation & Test in Europe Conference & Exhibition, DATE 2018, Dresden, Germany, 19–23 March 2018, pp. 563–568. IEEE (2018)
12. Wang, P., et al.: An efficient design and implementation of LSM-tree based key-value store on open-channel SSD. In: Ninth Eurosys Conference 2014, EuroSys 2014, Amsterdam, The Netherlands, 13–16 April 2014, pp. 16:1–16:14. ACM (2014)
13. Luo, C., Carey, M.J.: LSM-based storage techniques: a survey. VLDB J. **29**(1), 393–418 (2020)
14. Cao, Z., Dong, S., Vemuri, S., Du, D.H.C.: Characterizing, modeling, and benchmarking RocksDB key-value workloads at Facebook. In: 18th USENIX Conference on File and Storage Technologies, FAST 2020, Santa Clara, CA, USA, 24–27 February 2020, pp. 209–223. USENIX Association (2020)

15. Lee, H., Lee, M., Eom, Y.I.: SFM: mitigating read/write amplification problem of LSM-tree-based key-value stores. IEEE Access **9**, 103153–103166 (2021)
16. Raab, F.: TPC-C - the standard benchmark for online transaction processing (OLTP). In: The Benchmark Handbook for Database and Transaction Systems, 2nd edn. Morgan Kaufmann (1993)

Blink: Lightweight Sample Runs for Cost Optimization of Big Data Applications

Hani Al-Sayeh[1]([✉])(ID), Muhammad Attahir Jibril[1](ID), Bunjamin Memishi[2](ID), and Kai-Uwe Sattler[1](ID)

[1] TU Ilmenau, Ilmenau, Germany
{hani-bassam.al-sayeh,muhammad-attahir.jibril,kus}@tu-ilmenau.de
[2] Riinvest College, Pristina, Kosovo
bunjamin.memishi@riinvest.net

Abstract. Distributed in-memory data processing engines accelerate iterative applications by caching datasets in memory rather than recomputing them in each iteration. Selecting a suitable cluster size for caching these datasets plays an essential role in achieving optimal performance. We present BLINK, an autonomous sampling-based framework, which predicts sizes of cached datasets and selects optimal cluster size without relying on historical runs. We evaluate BLINK on iterative, real-world, machine learning applications. With an average sample runs cost of 4.6% compared to the cost of optimal runs, BLINK selects the optimal cluster size, saving up to 47.4% of execution cost compared to average cost.

1 Introduction

Modern distributed systems such as Spark [17] enhance the performance of iterative applications by caching crucial datasets in memory instead of recomputing or fetching them from slower storage (e.g., HDFS) in each iteration [16]. To measure the impact of repetitive re-computations on system performance, we run *Support Vector Machine* application (SVM in Spark MLLib 2.4.0 [11]) on an input dataset of 59.5 GB using different cluster sizes (1–12 machines) on our private cluster (cf. Sect. 6). We measure the execution time and the cost (#machines × time) of each run. As depicted in Fig. 1, we distinguish three areas:

- Area A : Increasing the cluster size decreases both execution time and cost.
- Area B : Increasing the cluster size decreases time but increases cost.
- Area C : The junction of A&B , where the highest cost efficiency is achieved.

In area A, the total memory capacity of the cluster machines is not enough for caching all partitions of a certain crucial dataset in SVM. As a result, many of its partitions do not fit in memory and are re-computed in all iterations, which is very expensive. A deeper dive into a single iteration shows that: 1. The percentage of cached data partitions in area A for 1 to 7 machines are 17%, 35%, 52%, 70%, 87%, 92% and 100% respectively. 2. On average, a task that reads an already cached partition runs 97× shorter than a task that recomputes a partition of equal size. In area B, increasing the cluster size reduces the execution

© Springer Nature Switzerland AG 2022
S. Chiusano et al. (Eds.): ADBIS 2022, CCIS 1652, pp. 144–154, 2022.
https://doi.org/10.1007/978-3-031-15743-1_14

time of the parallel part of the application but does not influence the serial part [6]. The data transfer overhead between machines also increases. These factors decrease cost efficiency.

Currently, optimal resource provisioning based on accurate prediction of the size of cached datasets remains an open challenge. In summary, we make the following contributions:

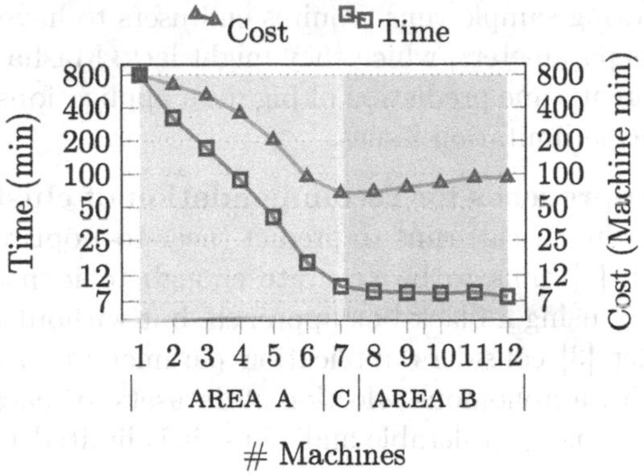

Fig. 1. Selection of cluster size (SVM)

- We introduce an efficient approach for minimizing the cost of sample runs.
- We present BLINK, a lightweight sampling-based framework that predicts the size of cached datasets and selects an optimal cluster size (area C).
- We perform an extensive analysis of machine learning applications and stress their minimal sampling requirements for an optimal cluster size selection.

We evaluate BLINK on 8 real-world applications. Relying on tiny sample datasets, BLINK selects the optimal cluster size for all 8 actual runs, which reduces execution cost to 52.6% compared to the average cost across all cluster sizes with an average sample runs cost of 4.6%.

2   Related Work

Caching decision support tools help application developers to determine which datasets to cache and when to purge them from memory [1,10]. However, these tools do not consider the size of the datasets and the required cluster configuration that guarantees eviction-free runs.

Cache eviction policies and **approaches to auto-tuning of memory configuration** tackle cache limitation in a best-effort manner but with penalties caused by cache eviction. This makes them suitable solutions if an inappropriate cluster size is selected (area A in Fig. 1). MRD [12] and LRC [16] are DAG-aware cache eviction policies in Spark that rank cached datasets based on their reference distance and reference count respectively. We apply both policies for the same SVM experiments (depicted in Fig. 1) and do not realize any performance improvement. This is because only one dataset is cached in SVM. MemTune [15] is a memory manager that re-adjusts memory regions during application run. RelM [9] introduces a safety factor to ensure error-free execution in resource-constrained clusters.

Runtime Prediction Approaches. Ernest [14] is a sampling-based framework that predicts the runtime of compute-intensive long-running Spark applications.

To reduce the overhead of sample runs, it decreases the number of iterations during these sample runs to make their overhead tolerable. This is not always practical because tuning an application parameter like the number of iterations during sample runs requires end users to have knowledge of the application and its parameters, which they might lack. Masha [4] is a sampling-based framework for runtime prediction of big data applications. Both frameworks do not address cache limitation issues.

Approaches for recommendation of cluster configuration rely on sample (or historical) runs to predict (near-to-) optimal cluster configuration. CherryPick [5] aims to be accurate enough to identify poor configurations and adaptive using a black-box approach, but without considering cache limitations. Juggler [3] considers application parameters to recommend cluster configurations with autonomous selection of datasets for caching. But, its offline-training overhead is not tolerable and, thus, it is limited to recurring applications.

3 Background

Spark runs applications on multiple *executors* that perform various parallel operations on partitioned data called Resilient Distributed Dataset or RDD [17]. A class of operations called *transformations* (e.g., filter, map) create new RDDs from existing ones while another class called *actions* (e.g., count, collect) return a value to the (*driver*) program after making computations on RDDs. An application is the highest level of computation and consists of one or more sequential *jobs*, each of which is triggered by an action. A job comprises of a sequence of transformations, represented by a DAG, followed by a single action. When a transformation is applied on an RDD, a new one is created. The parent-child dependency between RDDs is represented in a logical plan, by way of a lineage or DAG starting from an action up to either the root RDDs that are cached or original data blocks from the distributed file system.

As different jobs may consist of many transformations in common, we merge all their DAGs to represent an application in a single DAG of transformations, as illustrated with the *Logistic Regression* application in Fig. 2. The number of times a dataset is computed is determined by the number of its child branches in the resulting DAG.

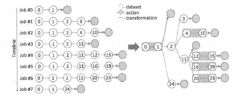

Fig. 2. Merging DAGs.

As depicted in Fig. 3, Spark splits memory into multiple regions. We focus on the storage and the execution regions, respectively used for caching datasets and computation [18]. Both regions share the same memory space

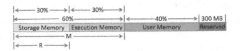

Fig. 3. Spark: Memory layout.

(i.e., the unified region M) such that if the execution memory is not utilized, all the available memory space can be used for caching, and vice versa. There is a minimum storage space R below which cached data is not evicted. That is, in each executor, at least R and at most M can be utilized to cache datasets.

4   Efficient Sample Runs

In this section, we explain how to minimize the cost of sample runs with empirical evaluations. Specifically, we show that the sample run phase required for predicting the size of the cached datasets is less challenging than that required for execution time prediction. As previous studies tackle data sampling challenges [7,8], we do not address them in this work, similar to Ernest [14] and Masha [4].

4.1   Size of Sample Runs

Few sample runs are sufficient to predict the size of the cached datasets. For example, if we conduct two short-running experiments of the same application using the same data and same cluster configuration, the sizes of datasets do not vary. However, this is not the case regarding execution time. To validate this, we select SVM, which caches one dataset, to run 10 experiments on 738.1 MB (data scale 1, 12 blocks), 10 experiments on 1501.6 MB (data scale 2, 24 blocks) and 10 experiments on 2.2 GB (data scale 3, 36 blocks). We conduct all runs on a single machine.

As illustrated in Fig. 4, we see that the size of the cached dataset remains constant in all runs of the same data scale. Also, we notice a considerable variance in execution time between the runs of the same data scale, which affects the construction and training of prediction models. To overcome this problem, we either run several experiments on the same data scale and obtain the statistical average (or median) or increase the size of sample

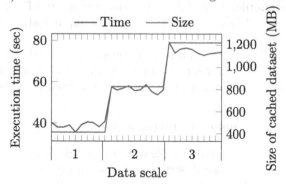

Fig. 4. Short-running experiments.

datasets to make sample runs longer and, thus, the execution time variance relatively lower. However, both solutions increase the cost of sample runs tremendously, which explains why runtime prediction approaches are limited to long-running applications.

To build size prediction models of the cached datasets, we carry out sample runs on tiny datasets within the range of 0.1%–0.3% of the original data.

4.2 Parallelism

Distributed file systems (e.g., HDFS) store original data by fragmenting it into equal chunks, namely blocks. The size of blocks is configurable (64 or 128 MB by default). In order to decrease the data size during sample runs, we either (1) reduce the size of each block (BLOCK-S), or (2) select few data blocks (BLOCK-N). For example, if the block size is configured to be 64 MB, 1 TB of data is stored in 16K blocks. Thus, 16 blocks out of them could be selected for a sample run of 0.1% of the original data.

BLOCK-N is less costly than BLOCK-S because it requires selecting data blocks from a distributed file system. BLOCK-S is more complicated and brings extra overhead in preparing the sample data. Since we are not expecting memory limitation during sample runs, increasing the parallelism increases the execution time of each sample run (i.e., data shuffling and cleaning).

In order to validate this, we conduct two runs of SVM with an input data of 1.2 GB on a single machine. The number of data blocks in the first run is 10 and it takes 41 s. In the second run, the number of data blocks is 1000 and it takes 3.5 min. In addition, during the first and second runs, the size of the cached dataset is 728.9 MB and 747.8 MB, respectively. This shows that the size of datasets is influenced by the parallelism level. Hence, in the case of BLOCK-N, if we reduce the number of tasks during sample runs, then predicting the size of the cached datasets might be affected. To tackle this problem, we keep the number of tasks proportional to the data scale by fixing the block size. For example, if the full-scale dataset consists of 16K blocks, then the sample runs with 0.1%, 0.2% and 0.3% of the input data scale will contain 16, 32, and 48 tasks respectively.

For some applications, the size of the original data is relatively small (as we will show in Sect. 6) and, thus, the number of its blocks is not enough to apply BLOCK-N. In such cases, BLOCK-S is used in spite of its costs.

4.3 Cluster Configuration

We carry out all sample runs on a single machine to reduce the cost of the sample runs. The serial part of a short-running experiment is relatively high compared with the parallel part and, hence, adding more machines during a sample run might not speed up the execution time. Rather, it leads to higher execution costs because of the increased overhead of negotiating resources (e.g., by YARN) and the increase in data transfer overhead with the addition of more machines. To validate this, we run SVM on 1.2 GB input data using a single machine and also using 12 machines. The execution cost on 12 machines is 13.9× higher than on a single machine. The exception that makes carrying out sample runs on a single machine too costly is when cached datasets do not fit in the memory of a single machine. However, this is unlikely for sample runs with tiny datasets.

4.4   Number of Sample Runs

Our experiments with all applications in HIBENCH 7.0 show that the prediction models for the size of the cached (and non-cached) datasets with respect to the input data scale are linear. Therefore, two sample runs are sufficient to construct a model. However, knowing that sample runs are lightweight, more sample runs could be conducted to apply *cross validation* to choose a well-fitting linear model.

5   Blink

We present BLINK, a sampling-based framework that performs optimal resource provisioning for big data applications. As depicted in Fig. 5, the *Sample runs manager* (Sect. 5.1) first carries out lightweight sample runs on 0.1%–0.3% data samples of the original data. Based on these runs, the *size predictor* (Sect. 5.2) and *execution memory predictor* (Sect. 5.3) train prediction models to predict the size of cached datasets and the required amount of execution memory per machine in the actual run respectively. Finally, based on these models and the allocated memory in each machine, the *cluster size selector* (Sect. 5.4) selects the optimal cluster size that guarantees eviction-free runs.

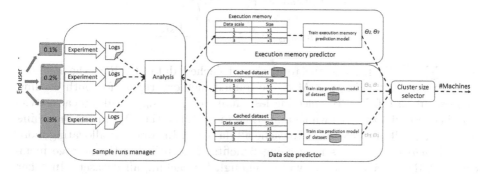

Fig. 5. Overview of BLINK.

5.1   Sample Runs Manager

The sample runs manager carries out three sample runs on tiny data samples (0.1%–0.3% of the original data) on a single machine and monitors the sample runs to make quick decisions regarding the following atypical cases:

- If there is no cached dataset in the application, the sample runs manager selects a single machine (the longest execution time but the cheapest cost).
- If there are cached datasets and eviction occurs, which is unusual with tiny datasets, it carries out new sample runs with smaller sampling scales.

While conducting sample runs, *SparkListener* collects runtime metrics and stores them as log files in the distributed file system (e.g., HDFS). The sample runs manager analyzes the logs and collects the size of each cached dataset.

5.2   Data Size Predictor

After carrying out sample runs, the data size predictor trains the following model
to predict the size of cached datasets in the actual runs:

$$D_{size} = \theta_0 + \theta_1 \times datascale \tag{1}$$

Our experiments show that the sizes of all cached datasets fit into this model. For
each cached dataset, the data size predictor takes the scale of the data sample
as a feature and its size as a label. Thus, the scales in sample runs are 1, 2, and
3; while in the actual run, the scale is 1000. We use the *curve_fit* solver with
enforced positive bounds to train the models while avoiding negative coefficients,
and Root Mean Square Error (RMSE) to evaluate the models.

5.3   Execution Memory Predictor

The minimum and the maximum amount of memory for caching in each machine
are known (M and R in Fig. 3) and, in turn, the minimum and the maximum
number of machines can be determined using the following equations:

$$Machines_{min} = \lceil \frac{\sum^{CachedDs} D_{size}}{M} \rceil$$

$$Machines_{max} = \lceil \frac{\sum^{CachedDs} D_{size}}{R} \rceil$$

where $\sum^{CachedDs} D_{size}$ is the total size of cached datasets, R is the memory
region used for caching and M is the unified memory region for both caching
and execution (cf. Fig. 3). Selecting less than $Machines_{min}$ leads to cache eviction because utilizing the whole unified memory space (i.e., M) in each machine
for caching will not be enough to cache all datasets. In contrast, allocating more
than $Machines_{max}$ gives no caching benefits since utilizing the storage memory (i.e., R) in each machine will be enough for caching all datasets. In other
words, $Machines_{max}$ is required to cache datasets without eviction when the
entire (M−R) memory region is utilized for execution. If M is not utilized at
all, then the entire region can be used for caching and, hence, $Machines_{min}$ is
required to cache datasets without evictions. Considering that the gap between
$Machines_{min}$ and $Machines_{max}$ may be quite wide and the execution memory
utilization differs from one application to another, there is a need for a precise
prediction of the amount of memory required for execution. Similar to the data
size predictor (cf. Sect. 5.2), the execution memory predictor analyzes the execution memory usage in sample runs and trains linear models to predict the total
amount of execution memory required for the actual runs. Our experiments show
that the relationship between the data sample scale and the amount of execution
memory fits into the following model, although the execution memory predictor
evaluates many other models:

$$Memory_{execution} = \theta_2 + \theta_3 \times datascale$$

5.4 Cluster Size Selector

Based on M and R in Fig. 3 (i.e., machine/instance specification), the cluster size selector calculates the required memory for execution per machine as follows:

$$MachineMemory_{execution} = \min(M - R, \frac{Memory_{execution}}{Machines})$$

Then, it selects the minimal number of machines that fulfills the condition below:

$$\frac{\sum^{CachedDs} D_{size}}{Machines} < (M - MachineMemory_{execution}) \times Machines$$

In multi-tenant environments, the recommended cluster configuration is not affected by concurrent application runs hosted on the same machines because they are deployed in isolated virtual machines, and cluster managers (e.g., YARN [13]) do not offer an occupied memory region (i.e., M) to newly submitted applications.

Table 1. Evaluated Spark MLlib applications. Recommended cluster size is shown in bold. Shadowed cells refer to cluster sizes that do not cause cache evictions. Time unit is represented in minutes. Cost unit is represented in machine minutes.

	#Machines	ALS		BAY		GBT		KM		LR		PCA		RFC		SVM	
		Time	Cost	Time	Cost	Time	Cost	Time	Cost	Time	Cost	Time	Cost	Time	Cost	Time	Cost
Sample runs	1	5.8	5.8	1.4	1.4	1.4	1.4	1.2	1.2	1.0	1.0	7.7	7.7	3.9	3.9	1.2	1.2
Approach		BLOCK-S		BLOCK-N		BLOCK-S		BLOCK-S		BLOCK-N		BLOCK-S		BLOCK-N		BLOCK-N	
Scale 100% (size)		5.6 GB		17.6 GB		30.6 MB		21.5 GB		22.4 GB		1.5 GB		29.8 GB		59.6 GB	
Scale 100% (#Blocks)		100		2K		100		200		2K		50		2K		2K	
Actual runs	1	**27.2**	**27.2**	63.3	63.3	**9.8**	**9.8**	137.2	137.2	337	337	**77.4**	**77.4**	361.6	361.6	804.8	804.8
(100% data scale)	2	14.5	29.0	29.1	58.2	6.3	12.6	45.4	90.9	133.5	266.9	41.9	83.9	125.4	250.7	325.6	651.2
	3	9.6	28.8	22.2	66.5	5.2	15.6	18.2	54.5	47.6	142.7	30.7	92	91.2	273.6	172.3	516.9
	4	8.7	34.9	14.3	57.1	8.7	34.9	**3.5**	**13.9**	17.3	69.3	28.8	115.3	**60.3**	**241**	88.5	354.1
	5	8.3	41.4	11	54.8	6.9	34.5	3.2	15.8	**8.6**	**42.9**	26.7	133.3	52.3	261.3	40.7	203.3
	6	7.5	45.2	10.1	60.8	5	29.9	2.7	16.5	7.7	46	25.2	151.2	51.4	308.4	15.7	94.4
	7	4.5	31.4	**4.1**	**28.5**	7.7	53.9	2.1	14.8	7.2	50.6	24.8	173.3	46.5	325.4	**9.6**	**67.2**
	8	4.1	33.1	3.8	30.4	4	32.2	2.3	18.8	6.9	55.6	22.4	179.5	47.2	377.8	8.6	68.9
	9	3.9	35.2	3.7	33.2	4.7	42	2.1	18.9	6.4	57.6	20.9	187.9	41.2	370.5	8.4	75.2
	10	3.6	36.3	3.5	35.3	6.2	62	1.9	19.3	6.3	63	19.5	194.6	39.8	397.5	8.3	83.5
	11	3.6	39.6	3.5	38.3	5.5	60.6	1.9	21.4	5.9	65.2	18.6	204.4	40.2	442.3	8.4	92.5
	12	3.2	38.9	3.4	41	6.1	72.9	1.9	23.2	5.5	66.2	18.3	219.1	36.7	440.6	7.7	92.9
	Avg	8.2	35.1	14.3	47.3	6.3	38.4	18.5	37.1	49.2	105.2	29.6	151	82.8	337.6	124.9	258.7

6 Evaluation

For evaluation, we use 8 applications from Spark MLlib 2.4.0: *Alternating Least Squares* (ALS), *Bayesian Classification* (BAY), *Gradient Boosted Trees* (GBT), *K-means clustering* (KM), *Logistic Regression* (LR), *Principal Components Analysis* (PCA), *Random Forest Classifier* (RFC), and *Support Vector Machine* (SVM).

Sample Runs. For conducting sample runs and measuring the robustness of the extracted models for re-usability on clusters with different machine types, we use a single node – Intel Core i3-2370M CPU running at 4x 2.40 GHz, 3.8 GB DDR3 RAM and 388 GB disk. For each application, we carried out 3 runs on sample data size in the range of 0.1%–0.3% of the complete input data scale.

Actual Runs. We made all actual runs on a private 12-node cluster equipped with Intel Core i5 CPU running at 4x 2.90 GHz, 16 GB DDR3 RAM, 1 TB disk, and 1 GBit/s LAN. All nodes used in the experiments run Hadoop MapReduce 2.7, Spark 2.4.0, Java 8u102 and Apache YARN on top of HDFS. In our extended evaluation [2], we show that the models extracted from the sample runs are reusable for larger data scales (up to $18 \times 10^4\%$) and are useful to determine the bounds on resource-constrained clusters (i.e., the maximum data scale of an application that a cluster can run without eviction).

6.1 Selected Cluster Size

As mentioned in Sect. 1, we consider an optimal cluster size as the minimum number of machines that fit all cached datasets in memory without cache eviction. The Shadowed cells in Table 1 show the cluster sizes where no eviction occur, while the bold numbers indicate the cluster sizes selected by BLINK for each application. Table 1 shows that for all applications, BLINK selects the optimal cluster size (see the first shadowed cell of each application actual run in bold).

To evaluate the efficiency of BLINK, we compare the sum of sample runs cost and actual run cost for the cluster size selected by BLINK to the average and worst costs of actual runs. Figure 6 shows that compared to the average and the worst costs, BLINK reduces the cost

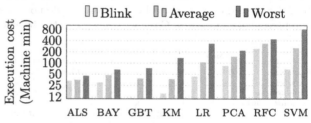

Fig. 6. BLINK cost optimization.

to 52.6% and 25.1%, respectively. In some cases, the worst cluster size (that leads to the highest cost) is a single machine due to lots of recomputations (SVM) and in other cases, it is the maximum cluster size because resources are wasted during data shuffling and processing of serial parts (RFC).

6.2 Overhead of Sample Runs

We compare the cost of sample runs with the cost of the corresponding actual run on optimal cluster configuration. Figure 7 shows that on average, sample runs cost 8.1% compared with the cost of the actual run on optimal cluster size. At worst, the overhead is 21.3% (ALS) while at best, it

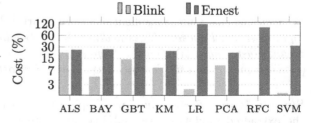

Fig. 7. Sample runs cost of BLINK and Ernest.

is 1.6% (RFC). Taking each sampling approach separately, we see that the average cost of sample runs of BLOCK-N is 2.7%, with a worst case of 5.1% (BAY) and a best case of 1.6% (RFC). For BLOCK-S, the average cost of sample runs is 13.3%, with a worst case of 21.3% (ALS) and a best case of 8.6% (KM). Altogether, BLOCK-S costs about 4.9× more than BLOCK-N. Nonetheless, the cost of BLOCK-S is still tolerable because we are comparing its cost with the costs of optimal actual runs. Note that all sample runs are carried out without changing any application parameter (e.g., number of iterations). Taking KM as a short-running application (3.5 min on the optimal cluster size; cf. Table 1), sample runs cost 8.6% of the cost of the actual run on the optimal cluster size. Hence, BLINK is also effective for short-running applications.

Even though Ernest (cf. Sect. 2) predicts application runtime rather than cluster size, we compare the cost of its sample runs with the cost of those carried out by the sample runs manager (cf. Sect. 5.1). We carry out 7 sample runs, as recommended by Ernest's optimal experiment design, on 1–2 machines with sample datasets (1% –10% of the original data). The sample runs of Ernest cost 16.4× more than those of BLINK (as depicted in Fig. 7).

7   Conclusion

BLINK is an autonomous sampling-based framework that selects an optimal cluster size with the highest cost efficiency for running big data applications. The evaluation of BLINK shows very good results in terms of selecting an optimal cluster size with high prediction accuracy.

Acknowledgement. This research was partially funded by the Thuringian Ministry for Economy, Science and Digital Society under the project thurAI and by the Carl-Zeiss-Stiftung under the project MemWerk.

References

1. Al-Sayeh, H., Jibril, M.A., Bin Saeed, M.W., Sattler, K.U.: SparkCAD: caching anomalies detector for spark applications. In: VLDB 2022 (2022)
2. Al-Sayeh, H., Jibril, M.A., Memishi, B., Sattler, K.U.: Blink: lightweight sample runs for cost optimization of big data applications. In: CoRR (2022). https://arxiv.org/abs/2207.02290
3. Al-Sayeh, H., Memishi, B., Jibril, M.A., Paradies, M., Sattler, K.U.: Juggler: autonomous cost optimization and performance prediction of big data applications. In: ACM SIGMOD 2022 (2022)
4. Al-Sayeh, H., Memishi, B., Paradies, M., Sattler, K.U.: Masha: sampling-based performance prediction of big data applications in resource-constrained clusters. In: VLDB DISPA 2020 (2020)
5. Alipourfard, O., Liu, H.H., Chen, J., Venkataraman, S., Yu, M., Zhang, M.: CherryPick: adaptively unearthing the best cloud configurations for big data analytics. In: USENIX NSDI 2017 (2017)

6. Amdahl, G.M.: Validity of the single processor approach to achieving large scale computing capabilities. In: AFIPS 1967 (Spring), Spring Joint Computer Conference (1967)
7. Chakaravarthy, V.T., Pandit, V., Sabharwal, Y.: Analysis of sampling techniques for association rule mining. In: ICDT 2009 (2009)
8. Hamidi, H., Mousavi, R.: Analysis and evaluation of a framework for sampling database in recommenders. J. Glob. Inf. Manag. **26**, 41–57 (2018)
9. Kunjir, M., Babu, S.: Black or white? How to develop an AutoTuner for memory-based analytics. In: ACM SIGMOD 2020 (2020)
10. Li, H., et al.: Detecting cache-related bugs in spark applications. In: ACM SIGSOFT 2020 (2020)
11. Meng, X., et al.: MLlib: machine learning in apache spark. J. Mach. Learn. Res. **17**, 1235–1241 (2016)
12. Perez, T.B.G., Zhou, X., Cheng, D.: Reference-distance eviction and prefetching for cache management in spark. In: ACM ICPP 2018 (2018)
13. Vavilapalli, V.K., et al.: Apache Hadoop YARN: yet another resource negotiator. In: ACM SoCC 2013 (2013)
14. Venkataraman, S., Yang, Z., Franklin, M., Recht, B., Stoica, I.: Ernest: efficient performance prediction for large-scale advanced analytics. In: USENIX NSDI 2016 (2016)
15. Xu, L., Li, M., Zhang, L., Butt, A.R., Wang, Y., Hu, Z.Z.: MEMTUNE: dynamic memory management for in-memory data analytic platforms. In: IEEE IPDPS 2016 (2016)
16. Yu, Y., Wang, W., Zhang, J., Letaief, K.B.: LRC: dependency-aware cache management for data analytics clusters. In: IEEE INFOCOM 2017 (2017)
17. Zaharia, M., et al.: Resilient distributed datasets: a fault-tolerant abstraction for in-memory cluster computing. In: USENIX NSDI 2012 (2012)
18. Zhu, Z., Shen, Q., Yang, Y., Wu, Z.: MCS: memory constraint strategy for unified memory manager in spark. In: IEEE ICPADS 2017 (2017)

ADBIS Short Papers: Data Pre-processing and Cleaning

A RoBERTa Based Approach for Address Validation

Yassine Guermazi(✉), Sana Sellami, and Omar Boucelma

Aix Marseille Univ, Université de Toulon, CNRS, LIS, Marseille, France
{yassine.guermazi,sana.sellami,omar.boucelma}@lis-lab.fr

Abstract. Address verification is becoming more and more mandatory for businesses involved in parcel or mail delivery. Situations where shipments are returned or delivered to the wrong person (or legal entity) are harmful and may incur several costs to the stakeholders. Indeed, addresses often carry incorrect information that need to be corrected prior to any shipment process. In this paper we propose a 2-step address validation approach consisting of (1) standardization and (2) classification, both steps are based on RoBERTa, a pre-trained language model. Experiments have been conducted on real datasets and demonstrate the effectiveness of the approach in comparison to other methods.

Keywords: Data quality · Address cleansing · Address classification · Natural language processing · Deep learning · Transformers · RoBERTa

1 Introduction

Bad address data severely impacts many industries such as postal services, e-commerce or transportation businesses. *"Shipment Returned because of Bad address"* or *"Package Delivered to Wrong Address"* are some of the situations that often happen in the general context of parcel (or surface mail) delivery.

In this paper we describe a solution for the *address verification* problem: we propose a cleansing and address validation process, in providing (1) a categorization of dirt (e.g. typos, misspelling, geographic inconsistencies), (2) an address standardization method, and finally (3) an address classification method.

Although the problem is a pretty old one, it recently received a lot of attention, both from industry and academia. In the industry, software vendors such as Experian[1] or Informatica[2] are expanding their businesses in providing address verification/validation solutions. In the academia, some research has focused on address cleansing solutions, including preprocessing and parsing, especially for structured addresses [1–3]. Address validation was usually performed based on geocoding solutions such as geocoding APIs (e.g. Google, Bing). However, these tools are not able to manage unstructured and dirty addresses (e.g. missing

---

[1] https://www.edq.com/demos/address-verification/.
[2] https://www.informatica.com/products/data-quality/data-as-a-service/address-verification.html.

© Springer Nature Switzerland AG 2022
S. Chiusano et al. (Eds.): ADBIS 2022, CCIS 1652, pp. 157–166, 2022.
https://doi.org/10.1007/978-3-031-15743-1_15

attributes, geographic inconsistencies) which are frequent in developing countries. In particular, geographic inconsistencies occur when at least two address attributes do not coexist in the same geographical area. The most frequent are those related to address elements (*city, district, road*) as illustrated in Table 1.

Table 1. Examples of *Invalid* addresses in Senegal

Inconsistency's type	Addresses	Description
City inconsistency (CI)	Sicap Amitie 2 Villa Numero 4030 **Louga**	The address does not really exist in **Louga** city
District inconsistency (DI)	**Sicap Amitie III** Vdn Numero 9982 Pres Auchan Dakar Senegal	**Vdn** road does not exist in **Sicap amitie III** district
Road inconsistency (RI)	**Route De Ngor** X Avenue Birago Diop Dakar Senegal	There is no intersection (X) between Route (i.e. Road) and Avenue

Table 2. An example of polysemy in Senegalese addresses

Polysemous word	Referring place	Example
Diourbel	Road name	Rue Saint Louis **Diourbel** Point E BP 116 Dakar Senegal
	City	Route De La Gare Face Pharmacie Baol **Diourbel** Senegal

To come up with the solution described in this paper, we consider the problem as a text classification one [6] after a standardization phase has been performed. However, in doing so, we had to face polysemous difficulties, e.g. place names that may refer to different places as illustrated in Table 2. Identifying and resolving polysemous situations is mandatory in order to avoid classification distortion.

The rest of the paper is organized as follows. Section 2 reviews some address parsing and classification work. In Sect. 3, we detail our solution while, in Sect. 4 we describe our experimental results. Finally, Sect. 5 concludes this paper.

2   Related Work

We describe in this section related works on address parsing and classification.

2.1   Address Parsing

In the field of NLP, parsing is considered as a sequence labeling task [3]. As far as parsing models, we looked at Hidden Markov models (HMM) [11] and Conditional Random Field (CRF) based models [12]. HMM does not take into account all possible address patterns, in particular those with low probabilities. CRFs perform better than HMM because they use a conditional probability instead of the independence assumption made in HMM. However, their performance is affected by the presence of non-standardized addresses and also polysemous words. Recently, deep learning models including Transformers, have been proposed for address parsing. In [13], authors proposed a BERT+CRF approach for parsing Chinese addresses: BERT is applied first for generating address contextual representation, then a CRF model is applied for predicting tags. The evaluation results performed on Chinese addresses show that the F1-score is better than approaches that combine Word2vec, BiLSTM and CRF.

Promising results of BERT applications in sequence labeling, particularly in address parsing [13] motivate us to apply RoBERTa [10] in parsing. Compared to BERT, RoBERTa allows to get rid of the next sentence prediction objective in model's pre-training which improve performance in some downstream tasks.

2.2 Address Classification

Recently, static or contextual word embedding models have been used to perform address classification. Seng et al. [4] proposed an approach that classifies addresses to its property type. It consists in applying Long Short-Term Memory Neural Networks (LSTM) on Word2Vec [5] address representations. However, static word embedding, such as Word2vec, cannot handle polysemy. To address this problem, contextual word embedding among which Pre-trained Language Models (PLM), such as BERT [7], have made it possible to strengthen the contextual modeling of texts. In the address classification context, Mangalgi et al. [6] propose a RoBERTa-based approach to classify Indian addresses according to the sub-regions to which they belong. A comparison with Word2vec and Bi-LSTM approaches shows that the RoBERTa approach outperforms the other ones in terms of accuracy. Indeed, Word2vec loses the sequential information by averaging the word vectors. RoBERTa better captures the context than Bi-LSTM. However, PLM, rarely consider incorporating structured semantic information which can provide rich semantics for language representation.

For better language understanding, some works have investigated the grounding of PLM with high quality (domain) knowledge, which are difficult to learn from raw texts. Indeed, incorporating external knowledge into PLM has proven effective in various NLP tasks [8,9].

Our address classification approach draws inspiration from these recent works. It consists on injecting knowledge in the form of address tag embedding into a PLM. These tags result from the address parsing step.

3 Address Validation Approach

In this section, we describe our RoBERTa-based approach for address validation which consists of two main steps: (1) address standardization in order to clean data and to obtain the different address tags and (2) a binary (valid, invalid) address classification.

3.1   Address Standardization

Address standardization refers to the transformation of an address into a normalized standard format. It involves two tasks: preprocessing and parsing.

Preprocessing. The purpose of this step is to normalize entities and to clean addresses in removing special characters and correcting different spelling errors. For that, we adopt a dictionary-based approach which provides the keywords that may be used to define the address components (city, road, etc.) as well as common abbreviations of these words. In addition, we use a spell checker *pyspellchecker*[3]/[4] in order to correct address keywords.

Parsing. Given an address $A = \{a_1, .., a_n\}$ where a_i is the i-th word and n represents the length of the address, the parsing of A aims to assign a label l to each word a_i of A among the corresponding list of address tags Y; $Y = \{IB, EB, P, Z, HN, RN, D, RS, PB, ZC, C, CO\}$. These tags are defined following the address model depicted in Fig. 1.

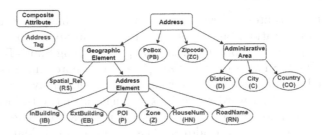

Fig. 1. Address model

We propose a parsing method (Fig. 2) which consists firstly in generating a contextual representation of an address A using pre-training RoBERTa model on a corpus of addresses (Sect. 4.1) by following these two sub-steps:

- RoBERTa calculates the input representations of A by summing over the token, position, and segment embedding. Token embedding for each token is generated using byte-level BPE tokenizer. Position embedding includes the positional information of each token in the address. Segment embedding provides the same label to the tokens that belong to the address.
- Input address representation goes through 12 transformer encoders which capture the contextual information for each token by self-attention and produces a sequence of contextual embeddings noted as H.

Then, the resulting representation is passed to a tagging layer to obtain address tags, using the IOB tagging scheme. A linear layer takes as input the last hidden state of the sequence $H = \{h_1, .., h_n\}$ and provides as result the prediction of the tags T.

---

[3] https://readthedocs.org/projects/pyspellchecker/downloads/pdf/latest/.
[4] https://norvig.com/spell-correct.html.

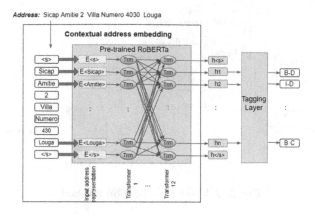

Fig. 2. Address parsing method

3.2 Address Classification

We propose a RoBERTa based classification method (Fig. 3) to classify addresses to *Valid* or *Invalid*. It consists of two steps: 1) generating a fusion of two vector representations which are the contextual vector representation of the address and the vector representation of the address tags, and 2) a classification of addresses according to resulted vectors.

Vectors Fusion. We use a concatenation function to fuse two embedding vectors as follows:

1. Contextual address embedding: we retrieve the contextual vector representations H of the address A, generated by the pre-trained RoBERTa model, in the address parsing step (see Sect. 3.1).
2. Address tags embedding: the output of the address parsing step is n tags denoted by $T = \{t_1, .., t_n\}$. Since these tags are at the word level, their length is equal to the length n of an address A. We use a look-up table to map these tags to $\{id_1, .., id_n\}$ and feed a linear layer (fully connected layer) in order to obtain the tags embedding, denoted as $W = \{w_1, .., w_n\}$, of A.

Address Classification. It is performed using a linear layer (fully connected layer). First, this layer takes as input the embedding fusion vector and generates as output the class logits (probabilities), knowing that the objective function of the training is the CrossEntropy. Then, the Argmax function is applied to these probabilities to obtain the predicted class.

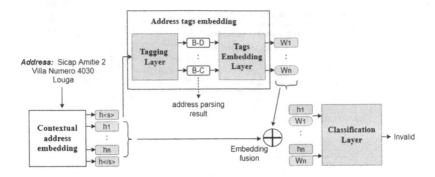

Fig. 3. Address classification method

4   Evaluation

In this section, we describe the experiments carried out in order to evaluate our approach.

4.1   Experimental Setting

Dataset Description. Parsing and classification have been performed in using two real-world datasets: (1) a French dataset J_f, which represents 10000 structured addresses extracted from the French Sirene directory[5] and (2) Senegalese dataset J_s, which contains 500 unstructured addresses collected from a Senegalese companies directory,[6] characterized by the presence of spatial operators and often the absence of keywords allowing the identification of address elements.

Evaluation Setup. GeLU activation is used in RoBERTa with the ADAM Optimizer. The dropout and learning rate are set respectively to 0.1 and 3e−5, in such a way to maximize the accuracy in the validation set. To avoid overfitting, we use the early stop technique based on loss validation by setting a maximum number of training epochs (= 12) and a batch size of 32.

The pretraining of RoBERTa is performed through the Pytorch framework.[7] We generated two pretrained RoBERTa models corresponding to each of the following corpora: (1) French corpora composed of 1,048,575 addresses[8] and (2)

---

[5] https://www.data.gouv.fr/fr/datasets/base-sirene-des-entreprises-et-de-leurs-etablissements-siren-siret.

[6] https://www.goafricaonline.com/.

[7] https://pytorch.org/.

[8] https://www.data.gouv.fr/fr/datasets/base-sirene-des-entreprises-et-de-leurs-etablissements-siren-siret.

Senegalese corpora composed of 31893 addresses collected from Web business directories.[9]/[10]/[11]/[12]

4.2 Address Parsing Evaluation

We describe in this section the experiments carried out for the parsing of French J_f and Senegalese J_s addresses. Each dataset is split into a training, validation, and test sets using the ratio of 3:1:1. Moreover, we perform a manual labeling of the datasets.

Baseline Methods. We compare our method with two known address parsing sequence models: (1) HMM, implemented with Febrl[13] and (2) CRF, implemented with python-crfsuite library.[14]

Results. Table 3 illustrates our results in terms of F-measure. First, it is worth noticing that RoBERTa outperforms the two other methods for all datasets. Second, HMM and CRF seem more accurate in case of J_f, the French dataset, given the structured nature of addresses, but less accurate in the case of J_s, the Senegalese dataset. Indeed, J_s addresses contains more polysemous words and are less structured than J_f ones. Finally, we note that RoBERTa better handles polysemous words as illustrated in Table 4 but fails in parsing addresses that lacks for some address elements and/or keywords: those addresses can be characterized as poorly contextualized addresses. Besides, the low frequency of some address elements in the pre-training corpus prevent RoBERTa from an efficient learning context.

Table 3. F-measure of address parsing methods

Method	J_f	J_s
HMM	0.973	0.931
CRF	0.984	0.947
RoBERTa	**0.988**	**0.956**

Table 4. Percentage of polysemous resolution

Method	J_f	J_s
HMM	71.1%	46.6%
CRF	82.1%	68.8%
RoBERTa	**91.2%**	**86.6%**

---

[9] https://creationdentreprise.sn/.
[10] http://pagesjaunesdusenegal.com/.
[11] https://www.goafricaonline.com/.
[12] https://www.yelu.sn/.
[13] http://users.cecs.anu.edu.au/~Peter.Christen/Febrl/febrl-0.3/febrldoc-0.3/node24.html.
[14] https://github.com/scrapinghub/python-crfsuite.

4.3   Address Classification Evaluation

We evaluate classification approach on J_f and J_s dataset. We assume that the addresses belonging to these 2 datasets are of class *Valid* because they come from 2 reliable data sources: (1) French government database and (2) an African company directory. To resolve the imbalance class problem in J_f and J_s, we propose a data augmentation technique that allows the generation of synthetic addresses labelled as *Invalid* by applying transformations to collected *Valid* addresses.

Data Augmentation: The proposed method is based on address attributes replacement and consists in: (1) Creating two subsets G_f and G_s, from the French and the Senegalese corpora, which represent a hierarchical arrangement of address elements per country and (2) Applying a sequence of attribute replacement, for each dataset J_f and J_s, using G_f and G_s, to create different types of geographic inconsistency (see examples in Table 1).

For each dataset $J \in \{J_f, J_s\}$, we have divided J into two subsets: J_v (70% of addresses) and J_v' (30% of addresses). The generation of *Invalid* addresses is performed on the J_v' dataset. For French addresses, we have injected two types of invalidity related to the most frequent address elements which are: city (CI) and road (RI). For the Senegalese dataset, we have also injected inconsistency for the district (DI) which is more frequent for this dataset than for French addresses. We denote the invalid dataset as J_{inv}'. The classification dataset, denoted as Jc (i.e. $J_f c$ or $J_s c$), is thus composed by J_v, representing *Valid* addresses, and J_{inv}' representing *Invalid* addresses such as $size(J_{inv}') = size(J_v)$.

Baseline Models. We compare our approach "AllRoBERTa" with the models used in address classification works which are based on (1) static word embedding (Word2vec) plus a SVM classifier and (2) RoBERTa with no knowledge injection. The idea here is to compare the effectiveness of static versus contextual word embedding and to outline the importance of knowledge injection in the proposed approach.

Results. Table 5 illustrates the classification results obtained with the different approaches. We notice that whatever the type of invalidity or the country, "AllRoBERTa" is more efficient. We note also that RoBERTa-based models are more efficient than a Word2vec one for both datasets. This can be explained by the highly contextualized representations offered by RoBERTa. Moreover, pre-training RoBERTa on a large corpus of business addresses allows the model to learn several geographical facts related to the context of each address element. Classification results show that our "AllRoBERTa" is a promising solution which can be useful, mainly when geographic databases are missing in some countries such as Senegal.

We evaluated the percentage of polysemy in the misclassified Senegalese addresses. As illustrated in Table 6, for all the tested approaches, more than 50% of the misclassified addresses are polysemous. This ratio can even reach 72.5%

Table 5. F-measure of different address classification approaches

Approach	J_fc		J_sc		
	CI	RI	CI	DI	RI
Without parsing					
Word2vec+ SVM	0.911	0.862	0.9	0.869	0.848
RoBERTa	0.949	0.928	0.942	0.919	0.912
With parsing					
AllRoBERTa	**0.981**	**0.957**	**0.971**	**0.948**	**0.938**

Table 6. Impact of polysemy in Senegalese addresses classification

Approach	Polysemy percentage in misclassified addresses
Word2vec + SVM	72.5%
RoBERTa	69.2%
AllRoBERTa	51.2%

Table 7. Impact of a "perfect" parsing on addresses classification

Approach	J_sc		
	CI	DI	RI
AllRoBERTa	0.971	0.948	0.938
Parsing "Ground Truth" + RoBERTa	**0.985**	**0.965**	**0.958**

in the case of a classification based on "Word2vec + SVM". For AllRoBERTa, classification errors come from cases of unresolved polysemous situations during parsing, with a percentage greater than 83%. We conclude that polysemous elements badly impacts the address classification process.

Finally, we analyzed the impact of the introduction of address tags in the classification. To this end, we first performed manual parsing (J_s) in order to perfectly identify addresses tags, then we carried out the classification (J_sc) with RoBERTa. The obtained results compared to AllRoBERTa (Table 7) show that the quality of parsing has an impact on the classification results.

5 Conclusion

In this paper, we described an address validation approach based on RoBERTa, a pre-trained transformer-based language model. Usage of RoBERTa is motivated by its ability to manage polysemy. We inject semantic address tags into the pre-trained RoBERTa model in order to improve semantic understanding of domain-specific data. Experimental evaluations, carried out on two real-world datasets involving French and Senegalese addresses, show the effectiveness of our solution. In the future, we intend to extend this work in at two directions: (1) explore an active learning method to minimize the efforts of manually labelling data sets and (2) make the approach usable through an address validation API.

References

1. Matci, D.K., Avdan, U.: Address standardization using the natural language process for improving geocoding results. Comput. Environ. Urban Syst. **70**, 1–8 (2018)
2. Xi, X.-F., Wang, L., Zou, E., Zeng, C., Fu, B.: Joint learning for non-standard Chinese building address standardization. In: 2018 IEEE International Smart Cities Conference (ISC2), pp. 1–8. IEEE (2018)
3. Abid, N., ul Hasan, A., Shafait, F.: DeepParse: a trainable postal address parser. In: 2018 Digital Image Computing: Techniques and Applications (DICTA), pp. 1–8. IEEE (2018)
4. Seng, L.: A two-stage text-based approach to postal delivery address classification using long short-term memory neural networks (2019)
5. Mikolov, T., Sutskever, I., Chen, K., Corrado, G., Dean J.: Distributed representations of words and phrases and their compositionality. In: Advances in Neural Information Processing Systems (NIPS), pp. 3111–3119 (2013)
6. Mangalgi, S., Kumar, L., Tallamraju, R.B.: Deep contextual embeddings for address classification in e-commerce. arXiv preprint arXiv:2007.03020 (2020)
7. Devlin, J., Chang, M.-W., Lee, K., Toutanova, K.: BERT: pre-training of deep bidirectional transformers for language understanding. arXiv preprint arXiv:1810.04805 (2018)
8. Zhang, Z., Han, X., Liu, Z., Jiang, X., Sun, M., Liu, Q.: ERNIE: enhanced language representation with informative entities. In: ACL, pp. 1441–1451 (2019)
9. Zhang, Z., et al.: Semantics-aware BERT for language understanding. In: Proceedings of the AAAI Conference on Artificial Intelligence, vol. 34, no. 05, pp. 9628–9635 (2020)
10. Liu, Y., et al.: RoBERTa: a robustly optimized BERT pretraining approach. arXiv preprint arXiv:1907.11692 (2019)
11. Christen, P., Belacic, D.: Automated probabilistic address standardisation and verification. In: Australasian Data Mining Conference (AusDM 2005), Sydney, pp. 53–67 (2005)
12. Wang, M., Haberland, V., Yeo, A., Martin, A., Howroyd, J., Bishop, J.M.: A probabilistic address parser using conditional random fields and stochastic regular grammar. In: 2016 IEEE 16th International Conference on Data Mining Workshops (ICDMW), pp. 225–232. IEEE (2016)
13. Zhang, H., Ren, F., Li, H., Yang, R., Zhang, S., Du, Q.: Recognition method of new address elements in Chinese address matching based on deep learning. ISPRS Int. J. Geo-Inf. **9**(12), 745 (2020)

Empirical Comparison of Semantic Similarity Measures for Technical Question Answering

Nabil Moncef Boukhatem[1,3](✉), Davide Buscaldi[2](✉), and Leo Liberti[3](✉)

[1] OneTeam, Paris, France
nboukhatem@oneteam.fr
[2] LIPN CNRS, Université de Paris-Nord, Villetaneuse, France
buscaldi@lipn.univ-paris13.fr
[3] LIX CNRS Ecole Polytechnique, Institut Polytechnique de Paris,
91128 Palaiseau, France
liberti@lix.polytechnique.fr

Abstract. We consider the task of looking for the answer to a given user question by means of identifying the most relevant document in a technical knowledge base. We briefly introduce the NLP fields related to this task, then discuss what we think are the most promising methods to accomplish the task. The main aim of the paper is to benchmark the chosen methods on two different knowledge bases (one proprietary, one public). Every document in each KB consists of a title and a text describing a solution to a technical problem. Our tests point out that the best method for the task at hand is the use of Sentence Transformers, a deep learning based method using pre-trained language models.

1 Introduction

This paper concerns a variant of a well known Natural Language Processing (NLP) task, namely Question Answering (QA) [10, Ch. 25]. QA aims at automatically answering questions posed by humans in natural language (NL). The variant we are interested in is actually a restriction of QA: our output is a relevant document in a Knowledge Base (KB) instead of an answer written in NL.

The reason why we look at this variant is that it represents an important need in industrial contexts. Typically, in our scenario of interest, the "user" who asks the question may be an employee of the firm, or a technically skilled client. This scenario is very different from those leading to open-text QA systems yielding NL answers, where the user may be any individual. In the latter case, users may employ more informal language to phrase their questions, and will expect answers in NL.

In our case, we expect the user to pose more precise questions, and, what's more important, we know that he or she will accept the pointer to documents containing an explanation as a valid answer. This effectively sets our task at the intersection of QA and Document Retrieval (DR) [11]. We note, however, that in

S. Chiusano et al. (Eds.): ADBIS 2022, CCIS 1652, pp. 167–177, 2022.
https://doi.org/10.1007/978-3-031-15743-1_16

DR the user queries need not be cast in NL. For later reference, we name our task of interest Natural Language Document Retrieval (NLDR). Methods that can be used to address NLDR have to match the meaning of the user question to relevant semantic indicators in the documents of the KB. This establishes a connection between NLDR and semantic similarity measures and techniques [4]. All NL tasks may either refer to a general language setting, or to a specific language domain [5,13]. The present research focuses on specific language domains: our motivation stems from the interest of a service firm with several industrial clients: each client is considered a specific domain.

The contribution of the paper is a computational comparison of known methods that can address the NLDR task, based on popular performance measures, on two KBs: a proprietary one belonging to the service firm that motivates this work, and the public IBM TechQA dataset [3]. In the rest of this section, we briefly survey the fields of NLP that are relevant to the NLDR task, from the point of view of the methods we benchmark. In the rest of the paper we review the benchmarked methods, and we comment our computational results.

1.1 Document Retrieval

Document retrieval methods are usually structured around a user query (which may be expressed in either formal or NL) and a KB (database, corpus, graph...). The goal is to return the KB entry that is the most relevant to the query. The relevance depends on the satisfaction of the user's information need. The document ranking can be obtained in different ways: using word frequency and co-occurrence (e.g. [16]); using several independent syntactical statistics and co-occurrence measures (e.g. [2]); using a weighted variant of TF-IDF called BM25F (e.g. [14]); using sentence transformers (ST) to compare the given query with sentences in the corpus (e.g. [15]).

1.2 Technical Question Answering

By *Technical* QA (TQA) we mean QA over a restricted domain, where user information needs are predominantly oriented to solving technical issues. The interest about these systems has been growing in recent years, as testified by the proliferation of technical forums, some dedicated to developers, such as Stack Exchange,[1] while others are related to a particular product or service. Early TQA systems were based on syntactical analysis of sentences, possibly with some elementary form of logical entailment, complemented by semantics stored in a specific-domain ontology [13]. More recently, ST have been used in TQA in [17], where an elaborate pipeline was used to train a neural network to match questions to answers, using training sets constructed from specific domains. In [18], the authors address the difficulty of forming large enough training sets for TQA related tasks. They propose a system relying on a general-purpose

---

[1] https://data.stackexchange.com/.

QA training set, before applying a transfer learning techniques using the IBM TechQA training set [3].

A TQA system can also be seen as an evolution towards an automatization of a classic technical support system, relieving technical experts from the burden of browsing the documentation necessary to answer users questions. A standard situation in a classic support system can be described as follows: a user encounters an issue with its system, gets in touch with the support team and describes the situation he or she is facing. The contact can occur by phone, by email or by filling a form directly in the support system.

The user usually tries to describe the situation to help the support agent identify the source of the issue and how to fix it, for example: *"Hello, I've been struggling all day with my internet connection without any improvement. Is it possible to help me please?"*

Depending on the root cause, issues may need to be escalated to qualified agents, investigated by means of technical user guides, and/or compared to known issues. Finding a solution often requires time and effort. Each issue and the corresponding solution is documented, and the information is archived for later reference. The documentation process helps creating the set of documents (or *corpus*) C that will be considered the knowledge base of a TQA system.

Typical users of TQA systems (usually clients or support agents) describe their issues in NL by typing a query Q. The system then compares Q to the documents in the set C so that the most fitting answer (or set K of answers) can be retrieved. If an answer is found, it is returned to the user, who can rate its pertinence by providing a score. In the above example, a typical TQA system might infer issues in the internet connection because of the words *"struggling"*, *"internet"* and *"connection"*. This would lead the system to retrieve a set of documents about internet connection issues.

1.3 Semantic Similarity Measures

Semantic similarity measures are functions mapping sentence pairs to a similarity measure (usually in $[0, 1]$), by means of semantic considerations. We refer the reader to the recent comprehensive survey [4]. In particular, with reference to [4, Fig. 1], the methods we benchmark are mostly based on edge counting (e.g. WordNet information), information content (e.g. word frequency), transformer models (based on BERT [7]).

2 Compared Methods

We chose a set of semantic similarity methods that cover the most important models for the semantic representation of documents: keyword-based similarity methods represent the content of a document by means of selected words, which are matched to the user question. Methods such as BM25, which is at the core of Whoosh and Elasticsearch, rely instead on the fact that all words in a document have some degree of importance. The document itself is represented by a (sparse) vector in the space of words. Finally, ST take advantage from deep learning to obtain a semantic representation of the document as dense vectors.

The TQA methods we test are organized around two pieces of input: Q, which represents a question, and C, which represents a corpus. The representation could be based on words, vectors, or the output of an artificial neural network. The general algorithmic scheme is two-phase: pre-processing (on C) and on-the-fly (on Q). The output is usually a ranked set of documents from C.

2.1 Keyword Extraction Based Methods

In this section, we summarize two of the methods we benchmark, where the representation of Q and C is based on keyword extraction (KE). The first phase (pre-processing) extracts keywords from C. The second phase (on-the-fly) extracts them from Q. The methods then match keywords from Q with those from C in order to obtain relevant documents. The difference between the methods is the computation of the weights assigned to keywords from C.

Algorithm 1. Weighting method 1

1: Corpus_KW, Tokens_L, Synonyms_L, Stemmed_Tokens, Scores, Orders : List
2: **for** D document in C corpus **do**
3:     Corpus_KW.append(KW_extraction(D["title"]))
4: Tokens_L ← (Tokenize(Q))
5: Remove Stopwords from Tokens_L
6: **for** T Token in Tokens_L **do**
7:     Synonyms_L.append(Syn(T))
8:     **for** X word in Synonyms_L **do**
9:         Stemmed_Tokens.append(Stem(X))
10:     Stemmed_Tokens.append(Stem(T))
11: **for** i from 0 to len(Corpus_KW) **do**
12:     **for** k Word in Corpus_KW[i] **do**
13:         **for** w Word in Stemmed_Tokens **do**
14:             **if** w == k **then**
15:                 Scores[i] ← Scores[i] + 1
16:     Scores[i] ← Scores[i] / len(Corpus_KW[i])
17: Return the documents of C with the highest scores

In the first method, the pre-processing phase scans C, and for each document D in C extracts three keywords from the title and three from the content of the document using a KE algorithm. Each keyword is simply assigned a unit weight. During the on-fly-phase, Q is cleaned from stopwords and tokenized. Each token is stemmed and listed along with its synonyms found in WordNet [8]. Each time a token (or one of its synonyms) is found, we increment the document score by the token weight. We divide the score by the number of the document keywords to obtain the final score of the document. The documents with maximum score value are returned to the user.

In the second method, the pre-processing phase scans C, and for each document D in C extracts three keywords from the title, which are given a weight of 2, and three other keywords from the text, which are given a weight of 1, using a KE algorithm. Identical keywords have their weights summed. The three keywords with highest weight values are selected. The on-the-fly phase Q is the same as for the first method. After finding keywords, for both methods we form a list of all corpus documents containing at least one keyword from Q. Documents are then ranked by decreasing keyword weight sums.

Algorithm 2. Weighting method 2

1: Document_KW,   Corpus_KW,   Query_Words,   Tokens_L,   Synonyms_L, Stemmed_Tokens, Scores, Orders : List
2: Title_Weight = 2
3: Body_Weight = 1
4: **for** D document in C corpus **do**
5:     // *If an element already exists, weights are summed*
6:     Corpus_KW.append(KW_extraction(D["title"]), Title_Weight)
7:     Corpus_KW.append(KW_extraction(D["body"]),Body_Weight)
8:     Corpus_KW.append(Max(Document KW,3))
9: Tokens_L.append(Split(Q))
10: Remove Stopwords from Tokens_L
11: **for** T Token in Tokens_L **do**
12:     Synonyms_L.append(Syn(T))
13:     **for** X word in Synonyms_L **do**
14:         Stemmed_Tokens.append(Stem(X))
15:     Stemmed_Tokens.append(Stem(T))
16: **for** i from 0 to len(Corpus_KW) **do**
17:     **for** k,l Word,weight in Corpus_KW[i] **do**
18:         **for** w Word in Stemmed_Tokens **do**
19:             **if** w == k **then**
20:                 Scores[i] = Scores[i] + 1
21:     Scores[i] = Scores[i] / len(Corpus_KW[i])
22: Return the documents of C with the highest scores

For KE purposes, we used two well-known systems, namely Rapid Automatic Keyword Extraction (RAKE) [16] and Yet Another Keyword Extraction (YAKE) [2]. Two unsupervised automatic KE methods that are domain, corpus, and language independent.

2.2   Word Vector Based Methods

Whoosh is an open-source Python search engine library for indexing and searching text based on the BM25F algorithm [14]. Whoosh is a native Python alternative to Lucene [1], which strongly inspired its development. Whoosh's index is *fielded*: the user defines a set of fields(D) of fields that represent the structure and content of the document D into the index. For instance, a classic fields referring to scientific papers involve indexing title, abstract and body into different fields. Keywords are selected by means of *analyzers*, which usually provide language-dependent stemming algorithms and stop-word lists. Given a query $Q = \{t_1^{(q)}, \ldots, t_n^{(q)}\}$ and a document $D = \{t_1^{(d)}, \ldots, t_m^{(d)}\}$, the matching score between them is calculated according to BM25F:

$$\mathsf{BM25F}(D, Q) = \sum_{t \in Q \cap D} \frac{\mathsf{TF}(t, D)}{k_1 + \mathsf{TF}(t, D)} \mathsf{IDF}(t),$$

where t is a term shared by both the query and the document, k_1 a constant set at 1.2, and $\mathsf{TF}(t, D)$ is the *normalized term frequency*: $\mathsf{TF}(t, D) = \sum_{c \in \mathsf{fields}(D)} w_c \frac{\mathsf{freq}_c(t, D)}{1 + b_c \frac{l_{(D, c)}}{\hat{l}_c}}$, where $\mathsf{freq}_c(t, D)$ are the occurrences of the term t in the field c of document d, $l_{(D, c)}$ is the length of the field c in document D, and $\hat{l}_c$ is the average length of field c. Moreover, b_c is a field-dependant parameter, usually set at 0.75, and w_c is a boost factor (by default set at 1.0) that can be specified by the user at query time. The inverse document frequency $\mathsf{IDF}(t)$ is usually calculated as $\mathsf{IDF}(t) = \frac{N - df(t) + 0.5}{DF(t) + 0.5}$, where N is the number of documents in the collection and $\mathsf{DF}(t)$ is the document frequency of term t.

ElasticSearch (ES) is another open-source search engine built on top of Lucene [1]. It allows users to store, index and search large sets of documents. Unlike Whoosh, explicit mapping schemata are not necessary. ES supports structured queries, full text queries, and complex queries that combine the two. Structured queries are similar to SQL queries, while full-text queries find and return all documents that match the query, sorted by relevance. In addition to searching for individual terms, ES supports phrase searches, similarity searches, and prefix searches. ES, like Lucene [1], also uses the Okapi BM25 algorithm as a default ranking function for documents relevance in the search process. BM25 is a bag-of-words retrieval function that ranks documents based on the terms $q_1, \ldots, q_n$ of a query Q appearing in each document D, regardless of their proximity within the document and given by:

$$\mathsf{BM25}(D, Q) = \sum_{i=1}^{n} \mathsf{IDF}(q_i) \frac{\mathsf{F}(q_i, D)(k_1 + 1)}{\mathsf{F}(q_i, D) + k_1(1 - b + b\frac{|D|}{\mathsf{avgdl}})},$$

where $\mathsf{F}(q_i, D)$ is q_i's term frequency in the document D, $|D|$ is the length of the document D in words, and avgdl is the average document length in the text collection from which documents are drawn. k_1 and b are free parameters.

2.3 Deep Learning Based Methods

ST are a Python framework for calculating sentence, paragraph and image embeddings. It is an implementation of Sentence-BERT (SBERT) [15], a modification of the pretrained BERT network [7] that uses Siamese network structures to derive semantically meaningful sentence embeddings. SBERT uses a pretrained BERT and RoBERTa networks, and adds a pooling operation to their output in order to derive a fixed sized sentence embedding. It can be described as a document processing method of mapping sentences to real-valued vectors such that sentences with similar meanings are close in vector space.

To achieve this goal, a semantic representation of a sentence is built by adapting a transformer model in a Siamese architecture: sentences are processed pair by pair, with the same neural network structure. A vector is produced by each network, on which a distance is calculated. The loss function for the complete network consists in minimizing the distance between semantically similar sentences, and maximizing the distance between semantically distant sentences.

The obtained word embeddings can be compared using similarity measurements, such as cosine similarity. Such scores can be exploited in different NLP tasks, including information and document retrieval. Cosine similarity is the cosine of the angle between two word vectors E and V (the dot product of the two vectors divided by the product of their lengths):

$$\cos(E, V) = \frac{E \cdot V}{\|E\|\|V\|} = \frac{\sum_{i=1}^{n} E_i V_i}{\sqrt{\sum_{i=1}^{n} (E_i)^2} \sqrt{\sum_{i=1}^{n} (V_i)^2}}.$$

The embeddings of the dataset are constructed using two different pre-trained models (RoBERTa and MiniLM-L6), and then stored in D. Cosine similarity ranges in $[0, 1]$, with 0 indicating dissimilarity and 1 identity. Cosine similarity is used in order to find the 3 most similar documents to return to the user.

The RoBERTa model [12] is based on Google's BERT model [7]. It has different key hyperparameters and training data size. It maps sentences and paragraphs to a 1024-dimensional vector space (with dense word vectors). It performs well in tasks like clustering or semantic search.

The MiniLM-L6 model is a sentence-transformer model mapping sentences and paragraphs to a 384-dimensional vector space (with dense word vectors). It can be used as a sentence and short paragraph encoder to capture the semantic information for information retrieval, clustering or sentence similarity tasks.

3   Computational Comparison

3.1   Evaluation Data

The different methods were evaluated using two different English datasets: one small dataset from the sponsoring company, and a large dataset from IBM.

The company's dataset, which we refer to as the "OT dataset" (from the name of the company), is composed of 76 technical documents, considered as a knowledge base, indicating how to fix common and less-common issues faced by users. In addition, it contains a test set composed of 35 questions, with 19 answerable and 16 non-answerable questions. The test set was collected from interactions between the company's clients and the TQA system we put in place at the company's site. Documents in the dataset, which consist of a title and a body, describe in which situation the solution was applied. The questions consist of a query, a boolean truth value determining if the question is answerable or not, and the ID of the corresponding answer if it is available.

The other dataset is called "TechQA" [3]. It consists of a collection of 28 481 technical documents, called *technotes*, that address specific technical questions, and of annotated questions with answers in the collection. Questions are divided into two sets, which the TechQA documentation describes as "training set", containing 450 answerable and 150 unanswerable questions, and "development set", containing 160 answerable and 150 unanswerable questions (by unanswerable we mean that there is no document in the collection containing an answer to the question). We chose the training set for testing purposes in this computational evaluation as the methods that we tested do not require training. Each question consists of a title and a body, and the answerable ones are paired with a set of answers consisting in the technote ID and the start and end offset of the answer within the document referenced by the ID.

3.2   Evaluation Measures

The measures selected for the evaluation of the methods are: *Precision, Recall, F1-Score* [9] and *Mean Reciprocal Rank (MRR)* [6]. We considered for the set Q only the answerable questions, to highlight the position in which the pertinent document is returned when an answer is available. In the following, we refer to this constrained measure as MRR_a.

3.3   Results

The results obtained are detailed in Table 1. It can be noted that, for the keyword extraction methods, RAKE is less accurate than YAKE: in particular, we observed that it is often unable to find appropriate keywords, and, conversely, repeats identified special characters (like "\\" or "&") as keywords. Both methods, however, were ineffective on the QA dataset, showing that these keyword extraction methods are not particularly fit to extract the technical keywords in that dataset.

Applying the evaluation methods to document titles (instead of contents) turned out to produce better results both with KE and ST methods. While titles contain less information, it appears to be more valuable or better exploitable (by current methods) than the information existing in the content.

ES gave good results on large datasets like TechQA, while Whoosh was better suited to smaller datasets like OT. The reason is related to the indexing technique of both methods: while Whoosh only indexes the two fields "title" and "content", ES makes use of all the fields of the document, including metadata.

ST gave the best results for both datasets, with a very high MRR score in each case. We can add that the use of MiniLM model is less time-consuming (about 3 times less) with a vector size 3 times smaller compared to the use

Table 1. Results obtained for the chosen evaluation methods, on the OT and TechQA datasets. When present, the labels (C) and (T) indicate that, respectively, only the content and the title have been indexed. For RAKE and YAKE, the −w and +w suffixes indicate the unweighted and weighted versions, respectively.

Method	OT dataset				TechQA dataset			
	Precision	Recall	F1-score	MRR_a	Precision	Recall	F1-score	MRR_a
RAKE-w (C)	12.86%	23.68%	16.66%	0.175	0.16%	0.22%	0.19%	0.001
RAKE-w (T)	17.14%	31.58%	22.22%	0.254	0.5%	0.67%	0.57%	0.001
RAKE+w	25.71%	47.37%	33.33%	0.228	0.22%	24.17%	0.44%	0.214
YAKE-w (C)	22.86%	42.11%	29.63%	0.333	1.78%	1.33%	1.52%	0.014
YAKE-w (T)	**45.71%**	**84.21%**	**59.26%**	**0.781**	0.5%	0.67%	0.57%	0.001
YAKE+w	42.86%	78.94%	55.55%	0.658	0.5%	0.67%	0.57%	0.006
ElasticSearch	11.76%	21.05%	15.09%	0.211	**26.83%**	**35.78%**	**30.67%**	**0.358**
Whoosh	31.43%	47.37%	37.79%	0.474	0.83%	1.02%	0.92%	0.011
ST-RoBERTa (C)	31.42%	57.89%	40.74%	0.697	12.83%	14.11%	14.67%	0.679
ST-RoBERTa (T)	**51.43%**	**94.74%**	**66.67%**	0.917	21.83%	29.11%	24.95%	**0.917**
ST-MiniLM (C)	37.14%	68.42%	48.15%	0.923	28.5%	38%	32.57%	0.812
ST-MiniLM (T)	**51.43%**	**94.74%**	**66.67%**	**0.963**	**33.83%**	**45.11%**	**38.66%**	0.889

of the RoBERTa model but with similar results. MiniLM even achieved better results on TechQA, as well as a higher MRR score on OT.

We carried out an analysis of the results obtained by ST, focusing on questions for which the answer was either right but with a low score or wrong with a high score. For the ST-MiniLM (T) model, we identified 49 questions of the 1st type and 11 questions of the 2nd type. We observed that in the second case, many questions had acronyms and uncommon words, so we carried out an evaluation to compare the number of words of the questions to the number of tokens extracted by the SBERT tokenizer. We could observe that the ratio token/words is higher in questions of the 2nd type (1.67) than for questions of the 1st type (1.53), which is also lower than the ratio for the other questions (1.58). This proved our intuition and it also shows that the SBERT model is less effective when dealing with words that are not in the vocabulary and it has to resort to sub-word token to build a semantic representation of the full word.

4 Conclusion and Further Works

In this paper, we presented an empirical comparison of various semantic similarity measures, based on both sparse and dense vector representations, on the technical QA task. The results confirm for this task the general behaviour that keyword-based and classic models may be useful in small scenarios but they are not particularly useful for large and complex corpora. For the dense representations based on neural models, we observed that the results obtained using only the title information exceed by a large margin those obtained with the content.

This may imply that further research on the representation of larger texts is required and SBERT semantic similarity measures are meaningful only at sentence level. We found out also that SBERT models have some problems with the acronyms and rare words that are characteristics of the TechQA task, and their results are worse when these words are not in the dictionary and they have to recur to sub-word tokens.

References

1. Białecki, A., Muir, R., Ingersoll, G.: Apache Lucene 4. In: Proceedings of the SIGIR 2012 Workshop on Open Source Information Retrieval, pp. 17–24 (2012)
2. Campos, R., Mangaravite, V., Pasquali, A., Jorge, A., Nunes, C., Jatowt, A.: YAKE! Keyword extraction from single documents using multiple local features. Inf. Sci. **509**, 257–289 (2020)
3. Castelli, V., et al.: The TechQA dataset. In: Proceedings of the 58th Annual Meeting of the Association for Computational Linguistics, pp. 1269–1278. ACL (2020)
4. Chandrasekaran, D., Mago, V.: Evolution of semantic similarity – a survey. ACM Comput. Surv. **54**(2), Art. 41 (2021)
5. Cimiano, P., Unger, C., McCrae, J.: Ontology-Based Interpretation of Natural Language. Morgan & Claypool, San Rafael (2014)
6. Craswell, N.: Mean reciprocal rank. In: Encyclopedia of Database Systems (2009)
7. Devlin, J., Chang, M.W., Lee, K., Toutanova, K.: BERT: pre-training of deep bidirectional transformers for language understanding. In: Proceedings of the 2019 Conference of the North American Chapter of the Association for Computational Linguistics: Human Language Technologies, Minneapolis, Minnesota, vol. 1, pp. 4171–4186. ACL (2019)
8. Fellbaum, C.: WordNet: An Electronic Lexical Database. Bradford Books, Bradford (1998)
9. Goutte, C., Gaussier, E.: A probabilistic interpretation of precision, recall and F-score, with implication for evaluation. In: Losada, D.E., Fernández-Luna, J.M. (eds.) ECIR 2005. LNCS, vol. 3408, pp. 345–359. Springer, Heidelberg (2005). https://doi.org/10.1007/978-3-540-31865-1_25
10. Jurafsky, D., Martin, J.: Speech and Language Processing. Stanford University, Stanford (Draft 191016)
11. Kruschwitz, U.: Intelligent Document Retrieval. Springer, Dordrecht (2005). https://doi.org/10.1007/1-4020-3768-6
12. Liu, Y., et al.: RoBERTa: a robustly optimized BERT pretraining approach. CoRR (2019)
13. Mollá, D., Vicedo, J.L.: Question answering in restricted domains: an overview. Comput. Linguist. **33**(1), 41–61 (2007)
14. Pérez-Agüera, J., Arroyo, J., Greenberg, J., Perez Iglesias, J., Fresno, V.: Using BM25F for semantic search. In: Proceedings of the 3rd International Semantic Search Workshop, pp. 1–8 (2010)
15. Reimers, N., Gurevych, I.: Sentence-BERT: sentence embeddings using Siamese BERT-networks. In: Proceedings of the 2019 Conference on Empirical Methods in Natural Language Processing. ACL (2019)
16. Rose, S., Engel, D., Cramer, N., Cowley, W.: Automatic keyword extraction from individual documents. In: Berry, M., Kogan, J. (eds.) Text Mining: Applications and Theory, pp. 1–20. Wiley, Hoboken (2010)

17. Yu, W., et al.: A technical question answering system with transfer learning. In: Proceedings of the 2020 Conference on Empirical Methods in Natural Language Processing (EMNLP): System Demonstrations, pp. 92–99. ACL (2020)
18. Yu, W., et al.: Technical question answering across tasks and domains. In: Proceedings of the North-American chapter of the Association for Computational Linguistics: Human Language Technologies: Industry Papers, pp. 178–186. ACL (2021)

Denoising Architecture for Unsupervised Anomaly Detection in Time-Series

Wadie Skaf$^{(\boxtimes)}$ and Tomáš Horváth

Telekom Innovation Laboratories, Data Science and Engineering Department
(DSED), Faculty of Informatics, Eötvös Loránd University, Pázmány Péter stny.
1/A, Budapest 1117, Hungary
{skaf,tomas.horvath}@inf.elte.hu

Abstract. Anomalies in time-series provide insights of critical scenarios across a range of industries, from banking and aerospace to information technology, security, and medicine. However, identifying anomalies in time-series data is particularly challenging due to the imprecise definition of anomalies, the frequent absence of labels, and the enormously complex temporal correlations present in such data. The LSTM Autoencoder is an Encoder-Decoder scheme for Anomaly Detection based on Long Short Term Memory Networks that learns to reconstruct time-series behavior and then uses reconstruction error to identify abnormalities. We introduce the Denoising Architecture as a complement to this LSTM Encoder-Decoder model and investigate its effect on real-world as well as artificially generated datasets. We demonstrate that the proposed architecture increases both the accuracy and the training speed, thereby, making the LSTM Autoencoder more efficient for unsupervised anomaly detection tasks.

Keywords: Anomaly detection · Time-series · Autoencoder

1 Introduction

An outlier or anomaly is a data point that differs dramatically from the rest of the data. Hawkins [5] defined an anomaly as an observation that deviates significantly from the rest of the observations, raising suspicions that it was generated by an unusual mechanism. Anomaly detection is used in a variety of industries, including network intrusion detection, credit card fraud detection, sensor network malfunction detection, and medical diagnosis [4].

In time-series data, an outlier or anomaly is a data point that deviates significantly from the overall trend, seasonal or cyclical pattern of the data. By significance, the majority of data scientists mean statistical significance, which indicates that the data point's statistical properties are out of phase with the rest of the series. Anomalies are classified into two broad categories: a point anomaly is a single data point that has reached an abnormal value, whereas a collective anomaly is a continuous sequence of data points that are considered anomalous collectively, even if the individual data points are not. Anomaly detection methods can be classified into the following two broad methodologies:

S. Chiusano et al. (Eds.): ADBIS 2022, CCIS 1652, pp. 178–187, 2022.
https://doi.org/10.1007/978-3-031-15743-1_17

1. Semi-supervised anomaly detection models are designed to be trained exclusively on data that does not contain anomalies and then tested on samples containing anomalies and non-anomalies to determine their accuracy.
2. Unsupervised anomaly detection models are designed to be trained on data containing a mixture of anomalous and non-anomalous data samples without specifying which are which, and then tested on samples containing both anomalies and non-anomalies to determine their accuracy.

An Encoder-Decoder scheme for anomaly detection (EncDec-AD) based on Long Short Term Memory Networks (LSTM) was proposed in [17], in which the encoder learns a vector representation of the input time-series, the decoder uses this representation to reconstruct the time-series, and the reconstruction error at any subsequent time instance is used to compute the likelihood of an anomaly at that point. However, since EncDec-AD trains on only regular sequences, it is a semi-supervised learning-based anomaly detection system. In this paper, we extend this model with a denoising architecture and put it to the test for unsupervised anomaly detection.

The paper is structured as follows. We begin by formalizing and discussing the topic of unsupervised time-series anomaly detection, delving into the details of the anomaly detection process using LSTM Autoencoders. After that, we describe the proposed denoising architecture and set up the experiments, before reporting and summarizing our major findings.

2   Related Work

Numerous anomaly detectors based on classic statistical models have been presented throughout the years (e.g., [10, 16, 19, 23], mainly time series models) for computing anomaly scores. Due to the fact that these algorithms often make simplistic assumptions about the application domain, expert assessment is required to choose an appropriate detector for every particular domain and then fine-tune the detector's parameters using the training data. According to [15], simple ensembles of these detectors, such as majority vote [8] and normalization [21], are also ineffective. As a result, these detectors are seldom used in practice.

To address the difficulties associated with algorithm/parameter tweaking for classic statistical anomaly detectors, supervised ensemble techniques such as EGADS [12] and Opprentice [15] have been developed. They train anomaly classifiers utilizing user feedback as labels and traditional detector output as features. Both EGADS and Opprentice demonstrated promising results, but they depend substantially on high-quality labels, which is often not practical in large-scale applications. Additionally, using numerous conventional detectors to extract features during detection adds significant computing overhead, which is a practical problem.

Recently, there has been an increase in the use of unsupervised machine learning algorithms for anomaly detection, including one-class SVM [1,6], clustering-based approaches such as K-Means [18] and Gaussian Mixture Model (GMM) [13], Kernel Density Estimation (KDE) [3], Auto-Encoder (AE) [2], Variational

Auto-Encoder (VAE) [2]. The aim is to place an emphasis on normal patterns rather than on anomalies. Generally speaking, these algorithms generate the anomaly score by first identifying "normal" regions in the original or some latent feature space, and then determining "how distant" an observation is from the normal regions.

Additionally, significant progress has been made lately in training generative models using deep learning approaches for the purpose of performing anomaly detection, such as Generative Adversarial Network (GANs) [9,14,24,25].

Despite the enormous potential of the aforementioned models and algorithms, SVM, KDE, AE, and VAE are not designed to handle time-series data, and using more complicated models such as GANs results in lengthy training times and high resource requirements. Given the aforementioned limitations, and given that deep learning architectures have exceptional learning abilities and are particularly adept at tolerating non-linearity in complicated temporal correlations [11], we enhance the LSTM Encoder-Decoder architecture [17] that has been used in supervised anomaly detection by introducing the Denoising Architecture, which enables it to perform unsupervised time-series anomaly detection in addition to significantly reducing the training time.

3   Unsupervised Time-Series Anomaly Detection

Given a time-series $X = (\mathbf{x}_1, \mathbf{x}_2, \ldots, \mathbf{x}_T)$, unsupervised time-series anomaly detection is the process of identifying segments $A^j = (\mathbf{x}_j, \mathbf{x}_{j+1}, \ldots, \mathbf{x}_{j+n_j})$ of anomalous points, where $\mathbf{x}_i \in \mathbb{R}^N$, $j \geq 1$, $n_j \geq 0$ such that $j + n_j \leq T$, and $N = 1$ in the case of a univariate time-series while $N > 1$ in the case of a multivariate time-series. Each A^j is a single data point (when $n_j = 0$) or a continuous series of data points (when $n_j > 0$) in time that exhibit(s) anomalous or unexpected behavior value(s) inside the segment that do not appear to conform to the signal's predicted temporal behavior.

This process is different from and is more complicated than time-series classification [7] and supervised time-series anomaly detection [20] in a few aspects:

1. Absence of prior knowledge of anomalies or prospective anomalies: unlike the supervised methods, which use previously recognized anomalies to train and optimize the model, the unsupervised methods use all the data to train the model to understand the time-series patterns, then ask it to find anomalies. The detector will then be checked to ensure that it recognized anything useful to end users. Additionally, n_j (the length of A^j) is variable and unknown in advance, complicating this method even further.
2. Unsupervised methods do not rely on baselines: for a large number of real-world systems, simulation engines can generate a signal that approximates normal contexts, providing a baseline against which models can be trained, with any deviations considered anomalies. Unsupervised time-series anomaly detection algorithms do not depend on such baselines but rather learn time-series patterns from real data that may contain anomalies or abnormal patterns.

3. Not all identified anomalies are cause for concern: detected anomalies may not necessarily indicate issues, but may be the result of an external phenomena such as a rapid change in ambient circumstances, additional information such as a test run, or other factors not evaluated by the system, such as regime settings changes. In this context, it is up to the end user or domain expert to determine whether the anomalies detected by the model are worrisome.

4   Anomaly Detection Using LSTM Autoencoder

Autoencoder is a generative unsupervised deep learning model that reconstructs high-dimensional input data by utilizing a neural network with a narrow bottleneck layer that contains the latent representation of the input data in between the Encoder and Decoder. The autoencoder attempts to minimize reconstruction error as part of its training procedure. As a result, the magnitude of the reconstruction loss can be used to detect anomalies.

During the training process, the data is transferred to the Encoder, which generates a fixed-length vector representation of the input time-series. This representation is then used by the LSTM decoder to reconstruct the time-series using the current hidden state and the estimated value at the previous time step. Given a time-series of length L as an input, $X = (\mathbf{x}_1, \mathbf{x}_2, \ldots, \mathbf{x}_T)$, where $\mathbf{x}_i \in \mathbb{R}^N$, $E_k^{(t)}$ is the hidden state of the kth layer of the encoder at time $t \in \{1, \ldots, L\}$, where $k \in \{1 \ldots H\}$; H denotes the number of hidden layers in each of the encoder and the decoder, and $E_k^{(t)} \in \mathbb{R}^u$; u is the number of LSTM units in hidden layer k of the encoder. The final state of the final hidden layer $E_H^{(L)}$ of the encoder outputs the data's latent representation, which is used as an initial state of the decoder. The decoder then reconstructs the original input by using the input $x^{(i)}$ to obtain the hidden state $D_1^{(i-1)}$ (The hidden state of the first layer of the decoder at time $(i-1)$) then proceed with all the hidden states of the first layer then outputs $z_1^{(i)}$ to the next hidden layer, and so on until the final layer, when the decoder utilizes $z_{H-1}^{(i)}$ to derive the hidden state $D_H^{(i-1)}$ then estimates $\mathbf{x}'_{i-1}$ corresponding to $\mathbf{x}_{i-1}$. The autoencoder is trained with the purpose of minimizing the following:

$$\sum_X \sum_{i=1}^{L} \|\mathbf{x}_i - \mathbf{x}'_i\|^2 \tag{1}$$

After training the autoencoder, time-series signals $I = \{X_1, X_2, \ldots., X_N\}$ of length L are passed to it in order to reconstruct them, then the reconstruction error of each point is calculated using Eq. 2

$$s_i = \|\mathbf{x}_i - \mathbf{x}'_i\| \tag{2}$$

where $\mathbf{x}_i \in X_i$; $X_i \in I$ can be used as an anomaly score, and by specifying a reconstruction error threshold, anomalous values can be flagged if the reconstruction error (Eq. 2) exceeds the value specified threshold.

5 The Denoising Architecture

Dropout [22] is a strategy for decreasing overfitting in neural networks. Back-propagation learning by itself accumulates brittle co-adaptations that work for the training data but do not generalize to unobserved data. By making the existence of any specific hidden unit unstable, random dropout disrupts these co-adaptations. This approach was discovered to significantly increase the performance of neural networks across a broad range of application fields. The term "dropout" refers to units that are dropped from a neural network (both hidden and visible). By dropping a unit from the network, we imply disconnecting it from all incoming and outgoing connections temporarily.

In our proposed architecture, we add a dropout layer after each LSTM layer in the LSTM Autoencoder. As a result, during the training phase, the output of each LSTM layer would be randomly set to 0 with a probability p, which is done by generating a random number r, and if this number is less than or equal to p, the output would be set to zero, otherwise it would pass with no changes, as shown in Eq. 3.

$$Output(x) = \begin{cases} 0 & \text{if } r \leq p \\ LSTM(x) & \text{Otherwise} \end{cases} \tag{3}$$

The proposed denoising architecture randomly exposes the model to extreme cases (zeroes) during training, allowing it to more accurately generalize the normal samples without being significantly affected by anomalous samples, so that its weights do not change significantly in the presence of an anomalous sample. As a result, after training, the model will be capable of efficiently reconstructing normal samples while struggling to reconstruct anomalous samples, resulting in a higher reconstruction errors (Eq. 2) for the anomalous samples, allowing for a more precise definition (threshold) of anomalies. The selection of probability p is based on a number of factors, which are discussed in Sect. 6.3.

Figure 1 shows an example of an LSTM Autoencoder with a denoising architecture.

6 Experimental Results

6.1 Datasets

To examine the denoising architecture's effect on the LSTM Autoencoder, we use the Yahoo S5 dataset[1], which is composed of four subsets: A1, A2, A3, and A4. The A1 dataset is based on real-world production traffic on Yahoo systems, whereas the remaining datasets are all made up of synthetic data. A2 lacks anomaly points and thus cannot be used to calculate the metrics described in Sect. 6.2; therefore, it is not used in the experiments. All values in the datasets are timestamped with a one-hour timestep.

---

[1] Yahoo S5 Dataset can be requested here: https://webscope.sandbox.yahoo.com/catalog.php?datatype=s&did=70.

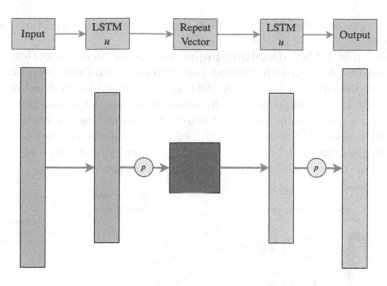

Fig. 1. An illustration of an LSTM autoencoder with two LSTM layers, each has u units and followed by a dropout layer with a probability of p

6.2 Experimental Setup

1. Data Preparation: We performed data normalization on each dataset to ensure that the data was within the range $[-1, 1]$. A sliding window with a window size of 24 (representing 24 h) and a step size of 1 was used to generate the training samples, resulting in a sequence of 24 consecutive data points for each training sample.
2. Architectures: We examine the effect of the denoising architecture on a variety of architectures, beginning with the simplest possible Autoencoder with two LSTM layers - one for the Encoder and one for the Decoder - and progressing deeper as shown in Table 1.
3. Evaluation Metrics: We use Recall, Precision, and F1-Score matrices to quantify the effect of the denoising architecture on model accuracy, as well as the number of epochs to quantify the training time speed.
4. Comparison Baseline: For each pair of (Dataset, Architecture), we will first train the model without any dropout layers and use that as a baseline for comparison, and then gradually add dropout layers with increasing probability up to 0.5 and compare the results to the baseline.

6.3 Benchmarking Results

We outline the results of the experiments in Table 1,where we list each architecture along with the obtained results. The notion for architectures used in the table is as follows: for the sake of simplicity, the architecture is represented by the layers of the encoder, so number 16 represents an autoencoder with two

LSTM layers, each of which has 16 units, one for the encoder and the other for the decoder.

As shown in Table 1, **denoising improved the accuracy metrics on the real dataset (A1) and shortened the training time on both real and synthetic datasets (A1 & A3 & A4)**. In comparison to the baseline (without dropout layers), improvements in recall, precision, f-1 score, and number of epochs are observed for the real dataset (A1), as well as the training speed for the all datasets (A1 & A3 & A4). All of these enhancements are outlined in Table 1 and illustrated in Fig. 2. In (A3 & A4), a very slight decrease of $\epsilon < 0.001$ happened in precision and f-1 score, which can be ignored.

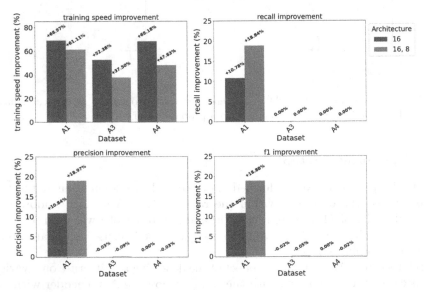

Fig. 2. A comparison of the denoising architecture to the baseline, demonstrating the improvement in accuracy metrics and training speeds.

The optimal probability of dropout p varies by dataset, being 0.4 for dataset A1 for all architectures, 0.2 for dataset A4 for all architectures, and 0.2 and 0.1 for dataset A3 for architectures (16), (16, 8), respectively. And to examine this, we investigate the ratio of anomaly points in each dataset presented to the model per epoch during training in the Table 2. As can be observed, there is a negative correlation between the number of anomaly samples and the optimal p, which sounds plausible because when there are more anomaly samples, the model will perceive more anomalies and will be more robust to them without requiring a higher dropout rate p—The more anomaly samples the model perceives, the less a single anomaly sample significantly alters the neural network's weights. Thus, in general, the amount of p should be determined by the number or ratio of anomaly samples, which in the case of unsupervised, can be determined by knowing the expected number of anomalies or by experimentation.

Table 1. Experiments results

Dataset	Arch	Dropout	Epochs	Recall	Precision	F1	Dataset	Arch	Dropout	Epochs	Recall	Precision	F1
A1	16	0.0	29	0.198542	0.442812	0.274159	A3	16	0.0	21	1.000000	0.597741	0.748233
		0.1	14	0.199482	0.444911	0.275459			0.1	9	1.000000	0.597222	0.747826
		0.2	18	0.205363	0.457787	0.283534			0.2	10	1.000000	0.597568	0.748097
		0.3	18	0.216184	0.481656	0.298425			0.3	18	1.000000	0.597568	0.748097
		0.4	9	0.219948	0.490814	0.303769			0.4	10	0.990795	0.591896	0.741076
		0.5	9	0.217596	0.485310	0.300471			0.5	9	0.946221	0.565268	0.707737
	16, 8	0.0	18	0.194778	0.434190	0.268918		16, 8	0.0	24	1.000000	0.597741	0.748233
		0.1	27	0.209598	0.466981	0.289333			0.1	15	1.000000	0.597222	0.747826
		0.2	11	0.214067	0.477189	0.295551			0.2	12	0.998062	0.596583	0.746783
		0.3	7	0.220654	0.492130	0.304694			0.3	9	0.928779	0.555169	0.694943
		0.4	7	0.231475	0.516535	0.319688			0.4	14	0.816376	0.487699	0.610618
		0.5	14	0.218772	0.487933	0.302095			0.5	14	0.837209	0.500000	0.626087

Dataset	Arch	Dropout	Epochs	Recall	Precision	F1
A4	16	0.0	22	1.000000	0.401042	0.572491
		0.1	11	1.000000	0.401042	0.572491
		0.2	7	1.000000	0.401042	0.572491
		0.3	7	1.000000	0.401158	0.572609
		0.4	10	1.000000	0.401042	0.572491
		0.5	11	1.000000	0.401158	0.572609
	16, 8	0.0	23	1.000000	0.401390	0.572846
		0.1	7	1.000000	0.401042	0.572491
		0.2	12	1.000000	0.401274	0.572727
		0.3	14	0.997114	0.400116	0.571074
		0.4	17	0.994228	0.398727	0.569186
		0.5	17	0.805195	0.323104	0.461157

Table 2. The percentage of anomaly points, the total number of samples, and the optimal $p(s)$ for each dataset

	A1	A3	A4
Total number of samples	2,238,624	3,974,400	3,974,400
Number of anomaly samples	6286	22,203	19,855
Anomaly samples percentage	0.280%	0.559%	0.499%
Optimal $p(s)$	0.4	0.1, 0.2	0.2

7   Conclusion

In this paper, we introduced the Denoising Architecture as an addition to the LSTM Autoencoder to extend its usage to unsupervised anomaly detection for point anomalies and evidenced that it resulted in noticeable improvements in accuracy metrics such as precision, recall, and f-1 score (up to 18%), as well as a remarkable increase in training speed (up to 68%), and we demonstrated that improvements in accuracy occur only when real-world datasets are used, whereas synthetic datasets only show improvements in training speed. Additionally, we addressed how to choose the appropriate dropout probability p, showing that the more anomalous samples present or expected in the data stream, the smaller p should be.

References

1. Amer, M., Goldstein, M., Abdennadher, S.: Enhancing one-class support vector machines for unsupervised anomaly detection. In: Proceedings of the ACM SIGKDD Workshop on Outlier Detection and Description, ODD 2013, pp. 8–15. Association for Computing Machinery, New York, NY, USA (2013). https://doi.org/10.1145/2500853.2500857
2. An, J., Cho, S.: Variational autoencoder based anomaly detection using reconstruction probability. Special Lecture on IE, vol. 2, pp. 1–18 (2015)
3. Cao, V.L., Nicolau, M., McDermott, J.: One-class classification for anomaly detection with kernel density estimation and genetic programming. In: Heywood, M.I., McDermott, J., Castelli, M., Costa, E., Sim, K. (eds.) EuroGP 2016. LNCS, vol. 9594, pp. 3–18. Springer, Cham (2016). https://doi.org/10.1007/978-3-319-30668-1_1
4. Chandola, V., Banerjee, A., Kumar, V.: Anomaly detection: a survey. ACM Comput. Surv. **41**, 1–58 (2009). https://doi.org/10.1145/1541880.1541882
5. Douglas, H.: Identification of Outliers, vol. 11. Springer, Dordrecht (1980). https://doi.org/10.1007/978-94-015-3994-4
6. Erfani, S.M., Rajasegarar, S., Karunasekera, S., Leckie, C.: High-dimensional and large-scale anomaly detection using a linear one-class svm with deep learning. Pattern Recog. **58**, 121–134 (2016). https://doi.org/10.1016/j.patcog.2016.03.028, https://www.sciencedirect.com/science/article/pii/S0031320316300267
7. Ismail Fawaz, H., Forestier, G., Weber, J., Idoumghar, L., Muller, P.-A.: Deep learning for time series classification: a review. Data Min. Knowl. Disc. **33**(4), 917–963 (2019). https://doi.org/10.1007/s10618-019-00619-1
8. Fontugne, R., Borgnat, P., Abry, P., Fukuda, K.: MAWILab: combining diverse anomaly detectors for automated anomaly labeling and performance benchmarking, p. 8, November 2010. https://doi.org/10.1145/1921168.1921179
9. Geiger, A., Liu, D., Alnegheimish, S., Cuesta-Infante, A., Veeramachaneni, K.: TadGAN: time series anomaly detection using generative adversarial networks, September 2020
10. Knorn, F., Leith, D.: Adaptive Kalman filtering for anomaly detection in software appliances, pp. 1–6, May 2008. https://doi.org/10.1109/INFOCOM.2008.4544581
11. Kwon, D., Kim, H., Kim, J., Suh, S.C., Kim, I., Kim, K.J.: A survey of deep learning-based network anomaly detection. Clust. Comput. **22**(1), 949–961 (2017). https://doi.org/10.1007/s10586-017-1117-8
12. Laptev, N., Amizadeh, S., Flint, I.: Generic and scalable framework for automated time-series anomaly detection. In: Proceedings of the 21th ACM SIGKDD International Conference on Knowledge Discovery and Data Mining, KDD 2015, pp. 1939–1947. Association for Computing Machinery, New York, NY, USA (2015). https://doi.org/10.1145/2783258.2788611
13. Laxhammar, R., Falkman, G., Sviestins, E.: Anomaly detection in sea traffic - a comparison of the gaussian mixture model and the kernel density estimator. In: 2009 12th International Conference on Information Fusion, pp. 756–763 (2009)
14. Li, D., Chen, D., Jin, B., Shi, L., Goh, J., Ng, S.-K.: MAD-GAN: multivariate anomaly detection for time series data with generative adversarial networks. In: Tetko, I.V., Kůrková, V., Karpov, P., Theis, F. (eds.) ICANN 2019. LNCS, vol. 11730, pp. 703–716. Springer, Cham (2019). https://doi.org/10.1007/978-3-030-30490-4_56

15. Liu, D., et al.: Opprentice: towards practical and automatic anomaly detection through machine learning. In: Proceedings of the 2015 Internet Measurement Conference (2015)

16. Lu, W., Ghorbani, A.A.: Network anomaly detection based on wavelet analysis. EURASIP J. Adv. Sig. Process. **2009**(1), 1–16 (2009). https://doi.org/10.1155/2009/837601

17. Malhotra, P., Ramakrishnan, A., Anand, G., Vig, L., Agarwal, P., Shroff, G.: LSTM-based encoder-decoder for multi-sensor anomaly detection, July 2016

18. Münz, G., Li, S., Carle, G.: Traffic anomaly detection using kmeans clustering. In: In GI/ITG Workshop MMBnet (2007)

19. Pincombe, B.: Anomaly detection in time series of graphs using ARMA processes. ASOR Bull. **24**, 2 (2005)

20. Qiu, J., Du, Q., Qian, C.: KPI-TSAD: a time-series anomaly detector for KPI monitoring in cloud applications. Symmetry **11**, 1350 (2019). https://doi.org/10.3390/sym11111350

21. Shanbhag, S., Wolf, T.: Accurate anomaly detection through parallelism. IEEE Netw. **23**(1), 22–28 (2009). https://doi.org/10.1109/MNET.2009.4804320

22. Srivastava, N., Hinton, G., Krizhevsky, A., Sutskever, I., Salakhutdinov, R.: Dropout: a simple way to prevent neural networks from overfitting. J. Mach. Learn. Res. **15**(1), 1929–1958 (2014)

23. Yaacob, A.H., Tan, I.K.T., Chien, S.F., Tan, H.: Arima based network anomaly detection. In: 2010 2nd International Conference on Communication Software and Networks, pp. 205–209 (2010)

24. Yoon, J., Jarrett, D., van der Schaar, M.: Time-series generative adversarial networks, vol. 32. Curran Associates, Inc. (2019). https://proceedings.neurips.cc/paper/2019/file/c9efe5f26cd17ba6216bbe2a7d26d490-Paper.pdf

25. Zhou, B., Liu, S., Hooi, B., Cheng, X., Ye, J.: BeatGAN: anomalous rhythm detection using adversarially generated time series. In: International Joint Conferences on Artificial Intelligence Organization, pp. 4433–4439, August 2019. https://doi.org/10.24963/ijcai.2019/616

Discretizing Numerical Attributes: An Analysis of Human Perceptions

Minakshi Kaushik[1]([✉]) [iD], Rahul Sharma[1] [iD], Ankit Vidyarthi[2] [iD],
and Dirk Draheim[1] [iD]

[1] Information Systems Group, Tallinn University of Technology, Akadeemia tee 15a,
12618 Tallinn, Estonia
{minakshi.kaushik,rahul.sharma,dirk.draheim}@taltech.ee
[2] Jaypee Institute of Information Technology, Noida, India

Abstract. To partition numerical attributes, machine learning (ML) has used a variety of discretization approaches that partition the numerical attribute into intervals. However, an effective method for discretization is still missing in various ML approaches, e.g., association rule mining. Moreover, the existing discretization techniques do not reflect best the impact of the independent numerical factor on the dependent numerical target factor. The main objective of this research is to develop a benchmark approach for partitioning numerical factors. We present an in-depth analysis of human perceptions of partitioning a numerical factor and compare it with one of our proposed measures. We also examine the perceptions of various experts in data science, statistics and engineering disciplines by using a series of graphs with numerical data. The analysis of the collected responses indicates that 68.7% of the human responses were approximately close to the values obtained by the proposed method. Based on this analysis, the proposed method may be used as one of the methods for discretizing the numerical attributes.

Keywords: Machine learning · Data mining · Discretization · Numerical attributes · Partitioning

1 Introduction

Various types of variables are available in real-world data. However, discrete values have explicit roles in statistics, machine learning, and data mining. Presently, there is no benchmark approach to find the optimum partitions for discretizing complex real-world datasets. Generally, if a factor impacts another factor, in that case, humans can easily perceive the compartments or partitions because the human brain can easily perceive the differences between the factors and detect the partitions. However, it is not easy for a human or even an expert to find the appropriate compartments in complex real-world datasets.

This work has been partially conducted in the project "ICT programme" which was supported by the European Union through the European Social Fund.

© Springer Nature Switzerland AG 2022
S. Chiusano et al. (Eds.): ADBIS 2022, CCIS 1652, pp. 188–197, 2022.
https://doi.org/10.1007/978-3-031-15743-1_18

Existing discretization techniques do not reflect best the impact of the independent numerical factor on the dependent numerical target factor. Moreover, no discretization approach uses numerical attributes as influencing and response factors. To find the cut-points for the cases of two-partitioning and three-partitioning, we have proposed two measures *Least Squared Ordinate-Directed Impact Measure* (LSQM) and *Least Absolute-Difference Ordinate-Directed Impact Measure* (LADM) [10]. These measures provide a simple way to find partitions of numerical attributes that reflect best the impact of one independent numerical attribute on a dependent numerical attribute.

In this paper, the outcome of *LSQM* measure is compared with the human perceived cut-points to assess the accuracy of the measure. We use numerical attributes as influencing and response factors to distinguish them from the existing approaches. A series of graphs with different data points are used to collect the human responses. Here, data scientists, machine learning experts and other non-expert persons are referred to as humans.

The idea of this research emerged from the research on partial conditionalization [5,6], association rule mining (ARM) [17,19] and numerical association rule mining (NARM) [11,12,20]. These papers discuss the discretization process as an essential step for NARM. Moreover, research on discretizing the numerical attributes is an essential step in frequent itemset mining, especially for quantitative association rule mining [20].

In the same sequence, we have also presented a tool named Grand report [16] and a framework [18] for unifying ARM, statistical reasoning, and online analytical processing. These paper strengthens the generalization of ARM by finding the partitions of numerical attributes that reflect best the impact of one independent numerical attribute on a dependent numerical attribute. Our vision is to develop an ecosystem to generalize the machine learning approaches by significantly improving the ARM from different dimensions.

The paper is organized as follows. In Sect. 2, we discuss related work. In Sect. 3, we explain the motivation for conducting this study. Section 4 describes the *LSQM* method. Then we discuss the design of the experiment in Sect. 5. In Sect. 6, analysis and results are given. The conclusion and future work are given in Sect. 7.

2   Related Work

Based on human perception evaluation and different discretization techniques, we discuss the related work in the direction of discretization, clustering techniques and human perception.

A variety of discretization methods are available in the literature [9,13,14]. Dougherty et al. [4] compared and analyzed discretization strategies along three dimensions: global versus local, supervised versus unsupervised, and static versus dynamic. Liu et al. [14] performed a systematic study of existing discretization methods and proposed a hierarchical framework for discretization methods from the perspective of splitting and merging. The unsupervised static discretization method, such as equal-width, uses the minimum and maximum values of the continuous attribute and then divides the range into equal-width intervals called bins. In contrast, the equal-frequency algorithm determines an equal number of continuous values and places them in each bin [2].

In state of the art, many studies have used human perception to evaluate the various techniques. However, they are not completely related to discretization. Etemadpour et al. [7] conducted a perception-based evaluation of high-dimensional data where humans were asked to identify clusters and analyze distances inside and across clusters. Demiralp et al. [3] used human judgments to estimate perceptual kernels for visual encoding variables such as shape, size, colour, and combinations. The experiment used Amazon's Mechanical Turk platform, with twenty Turkers completing thirty MTurk jobs. In [1] authors evaluated benchmarking clustering algorithms based on human perception of clusters in 2D scatter plots. The authors' main concern was how well existing clustering algorithms corresponded to human perceptions of clusters. Our work is also related to considering human perceptions for evaluating our proposed *LSQM* measure for discretizing numerical attributes.

3 Motivation

For years, obtaining discrete values from numerical values has been a complex and ongoing task. The main issue with the discretization process is obtaining the perfect intervals with specific ranges and numbers of intervals. In the state of the art, several discretization approaches such as equi-depth, equi-width [2], MDLP [8], Chi2 [15], D2 [2], etc. have been proposed. However, determining the most effective discretizer for each situation is still a challenging problem.

In [10], we presented an order-preserving partitioning method to find the partitions of numerical attributes that reflect best the impact of one independent numerical attribute on a dependent numerical attribute. In extreme cases (such as step-functions), humans can easily visualize the perfect partitions and even the number of compartments. However, in distinct cases, the ideal partition range depends on the perception of data experts. In state of the art, no investigation is available to understand the human perception of partitioning. Moreover, the current literature provides a comparison of discretization methods and compares their results. In this paper, we take a different approach to compare the human perception of discretization with the outcome of the proposed discretization method. We aim to visualize the differences between the outcomes of the proposed methods and the human perception of discretization.

4 The LSQM Method

In the *LSQM* method [10], we discretize the independent numerical attribute on the basis of order-preserving partitioning to learn the impact on the numerical target attribute. The number of cut-points is $k - 1$, where k is the number of partitions suggested by the user. The measure first calculates the squared difference between the y-value of each data point and the average of y-values of the current partition. The order of the independent variable is preserved using the value of data points. Therefore, the value of data points of one partition will always be less than the value of data points of the next partition. After summing up the squared differences of the several partitions, $LSQM$ retrieves the minimum values as cut-points.

Definition 1 (Least Squared Ordinate-Directed Impact Measure).
Given real-valued data points $(<x_i, y_i>)_{2 \leq i \leq n}$, *we define the* least squared ordinate-directed impact measure *for k-partitions as follows:*

$$\min_{i_0=0<i'_1<...<i'_{k-1}<i'_k=n} \sum_{j=1}^{k} \sum_{i'_{j-1}<i''\leq i'_j} (y_{i''} - \mu_{i'_{j-1}<\phi\leq i'_j})^2 \tag{1}$$

where the average of data values in a partition $\mu_{a<\phi\leq b}$ *between indexes* a *and* b *$(a < b \leq n)$ is defined as*

$$\mu_{a<\phi\leq b} = \frac{\sum_{a<\phi\leq b} y_\phi}{b - a} \tag{2}$$

In (1), we have that i'_j is the highest element in the j-th partition, where *highest element* means the data point with the highest index.

The definition of the $LSQM$ measure seems similar to the k-means clustering algorithm. The k-means clustering algorithm is a partitioning clustering algorithm to classify objects into k different clusters. The $LSQM$ measure is different from the k-means algorithm as k-means is based on the Euclidean distance metric between two vectors, X and Y. It also has the severe drawback that its efficiency is highly dependent on the initial random selection of cluster centres. However, the $LSQM$ measure is based on order-preserving partitioning for the independent variable. This measure also does not depend on the initial point chosen for starting.

5   Experimental Design

To understand how humans partition numerical factors, we designed a series of graphs and asked several experts to partition the data points given in the graphs. Initially, to produce a diverse collection of graphs with different data points, a set of graphs was shared and discussed with our own research team. These graphs include step functions, linear functions, and mixed data graphs. Finally, eight graphs were selected to be shared with humans (see Fig. 1). These graphs are obtained from eight synthetic datasets (D1 to D8). These synthetic datasets (D1 to D8) consist only two numerical attributes. A copy of all these datasets is available in the GitHub repository[1].

---

[1] https://github.com/minakshikaushik/LSQM-measure.git.

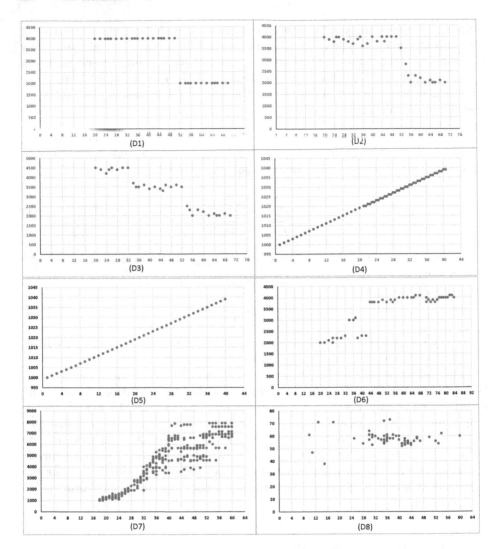

Fig. 1. Graphs for datasets D1 to D8.

We designed a Google form by providing a series of graphs containing different types of numerical data points and relevant questions to collect human responses and their perceptions about discretization. The google form was sent to fifty DS/ML experts and non-experts to estimate the number of partitions and the ranges of these partitions to obtain the cut-points.

Table 1. The comparison of human perception to identify the number of partitions based on their profile.

Responders	Partitions	Datasets							
		D1	D2	D3	D4	D5	D6	D7	D8
DS/ML-Experts (60%)	No				33.3%	93.3%			
	2	93.3%	73.3%	0%	53%	0%	13.3%	60%	33.3%
	3	6.67%	26.6%	93.3%	13.3%	6.6%	26.6%	20%	66.6%
	4			6.66%			26.6%	20%	0%
	5						33.3%		
Non-experts (40%)	No				20%	90%			
	2	90%	60%	0%	70%	0%	30%	40%	60%
	3	10%	40%	100%	10%	10%	0%	40%	30%
	4			0%			40%	20%	10%
	5						30%		

Table 2. The comparison of human perceived cut-points with the LSQM measure.

D	P	Human perception		LSQM
		R	Approx. near cut-points	Cut-points
D1	2	92%	50(91.3%), 48(8.6%)	50
	3	8%	(48,60)(50%), (20,50)(50%)	(20, 50)
D2	2	68%	50(88.2%), 52(11.7%)	52
	3	32%	(50,54)(37.5%), (20,53)(25%)	(52, 54)
D3	3	96%	(32,52)(62%), (30,52)(16.6%)	32,52
	4	4%	(20,32,52)(100%)	(32,52,55)
D4	0	28%	NA	NA
	2	60%	20(86.6%), 25(13.3%)	20
	3	12%	(20,45)(66.6%), (20,30)(33.3%)	(12, 24)
D5	0	92%	NA	NA
	2	0%	NA	20
	3	8%	(14,28)(100%)	(13, 26)
D6	2	20%	32(40%), 42(40%) 50(20%)	42
	3	16%	(42,68)(50%), (32,42)(25%)	(32, 42)
	4	32%	(32,37,42)(87.5%), (33,37,43)(12.5%)	(32, 37, 42)
	5	32%	(32,42,37,68)(87.5%), (17,32,38,42)(12.5%)	(32, 37, 42, 56)
D7	2	52%	40(84.6%), 50(7.6%), 36(7.6%)	35
	3	28%	(32,39)(57.1%)	(32, 39)
	4	20%	(32,39,50)(60%), (41,47,53)(40%)	(32,39,52)
D8	2	44%	18(36%), 30(27%)	40
	3	52%	(28,47)(53.8%), (18,47)(23%)	(13, 15)
	4	4%	(18,47,54)(100%)	(11, 13, 15)

D: Datasets; P: number of partitions; R: percentage of responses

The following data was gathered and compiled from the experiments: respondent identification (name), their email addresses, domain expertise (DS/ML expert or non-expert), number of partitions identified, and ranges of each partition.

6   Analysis and Result

Out of the fifty responses received via the Google form, two were incomplete; therefore, we did not consider them for the analysis. From the rest of the forty-eight responses, we divided the responses into two categories, expert responses and non-expert responses.

Table 3. Similarity between human perceived cut-points and LSQM cut-points.

	P	Datasets							
		D1	D2	D3	D4	D5	D6	D7	D8
LSQM	2	50	52		20		42	35	40
	3	(20,50)	(52,54)	(32,52)	(12,24)	(13,26)	(32,42)	(32,39)	(13,15)
	4			(32,52,55)			(32,37,42)	(32,39,52)	(11,13,15)
	5						(32,37,42,56)		
Human	2	50	52		20		42	36	30
percep.	3	(20,50)	(50,54)	(32,52)	(20,30)	(13,26)	(32,42)	(32,39)	(18,47)
	4			(20,32,52)			(32,37,42)	(32,39,52)	(18,47,54)
	5						(32,37,42,68)		
Matching%	2	91.3%	11.7%		80.6%		40%	0%	0%
	3	50%	19%	62%	0%	100%	25%	57%	0%
	4			59%			85.7%	60%	0%
	5						75%		
Matching	2	VH	L		VH		M	NM	NM
Status	3	M	L	H	NM	VH	L	H	NM
	4			M			VH	H	NM
	5						H		

P: Number of partitions, VH: 80–100%, H: 60–80%, M: 40–60%, L: 1–40%, NM: 0%

Table 1 illustrates the comparison of human perception to identify the number of partitions between the DS/ML experts' responses and non-expert people. We received 60% responses from DS/ML experts and 40% of answers from non-expert people. We analyzed that responses from both categories were opposite for graph D8. Out of the total responses for D8, 33.3% responses of DS/ML experts marked two partitions and 66.6% responses of experts marked three partitions; however, 60% of non-experts marked two partitions, and only 30% marked three partitions. In graphs D3 and D5, we analyzed that no contributor (experts or non-experts) marked two partitions. No non-expert contributors marked three partitions for graph D6 and four partitions for graph D3; whereas 26.6% of DS/ML experts identified three partitions for D6, and 6.66% experts marked

Table 4. Analysis of unmatched datasets in regard of number of partitions for the LSQM and human perceived cut-points.

Dataset	Partitions	LSQM method		Human perception		Remarks
		LSQM cut-points	Logical correctness	Human perceived cut-points	Logical correctness	
D8	2	40	Yes	30	Yes	Matter of perception
	3	(13,15)	No	(18,47)	Yes	LSQM to be improved
	4	(11,13,15)	No	(18,47,54)	Yes	LSQM to be improved
D4	3	(12,24)	Yes	(20,30)	Yes	Matter of perception
D7	2	35	Yes	36	Yes	Matter of perception

four partitions in the graph D3. Table 2 illustrates the comparison between the results of human perception and the *LSQM* measure. Table 3 describes the similarity percentage between cut-points provided by human perceived experiment outcome and the *LSQM* measure outputs. We have mentioned the cut-points from responses near the *LSQM* provided cut-points. We determine the matching status by distributing the matching percentage into the following categories: VH (Very High), H (High), M (Medium), L (Low) and NM (No match). The distribution of ranges is mentioned at the bottom of Table 3. It is clear from Table 3 that human perceived cut-points and the cut-points identified by the proposed measure *LSQM* do not match for the datasets D8, D4 and D7. In Table 4, we present an analysis and reason for not getting similar cut-points for the datasets D4, D8 and D7. If we look at Fig. 1(D8), then it seems logical to have cut-points at the data points of 40 (LSQM cut-point) and 30 (Human perceived cut-point) for two partitions on the X-axis. Humans divided the scattered points into first partition and dense data points into the second partition. In contrast, the *LSQM* measure calculated the cut-point in the middle of the dense data points. This case can be observed as a matter of perception for human perceived cut-points, while the cut-points marked by the LSQM measure seem analytically correct. For the cases of three partitions and four partitions, human perceived cut-points $(18, 47)$ and $(18, 47, 54)$ are good, but the cut-points provided by the *LSQM* measure are not satisfactory. The cut-points provided by the *LSQM* $(12, 24)$ and human perception experiment $(20, 30)$ for D4 are also the case of matter of perception. Similarly, cut-points 35 and 36 for D7 do not match exactly. However, as the data points in the graph are scattered; therefore, the difference between the cut-points of the proposed measure and the human perceived cut-points is negligible and both can be considered the best cut-points. This case can be observed as a matter of perception. Although these cut-points do not match the *LSQM* measure cut-points, the correctness of the measure is not affected due to non-similarity.

Out of the total responses for D1 to D7, we analyzed that 25% responses were matching *Very High*, 25% responses were matching *High*, 18.7% responses were matching *Medium* and 18.7% responses were matching *Low*. By aggregating all the matching status, 68.7% responses were similar to the responses marked by the proposed *LSQM* measure. By the overall analysis, it is clear that for initial datasets (D1 to D7), the proposed measure brought approximately equivalent results to human perception. The analysis is conducted for the datasets D1 to D7 because some random cut-points were observed by the human for the dataset D8 which are difficult to match with the analytically calculated cut-points by the *LSQM*. An analysis and reason for not getting similar cut-points for the dataset D8 are given in Table 4.

7 Conclusion

This paper is the first step toward understanding the human perception of partitioning numerical attributes. We first assessed the human perception of partitioning numerical attributes by examining a series of graphs with numerical data. Furthermore, we compared the human perceived cut-points of partition with the results of the proposed *LSQM* measure. The proposed measure produces cut-points mostly close to human perceived cut-points. The overall analysis shows that the proposed measure produced results that were approximately equivalent to human perception for the datasets (D1 to D7). The present results of the proposed measure are encouraging, and it is a significant step towards the generalization of ARM by finding the partitions of numerical attributes that reflect best the impact of one independent numerical attribute on a dependent numerical attribute. In future work, we plan to implement with *inter*-measures for comparing partitions of different numbers of k-partitions.

References

1. Aupetit, M., Sedlmair, M., Abbas, M.M., Baggag, A., Bensmail, H.: Toward perception-based evaluation of clustering techniques for visual analytics. In: IEEE Visualization Conference on Proceedings of the VIS 2019, pp. 141–145 (2019)
2. Catlett, J.: On changing continuous attributes into ordered discrete attributes. In: Kodratoff, Y. (ed.) EWSL 1991. LNCS, vol. 482, pp. 164–178. Springer, Heidelberg (1991). https://doi.org/10.1007/BFb0017012
3. Demiralp, Ç., Bernstein, M.S., Heer, J.: Learning perceptual kernels for visualization design. IEEE Trans. Vis. Comput. Graph. **20**(12), 1933–1942 (2014)
4. Dougherty, J., Kohavi, R., Sahami, M.: Supervised and unsupervised discretization of continuous features. In: Machine Learning Proceedings 1995, pp. 194–202. Elsevier (1995)
5. Draheim, D.: Generalized Jeffrey Conditionalization: A Frequentist Semantics of Partial Conditionalization. Springer, Cham (2017). https://doi.org/10.1007/978-3-319-69868-7
6. Draheim, D.: Future perspectives of association rule mining based on partial conditionalization. In: The 30th International Conference on Database and Expert Systems Applications, Proceedings of the DEXA 2019. LNCS, vol. 11706, p. xvi. Springer, Heidelberg (2019). https://doi.org/10.13140/RG.2.2.17763.48163

7. Etemadpour, R., da Motta, R.C., de Souza Paiva, J.G., Minghim, R., de Oliveira, M.C.F., Linsen, L.: Role of human perception in cluster-based visual analysis of multidimensional data projections. In: International Conference on Information Visualization Theory and Applications, Proceedings of IVAPP, pp. 276–283 (2014)

8. Fayyad, U., Irani, K.B.: Multi-interval discretization of continuous valued attributes for classification learning, 1993. In: The 13th International Joint Conference on Artificial Intelligence, Proceedings of IJCAI 1993 (1993)

9. Garcia, S., Luengo, J., Sáez, J.A., Lopez, V., Herrera, F.: A survey of discretization techniques: taxonomy and empirical analysis in supervised learning. IEEE Trans. Knowl. Data Eng. **25**(4), 734–750 (2012)

10. Kaushik, M., Sharma, R., Peious, S.A., Draheim, D.: Impact-driven discretization of numerical factors: case of two- and three-partitioning. In: Srirama, S.N., Lin, J.C.-W., Bhatnagar, R., Agarwal, S., Reddy, P.K. (eds.) BDA 2021. LNCS, vol. 13147, pp. 244–260. Springer, Cham (2021). https://doi.org/10.1007/978-3-030-93620-4_18

11. Kaushik, M., Sharma, R., Peious, S.A., Shahin, M., Ben Yahia, S., Draheim, D.: On the potential of numerical association rule mining. In: Dang, T.K., Küng, J., Takizawa, M., Chung, T.M. (eds.) FDSE 2020. CCIS, vol. 1306, pp. 3–20. Springer, Singapore (2020). https://doi.org/10.1007/978-981-33-4370-2_1

12. Kaushik, M., Sharma, R., Peious, S.A., Shahin, M., Yahia, S.B., Draheim, D.: A systematic assessment of numerical association rule mining methods. SN Comput. Sci. **2**(5), 1–13 (2021)

13. Kotsiantis, S., Kanellopoulos, D.: Discretization techniques: a recent survey. GESTS Int. Trans. Comput. Sci. Eng. **32**(1), 47–58 (2006)

14. Liu, H., Hussain, F., Tan, C.L., Dash, M.: Discretization: an enabling technique. Data Min. Knowl. Disc. **6**(4), 393–423 (2002)

15. Liu, H., Setiono, R.: Feature selection via discretization. IEEE Trans. Knowl. Data Eng. **9**(4), 642–645 (1997)

16. Arakkal Peious, S., Sharma, R., Kaushik, M., Shah, S.A., Yahia, S.B.: Grand reports: a tool for generalizing association rule mining to numeric target values. In: Song, M., Song, I.-Y., Kotsis, G., Tjoa, A.M., Khalil, I. (eds.) DaWaK 2020. LNCS, vol. 12393, pp. 28–37. Springer, Cham (2020). https://doi.org/10.1007/978-3-030-59065-9_3

17. Shahin, M., et al.: Big data analytics in association rule mining: a systematic literature review. In: International Conference on Big Data Engineering and Technology, Proceedings of the BDET 2021, pp. 40–49. ACM (2021)

18. Sharma, R., et al.: A novel framework for unification of association rule mining, online analytical processing and statistical reasoning. IEEE Access **10**, 12792–12813 (2022)

19. Sharma, R., Kaushik, M., Peious, S.A., Yahia, S.B., Draheim, D.: Expected vs. unexpected: selecting right measures of interestingness. In: Song, M., Song, I.-Y., Kotsis, G., Tjoa, A.M., Khalil, I. (eds.) DaWaK 2020. LNCS, vol. 12393, pp. 38–47. Springer, Cham (2020). https://doi.org/10.1007/978-3-030-59065-9_4

20. Srikant, R., Agrawal, R.: Mining quantitative association rules in large relational tables. In: International Conference on Management of Data, Proceedings of the ACM SIGMOD 1996, pp. 1–12 (1996)

ADBIS Short Papers: Modeling and Querying of Graph Databases

Interactive Knowledge Graph Querying Through Examples and Facets

Yael Amsterdamer$^{(\boxtimes)}$ and Laura Gáspár

Bar-Ilan University, Ramat Gan, Israel
yael.amsterdamer@biu.ac.il

Abstract. Knowledge graphs are a highly useful form of information representation. To assist end users in understanding the contents of a given graph, multiple lines of research have proposed and studied various data exploration tools. Despite major advancements, it remains highly non-trivial to find entities of interest in a large-scale graph where the user requirements may depend on the initially unknown contents and structure of the graph. We provide in this paper a formal approach for the problem, which combines in a novel way ideas from two approaches: query-by-example and faceted search. We first provide a novel model for user interaction that includes different formal semantics for interpreting the answers. The semantics correspond to natural interpretations of feedback in faceted search. We show that for each of these semantics, any sequence of user feedback may be encoded as a SPARQL query under standard closed-world semantics. We then turn to the problem of iteratively choosing which user feedback to prompt in order to optimize the expected length of interaction. We show that depending on the probabilities of user answers, the optimal choice of question may depend on the semantics; in contrast, we show that for a natural way of estimating the probabilities, the optimal choices coincide.

1 Introduction

The widespread adoption of knowledge graphs (KGs) as means for representing information calls for effective ways to allow users to query and explore them. SPARQL, the predominant query language for RDF graphs, allows specifying complex data selection criteria. Yet, writing formal queries requires the user not only to master the query language, but also to be familiar with the contents and structure of the queried KG, and to have a crisp notion of a question to be asked over this data. Due to the difficulty of these tasks in light of the increasing scale of KGs, assisting the uninformed user in identifying *relevant parts thereof* is a crucial need.

These challenges are well known and have been extensively studied, leading to the development of dedicated *data exploration tools*. There is a wide range of approaches and technologies for this task (see [9] for a survey). A prominent approach is that of *query-by-example* (e.g., [1,2,4,10,12,13,16,19,23,26,30]), which aims at the "reverse-engineering" of a query from output examples. The work of [2,4,12,19] has developed such solutions specifically for SPARQL.

© Springer Nature Switzerland AG 2022
S. Chiusano et al. (Eds.): ADBIS 2022, CCIS 1652, pp. 201–211, 2022.
https://doi.org/10.1007/978-3-031-15743-1_19

Another prominent approach, *faceted search*, allows users to browse through criteria that may be added to a gradually forming query, typically by providing a friendly interface for criteria selection, and dynamically updating the results (e.g., [5,6,11,17,27,28]).

Despite this great progress, the problem is far from being solved. In some cases, users may be unable to provide enough output examples (for query-by-example); and browsing the list of properties (as in faceted search) may be ineffective since properties are numerous, sparse, and not shown in context of entities and other properties. Faceted search may also lead to a "dead end" when the combination of properties does not capture the desired result [3]. Consider, for instance, a criminologist examining governmental data published in RDF, with the goal of studying properties of "interesting criminals". Since this user is not familiar with the contents of the repository, she may struggle in finding relevant information. Presenting the user example entities and properties may help her discover, e.g., that "a criminal" in this repository is identified through `convictedOf` properties, or realize that the data contains many historic convicts while she is interested in relatively recent convictions of living people (and what properties identify this relevant data).

We develop an approach for querying a given KG for entities based on their properties without prior knowledge of the KG contents. Our approach "marries" ideas from query-by-example and faceted search. Briefly, the framework selects example entities as well as example properties thereof to show the user. The display of entities with multiple properties provides context to understand their meaning and use in the KG. The user is then asked to provide feedback for the properties, in the spirit of faceted search. Following interactive variants of query-by-example, the entities and subsets of properties are chosen to maximize the expected information gain at each step. Then, feedback at the property level rather than entity level can greatly speed up the convergence to the user intention. This paper lays formal foundations for the proposed approach, making the following contributions.

Formal Model for User Interaction. Our first contribution is a formal model for user interactions. Questions are properties of entities in the KGs and answers take the form of "yes"/"no"/"Don't care". The novelty lies in the interpretations of answers: KBs are typically incomplete, e.g., when they are built through automated processes that may fail to cover all real-world facts, either due to recall issues or due to incomplete data sources. In such cases an entity may in fact match the user intention (e.g., be an "interesting criminal") even if a requested property (e.g., conviction date) is not present with respect to it. We provide four different semantics for interpreting user feedback and defining its effect on qualifying entities. We show that there is a chain of inclusion between the sets of qualifying entities according to the four semantics. Thus, the semantics may be viewed as means of balancing the precision and recall of queries, taking into account the incomplete nature of KGs.

Encoding in Standard Semantics. User answers provide an evolving specification of entity properties. We show that regardless of the semantics of user interaction, the specification may be encoded as a SPARQL query under standard closed-world semantics that SPARQL engines use, and that further the resulting SPARQL query is a member of a simple SPARQL fragment. Retrieving the relevant entities then simply involves invoking a SPARQL query engine.

Choosing the Right Questions. Given a choice of semantics for user interaction, the problem is then to choose questions in a way that minimizes the set of candidate entities. We provide a natural probabilistic formulation of the problem and show that, in general, the optimal choice of question may depend on the semantics assigned to user feedback. In contrast, we show a concrete way of inferring the probabilistic distribution of anticipated future answers based on statistics of past answers, and show that in this case, the optimal question to ask no longer depends on the semantics of user feedback.

2 Model

We next provide a model for KGs and modes user interaction.

2.1 Knowledge Graphs

We use here a simple KG model in the spirit of languages such as RDF and OWL, abstracting away details that are not necessary for our setting. Let $\mathcal{E}$ and $\mathcal{P}$ be sets of element ids and property ids respectively, standing for entities/concepts from the knowledge domain, and their properties. We further allow defining literal values from Σ^* assuming $\Sigma^* \cap \mathcal{E} = \Sigma^* \cap \mathcal{P} = \emptyset$.

Definition 1 (Facts and knowledge graphs). *A* fact *over $\mathcal{E}$, $\mathcal{P}$, Σ^* is a triple of the form* {subject property object} *where* subject $\in \mathcal{E}$, property $\in \mathcal{P}$ *and* object $\in \mathcal{E} \cup \Sigma^*$. *A KG is a set of facts.*

Example 1. The KG could include facts {Saddam_Hussein type Leader} and {Saddam_Hussein birthDate "1937-04-28"}. In this case, Saddam_Hussein, Leader$\in \mathcal{E}$, type,birthDate$\in \mathcal{P}$ and "1937-04-28"$\in \Sigma^*$.

A KG can be also viewed as a labelled directed graph with parallel edges and self-edges, where the vertices are elements in $\mathcal{E}$ or Σ^*, and every directed edge (subject, object) is labelled by some property $\in \mathcal{P}$. We interpret the meaning of a KG under *an open world assumption*: facts in the KG are asserted to be true and facts not in it may be true or false. This allows incompleteness, which typically holds in practice, as explained above.

2.2 Interaction Model

Let u be a user who seeks entities in the KG $\mathcal{G}$. Denote by E_u the set of all entities that are acceptable by u. To identify E_u, we ask u questions that can be interpreted as "Should entities in E_u have property p with object o?" e.g., "Should entities in E_u have the property type with object Leader?" Formally,

```
1  SELECT DISTINCT $e
2  WHERE {$e type Person. $e convictedOf []
3         MINUS {$e deathPlace []}
4         MINUS {$e convictedOf WarCrimes}.}
```

Fig. 1. Example selection query Q_{ex}.

Definition 2 (Question and answer model). *A user question in our framework is denoted by $q_{p,o}$ where $p \in \mathcal{P}$ and $o \in \mathcal{E} \cup \Sigma^* \cup \{[]\}$ ([] stands for an undistinguished variable, allowing any value to be assigned to it). In response a user u chooses an answer $a \in \{\Box p(o), \Box\neg p(o), \Diamond p(o)\}$, where for $o \neq []$, $\Box p(o)$ (resp., $\Box\neg p(o)$) implies that for every $e \in E_u$, the fact $\{e \ p \ o\}$ is true (resp., is false); $\Diamond p(o)$ implies that this fact may or may not hold for any $e \in E_u$ ("don't care"). If $o = []$, $\Box p([])$ (resp., $\Box\neg p([])$) implies that for every $e \in E_u$ there is some (no) value $o' \in \mathcal{E} \cup \Sigma^*$ such that the fact $\{e \ p \ o'\}$ is true (resp.,is false).*

We next recall the syntax of a simple fragment of SPARQL, which we refer to as sSPARQL, and which will be used to encode user answers.

Definition 3. *A fact pattern is a triple $e \ p \ o$ where e is a variable (fixed for a given query) and $p \in \mathcal{P}$ and $o \in \mathcal{E} \cup \Sigma^* \cup \{[]\}$. A fact $\{s \ \text{pred} \ \text{obj}\}$ matches the pattern if $p = \text{pred}$, if either $o = \text{obj}$ or $o = []$. s is assigned to e. We say that an entity s matches a pattern P with respect to a KG $\mathcal{G}$ if there exists a fact $\{s \ \text{pred} \ \text{obj}\} \in \mathcal{G}$ that matches P.*

An sSPARQL query is composed of a SELECT and WHERE clauses, as well as an optional MINUS operator. The WHERE clause consists of groups of fact patterns (in curly brackets), which may be nested in MINUS operators. Queries in our fragment always contain the clause SELECT DISTINCT $e, which means the result of a query Q over a KG $\mathcal{G}$, denoted $Q(\mathcal{G})$, includes every distinct assignment to $e that is consistent with the query contents (whose semantics are defined below).

Example 2. Figure 1 illustrates a query selecting all human convicts of who are still alive and have not been convicted of war crimes. This is done by selecting entities (assignments to $e) that e.g., do not appear in a fact with a deathPlace predicate (with any object), etc.

Let $\{\langle q_{p_1,o_1}, a_1 \rangle, \ldots, \langle q_{p_n,o_n}, a_n \rangle\}$ be a set of questions and respective answers after n interaction steps. We encode the answers in a query Q_n, where for each i, $a_i = \Box p_i(o_i)$ is encoded by $e p_i o_i and $a_i = \Box\neg p_i(o_i)$ is encoded by MINUS{$e p_i o_i}. Q_n does not encode $\Diamond p(o)$, as it has no effect on the selected entities (but we record the answer, to avoid repeating a question).

Example 3. The query in Fig. 1 is an encoding of the answer sequence $\Diamond$gender(male), $\Box\neg$deathPlace(Any), $\Box$convictedOf(Any), $\Box\neg$convictedOf(WarCrimes), $\Box$type(Person). Note that the order of answers in not important, and that the first, "Don't care" answer is not encoded.

2.3 Multiple Semantics for User Interactions

As explained in the Introduction, there are multiple reasonable semantics that could be assigned to the user feedback, and thereby to the sSPARQL query corresponding to it. We can also view the different semantics as different ways of balancing the precision and recall of queries over an incomplete KG with respect to a (complete) ground truth. We next overview these semantics.

Closed-World Semantics. Given a query over a KG $\mathcal{G}$, typical SPARQL engines interpret it as follows: an assignment s to e is consistent with a pattern group if for each pattern p in the group s matches p with respect to $\mathcal{G}/$ (This also holds if a matching fact could be *inferred* from $\mathcal{G}$'s contents, using logical inference rules as in RDF Schema and OWL.) MINUS has the usual meaning of difference. This semantics is closed-world in the sense that we regard missing facts as false.

Open-World Semantics. Closed-world semantics strictly excludes entities with missing information. An alternative semantics, which complies with the open-world assumption, ignores fact patterns in a "positive" context, i.e., not nested in negation. We thus cannot prune entities for which a matching fact does not occur in the graph; however, MINUS has the same semantics as in Closed-world.

Step-Wise Semantics. Closed world semantics may miss valuable results due to missing facts, while Open-world semantics misses important characterization of entities via positive fact patterns. As an intermediate solution, step-wise semantics has the same semantics as closed-world, as long as query result set is not empty. If it is empty, the query results contain all the entities that match a *maximal subset* of the fact patterns in positive context.

Weighted Semantics. Finally, we consider a semantics that accounts for *which* positive fact patterns are matched by each selected entity, and the importance of each constraint. For that, we associate with each fact pattern group G an importance weight $w(G)$. Let $G(\mathcal{G})$ denote the set of entities that match the pattern group under Closed-world Semantics, and let $\{G_1, \ldots, G_m\}$ denote the maximal subsets of positive constraints used in Step-wise semantics. We will then return the entities that match one of $\arg\max_{G_i} w(G_i)$ as well as the MINUS constraints. An example weight function, in the spirit of information theory is $w(G) = -\log\left(\frac{|G(\mathcal{G})|}{|\mathcal{G}|}\right)$, i.e., inverse to the number of entities matching G.

We next exemplify the different semantics.

Example 4. Reconsider the query Q_{ex} in Fig. 1, and now assume a small KG $\mathcal{G}_{ex}$ that contains the following facts: {Saddam_Hussein deathPlace Baghdad. Saddam_Hussein type Person. Silvio_Berlusconi convictedOf Fraud. Angela_Merkel type Person.} The entity of Saddam_Hussein matches the third pattern and hence does not match the first MINUS constraint, and will not appear in the query result under any of our semantics.

Under Closed-world semantics, $Q_{ex}(\mathcal{G}_{ex}) = \emptyset$, as each entity does not match some constraint: Silvio_Berlusconi and Angela_Merkel do not match the first pattern and the second patterns respectively.

Under Open-world semantics, we ignore the positive constraints and will return both of these entities.

Under Step-wise semantics, we have two maximal subsets of the patterns in positive context that match some entity. {$e type Person} and {$e convicted of []}, matched by Silvio_Berlusconi and Angela_Merkel respectively, Hence in this case, Step-wise semantics yields the same output as Open-world.

Finally, assuming a weight function inverse to number of matching entities, the second pattern will have a higher weight (having only one matching person, as opposed to two matching the first pattern). This means we will return the results adhering to the maximal subset of patterns containing only the second pattern, and adhering to the MINUS constraints - only Silvio_Berlusconi. This makes sense if, intuitively, the omission of Silvio_Berlusconi type Person is more likely than the omission of Angela_Merkel convictedOf o for any o.   □

Proposition 1. *For any* Q *and* $\mathcal{G}$, *it holds that* $Q_{\text{Closed-world}}(\mathcal{G}) \subseteq Q_{\text{Weighted}}(\mathcal{G}) \subseteq Q_{\text{Step-wise}}(\mathcal{G}) \subseteq Q_{\text{Open-world}}(\mathcal{G})$

Assume we start with the entire domain $\mathcal{E}$ as the set of candidate entities. We may also show that under Closed-world and Open-world semantics, a sequence of questions and answers yields queries whose set of answers monotonically decreases. This matches the intuition that query answers are equivalent to constraints that serve to narrow down the set of candidate entities.

Proposition 2. *Let* $\langle q_{p_1,o_1}, a_1 \rangle, \ldots, \langle q_{p_n,o_n}, a_n \rangle$ *be a sequence of questions and answers. For every* i, $1 \leq i < n$, *and any input graph* $\mathcal{G}$, $Q_{\text{sem}}^{i+1}(\mathcal{G}) \subseteq Q_{\text{sem}}^{i}(\mathcal{G})$, *where* Q^i *is the query encoding of* $\{\langle q_{p_1,o_1}, a_1 \rangle, \ldots, \langle q_{p_i,o_i}, a_i \rangle\}$, *and* $Q_{\text{sem}}^{i}(\mathcal{G})$ *is its evaluation on* $\mathcal{G}$ *under semantics* sem *which is one of Closed-world, Open-world.*

This monotonicity property does not hold for Step-wise and Weighted semantics, since as more constraints are added, new fact pattern subsets that select additional entities may be formed.

Encoding in Closed-World Semantics. Since standard SPARQL engines evaluate queries under Closed-world semantics, it may be useful to encode queries under other semantics by their Closed-world equivalents. Indeed, we may show:

Proposition 3. *For every* sSPARQL *query* Q *and a KG* G, *there exist* sSPARQL *queries* Q^1, Q^2, Q^3 *such that*

- $Q_{\text{Step-wise}}(\mathcal{G}) = Q_{\text{Closed-World}}^{1}(\mathcal{G})$
- $Q_{\text{Weighted}}(\mathcal{G}) = Q_{\text{Closed-World}}^{2}(\mathcal{G})$
- $Q_{\text{Open-World}}(\mathcal{G}) = Q_{\text{Closed-World}}^{3}(\mathcal{G})$.

Proof. We next provide the encoding for each semantics.

- *Open-world to Closed-world.* Given a query Q under Open-world semantics, a query Q' such that $Q_{\text{Open-world}} \equiv Q'_{\text{Closed-world}}$ can be obtained by omitting the positive constraints from Q.
- *Step-wise to Closed-world.* Given a query Q under Step-wise semantics, a query Q' such that $Q_{\text{Step-wise}}(\mathcal{G}) = Q'_{\text{Closed-world}}(\mathcal{G})$ for a given KG $\mathcal{G}$ can be obtained by using a UNION operation, which operates over two fact pattern groups and is supported by standard SPARQL engines under the standard (Closed-world) semantics: define a group for each maximal subset of the fact patterns, and replace the positive part of the query by this union.
- Given a query Q under Weighted Semantics, a query Q' s.t. $Q_{\text{Step-wise}}(\mathcal{G}) = Q'_{\text{Closed-world}}(\mathcal{G})$ for a given KG $\mathcal{G}$ and weight function w can be obtained similarly to Step-wise semantics, except that we take the union of the groups with the maximal weight only.

Note that for Step-wise and Weighted semantics, the encoding depends on the input graph, since these semantics depend on the (non-)emptiness of queries. The encoding of the latter further depends on the weight function. To bind $e when the positive part of the query is empty, we have added the fact pattern $e [] []. When a union consists of only one group, we replace it by that group.

3   Optimizing the Interaction

We next develop a framework for iteratively choosing the questions to users, namely combinations of property and object that should/not hold for the sought entities, conditional on the semantics of previous answers. In this framework, we focus on the following scenario.

(a) Questions of the form $q_{p,o}$ are the only means of interacting with users, as opposed to e.g., allowing to users to type input such as search strings.
(b) For a given query semantics sem, there exists a query Q in our fragment such that $Q_{\text{sem}}(\mathcal{G}) = E_u$.
(c) Users always give answers that are accurate to their needs by Definition 2, as opposed to users who may make mistakes.
(d) The process can halt as soon as an entity $e \in E_u$ is presented to the user, as opposed to only when the intended query Q is discovered.

These assumptions follow a "clean" model of exploration focusing on the type of questions that we consider (Assumptions (a)–(c)), and allowing for fast convergence (Assumption (d)). Under these assumptions, we consider the choice of questions that would maximize the expected number of eliminated entities, using a probabilistic model for answers.

Definition 4 (Problem definition). *Denote by $Pr_{p,o}(\square)$ and $Pr_{p,o}(\square\neg)$ the probability of "must" and "must not", respectively, for a specific question $q_{p,o}$ at a certain point of the interaction. The probability of "don't care" is then $Pr_{p,o}(\Diamond) = 1 - Pr_{p,o}(\square) - Pr_{p,o}(\square\neg)$. Let E_φ and $E_{\varphi'}$ denote, respectively, the*

*output of the current query φ, encoding the user answers thus far, and the output
of the next query φ'. Let $E_{p,o} \subseteq E_{\varphi}$ be the set of currently relevant entities e
(i.e., in the output of the current φ) with the property $\{e \; p \; o\}$. The expected
number of eliminated entities for a question $q_{p,o}$, under Closed-world semantics
is then*

$$\mathrm{E}[|E_{\varphi} - E_{\varphi'}|] = Pr_{p,o}(\square) |E_{\varphi} - E_{p,o}| + Pr_{p,o}(\square\neg) |E_{p,o}|$$

*For Open-world semantics we can replace $|E_{\varphi} - E_{p,o}|$ by 0. We seek the question
$q_{p,o}$ that maximizes this quantity.*

3.1 Assuming Known Probabilities

Problem Definition 4 gives rise to a simple greedy algorithm that computes the
expected probability for each combination of property and object, and then poses
the question that maximizes this value. We next describe two optimizations to
this algorithm. First, we observe that only questions in $\{q_{p,o} \mid \{s, p, o\} \in \mathcal{G} \land s \in E_{\varphi}\} \cup \{q_{p,\square} \mid \exists o \; \{s, p, o\} \in \mathcal{G} \land s \in E_{\varphi}\}$, i.e., questions on facts about candidate
entities, may lead to entity elimination greater than 0. We may thus only consider
questions in this set. Second, to compute the formula we need to compute $E_{p,o}$
for each question in the above mentioned set. To do so in a comparatively efficient
way, we can do a single pass over all the facts $\{s, p, o\}$ such that in $s \in E_{\varphi}$, and
with each such fact increment the counter for $E_{p,o}$ and $E_{p,\square}$, ignoring questions
already asked. The set of such facts may be retrieved by adding a pattern $e
$p $o to φ and executing a SPARQL query that returns the values of these
three variables. The number of eliminated entities may differ between different
semantics. For example, if the probability to a "must" answer is high, Closed-
world semantics will prefer a question corresponding to a pattern that matches
few entities. In contrast, Open-world semantics only eliminates entities given a
negative answer, and will thus prioritize patterns that match many entities.

3.2 Estimating the Probabilities

We next consider the estimation of $\mathrm{Pr}_{p,o}(\square)$ and $\mathrm{Pr}_{p,o}(\square\neg)$ at each step of the
interaction, assuming no external information (e.g. statistics for past user inter-
action). As preliminary results, we focus on the case where the user is interested
in finding a single entity, with each entity in E_{φ} having an a-priori equal probability of being the chosen one. Let the general probability of "don't care" be $\mathrm{Pr}_{\diamond}$,
then $\mathrm{Pr}_{p,o}(\square) = (1 - \mathrm{Pr}_{\diamond})\frac{|E_{p,o}|}{|E_{\varphi}|}$, and similarly $\mathrm{Pr}_{p,o}(\square\neg) = (1 - \mathrm{Pr}_{\diamond})\frac{|E_{\varphi} - E_{p,o}|}{|E_{\varphi}|}$.
By substituting these expressions in the expectation formula above, we get that
to maximize the expected number of eliminated entities with a single question
$q_{p,o}$ it suffices to choose one that maximizes $|E_{p,o}| \, |E_{\varphi} - E_{p,o}|$.

To extend this estimation to sets of properties, one can define the probability
of a combination of answers analogously to a single answer. A technical difficulty
here, beyond the increased computational complexity, is that the $\mathrm{Pr}_{\diamond}$ factor no
longer cancels out, as the number of "don't care" answers can vary. A possible
simplification adopted by our current preliminary prototype is assuming that

properties are roughly independent, namely, for any p, o, p', o' the fraction of entities in $\frac{|E_{p,o}|}{|E_\varphi|}$ resembles $\frac{|E_{p,o}-E_{p',o'}|}{|E_\varphi-E_{p',o'}|}$, and thus it suffices to choose the k questions $q_{p,o}$ with highest $|E_{p,o}| |E_\varphi - E_{p,o}|$ values.

For open-world semantics, the first term is equal to 0, but since the second term is proportional to $|E_{p,o}| |E_\varphi - E_{p,o}|$, it turns out that for this choice of probabilities, the same questions are selected for Closed-world and Open-world semantics, even though the entities they eliminate are different.

4 Related Work

The challenges faced by users interacting with huge Knowledge Bases are well known and there is a vast body of research on knowledge graph exploration [9]. For entity search, different techniques have been proposed for KB keyword search [14,15,22,29]. For query formation, one solution is using *query-by-example* to infer queries from positive/negative output examples. There is a broad line of work [1,10,13,16,26,30] on query by example in relational databases, but these techniques are not effective for properties of KBs such as sparsity, heterogeneity and incompleteness. Existing query-by-example frameworks for knowledge graphs (e.g., [2,4,12,19]) require the user to provide seed examples. *Faceted Search* is a popular technique for refining search results by proposing further constraints (e.g., [7,8,18,20,21,24,25,27,28]). Work on faceted search in knowledge graphs (e.g., [6,17,27]) focuses on facet computation, hierarchical browsing and visualization. Given a formal query, one may use similarity search [31] to find similar queries. To our knowledge, no previous work in this context has studied the alleviation of the exploration problem by combining query-by-example with principles of faceted search.

Acknowledgements. This work was partly funded by the Israel Science Foundation (grant No. 2015/21) and by the Israel Ministry of Science and Technology.

References

1. Abouzied, A., Angluin, D., Papadimitriou, C.H., Hellerstein, J.M., Silberschatz, A.: Learning and verifying quantified boolean queries by example. In: PODS (2013)
2. Abramovitz, E., Deutch, D., Gilad, A.: Interactive inference of SPARQL queries using provenance. In: ICDE (2018)
3. Amsterdamer, Y., Callen, Y.: Provenance-based SPARQL query formulation. In: Strauss, C., Cuzzocrea, A., Kotsis, G., Tjoa, A.M., Khalil, I. (eds.) Database and Expert Systems Applications: 33rd International Conference, DEXA 2022, Vienna, Austria, August 22–24, 2022, Proceedings, Part I, pp. 116–129. Springer, Cham (2022). https://doi.org/10.1007/978-3-031-12423-5_9
4. Arenas, M., Diaz, G.I., Kostylev, E.V.: Reverse engineering SPARQL queries. In: WWW (2016)

5. Arenas, M., Grau, B.C., Kharlamov, E., Marciuska, S., Zheleznyakov, D.: Faceted search over ontology-enhanced RDF data. In: CIKM (2014)
6. Arenas, M., Cuenca Grau, B., Kharlamov, E., Marciuška, Š., Zheleznyakov, D.: Faceted search over RDF-based knowledge graphs. J. Web Semant. **37–38**, 55–74 (2016). https://doi.org/10.1016/j.websem.2015.12.002
7. Atzori, M., Zaniolo, C.: Swipe: searching Wikipedia by example. In: WWW (2012)
8. Auer, S., Lehmann, J.: What have Innsbruck and Leipzig in common? Extracting semantics from wiki content (2007)
9. Bikakis, N., Sellis, T.K.: Exploration and visualization in the web of big linked data: a survey of the state of the art. In: LWDM (2016)
10. Bonifati, A., Ciucanu, R., Staworko, S.: Interactive inference of join queries. In: EDBT (2014)
11. Dakka, W., Ipeirotis, P.G., Wood, K.R.: Automatic construction of multifaceted browsing interfaces. In: CIKM (2005)
12. Diaz, G., Arenas, M., Benedikt, M.: SPARQLByE: querying RDF data by example. Proc. VLDB Endow. **9**(13), 1533–1536 (2016)
13. Dimitriadou, K., Papaemmanouil, O., Diao, Y.: Explore-by-example: an automatic query steering framework for interactive data exploration. In: SIGMOD (2014)
14. Elbassuoni, S., Blanco, R.: Keyword search over RDF graphs. In: CIKM (2011)
15. Elbassuoni, S., Ramanath, M., Schenkel, R., Weikum, G.: Searching RDF graphs with SPARQL and keywords. IEEE Data Eng. Bull. **33**(1), 16–24 (2010)
16. Fariha, A., Meliou, A.: Example-driven query intent discovery: abductive reasoning using semantic similarity. Proc. VLDB Endow. **12**(11), 1262–1275 (2019)
17. Fuenmayor, L., Collarana, D., Lohmman, S., Auer, S.: FaRBIE: a faceted reactive browsing interface for multi RDF knowledge graph exploration. In: ISWC (2017)
18. Hearst, M.A.: Clustering versus faceted categories for information exploration. Commun. ACM **49**(4), 59–61 (2006)
19. Jayaram, N., Khan, A., Li, C., Yan, X., Elmasri, R.: Querying knowledge graphs by example entity tuples. In: ICDE (2016)
20. Kamath, S., Manjunath, D., Mazumdar, R.: On distributed function computation in structure-free random wireless networks. IEEE Trans. Inf. Theor. **60**(1), 432–442 (2014)
21. Koren, J., Zhang, Y., Liu, X.: Personalized interactive faceted search. In: WWW (2008)
22. Le, W., Li, F., Kementsietsidis, A., Duan, S.: Scalable keyword search on large RDF data. IEEE Trans. Knowl. Data Eng. **26**(11), 2774–2788 (2014)
23. Li, Y., Yang, H., Jagadish, H.V.: NaLIX: an interactive natural language interface for querying XML. In: SIGMOD (2005)
24. Mass, Y., Ramanath, M., Sagiv, Y., Weikum, G.: IQ: the case for iterative querying for knowledge. In: CIDR (2011)
25. Mass, Y., Sagiv, Y.: Knowledge management for keyword search over data graphs. In: CIKM (2014)
26. Psallidas, F., Ding, B., Chakrabarti, K., Chaudhuri, S.: S4: top-k spreadsheet-style search for query discovery. In: SIGMOD (2015)
27. Qarabaqi, B., Riedewald, M.: User-driven refinement of imprecise queries. In: ICDE (2014)
28. Roy, S.B., Wang, H., Das, G., Nambiar, U., Mohania, M.K.: Minimum-effort driven dynamic faceted search in structured databases. In: CIKM (2008)
29. Tran, T., Wang, H., Rudolph, S., Cimiano, P.: Top-k exploration of query candidates for efficient keyword search on graph-shaped (RDF) data. In: ICDE (2009)

30. Weiss, Y.Y., Cohen, S.: Reverse engineering SPJ-queries from examples. In: Sallinger, E., den Bussche, J.V., Geerts, F. (eds.) PODS (2017)
31. Zheng, W., Zou, L., Peng, W., Yan, X., Song, S., Zhao, D.: Semantic SPARQL similarity search over RDF knowledge graphs. Proc. VLDB Endow. **9**(11), 840–851 (2016)

It's Too Noisy in Here: Using Projection to Improve Differential Privacy on RDF Graphs

Sara Taki, Cédric Eichler, and Benjamin Nguyen[(✉)]

INSA Centre Val de Loire, Laboratoire d'Informatique Fondamentale d'Orléans,
Bourges, France
{sara.taki,cedric.eichler,benjamin.nguyen}@insa-cvl.fr

Abstract. Differential privacy is one of the most popular and prevalent definitions of privacy, providing a robust and mathematically rigid definition of privacy. In the last decade, adaptation of DP to graph data has received growing attention. Most efforts have been dedicated to unlabeled homogeneous graphs, while labeled graphs with an underlying semantic (e.g. RDF) have been mildly addressed.

In this paper, we present a new approach based on graph projection to adapt differential privacy to RDF graphs, while reducing query sensitivity. We propose an edge-addition based graph projection method that transforms the original RDF graph into a graph with bounded typed-out-degree. We demonstrate that this projection preserves neighborhood, allowing to construct a differentially private mechanism on graphs given a similar mechanism on graphs with bounded typed-out degree. Experimental and analytical evaluation through a realistic twitter use-case show that this provide up to two orders of magnitude of utility improvement.

Keywords: Differential privacy · RDF · SPARQL · Graph projection

1 Introduction

RDF [11] is a standard way to model semantic (or linked) data. An RDF data set is a set of triples (subject-predicate-object) which form a labeled directed graph. The use of Linked Data is increasing, and thus privacy in such data sources is becoming an issue [3]. Indeed, directly publishing graph data may result in disclosure of sensitive information and therefore to privacy violations.

Differential privacy [4] (DP) is currently one of the most popular and prevalent definitions of privacy. In the last decade, adapting differential privacy to graphs has received growing attention. However, most efforts have been dedicated to unlabeled, homogeneous graphs, while labeled graphs with an underlying semantic have seldom been addressed. This original type of graph is our focus in this paper. It is important to note that many queries are highly sensitive to small modifications of the original graph, which means directly using differential privacy to perturb the query results is a bad option.

S. Chiusano et al. (Eds.): ADBIS 2022, CCIS 1652, pp. 212–221, 2022.
https://doi.org/10.1007/978-3-031-15743-1_20

Contribution. In this paper, we propose a new approach based on graph projection to adapt differential privacy to edge-labeled directed graphs, i.e. RDF graphs, while reducing the sensitivity of different kinds of queries. We take into account the semantic of the graph by adopting a QL-outedge privacy, QL being the set of sensitive relations. The main idea behind our approach is to use *graph projection* within the DP mechanism in order to reduce the sensitivity of queries. For projection to be adequate w.r.t. the privacy definition, we propose an edge-addition based graph projection method that transforms the original RDF graph into a graph of bounded QL-out-degree. We evaluate our contribution analytically and experimentally w.r.t. a real twitter use-case, showing significant improvement over a naive approach without projection. The rest of the paper is organized as follows. Fundamental concepts of differential privacy are introduced in Sect. 2. Sect. 3 surveys related work and introduces the neighborhood definition associated to the considered privacy model. Our approach and contributions are described in Sect. 4. Section 5 presents an analysis of the approach. We finally conclude and point some future works in Sect. 6.

2   Background: Differential Privacy

This section provides core background about DP, originally introduced by Dwork in 2006 [4,5].

2.1   Definition of Differential Privacy

An algorithm is differentially private if it is likely to yield the same output on neighboring (or adjacent) databases. The robustness of DP is quantified by a positive parameter ϵ, called privacy budget. The exact protection and the notion of individuals' contributions are defined based on the concept of neighboring (or adjacent) databases. Given a distance on databases, we say that two databases are neighbors if they are at distance 1. In this section, we note d the distance on the considered space. For a comprehensive overview of concepts and definitions, we refer to [4]. Formally, we model a database [9] as a vector $x \in D^n$, where x_i denotes the data provided by individual i. The distance between two databases $x, y \in D^n$ is $d(x, y) = |\{i | x_i \neq y_i\}|$.

Definition 1 (ϵ, δ-**differential Privacy**). *A randomized mechanism* $K: D^n \to \mathbb{R}^k$ *preserves* (ϵ, δ)-*differential privacy if for any pair of databases* $x, y \in D^n$ *such that* $d(x,y) = 1$, *and for all sets* S *of possible outputs:*

$$Pr[K(x) \in S] \leq e^\epsilon Pr[K(y) \in S] + \delta \qquad (1)$$

where the probability is taken over the randomness of K.

In what follows, we consider ϵ-DP which is (ϵ, δ) $- DP$ with $\delta = 0$.

2.2 Noise Calibration

One way of achieving DP is to add to an appropriate amount of noise to the query results, calibrated to its *global sensitivity*. *Global sensitivity* (*GS*) measures the maximal variation of the query result when evaluated upon any two neighboring databases. *GS* depends only on Q, d, and the considered space of databases.

Definition 2 (Global Sensitivity (*GS*) [4]). *For $f : D^n \to \mathbb{R}^k$ and all $x,y \in D^n$, the global sensitivity of f is*

$$\Delta f = \max_{x,y:d(x,y)=1} \| f(x) - f(y) \|_1 \tag{2}$$

where $\|\|_1$ denotes the L1 norm.

Theorem 1 (Laplace Mechanism [5]). *In the Laplace mechanism, in order to publish $f(x)$ where $f: D^n \to \mathbb{R}$ and $x \in D^n$ while satisfying ϵ-DP, one publishes*

$$K(x) = f(x) + Lap(\Delta f / \epsilon) \tag{3}$$

where Lap ($\Delta f / \epsilon$) represents a random draw from the Laplace distribution centered at 0 with scale $\Delta f / \epsilon$.

2.3 Sensitivity and Privacy on (Bounded QL-Out-Degree) Graphs

Applied to graphs and subspaces of graphs, these definitions become:

Notations. An edge-labeled directed graph is a graph G = (V, E) where V is a set of vertices, E is a set of edges such that E ⊆ V × L× V, with L the set of possible edge labels. We note $\mathcal{G}$ the set of such graphs.

Definition 3 (Restricted ϵ, δ-differential Privacy). *A randomized mechanism K: $\mathcal{G} \to S$ is $(\epsilon, \delta)_R$ differentially private over $R \subseteq \mathcal{G}$ w.r.t. a distance d over R, if for all pairs $(G_1, G_2) \in R^2$,*
$d(G_1, G_2) = 1 \implies Pr[K(G_1) \in S] \le e^\epsilon Pr[K(G_2) \in S] + \delta$

Definition 4 (Global sensitivity on Bounded Graphs). *For any $f : \mathcal{G} \to \mathbb{R}^k$, the global sensitivity of f on $R \subseteq \mathcal{G}$, w.r.t a distance d over R is:*

$$\Delta_d^R f = \max_{(G_1,G_2)\in R^2:d(G_1,G_2)=1} \| f(G_1) - f(G_2) \|_1 \tag{4}$$

By convention, $\Delta_d^{\mathcal{G}}$ is noted Δ_d. Considering the definitions it is trivial that for any f, R, and d, $\Delta_d^R f \le \Delta_d f$.

3 Related Work

In this paper, we propose a novel approach to construct DP-mechanism over RDF graphs. In this section, we provide an overview of DP over graphs and privacy preserving RDF querying.

3.1   Models and Distances for DP on Graphs

When using DP, the privacy model is tightly related to a distance on the considered database-space. On graph data, two distances are classically adopted: edge-DP and node-DP [7]. In node-DP, neighboring graphs are defined as graphs that differ by one node and all its incident edges. Node-DP represents the strongest privacy model for graphs. It protects the contribution of a node and all of its incident edges. In edge-DP, neighboring graphs are defined as graphs that differ by at most one edge. Edge-DP is the weakest graph privacy model. It only protects the contribution of a single edge. The sensitivity of many queries under node-DP is high and sometimes unbounded. This will degrade the utility (i.e. accuracy of the query answer), or even make it impossible to construct a DP-mechanism. In what follows, we consider other privacy models we believe to be reasonable in the context of RDF and should result in better utility.

Outedge DP [16] was introduced in the context of social networks. This privacy model protects *all* the outedges of a node. In the context of RDF, this means protecting all the triples a node is the subject of. Besides, QL-Outedge DP [14] was introduced for edge-labeled directed graphs. It is similar to outedge privacy but considers edges' semantics by only protecting edges of a given set QL (i.e. *sensitive* labels). In what follows, we adopt QL-outedge privacy and formalize the related distance as follows:

$$d_{QL}((V_1, E_1), (V_2, E_2)) = \begin{cases} \infty \ if \ V_1 \neq V_2 \vee \exists(u,v) \in (V_1 \cap V_2)^2, \exists l \in L \setminus QL \\ \quad such \ that \ (u,l,v) \in (E_1 \cup E_2) \setminus (E_1 \cap E_2) \\ |V| \ else, \ where \ V = \{v \in V_1 | \exists l \in QL, \exists u \in V_1 : \\ \quad (v,l,u) \in (E_1 \cup E_2) \wedge (v,l,u) \notin (E_1 \cap E_2)\} \end{cases}$$

3.2   Applying DP on Graphs

Previous studies described in [1,10,13,16] show different approaches to work with DP in graphs. [10] presents various techniques for designing node-DP algorithms for network data. The main idea is to *project* the input graph onto the set of graphs with maximum degree less than a specific threshold to bound the sensitivity of queries. It is based on a naive truncation that simply discards nodes of high degree. However, their techniques are designed for undirected labeled graphs, and not RDF, contrary to our proposed approach, which is also based on graph projection.

3.3   Privacy over RDF

Delanaux et al. [3] developed a declarative framework for anonymizing RDF graphs by using blank nodes to hide sensitive data. Anonymisation of RDF data inspired by the k-anonymity model was also studied in [8,12]. It does not however provide the formal privacy guarantees that DP does.

Most of the literature related to DP on RDF datasets appear to be theoretical. In fact, to the best of our knowledge, the only work investigating DP in the context of RDF that directly provides experiments is [15]. However, [15] gives a DP realisation via local sensitivity without the use of a smoothing function hence failing to comply with the privacy guarantees stated in [6].

4 Proposed Approach: From a Subspace with Low-Sensitivity Queries to $\mathcal{G}$

The main challenge when developing QL-outedge-DP algorithms is that the sensitivity of many queries can be very high, or even unbounded, in $\mathcal{G}$. Consider a query that computes the maximum out-degree in a graph. Under QL-outedge-DP, the global sensitivity of this query is roughly the number of nodes in the graph, which is unbounded.

Therefore, the main idea behind our approach is *graph projection*, in order to transform the original graph G into a graph of bounded QL-out-degree. We show that such projection can significantly reduce the sensitivity of a query and consequently the magnitude of the noise added to achieve DP. This reduction may compensate the data loss inherent to the projection, improving utility.

In this section, we introduce the projection method and show how to make the whole mechanism (i.e., a projection followed by a query over the projected space) differentially private.

4.1 Proposed Projection Method

In what follows, we propose an edge-addition based graph projection method named T_{QL}. Projection by edge-addition was introduced by [2] for unlabeled, undirected graphs and is herein expanded to edge-labeled directed graphs.

Notations. We note $\mathcal{G}_{QL}^{D}$ the set of graphs with maximum QL-out-degree D for a given QL $\subseteq L$; i.e. the set of graphs whose vertices are the source of at most D edges whose labels are in QL. Note that $\mathcal{G}_{QL}^{D} \subset \mathcal{G}$.

Projection Algorithm. Projection method $T_{QL} : \mathcal{G} \rightarrow \mathcal{G}_{QL}^{D}$ described in Algorithm 1 transforms the original graph G into one of its sub-graphs $\tilde{G}$, such that the maximum QL-out-degree of a node in $\tilde{G}$ is less than or equal to D. First, the projection creates a graph with the same nodes as G but without any edges. It then tries to insert each edge of G following an *edge ordering function* –noted A– that takes a graph and outputs an ordered lists of its edges. An edge $e = (v_1, \ell, v_2)$ is successfully inserted whenever its insertion preserves the constraint, i.e. it does not raise the *QL-out-degree* of v_1 over D.

Edge Ordering. As seen above, the algorithm attempts to insert the edge in some predetermined order. Using a different order may produce a different result. This edge ordering must be stable in the sense that given two neighboring graphs G_1 and G_2, if two edges appear in G_1 and G_2 then their relative order must be the same in $A(G_1)$ and $A(G_2)$. We can construct a stable edge ordering quite easily. Indeed, as $E \subseteq V \times L \times V$, a first intuition is to consider orders on the space of sources, labels, and destination (e.g. lexicographical order) and to define a total edge order by combining the three.

Algorithm 1: T_{QL}: projection by edge-addition, Bound QL-out-degree

---

Input: A graph G = (V, E) $\in \mathcal{G}$, a bound D, a stable edge ordering A, a set of labels QL $\subseteq$ L

Output: A D-QL-out-degree bounded graph

1 $\tilde{E} \leftarrow \emptyset$;

2 **foreach** $v \in V$ **do** toBound(v)$\leftarrow$0;

3 **foreach** $e=(v_1,l,v_2) \in A(G)$ *and following A's order* **do**

4 | **if** $l \in QL \wedge$ *toBound*$(v_1) < D$ **then**

5 | | $\tilde{E} \leftarrow \tilde{E} \cup \{e\}$;

6 | | toBound(v_1)++;

7 | **end if**

8 | **if** $l \notin QL$ **then** $\tilde{E} \leftarrow \tilde{E} \cup \{e\}$;

9 **end foreach**

10 **return** $\tilde{G} = (V, \tilde{E})$

4.2   Privacy and Projections

We study the privacy guarantees of the mechanism composed by a projection followed by a query. Its sensitivity depends on the sensitivity of the projection, i.e. the maximal distance between any two neighboring graphs after projection.

Definition 5 (Global sensitivity of a projection [10]). *The global sensitivity of a projection* $T: \mathcal{G} \to R$ *w.r.t. a distance d over $\mathcal{G}$ and d_R over R is:*

$$\Delta_{(d,d_R)} T = \max_{(G_1,G_2)\in\mathcal{G}^2:d(G_1,G_2)=1} d_R(T(G_1),T(G_2)) \tag{5}$$

In what follows, we assume that the same distance d is used in $\mathcal{G}$ and R, and note $\Delta_d T$ instead of $\Delta_{(d,d)} T$. The sensitivity of the composed function $f \circ T$ is bounded by the sensitivity of T times the sensitivity of f on the projected space.

Theorem 2 (Sensitivity of the composed mechanism [10]). *Given a projection* $T : \mathcal{G} \to R \subseteq \mathcal{G}$, *a function* $f : R \to \mathbb{R}^k$, *and a distance d over $\mathcal{G}$:*

$$\Delta_d(f \circ T) \leq \Delta_d^R f \times \Delta_d T \tag{6}$$

4.3   Privacy on Unbounded Graphs Through Projection

Our context involves QL-outedge privacy model related to the distance and projection introduced in Sect. 3.1 and 4.1, respectively.

Lemma 1 (Global sensitivity of T_{QL}). $\Delta_{d_{QL}} T_{QL} = 1$

We omit the proof due to space restriction. Intuitively, by removing or adding all the QL-out-edges of some node v, v is the only impacted node in term of Ql-outdegree. Since the constraint of T_{QL} concerns QL-outdegree, and since the edge ordering is stable, QL-outedges of other nodes are handled in the same way by T_{QL}, the sole difference between the two projected graphs being the QL-out-edges of v. Therefore, the projected graphs are still neighbors w.r.t. d_{QL}. Hence, the global sensitivity of T_{QL} w.r.t. its related distance d_{QL}, is 1. Therefore:

- It preserve neighborhood, i.e., the projection of two neighboring graphs through the use of T_{QL} results in two neighboring graphs w.r.t. d_{QL}.
- According to Theorem 2, for any function f, the global sensitivity of the composed mechanism is no greater than the global sensitivity of f over the projected space w.r.t. d_{QL}.

It directly follows that any DP algorithm on $\mathcal{G}^D_{QL}$ can be transformed into an DP algorithm on $\mathcal{G}$ without any extra privacy budget (i.e. while preserving ϵ).

Proposition 1. *Given any mechanism f whose domain is $\mathcal{G}^D_{QL}$, if f is ϵ-DP w.r.t. d_{QL} then $f \circ T_{QL}$ is ϵ-DP on $\mathcal{G}$ w.r.t. d_{QL}.*

5   Analytical and Experimental Evaluation

This section introduces a metric to analytically evaluate the approach and confront it to a real use-case. Analytical expectations are experimentally confirmed, demonstrating the feasibility and interest of the approach. The evaluation relies on the Sentiment140 dataset composed of 1.6 million tweets,[1] which we have parsed and serialized in RDF/XML format. Its schema is shown in Fig. 1a. Experiments are conducted using Apache Jena to run SPARQL queries and use our Java 1.8 implementation of the projection algorithms.

Figure 1b provides an overview of the considered scenario, functions and values w.r.t. a query Q, a projection T, and a distance d. To simplify, we consider that $Q : \mathcal{G} \to \mathbb{R}$ and the laplacian mechanism is used to achieve DP. We are interested in particular in evaluating the expected utility loss (noted E) due to privacy, defined as is the expected difference between $\bar{q}n$ and q. $E = \int_0^\infty xG(x)\,dx$ with $G(x)$ the probability of answering $\bar{q}n$ such as $|\bar{q}n - q| = x$. With $b = \frac{\Delta^R_d}{\epsilon}$ we have:

$$E = \int_0^\infty x(\frac{1}{2b}exp(\frac{-|q - x - \bar{q}|}{b}) + \frac{1}{2b}exp(\frac{-|q + x - \bar{q}|}{b}))$$
$$= b * exp(\frac{-(q - \bar{q})}{b}) + q - \bar{q} \tag{7}$$

5.1   Considered Query and Interest of the Approach

We consider here a query Q over the Sentiment140 dataset that counts the number of users user "Garythetwit" has referenced. This query leverage the dataset's semantic, refers to a path of size greater than 1, and showcase several interesting properties. On the original dataset, Q outputs 55, $q = 55$.

---

[1] https://www.kaggle.com/kazanova/sentiment140.

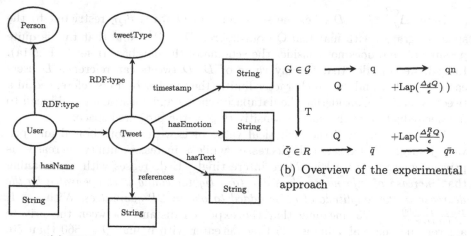

(a) RDF schema for Sentiment140

(b) Overview of the experimental approach

Fig. 1. Experimental context

Interest of the approach: Q considers solely *tweeted* and *references* outedges. If at least one of the two is sensitive, $\Delta_{d_{QL}} Q$ is infinite. Thus, in this case, it would not be possible to construct a DP mechanism directly from the original query without reducing its sensitivity. If neither *tweeted* nor *references* are considered sensitive, $\Delta_{d_{QL}} Q$ is 0. In what follows, we consider $QL = \{references, tweeted\}$.

5.2 Overall Utility of the Approach

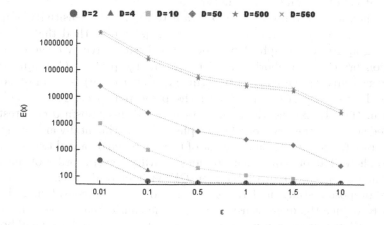

Fig. 2. Expected utility loss

Here, $\Delta_{d_{QL}}^{\mathcal{G}_{QL}^D} Q = D^2$, i.e. the sensitivity of Q w.r.t. d_{QL} restricted to the space of graphs with maximal QL-out-degree D is D^2. Note that this is quite pessimistic and does not consider the schema of the database (refer to Fig. 1a). Indeed, we consider that for any value of D, D tweets can reference D users each. In reality and due to character limits, the number of users referenced in a tweet is limited. Considering the database schema and constraints could lead to further reduction of the query's sensitivity over the projected space.

We report in Fig. 2 the analytical value of E with various bound and privacy budget. As expected, E decreases while ϵ increases: utility increases as privacy guarantees weaken. More interestingly, E decreases with D, meaning that *increase of information loss due to a tighter bound is compensated by the decrease in the amplitude of noise added to obtain DP guarantees*. With $\epsilon = 1$, $\frac{E with D = 560}{E with D = 50} \approx 125$, meaning that the expected distance between the private answer and the real value is 125 times greater with bound $D = 560$ than 50. Interestingly, as said before, $D = 560$ is an extremal case where the graph is not modified during projection. We have also seen that $\Delta_{d_{QL}} Q = \infty$ if at least tweeted or references is considered sensitive meaning that no DP mechanism can be trivially constructed over $\mathcal{G}$. A straightforward –but somewhat weak– approach would be to construct a restricted DP mechanism over some subspace of $\mathcal{G}$, typically $\mathcal{G}_{QL}^D$ with $D = 560$. This would provide exactly the same results as our approach with $D = 560$, which provides utility several orders of magnitude worse than a regular parameterization of our approach with a bound ≤ 50.

6   Conclusion

This paper presents a new approach based on *graph projection* to adapt differential privacy to edge-labeled directed graphs –e.g. RDF graphs– while reducing the amplitude of the randomized noise.

The main idea is to use *graph projection* to reduce the sensitivity of queries. We propose an edge-addition based graph projection method that transforms an RDF graph into a graph of bounded typed-out-degree. We show that this projection preserve neighborhood w.r.t. Ql-outedge privacy. Consequently, the global sensitivity of the composition (query o projection) is at most equal to the global sensitivity of the query over the projected space. Thus, we obtain a general method to expand the domain of any DP mechanism over a restricted projected space, to the space of RDF graphs. We experimentally and analytically demonstrate the feasibility and interest of the approach on a real twitter dataset. We also show that our approach provides a utility several order of magnitude better than a naive approach relying on DP without projection.

The proposed study underlines the importance of considering database schema to reduce the query sensitivity over the considered spaces. This implies studying DP in a space where a graph does not necessarily have neighbors and opens the possibility of the projected space to not be a subspace of the original space.

Acknowledgements. This work is supported by the French National Research Agency, under grant ANR-18-CE23-0010.

References

1. Blocki, J., Blum, A., Datta, A., Sheffet, O.: Differentially private data analysis of social networks via restricted sensitivity. In: Proceedings of the 4th Conference on Innovations in Theoretical Computer Science, pp. 87–96 (2013)
2. Day, W.Y., Li, N., Lyu, M.: Publishing graph degree distribution with node differential privacy. In: Proceedings of the 2016 International Conference on Management of Data, pp. 123–138 (2016)
3. Delanaux, R., Bonifati, A., Rousset, M.-C., Thion, R.: Query-based linked data anonymization. In: Vrandečić, D., et al. (eds.) ISWC 2018. LNCS, vol. 11136, pp. 530–546. Springer, Cham (2018). https://doi.org/10.1007/978-3-030-00671-6_31
4. Dwork, C.: Differential privacy. In: Bugliesi, M., Preneel, B., Sassone, V., Wegener, I. (eds.) ICALP 2006. LNCS, vol. 4052, pp. 1–12. Springer, Heidelberg (2006). https://doi.org/10.1007/11787006_1
5. Dwork, C., McSherry, F., Nissim, K., Smith, A.: Calibrating noise to sensitivity in private data analysis. In: Halevi, S., Rabin, T. (eds.) TCC 2006. LNCS, vol. 3876, pp. 265–284. Springer, Heidelberg (2006). https://doi.org/10.1007/11681878_14
6. Dwork, C., Roth, A., et al.: The algorithmic foundations of differential privacy. Found. Trends Theor. Comput. Sci. **9**(3–4), 211–407 (2014)
7. Hay, M., Li, C., Miklau, G., Jensen, D.: Accurate estimation of the degree distribution of private networks. In: 2009 Ninth IEEE International Conference on Data Mining, pp. 169–178. IEEE (2009)
8. Heitmann, B., Hermsen, F., Decker, S.: k-RDF-neighbourhood anonymity: combining structural and attribute-based anonymisation for linked data. PrivOn@ ISWC 1951 (2017)
9. Johnson, N., Near, J.P., Song, D.: Towards practical differential privacy for SQL queries. Proc. VLDB Endow. **11**(5), 526–539 (2018)
10. Kasiviswanathan, S.P., Nissim, K., Raskhodnikova, S., Smith, A.: Analyzing graphs with node differential privacy. In: Sahai, A. (ed.) TCC 2013. LNCS, vol. 7785, pp. 457–476. Springer, Heidelberg (2013). https://doi.org/10.1007/978-3-642-36594-2_26
11. Klyne, G., Carroll, J.J.: Resource description framework (RDF): concepts and abstract syntax. W3C Recommendation (2004). http://www.w3.org/TR/2004/REC-rdf-concepts-20040210/
12. Radulovic, F., García-Castro, R., Gómez-Pérez, A.: Towards the anonymisation of RDF data. In: SEKE, pp. 646–651 (2015)
13. Raskhodnikova, S., Smith, A.: Efficient Lipschitz extensions for high-dimensional graph statistics and node private degree distributions. arXiv preprint arXiv:1504.07912 (2015)
14. Reuben, J.: Towards a differential privacy theory for edge-labeled directed graphs. SICHERHEIT 2018 (2018)
15. Silva, R.R.C., Leal, B.C., Brito, F.T., Vidal, V.M., Machado, J.C.: A differentially private approach for querying RDF data of social networks. In: Proceedings of the 21st International Database Engineering & Applications Symposium, pp. 74–81 (2017)
16. Task, C., Clifton, C.: A guide to differential privacy theory in social network analysis. In: 2012 IEEE/ACM International Conference on Advances in Social Networks Analysis and Mining, pp. 411–417. IEEE (2012)

Modeling and Querying Sensor Networks Using Temporal Graph Databases

Bart Kuijpers[1] , Valeria Soliani[1,2] , and Alejandro Vaisman[2(✉)]

[1] Hasselt University, Hasselt, Belgium
bart.kuijpers@uhasselt.be
[2] Instituto Tecnológico de Buenos Aires, Buenos Aires, Argentina
{vsoliani,avaisman}@itba.edu.ar

Abstract. Transportation networks (e.g., river systems or road networks) equipped with sensors that collect data for several different purposes can be naturally modeled using graph databases. However, since networks can change over time, to represent these changes appropriately, a temporal graph data model is required. In this paper, we show that sensor-equipped transportation networks can be represented and queried using temporal graph databases and query languages. For this, we extend a recently introduced temporal graph data model and its high-level query language T-GQL to support time series in the nodes of the graph. We redefine temporal paths and study and implement a new kind of path, called Flow path. We take the Flanders' river system as a use case.

Keywords: Graph databases · Temporal databases · Sensor networks

1 Introduction and Related Work

A *sensor network* [1] is a collection of sensors that send their data to a central location for storage, viewing and analysis. These data can be used in various application areas, like traffic control and river monitoring. A sensor network through which a flow circulates (e.g., data, water, traffic) is called a *sensor-equipped transportation network*. These networks are rather *stable*, in the sense that the changes over time are minimal and occur occasionally. For example, the direction of the water flow in a river may change due to a flood or a branch may disappear due to long dry weather periods. Sensors attached to transportation networks produce time-series data, a problem studied in [3], where a formal model and a calculus are proposed. In that work, the network is modeled as a property graph (a graph whose nodes an edges are annotated with properties) [2] where nodes are associated with time series (see also [6]), obtained from the sensor measurements. One limitation of the work in [3], is that the model assumes that graphs are not *temporal*, that is, they do not keep track of their history. To address this problem, in this paper we propose to use the temporal graph data model proposed by Debrouvier et al. [4], denoted TGraph, where nodes and edges are labeled with temporal validity intervals telling the period when a node, an edge, or a property exists in the graph. Using this model we

© Springer Nature Switzerland AG 2022
S. Chiusano et al. (Eds.): ADBIS 2022, CCIS 1652, pp. 222–231, 2022.
https://doi.org/10.1007/978-3-031-15743-1_21

can query the existence or not of a graph object at a certain time instant, the values of a property measured by a sensor, and even the intervals where the sensor was working. In addition, the model comes with a high-level query language called T-GQL. TGraph builds upon three notions of paths: continuous, pairwise continuous, and consecutive. Intuitively, a *continuous path* (CP) is continuously valid during a certain time interval. A *pairwise continuous path* (PCP) is a path where consecutive edges overlap during a certain time interval. Finally *consecutive paths* (CSP) are paths where the temporal intervals between consecutive edges do not overlap (typically used for scheduling).

TGraph accounts mainly for connections between nodes, but do not address nodes associated with time series functions, like it is the case in sensor networks. For this, in this paper we extend TGraph and T-GQL, and redefine the temporal path notions to address queries like *"List the paths between two sensor nodes J and A were all temperature measurements are above a value τ, and the interval I when this occurred,"* which cannot be expressed by static graph models. We take the Flanders' river system as a use case and consider that some nodes, which represent river segments, are equipped with sensors while other ones are not. The model is introduced in Sect. 3. In addition, we redefine the three kinds of paths mentioned above and introduce the notion of *Flow Path* (Sect. 4), also showing how complex queries can be expressed using the extended T-GQL. Section 5 presents the algorithm to compute Flow Paths and describes how T-GQL queries are translated into Cypher using the underlying graph structure. We conclude in Sect. 6.

2   Background and Preliminary Definitions

A *transportation network* $\mathcal{TN}$ is a directed graph (N, E), where N is a finite set of nodes and $E \subseteq N \times N$ is a set of directed edges. Under this definition, we may model networks (e.g., rivers, roads, electrical) in at least two ways: (a) Segments represented by edges that connect two (geographic) points, modeled as nodes (illustrated on the left-hand side of Fig. 1); (b) Segments represented by nodes, and an edge between two nodes A and B indicates that the flow goes in the direction of the edge; we call FlowsTo the relationship (that is, the edge type) representing that the flow goes from A to B (Fig. 1, center). Following [3] we adopted the latter approach. Adding sensors to this network yields the notion of sensor-equipped transportation network.

Definition 1 (Sensor-equipped transportation network ([3])). *Consider a set $\mathbb{T}$ of (possible) time moments and a set $\mathbb{V}$ of (possible) measurement values. A* sensor-equipped transportation network SN, *is a 4-tuple* (N, E, S, TS), *such that* (N, E) *is a transportation network,* $S \subseteq N$ *is a set of sensor-equipped nodes (*sensor nodes, for short*), and* $\mathsf{TS} : S \to \mathcal{P}(\mathbb{T} \times \mathbb{V})$ *is a (*time-series*) function that maps sensors to a set (or sequence) of time-value pairs, ordered according with their time component (Fig. 1 (right)).*   □

In the sequel, we call the networks in Definition 1 *sensor networks*.

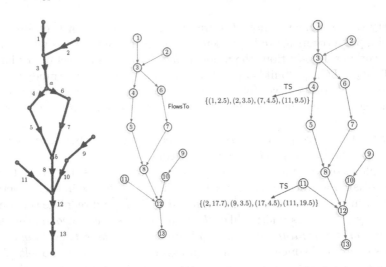

Fig. 1. Left: a physical transportation network for a river system (segments represented as edges); Center: representing segments as nodes; Right: a transportation network with the time series attached to sensor nodes 4 and 11 (segments represented as nodes)

Definition 2 (Temporal property graph [4]). *A temporal property graph is a structure $G(N_o, N_a, N_v, E)$ where G is the name of the graph, E is a set of edges, and N_o, N_a, and N_v are disjoint sets of nodes, called Object nodes, Attribute nodes, and Value nodes, respectively. Object and attribute nodes, as well as edges, are associated with a tuple (name, interval). The name represents the content of the node (or the type of the edge), and the interval the time period(s) when the node is (or was) valid. Analogously, value nodes are associated with a (name, interval) pair. For any node n, the elements in its associated pair are referred to as n.name, n.interval, and n.value. As usual in temporal databases, a special value Now tells that the node is valid at the current time. All nodes also have an (non-temporal) identifier denoted id.* □

A set of temporal constraints hold in Definition 2, and are intuitively explained next. First, all nodes with the same value associated with the same attribute node must be coalesced. Analogously, all edges with the same name between the same pair of nodes, must be coalesced. For nodes, it holds that: (a) An Object node can only be connected to an attribute node or to another object node; (b) Attribute nodes can only be connected to non-attribute nodes; (c) Value nodes can only receive edges from attribute nodes. Attribute nodes must be connected by only one edge to an object node, and value nodes must only be connected to one attribute node with one edge. Finally, for intervals: (d) The interval of an attribute (value) node must be included in the interval of its associated object (attribute) node; (e) Intervals associated with a value node must be disjoint; (f) The intervals of two edges between the same pair of nodes must be disjoint.

The model described above comes with a high-level query language denoted T-GQL. The language has a slight SQL flavor, although it is also based on Cypher [5], the query language of the Neo4j graph database. The implementation of T-GQL also extends Cypher with a collection of functions that allow handling the different kinds of temporal paths. T-GQL queries are translated into Cypher, hiding all the underlying structures that allow handling a temporal graph.

3 Temporal Graphs for Sensor Networks

The model in Definition 2 must be modified to handle sensor networks: We must distinguish Object nodes that hold a sensor from the ones which do not. We call the former *Segment* nodes. Also a list of time intervals indicates the periods of time where a segment had a working sensor on it. Properties that do not change across time are represented as usual in property graphs. We remark that we work with *categorical* variables. Also, we assume that there is at most one sensor per segment, which measures different variables, instead of many sensors that measure different variables.

Definition 3 (Sensor Network Temporal graph). *A Sensor Network Temporal Graph (SNGraph) is a structure $G(N_s, N_a, N_v, E)$ where G is the name of the graph, E a set of edges, and N_s, N_a, and N_v sets of nodes, denoted Segment, Attribute, and Value nodes, respectively. Nodes are associated with a tuple (name, interval), but in Segment nodes this tuple exists only if the segment contains (or ever contained) a sensor. In this case,* name *= Sensor, and* interval *represents the periods when a sensor worked. They may also have properties that do not change over the time (called static). An Attribute node represents a variable measured by the sensors, its* name *property is the name of such variable, and* interval *is its lifespan. A Value node is associated with an Attribute node, its* name *property contains the (categorical) values registered by the sensors, and* interval *the period when the measure was valid. The* name *property of the edges between Segment nodes represents the flow between two segments, and* interval *is the validity period of the edge. All nodes have a static identifier denoted* id. □

Temporal constraints in the TGraph model also hold for the model in Definition 3 (we omit them here). We use the Flemish river system in Belgium as a case study. Figure 2 shows a part of the Meuse river modeled as an SNGraph. There are five Segment nodes, three of which have sensors (the shaded ones, with id = 120, id = 345, and id = 1200), thus, name = Sensor. The static property riverName in Segment nodes contains the river's name. The Segment with id = 345 had a sensor between times 20 and 80 and measured two variables: Temperature and pH, thus, there are two Attribute nodes connected to it, one for each variable, with intervals [25–80] and [20–80], respectively. Note that time intervals in Attribute nodes are included in the interval of the Segment node, i.e., they satisfy the temporal constraints. There are two Value nodes for the Temperature Attribute node, such that between instants 20 and 25 the temperature was *Low*, between instants 25 and 27 it was *High*, and between instants 27 and 80 it went down to *Low* again. Finally, FlowsTo is the edge type.

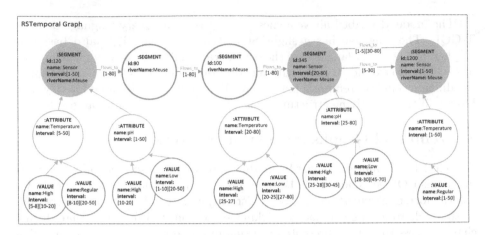

Fig. 2. A temporal graph for a river sensor network (sensor nodes are shaded).

4   Temporal Paths in Sensor Networks

We now redefine the path notions in [4], according with the model in Sect. 3.

Continuous Path in Sensor Networks. Many queries of interest can be answered using the model of Definition 3. For example: *"Starting from a segment, obtain all the paths and their corresponding time intervals T_i such that the temperature in the path has been simultaneously High for all nodes in the path during T_i"*. The original "Continuous Path" notion only accounts for the connections between nodes in a temporal graph, so it must be modified to compute a path restricted to a certain value of a variable measured by the sensors. The upper part of Fig. 3 shows, for each sensor node in a river, the temperatures registered during a certain period. Sensor nodes are denoted by a filled red square. Measures categorized as *High* (higher than 10) for the variable **Temperature** are denoted in red boxes over the registered measurement. The interval [10:30–11:00), where the value is *High* for all sensors, is framed in the figure. The lower part of the figure depicts the SNGraph, showing also non-sensor nodes. In the definitions next, the following notation is used: (a) An edge e between two nodes n_a and n_b is denoted $e\{n_a, n_b\}$; (b) An Attribute node is denoted $n_a\{n\}$ where n is the Object node connected to n_a; (c) A Value node is denoted $n_v\{n_a\}$ where n_a is the Attribute node connected to n_v.

Definition 4 (SN Continuous Path). *Let X be a variable that can take n possible values $x_1, x_2, \ldots, x_n$ during a certain time interval. Consider also an SNGraph G and a function $f(X)$. An SN continuous path for $f(X)$ (SNCP) with interval T from node s_1 to node s_k, traversing edges of type R, is a structure $P(S, R, f(X), T)$, where S is a sequence of k nodes $(s_1, \ldots, s_k)$, such that $s_i \in N_s$, s_i.name = Sensor, and T is an interval such that $\exists (a \in N_a, v \in N_v, v\{a\{s_i\}\})$ (a.name $= X$, v.name $= f(X)$, $T = \bigcap_{i=1,k} v_i$.interval $=$ and $T \neq \emptyset$). Between a pair (s_i, s_{i+1}) of sensor nodes, a path $e_1(s_i, n_1, R)$, $e_2(n_1, n_2, R)$, $\ldots$, $e_k(n_m, s_{i+1}, R)$ can exist, where $n_p \in N_s$ is a segment node with no sensor.* □

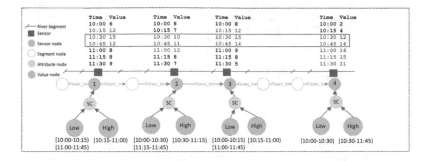

Fig. 3. SN Continuous Path with Temperature $= High$ in [10:30–11:00].

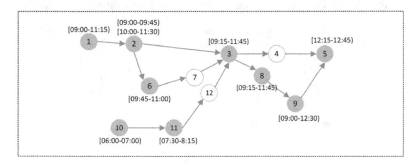

Fig. 4. Simplified SNGraph showing nodes with Temperature $= High$.

Example 1. Figure 4 depicts a simplified SNGraph where attributes and value nodes are not shown (R here is FlowsTo). The intervals tell when a *High* value of variable Temperature occurred. Filled nodes represent segments with a sensor and non-filled ones are non-sensor nodes. A query asking for all SNCPs between nodes 1 and 9, with *High* temperature values between 09:00 and 12:00, with a number of sensors between 5 and 7, returns (we use a concise notation, omitting the variable and the edge type): $Path_1 = [(1, 2, 3, 8, 9), [09:15–09:45]]$; $Path_2 = [(1, 2, 3, 8, 9), [10:00–11:15]]$; $Path_3 = [(1, 2, 6, 7, 3, 8, 9), [10:00–11:00]]$. □

Pairwise Continuous Path in Sensor Networks. Requiring a path to be valid throughout a time interval is a strong condition. A weaker notion of temporal path that asks for paths where there is an intersection in the intervals of every pair of consecutive sensor nodes may suffice. This is shown in Fig. 5. There is no SNCP with Temperature $= High$ that involves the four sensors but the value of Temperature of the first pair was *High* during the interval [10:30–11:00), for the next two segments during [11:15–11:45), and for the last two pairs during [11:30–12:00). That means, although there is no SNCP between the four sensors, there is a consecutive chain of pairwise temporal relationships between them, denoted an *SN pairwise continuous path* (SNPCP). We omit the formal definition here.

Analogously to the above, we define an *SN Consecutive Path* (SNCP) as paths composed of sensor nodes such that, for every pair of consecutive sensors,

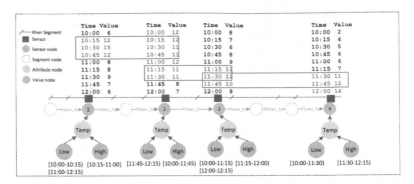

Fig. 5. SN Pairwise Continuous Path with **Temperature** $= High$.

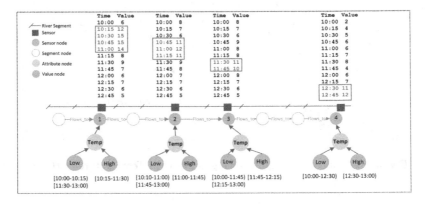

Fig. 6. Flow Path with **Temperature** $= High$.

the value of a function $f(X)$ is the same and the interval of the second period starts after the first one has finished. We omit the definition here, since SNCPs are included in the Flow Paths defined next.

Flow Paths in Sensor Networks. Sometimes, considering the paths above separately does not suffice to capture the characteristics of the flow. This is the case of an event that is detected by a sensor and may still be happening when is detected by the next sensor. Representing this situation requires a mixture of continuous and consecutive paths. Figure 6 depicts a *High* value of **Temperature** detected in one sensor earlier than the first time it is detected in the next one. Thus, the measurements overlap in the first pair of sensors but not in the other pairs. We call these paths as *Flow Paths*.

Definition 5 (Flow Path). *Let X be a temporal variable that can take n possible values $x_1, x_2, \ldots, x_n$ during a certain time interval, an SNGraph G, and a function $f(X)$. An SN Flow Path for $f(X)$ (SNFP) traversing edges of type R in G, is a structure $[S, R, f(X), T]$, where S is is a sequence of pairs $(s_1, [t_{s_1}, t_{e_1}]) \ldots, (s_k, [t_{e_k}, t_{s_k}]))$ and s_i is the i-th sensor node in S, for*

$1 \leq i \leq k$, and $\exists (a \in N_a, v \in N_v, v\{a\{s_i\}\})$ $(a.name = X, v.name = f(X)$, $[t_{s_i}, t_{e_i}] = v.interval$ and $T = \bigcup_{i=1,k} v_i.interval)$. For every pair $(s_i, [t_{s_i}, t_{e_i}])$, $(s_{i+1}, [t_{s_{i+1}}, t_{e_{i+1}}])$, $t_{s_{i+1}} > t_{s_i}$ holds. Between a pair of sensor nodes (s_i, s_{i+1}), a path $e(s_i, n_{i1}, R), e(n_{i_1}, n_{i_2}, R) \ldots e(n_{i_m}, s_{i+1}, R)$ can exist, where $n_{i_p} \in N_s$ is a segment node with no sensor.   □

Example 2. In Fig. 4, a query asking for all Flow Paths starting at node 2 such that the temperature was *High* between 09:00 and 13:00, with a minimum of 3 sensors, returns one SNFP (with Sensor nodes 2, 3, and 5): $Path_1 = [(2, 3, 4, 5, \{[09{:}00{-}9{:}45], [09{:}15{-}11{:}45], [12{:}15{-}12{:}45]\}]$ (node 4 is a non-sensor one). Intervals overlap in segments 2 and 3 but not in segments 3 and 5.   □

We extend T-GQL to address the new temporal paths (note the keywords `Variable` and `Value` below). An example of an SNCP query is: "Maximal time intervals (and the paths where they occurred) when temperature was *High* simultaneously, between '2022-03-10 05:00' and '2022-03-10 16:00', starting from the sensor located at segment 3. The number of sensors in the returned path must be between 3 and 5." The T-GQL expression for this query reads:

```
SELECT paths, interval MATCH (s1:Sensor), (s2:Sensor),
paths = SNCP((s1)-[:FlowsTo*3..5]-> (s2),
          '2022-03-10 05:00', '2022-03-10 16:00')
WHERE Variable = 'Temperature' AND Value = 'High'
          AND s1.id = 3;
```

5   Computing the Paths

The T-GQL language is implemented extending Cypher with a collection of procedures stored in the database's `Plugins` folder. T-GQL queries are translated into Cypher, and there is one procedure for each one of the temporal paths previously defined. Algorithm 1 describes the computation of the Flow Paths (Definition 5), the only one we include here for space reasons.

The algorithm receives a temporal graph G, the source and, optionally, the destination nodes (s and d, respectively), a variable X, a function f, a time interval I_q and a δ value that limits the time gaps between sensors. It returns a set of nodes S. To compute the solution, Algorithm 1 builds a transformed graph G_t, whose nodes contain either the interval when $f(X)$ was valid (if they are sensor nodes) or the interval of the previous sensor in the temporal graph, and the edges indicate the nodes reachable from that position. The nodes n in G_t have six attributes: a reference to the node in the original graph ($n.noderef$), a flag telling whether the node is a sensor or a non-sensor one ($isSensor$), a time interval when $f(X)$ was valid ($interval$), the number of sensors in the path $nbrOfSensors$), the number of sensors of a path that passes through that node ($length$), and a reference to the previous node in G_t, in order to allow rebuilding the paths after running the algorithm ($previous$).

After initialization, the algorithm adds the initially transformed graph node to a queue. This (sensor) node is a six-tuple that contains s, the interval time

---

Algorithm 1. Compute the Flow paths

Input: A graph G, a source node s, a destination node d, a variable X, a function $f(X)$, the maximum number of sensors in the path ns (optional), a query interval I_q and δ a period of time.
  Output: A set with the solutions S.
  Initialize the transformed graph G_t and Q (a queue of G_t nodes)
  if (s is sensor node) **then**
    $cInterval = f(s.X)$ nodes)
    if ($cInterval \cap I_q \neq \emptyset$) **then**
      $Q.enqueue((s, cInterval, true, 1, 1, null))$
      while not $Q.isEmpty$ **do**
        $curr = Q.dequeue()$
        for $(curr.node, interval, dest) \in \mathcal{G}.edgesFrom(curr.node)$ **do**
          if $not(G_t.containsNode(dest.id))$ **then**
            if $(dest.isSensor()$ and $dest.measures(X))$ **then**
              $dInterval = f(dest.X)$
              if $(dInterval \cap I_q == \emptyset)$ **then**
                $S.add(curr)$
                continue
              end if
              if $cInterval.start < dInterval.start$ **then**
                $newNode=(dest, dInterval, true, curr.nbrOfSensors + 1, curr.length + 1, curr)$
                if $(dest == d)$ or $(curr.nbrOfSensors == ns)$ **then**
                  $S.add(newNode)$
                 end if
              end if
           else
              $newNode=(dest, curr.interval, false, curr.nbrOfSensors, curr.length+1, curr)$
           end if
           $Q.insert(newNode)$
          end if
        end for
      end while
      return S
    end if
  end if

---

when $f(X)$ was valid for s, 1 as the number of sensors, 1 as the length of the path so far, and *null* as the reference to the previous node. An element *curr* is iteratively picked from the queue until the queue is empty. There is a node n_i in the temporal graph associated with *curr*. For each edge outgoing from n_i in G, there is a *dest* node associated with it. If the *dest* node is not in G_t and it is a sensor node, we obtain the interval time *dInterval* that corresponds to the times when $f(X)$ is valid for *dest*, and check that $dInterval \cap I_q \neq \emptyset$. We also check that the start time of *dInterval* is greater than the start time of *curr*. In that case, the path is expanded creating a sensor node *newNode* (the flag is true) whose interval is set as *dInterval*. If $dest = d$ or if we have reached the maximum number of sensors, *newNode* is added to S. In case *dest* is not a sensor node, the path is expanded with this node as a segment, and the interval will correspond to the previous sensor in that path (*cInterval*). When Q is emptied, the set of nodes in G_t is returned, and the algorithm reconstructs the paths following the link to the previous node until there is no such node.

Consider the following query, which computes the FPs between 06:00 and *Now* with at least five nodes.

```
SELECT paths MATCH (s1:Sensor), (s2:Sensor),
paths = SNFP((s1)-[:FlowsTo*]->(s2),'06:00','Now',5)
WHERE Variable = 'Temperature' AND Value='High' AND s1.id=10;
```

The query translated into Cypher using the underlying temporal graph structure (Fig. 2) is shown next. The SNFPs are computed using Algorithm 1.

```
MATCH (o1:Segment{name:'Sensor'}),(o2:Segment{name:'Sensor'})
WHERE o1.id = 10 AND o2.id = 8
```

```
CALL consecutive.flowSensor(o1,null,5,null,
 'Temperature','=','High', {edgesLabel:'FlowsTo',
 nodesLabel:'Segment',
 attributeLabel:'Temperature', valueLabel:'High',
 between:'06:00-Now', direction:'outgoing'})
YIELD path as internal_p1, intervals as internal_i1
WITH {path:internal_p1,intervals:internal_i1} as p
RETURN p.path as 'path', p.interval as 'intervals'
```

6   Conclusion and Future Work

We have shown how temporal graphs and temporal graph query languages can be used to model and query sensor networks. We have extended previous work in temporal graphs that allow supporting sensor networks. We have also extended the different notions of temporal paths, and added and implemented the notion of Flow path, that captures a wide variety of scenarios. The proposal has been implemented over the Neo4j graph database as a proof of concept, and our next step is to implement and test this model over large sensor network graphs using optimization techniques that we are developing, like temporal indices.

Acknowledgements. Valeria Soliani and Alejandro Vaisman were partially supported by Project PICT 2017-1054, from the Argentinian Scientific Agency.

References

1. Akyildiz, I.F., Su, W., Sankarasubramaniam, Y., Cayirci, E.: A survey on sensor networks. IEEE Commun. Mag. **40**(8), 102–114 (2002)
2. Angles, R.: The property graph database model. In: Proceedings of AMW, CEUR Workshop Proceedings, Cali, Colombia, 21–25 May 2018, vol. 2100. CEUR-WS.org (2018)
3. Bollen, E., Hendrix, R., Kuijpers, B., Vaisman, A.A.: Time-series-based queries on stable transportation networks equipped with sensors. ISPRS Int. J. Geo Inf. **10**(8), 531 (2021)
4. Debrouvier, A., Parodi, E., Perazzo, M., Soliani, V., Vaisman, A.: A model and query language for temporal graph databases. VLDB J. **30**(5), 825–858 (2021). https://doi.org/10.1007/s00778-021-00675-4
5. Francis, N., et al.: Cypher: an evolving query language for property graphs. In: Proceedings of SIGMOD, Houston, TX, USA, 10–15 June 2018, pp. 1433–1445. ACM (2018)
6. Zhang, S., Yao, Y., Hu, J., Zhao, Y., Li, S., Hu, J.: Deep autoencoder neural networks for short-term traffic congestion prediction of transportation networks. Sensors **19**(10), 2229 (2019)

Indexing Continuous Paths in Temporal Graphs

Bart Kuijpers[1], Ignacio Ribas[2], Valeria Soliani[1,2],
and Alejandro Vaisman[2]

[1] Hasselt University, Hasselt, Belgium
bart.kuijpers@uhasselt.be
[2] Instituto Tecnológico de Buenos Aires, Buenos Aires, Argentina
{iribas,vsoliani,avaisman}@itba.edu.ar

Abstract. Temporal property graph databases track the evolution over
time of nodes, properties, and edges in graphs. Computing temporal
paths in these graphs is hard. In this paper we focus on indexing Continuous Paths, defined as paths that exist continuously during a certain
time interval. We propose an index structure called TGIndex where index
nodes are defined as nodes in the graph database. Two different indexing
strategies are studied. We show how the index is used for querying and
also present different search strategies, that are compared and analyzed
using a large synthetic graph.

Keywords: Temporal graphs · Path indexing · Temporal graph index

1 Introduction

Property graphs [1], whose nodes and edges are annotated with properties, are
used in most graph databases in the marketplace. In practice, these graphs are
typically *static*, i.e., they do not change over time. However, in most real-world
applications, edges, nodes, and properties can be added, deleted, and updated
as needed. This is addressed in [2], where a model (TGraph) for temporal graph
databases is proposed. First-class citizens in this model are temporal paths of
three types: Continuous, Pairwise Continuous and Consecutive paths. Continuous Paths (CPs) are paths valid during a certain time interval. In Pairwise
Continuous Paths, every pair of adjacent edges has an overlapping time interval. In Consecutive Paths, for every pair of adjacent edges, the time validity of
one edge ends before the validity time interval of its consecutive one starts. The
model comes equipped with a high-level SQL-like query language called T-GQL,
which includes functions to compute the three kinds of paths above. In this work
we propose to index CPs to improve their computation.

Indexing paths in temporal and non-temporal graphs has been studied to a
limited extend. Pokorny et al. [7] index graph patterns in Neo4j, using a structure
stored in the same database as the graph (an approach we follow in this work).
Huo and Tsotras [5] study the problem of efficiently computing shortest-paths
on evolving social graphs. The authors use an extension of Dijkstra's algorithm

© Springer Nature Switzerland AG 2022
S. Chiusano et al. (Eds.): ADBIS 2022, CCIS 1652, pp. 232–242, 2022.
https://doi.org/10.1007/978-3-031-15743-1_22

to achieve this for a time-point or a time-interval. To improve performance of queries in temporal databases, Elmasri et al. [4] proposed a basic indexing technique for temporal data that can be combined with conventional attribute indexing schemes to process temporal selections and temporal join operations. In [6] an index structure for temporal attributes is proposed.

In this work we present two index structures for Continuous Paths, one that indexes all the paths and another one that indexes all paths of length two. In the latter case, computing the paths of length higher than two requires additional processing. We show how queries are evaluated using these indices. We also consider reducing the search space by limiting the time window to consider the one in which queries will most likely fit. As far as we are aware of, this is the first proposal for indexing temporal paths on graphs.

Section 2 briefly presents the concepts and definitions in which the present paper is based. Section 3 presents two index structures and in Sect. 4, we show how they are used to process T-GQL queries and updates over the temporal graph. In Sect. 5, we report experimental results. We conclude in Sect. 6.

2   Background

To make the paper self contained, we briefly present the TGraph model.

Definition 1 (Temporal property graph (TGraph) (cf. [2])). *A temporal property graph is a structure $G(N_o, N_a, N_v, E)$ where G is the name of the graph, E is a set of edges, and N_o, N_a, and N_v are disjoint sets of nodes, denoted object nodes, attribute nodes, and value nodes, respectively. Every object and attribute node and every edge are associated with a tuple (name, interval). The name represents the content of the node or the type of the edge, and the interval represents the period(s) when a node or edge is (was) valid. Analogously, value nodes are associated with a (name, interval) pair. For any node n, the elements in its associated pair are referred to as n.name, n.interval, and n.value. As usual in temporal databases, a special value Now tells that the node is valid at the current time. All nodes also have an identifier denoted id.* □

Nodes and edges in G satisfy a collection of temporal constraints. For the sake of space, we refer the reader to [2]. Figure 1 shows a social network represented using the model in Definition 1. There are three kinds of object nodes: Person, City, and Brand. There are also three types of edges: LivedIn, Friend, and Fan. The first one is labeled with the periods when someone lived somewhere, the second one with the periods when two people were friends. An edge of type Fan tells that Peggy Sue-Jones is a Samsung fan since 2005. The temporal attribute node Name represents the name associated with a Person node. Finally, for clarity, if a node is valid throughout the complete history, the temporal labels are omitted.

Definition 2 (Continuous Path). *Given a temporal property graph G, a continuous path (CP) with interval T from node n_1 to node n_k, traversing a relationship r, is a sequence $(n_1, \ldots, n_k, r, T)$ of k nodes and an interval T such that there is a sequence of consecutive edges of the form $e_1(n_1, n_2, r, T_1)$, $e_2(n_2, n_3, r, T_2)$, $\ldots$, $e_k(n_{k-1}, n_k, r, T_k)$, $T = \bigcap_{i=1,k} T_i$.* □

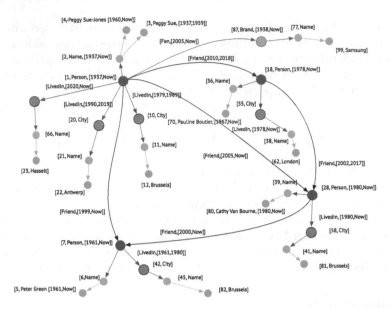

Fig. 1. A temporal property graph.

Continuous paths capture queries like: "Compute the friends of the friends of each person, and the period such that the relationship occurred through all the path." In Fig. 1, for example, Pauline (person node 18) was a friend of Cathy (person node 28) between 2002 and 2017. Also, Peggy Sue (person node 1) was a friend of Pauline between 2010 and 2018. Thus, the path ($PeggySue \xrightarrow{Friend} Pauline \xrightarrow{Friend} Cathy, [2010, 2017]$) will be in the answer.

3   An Index for Continuous Paths: TGIndex

Given a TGraph, the *Temporal Out Degree (tod)* of an Object Node $n \in N_o$, with relationship r is the number of intervals associated with r coming out from n. The *tod* of a node is always equal or greater than its out degree. Given an Object node n and a length L, the upper bound on the number of CPs is $O(tod_{max}^L)$. If N is the number of Object nodes then the TGraph may have a maximum number of CP bounded by $O(N * tod_{max}^L)$. Thus, computing the CPs in a temporal graph is an expensive operation. To improve performance when computing CPs, we consider indexing CPs. The idea proposed in this paper builds on [7] together with typical methods for time indexing [3]. We may reduce the search space if we knew in advance that queries ask for paths within a time window $[t_1, t_2]$, with $t_1 \geq t_0$ (the minimum timestamp in the graph) and $t_2 \leq Now$. This way we could index just those CPs. The CPs that exist outside the indexed time interval could be retrieved combining the Index and the TGraph. We describe next two indexing approaches denoted **TGIndexL** and **TGIndex2**.

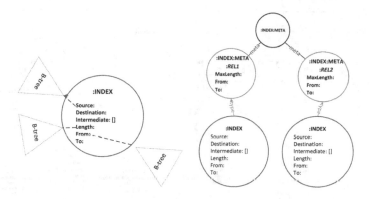

Fig. 2. Left: Ix node; Right: TGIndex nodes Meta and Ix.

The TGIndexL Approach. We denote $cPath_r^*$ the set of all the CPs in G with relationship r. Let $cPath_r^*(L, t_1, t_2)$ be a subset of $cPath_r^*$ such that the CPs have a maximum length L and are valid between t_1 and t_2. We call **TGIndexL** the index $I_{rL}[t_1, t_2]$ containing those cPaths. Each $I_{rL}[t_1, t_2]$ has N index nodes, each one corresponding to a $cPath_{rL}^i$, denoted $Ix(source, destination, length, from, to, intermediate)$ where: (a) *source* contains the *id* of the starting Object node of $cPath_{rL}^i$; (b) *destination* contains the *id* of the ending Object node of $cPath_{rL}^i$; (c) *length* is the number of hops involved in $cPath_{rL}^i$, where $length \leq L$; (d) *from* is the starting time of $cPath_{rL}^i$; (e) *to* is the ending time of $cPath_r^i$; (f) *intermediate* is an ordered list of the of the id's of the intermediate Object nodes $cPath_{rL}^i$. Each Ix node also has two outgoing edges labeled *start* and *end* that connect it with the starting and ending Object nodes of its corresponding $cPath_{rL}^i$, respectively. The Ix nodes are ordered by their starting time *from*, their *length* and by the *source* attribute. Following [4], Neo4j B-Tree indices on those properties are created for this. Figure 2 shows a sketch of the proposed index node. We can see that the TGIndex has two types of nodes: the Ix nodes (leaves), and the nodes containing metadata (indicated by the META label). When creating a $I_{rL}[t_1, t_2]$ we must indicate the time window $[t_1, t_2]$ to calculate only $cPath_r$ existing between t_1 and t_2. Figure 3 shows a TGIndex stored in the same TGraph database (GDB). For clarity, the TGraph shown is an abstraction of the actual graph, since each node of the graph is actually composed by the Object, Attribute and Value nodes.

The TGIndex2 Approach. For very large graphs, our experiments showed that it may take a long time to build TGIndexL, and the required storage space may be large. Therefore, we propose a second approach, denoted **TGIndex2**, that indexes only the CPs of length 2 and then answers the queries by rebuilding the paths from there. This way, the storage space is reduced and the query time, although higher than when indexing every possible CP, is lower than when we compute the paths over the original TGraph. The correctness of the procedure is based on two CPs properties: (a) Every CP of even length k can be decomposed

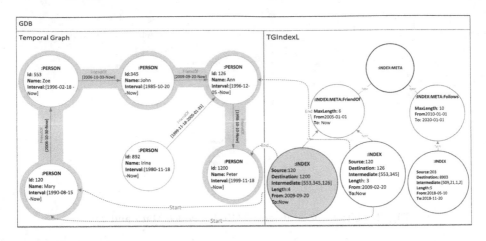

Fig. 3. Indexing paths in a temporal graph: GDB and TGIndexL.

in $k/2$ CPs of length $l = 2$; (b) Every CP of odd length k $cPath_r(n_1, \ldots, n_{k+1}, I)$ can be decomposed in a CP of (even) length $k-1$, $cPath'_r(n_1, \ldots, n_k, I')$ and the last edge e_k between n_k and n_{k+1}, $e_k\{n_k, n_{k+1}\}$. Here, the intersection between the intervals of the $cPath'_r$ and the edge interval equals the interval of the original $cPath$, $e_k.I \cap I' = I$. Every pair of consecutive edges can be seen as $k-1$ CPs of length 2 of the form $cPath^1_r(n_1, n_2, n_3, I_1)$, $cPath^2_r(n_2, n_3, n_4, I_2)$, $cPath^3_r(n_3, n_4, n_5, I_2)$, $\ldots$, $cPath^{k-1}_r(n_{k-1}, n_k, n_{k+1}, I_{k-1})$. Moreover, when k is even, to rebuild the original path we only need $cPath^1_r$, $cPath^3_r$, $\ldots$, which are the CPs whose destination matches the source of another one. If k were odd, we can split the CP into two: the one from n_1 to n_k (of length $k-1$), $cPath'_r(n_1, \ldots, n_k, I)$ and the last edge of the CP, which goes from n_k to n_{k+1}. Since $cPath'_r$ has even length, it can be decomposed into several CPs of length 2. The edge to complete the original path is obtained from the last CP of the decomposition $cPath^{k-1}_r$.

Figure 4 illustrates the above. There is an index node for every CP of length 2 valid in a certain time window. Retrieving a CP of length 2 from the index is straightforward: the index nodes themselves are the answer. To retrieve a path of even length k, $k \geq 4$ we obtain the $k/2$ index nodes Ix where $Ix_i.destination = Ix_{i+1}.source$. To improve the performance of queries asking for paths of length greater than 2, the indexed paths of length 2 are connected to each other by edges of type :concat, as Fig. 4 shows. An edge between two index nodes Ix_1 and Ix_2 exists if there is a CP of length 4 in G starting at $Ix_1.source$, ending at $Ix_2.destination$, with $\{Ix_1.destination, Ix_2.source\}$ being the intermediate nodes of the path, and a validity interval I. This interval, which can be obtained by the intersection of $[Ix_1.from, Ix_1.to]$ and $[Ix_2.from, Ix_2.to]$, is a property of the :concat edge. The structure of the index node Ix is the same as in TGIndexL except for the *length* property, which is no longer required.

To estimate the number of index nodes we must calculate the number of CPs of length 2. Let $M = tod_{max}$ and $N = |N_o|$ be the number of Objects nodes in

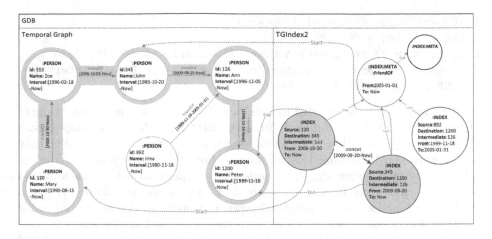

Fig. 4. Indexing paths in a temporal graph: GDB and TGIndex2

G, then the maximum number of paths of length 2 from a node x is $O(M^2)$. Then, there could be a maximum number of $O(N * M^2)$ of index nodes.

4   Using the Index

We explain now how a T-GQL query is executed using the two indexing approaches in the previous section. We start with TGIndexL and the T-GQL query:

```
SELECT paths
MATCH (p1:Person),(p2:Person),
paths = cPath((p1)-[:Friend*4]-> (p2),'2010','2020')
WHERE p1.Name = "Mary"
```

The query processor first checks if there is an index for the relationship Friend whose indexing interval includes the query interval. In the example, the query window time is completely included in the indexed window time. The query is translated into Cypher as:

```
MATCH (v1:Value)<--()<--(o1:Object)
WHERE v1.value = "Mary"
CALL graphindex.retrievePaths(o1,null,4,4,'2010','2020',
{EdgeLabel:'Friend', direction:'outgoing'}) YIELD path,interval
RETURN path, interval;
```

The function `graphindex.retrievePaths` is added to the Neo4j library, so it can be included in a Cypher expression.

Using the TGIndex2 approach for the same query, yields the translation:

```
MATCH (v1:Value)<--()<--(o1:Object)
WHERE v1.value = "Mary"
CALL graphindex.retrievePathsConcat( o1,null,4,4,'2010','2020',
{EdgeLabel:'Friend', direction:'outgoing'} ) YIELD path,interval
RETURN path, interval;
```

Two index nodes are needed to build a CP of length 4. The validity time of such path can be obtained directly from the interval in the relationship :concat (Fig. 4). For queries asking for CPs of length greater than 5, the validity time of the path can be obtained as the intersection of the intervals of the edges labelled :concat between the index nodes. If the query asks for a path of odd length, the procedure is similar but the last edge is obtained from the graph itself. The function **retrievePathsConcat** concatenates the connected index nodes and keeps only those paths whose intersection exist all the way. The time for retrieving CPs using this index depends not only on the number of indexed nodes but also on the execution times of the Neo4j pattern matching strategies. However, navigation through the :concat edges of the index can also be done in Breadth First Search (BFS) fashion. The BFS strategy starts from all the index nodes whose source matches the required id. We developed an algorithm for this second strategy to retrieve paths, denoted **retrievePathsBFSConcat**, not shown here for the sake of space.

Strategies for Querying a TGraph using TGIndices. Consider a TGraph G and a TGIndex over a relationship r with interval $I_x = [t_1, t_2]$, in any of its implementations. Given the query $QcPath_r(source = x, dest = y, L_{min} = l_1, L_{max} = l_2, I_q)$, there are three possible scenarios: (a) $I_q \subseteq I_x$: I_q **during** I_x; (b) $I_q \cap I_x = \emptyset$: I_q **before** I_x or I_x **before** I_q; (c) $I_q \cap I_x \neq \emptyset$: I_q **overlaps** I_x or I_x **during** I_q. In the first case, as explained before, the result set R will be fully provided by the index. In the second, the index could not answer the query at all and the standard procedure **consecutive.continuous** described in [2] must be used. In the last case, the result can be obtained combining the use of the index and G itself. Depending on whether or not I_x is fully included in the query interval, it will be necessary to divide I_q into two or, at most, three separate time intervals. If $I_x \subset I_q$ there will be three sub-intervals: the interval before I_x, I_x itself, and the interval after I_x. Otherwise there will be two sub-intervals.

5   Experimental Evaluation

We compare the performance of the *cPath* function built-in T-GQL against both TGindex2 strategies which we called IndexConcat (the one that uses Neo4j's pattern matching) and IndexBFSConcat (our BFS algorithm), using two kind of queries: (a) SO: Source-only CP queries, which obtain all the CPs of a certain length starting from a given node; (b) P2P: Point-to-Point CP queries, which, given a source and a destination node, obtain all the CPs between them. We discarded the TGIndexL strategy since the index sizes and building times were very high for our graph. We created a synthetic TGraph representing a social

Table 1. TGraph: *tod* & #CPs starting from selected Object Nodes.

id	172	307	19	13	448	409	65779
tod	1396	55	57	50	37	29	28
#cPaths by length 2	151	136	108	64	61	6	1
3	1747	701	1209	703	722	10	23
4	4212	3375	2683	1509	1436	19	59
5	19464	15835	12419	0	6590	71	317
6	89037	73353	57067	0	30775	291	1499
7	409741	337693	261294	0	142791	1294	6560
8	1880151	1558243	1201237	0	660440	5916	30502

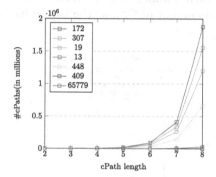

Length	Class 1	Class 2
2	1	301
3	2	2263
4	6	3375
5	28	11376
6	156	54238
7	379	260262
8	1800	111111

Fig. 5. Left: CPs growth vs. length in SN TGraph; Right: Social network TGraph: Number of cPaths between two nodes classified as Class 1 and Class 2

network (SN) with 38,500 Object nodes of type Person and the same number of Attribute and Value nodes, resulting in 115,500 nodes in total. The Object nodes are connected through a relationship Friend. The graph has two connected components C1 and C2. In C1, for the relationship Friend, the minimum out degree of the Object nodes is 25; there is a Follower Object node whose out-degree is 1,394 and its *tod* is 1396. There is also an Influencer Object node whose in-degree is 15,000. In C2 nodes are less connected than in C1 but there is also a large number of CPs. For this graph, a TGIndex2 is created with a time window of 3 months obtaining 178,436 Index Nodes representing CPs of length 2. We selected representative Object nodes from C_1 to run the SO tests based on their *tod*, the degree and the number of indexed CPs of length 2 that start from them during the indexed period. We denote the nodes by their Object id. Table 1 shows those nodes, their *tod*, and the number of CPs from length 2 to length 8.

The selected time window impacts on the number and length of the CPs. Based on this, we classify the nodes as *Big* (172, 307 and 19), *Medium* (13 and 448) and *Low* (409 and 65779). We run the tests five times for each node and then average the resulting time. To run the P2P queries we chose eight pair of nodes from C_1 and C_2 and observed the number of CPs from length 2 to length

6. On the right-hand side of Fig. 5 we classify the paths as "Class 1" when the resulting number of CPs is low and "Class 2" when that number is high.

Figure 6 shows *the execution times of SO queries* (results are expressed in milliseconds and the scale is logarithmic). We can see that the BFS concat strategy is always better than the other two ones. When the number of CPs is high, both BFS strategies outperform IndexConcat. Also, the difference is higher for paths of even length than for the ones of odd length. On the right-down portion of the figure we show the averaged results of the three kinds of nodes for CPs of even length. We can see that both index strategies are better than the cPath strategy. Classifying the nodes according to the number of CPs retrieved suggests that the BFS strategy should be considered when the source nodes have a high *tod* and produce a high number of CPs of length 2, which in turn will produce a high growth in the number of CPs as the length increases.

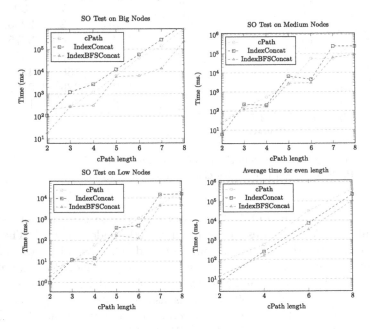

Fig. 6. Execution times vs. path length for SO tests.

Figure 7 compares the *execution times of P2P queries*. In this case, the *index-Concat* strategy clearly outperforms the others when the number of CPs is low (below 1000 approximately). This is because the Neo4j strategy over the index nodes is fast when the search space is small. The number of CPs retrieved when the query mentions a source and a destination node are less than when the source is not mentioned. The *BFSconcat* strategy on the other hand, presents a better performance as long as the number of CPs increases to a very high number of paths, since it prunes the results and avoids evaluating non-useful paths. For higher numbers, the three strategies perform similarly.

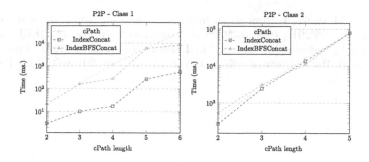

Fig. 7. P2P Test execution time vs. length. Left: Class 1; Right: Class 2.

6   Conclusion

We studied strategies for indexing CPs in temporal graph databases. These indices are defined within the database and point to the start of the CPs being indexed. Two kinds of indices were defined: one that indexes paths of any length (called TGIndexL) and another one that indexes paths of length two (called TGIndex2) and builds paths of any length from there. Since the latter proved to be more effective, two query evaluation strategies are defined for it. One of them takes advantage of the underlying graph database search strategies, and the other one using a BFS strategy. We carried out tests over synthetic temporal graphs with good results since both strategies outperformed the non-indexed evaluation algorithm. We believe that the ideas developed in this paper set the basis for future work that includes developing new evaluation and implementation strategies and performing further experiments.

Acknowledgements. Valeria Soliani and Alejandro Vaisman were partially supported by Project PICT 2017-1054, from the Argentinian Scientific Agency.

References

1. Angles, R.: The property graph database model. In: Proceedings of AMW 2018, CEUR Workshop Proceedings, Cali, Colombia, 21–25 May 2018, vol. 2100. CEUR-WS.org (2018)
2. Debrouvier, A., Parodi, E., Perazzo, M., Soliani, V., Vaisman, A.: A model and query language for temporal graph databases. VLDB J. **30**(5), 825–858 (2021). https://doi.org/10.1007/s00778-021-00675-4
3. Elmasri, R., Kim, Y., Wuu, G.T.J.: Efficient implementation techniques for the time index. In: Proceedings of ICDE 1991, 8–12 April 1991, Kobe, Japan, pp. 102–111. IEEE Computer Society (1991)
4. Elmasri, R., Wuu, G.T.J., Kim, Y.: The time index: an access structure for temporal data. In: Proceedings of VLDB 1990, Brisbane, Queensland, Australia, 13–16 August 1990, pp. 1–12. Morgan Kaufmann (1990)
5. Huo, W., Tsotras, V.J.: Efficient temporal shortest path queries on evolving social graphs. In: SSDBM, Aalborg, Denmark, 30 June–2 July 2014, pp. 38:1–38:4 (2014)

6. Kvet, M., Matiasko, K.: Impact of index structures on temporal database performance. In: EMS 2016, Pisa, Italy, 28–30 November 2016, pp. 3–9. IEEE (2016)
7. Pokorný, J., Valenta, M., Troup, M.: Indexing patterns in graph databases. In: DATA 2018, Porto, Portugal, 26–28 July 2018, pp. 313–321. SciTePress (2018)

ADBIS Short Papers: Data Science and Machine Learning

Unsupervised Extractive Text Summarization Using Frequency-Based Sentence Clustering

Ali Hajjar and Joe Tekli[✉] [iD]

School of Engineering, Lebanese American University (LAU), 36, Byblos, Lebanon
ali.hajjar@lau.edu, joe.tekli@lau.edu.lb

Abstract. Large texts are not always entirely meaningful: they might include repetitions and useless details, and might not be easy to interpret by humans. Automatic text summarization aims to simplify text by making it shorter and (possibly) more informative. This paper describes a new solution for extractive text summarization, designed to efficiently process flat (unstructured) text. It performs unsupervised frequency-based document processing to identify the candidate sentences having the highest potential to represent informative content in the document. It introduces a dedicated feature vector representation for sentences to evaluate the relative impact of different sentence terms. The sentence feature vectors are run through a partitional k-means clustering process, to build the extractive summary based on the cluster representatives. Experimental results highlight the quality and efficiency of our approach.

Keywords: Automatic text summarization · Extractive summaries · Word space model · Feature representation · k-means clustering

1 Introduction

The exponential increase of data published on the Web has reignited interest in automatic text summarization, aiming to save data storage space and allow faster access to the most informative data. In this study, we introduce a new solution for extractive summarization of flat (unstructured) text, by integrating and adapting different existing techniques in a novel way, aiming to provide a simple, flexible, and computationally efficient solution. While most extractive solutions utilize term-based feature vectors with heuristic or linear optimization solutions, our solution performs sentence feature vector extraction, followed by sentence clustering using their feature vector similarity, and summary building based on the cluster representatives. The user selects the input text document, and the size of the final summary expressed in number of sentences. The input text is processed for term frequency computation in order to produce a co-occurrence matrix describing the feature vector of each sentence in the original text. The sentence feature vectors are utilized to compute pair-wise sentence similarities, in order to perform partitional k-means clustering to group similar sentences together, where the number of k clusters corresponds to the size of the summary provided initially by the user. Different strategies are suggested to select the most representative sentences from the clusters,

S. Chiusano et al. (Eds.): ADBIS 2022, CCIS 1652, pp. 245–255, 2022.
https://doi.org/10.1007/978-3-031-15743-1_23

to form the output summary. Experiments highlight our solution's quality and almost linear computation time.

The remainder of the paper is organized as follows. Section 2 reviews the related works. Section 3 described our proposal. Section 4 describes our experimental evaluation, before concluding in Sect. 5 with ongoing works and future directions.

2 Related Works

Automatic text summarization techniques can be grouped in two main categories: abstractive and extractive. Abstractive summarization aims to generate summaries the way humans perform summarization, by transforming the original text to generate a new text summary. They usually follow a predefined schema describing a certain ordered pattern of content organization [1, 2]. The schema can be expressed using knowledge-based or rule-based representations. Knowledge-based approaches use of a machine-readable semantic graph made of a set of concepts representing word senses, and a set of links representing semantic relations (synonymy, hyponymy, etc., [3, 4]). A text document is represented as a semantic sub-graph, and the process of summarization consists in generating a reduced semantic sub-graph using some heuristic rules, in order to generate the reduced output summary. Rule-based approaches describe text relationships using typed dependencies between pairs of words. Rule-based information extraction and content selection heuristics [5–7] are then used to generate new texts, based on training data consisting of sets of input texts and expected (summarized) output texts. Text chunks from the original text are matched against the rules, and sent to a generation module trained based on the training data, in order to produce the output summary. Recent approaches have utilized deep learning transformer based encoder-decoder architectures like BERT to produced trained summarization models, e.g., [8–10]. Deep learning solutions have shown promising results compared with their counterparts [8], albeit requiring training data and training time which are not always available.

Extractive summarization promotes a less complex process of identifying and extracting the most informative text tokens, without content re-writing or generation. Most techniques in this context use unsupervised term frequency computation, e.g., [11, 12], identifying and combining the sentences or paragraphs including the most frequent terms to form the summary. This is based on the assumption that the high frequency of specific words in a text may be a good indicator of its significance. As a result, text chunks or sentences are compared based on their most common terms. Heuristic or linear optimization solutions can be used to identify the most representative text chunks or sentences, to be combined into the output summary. The authors in [13] use an adapted quantum-inspired genetic algorithm, using a modified quantum measurement and a self-adaptive quantum rotation gate based on the quality and length of the summary. The authors in [14] use a swarm optimization solution, using word mover distance and normalized Google distance to evaluate text similarity. The authors in [15] use a projected gradient descent algorithm to perform summarization through convex optimization. In [16], the authors represent sentences as nodes in an undirected graph, where every distinct term is represented as a vertex, regardless of the number of term occurrences. The process extracts the most connected nodes in the graph, to form the output summary. Some approaches, e.g., [2, 17], perform feature vector transformation using singular value decomposition (SVD) or latent derelict analysis (LDA), to represent terms or sentences in a latent semantic space. Sentence selection is then conducted in the latent semantic space, before compiling the sentences in their original form to construct the output summary.

While most extractive solutions utilize term-based feature vectors combined with heuristic or linear optimization solutions, we introduce a new sentence-based feature vector representation combined with a partitional clustering algorithm, aiming to improve summarization efficiency and quality.

3 Proposal

Our solution consists of four main components (cf. Fig. 1): i) linguistic pre-processing, ii) sentence feature vector representation, iii) sentence clustering, and iv) summary building.

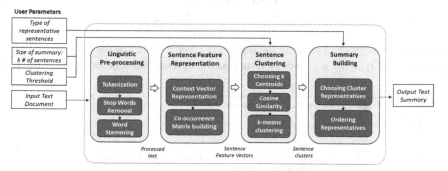

Fig. 1. Simplified activity diagram describing our approach

3.1 Linguistic Pre-processing

A sequence of preprocessing tasks is first executed before the documents can be processed for sentence feature extraction, clustering, and summarization. First, this component converts all words to their lowercase form, performs tokenization to distinguish separate terms, and removes stop-words from the obtained term sequences. Second, it performs stemming or lemmatization, following the user's preference: i) stemming converts all the words to their original syntactic forms (stems) using syntactic stemming rules[1]; ii) lemmatization transforming words into their original lexical forms using a lexical reference[2]. Third, it performs sentence extraction based on text punctuations, where each sentence is represented as a sequence of stemmed/lemmatized terms.

---

[1] We use the Porter Stemmer in our approach since it is one the most effective in the literature.

[2] We use the WordNet lexical dictionary [3] to perform lemmatization, due to its common usage in the literature.

3.2 Sentence Feature Vector Representation

Different from most existing approaches which rely on term frequencies in processing text documents (cf. Sect. 2), we introduce a sentence-based feature vector representation to capture the syntactic similarities between sentences. We adopt sentences as our base extractive summarization unit, and aim to identify the most informative sentences to put in the output summary. The pseudocode for our sentence feature representation component is shown in Fig. 2. It accepts as input: a text document to be summarized, and produces as output: the set of sentence feature vectors for all sentences in the document. For each sentence s_i in the document, the algorithm builds a square matrix M_i designed to store the co-occurrence scores of all terms in s_i (cf. Fig. 2, lines 1–5). Each line k in M_i represents the context vector of term t_k, denoted as the centroid of line k. Each column ℓ represents a term t_ℓ in the context of centroid t_k. The term co-occurrence weight of t_ℓ in the context of t_k is computed according to the relative distance of t_ℓ from t_k:

$$
w_\ell^k = \begin{cases} \text{if}(\ell < k) \ \frac{\ell}{k-1} \\ \text{if}(\ell = k) \ 0 \\ \text{else} \quad 1 - \left(\frac{\ell-k-1}{m-k}\right) \end{cases} \tag{1}
$$

where m is the number of terms in the sentence, ℓ and k represent the occurrence indices of terms t_ℓ and t_k in the sequence of term tokens representing sentence s_i (cf. Linguistic

```
Algorithm Sentence_Feature_Vector_Representation
Input: D        // Text document
Output: Vs      // Set of sentence feature vectors

Begin
S = {s₁, s₂, ..., sₘ}   // set of sentences in D, each represented as a sequence of term tokens    1
Vs = φ                  // set of sentence feature vectors in D                                   2
For each sentence sᵢ in S
  Tᵢ = {t₁, t₂, ..., tₘ}                        // set of terms in sᵢ                              3
  Initialize co-occurrence matrix Mᵢ of dimensions m²   // term co-occurrence matrix              4
                                                                                                  5
  For each line k in M                                                                            6
    Consider term tₖ as sentence centroid                                                         7
    For each column ℓ in Mᵢ                     // weight of tℓ in context of tₖ                   8
      Compute wℓᵏ following formula (1)                                                            9
    Compute context(tₖ) = [w₁ᵏ, ..., wₖ₌₀ᵏ, ..., wₘᵏ]   // context vector of centroid tₖ          10
  Initialize reduced co-occurrence matrix Mᵢ* of dimension n²  // n is number of distinct terms in S  11
  For each line ℓ in Mᵢ*                                                                          12
    Compute aggregate weight wₗ = Σ∀tⱼ∈occurrence of tℓ (wⱼ,Mᵢ)                                    13
  For each column k in Mᵢ*                                                                        14
    Compute aggregate weight wₖ = Σ∀tⱼ∈occurrence of tₖ (wⱼ,Mᵢ)                                    15
  Initialize sentence feature vector Vᵢ of dimension n                                            16
  For each column k in Vᵢ                                                                         17
    Compute aggregate weight agg(tₖ,Mᵢ*) = Σ∀tⱼ∈context(tₖ) wⱼᵏ                                    18
  Normalize weights in Vᵢ          // divide by sum of aggregate weights                          19
  Add Vᵢ to Vs                     // other normalizations can be used                            20
  Return Vs                                                                                       21
End
```

Fig. 2. Pseudocode of the sentence feature vector representation component

preprocessing in Sect. 3.1). Once weights in the co-occurrence matrix are computed (lines 6–10), the algorithm aggregates the weights of identical terms occurring multiple times in the sentence (lines 11–15). It then creates a reduced feature vector including only the distinct terms as vector dimensions (lines 16–18), and aggregates the weights from all centroid context vectors (i.e., from all lines in the co-occurrence matrix) to from the output sentence feature vector (lines 19–21).

Consider the following sentence, extracted from one of our test documents (cf. Sect. 5): *A solar eclipse occurs when the Moon passes between Earth and the Sun, thereby totally or partially obscuring Earth's view of the Sun.* The tokenized representation of the sentence (following linguistic pre-processing, cf. Sect. 3.1) is:

Term indices	1	2	3	4	5	6	7	8	9	10
	solar	eclipse	occurrence	Moon	passage	Earth	Sun	obscurity	Earth	Sun

where the number of terms $m = 10$. Consider the first term *solar* as centroid, i.e., $k = 1$:

Term indices	1	2	3	4	5	6	7	8	9	10
	solar	eclipse	occurrence	Moon	passage	Earth	Sun	obscurity	Earth	Sun
Context weights	0	9/9	8/9	7/9	6/9	5/9	4/9	3/9	2/9	1/9

centroid $\ell=k$ $\ell>k$

The weight of context term *eclipse* having $\ell = 2\,(>k)$ is $1-\frac{\ell-k-1}{m-k} = 1-\frac{2-1-1}{10-1} = 1$. It is the highest weight in this context vector since it represents the closest term to the centroid. The weights decrease gradually as the terms occur farther away form the centroid, reaching minimum weight $= \frac{1}{9}$ for the last term *Sun*.

Consider the forth term *Moon* as centroid, i.e., $k = 4$:

Term indices	1	2	3	4	5	6	7	8	9	10
	solar	eclipse	occurrence	Moon	passage	Earth	Sun	obscurity	Earth	Sun
Context weights	1/3	2/3	3/3	0	6/6	5/6	4/6	3/6	2/6	1/6

$\ell<k$ centroid $\ell=k$ $\ell>k$

Here, the weight of term *eclipse* having $\ell = 2\,(<k)$ is $\frac{\ell}{k-1} = \frac{2}{4-1} = \frac{2}{3}$. The weight of term *passage* having $\ell = 5\,(>k)$ is $1-\frac{\ell-k-1}{m-k} = 1-\frac{5-4-1}{10-4} = 1$. The weights decrease gradually as the terms occur farther away from the centroid, reaching minimum weight $= \frac{1}{3}$ for the last term to the left, *solar*, and minimum weight $= \frac{1}{6}$ for the last term to the right, *Sun*. The complete co-occurrence matrix for all centroids is shown in Table 1. Table 2a shows the reduced co-occurrence where term repetitions (highlighted in color in Table 1) have been aggregated. Table 2b shows the output sentence feature vector, normalized w.r.t.[3] the sum of the aggregate scores (other normalization functions can be used, following user preferences). The latter highlights the weight of every term in

---

[3] With respect to.

the sentence, considering its relative potion w.r.t. all other terms as well as its number of repetitions in the sentence.

3.3 Sentence Clustering

Once the sentence feature vectors are computed, we utilize k-means partitional clustering to group similar sentences together, where the sentence clusters serve as seeds for extractive summarization. We adopt k-means as one of the most prominent and efficient algorithms, allowing the user to choose and control the size of the summary, where the number of clusters k represent the number of sentences in the output summary. Yet, other clustering algorithms can be used (e.g., k-medians, k-medoids, and constrained partitional clustering where the number of output clusters is chosen by the user [18]).

Table 1. Sample square co-occurrence matrix

Centroid terms	Context terms									
	solar	eclipse	occurrence	Moon	passage	Earth	Sun	obscurity	Earth	Sun
solar	0	9/9	8/9	7/9	6/9	5/9	4/9	3/9	2/9	1/9
eclipse	1/1	0	8/8	7/8	6/8	5/8	4/8	3/8	2/8	1/8
occurrence	1/2	2/2	0	7/7	6/7	5/7	4/7	3/7	2/7	1/7
Moon	1/3	2/3	3/3	0	6/6	5/6	4/6	3/6	2/6	1/6
passage	1/4	2/4	3/4	4/4	0	5/5	4/5	3/5	2/5	1/5
Earth	1/5	2/5	3/5	4/5	5/5	0	4/4	3/4	2/4	1/4
Sun	1/6	2/6	3/6	4/6	5/6	6/6	0	3/3	2/3	1/3
obscurity	1/7	2/7	3/7	4/7	5/7	6/7	7/7	0	2/2	1/2
Earth	1/8	2/8	3/8	4/8	5/8	6/8	7/8	8/8	0	1/1
Sun	1/9	2/9	3/9	4/9	5/9	6/9	7/9	8/9	9/9	0

Context vectors of all centroid terms (row group label for Table 1)

Table 2. Reduced co-occurrence matrix, and resulting sentence feature vector

a. Reduced co-occurrence matrix

Centroid terms	Aggregate context terms								Aggregate weights
	solar	eclipse	occurrence	Moon	passage	Earth	Sun	obscurity	
solar	0	9/9	8/9	7/9	6/9	5/9+2/9	4/9+1/9	3/9	9/9+…+3/9
eclipse	1/1	0	8/8	7/8	6/8	5/8+2/8	4/8+1/8	3/8	1/1+…+3/8
occurrence	1/2	2/2	0	7/7	6/7	5/7+2/7	4/7+1/7	3/7	…
Moon	1/3	2/3	3/3	0	6/6	5/6+2/6	4/6+1/6	3/6	…
passage	1/4	2/4	3/4	4/4	0	5/5+2/5	4/5+1/5	3/5	…
Earth	1/5+1/8	2/5+2/8	3/5+3/8	4/5+4/8	5/5+5/8	0+6/8	4/4+7/8	3/4+8/8	…
Sun	1/6+1/9	2/6+2/9	3/6+3/9	4/6+4/9	5/6+5/9	6/6+6/9	0+7/9	3/3+8/9	…
obscurity	1/7	2/7	3/7	4/7	5/7	6/7+2/2	7/7+1/2	0	1/7+…+0

Reduced context vectors (row group label for Table 2a)

b. Sentence feature vector

	solar	eclipse	occurrence	Moon	passage	Earth	Sun	obscurity
Normalized weights	0.104	0.109	0.109	0.109	0.106	0.184	0.169	0.109

/ Σ

In brief, k-means [19, 20] attempts to divide data objects (e.g., sentences in our case) into non-overlapping subsets, i.e., the clusters, such that each sentence is in exactly one cluster, by maximizing intra-cluster similarity and minimizing inter-cluster similarity. It first chooses (randomly, or heuristically) k sentences as initial centroids, and computes one cluster around each centroid by associating sentences to the clusters sharing minimum distances with their centroids. The centroids are re-computed recursively, and the clusters are adjusted around them, until reaching convergence where the cluster centroids stabilize.

We utilize the cosine measure to compute sentence feature vector similarity, yet other similarity measures can be used (e.g., Jaccard, Dice, Euclidian, e.g., [18, 21]).

3.4 Summary Building

Our extractive summary building process consists in choosing the best representative sentence from each sentence cluster, and combining them together to form the output summary. Here, we consider multiple approaches (other approaches can be added according to the user's needs): i) *Longest Sentence* (LS) – from each cluster, the system extracts the sentence with the maximum number of terms (based on the assumption that the longest sentence is likely the most elaborate/descriptive of the cluster), ii) *Shortest Sentence* (SS) – from each cluster, the system extracts the sentence with the minimum number of terms (based on the assumption that the user is interested in the most concise information from the cluster), and iii) *Most Similar Sentence* (MSS) – from each cluster, the system evaluates the pairwise sentence similarities and chooses the sentence with the highest average similarity with all others (based on the assumption the sentence which shares the maximum amount of similarity would best reflect their information content). Once the representative sentences are identified, the system combines them to form the output summary, by placing them one after the other according to their relative order in the original text, to preserve logical content ordering.

The time complexity of our solution comes down to the complexity of the sentence clustering algorithm, requiring $O(N \times k \times |s|)$ where N is the size of the input text document in number of sentences, k is the number of clusters (i.e., the size of the output summary in number of sentences), and $|s|$ is the maximum size of a sentence in number of terms. It simplifies to $O(N \times |s|)$ since $k \ll N$.

4 Experimental Evaluation

4.1 Prototype Implementation

We have implemented our topical organization solution using the Python programming language, to test and evaluate its performance. We perform the data serialization using Python's *PDFToText* library. We remove the stop-words and punctuations using Python's *NLTK* library. Later we convert the remaining text to lower-case and we stem each word using the *NLTK Porter Stemmer*. We utilize the WordNet API to perform lemmatization.

4.2 Experimental Metrics

We make use of the compression ratio (*CR*) metric [22] to evaluate the compactness of the produced summaries, compared with the size of the initial text document:

$$CR = \frac{Length\ of\ Symmary}{Length\ of\ Full\ Text} \in [0, 1] \qquad (2)$$

We also make use of the precision (*PR*) and recall (*R*) metrics [18, 23] to evaluate the quality of the system produced summaries w.r.t. their human generated counterparts. High *precision* denotes that sentences are actually in the right summary. High *recall* means that very few sentences are not in the summary where they should have been. High *precision* and *recall*, and thus high *F-value* indicates excellent summarization quality.

4.3 Experimental Data

We collected 18 news articles from the renounced online newspapers, consisting of an average 42 sentences and 855 terms per article[4]. We grouped the articles according to the desired (output) summary size, and requested the assistance of three human testers (graduates students) to generate the corresponding summaries. The testers were provided each an excel sheet that includes a header describing briefly the research and the experiment, and a body that includes an array containing as columns: i) reference number of source texts (articles), ii) summary size in number of sentences (e.g. $k = 2, k = 4, ..., k = 20$), and iii) a link to the source articles. The testers completed the summarization tasks together, and compiled the summaries into the reference dataset in our experiments.

4.4 Experimental Results

Figure 3 shows the precision, recall, and f-value results obtained for our experimental dataset, considering the size of the input text, the size of the produced summary, the compression ratio, and each of the three representative options (longest sentence - LS, shortest sentence - SS, and most similar sentence - MSS).

On the one hand, results in Fig. 3 show that the quality levels of our summarization solution, considering all three precision, recall, and f-value metrics, remain more or less steady w.r.t. the varying size of the input text documents, the varying size of the output summaries, and the varying compression ratio. This shows that our solution produces consistent results regardless of text size, summary size, and compression ratio. On the other hand, results show that summarization quality is affected by the cluster representative selection approach, where the longest sentence (LS) method consistently produces the best quality levels (with average f-value = 0.51), followed by the most similar sentence (MSS) method (with average f-value = 0.33), whereas the shortest sentence (SS) method usually ranks last (with average f-value = 0.14). Following our discussions with human testers, they usually consider the longest sentences to be the most informative and thus favor their presence in extractive summaries. Results also show that testers sometimes select sentences that are most related (i.e., similar) to others in the text. Nonetheless, testers very rarely allocate the shortest sentences in the summary, since they consider short sentences to be the least informative in the text.

---

[4] Available online: https://bit.ly/3FiaMLu.

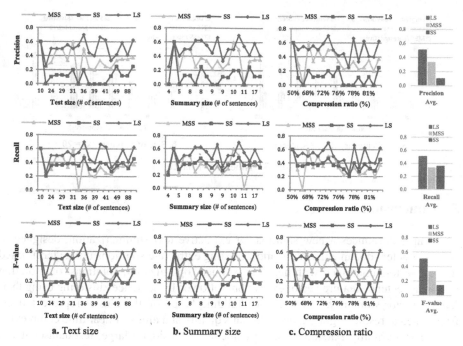

Fig. 3. Precision, recall, and f-value results obtained for our experimental dataset

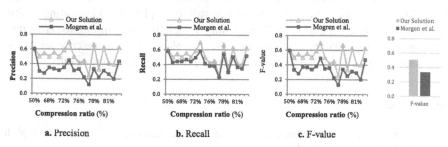

Fig. 4. Comparative evaluation

We compare our approach with an online extractive summarization solution by Morgen et al. [24, 25]. Figure 4 shows the best results produced by our approach, i.e., considering the longest sentence (LS) method. Results show that our approach produces summaries which are more accurate, highlighting a higher average precision level = 0.51 (compared with 0.32 for Morgen et al.). Both solutions produce comparable recall levels, with a slight improvement in favor of our solution. Average f-value results highlight the quality of our approach, producing average 0.51 (compared with 0.34 for Morgen et al.).

Time performance results in Fig. 5 highlight the polynomial (almost linear) complexity of our approach, which comes down to the sentence clustering algorithm.

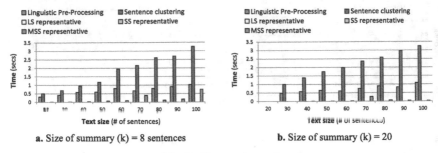

a. Size of summary (k) = 8 sentences   b. Size of summary (k) = 20

Fig. 5. Time performance

5 Conclusion

This paper introduces a new solution for extractive text summarization, designed to process flat text documents. While most existing extractive solutions utilize term-based feature vectors combined with heuristic or linear optimization solutions, our solution performs sentence feature vector extraction, followed by sentence clustering using their feature vector similarity, and summary building based on the cluster representatives. Experimental results highlight our solution's quality and almost linear time complexity.

We are currently extending our experiments to consider a larger test dataset and to compare with multiple existing solutions. We are also investigating the selection process in the k-means algorithm and how to better consider the algorithm's convergence threshold in fine-tuning the summarization result, e.g., [19, 26]. Generating multiple representatives for each cluster and performing associative rule mining to select representatives [11] are also ongoing directions. We also plan to investigate encoder-decoder architectures, e.g., [27, 28], which have recently produced promising results.

References

1. Khan, A., Salim, N.: A review on abstractive summarization methods. J. Theor. Appl. Inf. Technol. **59**(1), 64–72 (2014)
2. Tanaka, H., et al.: Syntax-driven sentence revision for broadcast news summarization. In: Workshop on Language Generation and Summarization, pp. 39–47 (2009)
3. Miller, G., Fellbaum, C.: WordNet then and now. Lang. Resour. Eval. **41**(2), 209–214 (2007)
4. Getahun, F., Tekli, J., Chbeir, R., Viviani, M., Yetongnon, K.: Relating RSS news/items. In: Gaedke, M., Grossniklaus, M., Díaz, O. (eds.) ICWE 2009. LNCS, vol. 5648, pp. 442–452. Springer, Heidelberg (2009). https://doi.org/10.1007/978-3-642-02818-2_36
5. Genest, P., Lapalme, G.: Fully abstractive approach to guided summarization. In: Annual Meeting of the Association for Computational Linguistics (ACL), pp. 354–358 (2012)
6. Lau, R., et al.: Toward a fuzzy domain ontology extraction method for adaptive e-learning. IEEE Trans. Knowl. Data Eng. **21**(6), 800–813 (2009)
7. Özates S., et al.: Sentence similarity based on dependency tree kernels for multi-document summarization. In: International Conference on Language Resources and Evaluation (LREC) (2016)
8. Abdel-Salam, S., Rafea, A.: Performance study on extractive text summarization using BERT models. Information **13**(2), 67 (2022)

9. Cao, S., Yang, Y.: DP-BERT: dynamic programming BERT for text summarization. In: Fang, Lu., Chen, Y., Zhai, G., Wang, J., Wang, R., Dong, W. (eds.) Artificial Intelligence: First CAAI International Conference, CICAI 2021, Hangzhou, China, June 5–6, 2021, Proceedings, Part II, pp. 285–296. Springer, Cham (2021). https://doi.org/10.1007/978-3-030-93049-3_24

10. Aaditya M., et al.: Layer freezing for regulating fine-tuning in BERT for extractive text summarization. In: Pacific Asia Conference on Information Systems (PACIS), p. 182 (2021)

11. Haraty, R., Nasrallah, R.: Indexing Arabic texts using association rule data mining. Librar. Hi Tech **37**(1), 101–117 (2019)

12. Mansour, N., et al.: An auto-indexing method for Arabic text. Inf. Process. Manage. J. **44**(4), 1538–1545 (2008)

13. Mojrian, M., Mirroshandel, S.A.: A novel extractive multi-document text summarization system using quantum-inspired genetic algorithm: MTSQIGA. Exp. Syst. Appl. **171**, 114555 (2021)

14. Srivastava, A.K., Pandey, D., Agarwal, A.: Extractive multi-document text summarization using dolphin swarm optimization approach. Multimedia Tools Appl. **80**(7), 11273–11290 (2021)

15. Popescu, M., et al.: A highly scalable method for extractive text summarization using convex optimization. Symmetry **13**(10), 1824 (2021)

16. Kruengkrai, C., Jaruskulchai, C.: Generic text summarization using local and global properties of sentences. In: Web Intelligence, pp. 201–206 (2003)

17. Rani, R., Lobiyal, D.: An extractive text summarization approach using tagged-LDA based topic modeling. Multim. Tools Appl. **80**(3), 3275–3305 (2021)

18. Tekli, J.: An overview of cluster-based image search result organization: background, techniques, and ongoing challenges. Knowl. Inf. Syst. **64**(3), 589–642 (2022)

19. Haraty, R.A., Dimishkieh, M., Masud, M.: An enhanced k-means clustering algorithm for pattern discovery in healthcare data. Int. J. Distrib. Sensor Netw. **11**(6), 615740:1-615740:11 (2015)

20. Lloyd, S.: Least squares quantization in PCM. IEEE Trans. Info. Theor. **28**(2), 129–137 (1982)

21. Haraty, R., Hamdoun, M.: Iterative querying in web-based database applications. In: ACM Symposium on Applied Computing (SAC), pp. 458–462 (2002)

22. Mridha, M., et al.: A survey of automatic text summarization: progress, process and challenges. IEEE Access **2021**(9), 156043–156070 (2021)

23. Tekli, J., Chbeir, R., Yetongnon, K.: Structural similarity evaluation between XML documents and DTDs. In: Benatallah, B., Casati, F., Georgakopoulos, D., Bartolini, C., Sadiq, W., Godart, C. (eds.) WISE 2007. LNCS, vol. 4831, pp. 196–211. Springer, Heidelberg (2007). https://doi.org/10.1007/978-3-540-76993-4_17

24. Mogren, O., et al.: Extractive summarization by aggregating multiple similarities. In: Recent Advances in Natural Language Processing (RANLP), pp. 451–457 (2015)

25. Kågebäck, M., et al.: Extractive summarization using continuous vector space models. In: Workshop on Continuous Vector Space Models and their Compositionality (CVSC), pp. 31–39 (2014)

26. Tekli, J., Al Bouna, B., Bou Issa, Y., Kamradt, M., Haraty, R.: (k, l)-clustering for transactional data streams anonymization. In: Su, C., Kikuchi, H. (eds.) ISPEC 2018. LNCS, vol. 11125, pp. 544–556. Springer, Cham (2018). https://doi.org/10.1007/978-3-319-99807-7_35

27. Maziad, H., Rammouz, J.-A., Asmar, B.E., Tekli, J.: Preprocessing techniques for end-to-end trainable RNN-based conversational system. In: Brambilla, M., Chbeir, R., Frasincar, F., Manolescu, I. (eds.) ICWE 2021. LNCS, vol. 12706, pp. 255–270. Springer, Cham (2021). https://doi.org/10.1007/978-3-030-74296-6_20

28. Chakar, J., Sobbahi, R.A., Tekli, J.: Depthwise separable convolutions and variational dropout within the context of YOLOv3. In: Bebis, G., et al. (eds.) ISVC 2020. LNCS, vol. 12509, pp. 107–120. Springer, Cham (2020). https://doi.org/10.1007/978-3-030-64556-4_9

Experimental Comparison of Theory-Guided Deep Learning Algorithms

Simone Monaco$^{(\boxtimes)}$ and Daniele Apiletti

Department of Control and Computer Engineering, Politecnico di Torino,
Turin, Italy
{simone.monaco,daniele.apiletti}@polito.it

Abstract. The enrichment of machine learning models with domain knowledge has a growing impact on modern engineering and physics problems. This trend stems from the fact that the rise of deep learning algorithms is closely associated with an increasing demand for data that is not acceptable or available in many use cases. In this context, the incorporation of physical knowledge or a-priori constraints has been shown to be beneficial in many tasks. On the other hand, this collection of approaches is context-specific, and it is difficult to generalize them to new problems. In this paper, we experimentally compare some of the most widely used theory injection strategies to perform a systematic analysis of their advantages. Selected state-of-the-art algorithms have been reproduced for different use cases to evaluate their effectiveness with smaller training data and to discuss how the underlined strategies can fit into new application contexts.

Keywords: Theory-based machine learning · Domain knowledge in data-driven modeling · Physics-informed neural networks

1 Introduction

Machine learning and deep learning (DL) are obtaining extremely promising results in various applications, from research to industry, extending to many aspects of our everyday life.

We are currently living under a data deluge, with petabytes of data being produced every year in the world. This is precisely the ideal terrain for the growth of a great variety of techniques capable of extracting patterns, or more generally *learning* knowledge from data. The Data Science paradigm stems from the idea of producing knowledge gained through this tremendous amount of data samples [9]. However, scientific progress is historically related to the process of generation and validation of theories, whose observations should undergo. This opposite paradigm has been known since the 17th century as the Scientific Method. In the Big Data era, samples are continuously collected without a specific theoretical basis. This can lead, on the one hand, to opportunities to

S. Chiusano et al. (Eds.): ADBIS 2022, CCIS 1652, pp. 256–265, 2022.
https://doi.org/10.1007/978-3-031-15743-1_24

produce new knowledge discovery frameworks in many applications [2,6], but, on the other hand, to a systematic way of neglecting scientific theories. In fact, the success of the black-box applications of data science can be seen as "the end of the theory" [1] since they generally do not need scientific prior hypotheses.

Between these two extremes, in recent years new strategies have been developed to take advantage of both sides. All techniques that attempt to incorporate theory-based (e.g., physical laws) domain knowledge into otherwise blind data-driven models can be collected under the paradigm of Theory-guided Data Science (TGDS) [12].

In recent years, some surveys [12,21,24] tried to collect the different techniques blending theory-guided and data-driven modeling. The main difficulty is that each state-of-the-art solution is by design highly domain specific. Then it is very hard to compare the contribution that a selected strategy can apport when solving a specific problem. The contribution of this paper is to perform an experimental comparison by highlighting the main building blocks that different state-of-the-art solutions provide in two popular application contexts. The contribution is then twofold:

1. Experimentally reproducing, in a comparable way, different state-of-the-art algorithms applied to two scientific contexts.
2. Providing a unified formalism that groups the theory-driven elements proposed in the state-of-the-art, and their corresponding experimental evaluation to measure their contribution to the final model performance.

The paper is organized as follows. Section 2 presents the related works, among which the solutions to be experimentally compared have been selected. Section 3 describes the selected solutions, highlighting differences in the context and in the adopted strategies. Section 4 shows the experimental results of the comparative analysis, measuring the effects of each theory-guided strategy on the overall model performance. Finally, Sect. 5 draws conclusions and presents future works.

2 Related Works

The TGDS paradigm has been successfully applied in various scientific domains, from climate science [11,17], turbulence modeling [15], biological sciences [18], and quantum chemistry [16].

Von Rueden et al. [21] categorized each approach according to (i) the source of the integrated knowledge, (ii) the representation of the knowledge, and (iii) its integration, i.e., where it is integrated into the learning pipeline.

2.1 Knowledge Source

The *Source* of the knowledge and whether it is formalized into a theoretical set of equations or rules can differentiate the kind of source in one of the following cases.

Scientific Knowledge. Typically formalized and validated through scientific experiments and/or analytical demonstrations. In this category, all subjects of science, technology, engineering, and mathematics can be considered.

World Knowledge. General information from everyday life, without formal validation, can be considered an important element in enriching the learning procedure. This kind of knowledge is generally more intuitive and refers to human reasoning on the perceived world, for instance, the fact that a cat has two ears and can meow. Within this class, we can also consider linguistics, with syntax and semantics as examples.

2.2 Knowledge Representation

The *Knowledge Representation* category effectively corresponds to the formalized element of the prior information. Depending on the knowledge available for each specific task, different representations can be adopted. The most widespread alternatives, and the more interesting for this discussion, are briefly reported in the following.

Equations. Being them differential or algebraic, the final solution could follow a partially known behavior or be subject to some constraints, which can be formalized into equations. Constraints are generally associated with algebraic equations or inequalities. Notable examples can be the energy-mass equivalence (i.e., $E = mc^2$) or the mass invariance reflected by the Minkowski metric, which, for example, has been integrated through a *Lorentz layer* in [5]. Analogous reasoning can be performed for differential equations, which rule dynamical behavior of state variables, inputs, and outputs. This background is sometimes known but eventually unfeasible, partially known but not completely representative of the real solution, or completely unknown [23]. For these scenarios, machine learning algorithms can be applied to solve differential equations as in [20], to learn the residual dynamics with known priors of the behavior of the solution as in [22], or finally to learn the spatiotemporal dynamics itself as in [13].

Domain-Specific Invariances. Input data might have a peculiar invariance due to their structure, i.e., images to be classified can preserve some properties even with translations or rotations. Other kinds of data can be permutationally invariant or even time-invariant, or subject to periodicity. For each of these cases, some particular model architectures can better express these features [3].

2.3 Knowledge Integration

Concerning the *Knowledge integration* into machine learning algorithms, it can be applied either at the beginning of the pipeline, e.g., within the training dataset, or in the middle, defining customized architectures for learning strategies, or at the end, driving the model outcomes.

The classic approach of embedding prior information into data is provided by feature engineering, in which secondary data is produced out of the sampled one in order to emphasize something already known in the specific domain. One known alternative is to add synthetic information obtained from simulated data, as in [7], hence having a final algorithm devoted to finding the residuals of such approximated solutions.

Prior information can then be introduced constraining the learning procedure by adding a physical-aware loss function to the ordinary supervised ones. This general strategy can be summarized as follows [24]:

$$\mathcal{L} = \mathcal{L}_{\text{SUP}}(Y_{\text{TRUE}}, Y_{\text{PRED}}) + \gamma \mathcal{L}_{\text{PHY}}(Y_{\text{PRED}}) + \lambda R(W); \tag{1}$$

where $\mathcal{L}_{\text{SUP}}$ is the measure of the supervised error (e.g., MSE, cross-entropy), R is an eventual additional regularization term to limit the complexity of the model, and $\mathcal{L}_{\text{PHY}}$ is the theory-guided contribution. The last term can either incorporate algebraic, differential, or logical equations. In this scenario, the work of Willard et al. in [24], when predicting the lake temperature over the variation of the depth, introduces a penalty for predictions leading to water density not respecting the theory-bounded increase with depth.

Beucler et al. [4] enforced conservation laws instead in the context of climate modeling. Starting from these constraints, they both apply them as *soft constraints* in the loss function, and as *hard constraints* reducing the cardinality of the prediction from optimizable neural networks and calculating the residual ones through fixed layers. An analogous idea applied to AC optimal power flow can be found in [10].

A more challenging strategy to incorporate physical information into DL models can be provided in the design itself of the model architectures or, generally speaking, the hypothesis set [21]. State of the art DL can be in fact designed to naturally express *relational inductive biases* [3] even before the training procedure. In this sense, convolutional layers are naturally devoted to dealing with spatial invariances on images, while recurrent layers can track sequential features, e.g., time series. The DL component capable of operating this reasoning on arbitrary relational structures is the Graph Neural Network (GNN). Graph layers can be the counterpart of the CNN to graph-structured data [14] or even enhance representation via knowledge graphs in imaging or natural language processing [19].

In the context of TGDS, we selected our top picks for this experimental comparison among those works providing both code and datasets and being representative of the aforementioned strategies. They are collected in the following:

1. Enforcing domain-specific constraints within the *loss function* (**LF**)
2. Reduce the network output space in order to make the solution exactly fulfill the known *hard constraints* (**HC**)
3. Incorporate the semantics of the problem to build a use-case specific *model architecture* in order to express the prior knowledge (**MA**)

3 Datasets and Methods

For each of the proposed strategies, we experimentally compare state-of-the-art solutions in two application contexts defined by a specific dataset each, whose main characteristics, e.g., cardinality, and size, are reported in Table 1.

Table 1. Dataset and characterization of the use cases under comparison.

	Cardinality	Dimensions	Disk space
Lake temperature	76050	11	12 Mb
Convective movements	84 M	304	171 Gb

3.1   Lake Temperature

The first study concerns the problem of modeling the water temperature in a lake on the variation of depth and weather conditions. We selected two works of Daw et al., first introducing the Physics-guided Neural Network (PGNN) [7], and later addressing the same task with a recurrent neural network named Physics-Guided Architecture (PGA-LSTM from now on) [8]. We compared the proposed architectures on the same dataset, provided in [8], whereas previous works were conducted on different datasets. The available data are a collection of lake Mendota weather measurements in Wisconsin (USA), between April 2009 and December 2017. Extra information is then provided by simulated prediction of the temperature.

PGNN consists of a multilayer perceptron (MLP) trained with a loss function with a physical term, following the approach of Eq. 1. Such a loss formalizes the fact that water density monotonically increases with depth. Given the known analytic relation between density ρ and temperature t, the difference between consecutive densities (i.e., from depth i to $i + 1$), which are functions of the network predictions, have to be negative to be consistent with this property.

We can then classify the PGNN algorithm according to the source, representation, and integration criteria as follows: source is scientific knowledge, representation is equations and simulation results, and integration is obtained by means of enrichment of training data and constraints in the learning procedure. The physical injection is provided here according to the **LF** strategy.

The same physical information, but with a different strategy, is introduced in the PGA-LSTM approach. Its architecture consists of an LSTM-based autoencoder extracting temporal features from input data and exploiting the recurrence in the time dimension. The output of this first network is then appended and passed through a second recurrent network, now working on the depth dimension, specifically designed to preserve the monotonicity of the density over depth. This has been done by giving an extra recurrent connection within the base-LSTM architecture between the density, enforced as a physical intermediate in a loss function penalizing the mean squared error between the ground truth and the predicted value of both temperature and density. The PGA-LSTM architecture, while having the same source of knowledge as the PGNN, represents the knowledge through spatial and time invariances instead of equations, and integrates the knowledge into the model architecture itself. Hence, we classify the PGA-LSTM as an example of the **MA** strategy.

3.2 Convective Movements in Climate Modeling

The second experimental comparison context is the application of neural networks to climate modeling as provided in [4]. The objective of the network is to predict the rate at which heat and water are redistributed due to convective movements. The local climate is described through a set of over 300 variables both regarding thermodynamic properties over varying level profiles, and large-scale conditions not dependent on height. Then the network goal is to predict the associated time-tendencies from convection and additional variables from system conservation laws. The whole data is the simulated climate for 2 years using a parametrized atmosphere model. Among the many variables, the authors pointed out 4 quantity conservations, namely: column-integrated energy, mass, long-wave radiation, and short-wave radiation. These physical laws are translated into equations relating input and output. Since they are derived to be linear relations, they can be inserted into a matrix of coefficients C producing zero when multiplied by the vector of inputs and outputs.

These constraints are then added either as *soft constrains*, i.e., the squared norm of $\mathcal{H}$ as an additional loss function term, leading to an approach named LCNet; or as *hard constraints*, generating an ACnet. The latter strategy is carried out by developing an MLP with trainable parameters and a number of output features equal to the cardinality of $\bar{y}$ minus the number of constraints, concatenating the others obtained in a deterministic way.

For both LCNet and ACNet, the information is provided by *scientific knowledge* as source and *equations* as representation. Instead, we can identify the integration to be provided by the loss function for the former, and as the *hypotesis set* for the latter. For this reason, the two strategies will be compared as representatives of the solutions applying the **LF** and **HC**, respectively.

4 Experimental Results

Since the main contribution of a theory-guided algorithm is its capability of obtaining a reliable and more physical-consistent solution with reduced amounts of data, we experimentally evaluated all the strategies by measuring their predictive power over variations of the training dataset size. Each model has been trained on a subset of the data, starting from the initial timestamp. The test set, instead, has been fixed as a separate subset. The whole process has been repeated 10 times for each experiment, averaging the results to reduce uncertainty.

A limit of the current work is due to the seasonal effect of the starting date: we plan to address this issue in future work. Currently, all comparisons have been made under the assumption that the potential advantages of choosing different periods of the year for training are common to all models.

Both the code and the datasets are available upon request. However, we plan to make them directly available to the public in the near future.

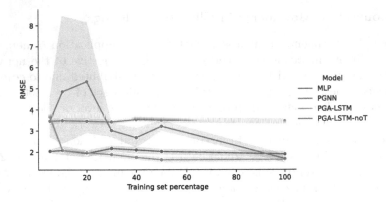

Fig. 1. Lake temperature, RMSE of the models varying the training set size.

Lake Temperature. In Fig. 1 we compare the performance over varying training set percentages of the PGNN algorithm, its counterpart without the physical loss function (labeled as MLP), and the PGA-LSTM, besides an additional experiment for PGA-LSTM without the temporal features extracted by the encoder (labeled PGA-LSTM-noT). The first remarkable observation is that the physical loss function of the PGNN does not provide a beneficial contribution to any subset of the training set. On the other hand, the plain MLP obtains a valuable performance with respect to other models, being the second best after PGA-LSTM, and even being on par for small percentages of the training set (10–20%). This suggests that, for this simple problem, the theory-guided contribution might be limited. The use of the encoder to extract temporal features in the PGA-LSTM, instead, gives a large contribution in improving the results with respect to the PGA-LSTM-noT, in particular for small training sizes.

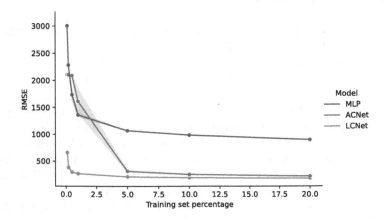

Fig. 2. Convective modeling, RMSE of the models varying the training set size.

Convective Movements in Climate Modeling. In Fig. 2 we compare a purely data-driven MLP architecture and the two theory-guided solutions LCNet and ACNet. Due to the much larger dimension of the dataset (171 Gb vs few Mb of the other use cases), experiments are reported for 1–20% of the training set, with a clear asymptotic trend suggesting unsurprising behaviors for larger sizes. With the largest training subset (20%), the MLP obtains an RMSE of 888 W^2/m^4, around 4 times higher than both the theory-guided approache, with LCNet and ACNet at 176 and 208 W^2/m^4, respectively.

The **HC** strategy (ACNet) strongly depends on the training set size: when the training set increases from 1 to 5% of the complete dataset, the ACNet RMSE abruptly decreases from 1612 W^2/m^4 to 310 W^2/m^4. Also the MLP and the LCNet exhibit the same trend for smaller training set sizes, below 1%. In this use case, the LCNet with the **LF** approach is the best performing solution, and its theory-guided injection yields a large improvement over the pure data-driven MLP for all training sizes, and in particular reaching a top performance asymptote by being trained on 1% of the dataset size only.

5   Conclusions

The paper aims at providing an experimental comparison of selected state-of-the-art theory-guided approaches to evaluate the contribution of the different injection techniques of a priori knowledge. We classified the injection of knowledge into loss function (LF), hard constraints (HC), and model architecture (MA) building blocks. Among the two selected use cases, for which datasets and code were made available by the authors, we compared the contribution of different theory-guided injection techniques.

Even though the results are preliminary with respect to the width of the field, we observed that an architecture specifically designed for a physical phenomenon (MA) brings better performance even with smaller dataset sizes, with higher benefits for more complex problems. However, the theory-guided injection for less complex problems is not always beneficial, as in the Lake temperature use case.

Enforcing hard constraints (HC) is shown to be effective in the convective movement use case, with comparable results with respect to the approach with a soft loss function penalty. Nevertheless, it is difficult to be applied, since the strict equality constraints between input and outputs are not necessarily present in all use cases. In literature [4] the extension of this approach to inequality constraints, as upper bounds in parts of the solution, is proposed. We plan to investigate this possibility or enforce constraints even in situations in which they are not strict, as in our third use case.

Finally, the domain-driven loss function (LF) seems the most promising one. Although it does not obtain when kept alone the best results in all the comparisons, it is more easily applicable to most contexts. In facts, it is in both the proposed use cases of the paper, and eventually, it can be integrated with other strategies.

In future work, we plan to extend the experimental comparison to new use cases and possibly provide more strategies for each use case, with the goal to identify suggestions on the most promising theory-guided approaches for different classes of scientific problems.

Acknowledgment. The research leading to these results has been partially supported by the SmartData@PoliTO center for Big Data and Machine Learning technologies.

References

1. Anderson, C.: The end of theory: the data deluge makes the scientific method obsolete. Wired Mag. **16**(7), 7–16 (2008)
2. Baldi, P., Brunak, S.: Bioinformatics: the Machine Learning Approach. MIT Press, Cambridge (2001)
3. Battaglia, P.W., et al.: Relational inductive biases, deep learning, and graph networks. arXiv:1806.01261 [cs, stat], October 2018
4. Beucler, T., Pritchard, M., Rasp, S., Ott, J., Baldi, P., Gentine, P.: Enforcing analytic constraints in neural networks emulating physical systems. Phys. Rev. Lett. **126**(9), 098302 (2021)
5. Butter, A., Kasieczka, G., Plehn, T., Russell, M.: Deep-learned top tagging with a Lorentz layer. SciPost Phys. **5**(3), 028 (2018)
6. Castelvecchi, D., et al.: Artificial intelligence called in to tackle LHC data deluge. Nature **528**(7580), 18–19 (2015)
7. Daw, A., Karpatne, A., Watkins, W., Read, J., Kumar, V.: Physics-guided neural networks (PGNN): an application in Lake Temperature Modeling. arXiv:1710.11431 [physics, stat] (2017)
8. Daw, A., Thomas, R.Q., Carey, C.C., Read, J.S., Appling, A.P., Karpatne, A.: Physics-guided architecture (PGA) of neural networks for quantifying uncertainty in lake temperature modeling. arXiv:1911.02682 [physics, stat], November 2019
9. Dhar, V.: Data science and prediction. Commun. ACM **56**(12), 64–73 (2013)
10. Donti, P.L., Rolnick, D., Kolter, J.Z.: DC3: a learning method for optimization with hard constraints. ICLR 21 + arXiv:2104.12225 [cs, math, stat], April 2021
11. Faghmous, J.H., Kumar, V.: A big data guide to understanding climate change: the case for theory-guided data science. Big Data **2**(3), 155–163 (2014)
12. Karpatne, A., et al.: Theory-guided data science: a new paradigm for scientific discovery from data. IEEE Trans. Knowl. Data Eng. **29**(10), 2318–2331 (2017). https://doi.org/10.1109/TKDE.2017.2720168, arXiv:1612.08544
13. Kashinath, K., et al.: Physics-informed machine learning: case studies for weather and climate modelling. Phil. Trans. R. Soc. A **379**(2194), 20200093 (2021)
14. Liang, X., Hu, Z., Zhang, H., Lin, L., Xing, E.P.: Symbolic graph reasoning meets convolutions. In: Advances in Neural Information Processing Systems, vol. 31 (2018)
15. Mohan, A.T., Gaitonde, D.V.: A deep learning based approach to reduced order modeling for turbulent flow control using LSTM neural networks. arXiv preprint arXiv:1804.09269 (2018)
16. Muralidhar, N., et al.: PhyNet: physics guided neural networks for particle drag force prediction in assembly. In: Proceedings of the 2020 SIAM International Conference on Data Mining, pp. 559–567. SIAM (2020)

17. O'Gorman, P.A., Dwyer, J.G.: Using machine learning to parameterize moist convection: potential for modeling of climate, climate change, and extreme events. J. Adv. Model. Earth Syst. **10**(10), 2548–2563 (2018)
18. Peng, G.C., et al.: Multiscale modeling meets machine learning: what can we learn? Arch. Comput. Methods Eng. **28**(3), 1017–1037 (2021)
19. Peters, M.E., et al.: Knowledge enhanced contextual word representations. arXiv preprint arXiv:1909.04164 (2019)
20. Raissi, M., Perdikaris, P., Karniadakis, G.E.: Physics informed deep learning (Part I): data-driven solutions of nonlinear partial differential equations. arXiv:1711.10561 [cs, math, stat], November 2017. Version: 1
21. von Rueden, L., et al.: Informed machine learning - a taxonomy and survey of integrating knowledge into learning systems. IEEE Trans. Knowl. Data Eng. 1 (2021). https://doi.org/10.1109/TKDE.2021.3079836, arXiv:1903.12394
22. Seo, S., Liu, Y.: Differentiable physics-informed graph networks. Technical report, February 2019. https://ui.adsabs.harvard.edu/abs/2019arXiv190202950S, Publication Title: arXiv e-prints ADS Bibcode: 2019arXiv190202950S Type: Article
23. Wang, R., Yu, R.: Physics-guided deep learning for dynamical systems: a survey. arXiv:2107.01272 [cs], March 2022
24. Willard, J., Jia, X., Xu, S., Steinbach, M., Kumar, V.: Integrating scientific knowledge with machine learning for engineering and environmental systems. arXiv:2003.04919 [physics, stat], March 2022

Hyper-parameters Tuning of Artificial Neural Networks: An Application in the Field of Recommender Systems

Vaios Stergiopoulos[1]([✉])[iD], Michael Vassilakopoulos[1][iD], Eleni Tousidou[1][iD], and Antonio Corral[2][iD]

[1] Data Structuring and Engineering Laboratory, Department of Electrical and Computer Engineering, University of Thessaly, Volos, Greece
{vstergiop,mvasilako,etousido}@uth.gr
[2] Department of Informatics, University of Almeria, Almeria, Spain
acorral@ual.es

Abstract. In this work, we carry out the hyper-parameters tuning of a Machine Learning (ML) Recommender Systems (RS) which utilizes an Artificial Neural Network (ANN), called CATA++. We have performed tuning of the activation function, weight initialization and training epochs of CATA++ in order to improve both training and performance. During the experiments, a variety of state-of-the-art activation functions have been tested: ReLU, LeakyReLU, ELU, SineReLU, GELU, Mish, Swish and Flatten-T Swish. Additionally, various weight initializers have been tested, such as: XavierGlorot, Orthogonal, He, Lecun. Moreover, we ran experiments with different epochs number from 10 to 150. We have used data from CiteULike and AMiner Citation Network. The recorded metrics (Recall, nDCG) indicate that hyper-parameters tuning can reduce notably the necessary training time, while the recommendation performance is significantly improved (up to +44.2% Recall).

Keywords: Hyper-parameters tuning · Neural networks · Activation function · Weight initialization · Training epochs · Recommender systems

1 Introduction

Recently, hyper-parameters (HP) tuning of Deep Neural Networks (DNN) has emerged as an important topic. DNN performance, however, is known to be highly sensitive to the HP setting.

The activation function is an important component of ANN because they turn an otherwise linear classifier into a non-linear one, which has proven key to the high performances witnessed across a wide range of tasks in recent years. While different activation functions seem equivalent on a theoretical level, they often show very diverse behavior in practice. Moreover, they are characterized by a variety of properties, such as ones relating to their derivatives, monotonicity,

© Springer Nature Switzerland AG 2022
S. Chiusano et al. (Eds.): ADBIS 2022, CCIS 1652, pp. 266–276, 2022.
https://doi.org/10.1007/978-3-031-15743-1_25

and whether their range is finite or not [18]. A proper initialization of the weights in an ANN is critical to its convergence [19]. The training epochs optimization mainly consists of two problems, namely under-fitting and over-fitting.

In this work, we aim to optimize the above-mentioned HP of the ANN utilized in CATA++: A Collaborative Dual Attentive Auto-encoder Method for Recommending Scientific Articles [1]. A variety of state-of-the-art activation functions and weight initializers have been tested with different numbers of training epochs in the range [10,150].

The remainder of this paper is organized in the following order. Firstly, all related work is presented in Sect. 2. Secondly, the essential preliminaries and theoretical background are explained in Sect. 3. Next, our experimental results are demonstrated thoroughly in Sect. 4. In Sect. 5, we discuss about the experimental results and the discovered good practices for HP tuning. Lastly, in Sect. 6 we conclude this work and discuss related future work directions.

2   Related Work

There are numerous scientific publications and ongoing research in the subject of HP tuning in order to improve ANNs performance. The aim of HP optimization is to choose the HP values that return the best results in the validation phase.

To begin with, Alfarhood and Cheng [1] introduce a Collaborative Dual Attentive Autoencoder (CATA++) RS method that utilizes an item's content and learns its latent space via two parallel autoencoders. They employ the attention mechanism in the middle of the autoencoders to capture the most significant segments of contextual information, which leads to a better representation of the items in the latent space. They have utilized Matrix Factorization, Collaborative Filtering and an ANN in order to improve the recommendation performance (we did not modify the structure of CATA++). In this work we have chosen to use CATA++ RS for our experiments, as it outperformed other RS, as described in [1] and in experiments we have run (due to space limitation cannot include them here). More details on work related to this paper are included in [4].

The lack of optimization for some HP in CATA++ motivated us to optimize its recommendation performance by HP tuning.

3   Background

Major gains have been recently made in RS due to advances in deep neural networks, the most critical of them are described in the following.

3.1   Activation Functions

Activation functions play a key role in neural networks; therefore it becomes fundamental to understand their advantages and disadvantages in order to achieve reduced training time and better recommendation performance. In Table 1, we present the most widely used activation functions (more details are given in [4]).

Table 1. Mathematical expression of activation functions

Activation function	Mathematical expression
Rectifier Linear Unit (ReLU) [5,6]	$relu(x) = max(0, x)$
Leaky-ReLU (LReLU) [7]	$lrelu(x) = ax$, if $x < 0$ and $lrelu(x) = x$, if $x \geq 0$
Exponential LU (ELU) [8]	$elu(x) = a \cdot (\exp(x) - 1)$, if $x \leq 0$ and $elu(x) = x$, if $x > 0$
SineReLU [9]	$sinerelu(x) = \varepsilon \cdot (sin(x) - cos(x))$, if $x \leq 0$ and $sinerelu(x) = x$, if $x > 0$
Gaussian Error LU (GELU) [10]	$gelu(x) = x \cdot \frac{1}{2} \left[1 + erf \left(\frac{x}{\sqrt{2}} \right) \right]$
Mish [11]	$mish(x) = x \cdot \tanh(\text{softplus(x)})$, where: $\text{softplus}(x) = ln(1 + e^x)$
Swish [12]	$swish(x) = x \cdot sigmoid(x)$, where: $sigmoid(x) = (1 + exp(-x))^{-1}$
[0.1cm] Flatten-T Swish (FTS) [13]	$fts(x) = x/(1 + e^{-x}) + T$, if $x \geq 0$ and $fts(x) = T$, if $x < 0$

After many experiments we selected the value of the following HP variables. The ELU HP α controls the value to which an ELU saturates for negative net inputs. The default value of $\alpha = 1.0$ was used in our experiments. Since SineReLU is a sinusoidal wave when $x \leq 0$, it is a differentiable function for all input values of x. Also, ε works as a HP, used to control the wave amplitude; usually $\varepsilon \in [0.002, 0.025]$ and we set $\varepsilon = 0.0025$. The default value of T = 0 was used in our FTS activation experiments.

3.2   Weight Initialization

This section covers various weight initialization techniques that determine the algorithm by which weights are initially selected for a neural network. We have used various state-of-the-art weight initialization algorithms, that are available online, in the official Keras library[1].

Glorot and Bengio [14] have examined the effect of different HP on training and have also studied the propagation of gradients. The authors commented that the variance of the gradients decreases during backpropagation. In our experiments we test both versions: Glorot-Normal (Gaussian distribution) and Glorot-Uniform (Uniform distribution). In the remainder of this article these initializations are referred as GloN and GloU, respectively.

He et al. [15] have modified the scaling factor for weights given by Glorot and Bengio [14] to consider the rectifier non-linearities. In this work we test both

---

[1] https://keras.io/api/layers/initializers/.

versions: He-Normal (Gaussian distribution) and He-Uniform (Uniform distribution). These initializations are referred as HeN and HeU, respectively.

Orthogonal weight initialization [16] is a new class of random orthogonal initial conditions on weights that, like unsupervised pre-training, enjoys depth independent learning times. Moreover, Lecun Normal and Lecun Uniform [17] (referred as LecN and LecU respectively) have been used during our experiments.

Finally, we have experimented with some of the weight initialization classes implemented in the Keras library: Random-Normal, Random-Uniform, Identity, Zeros, Ones. However, these classes are not included in this work because of their significantly poor overall performance.

4 Experiments

In this section, we will try to optimize a neural network by performing HP tuning in order to obtain a high-performing model of CATA++². In the experiments we compare different CATA++ versions (changing: activation function, weight initialization and epochs number) against the default CATA++ (the paper's original version: ReLU-150 epochs).

4.1 Datasets

Two scientific article datasets are used for experiments. The first dataset, called citeulike-a, was gathered from the CiteULike website which is currently unavailable. CiteULike was a web service that let users create their own library of academic publications.

Secondly, experiments were conducted using the DBLP-Citation-network-V13 (2021-05-14)³ from AMiner [2] available at the time of authoring this work, consisting of 5,354,309 papers and 48,227,950 citation relationships. Using this enormous dataset and executing the algorithms for data preprocessing described in [3], we created three datasets and the necessary input files for CATA++.

Data sparsity for the datasets is calculated based on the user-item interaction. The datasets we used with their characteristics are described in Table 2.

Table 2. Datasets

Dataset	#users	#items	#tags	#sparsity	Recommendation type (item)
citeulike-a	5,551	16,980	7,386	99.78%	Scientific publications
dblp13_collection1	6,959	27,025	6,137	99.95%	Scientific publications
dblp13_venues	14,687	2,389	7,187	99.92%	Publication venues
dblp13_people	16,213	21,630	8,144	99.99%	People (collaboration)

---

² Code available at https://www.github.com/jianlin-cheng/CATA.
³ https://www.aminer.cn/citation.

4.2   Evaluation Metrics

The evaluation of our experiments is accomplished using two metrics: Recall & normalized Discounted Cumulative Gain (nDCG). Recall (Sensitivity or True Positive Rate) is the number of relevant documents retrieved by a search divided by the total number of existing relevant documents. However, Recall does not measure the ranking quality within the top-K list. Therefore, nDCG is used to show the ability of a model to recommend articles at the upper part of the recommendation list. Recall and nDCG are computed as described in [1]. More details on the metrics we used are presented in [4].

4.3   CiteULike Experiments in Recommendation of Scientific Publications

Firstly, using the citeulike-a dataset we ran experiments with the default parameters of CATA++, followed by experiments where different weight initialization techniques where used. During this first experimental phase no other HP was modified. The observed Recall results are recorded in Table 3, where we can compare the performance of the different weight initializers used in the tuned CATA++, to the Default CATA++ performance. We present the Recall values for a different number of recommendations, from K = 10 to K = 300. The best performance for each column (number K) is highlighted. In Table 3, the best performance is recorded when we are using CATA++ with Glorot-Uniform or He-Normal weight initializations; for the case of K = 300, the achieved Recall performance for CATA++ with He-Normal is more than 2% improved.

Table 3. Recall performance @K = 10, 50, 100, 150, 200, 250, 300 recommendations

Weight Init	10	50	100	150	200	250	300
GlorotN	0.0399	0.1245	0.1848	0.2275	0.2606	0.2876	0.3111
GlorotU	**0.0415**	0.1277	0.1905	0.2339	0.2675	0.2945	0.318
Orthogonal	0.041	0.1265	0.188	0.2309	0.2638	0.2902	0.3129
HeNormal	0.0408	**0.1284**	**0.1917**	**0.2351**	**0.2696**	**0.2963**	**0.3198**
HeUniform	0.0403	0.1264	0.1873	0.2302	0.2637	0.2905	0.3137
LecunN	0.0402	0.1265	0.1886	0.2321	0.2655	0.2922	0.3153
LecunU	0.0399	0.1257	0.1875	0.2309	0.2645	0.2918	0.3153
VarianceScaling	0.0379	0.1221	0.1834	0.2272	0.2603	0.2879	0.311
Default CATA++	0.0409	0.1274	0.1892	0.2318	0.2655	0.2936	0.3134

The full Recall results for all tested activation functions in CATA++ and for K=300 recommendations, are presented in Table 4. The best performance for each activation is highlighted. In Table 4, we can see that Swish, Mish, GELU, SineReLU, ELU and LeakyReLU perform best for a number of epochs in the

range [40,80], as they all utilize the negative values to influence the network training. On the contrary, FTS and ReLU need 150 epochs to achieve their best score
(due to the "dying ReLU problem"). The slow training of FTS was expected, due
to the default value of T (=0); so the representations in the negative form could
not benefit the network's training. Therefore, the top 3 recorded performances
on Recall @K = 300 are: FTS-150, ReLU-150, GELU-40 & SineReLU-60 (both
at 3rd position). In [4], CiteULike experiments are further studied.

Table 4. Recall performance for the tested activation @K = 300 (citeulike-a dataset)

Activation vs Epochs	10	20	40	60	80	100	120	150
FTS	0.2898	0.2907	0.2926	0.3069	0.31	0.314	0.3144	**0.3196**
Swish	0.286	0.2871	0.2951	**0.3001**	0.2959	0.285	0.2846	0.2835
Mish	0.284	0.2907	0.2963	0.297	**0.2984**	0.293	0.2922	0.2913
GELU	0.2995	0.301	**0.3019**	0.3001	0.3014	0.2952	0.2941	0.2933
SineReLU	0.2892	0.2933	0.2975	**0.3018**	0.2994	0.2981	0.297	0.2952
ELU	0.2865	0.2901	**0.2925**	0.2916	0.2892	0.2873	0.286	0.2846
LeakyReLU	0.2918	0.293	0.2941	0.2953	**0.3012**	0.2946	0.2893	0.2887
ReLU (default)	0.2843	0.2898	0.3031	0.3049	0.3087	0.3098	0.3101	**0.3134**

After the above-mentioned experiments with citeulike-a, we have created
seven tuned versions of CATA++: one version for each activation, along with the
most suitable weight initialization and number of epochs. The tuned-CATA++
models are presented in Table 5; xx is the epochs number in the [40,80] range.

4.4 AMiner Citation Network Experiments

In the present section we continue the experiments with the datasets created by
AMiner, in order to further tune the versions of CATA++ described in Table 5.

Table 5. The hyper-tuned versions of CATA++

CATA++ Version	Activation	Weight init.	Training epochs
FTS-heN-150	Flatten-T Swish	He-Normal	150
Swish-gloU-xx	Swish	Glorot-Uniform	40–80
Mish-heN-xx	Mish	He-Normal	40–80
GELU-gloU-xx	GELU	Glorot-Uniform	40–80
SReLU-heN-xx	SineReLU	He-Normal	40–80
ELU-gloU-xx	ELU	Glorot-Uniform	40–80
LReLU-heN-xx	LeakyReLU	He-Normal	40–80

Recommendation of Scientific Publications. Using the dblp13_collection1 and tuned-CATA++ as a RS for scientific publications we ran the tests of Table 6. The best performance for each K (column in Table 6) is highlighted. We ran many tests in order to select the training epochs number for each model version. In Table 6, each version is recorded with the epochs number that provided the best Recall. Therefore, the top 3 recorded performances on Recall@K = 300 are: SReLU-heN-20, Mish-heN-20 and GELU-gloU-40. We can note here that SineReLU and Mish have recorded their best performance at only 20 epochs of training. We had similar results regarding nDCG (for all datasets) for the different CATA++ versions, but due to space limitations we cannot include them here.

Table 6. Recall performance for the different CATA++ versions (Paper RS)

CATA++ version	10	50	100	150	200	250	300
FTS-heN-150	0.0582	0.1429	0.189	0.2203	0.2461	0.2664	0.2837
Swish-gloU-60	0.0719	0.1757	0.2301	0.2659	0.2921	0.3144	0.3325
Mish-heN-20	0.1135	**0.2393**	**0.2995**	0.3368	0.364	0.386	0.4045
GELU-gloU-40	0.0724	0.1767	0.2329	0.2707	0.2985	0.3223	0.3417
SReLU-heN-20	**0.1144**	0.2391	**0.2995**	**0.3379**	**0.3656**	**0.3884**	**0.407**
ELU-gloU-40	0.0582	0.1559	0.2107	0.2475	0.2758	0.2975	0.3163
LReLU-heN-80	0.0454	0.1303	0.1805	0.2134	0.2384	0.26	0.2783
Default CATA++	0.0495	0.1338	0.1832	0.2166	0.2431	0.2643	0.2823

Recommendation of Publishing Venues. The same experiments for the tuned CATA++ versions and the default one, were executed with the dblp13_venues dataset to recommend publishing venues. The Recall performance for all alternatives is presented in Table 7. Therefore, the top 3 recorded performances on Recall @K = 300 are: LReLU-heN-80, SReLU-heN-40 and Mish-heN-20.

Table 7. Recall performance for the different CATA++ versions (Venue RS)

CATA++ version	10	50	100	150	200	250	300
FTS-heN-150	0.2375	0.45	0.5389	0.585	0.6159	0.6393	0.6578
Swish-gloU-40	0.2247	0.4497	0.54	0.5916	0.6237	0.6485	0.6689
Mish-heN-20	0.2153	0.4481	0.5458	0.5947	0.6273	0.6528	0.6736
GELU-gloU-40	0.2234	0.4481	0.5387	0.5872	0.6173	0.64	0.6605
SReLU-heN-40	0.2304	**0.4686**	**0.5675**	0.6174	0.6506	0.6779	0.6976
ELU-gloU-40	0.197	0.4414	0.5407	0.5897	0.6243	0.65	0.6707
LReLU-heN-80	0.2361	0.4645	0.5671	**0.6197**	**0.6543**	**0.6787**	**0.6991**
Default CATA++	**0.2397**	0.4635	0.5555	0.5977	0.6292	0.6515	0.6709

Recommendation of People (collaboration RS). Finally, the same experiments for the tuned CATA++ versions and the default one, were executed with the dblp13_people dataset to recommend researchers for collaboration. The Recall performance for all the alternatives is presented in Table 8. Therefore, the top 3 recorded performances on Recall @K = 300 are: FTS-heN-150, LReLU-heN-80, SReLU-heN-60 and ELU-gloU-60 (both at 3rd position).

In Fig. 1 (paper RS), Fig. 2 (venue RS) and Fig. 3 (people RS) we present the performance of the top-3 variants compared to default CATA++. As someone can see, there is a significant difference in Recall performance to default CATA++, as HeN and GloU offer better results in these bigger datasets with increased sparsity (compared to citeulike-a) and at the same time SReLU, LReLU, Mish, GELU and ELU use the negative values in order to improve the network's training. Also, FTS benefits from HeN initialization and outperforms all other activations in Fig. 3.

Table 8. Recall performance for the different CATA++ versions (People RS)

CATA++ version	10	50	100	150	200	250	300
FTS-heN-150	**0.627**	**0.6982**	**0.7259**	**0.7434**	**0.7558**	**0.7636**	**0.7708**
Swish-gloU-60	0.5499	0.6606	0.6978	0.7178	0.7323	0.7431	0.7528
Mish-heN-60	0.5406	0.6639	0.6974	0.7183	0.7322	0.7427	0.7513
GELU-gloU-60	0.521	0.6567	0.697	0.7175	0.7312	0.7421	0.7518
SReLU-heN-60	0.5371	0.6621	0.701	0.7231	0.743	0.7516	0.761
ELU-gloU-60	0.5913	0.6794	0.711	0.7286	0.7421	0.7511	0.76
LReLU-heN-80	0.581	0.6747	0.7101	0.7307	0.745	0.7557	0.7647
Default CATA++	0.5814	0.6622	0.6931	0.712	0.7249	0.7368	0.7459

5 Discussion of the Experimental Results

Our findings suggest that the weight initializations that provide the best performance are He-Normal and Glorot-Uniform, offering an improvement of more than 2% in the citeulike-a dataset experiments.

In Table 9, the CATA++ versions with the top-3 Recall performances for each experiment are shown, along with the improvement (maximized for the second dataset, since its size optimized training) compared to the default CATA++.

One of our major findings is that SineReLU is a really stable activation function, performing really well in different datasets (always in the top-3 performances in Table 9). A very important new activation is FTS which gave the top performance in two out of four datasets-experiments. While ReLU remains competitive, we have found other activations that usually outperform it; SineReLU, FTS, Leaky-ReLU and Mish are definitely in this category.

Finally, we found that except for ReLU and FTS, the other activation functions produce their highest performance in fewer training epochs, namely in the range [20,80]; the exact number depends on the dataset's and model's characteristics. If FTS is used with $T < 0$, it will probably need fewer epochs too.

Table 9. Top-3 Recall performances for the different experiments (datasets)

Dataset-experiment	CATA++ version (Top-3)	Improvement
citeulike-a (Table 4)	**FTS-150**	**2%**
	Default CATA++	0
	GELU-40 & SReLU-60	0
dblp13_collection1 (Table 6)	**SReLU-heN-20**	**44.2%**
	Mish-heN-20	43.3%
	GELU-gloU-40	21%
dblp13_venues (Table 7)	**LReLU-heN-80**	**4.2%**
	SReLU-heN-40	4%
	Mish-heN-20	0.4%
dblp13_people (Table 8)	**FTS-heN-150**	**3.3%**
	LReLU-heN-80	2.5%
	SReLU-heN-60 & ELU-gloU-60	2%

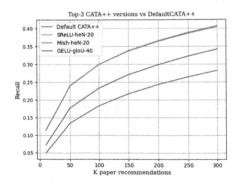

Fig. 1. Top-3 paper RS

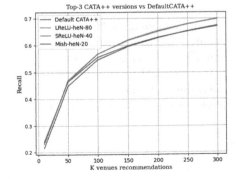

Fig. 2. Top-3 venue RS

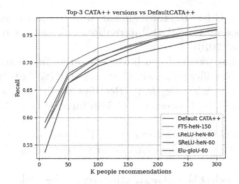

Fig. 3. Top-3 people RS

6   Conclusion and Future Work

We have conducted a large-scale comparison of activation functions and weight initializations using CATA++, a ML RS, and different datasets. Activation function, weight initialization and training epochs are among the most important and determinant factors. An activation function turns an otherwise linear classifier into a non-linear one and affects its training, a proper weight initializer is critical for the convergence of the ANN, while the proper number of training epochs helps to avoid under-fitting and over-fitting. We showed that the training time (epochs number) can be reduced, while at the same time the recommendation performance can increase up to 44.2%. Regarding future work, our experiments could be expanded to cover other HP, like hidden layers number or latent factors dimension, or other neural networks RS.

Acknowledgements. The work of M. Vassilakopoulos and A. Corral was funded by the MINECO research project [TIN2017-83964-R] and the Junta de Andalucia research project [P20_00809].

References

1. Alfarhood, M., Cheng, J.: CATA++: a collaborative dual attentive autoencoder method for recommending scientific articles. IEEE Access **8**, 183633–183648 (2020)
2. Tang, J., et al.: ArnetMiner: extraction and mining of academic social networks. In: Proceedings of 14th ACM SIGKDD International Conference on Knowledge Discovery and Data Mining, pp. 990–998 (2008)
3. Stergiopoulos, V., Tsianaka, T., Tousidou, E.: AMiner citation-data preprocessing for recommender systems on Scientific Publications. In: Proceedings of 25th Pan-Hellenic Conference on Informatics, pp. 23–27. ACM (2021)
4. Stergiopoulos, V., Vassilakopoulos, M., Tousidou, E., Corral, A.: An application of ANN hyper-parameters tuning in the field of recommender systems. Technical report, Data Structuring & Engineering Laboratory, University of Thessaly, Volos, Greece (2022). https://faculty.e-ce.uth.gr/mvasilako/techrep2022.pdf

5. Nair, V., Hinton, G.E.: Rectified linear units improve restricted Boltzmann machines. In: Proceedings of 27th International Conference on Machine Learning, pp. 807–814, Haifa (2010)

6. Pedamonti, D.: Comparison of non-linear activation functions for deep neural networks on MNIST classification task. CoRR:1804.02763 (2018)

7. Maas, A.L., Hannun, A.Y., Ng, A.Y.: Rectifier nonlinearities improve neural network acoustic models. In: Proceedings of 30th International Conference on Machine Learning (2013)

8. Clevert, D., Unterthiner, T., Hochreiter, S.: Fast and accurate deep network learning by exponential linear units (ELUs). arXiv:1511.07289v5 and ICLR (Poster) (2016)

9. Rodrigues, W.: SineReLU - An Alternative to the ReLU Activation Function (2018). https://wilder-rodrigues.medium.com/sinerelu-an-alternative-to-the-relu-activation-function-e46a6199997d. Accessed 5 Mar 2022

10. Hendrycks, D., Gimpel, K.: Gaussian error linear units (GELUS). arXiv:1606.08415v4 (2020)

11. Misra, D.: Mish: a self regularized non-monotonic neural activation function. CoRR:1908.08681 (2019)

12. Ramachandran, P., Zoph, B., Le, Q.: Searching for activation functions. CoRR:1710.05941 (2017) and ICLR' 2018 (Workshop) (2018)

13. Chieng, H., Wahid, N., Pauline, O., Perla, S.: Flatten-T Swish: a thresholded ReLU-Swish-like activation function for deep learning. Int. J. Adv. Intell. Inform. 4(2), 76–86 (2018)

14. Glorot, X., Bengio, Y.: Understanding the difficulty of training deep feedforward neural networks. In: Proceedings of the 13th International Conference on Artificial Intelligence and Statistics, Chia Laguna Resort, Sardinia, Italy, vol. 9 of JMLR: W&CP 9 (2010)

15. He, K., et al.: Delving deep into rectifiers: surpassing human-level performance on ImageNet classification. In: IEEE International Conference on Computer Vision, pp. 1026–1034 (2015)

16. Saxe, A.M., McClelland, J.L., Ganguli, S.: Exact solutions to the nonlinear dynamics of learning in deep linear neural networks. arXiv:1312.6120 and ICLR 2014 (2014)

17. LeCun, Y.A., Bottou, L., Orr, G.B., Müller, K.-R.: Efficient BackProp. In: Montavon, G., Orr, G.B., Müller, K.-R. (eds.) Neural Networks: Tricks of the Trade. LNCS, vol. 7700, pp. 9–48. Springer, Heidelberg (2012). https://doi.org/10.1007/978-3-642-35289-8_3

18. Eger, S., Youssef, P., Gurevych, I.: Is it time to swish? Comparing deep learning activation functions across NLP tasks. CoRR:1901.02671 (2019) and EMNLP 2018, pp. 4415–4424 (2018)

19. Kumar, S.K.: On weight initialization in deep neural networks. CoRR:1704.08863 (2017)

Exploring Latent Space Using a Non-linear Dimensionality Reduction Algorithm for Style Transfer Application

Doaa Almhaithawi[1]([✉]), Alessandro Bellini[2], and Stefano Cuomo[2]

[1] Department of Control and Computer Engineering,
Politecnico di Torino, Turin, Italy
doaa.almhaithawi@polito.it
[2] Mathema, Firenze, Italy
{abel,stefano.cuomo}@mathema.com

Abstract. A latent space represents data by embedding them in a multidimensional vector space. In this way an abstract estimation of any complex domain could be created. Empirical approach for exploring the latent space generated by known pre-trained model of human face images using a nonlinear dimensionality reduction algorithm is presented in this paper. One aim was to find more detailed entangled features (beard and hair color) between the real images and their representation, in artistic face portrait application. Experimental results showed that sparse vectors in the latent space could be useful to obtain optimal results with relatively low effort. To evaluate our work, we present the results of a survey that was sent to 25 thousand subscribers of the real world application and got around 360 responses. The main goal of the survey was to find some quantitative measurements that can be used in our research.

Keywords: Latent space exploration · Nonlinear dimensionality reduction · Generative adversarial networks (GANs) · Industrial survey

1 Introduction

Image processing is one of the matured fields of study. An image latent space is usually characterized by high dimensionality. However, style transfer is not the only application for which it is used it includes image generation, manipulation, object and face recognition and more. We started our research with a real life application[1], based on images latent space for portraying human faces in Leonardo Davinci style [1]. One of the downfalls for such applications is the subjective evaluation, based only on the human artistic or technical knowledge. In general, we can say that style mixing applications have two major limitations: a) *Identity preservation.* Mixing two vectors (subject and style) is a trade-off between the subject image identity and the style image. We need to find the

---

[1] DaVinciFace is a software registered at SIAE (Italian Authors' and Publishers' Association) - www.davinciface.com.

© Springer Nature Switzerland AG 2022
S. Chiusano et al. (Eds.): ADBIS 2022, CCIS 1652, pp. 277–286, 2022.
https://doi.org/10.1007/978-3-031-15743-1_26

most suitable settings to maximize both. b) *Evaluate the results.* Results are subjective and personal, thus we need quantitative measurements to know which settings are more appreciated and accepted by the users of the system.

Latent spaces can be explained as numerical representations of real data into a multi-dimensional space, with high dimensionality. To determine the shape and properties of a latent space is essential for understanding the underlying data structure and for developing and improving applications such as data augmentation, pattern recognition, anomaly detection, and more. Style transfer is one of the research areas where latent spaces are used in images [2,3], text [4], and videos [5].

This paper presents a methodology to explore images latent space using nonlinear dimensionality reduction algorithms. Searching for information about disentanglement of the latent representation and the original image's features. These help solve the above mentioned limitations in identity preservation and include the results in a survey to evaluate our work. The paper is organized as follows. Section 2 discusses related work, Sect. 3 introduces the used methodology, Sect. 4 presents the experiment and the results, Sect. 5 presents the evaluation and the survey design, finally Sect. 6 discusses future directions of this research activity.

2   Related Work

Recently, several research studies have been devoted to the exploration of latent spaces, and nonlinear dimensionality reduction.

Latent representation is an important step in most real-world applications, and learning this representation facilitates the extraction of information for downstream tasks (e.g., classification or prediction) [6]. Although the data represented may differ depending on the domain, the transformation and exploration of the latent space are the analogous.

Dimensionality reduction is mainly divided into linear and nonlinear. Nonlinear projection-based techniques can be used for various purposes, such as feature extraction, visualization [7], pattern recognition [8], or even as a pre-processing step [9].

Style transfer is an active area of research in various domains, such as text processing [4], where adversarial networks have been used. It has also been used in images with various applications, such as artistic style transfer in [2], medical cancer classification [3], and videos in [5].

In the following we introduce the main components used in this paper, Sect. 2.1 presents the StyleGAN2 human face images latent space, while Sect. 2.2 presents ISOMAP, the nonlinear dimensionality reduction algorithm and Sect. 2.3 presents the application DaVinciFace.

StyleGAN2 latent space, proposed in [10], maps the input to a latent intermediate space W; in contrast, earlier generators use only the input layer. Therefore, the mapping network to this intermediate space consists of 8 layers, and the synthesis network (that generate the final image) consists of 18 layers, two for each resolution $(4^2 - 1024^2)$. The output of the last layer is converted to RGB with a separate convolution. As described in [11], the coarse spatial resolutions (the first 4 vectors) correspond to high-level aspects such as pose, general hairstyle, face shape, and glasses, but not the colors (eyes, hair, lighting) or finer facial features. The medium resolution styles (the second 4 vectors) adopt minor facial features, hairstyle, open/closed eyes and the fine styles (the remaining 10 vectors) mainly bring in the color scheme and microstructure.

As mentioned in [12], StyleGAN usually requires task-specific training for various tasks. However, it has enabled various image manipulation and editing tasks due to its high-quality generation and disentangled latent space.

Dimensionality Reduction. Since visualization is used to achieve a better understanding of the data, dimensionality reduction methods are considered effective and easy to use, especially techniques based on nonlinear projections. ISOMAP, discussed in [13], differs from other dimensionality reduction algorithms such as t-SNE and UMAP in that it focuses on the global structure, while the other algorithms give better results where the local geometry is close to an Euclidean geometry.

DaVinciFace application is a style transfer system developed by Mathema - an Innovative SME[2] based in Florence, Italy - that aims to create a portrait in the style of Leonardo Da Vinci from a photograph of a human face as one of its main projects.

3   Methodology

In our work, we use the latent space generated by a StyleGAN2, and we explore and visualize this space using nonlinear projection-based technique (ISOMAP) to improve the results of machine learning applications for real human faces portrait application (DaVinciFace). Figure 1 shows the main building blocks of the proposed methodology

First, the human representation of the image is projected into the latent space using a pre-trained model (StyleGAN2) after some pre-processing activities which are the a) face detection uses a pre-trained model on the FFHQ dataset[3] that extracts the most evident face in the image and b) centering and cropping the detected face in a square frame with the dimensions 1024*1024. The output of this step for each image consists of 18 vectors with a length of 512 each.

Next, we visualized a dataset of latent vectors which are considered a high dimensional space vectors (18 * 512) in a 2-dimensional space using three different nonlinear dimensionality reduction algorithms (the dataset is described below). While dimensionality reduction methods have no variation in density but mostly symmetric and oval distribution, ISOMAP representation shows higher variation in distributions. Some vectors were sparse while others were dense or sometimes highly dense (most of the points are focused around zero), that gives

---

[2] SME: small-to-medium enterprise.
[3] paperswithcode.com/dataset/ffhq, last visite 14/05/2022.

us the motivation to use ISOMAP results in our study to find the clue to build the resulting image.

We then performed an experiment to optimize the application for style transfer and compared the different results by changing the vectors used from the subject image and the style image (from the 18-vector we focused on the middle vectors from 8th till 11th because led by [11] the first 8 are essential to be from the subject and the last 4 to be from the style image).

To evaluate the results, we started a survey to collect subjective feedback from the users. The survey was sent to 25 thousand subscribers and got around 360 responses. The content of the survey mainly exposes different resulting images with different settings and asks the participant to select the best regarding style and identity preservation, thus, the responses would be used as a quantitative measurement in our research.

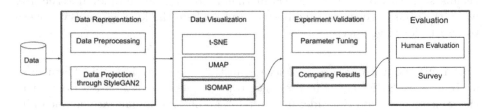

Fig. 1. The main steps of our methodology

4   Experimental Validation

In this section we present the dataset used in our experiment. Then the results of ISOMAP with a brief comparison, and finally the output images with the new adjustment sets compared to the old ones.

4.1   The Dataset

We used a dataset of 1,158 input images from the application test environment with the corresponding latent vectors (18 * 512 each). The dataset is copyrighted and therefore cannot be published. However, the photos are images of faces from the Internet or from company employees used in the initial steps of building and testing the application.

4.2   Visualization

In this experiment we used ISOMAP to visualize each vector distribution in 2-dimensional space to distinguish the effect of each vector on the output with respect to its density, as a way to explore entanglements between the real images and their representation in the latent space. The ISOMAP representation of the 18 vectors of StyleGAN2 showed high variation in distributions, some vectors were sparse while others were dense or sometimes highly dense (most of the points are focused around zero), that gives us the motivation to use the results in our study to find the effect of sparse versus the dense vectors on building the resulted image. We focused on the middle vectors (from the 8th till 11th) which hold the most controversy about the trade-off between the identity and style. Figure 2, shows 3 vectors 7th, 8th and 9th as examples, the 8th vector is sparse while the others are more dense. That could also give us some indications that the StyleGAN2 space features do not only have local but global geometry that could be close to Euclidean geometry. We observed in the 18 vectors representation, that the first (coarse features) and the last (fine style) vectors, are more sparse while the middle vectors are relatively dense (except 8th), that could point that, the more the feature is distinguish between the samples the more the sparse its distribution. As a consequence, even a slight change (or transaction) in such features will produce the most change in the original space.

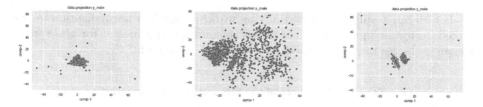

Fig. 2. Example of ISOMAP 2-dimension representation of 3 vectors (7th, 8th and 9th)

4.3   Comparing the Results

Two examples are shown in Table 1, where we compared between the old settings on the left and the effect of adding each vector of the 8th, 9th, 10th, and 11th respectively to be from the subject image instead of the style image, the 8th vector effect is shown on the identity also on features like beard in the top example and hair color on the bottom example and its representation was sparse in the ISOMAP. While adding vectors improve the identity slightly, especially the eye color as in the bottom example, the style is progressively lost, and we need to find suitable combinations to have as much style while preserving the identity.

Table 2 focuses on the effect of the 8th vector and shows the comparison between the first 12 vectors of the original image with and without the 8th vector. We observed the same effect on 3 other bearded men images and 3 other blonde-haired images. Based on the obtained results, we can conclude that this sparse vector has a greater effect on distinguishing patterns than the other dense vectors.

Table 1. Comparing the results of the style transfer from left to right, the original image, the old settings, first 9 vectors, first 10 vectors, first 11 vectors, first 12 vectors from original image while the rest are from the style image

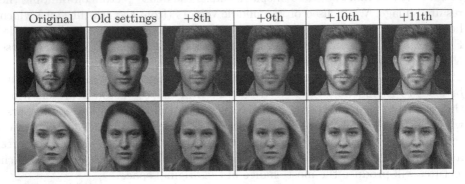

Original	Old settings	+8th	+9th	+10th	+11th

For all our results mentioned in this work, we applied normalization of the mixed vector in the latent space before generating the image, for the results shown in Tables 1 and 2 we used a variance percentage of 0.4%.

Table 2. The effect of the 8th vector on the identity.

With 8th vector	Without 8th vector	With 8th vector	Without 8th vector

5   Evaluation

In this section we present our evaluation of the results where we conducted a survey with different settings. The survey was advertised and conducted online and was sent to all users (almost 25 thousand) on batches (2,000 each). We opened the survey during a month, from a total of 525 views 370 completed the survey and 9 started without finishing (the participation rate is 72.2%, the completion rate is 97.6% and the average time for completion is 3:44 min). For designing the survey and collecting responses we used SurveyHero[4]. The survey included 18 questions, both quantitative and qualitative data were gathered.

The survey was designed mainly to help finding insights to answer the following questions:

---

[4] www.surveyhero.com is a software to design, collect and analyze responses of surveys.

RQ1. Is there a better trade-off (between style and identity preservation) than the current version?
RQ2. Does the current version preserve the identity of the subject?

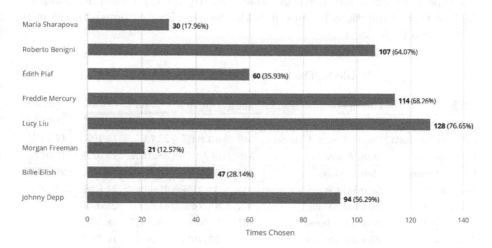

Fig. 3. The second type of questions results. The percentage indicates the number of users choose the celebrity divided by the number of total responses to this question (the total responses were 167)

To address the questions mentioned before we designed 9 questions that related to the comparison in two types:

1. Type 1 addresses RQ1, which has 8 questions, present four different celebrities (Monica Bellucci, Luciano Pavarotti, G-Dragon and Barack Obama) and four other test cases with the following four settings each and the participant can choose the most convenient: (a) Old settings; (b)The first 9 vectors from the original image; (c) The first 11 vectors from the original image; (d) The first 12 vectors from the original
2. Type 2 addresses RQ2, one question presents 8 other celebrities (Maria Sharapova, Roberto Benigni, Edith Piaf, Freddie Mercury, Lucy Liu, Morgan Freeman, Billie Eilish, Johnny Depp) with the old settings and asks the user to choose all that are recognized, to measure the identity preservation in the old settings.

5.1 Survey Results

The demographic questions show the followings:

- The Gender, from 360 responses, 224 (62.22%) are men, 129 (35.83%) are women and 7(1.94%) prefer not to declare

- The Age, from 360 responses, most of the participants are between 31 and 45 years old (39.44%)
- The Art-Related background, mostly interested in art 56.82%

The type 1 responses are shown in Table 3, mostly the old version is chosen for the first type of questions because it holds the most style although the identity preservation is the least.

Table 3. The first type of questions results.

Image	Description	Total	Option a	Option b	Option c	Option d
Monica Bellucci	Woman, Brunette	343	**67.93%**	10.79%	9.62%	11.66%
Luciano Pavarotti	Bearded man, Caucasian	349	**43.84%**	23.78%	15.19%	17.19%
G-Dragon	Man, Asian	346	35.84%	**38.15%**	13.01%	13.01%
Barack Obama	Man, African	350	26%	16.57%	28.57%	**28.86%**
Test case 1	Bearded man, Caucasian	346	**31.21%**	25.43%	22.83%	20.52%
Test case 2	Woman, blonde	352	**45.74%**	11.65%	23.30%	19.32%
Test case 3	Woman, African	167	30.54%	**31.14%**	18.56%	19.76%
Test case 4	Woman, Brunette	167	**57.49%**	20.96%	10.78%	10.78%

The type 2 responses are shown in Fig. 3, Lucy Liu was the most recognized. Then Freddie Mercury and Roberto Benigni.

5.2 Survey Validation

We discuss the threats to internal validity which are the instrumentation, the selection of the subjects, and the maturation. The instrumentation, while designing this survey we conducted a pilot survey with 19 participants, which helped in determining to what extent the questionnaire was understandable and complete. The participants have the chance to give their feedback to the questionnaire regarding wording, clarity, and presentation. Subject selection, participants took part in the survey voluntarily. The maturation threats from fatigue or boredom were not taken into consideration due the fact that the average time for completion is 3:44 min. Concerning the external validity (that limit the ability to generalize the experiment results outside the experiment setting) had to do with the representativeness of the subject population. To guarantee the representativeness of the subject population, the survey was sent to all the users of the DaVinciFace.

6 Discussion and Future Directions

The results of our research can be summarized in the following points:

Relevance of the 8th Vector from the StyleGAN2 Structure. This vector shows a large impact on identity preservation and some detailed features such as beard and hair color. This vector was represented as a sparse distribution using ISOMAP.

The Survey Results. The results for the type 1 questions shows that mostly the old version is preferable because it holds the most style while the identity preservation is the least. In exception are the Asian and African races, where more identity was needed. Also, the type 2 question supports the previous findings. The examples of Asian and African races have the most conflict in opinion, while Women examples the old version is mainly selected. Thus, this work was practically useful to improve a real and existing product while addressing some users comments, such as: (i) Obtaining more demographic information about the users; (ii) Obtaining artistic reviews; (iii) quantifying the results of the application which are subjective (qualitative); and (iv) Measuring the importance of identity preservation. Although the main purpose of this work was to improve the resemblance of the portrait to the subject image, the survey showed that the users' opinion is more drift towards keeping the most style more than having more identity. This could be due to the fact that this application is artistic and this is an artistical point of view. However, this is not the case in special subjects with races and features different from Leonardo's original portrait like Africans, Asiana and bearded men which needs more investigation. The research activities presented can be significantly extended in the following research directions.

Defining a New Set of Quantitative Metrics to Objectively Evaluate Dense Vectors Compared to Sparse Vectors: To this end, we will use the results of the survey presented in this article to answer more questions such as:

- How does demographics (race, gender) of the subject affect the result;
- Are the results of the questions different in case the participant had a personal experience with the system;
- Are the results of the questions different if the participant is professionally related to art (artist or art-student).

The Influence of Density on the Discrimination of Samples. We can anticipate that the density is an important factor in exploring the latent space, and it could play a role in the creation of the space. However, these preliminary results are promising and represent a first attempt at a more comprehensive and robust study looking into the impact of sparse vectors compared to dense vectors on the construction of the resulting image.

Investigate the Applicability of the Proposed Method to Support the Data Labeling Phase: Any information about the data structure helps in later processing and one of them is data labeling. The used dataset was manually labeled, which will give us the opportunity to explore these labels, especially the gender.

References

1. Davinciface homepage. https://www.davinciface.com/. Accessed 6 July 2022
2. Xu, Z., Wilber, M., Fang, C., Hertzmann, A., Jin, H.: Learning from multi-domain artistic images for arbitrary style transfer. arXiv preprint arXiv:1805.09987 (2018)
3. Shaban, M.T., Baur, C., Navab, N., Albarqouni, S.: StainGAN: stain style transfer for digital histological images. In: 2019 IEEE 16th International Symposium on Biomedical Imaging (ISBI 2019), pp. 953–956. IEEE (2019)
4. Prabhumoye, S., Tsvetkov, Y., Salakhutdinov, R., Black, A.W.: Style transfer through back-translation. arXiv preprint arXiv:1804.09000 (2018)
5. Ruder, M., Dosovitskiy, A., Brox, T.: Artistic style transfer for videos and spherical images. Int. J. Comput. Vision 126(11), 1199–1219 (2018)
6. Bengio, Y., Courville, A., Vincent, P.: Representation learning: a review and new perspectives. IEEE Trans. Pattern Anal. Mach. Intell. 35(8), 1798–1828 (2013)
7. Shen, J., Wang, R., Shen, H.-W.: Visual exploration of latent space for traditional Chinese music. Vis. Inform. 4(2), 99–108 (2020)
8. Crecchi, F., Bacciu, D., Biggio, B.: Detecting adversarial examples through non-linear dimensionality reduction. arXiv preprint arXiv:1904.13094 (2019)
9. Tasoulis, S., Pavlidis, N.G., Roos, T.: Nonlinear dimensionality reduction for clustering. Pattern Recogn. 107, 107508 (2020)
10. Karras, T., Laine, S., Aittala, M., Hellsten, J., Lehtinen, J., Aila, T.: Analyzing and improving the image quality of StyleGAN. In: Proceedings of the IEEE/CVF Conference on Computer Vision and Pattern Recognition, pp. 8110–8119 (2020)
11. Karras, T., Laine, S., Aila, T.: A style-based generator architecture for generative adversarial networks. In: Proceedings of the IEEE/CVF Conference on Computer Vision and Pattern Recognition, pp. 4401–4410 (2019)
12. Chong, M.J., Lee, H.-Y., Forsyth, D.: StyleGAN of all trades: image manipulation with only pretrained StyleGAN. arXiv preprint arXiv:2111.01619 (2021)
13. Balasubramanian, M., Schwartz, E.L.: The isomap algorithm and topological stability. Science 295(5552), 7–7 (2002)

A Novel Method for University Ranking: Bounded Skyline

Georgios Stoupas[1]([envelope])[iD] and Antonis Sidiropoulos[2][iD]

[1] Aristotle University of Thessaloniki, Thessaloniki, Greece
grgstoupas@csd.auth.gr
[2] International Hellenic University, Thessaloniki, Greece
asidirop@ihu.gr

Abstract. University rankings are not only used as tools in academia and funding agencies, but they are popular readings in newspapers and magazines as well, plus they are of major importance for decision-making in academia. University rankings have been heavily criticized for a number of reasons, including that they are based on arbitrary choices about the key performance indicators, their special weights and so on. For these reasons, we argue that there is no meaning in ranking universities in an absolute ordered manner; instead universities should be ranked in "sets of equivalence". To this end, here we present a modified version of Skyline and Rainbow Ranking, named *Bounded Skyline* and *Bounded Rainbow Ranking*, respectively. The results are analyzed and discussed in comparison to traditional university rankings.

Keywords: Dominance · Skyline · Rainbow Ranking · University ranking

1 Introduction

University rankings are of major importance for decision-making by prospective students, academic staff, and funding agencies. The placement of universities in these lists is a crucial factor and academic institutions adapt their strategy according to the particular criteria of each evaluation system.

Probably, the most popular global rankings are: ARWU[1] (Academic Ranking of World Universities) of Shanghai, THE[2] (Times Higher Education) World University Rank and QS[3] (Quacquarelli Symonds) Rank. All these lists base their respective ranking on some set of key performance indicators, which differ from one organization to the other. The reader can retrieve these indicators from the respective sites.

In [3,4] we have used the concept of Skyline and the Rainbow Ranking to overcome the most significant drawback of all these rankings: the arbitrary

---

[1] https://www.shanghairanking.com/.
[2] https://www.timeshighereducation.com/.
[3] https://www.topuniversities.com/.

© Springer Nature Switzerland AG 2022
S. Chiusano et al. (Eds.): ADBIS 2022, CCIS 1652, pp. 287–298, 2022.
https://doi.org/10.1007/978-3-031-15743-1_27

weight selection. We have used the standard dominance definition [3], as well as a dominance definition based on the mathematical concept of Majorization [4]. In this work we elaborate the Skyline definitions by adding the ability to define bounds. This concept can be used in combination with any dominance type. We notice, however, that here we use only the standard dominance type.

2   Skyline and Rainbow Ranking

The Skyline operator [1] has been utilized in the Databases field for decades and dates back to the definition of the Pareto frontier in Economics [7]. However, the Skyline approach does not refer to efficient resource allocation; rather it provides a multi-criteria selection of "objects". In particular, the Skyline operator is used as a database query to filter only those objects that are not worse than any other (they are not dominated) [1]. The Skyline operator takes advantage of the concept of "dominance", which is symbolized with $\prec$ and $\succ$. There are various definitions of the dominance relationship:

Definition 1 (Dominance [2]**).** *Point s dominates point t, denoted as $s \succ t$, iff $\forall i \in [1, d]$, $s[i] \geq t[i]$ and $\exists j \in [1, d]$, such that $s[j] > t[j]$. If neither s dominates t, nor t dominates s, then s and t are incomparable, denoted $s \prec\succ t$. The relation $s \succeq t$ denotes that either $s \succ t$ or $\forall i \in [1, d], s[i] = t[i]$.*

Definition 2 (Strict Dominance [2]**).** *Point p strictly dominates point q, denoted as $p \ggg q$, iff $\forall i \in [1, d], p[i] > q[i]$.*

Here, we note that there are two opposite views in the literature about dominance, depending on the application. In some applications we may need positive domination, in the sense that the best value is the greater one. Elsewhere, we may need negative domination, i.e. the best value is the smallest one. For example, an attribute as *h-index* needs positive domination but the attribute *rank position* needs negative domination.

Definition 3 (Skyline [6]**).** *Given a dominance relationship in a dataset, a Skyline query returns the objects that cannot be dominated by any other object.*

$$SKY(S) = \{s \in S : \nexists t \in S, \ t \succ s\} \tag{1}$$

Apparently, we can apply the Skyline concept in Scientometrics. For example, given a set of attributes that characterize scientific performance, the Skyline operator outputs the entities (e.g. authors, institutions etc.) that cannot be surpassed by any other entity in the dataset [3].

An extension of the Skyline operator, the Rainbow Ranking [5], applies iteratively the Skyline operator until all entities (i.e., scientists) of a dataset have been classified into a Skyline level. More specifically, given a set of scientists X_1, the first call of the Skyline operator produces the first Skyline level, which is denoted as S_1. Next, the Skyline operator is applied on the dataset $X_1 - S_1$, to derive the second Skyline layer, denoted as S_2. This process continues until all the scientists of the dataset have been assigned to a particular Skyline level S_i.

To give more semantics to the method, a particular value should characterize the Skyline levels. Should this value be the iteration number, then this would convey limited interpretability since the relativeness would be lost. It is crucial to designate the position of a scientist among their peers. Therefore, a normalization of this value is necessary.

Definition 4 (Rainbow Ranking [5]**).** *Given a dominance relationship in a dataset, RainbowRanking is the process of iteratively applying the Skyline operator. Each Skyline operation produces a Skyline level (i.e. 1, 2, 3, ...). Thus, the RR-index of an object a is defined as:*

$$RR(a) = 100 - 100 \times \frac{|A_{above}(a)| + |A_{tie}(a)|/2}{|A|} \tag{2}$$

where A is the set of objects, $A_{above}(a)$ is the number of objects at higher Skyline levels than object a, and $A_{tie}(a)$ is the number of objects at the same Skyline level with object a, excluding object a.

Apparently, it holds that: $0 < RR(a) \le 100$, and consequently the best object has the greatest value.

Table 1. A synthetic 2D example.

Researcher	C	P	Rank order by C	Rank order by P	RR layer	Bounded RR layer
R1	32	51	1	19	1	4
R2	32	40	1	23	1	4
R3	32	20	1	26	1	4
R4	31	63	4	18	1	1
R5	30	72	5	13	1	1
R6	30	43	5	21	2	5
R7	30	32	5	25	2	5
R8	26	70	8	14	2	2
R9	25	65	9	16	3	3
...						
R20	14	90	20	8	3	5
R21	14	89	20	9	3	5
R22	13	99	22	1	1	4
R23	13	94	22	3	2	5
R24	13	75	22	12	4	6
R25	12	91	25	6	3	6
R26	10	92	26	5	3	6

3   Motivation

In the sequel, we present an example. Let's assume that we have a set R of researchers $\{R1, R2, \ldots, R26\}$. We have computed their average number of citations per publication C and number of publications P and we need to generate a unique ranking based on both attributes. Columns C and P of Table 1 show the particular values of the example. The first examined method is Rainbow Ranking, which is based on Skylines. In Fig. 1a we show the plot, whereas in column "RR Layer" of Table 1 we show for each entity the Skyline layer that it belongs. In Fig. 1b we focus on the first Skyline layer. It can be seen that researchers $R1, R2, R3, R4, R5, R11, R17, R22$ belong in the first Skyline layer. At this point, we notice that the Researcher $R3$ is ranked last by the number of publications P (y-axis). Is it fair, the last ranked researchers by one of the metrics to be clustered within the first class of researchers? In most cases, this is no fair. Let's recall the classic Skyline example from [1], where we seek for cheap hotels nearby the sea, and extend our search under the constraint that the hotels should not be very expensive neither they should be too far from the sea. This means that in every attribute we must apply a lower (upper) limit.

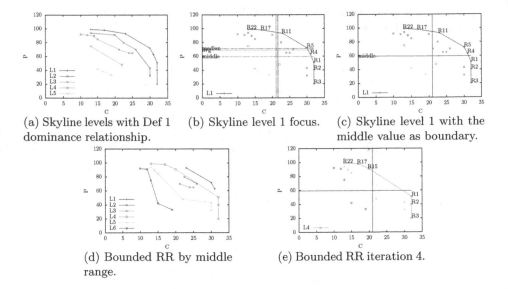

(a) Skyline levels with Def 1 dominance relationship.

(b) Skyline level 1 focus.

(c) Skyline level 1 with the middle value as boundary.

(d) Bounded RR by middle range.

(e) Bounded RR iteration 4.

Fig. 1. Skylines under Strict dominance.

Returning to our example with the researchers, rephrase the query as *find the researchers who are not dominated by any other by either C and P metric, as well as they are ranked in a acceptable position by any metric*. What is an acceptable position? In other words: which should be the boundary that will allow the researchers (or not) to become members of the Skyline? There are several choices to use as a boundary:

- the average value,
- the median value, or
- the range middle value.

It is obvious that any value from the ones mentioned earlier could be used, either computed as a percentage of each dimension range, or computed as a population limit. In this work, we will use only the above-mentioned values: average, median and range middle, and we will try to investigate how each one affects the final ranking. In Fig. 1b we show the above possible bounds for both P and C metrics. It is obvious that depending on the data distributions, the selection of the boundary will affect more or less the final Skyline selection. Assuming that we use the range middle value as a boundary for all dimensions (Fig. 1c), then the selected elements from the Skyline will be only the ones that reside in the top-right quadrant of the plot.

Definition 5 (Bounded Skyline). *Given a dominance relationship in a dataset and a boundary function B, a Bounded Skyline query returns the objects that cannot be dominated by any other object and simultaneously all its attributes are better than the computed B[i].*

$$SKY(S) = \{s \in S : \nexists t \in S, \ t \succ s \wedge \forall i \in [1, d], s[i] >= B[i]\} \qquad (3)$$

By using the Definition 4 we can define the RainbowRanking (RR) based on the Bounded Skyline. In every iteration, the boundary for each dimension will change[4], since it will be recomputed based only on the remaining elements. So the cross shown in Fig. 1c may be moved during each iteration.

At this point, we notice that given a set of elements in a d-dimensional space, the Bounded Skyline may be an empty set. This situation may occur when, in the hypothetical 2D example presented in Fig. 1, the upper right quadrant of the plot has no points of the unbounded Skyline. If this situation occurs, then in the specific iteration of the RR computation, the unbounded Skyline should be used instead of the bounded one. In Fig. 1d we show the resulting Skyline layers by applying the RR iterations. In iterations 4, 5, and 6 the resulting Bounded Skyline is empty, and thus the unbounded one is used. In Fig. 1e we show in detail the computation of layer 4. The unbounded Skyline does not contain any points in the upper right quadrant. Therefore, the Bounded Skyline set is empty. This leads that the unbounded one will be used.

Finally, it is anticipated that the Bounded RainbowRanking (RR) will produce:

- better refined and smaller Skyline sets than the unbounded algorithm, especially in the top layers,
- more or equal number of layers than the unbounded algorithm.

---

[4] There are some cases that it will not change, like the case of Fig. 1 from iteration 1 to 4, since the minimum and maximum values remain the same. In iterations 5 and 6, it will change.

4   Experiments

In this section, we present the experiments performed. The first set of experiments is based on the QS ranking. They rank the universities according to six criteria. Each criterion is assigned a specific weight. Our first claim is that given a different set of weights, the rank order will change. Especially in the high ranked universities. In Table 2 we present the original QS rank table for the top-30 universities. Our second claim is that a significant set of universities get very similar scores. Thus, these universities should be ranked at the same position. In [3,4] we have proposed the use of Skyline and the Rainbow Ranking to overcome these drawbacks. The simple Skyline method still has some drawbacks as explained in Sect. 3.

It can be seen from Table 3 that a significant number of universities are grouped into the layer #1 Skyline. Actually, 103 universities are ranked into the first layer by the unbounded method. This size is considered to be too excessive. Here, we notice that the Skyline produces large layers in the case that there are a lot of dimensions which are uncorrelated to each other. Thus, since there are six uncorrelated criteria, it is almost obvious that the groups will be rather large. The Bounded Skyline adds another filter in the Skyline iterations. In Table 3 we can see that by using a bounded variation of Skyline, after the 16th positioned university by QS, a lot of universities are sunk to a lower Skyline layer by almost

Table 2. Top 20 Universities by QS 2022.

University	# by QS	QS overall	Academic reput.	Employer reput.	Faculty student ratio	Cit. per faculty	Intern faculty ratio	Intern students ratio
MIT	1	100	100	100	100	100	100	91.4
U. of Oxford	2	99.5	100	100	100	96	99.5	98.5
Stanford U.	3	98.7	100	100	100	99.9	99.8	67
U. of Cambridge	4	98.7	100	100	100	92.1	100	97.7
Harvard U.	5	98	100	100	99.1	100	84.2	70.1
California I. of Tech. (Caltech)	6	97.4	96.7	89.9	100	100	99.4	87.7
Imperial College London	7	97.3	98.4	99.8	99.8	88.1	100	100
ETH Zurich	8	95.4	98.7	95.3	80.4	99.8	100	98.2
UCL	9	95.4	99.4	98.9	99	78	99.5	100
U. of Chicago	10	94.5	99.2	93.5	95.5	90.6	71.9	84.9
National U. of Singapore (NUS)	11	93.9	99.6	97.1	86.8	90.6	100	70.3
Nanyang Tech. U. Singapore (NTU)	12	90.8	91.1	85	89.5	95.5	100	69.6
U. of Pennsylvania	13	90.7	96.2	93.8	100	74	94.1	58.1
EPFL	14	90.2	82.4	79.7	94.8	99.8	100	100
Yale U.	15	90.2	99.9	100	100	59.9	87.9	69.8
The U. of Edinburgh	16	89.9	98	97.3	83.5	69.5	99.3	99.6
Tsinghua U.	17	89	98.6	97.6	90.6	96	15.3	26.3
Peking U.	18	88.8	99.4	98.2	83.5	89.5	45.6	38.5
Columbia U.	19	88.7	99.6	98.4	100	55.6	52.7	98.4
Princeton U.	20	88.6	99.9	99.2	68.7	100	28.1	63.2

all the variations. The first sank university is Tsinghua University, which is ranked at the 17th position by QS and in the 2nd unbounded Skyline layer. By two of the three bounded variations, it is downgraded to positions five and seven. We claim that this improves the clustering results. This is more obvious for the University of Tokyo case. Its score is only 3.3/100 for the criterion International Faculty Ratio. Due to the high score on the other indicators, the University of Tokyo is ranked high by both QS Rank and the unbounded Skyline. The Bounded Skyline variations downgrade it to layers 5 to 7.

Table 3. Top 30 Universities by QS 2022.

University	QS rank	Simple Skyl layer	Middle BSkyl layer	Avg BSkyl layer	Median BSkyl layer	QS Sc.	Simple RR Sc.	Middle RR Sc.	Avg RR Sc.	Median RR Sc.
MIT	1	1	1	1	1	100	95.8	99.2	98.6	98.1
U. of Oxford	2	1	1	1	1	99.5	95.8	99.2	98.6	98.1
Stanford U.	3	1	1	1	1	98.7	95.8	99.2	98.6	98.1
U. of Cambridge	4	1	1	1	1	98.7	95.8	99.2	98.6	98.1
Harvard U.	5	1	1	1	1	98	95.8	99.2	98.6	98.1
CalTech	6	1	1	1	1	97.4	95.8	99.2	98.6	98.1
Imperial College London	7	1	1	1	1	97.3	95.8	99.2	98.6	98.1
ETH Zurich	8	1	1	1	1	95.4	95.8	99.2	98.6	98.1
UCL	9	1	1	1	1	95.4	95.8	99.2	98.6	98.1
U. of Chicago	10	2	2	2	2	94.5	89.5	98.0	96.0	94.6
NUS Singapore	11	1	1	1	1	93.9	95.8	99.2	98.6	98.1
NTU Singapore	12	1	1	1	1	90.8	95.8	99.2	98.6	98.1
U. of Pennsylvania	13	1	1	1	1	90.7	95.8	99.2	98.6	98.1
EPFL	14	1	1	1	1	90.2	95.8	99.2	98.6	98.1
Yale U.	15	1	1	1	1	90.2	95.8	99.2	98.6	98.1
The U. of Edinburgh	16	2	2	2	2	89.9	89.5	98.0	96.0	94.6
Tsinghua U.	17	2	5	7	2	89	89.5	92.3	85.5	94.6
Peking U.	18	2	5	2	2	88.8	89.5	92.3	96.0	94.6
Columbia U.	19	1	1	1	1	88.7	95.8	99.2	98.6	98.1
Princeton U.	20	1	5	1	1	88.6	95.8	92.3	98.6	98.1
Cornell U.	21	2	2	2	2	88.3	89.5	98.0	96.0	94.6
The U. of Hong Kong	22	1	1	1	1	86.3	95.8	99.2	98.6	98.1
The U. of Tokyo	23	1	5	7	6	86.2	95.8	92.3	85.5	81.6
U. of Michigan-Ann Arbor	24	3	5	3	3	86.2	83.5	92.3	94.1	91.2
Johns Hopkins U.	25	1	1	1	1	85.9	95.8	99.2	98.6	98.1
U. of Toronto	26	2	2	2	2	85.3	89.5	98.0	96.0	94.6
McGill U.	27	3	3	3	3	84	83.5	97.5	94.1	91.2
The Australian National U.	28	1	5	1	1	84	95.8	92.3	98.6	98.1
The U. of Manchester	29	2	2	2	2	84	89.5	98.0	96.0	94.6
Northwestern U.	30	2	5	2	2	82.8	89.5	92.3	96.0	94.6

In Fig. 2 we show the group sizes by each ranking method. The QS produces one-member clusters up to position 500 (Fig. 2b). After this position, universities are clustered by an unknown methodology. On the other hand, in Fig. 2a the Skyline layer sizes are shown. The unbounded Skyline produces 12 layers and their sizes are varying around 100. The Middle range Bounded Skyline produces the smallest high ranked sets (about 10–20 members) but finally generates a total number of only 14 layers. The averaged value Bounded Skyline produces bigger high ranked sets (about 20–30 members) and generates 17 layers. Finally, the median value Bounded Skyline produces even bigger high ranked sets (about 40–50 members) and a total number of 27 layers. For the cases of median and average Bounded Skylines we must notice the sharp increases of layer sizes at the lower ranked layers. This is caused by the phenomenon that the algorithm reaches a case that there are no universities that fulfill the bounding criteria, and therefore the unbounded variation is used for the specific layer.

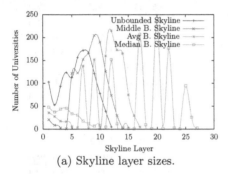
(a) Skyline layer sizes.

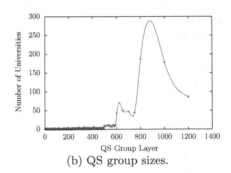
(b) QS group sizes.

Fig. 2. Rank groups' sizes.

Table 4 shows the Kendall τ, Spearman and Pearson coefficients of the resulting scores. It is self-evident that any comparison of RR scoring will have the same exactly Pearson and Spearman coefficients. All the RR variations have very high coefficients compared to each other. It is worth noticing that the middle range Bounded RR has a very high coefficient with the unbounded one. Also, it is worth noticing that the bounded variations are closer to the QS score than the unbounded RR one. The unbounded and the original QS are the most dissimilar methods. The Middle range RR is the most similar to both QS and unbounded RR, and thus combines the advantages of the QS scoring and the Skyline (and RR) notion of clustered ranking.

Table 4. Kendall τ, Spearman ρ, and Pearson r coefficients

		QS	RR	Middle RR	Avg RR	Median RR
Kendall τ	QS	–	0.65	0.68	0.67	0.66
	RR	–	–	0.92	0.84	0.72
	Middle RR	–	–	–	0.90	0.77
	Avg RR	–	–	–	–	0.81
	Median RR	–	–	–	–	–
Pearson r	QS	–	0.76	0.78	0.77	0.76
	RR	–	–	0.97	0.93	0.85
	Middle RR	–	–	–	0.96	0.88
	Avg RR	–	–	–	–	0.91
	Median RR	–	–	–	–	–
Spearman ρ	QS	–	0.78	0.79	0.79	0.78
	RR	–	–	0.97	0.93	0.85
	Middle RR	–	–	–	0.96	0.88
	Avg RR	–	–	–	–	0.91
	Median RR	–	–	–	–	–

Table 5. Top 20 Universities by NTU 2021.

Universities	Final Pos.	Score	11y arts	Cur. arts	11y Cits	Cur. Cits.	Avg Cits	H	HiCi papers	High J arts
Harvard U.	1	97.8	100	100	100	100	77.5	100	100	100
Stanford U.	2	64.3	55.7	66.7	53.3	65.6	77.3	87.9	55.8	61.6
U. of Toronto	3	61.2	65.4	71.2	52.8	60.9	63.3	81.2	48.9	54.4
U. of Oxford	4	60.1	56.4	66.6	50.7	62	71.7	79.8	47.9	55.4
Johns Hopkins U.	5	59.7	57.8	67.9	50.6	60.5	69.3	80.3	46.2	54.4
U. of London	6	59.2	59.2	67.5	48.6	61.1	64.4	80.3	46.8	54.9
U. of Washington, Seattle	7	57.8	53.4	62	47.6	59.6	71.1	82.9	46.6	50.9
MIT	8	57.6	46.8	53.5	49.8	54.8	90.9	78	50	50.3
U. of Michigan, Ann Arbor	9	56.8	57.3	66.9	46.6	57.4	63.6	78.4	42.5	51.7
U. of Cambridge	10	55.1	52.3	59.3	46.4	55.3	70.7	75.7	43.8	48.8
Imperial College London	11	54.7	50.5	59.6	42.9	56.7	67	79.8	41.8	51.1
Columbia U.	12	54.7	50.3	59.5	44.1	53.9	69.7	75.7	43.3	51.2
U. of Pennsylvania	13	54.6	52.4	60.7	45.4	52.4	68.5	71.7	43	51.6
Tsinghua U.	14	54.4	52.2	69.2	36.6	64.3	52	77.1	36.4	56.9
U. of California, Los Angeles	15	53.6	52.4	58.5	45.6	51.6	68.9	73.5	42.1	46.7
U. of California, San Francisco	16	52.8	45.4	53.2	43.1	49.7	78.8	73.1	40.2	50.6
Shanghai Jiao Tong U.	17	52.6	56.3	78.3	35.1	59.6	45.5	72.1	30.8	51
U. of California, San Diego	17	52.6	48	56.1	42.5	51.8	70.9	77.6	40.6	46.1
Zhejiang U.	19	52.3	55.7	79.3	34.9	58.6	45.6	67.7	31	51.5
U. of California, Berkeley	20	52.1	47.9	51.5	46	47.4	79.2	71.3	43.3	42.6

Table 6. Top 30 Universities by NTU 2021 ordered by Avg Bounded RR.

Universities	NTU Pos.	Simple Sky layer	Mid. BSky layer	Avg BSky layer	Med. BSky layer	Simple RR pos.	Middle RR pos.	Avg RR pos.	Median RR pos.
Harvard U.	1	1	1	1	1	1	1	1	1
Stanford U.	2	2	2	2	2	6	2	5	5
U. of Toronto	3	2	5	2	2	6	11	5	5
U. of Oxford	4	2	2	2	2	6	2	5	5
Johns Hopkins U.	5	2	2	2	2	6	2	5	5
U. of London, U. College London	6	2	5	2	2	6	11	5	5
U. of Washington, Seattle	7	3	3	3	3	17	5	11	12
Massachusetts Institute of T.	8	1	5	1	1	1	11	1	1
U. of Michigan, Ann Arbor	9	3	6	3	3	17	24	11	12
U. of Cambridge	10	4	4	4	4	32	8	20	21
Imperial College London	11	3	5	3	3	17	11	11	12
Columbia U.	12	3	3	3	3	17	5	11	12
U. of Pennsylvania	13	3	3	3	3	17	5	11	12
Tsinghua U.	14	2	5	16	13	6	11	143	141
U. of California, Los Angeles	15	4	4	4	4	32	8	20	21
U. of California, San Francisco	16	1	5	1	1	1	11	1	1
Shanghai Jiao Tong U.	17	2	5	16	21	6	11	143	214
U. of California, San Diego	17	4	4	4	4	32	8	20	21
Zhejiang U.	19	2	5	16	21	6	11	143	214
U. of California, Berkeley	20	1	5	1	1	1	11	1	1
Yale U.	21	5	5	5	5	46	11	27	30
Cornell U.	22	5	6	5	5	46	24	27	30
Peking U.	23	3	9	16	15	17	35	143	168
U. of Melbourne	24	4	9	4	4	32	35	20	21
Duke U.	25	6	6	6	6	63	24	36	39
U. of Copenhagen	26	5	7	5	5	46	29	27	30
Swiss Federal Inst of T. - Zurich	27	3	5	3	3	17	11	11	12
Paris-Saclay U.	28	4	9	4	4	32	35	20	21
Northwestern U.	29	5	6	5	5	46	24	27	30
Sorbonne U.	30	4	9	4	4	32	35	20	21

Tables 5 and 6 present the rank results of the NTU (National Taiwan University) Ranking. Table 5 presents the top 20 universities by NTU and their corresponding scores on NTU criteria. Table 6 presents the results of our algorithms. Column *simple Skyline Layer* shows the rank layer each university is clustered and column *simple RR pos.* shows the absolute position of the cluster's first university. These two columns are repeated for every Bounded RR variation. Harvard, MIT, San Francisco and Berkeley are clustered into layer 1 by Average Bounded RR and by Median Bounded RR as well. MIT, San Francisco and Berkeley are upgraded related to the original NTU rank. Stanford, Toronto, Oxford, Johns Hopkins and UCL, are clustered into layer 2 by the same methods.

We must notice the Tsinghua University, Shanghai Jiao Tong, Zhejiang Zhejiang and Peking University, who are ranked highly by the NTU applied weighted aggregate, they are all downgraded by both Average and Median Bounded RR.

Finally, as it was expected, the NTU experiment produces much smaller layers than the QS one. This occurs because the QS criteria are uncorrelated but the NTU ones are all scientometric indicators, so they are somehow correlated to each other.

5   Conclusion

University ranking have been heavily criticized for a number of reasons, including that they are based on arbitrary choices about the key indicators, their special weights and so on. For these reasons, we argue that there is no meaning in ranking universities in an absolute ordered manner; instead universities should be ranked in "sets of equivalence". To this end, we have used the Skyline notion and the Rainbow Ranking algorithm to perform experiments on a dataset extracted from QS Ranking.

Also, we have defined the Bounded Skyline concept. This can be used in collaboration with any dominance relationship. Also, any statistical concept can be used as a boundary function to split a set of values (mean, middle, average etc.). We proved that the use of a boundary function improves the ranking produced by the application of Rainbow Ranking, especially for the higher ranked elements, by producing smaller sets. As a future work, we plan to improve the behavior of Bounded Rainbow Ranking for the cases that an iteration produces an empty Bounded Skyline. An alternative policy could be, that in the case a Bounded Skyline set is empty, downgrade the boundary for all dimensions by 50%. For example, while using as bound the middle range (50% of the range values), after meeting an empty Skyline set, set the boundary to 25% of the range values. After meeting another empty set, set the boundary to 12.5% and so on, instead of eliminating it. That is a dynamically adapted Bounded Skyline. Another extension could be the reduction of bound strictness. For example, if none of the elements fulfills the boundaries in all dimensions, let's accept the elements that fail in the minimum number of dimensions. These possible improvements will be examined in a future work.

The Bounded RR is applicable to other entities, such as authors or other higher order multi-dimensional entities. The performance study is left as a future work, too.

References

1. Borzsony, S., Kossmann, D., Stocker, K.: The skyline operator. In: Proceedings of the 17th International Conference on Data Engineering (ICDE), pp. 421–430, Singapore (2001)
2. Chester, S., Assent, I.: Explanations for skyline query results. In: Proceedings of the 18th International Conference on Extending Database Technology (EDBT), Brussels, Belgium (2015)

3. Sidiropoulos, A., Gogoglou, A., Katsaros, D., Manolopoulos, Y.: Gazing at the skyline for star scientists. J. Informet. **10**(3), 789–813 (2016)
4. Stoupas, G., Sidiropoulos, A.: Multi-dimensional ranking via majorization. In: Proceedings of the 25th European Conference on Advances in Databases & Information Systems (ADBIS), pp. 276–286. Tartu, Estonia (2021)
5. Stoupas, G., Sidiropoulos, A., Gogoglou, A., Katsaros, D., Manolopoulos, Y.: Rainbow ranking: an adaptable, multidimensional ranking method for publication sets. Scientometrics **116**(1), 147–160 (2018)
6. Tiakas, E., Papadopoulos, A.N., Manolopoulos, Y.: Skyline queries: an introduction. In: Proceedings 6th International Conference on Information, Intelligence, Systems & Applications (IISA), Corfu, Greece (2016)
7. Voorneveld, M.: Characterization of Pareto dominance. Oper. Res. Lett. **31**(1), 7–11 (2003)

QLDT: A Decision Tree Based on Quantum Logic

Ingo Schmitt$^{(\boxtimes)}$ ⓘ

Brandenburgische Technische Universität Cottbus-Senftenberg, Cottbus, Germany
schmitt@b-tu.de
https://www.b-tu.de/fg-dbis

Abstract. Besides a good prediction a classifier is to give an explanation how input data is related to the classification result. Decision trees are very popular classifiers and provide a good trade-off between accuracy and explainability for many scenarios. Its split decisions correspond to Boolean conditions on single attributes. In cases when for a class decision several attribute values interact gradually with each other, Boolean-logic-based decision trees are not appropriate. For such cases we propose a quantum-logic inspired decision tree (QLDT) which is based on sums and products on normalized attribute values. In contrast to decision trees based on fuzzy logic a QLDT obeys the rules of the Boolean algebra.

Keywords: Quantum logic · Decision tree · Interpretable AI

1 Introduction and Related Work

Data mining is the task of finding patterns in a large data collection. Methods of supervised learning find a mapping, called a model, between input objects and a target property. For the classification task the target property is categorical. Without loss of generality, we discuss classification tasks where the target property distinguishes between two classes denoted by the values 0 and 1, respectively. Let D be a set of objects (tuples of reals). Every object is characterized by its values for n given attributes: $o = (o_1, \ldots, o_n)$. Furthermore, we assume the existence of a hidden mapping m from D to the classes, that is, $m : D \to \{0, 1\}$. We explicitly know the mapping only for a subset $O \subset D$. That is, we hold $M = \{(o, m(o)) | o \in O\}$. Let $TR \subset M$ be the set of training data and $TE = M \setminus TR$ be test data.

Solving a classification problem means to construct a mapping function $cl : D \to \{0, 1\}$, called a *classifier*, from given training data TR. The classifier should approximate m and should provide a prediction on TE with high accuracy. The *accuracy* of a classifier is quantified as the fraction of correctly classified objects of all test objects: $accuracy = |\{(o, m(o)) \in TE | m(o) = cl(o)\}| / |TE|$. In addition to a good accuracy a classifier should explain to users the connection between object attributes and the corresponding class, see [2]. A very popular

ⓒ Springer Nature Switzerland AG 2022
S. Chiusano et al. (Eds.): ADBIS 2022, CCIS 1652, pp. 299–308, 2022.
https://doi.org/10.1007/978-3-031-15743-1_28

classification method is the decision tree (DT), see [1]. The DT is based on rules of Boolean logic and can be seen as a good trade-off between accuracy and power in order to explain the classifier [3]. That is, in contrast to works like [7,15] we use logic as a means for explanation.

For finding the class of an object using a DT we navigate from the root to a leaf. Following such a path means to check conjunctively combined conditions on object attribute values. If we regard objects as points in $[0,1]^n$ then every tree node split on an attribute corresponds to one or more hyperplanes being parallel to $n-1$ axes.

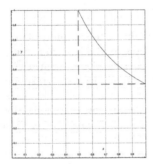 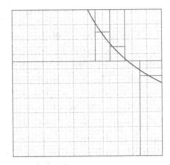

Fig. 1. (left) Class decision lines for $(x > 0.5) \wedge (y > 0.5)$ (dashed) and $x * y > 0.5$ (solid) and (right) space decomposition by axis-parallel decisions for $x * y > 0.5$

See Fig. 1 (left) for a two-dimensional case where the dashed line refers to the class separation for $(x > 0.5) \wedge (y > 0.5)$. For that class decision the attribute values interact conjunctively on the level of Boolean truth values. But what about scenarios where the interaction takes place on object values directly? See for example the solid class separation line in Fig. 1 (left) for $x * y > 0.5$. Let, for example, $x \in [0,1]$ encodes age and $y \in [0,1]$ encodes continuously the BMI of a person. Furthermore, the risk of severe health damage from COVID may increase gradually in the shape of a product of age and BMI. In that and similar cases, decision trees based on axis parallel decisions can only roughly approximate non-parallel decision lines and deteriorate, see Fig. 1 (right). A tighter approximation would lead to even more deterioration.

In contrast to the traditional decision tree based on Boolean decisions we develop a quantum-logic inspired decision tree (QLDT). Instead of combining Boolean values we regard attribute values from $[0,1]$ as results from quantum measurements and combine them directly by using negation, conjunction and disjunction following the concepts of quantum logic [9,12]. Different from fuzzy logic, quantum logic based on mutually commuting conditions obeys the rules of a Boolean algebra. Therefore, in contrast to decision trees based on fuzzy logic [5,6,10], every logical formula can be represented as a set of disjunctively combined minterms which are themselves conjunctions of positive or negated conditions (disjunctive normal form) [4]. After deriving a logic expression e in disjunctive normal form we generate a QLDT $(qldt(e))$ from it.

Referring to the solid decision line (left) in Fig. 1 we obtain the quantum logical expression $x \wedge y$ with $x, y \in [0, 1]$.

The evaluation of a traditional decision tree against an input object differs from the evaluation $[qldt(e)]$ of a QLDT. Starting from the QLDT root we navigate in a parallel manner to all leaves where for each leaf we obtain a leaf-specific evaluation value from $[0, 1]$. All evaluation values of class-1-leaves are summed up to a class value from $[0, 1]$. A final threshold τ is applied to the class value for a discrete class decision. The class decision can be written as $[qldt(e)] > \tau$ (in the example we yield $x * y > 0.5$).

For our quantum logic decision tree approach we identify the following advantages:

1. Quantum logic deals directly with continuous truth values;
2. In contrast to fuzzy logic our quantum-logic inspired approach obeys the rules of the Boolean algebra [8], for example $[e \wedge \neg e] = 0$ and $[e \wedge e] = [e]$;
3. Class separation lines are not restricted to be axis-parallel.[1]

In following sections we will develop the QLDT. It is based on the concepts of CQQL. The quantum-logic inspired language CQQL (commuting quantum query language) was introduced in [12, 14].

2   Commuting Quantum Query Language (CQQL)

Syntactically, a CQQL expression is an expression of propositional logic based on conjunction, disjunction, and negation. We assume n atomic, unary conditions on the n values of an object o. Such a condition expresses *gradually* whether an input value is a high value, e.g. a high BMI value. Each of the conditions returns a value from $[0, 1]$. The upper bound 1 is interpreted as *true* and the lower bound 0 as *false*. In [11] we prove that a quantum logic expression based on atomic conditions on different attributes form a Boolean (orthomodular, distributive) lattice.

A CQQL expression e in a specific syntactical normal form (CQQL normal form, see [11, 12, 14]) can be evaluated arithmetically. Each CQQL expression can be transformed into that normal form, see [12, 14].

Let the function $atoms(e)$ return the set of atomic conditions involved by a possibly nested condition e. The CQQL normal form requires that for each conjunction $e_1 \wedge e_2$ and for each disjunction $e_1 \vee e_2$ (but not for the special case of an exclusive disjunction) the atom sets are disjoint: $atoms(e_1) \cap atoms(e_2) = \emptyset$. If for $e_1 \vee e_2$ the conjunction $e_1 \wedge e_2$ is unsatisfiable in propositional logic then the disjunction is exclusive. We mark each exclusive disjunction by $\dot{\vee}$.

The evaluation of a CQQL expression e in the required normal form against an object o is written as $[\cdot]^o$. For brevity, we drop the object o and write just $[\cdot]$. In the following recursive definition of $[e]$, we distinguish five cases:

---

[1] Of course, there exist classification problems for which the decision tree based on Boolean logic fits perfectly.

1. Atomic condition: If e is an atomic condition then $[e] \in [0, 1]$ returns the result from applying the corresponding condition on o.
2. Negation: $[\neg e] = 1 - [e]$;
3. Conjunction: $[e_1 \wedge e_2] = [e_1] * [e_2]$;
4. Non-exclusive disjunction: $[e_1 \vee e_2] = [e_1] + [e_2] - [e_1] * [e_2]$; and
5. Exclusive disjunction: $[e_1 \veebar e_2] = [e_1] + [e_2]$.

We now extend the expressive power of a CQQL condition by introducing *weighted conjunction* ($e_1 \wedge_{\theta_1, \theta_2} e_2$) and *weighted disjunction* ($e_1 \vee_{\theta_1, \theta_2} e_2$). The work [13] develops the concept of weights in CQQL from quantum mechanics and quantum logic. Weight variables θ_1, θ_2 stand for values out of $[0, 1]$. A weight $[\theta_i] = 0$ means that the corresponding argument has no impact and a weight $[\theta_i] = 1$ equals the unweighted case (full impact). We regard every weight variable θ_i as a 0-ary atomic condition. Before we evaluate a condition with weights we map every weighted conjunction and weighted disjunction in e to an unweighted condition:

$$(e_1 \wedge_{\theta_1, \theta_2} e_2) \rightarrow ((e_1 \vee \neg \theta_1) \wedge (e_2 \vee \neg \theta_2))$$
$$(e_1 \vee_{\theta_1, \theta_2} e_2) \rightarrow ((e_1 \wedge \theta_1) \vee (e_2 \wedge \theta_2)).$$

For a certain classification problem we want to find a matching CQQL expression e together with a well-chosen output threshold value τ for cl: $cl_e^\tau(o) = th_\tau([e]^o)$ with $th_\tau(x) = \begin{cases} 1 \text{ if } x > \tau \\ 0 \text{ otherwise.} \end{cases}$

From the laws of the Boolean algebra we know that every expression e can be expressed in the complete disjunctive normal form, that is, every expression is equivalent to a subset of 2^n minterms. We implicitly assume for each of the n object attributes exactly one atomic condition c_j for $j = 1, \ldots, n$ and for an object $o = (o_1, \ldots, o_n)$ the equivalence $o_j = [c_j]^o \in [0, 1]$. The minterm subset relation for any logic expression can be expressed by use of minterm weights $\theta_i \in \{0, 1\}$:

$$e = \overset{\cdot 2^n}{\underset{i=1}{\bigvee}} minterm_{i, \theta_i} \text{ mapped to } \overset{\cdot 2^n}{\underset{i=1}{\bigvee}} minterm_i \wedge \theta_i = e \qquad (1)$$

and $minterm_i = \bigwedge_{j=1}^n c_{ij}$ with $c_{ij} = \begin{cases} c_j & \text{if } (i-1) \And 2^{j-1} > 0 \\ \neg c_j & \text{otherwise} \end{cases}$.

That is, the value $i - 1$ is considered as a bitcode and identifies a minterm uniquely and $j - 1$ stands for a bit position. The symbol '$\And$' stands for the bitwise and.

Please note that the disjunction of any two different complete minterms is always exclusive. Thus, e is in CQQL normal form and its evaluation against object o yields

$$[e]^o = \sum_{i=1}^{2^n} \theta_i \prod_{j=1}^n [c_{ij}]^o. \qquad (2)$$

3 Extraction of Minterms and Finding the Output Threshold

Next, we will extract a CQQL expression e in complete disjunctive normal form from training data. We have to find the weight θ_i for every minterm i. The starting point is the training set $TR = \{(x, y)\} = \{(o, m(o))\}$ with the input tuples (objects) $x = o \in [0, 1]^n$ and $y = m(o) \in \{0, 1\}$.

One important requirement for a classifier is high accuracy. Therefore, we maximize the accuracy of expression (1) depending on the minterm weights θ_i based on $TR = \{(x, y)\}$.

Accuracy acc for a continuous evaluation can be measured as sum over the two correct cases $(y = 1) \wedge [e]^x$ and $(y = 0) \wedge [\neg e]^x$ over all pairs $(x, y) \in TR$:

$$acc = \sum_{(x,y) \in TR} (y * [e]^x + (1-y)(1 - [e]^x))$$

$$= \sum_{i=1}^{2^n} \theta_i \sum_{(x,y) \in TR} \left((2y-1) \cdot \prod_{j=1}^{n} [c_{ij}]^x \right) + \sum_{(x,y) \in TR} (1-y).$$

We see after applying Eq. (2) and some reformulations, that accuracy shows a linear dependence on the minterm weight θ_i for fixed TR-pairs. The first derivative yields a constant gradient on θ_i:

$$\frac{\partial acc}{\partial \theta_i} = \sum_{(x,y) \in TR} (2y-1) \cdot \prod_{j=1}^{n} [c_{ij}]^x = \sum_{(x,1) \in TR} \prod_{j=1}^{n} [c_{ij}]^x - \sum_{(x,0) \in TR} \prod_{j=1}^{n} [c_{ij}]^x.$$

For maximizing accuracy a minterm weight θ_i should have the value 1 if $\frac{\partial acc}{\partial \theta_i} > 0$ and 0 otherwise. In other words, for the decision whether a minterm should be active or not it is sufficient to compare the impact of positive training data E_i against the impact of the negative training data N_i with

$$E_i = \sum_{(x,1) \in TR} \prod_{j=1}^{n} [c_{ij}]^x \text{ and } N_i = \sum_{(x,0) \in TR} \prod_{j=1}^{n} [c_{ij}]^x.$$

Please note that the decision depends on the relative number of positive training objects in TR. Therefore, let $\gamma_1 = |\{(x, 1) \in TR\}|$ and $\gamma_0 = |\{(x, 0) \in TR\}|$ be the number of positive and negative training objects, respectively. The fraction of negative objects is then given by $\gamma = \frac{\gamma_0}{\gamma_1 + \gamma_0}$. In the unbalanced case $(\gamma \neq 1/2)$ we compensate the effect on the minterm weight decision by:

$$[\theta_i] = \begin{cases} 1 & \text{if } \gamma \cdot E_i > (1-\gamma) \cdot N_i \\ 0 & otherwise \end{cases}. \tag{3}$$

Following that minterm decision rule we can decide for every minterm whether it is active or inactive. In case of $\gamma \cdot E_i \approx (1-\gamma) \cdot N_i$ the decision is not clear. We call such kind of minterms *unstable* because adding a single new training object may change the decision. Instable minterms are less expressive then stable ones and have a low impact on accuracy. We are interested in stable minterms. For measuring stability we compute the ratio ρ_i of $\gamma \cdot E_i$ to the sum of $\gamma \cdot E_1$ and $(1 - \gamma) \cdot N_i$ of a minterm i. A value for ρ_i close to $1/2$ indicates an unstable minterm i, a value near 1 means a stable active minterm i, and a value near 0 means a stable inactive minterm i.

The question arises: should instable minterms be active or inactive? We propose to sort all minterms by their values ρ_i and choose a ρ-threshold θ_ρ from them that provides a good trade-off between accuracy and compactness of expression e. The modified minterm decision rule is:

$$[\theta_i] = \begin{cases} 1 & \text{if } \frac{\gamma E_i}{\gamma E_i + (1 - \gamma) N_i} > \theta_\rho \\ 0 & otherwise \end{cases}. \tag{4}$$

After applying our minterm decision rule (4) we obtain the expression:

$$e = \bigvee_{i:[\theta_i]=1} \bigwedge_{j=1}^{n} c_{ij} \text{ with } [e]^x = \sum_{i:[\theta_i]=1} \prod_{j=1}^{n} [c_{ij}]^x.$$

Next we have to find the output threshold value τ for $cl_e^\tau(x) = th_\tau([e]^x)$. Let $\min_1 = \min_{(x,1)\in TR}[e]^x$ be the smallest evaluation result of the positive training objects and $\max_0 = \max_{(x,0)\in TR}[e]^x$ be the highest result of the negative training objects. In case of $\max_0 < \min_1$, positive objects and negative objects are well separated and we set τ to $(\max_0 + \min_1)/2$.

Otherwise, we have to choose a value τ from the interval $[\min_1, \max_0]$. In order to find a threshold which maximizes discrete accuracy we use the training objects from TR:

$$\tau = \operatorname*{arg\,max}_{\substack{(x,\_) \in TR \\ \tau_x := [e]^x \\ \tau_x \in [\min_1, \max_0]}} accuracy(e, \tau_x, TR)$$

where $accuracy(e, \tau_x, TR) = |\{(x,y) \in TR | y = cl_e^{\tau_x}(x)\}|/|TR|$.

4   Quantum Logic Decision Tree

Our expression e has so far been a disjunction of active minterms. The last step is to generate a quantum-logic inspired decision tree from e. The QLDT is just a compact presentation of the CQQL expression e based on active minterms. The basic idea is to regard the derived minterm weights as training data and to use a traditional DT algorithm for constructing the QLDT. The training data for the decision tree construction are just the bit values 0 and 1 from the binary code $bitcode(i - 1) = x_1 \ldots x_n$ of all minterm identifiers i: The bit values of the bitcode are regarded as attribute values and the minterm weight as target for the DT algorithm:

$$TR' = \{(bitcode(i - 1), [\theta_i]) \mid i = 1, \ldots, 2^n\}.$$

For example, following minterm weights for $n = 3$ and $e = (x_1 \wedge x_2 \wedge x_3) \vee (\neg x_1 \wedge x_2 \wedge x_3) = x_2 \wedge x_3$ with evaluation $[e]^x = x_2 * x_3$ produce the QLDT shown on the right hand side. The leaves correspond to the minterm weights:

i	x_1	x_2	x_3	$[\theta]$
1	0	0	0	0
2	0	0	1	0
3	0	1	0	0
4	0	1	1	1
5	1	0	0	0
6	1	0	1	0
7	1	1	0	0
8	1	1	1	1

But what about the evaluation of a QLDT? The generated decision tree looks like a traditional decision tree. However, its evaluation differs. The evaluation result of a QLDT should be the same as the evaluation result of e in disjunctive normal form:

$$[e]^x = \left[\dot{\bigvee}_{i:[\theta_i]=1} \bigwedge_{j=1}^{n} c_{ij} \right]^x = \sum_{i:[\theta_i]=1} \prod_{j=1}^{n} [c_{ij}]^x. \tag{5}$$

Please note, that the disjunction of minterms is always exclusive and, that is why, its evaluation leads to a simple sum. The same holds for a split node of a QLDT. A QLDT assigns implicitly to every leaf a set of minterms sharing the same path conditions (split attributes). Let L_A refers to the set of all active leaves where every leaf is represented by the identifier i of one of its assigned minterms and let $path_i$ be the path from the root to leaf i. Then the evaluation of a QLDT is given by:

$$[qldt(e)]^x = \sum_{i \in L_A} \prod_{j \in path_i} [c_{ij}]^x = [e]^x = \sum_{i:[\theta_i]=1} \prod_{j=1}^{n} [c_{ij}]^x.$$

Actually, the evaluation of a QLDT takes into account only active leaves. The inactive leaves correspond to $\neg e$ and are connected to the active leaves by $[e]^x + [\neg e]^x = 1$. Therefore, the inactive leaves are removed from the QLDT without any loss of semantics and as result the QLDT becomes more compact.

So, we end up with the QLDT classifier:

$$cl_e^\tau(x) = th_\tau([qldt(e)]^x) = \begin{cases} 1 \text{ if } \sum_{i \in L_A} \prod_{j \in path_i} [c_{ij}]^x > \tau \\ 0 \text{ otherwise.} \end{cases}$$

5   Experiment

Next, we shall apply our approach to an example scenario and compare the resulting traditional decision tree with the generated quantum-logic inspired decision tree.

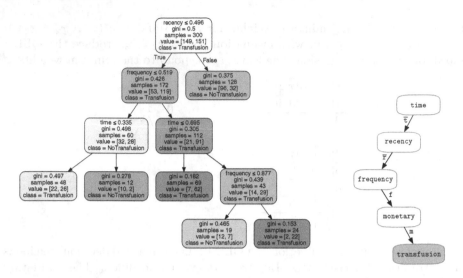

Fig. 2. Traditional decision tree for blood transfusion with 6 leaves (left) and quantum-logic inspired decision trees with best accuracy (right)

Our experimental dataset is the blood transfusion service center dataset.[2] A classifier needs to be found which predicts whether a person donates blood or not. To make that decision for every person we know `recency` (months since last donation), `frequency` (total number of donation), `monetary` (total blood donated in c.c.), and `time` (months since first donation). In the balanced case 178 people donated blood and 178 did not. 300 of the 356 people belong to the training set and the remaining ones to the test set. For further processing the attribute values of the four attributes are mapped to the unit cube $[0, 1]^4$ using a normalized rank position mapping. For that mapping all values of an attribute are sorted. Then, every value is mapped to its rank position divided by the total number of values. For tied values the rank maximum is taken.

We obtain a traditional decision tree with highest accuracy of 64%, 6 leaves, and 5 levels, see Fig. 2 (left). The QLDT, see Fig. 2 (right), with its highest accuracy of 70% is achieved when we choose $\theta_\rho = 0.71$ and $\tau = 0.045$. That QLDT corresponds to the expression

$$e := \neg t \wedge \neg r \wedge f \wedge m$$

and its evaluation is

$$[e] = [\neg t] \cdot [\neg r] \cdot [f] \cdot [m].$$

---

[2] https://archive.ics.uci.edu/ml/datasets.php.

Table 1. Differences between Boolean-logic-based decision tree and quantum-logic inspired decision tree

Criterion	DT	QLDT
Logic	Boolean logic	Quantum logic
Thresholds	On node level	On tree level
No. of classes	Many-class-classifier	One-class-classifier
No. of node children	≥ 2	<2
Type of attributes	Categorical, ordinal, metric	Ordinal, metric
Class decision	Based on one single leaf	Based on all leaves
Complexity	At most exponential	Exponential

6 Conclusion

In our paper we suggest a decision tree classifier based on quantum logic. In contrast to Boolean logic, quantum logic can directly deal with continuous data, which is beneficial for many classification scenarios. Other than fuzzy logic, our quantum logic approach obeys the rules of the Boolean algebra. Thus, Boolean expressions can be transformed accordingly and a check of hypotheses against a Boolean expression becomes feasible.

In Table 1 we compare Boolean-logic-based decision trees with our quantum-logic inspired decision tree with respect to several criteria.

In a Boolean-logic-based decision tree every split decision at node level represents Boolean conditions and can be regarded as axis-parallel decision lines within the input space. For a QLDT the restriction on axis-parallel lines does not hold. A QLDT is appropriate in scenarios where the classification decision relies on sums and products rather than on a combination of Boolean values. The input for the threshold-based class decision is the evaluation result of the QLDT.

A QLDT represents syntactically a Boolean expression for one class. Thus, a QLDT classifier is a one-class classifier. Many-class classification problems can be transformed to multiple one-class classifier decisions.

Every inner QLDT node corresponds to exactly one logical expression with two outcomes. Since we drop 0-class leaves, some inner nodes have only one child. In contrast, a split rule in a traditional decision tree leads to more than or exactly two children.

For our QLDT we assume attribute values from the unit interval. The normalized rank position mapping maps ordinal and metric attribute values to the unit interval. But what about categorical attributes? We did not discuss this aspect but it can be easily solved by introducing an artificial attribute for each category.

As discussed above, the class decision in a traditional decision tree depends on exactly one leaf in contrast to the sum of all leave scores in a QLDT.

The complexity of a Boolean-logic-based decision tree for a number n of attributes corresponds to the number of leaves. In the worst case, that number is exponential. However, in many real-case-scenarios the number of leaves is much lower. In contrast, the generation of the QLDT requires to compute the weight for all 2^n minterms. Thus, too many attributes cause a complexity problem more for the QLDT rather than for a Boolean-logic-based decision tree.

References

1. Aggarwal, C.C.: Data Mining: The Textbook. Springer, Cham (2015). https://doi.org/10.1007/978-3-319-14142-8
2. Doshi-Velez, F., Kim, B.: Towards a rigorous science of interpretable machine learning. arXiv preprint arXiv:1702.08608 (2017)
3. Freitas, A.A.: Comprehensible classification models: a position paper. ACM SIGKDD Explor. Newsl. **15**(1), 1–10 (2014)
4. Hüllermeier, E., Schmitt, I.: Non-additive utility functions: Choquet integral versus weighted DNF formulas. In: Gaul, W., Geyer-Schulz, A., Baba, Y., Okada, A. (eds.) German-Japanese Interchange of Data Analysis Results. SCDAKO, pp. 115–123. Springer, Cham (2014). https://doi.org/10.1007/978-3-319-01264-3_10
5. Janikow, C.Z.: Fuzzy decision trees: issues and methods. IEEE Trans. Syst. Man Cybern. Part B (Cybernetics) **28**(1), 1–14 (1998)
6. Jiménez, F., Martínez, C., Marzano, E., Palma, J.T., Sánchez, G., Sciavicco, G.: Multiobjective evolutionary feature selection for fuzzy classification. IEEE Trans. Fuzzy Syst. **27**(5), 1085–1099 (2019)
7. Lundberg, S.M., Lee, S.I.: A unified approach to interpreting model predictions. In: Advances in Neural Information Processing Systems, vol. 30 (2017)
8. Mittelstaedt, P.: Quantum logic. In: PSA 1974, pp. 501–514. Springer, Cham (1976). https://doi.org/10.1007/978-94-010-1449-6_28
9. Mittelstaedt, P.: Quantum Logic. D. Reidel Publishing Company, Dordrecht (1978)
10. Olaru, C., Wehenkel, L.: A complete fuzzy decision tree technique. Fuzzy Sets Syst. **138**(2), 221–254 (2003)
11. Schmitt, I.: Quantum query processing: unifying database querying and information retrieval. Citeseer (2006)
12. Schmitt, I.: QQL: a DB&IR query language. VLDB J. **17**(1), 39–56 (2008)
13. Schmitt, I.: Incorporating weights into a quantum-logic-based query language. In: Aerts, D., Khrennikov, A., Melucci, M., Toni, B. (eds.) Quantum-Like Models for Information Retrieval and Decision-Making. SSTEAMH, pp. 129–143. Springer, Cham (2019). https://doi.org/10.1007/978-3-030-25913-6_7
14. Schmitt, I., Baier, D.: Logic based conjoint analysis using the commuting quantum query language. In: Algorithms from and for Nature and Life, pp. 481–489. Springer, Cham (2013). https://doi.org/10.1007/978-3-319-00035-0_49
15. Strumbelj, E., Kononenko, I.: An efficient explanation of individual classifications using game theory. J. Mach. Learn. Res. **11**, 1–18 (2010)

DOING: 3rd Workshop on Intelligent Data – From Data to Knowledge

DOING: 3rd Workshop on Intelligent Data – From Data to Knowledge

Workshop Chairs

Mirian Halfeld Ferrari   University of Orleans, LIFO, France
Carmem Satie Hara   Federal University of Parana, Curitiba, Brazil

Program Committee Members

Ahmed Awad   University of Tartu, Estonia
Cheikh Ba   University of Gaston Berger, Senegal
Davide Buscaldi   Sorbonne University, France
Karin Becker   Federal University of Rio Grande do Sul, Brazil
Javam de Castro Machado   Federal University of Ceara, Brazil
Laurent d'Orazio   University of Rennes, France
Danilo Giordano   Polytechnic University of Turin, Italy
Sven Groppe   University of Lubeck, Germany
Nicolas Hiot   University of Orleans, France
Jixue Liu   University of South Australia, Australia
Wagner M. N. Zola   Federal University of Parana, Brazil
Anne-Lyse Minard-Forst   University of Orleans, France
Dung V. N. Nghiem   University of Orleans, France
Agata Savary   University of Tours, France
Rebeca Schroeder Freitas   State University of Santa Catarina, Brazil
Aurora Trinidad Ramirez Pozo   Federal University of Parana, Brazil
Domagoj Vrgoc   Pontifical Catholic University of Chile, Chile

Additional Reviewer

Jose Maria da Silva Monteiro Filho

Using Provenance in Data Analytics for Seismology: Challenges and Directions

Umberto Souza da Costa[4], Javier Alfonso Espinosa-Oviedo[3],
Martin A. Musicante[4], Genoveva Vargas-Solar[1]($\boxtimes$),
and José-Luis Zechinelli-Martini[2]

[1] CNRS, Univ Lyon, INSA Lyon, UCBL, LIRIS, UMR5205,
69622 Villeurbanne, France
`genoveva.vargas-solar@cnrs.fr`
[2] Fundación Universidad de las Américas Puebla, 72820 San Andrés Cholula, Mexico
`joseluis.zechinelli@udlap.mx`
[3] CPE Lyon, LIRIS, Université de Lyon, Villeurbanne, France
`javier.espinosa-oviedo@cpe.fr`
[4] Universidade Federal Rio Grande do Norte, DIMAp, Natal, Brazil
`{umberto,mam}@dimap.ufrn.br`

Abstract. We analyze data and meta-data modeling challenges to provide curated collections that can be easy to explore. Data exploration can use provenance tools to give insight into the conditions in which data are collected. We are concerned with data curation and exploration in seismic geophysics. We believe that the tasks involved in graph exploration depend highly on the knowledge domain. The discussion about possible solutions is driven by the hypothesis that graphs can be well-adapted data models to represent, explore and analyze seismic data. Given that data curation is done by human agents, it is essential to provide automatic tools to add provenance to seismic data.

Keywords: Data curation · Metadata extraction · Graph database design · Data exploration · Graph analytics and querying

1 Introduction

With the unprecedented volumes of heterogeneous data automatically collected and available, data collections must be heavily transformed before consumers can use them. Data curation approaches [1] have defined strategies and protocols to maintain data collections and contribute to preparing and integrating datasets to perform analytics tasks. Curation tasks include extracting meta-data and integrating semantic information. Data-driven analytics and experimentation have quality requirements concerning the validity of the data and results. For example, Geophysics experiments and analyses to study the seismic behavior

---

This work is partially funded by the project ADAGEO funded by the CNRS EDI program. Authors are listed in alphabetical order.

S. Chiusano et al. (Eds.): ADBIS 2022, CCIS 1652, pp. 311–322, 2022.
https://doi.org/10.1007/978-3-031-15743-1_29

of specific zones rely on data collections produced by observation stations (seismographs) and labeled manually by experts. Observation stations are located in different environments and submitted to conditions that can perturb sensed data. Their location determines the signals detected, for instance, near a mine, the sea, or a crowded urban space. Studying seismic behavior using the data collected from these sensors must consider this provenance meta-data to better drive conclusions. For instance, there was an extreme seismic event in Puebla city, located near the seismic zone along a geological fault line of the *Sierra Madre Occidental*, and it is a mine zone close to a risk zone of the Popocatepetl volcano. Geophysicists must compare data from different observation stations to classify an observation as an earthquake or a human-generated seismic event.

Exploratory search allows us to discover and understand relevant data to answer the informational needs of users. This data exploration task is critical for driving conclusions. Therefore, data and provenance meta-data should be associated to guide the construction of datasets. Next-generation data management engines should aid the user in understanding the data collections' content and guide to exploring data.

This position paper analyses data and meta-data modeling challenge to provide curated collections that can be easily explored. Data exploration can be done considering data provenance to give insight into the conditions in which data are collected. For instance, the device used to collect data, the human/synthetic agents that analyse data, etc. We focus on geophysics data curation and exploration because we believe these tasks depend highly on the knowledge domain. We postulate that:

- Graph data models can be well adapted for representing and storing the data and (provenance) meta-data mesh.
- Curation processes can consider enriching/completing graphs (data and their provenance) using link discovery techniques.
- Exploration can be done (semi) declaratively using domain-specific languages that can traverse and analyze graphs to extract the portions of data pertinent to given analytics tasks.

These postulates call for addressing graph database design, processing, and querying challenges discussed in this paper and put in perspective with graph modeling, processing, and querying existing work. Accordingly, the remainder of this paper is organized as follows. Section 2 introduces related work about data provenance and quality and data exploration techniques. Section 3 describes a motivation example that shows the need for a provenance-guided exploration for selecting data for performing geophysics analytics processes. Section 4 introduces our vision for modeling geophysics data collections and provenance using graphs. It proposes the general lines for defining a domain-specific graph exploration language for geophysics data collections. Finally, Sect. 5 concludes the paper and discusses research directions and opportunities.

2 Related Work

Provenance refers to sources of information involved in producing or delivering a product. The provenance role is crucial in deciding whether the information is reliable, integrating it with other diverse sources of information, and giving credit to its originators when using it. Provenance, therefore, provides a critical basis for assessing authenticity, enabling booth trust and reproducibility.

Despite its importance for the usability of information, tracking provenance, sharing data, and integrating them from multiple sources according to their provenance remains an open issue [13]. Several workflow management systems and Semantic Web systems have been developed and are actively used by communities of scientists. Many of these systems implement some form of provenance tracking internally and have begun standardising some common representations for provenance data to allow for the exchange and integration [5].

Provenance Models for Databases. A database can be broadly understood as a data repository that enables queries. Often, information about the data itself is also stored, indicating, in addition to its value, its type, and other restriction rules [2]. Data collected about data routinely may fall into the category of provenance information, *e.g.* creation date, creator, instrument or software used, data processing methods, etc. In this way, good data management practices are the basis for accurately recording provenance. Provenance is recorded as metadata that may include items used to compile provenance information: plain text files, spreadsheets, file names, databases, etc. Provenance data can be represented using different data models (relational, graph, documents) and are usually associated with the data items to which they refer.

The PROV-DM model [5] is an extension of the PROV model [8] released by the Provenance Interchange Working Group and recommended by the W3C. PROV-DM promotes the interoperable exchange of provenance information in heterogeneous environments. PROV-DM is defined using an abstract relational model and an OWL ontology, with various serializations including RDF and XML. PROV is generic and domain-independent. It provides extension points through which such systems and applications can extend PROV for their purposes [4].

Provenance-Based Querying on Graphs. Graph databases are a specific database type that falls under the NoSQL (Not Only SQL) category. Graph databases are composed of data items (the nodes of the graph) and links between them (the edges of the graph). Properties may be associated with the nodes. Popular graph database systems include RDF/SPARQL [12] and Neo4J [6]. The aspect of providing provenance explanations for query results has been mostly neglected. Based on query rewriting, the method, SPARQLprov [3] computes "how-provenance polynomials" for SPARQL queries over knowledge graphs. SPARQLprov has been evaluated on "real" and synthetic data. Results show that it incurs good runtime overhead w.r.t. the original query, competing with state-of-the-art solutions for how-provenance computation. The work in [9] establishes a translation between a formalism for dynamic programming over hypergraphs and the computation of semiring-based provenance for Datalog programs. The approach proposes a new method for computing provenance for a specific class of semirings.

Provenance in Scientific Workflows. Scientific workflows implement online data-driven experiments of specific experimental sciences (biology, geosciences, physics, etc.). Scientific workflow provenance, both for the data they derive and their specification, is essential to allow for reproducibility, sharing, and knowledge re-use in the scientific community. Harvesting provenance for streaming workflows introduces challenges related to the characteristics of scientific workflows. Executing these workflows produces a high rate of updates and promotes a large distribution of the tasks spread across several institutional infrastructures. Since activities are often externalized can be an obstacle to enabling provenance metadata extraction procedures. According to the target experimental science, maintaining and managing provenance in scientific workflows must be often specialized. For example, the work proposed by A. Spinuso et al. [11] is an example of an approach dealing with provenance in seismological processing workflows. The work by S. Bowers et al. [1] captures the dependencies of data and collection creation events on preexisting data and collections and embeds these provenance records within the data stream. A provenance query engine operates on self-contained workflow traces representing serializations of the output data stream for particular workflow runs. The bottom line is analyzing the large amounts of provenance data generated by workflow executions and extracting valuable knowledge of this data [7].

Discussion. According to the taxonomy proposed in [10] provenance can be used for describing processes and data under a fine or coarse-grained approach. It can be represented by syntactic and semantic information by annotating entities. Provenance data can be disseminated through visual graphs, queries, and services (providers). The work argues about provenance modeling for tagging data and processes using syntactic and numerical meta-data. We believe that provenance models must be close to the knowledge domain of data and processes. Therefore, the perimeter of our proposal is defined by seismologic data and provenance meta-data.

This work addresses the questions and open issues associated with the provenance of data-driven geophysics and seismology scientific workflows -our work models provenance regarding the data, activities, and agents participating in the workflow. We focus on modeling the meta-data used to deal with these components' provenance to curate seismic data and guide their exploration.

3   Motivation Use Case

Let us now present a motivational example. We focus on the data produced by a set of seismographs distributed across a large geographic region. Our data originated from seismographs in the Brazilian northeast in our case. The data produced by the seismographs are used to generate a weekly bulletin identifying seismic activity during a period of interest and create a database of seismic data to be used by researchers and students. The database should be tagged with provenance metadata.

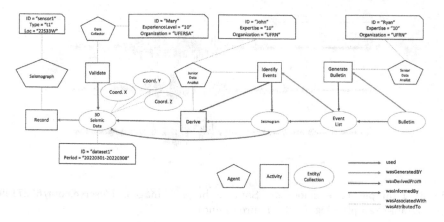

Fig. 1. Provenance graph for seismic bulletins.

Our provenance graph describes activities and the entities they produce. Several agents may affect this process. The provenance model introduces types and their relationships. Figure 1 shows a provenance graph for our scenario. The agents in charge of activities in this example are:

Seismographs: Devices that produce raw data.

Data Collector: Human agent, responsible for retrieving and validating the data from seismographs.

Junior Data Analyst: Human agent that identifies seismic events from the collected data. She produces a list of the seismic events in the bulletin.

Senior Data Analyst: Human-agent, responsible for checking the list of events produced by the Junior Analyst and generating the bulletin to be published.

Entities in our context correspond to data items and data collections. They are depicted as ovals in Fig. 1. In our example, we have the following:

3D Seismic Data: This collection comprises files that record seismic movements for a given period and location. One file for each dimension (North-South, East-West, and Up-Down). Each file contains a list of pairs formed by a timestamp and an integer value. Typical sampling rates range between 50 and 4000 records per second.

Seismogram: Corresponds to graphical representations of ground movements. These graphics depict the data acquired by seismographs of a given station, being analysed by a Junior Data Analyst to identify geological events, like earthquake and mining explosions, and their magnitudes (see Fig. 2).

Figure 2 shows two vertical lines marked as P wave (in red) and S wave (in green). The analyst creates these lines to indicate, respectively, the arrival to the sensor of the *primary* and *secondary* waves of a seismic event. Both waves are created at the same time by the seism. The primary wave travels through any

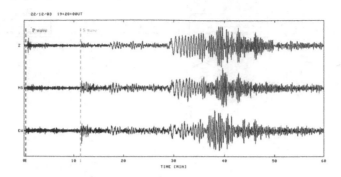

Fig. 2. Example of seismogram. Source: https://image3.slideserve.com/6627149/seismogram-example-1.jpg. (Color figure online)

media and arrives faster at the sensor. The secondary wave travels only through solid media and arrives later. The seismic analyst is in charge of identifying the seismic event by identifying the moments of arrival o these waves and the end of the event.

Event List: This is the result of the analysis of the Junior Data Analyst, reporting the events identified on Seismograms. The event list also includes the location of the seism epicenter, calculated by triangulation from the data registered by several seismographs.

Bulletin: The bulletin is the final document reporting the detected events and their properties, including locations and magnitudes, along a given period. This document needs to be certified by a Senior Data Analysts before its release to the general public.

So far, the data has been used to generate a bulletin from seismic activity. However, other problems can be solved using the collected data by seismographs. We describe an example in the following lines.

Using One Sensor Data to Locate the Epicenter of a Seism. This process uses the data collected by just one seismograph. The direction from where the seism originated is calculated by considering the initial movement detected by the hardware on each dimension (north-sud, east-west, and up-down). The distance from the epicenter to the sensor is given by the difference (in time) among the waves P and S, and the soil class around the seismograph. Notice that the P and S waves originated simultaneously at the epicenter. The P wave is faster than the S wave.

4   Exploring Geophysics Data Guided by Provenance

Geophysics data exploration is an analytics process that seeks to process data collection content (produced by metrology devices) for answering factual and analytics queries. Since the exploration is intended to support the manual or

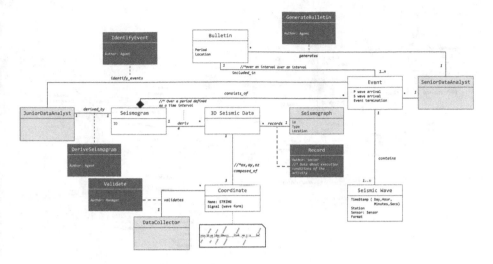

Fig. 3. Provenance based seismic events data (Color figure online)

semi-automatic identification of geophysics events, it is critical to produce results that answer queries, but that report the provenance of identified events and the conditions in which data and events are observed, detected, validated, and disseminated.

The first challenge is identifying and modeling the seismic and provenance concepts. We adopt a database-oriented approach for building a database of geophysical data and metadata with the following steps: (1) Designing the data and meta-data according to a real case described in the previous section (UML class diagram); (2) Transform the design into seismic and provenance property graphs; (3) Identifying exploration query types that can be asked on top of the seismic and provenance graphs.

Seismic and Provenance Data Design. Figure 3 illustrates the UML class diagram that combines the concepts of the seismic data (in white) and the associated provenance concepts (in violet and green). The diagram models the concepts representing seismic events detection, including the type of agents and actions that produce, validate, and process the data. These entities represent the meta-data used for tagging the seismic data with provenance.

Regarding data, in the central concept is an 3D Seismic Wave, that is built by processing 3 Coordinates (x,y,z). A Seismogram is derived with a set of 3D Seismic Waves. It contains the information an analyst uses to identify seismic events and build an Events list. An Event refers to a Seism or other earth movement. Each event has a duration and contains timestamps for its P and S waves and Termination. Events produced in an interval are reported in a Bulletin.

Provenance meta-data are of three types agents Data Collector, Junior Data Scientist and Senior Data Scientist (green), and Actions (violet). In UML, verbs (i.e., actions) are modelled by relations between classes, in consequence, as shown in

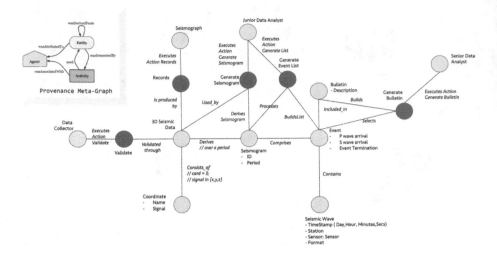

Fig. 4. Provenance and seismic graphs.

the diagram, Action's are associated to relations and Agents's are associated with (perform) actions. For example, a Seismogram is derived from a 3D Seismic Data by an agent of type Junior Scientist.

Seismic and Provenance Property Graphs. Once we have modeled the concepts representing the data and meta-data of seismic events detection and dissemination, we assume that graphs are the most adapted data model for storing these geophysics (meta)-data. Graphs can be then explored with factual and analytics queries that can produce results and associated explanations with provenance. We adopt a properties graph model to model seismic data and provenance meta-data for a first approach. Then it is possible to identify the type of queries that can be asked and evaluated on top of this type of graph.

Figure 4 shows a sample of the property graph schema of two interconnected graphs representing seismic data and provenance meta-data. The principle of our design is to separate data and meta-data to open the possibility of associating different meta-data "spaces" with the same data.

Notice that an alternative design strategy would be to add properties to the relations and nodes of the data graph. This second strategy is closer to relational provenance approaches, where attributes are added to the relational schema to tag tuples with provenance meta-data. The seismic data graph (nodes in grey in Fig. 4) is defined as follows:

```
consists_of(3DSeismicData, Coordinate)
derives(Seismogram, 3DSeismicData)
comprises(Seismogram, Event)
included_in(Event, Bulletin)
contains(SeismicWave, Event)
```

The provenance meta-data graph represents five action types performed on data during the event detection process (done manually by geophysics technicians or scientists). The provenance meta-data graph (nodes in green and violet in the figure) is connected with the seismic graph, and it is defined as follows:

```
Builds(GenerateBulletin, Bulletin)
Selects(GenerateBulletin, Event)
ExecutesActionGenerateBulletin(SeniorDataScientist, GenerateBulletin)

BuildsList(GenerateEventList, Event)
Processes(GenerateEventList, Sesimogram)
ExecutesActionGenerateList(JuniorDataAnalyst, GenerateEventList)

ExecutesActionGenerateSeismogram(JuniorDataAnalyst, GenerateSeismogram)
ExecutesActionRecords(Seismogram,Records)
ExecutesActionValidate(DataCollector,Validate)

DerivesSeismogram(GenerateSeismogram, Seismogram)
UsedBy(GenerateSeismogram, 3DSeismicDiagram)
ValidatedThrough(Validate, 3DSeismicDiagram)
```

It is through actions that the graph is connected with the seismic graph.

Expressing Provenance Based Queries for Exploring Seismic Graphs. The event detection and analysis process within the datasets collected by seismic stations is exploratory in geophysics. Collected data are periodically downloaded from stations and archived so geophysicists can explore signals and detect events produced in specific intervals and locations. Analytics results are plots, event histories, and bulletins. The manual and meticulous process performed on independent data batches makes it challenging to correlate the observations on different data collections and prevents discovering hidden knowledge. Since the detection is manual, it is essential to know who processes data, detects events, and produces bulletins. In general, examples of exploration queries are:

– Graph traversal, for example, *Which seismograms were used to detect the events produced in Natal reported in the bulletin in January? How many data collectors are located in Natal, and which junior analysts processed their collected coordinates to detect events?*
– Graph analytics can be used to answer the following types of queries:

Community Detection: *Which locations have the highest number of detected events? Who are the junior scientists participating in that detection? Which data collectors produced the noisiest readings?*

Centrality: *Which are the seismograms with 3D seismic data where junior scientists detected the most number of events?*

Similarity: *Which events have similar intensity and are located in the same region in subsequent intervals produced by the same data collector?*

Heuristic Link Prediction: *Are events reported in the march bulletin in a particular region related to those detected in a close area by other data collectors?*

Pathfinding and Search: *Who are the junior data scientists that have detected events from waves sensed by data collectors in the same region?*

Towards a Curation and Exploration Environment for Seismic Graphs with Provenance. Figure 5 illustrates the functional architecture of a possible environment that can provide tools to curate and explore seismic data. The environment must provide services to curate data and processes performed to process it, like extracting content, generating plots, observing and detecting events at intervals, and producing bulletins. From a curation perspective, the environment must archive, process data to extract content, allow agents to explore them, and keeps track of their actions. For example, who performs which action on which data? The system must coordinate a cyclic process that collects, archives, processes, and produces new data and provenance meta-data. Since some curation tasks cannot be completely automatic, junior scientists execute seismogram production and event detection; the environment must provide simple, well-adapted tools with well-adapted human-in-the-loop strategies. The environment is a library with seismic, experiments data, and meta-data that can be continuously curated (with new metadata) and where exploration and analytics pipelines can be performed. Thanks to the choice of graphs, exploration can lead to discovering patterns that can improve the knowledge about seismicity in different regions and its implications. We believe that the environment must rely on a domain-specific exploration queries engine. Thereby, queries can be expressed by geophysicists and then evaluated to produce results explained with provenance meta-data.

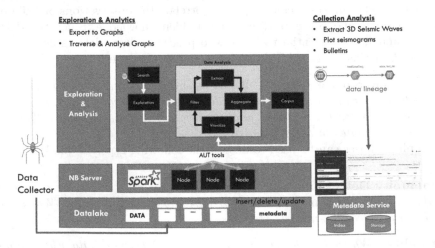

Fig. 5. Curation and exploration environment for seismic graphs with provenance

Challenges and Open Issues. Associating provenance meta-data to seismic data corresponds to a **curation process**. It focuses on activities and agents that act on data during the event detection process. The curation process keeps track of the conditions in which seismic events are detected and disseminated in bulletins.

First Challenge: The design of the graphs is the first challenge to address. Therefore, we consider two types of open issues. First, formally expressing the graphs to have a solid and sound representation. Then profile and study their properties to estimate storage, querying, and processing implications.

Second Challenge: Junior (senior) data scientists do the curation process manually, and some aspects like the **quality of data and meta-data** deserve to be modeled and assessed.

Third Challenge: Storing data and meta-data in a graph database enables the **exploration** of raw and processed data reported in plots and bulletins. In seismology, this feature is essential because it can establish connections across the data reported within bulletins. Storing and maintaining bulletins can build a history and perform analytics to **understand and discover seismic patterns**. Identifying and characterizing the exploration process and queries ad hoc for seismology is an open issue. Geophysicists have not formally expressed how they operate on data to solve questions and study different phenomena and implications. As questions are characterized, it will be possible to define exploration and processing operators and how provenance can contribute to implementing specialized exploration processes.

Fourth Challenge: The expression of exploration queries can combine path traversal, aggregation, and analytics operations. Query languages like Cypher enable the expression path traversal queries for property graphs. Extensions with data science cartridges like the one proposed by Neo4J allow to "declaratively" define pipelines that can apply machine learning models on graphs. However, we believe that for the seismic and, in general, the geophysics sciences with particular exploration requirements, it can be interesting to specify domain-specific query languages that integrate models and operations used in the discipline.

5   Conclusion and Future Work

This paper introduced problems, challenges, and open issues regarding the provenance-guided curation and exploration of geophysics data collections. We motivated the problem through a use case regarding seismic data collected by observation stations in the northeast region of Brazil. We exhibited the requirements regarding curation and exploration and highlighted the importance of provenance to ensure seismic events detection, validation, archival, and dissemination. Graphs can provide an intuitive, rich and mathematical way of curating and exploring data. The curation and exploration processes must be specialized for seismic data and analysis. We showed the general architecture of an environment that can implement our approach. As future work, we have to improve the provenance graph in order to broaden the scope, as well as include more details. Also, we count on proposing and implementing a Domain Specific Language for the curation and exploration of geophysical data sets. Other open issues like dealing with data volume, velocity, and variety will be experimented with in the context of the project ADAGEO (https://adageo.github.io/).

References

1. Bowers, S., McPhillips, T.M., Ludäscher, B.: Provenance in collection-oriented scientific workflows. Concurr. Comput. Pract. Experience **20**(5), 519–529 (2008)
2. Cheney, J., Chiticariu, L., Tan, W.C.: Provenance in databases: why, how, and where. Found. Trends Databases **1**, 379–474 (2009). https://doi.org/10.1561/1900000006
3. Hernández, D., Galárraga, L., Hose, K.: Computing how-provenance for SparQL queries via query rewriting. Proc. VLDB Endow. **14**(13), 3389–3401 (2021)
4. Missier, P., Dey, S., Belhajjame, K., Cuevas-Vicenttin, V., Ludäscher, B.: D-PROV: Extending the PROV provenance model with Workflow structure. In: 5th USENIX Workshop on the Theory and Practice of Provenance (TaPP 13). USENIX Association, Lombard, IL (Apr 2013), https://www.usenix.org/conference/tapp13/technical-sessions/presentation/missier
5. Moreau, L., et al.: Prov-DM: The PROV data model (2013). http://www.w3.org/TR/prov-dm/, World Wide Web Consortium (W3C)
6. Neo4j: Neo4j - The World's Leading Graph Database (2012). http://neo4j.org/
7. Oliveira, W., Oliveira, D.D., Braganholo, V.: Provenance analytics for workflow-based computational experiments: a survey. ACM Comput. Surv. (CSUR) **51**(3), 1–25 (2018)
8. Paul Groth, L.M.: PROV-overview (2013). https://www.w3.org/TR/prov-overview/, World Wide Web Consortium (W3C)
9. Ramusat, Y., Maniu, S., Senellart, P.: A practical dynamic programming approach to datalog provenance computation. arXiv preprint arXiv:2112.01132 (2021)
10. Simmhan, Y.L., Plale, B., Gannon, D.: A survey of data provenance in e-science. ACM SIGMOD Rec. **34**(3), 31–36 (2005)
11. Spinuso, A., Cheney, J., Atkinson, M.: Provenance for seismological processing pipelines in a distributed streaming workflow. In: Proceedings of the Joint EDBT/ICDT 2013 Workshops, pp. 307–312 (2013)
12. W3C: SPARQL 1.1 query language (2012). https://www.w3.org/TR/2012/PR-sparql11-query-20121108/
13. Wang, J., Crawl, D., Purawat, S., Nguyen, M., Altintas, I.: Big data provenance: challenges, state of the art and opportunities. In: 2015 IEEE International Conference on Big Data (Big Data), pp. 2509–2516 (2015). https://doi.org/10.1109/BigData.2015.7364047

Storing Feature Vectors in Relational Image Data Warehouses

Guilherme Muzzi Rocha[1], Piero Lima Capelo[1],
Anderson Chaves Carniel[2], and Cristina Dutra Aguiar[1]($\boxtimes$)

[1] Department of Computer Science, University of São Paulo, São Paulo, Brazil
cdac@icmc.usp.br
[2] Department of Computer Science, Federal University of São Carlos,
São Carlos, Brazil
accarniel@ufscar.br

Abstract. An image data warehousing also manipulates image data represented by feature vectors and attributes for similarity search. In this paper, we study the impact of storing feature vectors on the performance of analytical queries extended with a similarity search predicate over images. We consider the management of huge data volumes. Thus, we use Spark as support. Experimental results showed that depending on the query characteristics and the data warehouse design, the difference in performance was up to 86.23%. Based on these results, we propose guidelines for storing feature vectors in relational image data warehouses.

Keywords: Data warehouse design · Image data · Analytical queries

1 Introduction

An *image data warehousing* handles complex image data and conventional data. These complex data are the intrinsic characteristics of images represented by feature vectors and attributes for similarity search. The extract, transform, and load (ETL) process captures the intrinsic characteristics of images. The relational schema of the *image data warehouse* (IDW) contains fact and dimension tables designed to store image data. The *online analytical processing* (OLAP) supports queries extended with a similarity search predicate over images [9].

Using an image data warehousing enables a new range of analyses, empowering applications that require image manipulation aimed at decision-making. In this paper, we are interested in healthcare decision-making, as described in the application introduced in Example 1 and used throughout the paper.

Example 1. Consider a healthcare data warehousing application that manipulates complex data of *exam images* and conventional data related to *patients*, *dates*, *hospitals*, and exam *descriptions*. The objective is to record the *quantity* of exams considering these complex and conventional data as analysis perspectives. An extended OLAP query aimed at healthcare decision-making is *"How many exams have similar images to a given disease image?"*. A wide range of analyses

S. Chiusano et al. (Eds.): ADBIS 2022, CCIS 1652, pp. 323–331, 2022.
https://doi.org/10.1007/978-3-031-15743-1_30

can use this example query to evaluate pathologies prevalence based on similarity over images. Distinct perspectives may explore similar images considering historical data, geographical area, and age group. □

Different features can represent images, revealing particular perceptions to define similarity. Thus, an image data warehousing supports the definition of perceptual layers [9]. Each perceptual layer is an abstraction that represents a feature vector set and attributes for similarity search according to a given feature descriptor. Example 2 shows the specialists' perception of this concept.

Example 2. Consider that three healthcare specialists are submitting extended OLAP queries to the application described in Example 1. The first specialist analyzes *the texture feature of images using a Haralick Descriptor* [4], while the second specialist examines *the color feature of images using the Color Histogram* [4]. The third specialist explores *the features of color and texture using a Haralick Descriptor and the Color Histogram*. In fact, there are several Haralick Descriptors, such as Variance, Entropy, and Uniformity, increasing the possibility of analysis. Therefore, the precision of the similarity comparisons depends on the decision-making requirements and the healthcare specialist's perceptions. □

In an image data warehousing, the IDW is voluminous since it stores conventional data and intrinsic characteristics of images, as well as supports the storage of several perceptual layers. Furthermore, OLAP queries extended with a similarity search predicate over images demand high computational costs, requiring the processing of the operations of star-join and distance calculations. Another aspect is that the management of voluminous IDWs can benefit from using parallel and distributed data processing frameworks, such as Spark [11].

Processing extended OLAP queries in Spark introduces several issues. In this paper, we explore the impact of storing feature vectors on the performance of OLAP queries extended with a similarity search predicate over images. In the literature, most approaches explore how to store data in relational warehouses considering other data types, i.e., conventional, geographical, and socioeconomic [2,5,7]. The work of [1] investigates image data, but it uses a centralized computing environment and a small image dataset (131,656 images).

The contributions of this paper are threefold. First, we analyze four designs for storing feature vectors in relational IDWs. Second, we focus specifically on the processing of the similarity search predicate over images and study different query characteristics that can impact performance. Third, we propose guidelines for helping design an IDW regarding the storage of feature vectors.

This paper is organized as follows. Section 2 describes theoretical foundation. Section 3 details the IDW designs. Section 4 discusses the experimental evaluation. Section 5 proposes the guidelines. Section 6 concludes the paper.

2 Processing the Similarity Search Predicate in Spark

The processing of the similarity search predicate over images in Spark requires image comparisons, the definition of a similarity query, and the use of a suitable method to execute the extended OLAP query. We detail these aspects as follows.

Image comparisons require two steps [10]. First, a feature descriptor tracks the images in a dataset, processes their pixel values, and generates numerical representations, storing them in a feature vector set. The set contains the feature vectors of all images in the dataset generated by a given feature descriptor. Feature vectors have a specific dimensionality. For instance, Color Histogram and Haralick Descriptors (e.g., Variance, Entropy, and Uniformity) generate vectors with the dimensionality of 256 and 4, respectively. Second, a distance function compares the similarity between any two images using their feature vectors.

One of the most important similarity queries is the range query. Given a set S (e.g., feature vectors generated by the Color Histogram), a query center element s_q, a distance function d (e.g., the Euclidean distance L_2), and a query radius r_q, this query returns all elements $s_i \in S$ satisfying the condition $d(s_i, s_q) \leq r_q$.

The range query calculates the distance $d(s_i, s_q)$ between each element s_i and the query center s_q. In our work, we use the Omni technique [10] to reduce the number of distance calculations. It uses representative elements $f_i \in S$ and calculates in advance the distances $d(f_i, s_i)$ between these elements and all other elements in the dataset. The Omni filtering step defines candidate elements, such that each element s_i is a candidate for the answer if the condition $d(f_i, s_q) - r_q \leq d(f_i, s_i) \leq d(f_i, s_q) + r_q$ is satisfied. The Omni refinement step eliminates false positives by calculating distances for only the candidate elements.

BrOmnImg [8] is an efficient method that employs the Omni technique to process extended OLAP queries using Spark. *BrOmnImg* stores the filtering and refinement results in hash structures and broadcasts them to all cluster nodes to perform the star-join operation. To this end, *BrOmnImg* assumes that data files are small enough to be stored in the main memory. As a result, they can be transmitted and replicated to all cluster nodes during query processing.

3 Designing Image Data Warehouses

In the star schema of a relational IDW [9], the fact table stores numeric measures of interest and primary keys for dimension tables. A conventional dimension table has a primary key and descriptive attributes. An image dimension table has a primary key and intrinsic characteristics of images. Example 3 illustrates conventional tables using the healthcare application described in Example 1.

Example 3. Exam is a conventional fact table that contains the *quantity* measure and a primary key composed of references to all dimension tables. *Patient*, *Date*, *Hospital*, and *Description* are conventional dimension tables. Examples of attributes are: *Patient*: a primary key, gender, ethnicity, state; *Date*: a primary key, day, month, year; *Hospital*: a primary key, address, city, state, country; and *Description*: a primary key, exam type, equipment.   □

Furthermore, an IDW supports several perceptual layers, each referring to a feature vector set and its attributes for similarity search. These attributes are the distances between each Omni representative element and each image in the dataset. There are four star schemas described in [1] to support perceptual layers, which are depicted in Fig. 1 considering our healthcare application.

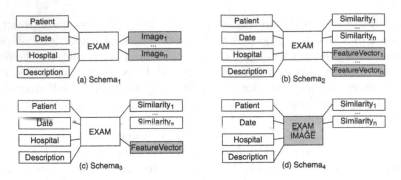

Fig. 1. Different designs for storing feature vectors in an IDW.

The characteristics of the schemas are detailed as follows, for $1 \leq i \leq n$ ($n =$ number of perceptual layers). In $Schema_1$ (Fig. 1a), each feature vector set i is stored along with its attributes for similarity search in an $Image_i$ table. Regarding $Schema_2$ (Fig. 1b), each feature vector set i is stored in a $FeatureVector_i$ table and its attributes for similarity search are stored in a $Similarity_i$ table. As for $Schema_3$ (Fig. 1c), all feature vector sets are stored together in one $FeatureVector$ table, while the attributes for similarity search of each set i are stored in a specific $Similarity_i$ table. Finally, in $Schema_4$ (Fig. 1d), all feature vector sets are stored together in the fact table, while the attributes for similarity search of each set are stored in a specific $Similarity_i$ table. In this case, the $ExamImage$ fact table contains the conventional numeric measures and the feature vector sets as facts. Example 4 shows instances for the image tables.

Example 4. Our application defines 4 perceptual layers. Consider the Color Histogram layer, a feature vector set produced by this extractor, and attributes for similarity search produce by Omni. In Fig. 1a, $Image_{ColorHis}$ stores a primary key, the feature vector set, and the attributes for similarity search. In Fig. 1b, $FeatureVector_{ColorHis}$ contains a primary key and the feature vector set, and $Similarity_{ColorHis}$ stores a primary key and the attributes for similarity search. The contents of $Similarity_{ColorHis}$ in Figs. 1c and 1d remains the same. But, in Fig. 1c, $FeatureVector$ contains a primary key and 4 feature vector sets. In Fig. 1d, $ExamImage$ stores a primary key composed of references to all dimension tables, the $quantity$ measure, and 4 feature vector sets. □

4 Experimental Evaluation

We conducted an experimental evaluation considering the application detailed in Examples 1 to 4. We used the ImgDW Generator tool [6], which populated the *Hospital* and *Date* conventional dimension tables with real data obtained from SUS (www2.datasus.gov.br/datasus/index.php) and days from 1970 to 2020, respectively. Due to privacy, the tool stored synthetic data in the *Patient* and

Table 1. Configurations of the similarity search predicate.

Feature descriptors	1% Selectivity	20% Selectivity
Haralick Variance	*LD1*	*LD2*
Haralick Variance, Entropy, and Uniformity	*LD3*	*LD4*
Color Histogram	*HD1*	*HD2*
Color Histogram and Haralick Variance	*HD3*	*HD4*
Color Histogram and Haralick Variance, Entropy, and Uniformity	*HD5*	*HD6*

Description conventional dimension tables and in the *Exam* conventional fact table. The tool set the *quantity* measure to 1 to record each exam occurrence.

The tool used real images from the Hospital das Clínicas da Faculdade de Medicina de Ribeirão Preto. We created the perceptual layers of Color Histogram and Haralick Variance, Entropy, and Uniformity, and obtained 3 elements as attributes for similarity search. The *Image*, *FeatureVector*, *Similarity*, and *ExamImage* tables were populated according to the schemas in Fig. 1. The tool produced 3 million exams related to 300 thousand patients, 6 thousand hospitals, 18,627 dates, 3 million images, and 3 million exam descriptions.

We adapted *BrOmnImg* to process the queries. We defined two dimensionalities of interest. The low dimensionality of 4 refers to the Haralick Variance, Entropy, and Uniformity feature vectors. The high dimensionality of 256 indicates the Color Histogram feature vectors. We also defined two selectivity values. The high selectivity of 1% returns few images, while the low selectivity of 20% returns several images. Combining these aspects, we produced the configurations shown in Table 1. *LD* and *HD* refer to low and high dimensionality, respectively.

The experiments were performed in a cluster with 5 nodes. Each node had, at least, 3 GB of RAM. We executed each query 5 times and calculated the average elapsed time in seconds and the standard deviation, after removing outliers. We flushed the cache and buffers after each query.

Figure 2a depicts the time spent to process the configurations with only low-dimensionality feature vectors. $Schema_1$ and $Schema_2$ guaranteed better performance than $Schema_3$ and $Schema_4$ since they store each feature vector set in a different table, handling smaller data volumes. The gains ranged from 46.62% to 86.23%. Comparing $Schema_1$ to $Schema_2$, the first provided better results for *LD3* (49.80%) and *LD4* (35.37%) since it stores the feature vector set and its attributes for similarity search together, decreasing the number of joins.

Figure 2b depicts the time spent to process *HD1*, *HD3*, and *HD5*. Evaluating the high-dimensionality feature vectors is the most expensive piece. Although *HD3* analyzes one low-dimensionality feature vector set, there is no significant performance difference between it and *HD1*. *HD5* increases query time concerning $Schema_3$ and $Schema_4$ since the storage of the feature vector sets together in the same table impaired these schemas. Regarding Fig. 2c, it depicts the time spent to process configurations *HD2*, *HD4*, and *HD6*. Considering *HD2*, there is no significant difference in the time spent because all the schemas require

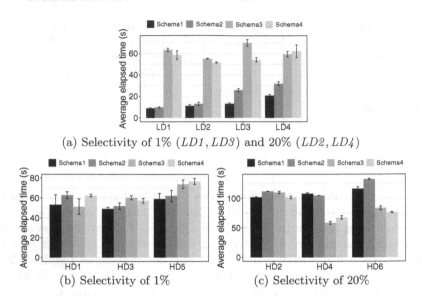

(a) Selectivity of 1% (*LD1*, *LD3*) and 20% (*LD2*, *LD4*)

(b) Selectivity of 1%   (c) Selectivity of 20%

Fig. 2. Performance results for the configurations detailed in Table 1.

the processing of several images. As for *HD4* and *HD6*, *Schema3* and *Schema4* improved query performance because they store all feature vectors in the same table. The gains varied from 27.94% to 45.49%. Here, *BrOmnImg* processed the low-dimensionality feature vectors first and used the results to filter the high-dimensionality feature vectors.

5 Guidelines for Designing Image Data Warehouses

From the experimental evaluation discussed in Sect. 4, we propose a set of guidelines for helping the design of a relational IDW. We also introduce the following taxonomy of tables based on the discussions in Sect. 3: conventional fact table (*ConFact*), image fact table (*ImgFact*), conventional dimension table (*ConDi*), feature vector dimension table (*FeatureVectorDi*), similarity dimension table (*SimilarityDi*), and image dimension table (*ImageDi*).

Guideline 1: *An IDW stores conventional data and characteristics of images (i.e., feature vectors and attributes for similarity search) and can be designed using the following types of tables: ConFact, ImgFact, ConDi, FeatureVectorDi, SimilarityDi, and ImageDi.*

The IDW can contain conventional data and image data represented by feature vectors and attributes for similarity search. According to the proposed types of tables introduced in Table 2, the image data can be modelled as: (i) several *ImageDi* tables; (ii) several *FeatureVectorDi* and *SimilarityDi* tables; (iii) a *FeatureVectorDi* table containing all feature vector sets and several *SimilarityDi* tables; or (iv) several *SimilarityDi* tables and feature vector sets stored as numerical measures in one *ImgFact* table.

Table 2. Taxonomy of tables in a relational IDW.

Name	Description	Contents
ConFact	Fact table with conventional numeric measures	– Primary key composed of references to all dimension tables – Conventional numeric measures
ImgFact	Fact table with conventional numeric measures and feature vectors	– Primary key composed of references to all dimension tables – Conventional numeric measures – One or more feature vector sets
ConDi	Dimension table with conventional attributes	– Primary key – Conventional descriptive attributes
FeatureVectorDi	Image dimension table with feature vectors	– Primary key – One or more feature vector sets
SimilarityDi	Image dimension table with attributes for similarity search	– Primary key – Distances for the Omni representative elements
ImageDi	Image dimension table with intrinsic features of images	– Primary key – Distances for the Omni representative elements – A specific feature vector set

Guideline 2: *Applications with high demand for analytical queries extended with a similarity search predicate that involves only low-dimensionality feature vectors should store each set separately in a specific table containing the intrinsic characteristics of images.*

Guideline 3: *Applications with high demand for analytical queries extended with a similarity search predicate that involves one high-dimensionality feature vector set and returns few images should store this set in a specific table containing the intrinsic characteristics of images. These applications should also store each low-dimensionality feature vector set that may compose the queries in a specific table containing the intrinsic features of the images.*

Each feature vector set i must be stored in an *ImageDi$_i$* table containing the respective feature vectors and attributes for similarity search. Keeping these data together reduces the need to join tables. Maintaining each feature vector set in a particular table reduces the data volume manipulated in query processing.

Guideline 4: *Applications with high demand for analytical queries extended with a similarity search predicate that involves one high-dimensionality feature vector set, returns several images and may contain low-dimensionality feature vector sets should store all the sets in the same table.*

The attributes for similarity search related to a feature vector set i must be stored in a *SimilarityDi$_i$* table. In contrast, all feature vector sets must be stored together in a *FeatureVectorDi* table or a *ImgFact* table. The storage of all feature vector sets together reduces the volume of images manipulated. Extended OLAP queries should first process the low-dimensionality feature vectors and then apply the results as a filter to process the high-dimensionality feature vectors.

6   Conclusions and Future Work

In this paper, we have explored different designs for storing feature vectors in relational image data warehouses manipulated in parallel and distributed computing environments. To comply with this goal, we have carried out an experimental evaluation to analyze how the designs impact the processing of analytical queries extended with a similarity search predicate over images. Based on the results indicating a difference in the performance of up to 86.23%, we propose guidelines to store image data in relational warehouses. As future work, we plan to extend our guidelines for designing relational warehouses that stores sound and patient records as complex data types. We also intend to employ spatial index structures to extend our experiments [3].

Acknowledgments. This work was supported by the São Paulo Research Foundation (FAPESP), the Brazilian Federal Research Agency CNPq, and the Coordenação de Aperfeiçoamento de Pessoal de Nível Superior, Brasil (CAPES), Finance Code 001. G. M. Rocha and C. D. Aguiar were supported by FAPESP, grants #2018/10607-3 and #2018/22277-8, respectively. A. C. Carniel was supported by Google as a recipient of the 2022 Google Research Scholar Program.

References

1. Annibal, L.P.: iStar: a star schema optimized for image data warehouses based on similarity. Master thesis, UFSCar (2011). In Portuguese
2. Boussahoua, M., Boussaid, O., Bentayeb, F.: Logical schema for data warehouse on column-oriented NoSQL databases. In: Proceedings of DEXA Conference, pp. 247–256 (2017)
3. Carniel, A.C., Ciferri, R.R., Ciferri, C.D.A.: FESTIval: a versatile framework for conducting experimental evaluations of spatial indices. MethodsX **7**, 100695 (2020)
4. Gonzalez, R., Woods, R.: Digital Image Processing, 3rd edn. Prentice-Hall, Upper Saddle river (2006)
5. Mateus, R.C., Siqueira, T.L.L., Times, V.C., Ciferri, R.R., de Aguiar Ciferri, C.D.: Spatial data warehouses and spatial OLAP come towards the cloud: design and performance. Distrib. Parallel Databases **34**(3), 425–461 (2015). https://doi.org/10.1007/s10619-015-7176-z
6. Rocha, G., Ciferri, C.D.A.: ImgDW generator: a tool for generating data for medical image data warehouses. In: Proceedings of SBBD Demos e WTDBD, pp. 23–28 (2018)
7. Rocha, G.M., Capelo, P.L., Ciferri, C.D.A.: Healthcare decision-making over a geographic, socioeconomic, and image data warehouse. In: Proceedings of ADBIS, TPDL and EDA Common Workshops and Doctoral Consortium, pp. 85–97 (2020)
8. Rocha, G.M., Ciferri, C.D.A.: Efficient processing of analytical queries extended with similarity search predicates over images in Spark. J. Inf. Data Manag. **11**(3), 209–227 (2020)
9. Teixeira, J.W., Annibal, L.P., Felipe, J.C., Ciferri, R.R., Ciferri, C.D.A.: A similarity-based data warehousing environment for medical images. Comput. Biol. Med. **66**, 190–208 (2015)

10. Traina-Jr, C., Filho, R.F.S., Traina, A.J.M., Vieira, M.R., Faloutsos, C.: The Omni-family of all-purpose access methods: a simple and effective way to make similarity search more efficient. VLDB J. **16**(4), 483–505 (2007)
11. Zaharia, M., Chowdhury, M., Franklin, M.J., Shenker, S., Stoica, I.: Spark: cluster computing with working sets. In: Proceedings of HotCloud Conference, p. 10 (2010)

Automatic Classification of Stigmatizing Articles of Mental Illness: The Case of Portuguese Online Newspapers

Alina Yanchuk[1]($\boxtimes$), Alina Trifan[1,2], Olga Fajarda[1,2],
and José Luís Oliveira[1,2]

[1] IEETA, DETI, University of Aveiro, Aveiro, Portugal
{alinayanchuk,alina.trifan,olga.oliveira,jlo}@ua.pt
[2] LASI - Intelligent Systems Associate Laboratory, Guimarães, Portugal

Abstract. The stigma related to mental health continues to be present in online newspapers, where mental diseases are often used metaphorically to refer to entities or situations outside the clinical of mental health. This project explores the implementation of Artificial Intelligence and Natural Language Processing techniques for the task of automatically classifying stigmatizing articles with references to the mental disorders of schizophrenia and psychosis. This work is implemented in Portuguese online news articles, collected from the *Arquivo.pt* repository, a public repository of archived Portuguese web pages, and can be adapted to other languages or similar problems. Nine machine and deep learning algorithms were implemented, most of them yielding results with a precision above 90%. In addition, the automatic detection of articles topics was also performed, through topic modeling with the *top2vec* model, which allowed concluding that the stigmatization of mental health occurs, essentially, in Economics and Politics related news. The results confirm the existence of stigma in Portuguese newspapers (52% of the 978 articles collected) and the effectiveness of the use of Artificial Intelligence models to detect it. Additionally, a set of 978 articles collected and manually classified with the classes ["stigmatizing", "literal"] is obtained.

Keywords: Artificial Intelligence · Deep learning · Machine learning · Natural Language Processing · Newspapers · Text classification · Topic modeling

1 Introduction

The presence of stigma in our society is still frequent. When associated with mental illness, it has negative implications on patients and their treatments. Stigmatization occurs when terms referring to mental illness are used in a figurative/metaphorical sense to describe entities or situations outside the health context. Within this scope, there is a need to fight the stigma present in the media outlets, where the use of stigmatizing expressions is still quite common. On the other hand, the analysis of journalistic news has shown a great growth

S. Chiusano et al. (Eds.): ADBIS 2022, CCIS 1652, pp. 332–343, 2022.
https://doi.org/10.1007/978-3-031-15743-1_31

in the research area and more and more computational approaches have been adopted to perform it, in contrast to traditional manual methods. Manual methods are characterized by the annotation, by humans, of the texts to be classified, while computational methods also use Artificial Intelligence (AI). The subfields of AI most relevant to this process are Machine Learning, Natural Language Processing (NLP) and Text Mining.

Thus, the result of this work consists of a set of automatic text classification models, which allow classifying the sense of articles, present in online Portuguese newspapers and holding references to the mental disorders of schizophrenia and psychosis, as stigmatizing or literal. Consequently, a set of 978 Portuguese articles annotated with the two classes is also created and made available as open source (along with the project code)[1]. Additionally, an automatic detection of topics present in the articles was also performed. All articles were retrieved from *Arquivo.pt*[2], the official data source.

The remainder of this manuscript is organized as follows: Section *Background* introduces the state of the art of the topics relevant to this work, Section *Methodology* describes the different steps performed in the development process, Section *Results and Discussion* presents the main results obtained and Section *Conclusion* portrays the main conclusions made.

2 Background

A document from the Portuguese Society of Psychiatry and Mental Health, released in 2021, reveals that mental diseases affect one in five Portuguese people (23%), being the COVID-19 pandemic a factor that has contributed to the increase of this number [24]. In Portugal, as in other countries, the reality of stigma exists, and mental disorders continue to be used metaphorically and in contexts that do not relate to the health field. The terms "schizophrenic," "bipolar", "depressive" and others are often used as adjectives to refer, figuratively, to situations or entities negatively, as for example when the word "schizophrenic" is used to refer to a ridiculous or contradictory situation.

In Europe, a study [5] analyzed 695 news items, featuring mental health-related terms, from 20 popular newspapers in Spain in the year 2010, and found the presence of 47.9% stigmatizing news items that used mental illness as metaphors. In Greece, 150 news items, referring only to schizophrenia, were analyzed, and a 34.9% presence of stigma in the metaphorical sense was found [6]. In the UK, this number constituted 11% of a total of 600 news items [10]. In the United States of America, 1740 articles were analyzed and the percentage of those using schizophrenia as a metaphor was 28% [13]. In Brazil, a study [7], which also focused only on schizophrenia, found a percentage of 34%, out of a total of 229 news stories evaluated, of stigma in the metaphorical sense and concluded that it is more present in the fields of Politics, Economics and Entertainment.

---

[1] https://github.com/alina-yanchuk02/stigmaClassification.
[2] https://www.arquivo.pt/.

Currently, only two Portuguese studies in this field were found. The first [26] conducted a content analysis of Portuguese news about mental health, published between January and June 2015, and revealed that depression and schizophrenia tend to be the most stigmatized diseases in the Portuguese press. The second [25] evaluated the use of the word "schizophrenia" in a total of 1058 Portuguese news stories, published between 2007 and 2013, and found that 40% of the news stories were stigmatizing, with the prominent area being Politics. However, these studies consisted only of an exploratory study, which used the typical manual approach of classification. Thus, the need arises to explore the automatic approach. Automatic text classification is a problem that belongs to the category of supervised learning, and it consists, in general, of a text cleaning and preprocessing step, a feature extraction step, a classification step, and a final step of models evaluation.

In the preprocessing of texts, NLP techniques are applied, the most common being tokenization, which is the breakdown of the text into tokens/terms, the removal of stop words, words with high frequency and no semantic importance for the text, and the lemmatization and stemming techniques, which consist of the transformation of derived words to their root form, taking into account the context and ignoring it, respectively. There are very few tools that implement these techniques accurately for the Portuguese language, most being applicable to English.

As for feature extraction, this is often translated into the bag-of-words model, a numerical representation that only takes into account the frequency of words and not the order in which they appear, and the Term Frequency-Inverse Document Frequency (TF-IDF) model, which also takes into account their importance in the text. More complex representations involve mapping words to certain established categories. One example is the word embeddings model, which consists of mapping words to dense and small vectors, allowing to better capture the semantics of words and making similar words have similar vectors. Other approaches consist of using dictionaries, such as Linguistic Inquiry and Word Count (LIWC), a dictionary, also available in Brazilian Portuguese, which maps words to four main types of categories: basic linguistic processes, psychological processes, relative expressions, and personal concerns [22]. Onan and Tocoglu [18] used attributes extracted from LIWC to identify satire in Turkish news and concluded that they generate better classification results than typical bag-of--words models.

As for text classification, the most commonly used algorithms are Decision Trees, Naive Bayes, Support Vector Machine (SVM), K-Nearest Neighbors (KNN) and Logistic Regression [2,3,16]. SVM with linear kernels and Naive Bayes are the algorithms that have shown the best results on this problem [3,16]. Deep learning, a type of machine learning that uses neural network-based learning, has also proven to be quite effective in the textual classification process, particularly in the detection of metaphors. Gao et al. [15] demonstrated that architectures based on Bidirectional Long Short-Term Memory (Bi-LSTM), combined with the representation of word embeddings, present state-of-the-art results in the identification and classification of text with metaphorical expressions. However, all these algorithms often rely on a large amount of training data. One approach that tries to overcome this limitation is transfer learning, where models are pre-trained and the knowledge learned in one domain is applied to another similar domain. One model that is part of this approach and has shown state-of-the-art results in the area of NLP is Bidirectional Encoder Representations from Transformers (BERT), a pre-trained model that is able to better understand the meaning and relationships of words in a sentence, as well as the context in which they are embedded [12]. BERTimbau [27] is the BERT model trained on the Brazilian Portuguese language.

In topic modeling field, the fundamental algorithms correspond essentially to the Linear Discriminant Analysis (LDA) and Probabilistic Latent Semantic Analysis (PLSA) algorithms, based on word distributions. However, these models have difficulty in capturing the text context and semantics of words, and the need to predefine the number of topics to be discovered. Angelov [4] presents the top2vec algorithm, which automatically detects topics present in a document, without the need for preprocessing, and generates representations that take into account the semantic content of the text. This algorithm tends to generate better results than traditional topic modeling algorithms.

Despite the great advances in the areas of NLP and Text Mining, these still have several gaps and limitations, particularly in the processing of texts where irony, metaphors, idioms and word ambiguities are present. Moreover, the area of automatic text classification itself is still underdeveloped in Portugal.

3   Methodology

The adopted methodology is characterized by four main steps: (i) data collection; (ii) filtering and manual data annotation; (iii) preprocessing; (iv) automatic classification and topic modeling.

3.1   Data Collection

The official data source, consisting of Portuguese newspaper articles and their metadata, was the repository *Arquivo.pt*, a repository of Portuguese web pages archived from 1996 to the present day. *Arquivo.pt API*[3] is the public Application Programming Interface (API) of the *Arquivo.pt* repository, which enables the collection of the preserved web pages and their metadata through text searches.

Starting with the search terms, it was decided to focus on articles that stigmatize the mental illness of schizophrenia, as previous studies have shown it to be one of the most commonly used illnesses in the press in a metaphorical sense. In addition, to increase the number of articles collected, terms referring to psychosis were also taken into account, since this is a condition that is part of the symptoms of schizophrenia and both disorders are often used

---

[3] https://github.com/arquivo/pwa-technologies/wiki/Arquivo.pt-API.

in a related way. Thus, taking into account all the words that can be derived from the words "schizophrenia" and "psychosis", through the use of derivation suffixes and without losing their original meaning, all the articles that had at least one of the following terms were collected from *Arquivo.pt*: ["esquizofrenia", "esquizofrénico", "esquizofrenico", "esquizofrénica", "esquizofrenica", "esquizofrénicas", "esquizofrenicas", "esquizofrénicos", "esquizofrenicos", "esquizofronicamente", "esquizofrenizar", "psicose", "psicótica", "psicotica", "psicóticas", "psicoticas", "psicótico", "psicotico", "psicóticos", "psicoticos"]/["schizophrenia", "schizophrenic", "schizophrenics", "schizophrenically", "schizophrenize", "psychosis", "psychose", "psychotic", "psychotics"]. Given the large number of existing Portuguese newspapers, only nine electronic journals were selected. The selection criteria were the popularity of the journal on the Internet [11], its relevance to the scope of the project, and its longevity. The journals used were Público[4], Observador[5], Diário de Notícias[6], Expresso[7], Correio da Manhã[8], Jornal de Notícias[9], Sábado[10], Visão[11] and A Bola[12]. Since *Arquivo.pt* allows searching only from the year 1996 and has an embargo period on access corresponding to one year [14], we defined the search time interval between 1996 and 2021. The total number of data returned by the API was 8235 web pages.

Next, the process of web scraping the HyperText Markup Language (HTML) of each page was performed. The web scraping process was done using the *newspaper* library [19], since it was able to quickly and efficiently perform the process for all the returned pages, even the older ones. In addition, all articles that did not contain in their title or content at least one of the terms from the search term set were removed (although they were returned by the API, not all had the terms located in the article text), and also the duplicate articles, by comparison of their content. The total number of structured data obtained was 1111.

3.2 Filtering and Manual Data Annotation

Given that automatic text classification implies the existence of already classified data, to train and test the models, manual annotation of all articles was performed. It was divided by a set of different human annotators and the same instruction was made available to each of them, stating under what circumstances, based on those presented in previous studies, each of the categories should be assigned. Each article was classified by at least two different annotators. An example of excerpts from classified articles can be seen in Table 1. The categories assigned were:

---

[4] https://www.publico.pt/.
[5] https://observador.pt/.
[6] https://www.dn.pt/.
[7] https://expresso.pt/.
[8] https://www.cmjornal.pt/.
[9] https://www.jn.pt/.
[10] https://www.sabado.pt/.
[11] https://visao.sapo.pt/.
[12] https://www.abola.pt/.

- Stigmatizing: the article is stigmatizing - it uses the disease reference in any metaphorical sense and within an inappropriate context, to reveal an idea that goes beyond the literal meaning of the term;
- Literal: the article is not stigmatizing - it uses the disease reference in its literal sense and within an appropriate context.

Table 1. Examples of article excerpts manually classified (with translation).

Article excerpt	Label
Os adeptos do Sporting estão a viver uma espécie de **"esquizofrenia"** coletiva./Sporting fans are experiencing a kind of collective **"schizophrenia"**	Stigmatizing
Talvez seja mesmo um thriller **psicótico**. Ou uma reflexão sobre o bem, o mal, o amor materno e as suas contradições. /Maybe it really is a **psychotic** thriller. Or a reflection on good, evil, maternal love and its contradictions	Stigmatizing
Eli, que tem agora 19 anos e está preso, sofre de **esquizofrenia** desde os 14 anos./Eli, who is now 19 and in prison, has suffered from **schizophrenia** since he was 14	Literal
O jovem de 20 anos (...) estava em "surto **psicótico**", segundo a psicóloga que acompanhou a missão no local./The 20-year-old (...) was in a "**psychotic** outburst," according to the psychologist who accompanied the mission to the scene	Literal

In the end, 978 articles were manually annotated with the labels/classes ["stigmatizing", "literal"]. It should also be noted that during this process some articles were discarded because they had structural problems, were duplicates or were not relevant to the problem.

3.3 Preprocessing

During the preprocessing phase, the documents (N = 978) were cleaned and NLP techniques were used to prepare the texts for the subsequent steps. Each document corresponds to the text obtained from the concatenation of the title and content of each article. This phase is quite important and is intended to allow the models to better understand the texts and generate more accurate and consistent results. The techniques applied were:

- Tokenization: breaking down the text of each document into a sequence of terms;
- Conversion of the text to all lowercase letters;
- Removal of stop words[13], obtained from NLTK [8];
- Removal of all URLs, text within brackets and square brackets, all punctuation marks, all terms containing numbers, all terms smaller than three characters in length, some irrelevant terms, and all personal pronouns connected to verbs.

---

[13] https://www.nltk.org/howto/portuguese_en.html.

Other techniques, such as lemmatization and stemming, were not applied because of the scarcity of precise tools for the Portuguese language, and also in order not to reduce the vocabulary of the documents too much further.

3.4 Automatic Classification and Topic Modeling

The classification step implied the occurrence of two phases: feature extraction and implementation of the classification models. In the extraction of features, four different models were used to generate numerical representations of the documents, namely the bag-of-words model, the TF-IDF model, the word embeddings model, using 300-dimensional vectors pre-trained for the Portuguese language with the GloVe [23] algorithm and obtained from the *NILC-Embeddings* repository[14], and the mapping of the terms from the texts to the 464 categories of the *Brazilian Portuguese LIWC 2007* Dictionary[15].

The classification process consisted of implementing the algorithms, training them using the training data, and then evaluating and comparing the results obtained using the test data. The training data corresponded to 80% (N = 782) of the total data (documents and their classes) and the test data to 20% (N = 196), both of which were selected randomly. Five traditional machine learning algorithms and four deep learning algorithms were used. The machine learning algorithms were implemented using the *scikit-learn* [21] library and consisted of:

- Logistic Regression: algorithm that uses a logistic function to model the probability of the given classes;
- SVM: algorithm that aims to find a hyperplane in a space of X dimensions (X - number of features) that distinguishes the given classes. The "Linear Support Vector Classification" (linear kernel) class was used;
- Naive Bayes: probabilistic algorithm based on Bayes' Theorem and the assumption of conditional independence of the attributes given a class. The "Multinomial Naive Bayes" class was used;
- KNN: algorithm that tries to find a predefined number of training samples closest in distance to the new point and predict the class from them;
- Random Forest: algorithm that sets a number of decision tree classifiers on several subsamples of the training data set and combines the results of each classifier to determine the final class.

The hyperparameters used in each of the models were obtained through an optimization process using the 5-Fold Cross Validation strategy. This process was implemented using the *scikit-optimize* library, which uses a sequential model-based optimization algorithm (using Gaussian processes) to find optimal solutions in less time. All the models were trained with the bag-of-words representation, TF-IDF and the one generated by the Portuguese LIWC dictionary.

The deep learning algorithms were implemented using the *TensorFlow* [1] library, the *Keras API* [9], the *PyTorch* [20] framework and the *transformers* [28] library, and consisted of:

---

[14] http://nilc.icmc.usp.br/embeddings.
[15] http://143.107.183.175:21380/portlex/index.php/pt/projetos/liwc.

- Convolutional Neural Network (CNN): a type of neural network that is typically used in image recognition but has also been used in NLP tasks. The CNN was implemented, sequentially, with an embedding layer, which generates 300-dimensional vectors using word embeddings from the GloVe model, a dropout layer, two 1D convolutional layers (with activation function "relu", and followed by a max-pooling layer), a flatten layer, a dense layer (with activation function "relu"), a dropout layer, and another dense layer (with activation function "sigmoid"). It was trained using a batch size of 32 and epochs value of 10;
- Long Short-Term Memory (LSTM): A type of recurrent neural network that retains only information necessary or useful for prediction. The LSTM network was implemented sequentially with an embedding layer, which generates 300-dimensional vectors using word embeddings from the GloVe model, an LSTM layer and a dense layer (with a "sigmoid" activation function). It was trained using a batch size of 32 and epochs value of 10;
- Bi-LSTM: a type of recurrent neuronal network similar to the LSTM network, but which processes information in both directions. It was implemented sequentially with an embedding layer, which generates 300-dimensional vectors using word embeddings from the GloVe model, an LSTM layer (embedded in a bidirectional layer) and a dense layer (with a "sigmoid" activation function). It was trained using a batch size of 32 and epochs value of 10;
- BERTimbau: We used the "Base" size of the pre-trained BERTimbau model (which has 12 layers/blocks of transformers, 12 attention heads and 110 million parameters), returned through the "AutoModelForSequenceClassification" class, which already has a classification layer implemented on top. It was trained using a batch size of 8 and epochs value of 4.

The optimization of the remaining hyperparameters of the first three models mentioned was done using the *KerasTuner* library [17], with the tuner "Hyperband", which uses the random search algorithm and tries to speed it up by adaptive resource allocation and early stopping.

The automatic topic detection was performed using the *top2vec* algorithm, which was trained on the 978 documents (using only the documents and ignoring their classes) and allowed to obtain a set of 50 descriptive terms of the discovered topics, the scores of their similarity to the topic and the most semantically similar documents to each topic.

4 Results and Discussion

As for the results of the manual annotation of the articles, it was found that 52% of the articles (N = 509) have a stigmatizing meaning and 48% (N = 469) have a literal meaning. As for the results obtained by the developed classification models, the values of the metrics used to evaluate their performance can be visualized and compared in Table 2. All the implemented combinations of classification algorithm and representation model are present in it.

Table 2. Evaluation metric values for each implemented combination of classification model and representation model.

Classification model	Representation model	Accuracy	Precision	Recall	F1
Logistic Regression	Bag-of-words	92.35	0.92	0.93	0.93
	TF-IDF	**93.37**	0.93	0.94	**0.94**
	LIWC	70.41	0.73	0.69	0.71
SVM	Bag-of-words	90.31	0.92	0.90	0.91
	TF-IDF	90.82	0.93	0.90	0.91
	LIWC	80.10	0.81	0.80	0.81
Naive Bayes	Bag-of-words	91.33	0.91	0.92	0.92
	TF-IDF	**93.37**	0.91	0.97	**0.94**
	LIWC	52.04	0.52	**1.00**	0.69
KNN	Bag-of-words	65.82	0.89	0.40	0.54
	TF-IDF	91.33	0.92	0.91	0.92
	LIWC	70.41	0.72	0.70	0.71
Random Forest	Bag-of-words	92.86	0.90	0.97	0.93
	TF-IDF	91.84	0.88	0.97	0.93
	LIWC	79.08	0.77	0.86	0.81
CNN	Word embeddings	87.76	0.92	0.83	0.88
LSTM	Word embeddings	87.24	**0.96**	0.78	0.87
Bi-LSTM	Word embeddings	91.33	0.90	0.94	0.92
BERTimbau	BERTimbau Tokenizer	91.84	0.93	0.91	0.92

Most of the models achieved good results, with accuracy above 90%, and with the Naive Bayes (93.37%) and Logistic Regression (93.37%) classification algorithms standing out at the top, both combined with the TF-IDF representation. In the feature representation, the TF-IDF, bag-of-words and word embeddings models stand out, with the Portuguese LIWC model presenting the worst results with quite significant differences. In the deep learning field, the models with the best accuracy were BERTimbau (91.84%) and Bi-LSTM (91.33%). Despite the good results, the deep learning algorithms combined with word embeddings did not, in general, outperform the traditional machine learning algorithms combined with simpler feature representations, which may suggest the need to experiment with other word embedding algorithms or generate new ones by training with a larger volume of Portuguese texts.

Ten topics were automatically detected, each defined by a set of 50 terms most descriptive of it. Table 3 shows the five most descriptive terms returned, sorted in descending order of semantic similarity to the topic, the overall label manually assigned to each topic, and the number of articles belonging to each topic. It is possible to verify that mental disorders are essentially portrayed in health topics and when associated with criminal actions, and that the highest percentage of stigmatizing articles, relative to the total of articles in this topic, is present in the Economics (97%) and Politics (96%) topics.

Table 3. Five most descriptive terms (with translation), overall label assigned, number of total and stigmatizing articles, along with the percentage of stigmatizing articles.

Descriptive terms	Topic	Total #	Stigma #	Stigma (%)
[doencas, estudo, doenca, medicamentos, ansiedade] /[disease, study, illness, medication, anxiety]	Health	232	13	6%
[homicidio, prisao, policia, crime, encontrado] /[murder, arrest, police, crime, found]	Crimes	158	13	8%
[filme, comedia, realizador, personagens, cinema] /[film, comedy, director, characters, cinema]	Cinema	112	61	54%
[europeia, austeridade, divida, euro, mercados] /[europe, austerity, debt, euro, markets]	Economics	92	89	97%
[eua, russia, militar, armas, washington] /[usa, russia, military, weapons, washington]	Armed Conflicts	85	79	93%
[partido, governo, psd, parlamentar, mocao] /[party, government, psd, parliament, motion]	Politics	80	77	96%
[livros, escritor, literatura, escritores, escrita] /[books, writer, literature, writers, writing]	Literature	70	44	63%
[banda, album, disco, pop, rock] /[band, album, record, pop, rock]	Music	70	63	90%
[desporto, futebol, jogo, lideranca, dirigentes] /[sport, soccer, game, leadership, managers]	Sports	41	37	90%
[magistrados, justica, judicial, tribunais, ministerio] /[magistrates, justice, judicial, courts, ministry]	Justice	38	34	89%

5   Conclusion

In this work, the collection and manual classification of articles, with references to mental disorders, from Portuguese newspapers present on the Internet and archived in the *Arquivo.pt* repository was performed, as well as the exploration of Artificial Intelligence techniques for the automatic classification and topic modeling tasks. Nine different machine learning and deep learning models were proposed for the classification task, and the *top2vec* algorithm was used for topic detection. In addition, a set of 978 articles manually annotated with the classes "stigmatizing" and "literal" was obtained, allowing us to explore how mental health, and more specifically the disorders of schizophrenia and psychosis, are portrayed in the Portuguese media. It was concluded that 52% of the articles were stigmatizing, and predominate in the areas of Economics and Politics.

To our knowledge, this is the first work that explores the classification of Portuguese texts containing metaphorical references through the use of computational techniques, and the major conclusions drawn are that most traditional machine learning algorithms yield good results and that the use of neural networks also suggests being quite promising. However, the field of Portuguese NLP is still very little explored, which is also revealed by the small number of models trained for the European Portuguese language.

Acknowledgment. This work was supported by FCT - Fundação para a Ciência e Tecnologia within project DSAIPA/AI/0088/2020.

References

1. Abadi, M., et al.: Tensorflow: large-scale machine learning on heterogeneous distributed systems. ArXiv (2016). https://doi.org/10.48550/ARXIV.1603.04467, https://tensorflow.org/
2. Aggarwal, C.C., Zhai, C.: A Survey of text classification algorithms. In: Aggarwal, C.C., Zhai, C. (eds.) Mining Text Data, pp. 163–222. Springer, New York (2012). https://doi.org/10.1007/978-1-4614-3223-4_6
3. Ahmed, J., Ahmed, M.: Online news classification using machine learning techniques. IIUM Eng. J. **22**(2), 210–225 (2021). https://doi.org/10.31436/iiumej.v22i2.1662
4. Angelov, D.: Top2vec: listributed representations of topics. ArXiv abs/2008.09470 (2020). https://doi.org/10.48550/ARXIV.2008.09470
5. Aragonès, E., López-Muntaner, J., Ceruelo, S., Basora, J.: Reinforcing stigmatization: lover. age of mental illness in Spanish newspapers. J. Health Commun. **19**(11), 1248–1258 (2014). https://doi.org/10.1080/10810730.2013.872726
6. Athanasopoulou, C., Välimäki, M.: 'Schizophrenia' as a metaphor in Greek newspaper websites. Stud. Health Technol. Inform. **202**, pp. 275–278. (2014). https://doi.org/10.3233/978-1-61499-423-7-275
7. Bevilacqua Guarniero, F., Bellinghini, R.H., Gattaz, W.F.: The schizophrenia stigma and mass media: l search for news published by wide circulation media in Brazil. Int. Rev. Psychiatr. (Abingdon, England) **29**(3), 241–247 (2017). https://doi.org/10.1080/09540261.2017.1285976
8. Bird, Steven, E.L., Klein, E.: Natural Language Processing with Python. O'Reilly Media Inc (2009). https://www.nltk.org/
9. Chollet, F., et al.: Keras (2015). https://keras.io
10. Chopra, A., Doody, G.: Schizophrenia, an illness and a metaphor: analysis of the use of the term 'schizophrenia' in the UK national newspapers. J. R Soc. Med. **100**, 423–426 (2007). https://doi.org/10.1258/jrsm.100.9.423
11. para a Comunicação Social, E.R.: Públicos e consumos de média - o consumo de notícias e as plataformas digitais em portugal e em mais dez países (2014). https://www.erc.pt/pt/estudos-e-publicacoes/consumos-de-media/estudo-publicos-e-consumos-de-media
12. Devlin, J., Chang, M.W., Lee, K., Toutanova, K.: BERT: pre-training of deep bidirectional transformers for language understanding. In: Proceedings of the 2019 Conference of the North American Chapter of the Association for Computational Linguistics: Human Language Technologies, Vol. 1 (Long and Short Papers), pp. 4171–4186. Association for Computational Linguistics (2019). https://doi.org/10.18653/v1/N19-1423
13. Duckworth, K., Halpern, J.H., Schutt, R.K., Gillespie, C.: Use of schizophrenia as a metaphor in US newspapers. Psychiatr. Serv. (Washington, D.C.) **54**(10), 1402–1404 (2003). https://doi.org/10.1176/appi.ps.54.10.1402

14. Fundação para a Ciência e Tecnologia: Recolha de conteúdos - sobre.arquivo.pt, https://sobre.arquivo.pt/pt/ajuda/recolha-e-arquivo-de-conteudos/
15. Gao, G., Choi, E., Choi, Y., Zettlemoyer, L.: Neural metaphor detection in context. In: Proceedings of the 2018 Conference on Empirical Methods in Natural Language Processing, pp. 607–613. Association for Computational Linguistics (2018). https://doi.org/10.18653/v1/D18-1060
16. Hsu, B.M.: Comparison of supervised classification models on textual data. Mathematics **8**(5) (2020). https://doi.org/10.3390/math8050851
17. O'Malley, T., et al.: Kerastuner (2019). https://github.com/keras-team/keras-tuner
18. Onan, A., Togoclu, M.: Satire identification in Turkish news articles based on ensemble of classifiers. Turk. J. Electr. Eng. Comput. Sci. **28**, 1086–1106 (2020). https://doi.org/10.3906/elk-1907-11
19. Ou-Yang, L.: Newspaper3k: aArticle scraping & curation. https://newspaper.readthedocs.io/en/latest/
20. Paszke, A., et al.: Pytorch: An imperative style, high-performance deep learning library. In: Wallach, H., Larochelle, H., Beygelzimer, A., d'Alché-Buc, F., Fox, E., Garnett, R. (eds.) Advances in Neural Information Processing Systems 32, pp. 8024–8035. Curran Associates, Inc. (2019). https://papers.neurips.cc/paper/9015-pytorch-an-imperative-style-high-performance-deep-learning-library.pdf
21. Pedregosa, F., et al.: Scikit-learn: machine learning in Python. J. Mach. Learn. Res. **12**, 2825–2830 (2011). https://scikit-learn.org
22. Pennebaker, J., Francis, M.: Linguistic Inquiry and Word Count. Lawrence Erlbaum Associates, Incorporated (1999). https://books.google.pt/books?id=6FnuAAAACAAJ
23. Pennington, J., Socher, R., Manning, C.: GloVe: global vectors for word representation. In: Proceedings of the 2014 Conference on Empirical Methods in Natural Language Processing (EMNLP), pp. 1532–1543. Association for Computational Linguistics, Doha, Qatar, October 2014. https://doi.org/10.3115/v1/D14-1162
24. dos Psicólogos Portugueses, O.: Desenvolvimento sustentável e sustentabilidade dos cuidados de saúde primários (2021)
25. Rodrigues-Silva, N., Falcão de Almeida, T., Araújo, F., Molodynski, A., Venâncio, Bouça, J.: Use of the word schizophrenia in Portuguese newspapers. J. Mental Health (Abingdon, England) **26**(5), 426–430 (2017). https://doi.org/10.1080/09638237.2016.1207231
26. Sociedade Portuguesa de Psiquiatria e Saúde Mental: Os media e a saúde mental - análise de conteúdo de notícias publicadas por meios de comunicação social portugueses (2016). https://www.sppsm.org/informemente/apresentacao/
27. Souza, F., Nogueira, R., Lotufo, R.: BERTimbau: pretrained BERT models for Brazilian Portuguese. In: 9th Brazilian Conference on Intelligent Systems, BRACIS, Rio Grande do Sul, Brazil, October 20–23 (to appear 2020)
28. Wolf, T., et al.: Transformers: State-of-the-Art Natural Language Processing, pp. 38–45. Association for Computational Linguistics, October 2020

A Multiplex Network Framework Based Recommendation Systems for Technology Intelligence

Foutse Yuehgoh[1,2,0](□) ⓘ, Sonia Djebali[1](✉) ⓘ, and Nicolas Travers[1,2](✉) ⓘ

[1] Léonard de Vinci Pôle Universitaire, Research Center, Paris La Défense, France
{foutse.yuehgoh,sonia.djebali,nicolas.travers}@devinci.fr
[2] CEDRIC Laboratory, Conservatoire National des Arts et Métiers, Paris, France
[3] Coexel, Toulon, France
foutse.yuehgoh@coexel.com
https://cedric.cnam.fr/lab/

Abstract. Network Science has become a flourishing interest in the last decades as we witness the Big Data explosion in many fields including, social science, biology and engineering. *Technology Intelligence* aims at surveying the prolific production of information and recent studies have identified their multiplex nature as a very important aspect to understand various aspects of information. However, the interplay between multiplexity and controllability of these networks is challenging. This paper aims to describe a flexible framework for a recommendation system, based on multiplex networks in the context of technological development. We detail the important characteristics of the multiplex network that is of interest to us, and show the advantages of this approach over the matrix approaches common in the literature.

Keywords: Complex systems · Multiplex networks · Neo4j · Recommendation systems · Framework

1 Introduction

The current economic context, marked by numerous economic tensions, pushes companies to explore novel strategies continuously to keep their business moving forward and to optimize their innovation processes. To identify technological opportunities and threats that may affect the future development and survival of companies, scientific experts of these companies have to scan and monitor developments in the external environment through a structured process called "*Technology Intelligence*".

Technology intelligence can be defined as the capture and delivery of technological information as part of a structured process through which an organization/company develops an awareness of technological threats and opportunities [4] for their experts. This involves constant search, through the experts, for

Supported by COEXEL.

key information from the internet provided by diverse data sources in order to stay up to date of current evolution in their field of interest with the aim of remaining competitive. These pieces of information produced by data sources from the internet are stored as structured *"documents"*. One issue that *Technology Intelligence* is faced with lies in real-time relevant recommendation of documents to experts. Indeed, Digitalization has led to the availability of very large amount of information, but this also brings an overload of information. Thus, it results in deep challenges in recommending relevant documents where the mass of data must be searched, matched and assessed *wrt.* experts' needs.

An interesting aspect of experts that the recommendation system can take into account is their multidimensional profile. In this context, experts actions characterize its profile. Furthermore, combining several profiles by relying on common behaviors, interactions, content... on documents could bring new knowledge, thus recommendations. This leads us to a graph model representation based on experts and documents.

Thanks to the multiplicity of information, such a graph integrates several dimensions that link documents together. A multi-dimensional graph can be exploited and updated in real time. This multi-dimensional graph, called a *multilayer graph*, helps to produce several kinds of recommendation dynamically depending on the combination of those dimensions.

More specifically, we are interested in studying homogeneous multilayer graphs where entities in each layer are the same while relationships in each layer is different, the so called as `multiplex` networks. It was noticed that modeling real-world complex systems as a multiplex network has improved understanding about their structures and dynamics. Multiplex networks are particular in that they have a highly correlated structure, as such, it is possible to extract more information when a system is represented this way than when it is represented as a single layer taken separately.

Our approach proposes a framework which aims at making recommendations based on a multiplex network of documents. This multiplex relies on the split-up of information, that connects documents with each other in different ways (*e.g.,* interactions, semantic, categories). Then each layer can be the basis of recommendations and the combination of those layers by a graph database paradigm will bring new ways of recommending documents. The article describes how the multiplex is built, its integration to a `Neo4j` database for dynamic manipulations, and thus specialized recommendations.

This paper is organized as follows. We first detail in Sect. 2 the related work on framework modeling with multiplex graph. Then, we formalize our multiplex data model and explain how to make recommendations in Sect. 3.

2   Related Work

In the literature, several works have been proposed to tackle the issue of information overload by rather tackling the problem at data crawling levels, *via* tag recommendation algorithms [5]. This approach is not sufficient, as we still get overwhelmed with the large volumes of data that has been crawled from recommended tags. Thus, with the availability and popularity of large complex data sets in the world, it is essential to model them as graphs in order to analyze them in different ways with different methods and to make recommendations, clustering, centralities, etc. Simple graphs are not adequate for modeling complex data sets with multiple entities and relationships. Multi-layer networks have proven to be more suitable, and especially multiplex networks when we want to deal with a single type of nodes. To systematically study multiplex networks, a general multiplex network structure is necessary [11].

Wu et al. [10], introduce an algorithm by which they could generate a multiplex network which can be tuned with layers degree correlation functions, the nodal degree correlation function between the layers, and the size of the resulting network. However, a legitimate understanding of such networks can be achieved only when both structural and functional aspects of the multiplexity are correctly taken into account. Indeed, depending on the type of multiplexity, same multiplex structures can behave quite differently.

Berner et al. [1], propose a multiplex decomposition framework which enables a rigorous description of the spectral properties of multiplex networks. They achieve this by establishing a connection between the eigenvalues of multiplex networks and the one of the individual layers for the special case when the adjacency matrices of individual layers are simultaneously triagonalizable. It allows them to greatly simplify the study of synchronization on multiplex networks. This approach is not adapted for a recommendation system as it does not take into account node updates on each layer (insert and delete).

Computing layer similarity is an important way of characterizing multiplex networks because various static properties and dynamic processes depend on the relationships between layers. To model both within-layer connections and cross-layer network dependencies simultaneously, *Li* et al. [6] designed a unified optimization framework called *MANE*, for multilayer network embedding representation learning. Network embeddings make us lose the explainability and flexibility of our model as such we prefer not to use the embedding approach.

Zitnik et al. [12] proposed *OhmNet*, an algorithm for hierarchy-aware unsupervised feature learning in multi-layer networks. The focus of their framework is to learn protein features in different tissues, by applying *Node2vec* to construct network neighbors for each node in each layer. These methods rely on factorization based methods or sampling based methods which are often: infeasible, not scalable, computation and memory intensive, or difficult in balancing relationship type for cross-layer sampling, biased towards high degree nodes, and have a limited capacity to capture complexity, respectively. In response to the above difficulties, *Nianwen Ning* et al. [9] propose an Adaptive Node Embedding Framework, *ANEF*, for multiplex networks. This framework samples the cross-layer context of a node by *Metropolis Hastings Random Walk and Forest Fire Sampling*, methods to avoid bias towards high degree nodes. However, our goal is to build a model that one can understand fully how it works, thus avoiding the use of embeddings.

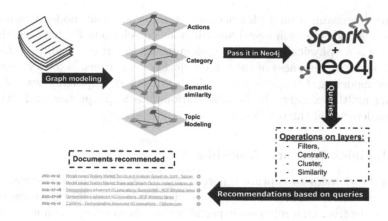

Fig. 1. The model architecture of the proposed multiplex framework

3   Our Approach

In our work, we want to make use of the advantages of multiplex networks in the context of document recommendations. As such, we propose a multiplex network framework, a recommendation system that can easily be tunable and adapted according to different contexts.

Figure 1 illustrates the process by starting with the document modeling in a graph with the respective relationships that exist between documents *wrt.* each target layer described above. Those documents (nodes) and relationships are passed into a *Neo4j* graph database to maintain the multiplex dynamically and manipulate it with graph operations (centrality, clustering, similarities, etc.) and pattern queries (filters, propagation, etc.). *Spark* helps to combined graph operations, leading to different recommendations.

3.1   Multiplex Network

A multiplex network [3] is a homogeneous multilayer network where nodes type in each layer is the same with a one-to-one mapping of the nodes in different layers, usually called replica nodes, where interlinks can exclusively connect to corresponding replica nodes. Usually, each layer turns to characterize a different type of interaction that exists amongst the nodes.

Definition 1. *A multilayer network is a graph* $\mathcal{M} = (Y, \mathcal{G})$ *where* Y *is the set of layers* $Y = \{\alpha | \alpha \in \{1, 2, ..., |Y|\}\}$*, and* $\mathcal{G}$ *the ordered list of networks characterizing relationships within each layer* α *where* $\mathcal{G} = \{G_1, ..., G_\alpha, ..., G_{|Y|}\}$*. The network* G_α *of layer* α *is defined by* $G_\alpha = (V_\alpha, E_\alpha)$*, where* V_α *is the set of nodes of layer* α*, and* E_α *is the set of links connecting nodes within layer* α*.*

Definition 2. *A multiplex network is a multilayer network* $\mathcal{M} = (Y, \mathcal{G})$ *where all vertices* V_α *from* $G_\alpha \in \mathcal{G}$ *are identical, s.t.:* $V_\alpha = V = \{i | i \in \{1, 2, ..., n\}\}$

Figure 1 contains a multiplex network of 4 layers, where nodes V are documents (same nodes in each layer) and different correlations E_α between them.

A node is a "physical object", *e.g.*, documents in our case, while node-layer pairs are different instances of the same object, *e.g.*, actions, category, semantic and topic modeling. Computing layer similarities is an important way of characterizing multiplex networks because various static properties and dynamic processes depend on the relationships between layers [2].

3.2   Multiplex Networks Modeling

Our framework tries to capture different semantics of relation between documents. The choice of these semantics is made based on its relevance to the targeted objective. One relies on experts' action showcasing documents attractiveness, others focus on content similarity, or experts' classification. We will consider actions that reveals to us whether a document was of interest to the expert or not. This leads us to have 4 main different types of links that will model interactions between the documents. These 4 different links represent the different layers in our multiplex network.

Action Layer G_{action}. Since many actions are applied on documents, each expressing a different experts actions' degree of interest. The sum of each kind of action on a document gives experts' interested on it. Thus, each node will be associated with an action vector $A = (a_0 \cdot c_0, ..., a_i \cdot c_i, ..., a_n \cdot c_n)$ with $c_i \in n$ count the number of actions[1] and a_i corresponds to actions' weight where $\sum_{i=1}^n a_i = 1$. Consequently, links $E_{action} \in G_{action}$ and their weight w correspond to pairwise distances (*e.g.*, Euclidean) between action vectors of nodes V, s.t.:

$$r_{i,j} \in E_{action} | v_i, v_j \in V \wedge w(r_{i,j}) = 1 - \frac{dist(A_i, A_j)}{dist_{max}(E_{action})}$$

The more experts act the same way on documents, greater the weight is between them. To simplify the layer:

- Documents with low action rates (*i.e.*, new documents) are not considered,
- If $w(r_{i,j}) < \tau_a$ the relationship is ignored with a given threshold τ_a.

Semantic Layer G_{sem}. We use the cosine similarity measure, to measure how similar documents are to each other. Consider two documents A and B representing the respective vectors of the documents, we obtain:

$$\cos \varphi = \frac{\sum_{i=1}^n A_i B_i}{\sqrt{\sum_{i=1}^n A_i^2} \sqrt{\sum_{i=1}^n B_i^2}}.$$

Two documents are linked if $\cos \varphi$ is greater than a threshold τ_{sem}, otherwise we consider that the documents are not similar enough. We then give a weight w_{AB} to their relationship based on the value of $\cos \varphi$. Thus, documents with $\cos \varphi$ of 0.8 and above are strongly connected.

---

[1] n types of actions like: reading the document, sharing it, rating it...

Table 1. Graph database operations on the multiplex

Operation	Neo4j operation	Effect on the multiplex
σ	Cypher: *MATCH* & *WHERE*	Filtering (nodes/edges)
κ	Clustering (GDS)	Grouping nodes
θ	Centrality (GDS)	Gives a score for each node based on topological centralities
ν	Similarity (GDS)	Gives scores for most similar nodes based on common properties
$\cap$	Cypher: `WITH` & Spark	Intersection of graphs
$\times$	Cypher: `WITH` & Spark	Combination of scores between graphs

Category Layer G_{cat}. Given that we have predefined categories to which each document belongs to (classified by experts), and that documents could belong to one or more categories. We use the *Jaccard Similarity index* to define how much two documents are linked together.

Given A and B, two sets containing the categories to which two documents belongs to respectively, the *Jaccard Similarity* between these documents will be given by, $J(A, B) = \frac{|A \cap B|}{|A \cup B|}$, which gives the weight of the link between the two documents w_{AB}. Thus two documents are linked if $w_{AB} > \tau_{category}$ as the value of $J(A, B)$ ranges between 0 and 1.

Topic Layer G_{topic}. In this layer, we want to cluster the documents into different topics in an unsupervised way. Topic modeling is first applied to the corpus of selected documents within a period of interest to extract a defined number of topics, such that each document will belong to particular topics. Then documents shall be linked based on whether they belong to common topics or not.

To measure the level of similarity here, i.e. w_{AB}, one can still make use of the Jaccard similarity $J(A, B)$ where A and B represent topics sets that two distinct documents belong to. This layer is an unsupervised version of the category layer.

3.3 Manipulations with Neo4j

Recommending documents from a multiplex in real time using various queries is challenging. To achieve this, we rely on a graph database `Neo4j` [8] which allows expressing online queries with various operations, either on pattern matching [7] or graph algorithms (GDS library[2]).

Applying a query on a layer of the multiplex produces filtering (*e.g.,* pattern matching, clustering), ordering (*e.g.,* centrality), and scoring (*e.g.,* similarity). Thus, we propose a graph database algebra $\mathcal{O}$ to represent those manipulations as summarized in Table 1. Any operation $o \in \mathcal{O}$ remains in a closed form, it takes a graph $G(V, E)$ and produces a new graph $G'(V', E')$ s.t. $V' \subset V$.

---

[2] Neo4j GDS Library: https://neo4j.com/docs/graph-data-science/current/.

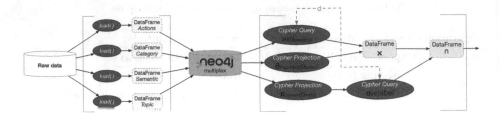

Fig. 2. Interactions with the multiplex network with Neo4j & Spark

Our approach benefits from the multiplex graph data model by combining results from several layers. In fact, since nodes V are identical in each layer we can combine the results of operations on different layers to provide different semantics of recommendation. We can see 1) unary operators which transform the graph with filtering (σ), grouping (κ), scoring (θ, ν) and 2) binary operators which combines two graphs for nodes selection ($\cap$) and scores merge ($\times$).

Query q below gives a composition example which recommends for a given input document $d \in V$ with: 1) the subgraph of layer G_{topic} with documents belonging to the community as d (*Louvain*), 2) a score for similar nodes with d in G_{cat}, and 3) scores for each node in G_{action} using the PageRank centrality. Nodes' scores from G_{cat} and G_{action} are combined together to produce recommendations and restricted to the same community in the G_{topic} layer.

$$q = \sigma_{d \in label}(\kappa_{Louvain}(G_{topic})) \cap \nu_d(G_{cat}) \times \theta_{PageRank}(G_{action})$$

3.4   Workflow of Recommendations

In order to manipulate the multiplex with *Neo4j*, we need to optimize operations, their combination and their usage in real time. We present here the integration of the multiplex in *Neo4j* and its interaction with *Spark*.

A first Spark workflow takes raw data and translates it into the multiplex according to layers' specification (Sect. 3.2). For new documents a node is added to the system, producing relationships in the semantic layer *wrt.* the content. Then, further experts' interactions generate new relationships with actions, categorization or topic clusters. Thanks to *Neo4j* typing, the multiplex network can be built as a unique graph with different relationships type. Queries will just focus on the corresponding layer (*i.e.*, type).

A second Spark workflow corresponds to the recommendation step. Operators presented in Sect. 3.3 are templated by integrating configurable Cypher queries. Some precomputed queries (*e.g.*, PageRank) are stored as *Cypher Projections* for further executions, and other combinations of operators as persisted *Spark DataFrames*. Cypher projections and persistent DataFrames can be updated according to several triggering strategies: data, timely, on-demand.

Binary operators like $\cap$ and $\times$ requires a specific manipulation. When the query allows it, we pipe results from one operator to another through WITH clauses. Otherwise, a merge between two *Spark* DataFrames are applied, thanks to the multiplex model node ids which are identical on both sides to ease the merge. An example of Spark workflows is illustrated in Fig. 2. We can notice that document d is used as a parameter of the workflow (ν and σ) to produce recommendations. It relies on Cypher projections previously computed and Cypher queries directly on the corresponding layers (*i.e.,* relationship type).

4 Conclusion

This paper proposes a framework for a recommender system in the *Technology Intelligence* field, using a multiplex graph by making use of the advantages of the graph database *Neo4j*. The main objective is to exploit data multidimensionality via a multiplex graph. We examine the respective instances of a document in our case to define the corresponding layers of our multiplex network. Then we show how this helps us easily apply filters/functions on this multiplex to obtain different kinds of recommendations. The flexibility of *Neo4j* proves that this is a promising approach and could help us handle important aspects such as real-time recommendations. For future work, we will experience the framework to showcase the flexibility and its efficiency. Furthermore, a formal specification of our algebra will ease the workflow optimization.

References

1. Berner, R., Mehrmann, V., Schöll, E., Yanchuk, S.: The multiplex decomposition: an analytic framework for multilayer dynamical networks. SIAM J. Appl. Dyn. Syst. **20**(4), 1752–1772 (2021)
2. Bródka, P., et al.: Quantifying layer similarity in multiplex networks: a systematic study. R. Soc. Open Sci. **5**(8), 171747 (2018)
3. Cozzo, E., De Arruda, G.F., Rodrigues, F.A., Moreno, Y.: Multiplex Networks: Basic Formalism and Structural Properties. Springer, Cham (2018). https://doi.org/10.1007/978-3-319-92255-3
4. Egger, M., Schoder, D.: Consumer-oriented tech mining: integrating the consumer perspective into organizational technology intelligence-the case of autonomous driving. In: HICSS (2017)
5. Krestel, R., Fankhauser, P.: Personalized topic-based tag recommendation. Neurocomputing **76**(1), 61–70 (2012)
6. Li, J., Chen, C., Tong, H., Liu, H.: Multi-layered network embedding. In: ICDM, pp. 684–692 (2018)
7. Medeiros, C., Costa, U., Musicante, M.: Standard matching-choice expressions for defining path queries in graph databases. In: DOING@ADBIS, pp. 97–108 (2021)
8. Negro, A.: Graph-Powered Machine Learning. Simon and Schuster, New York (2021)
9. Ning, N., Yang, Y., Song, C., Wu, B.: An adaptive node embedding framework for multiplex networks. Intell. Data Anal. **25**(2), 483–503 (2021)
10. Wu, K., Zou, W., Yao, Y., Zhou, Y.: An algorithm for multiplex network generation. In: CCC, pp. 1230–1235. IEEE (2016)

11. Zhou, Y., Zhou, J.: Algorithm for multiplex network generation with shared links. Phys. A Stat. Mech. Appl. **509**, 945–954 (2018)
12. Zitnik, M., Leskovec, J.: Predicting multicellular function through multi-layer tissue networks. Bioinformatics **33**(14), i190–i198 (2017)

Relation Extraction from Clinical Cases
for a Knowledge Graph

Agata Savary[1]($\boxtimes$) (iD), Alena Silvanovich[2], Anne-Lyse Minard[3] (iD), Nicolas Hiot[2,4] (iD),
and Mirian Halfeld Ferrari[2] (iD)

[1] LISN, Paris-Saclay University, CNRS, Orsay, France
agata.savary@universite-paris-saclay.fr
[2] LIFO, Université d'Orléans, INSA CVL, Orléans, France
mirian@univ-orleans.fr
[3] LLL, Université d'Orléans, Orléans, France
anne-lyse.minard@univ-orleans.fr
[4] EnnovLabs – Ennov, Paris, France
nhiot@ennov.com

Abstract. We describe a system for automatic extraction of semantic relations between entities in a medical corpus of clinical cases. It builds upon a previously developed module for entity extraction and upon a morphosyntactic parser. It uses experimentally designed rules based on syntactic dependencies and trigger words, as well as on sequencing and nesting of entities of particular types. The results obtained on a small corpus are promising. Our larger perspective is transforming information extracted from medical texts into knowledge graphs.

1 Introduction

Transforming data into information and then into knowledge is the focus of our action. In there, one of the main challenges concerns the construction of a graph database instance from a set of textual documents, sometimes referred to as the construction of a *knowledge graph*. This paper deals with a step towards this goal.

Nowadays, knowledge graphs are considered essential to allow smart data exploitation. They are intended to use an *organizing principle* so that a user (or a computer system) can reason about the underlying data. This organization principle puts in place a meta-data level (a schema) that adds context to knowledge discovery.

Our goal is to work on attributed graph databases which become very popular both in industry and academia [9]. This comprises nodes, representing entities (such as people, drugs, and exams), and edges, representing relationships between the entities. Graph databases are to be used when the relationships are as important as the entities themselves. Any number of attributes (also called properties), in the form of key-value pairs, may be associated with the nodes or the edges.

The ultimate goal of our work is to automatically map text to a given schema, building in this way a database instance that respects that schema. The schema here can be built as a collaborative task where techniques from natural language and database model design interact. Our corpus is a collection of clinical cases from which we would like to extract entities (classes) and relationships among them. The following example illustrates our general propose.

S. Chiusano et al. (Eds.): ADBIS 2022, CCIS 1652, pp. 353–365, 2022.
https://doi.org/10.1007/978-3-031-15743-1_33

Example 1. Let us consider the following clinical case extract[1]: "*A female patient in the age group 55–60 years presented to us with blurring of vision in both eyes. On slit-lamp examination, numerous circular to oval fleck-like discrete blue opacities at the level of deep corneal stroma and Descemet's membrane was observed.*" In this example, we want a database instance to represent patients having some symptoms and examinations they pass. Let us consider the representation of Fig. 1 where *Patient, Anatomy, Symptom* and *Examination* are types of nodes and edges represent relationships *PassExam, HasSymp, GivesRes* and *ConcernsAnat.* To structure the information, we can consider a schema designed by a database designer from the information obtained through natural language processing. This conception is a big challenge. One should consider many different aspects of the problem, starting with questioning what information is really important to store. The solution also depends on the (analytical) queries we want to consider later on. Then, for a graph database, we should also decide what information is represented as a node, a property or a relationship. Usually entities give rise to nodes, but the distinction between properties and relationships is not evident. Choices may impact the efficiency of query processing.   ☐

It should be clear that we do not want just to transform textual structures into a graph – some tools exist to represent a text corpus as a graph[2] [10, 17], but they do not go further, trying to build a higher-level model. Here, the idea is to add or infer metadata (the schema) and organize the information originally available in texts according to this higher abstraction. Different proposals exist as, for instance, to use generic taxonomies and ontologies as the property graph model. Our work focuses on custom data models: a public standard can be used as a starting point, but we let the database designer introduce her own organisation principles. In this scenario, the information extraction pipeline is usually composed of the following steps: named entity recognition and classification, co-reference resolution, and relationship extraction. These steps should deal with well known challenges in Natural Language Processing (NLP), such as disambiguation or temporal representations.

In this paper, we focus on relation extraction in NLP to contribute to the construction of a knowledge graph. In our previous work, we have concentrated our attention on entities: in [19] we propose a method for the extraction of nested entities that uses a cascade of Conditional Random Fields (CRF);[3] in [2] we propose entity enrichment in order to translate natural language queries into database queries. In fact, from the enrichment of an entity some relationships can be detected. In the current paper, we consider the extraction of entities as done in [19], and *we investigate the extraction of binary relations between them, broadening the initial ideas we used in* [2].

It is worth noticing that the gap between the mentioned NLP steps and the wanted data model is still big. The construction of such a model needs 'external' semantic information (standard or customized) and should always be driven by its intended use. The final efficiency of the model is essential.

---

[1] Extract from PubMed: https://pubmed.ncbi.nlm.nih.gov/35365471/.

[2] https://www.slideshare.net/lyonwj/natural-language-processing-with-graph-databases-and-neo4j.

[3] Conditional Random Fields [15] are probabilistic models often used in NLP for sequence labelling tasks as they take into account the context of the samples to label.

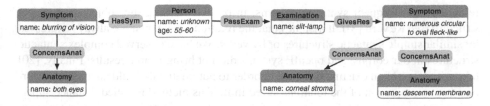

Fig. 1. Example of a Neo4J graph database instance we would like to obtain from the given text. Most properties of edges and nodes are not shown.

Paper Organization. Section 2 presents related works. Section 3 describes the medical corpus used in the experiments and summarizes the entity extraction method developed in previous work. Section 4 addresses the typology of relations between entities and Sect. 5 presents the rule-based method for extracting these relations from the corpus. Experimental results are shown in Sects. 6 and 7 closes the paper.

2 Related Work

Relation extraction is a task of information extraction which consists in detecting if two or more entities are linked and in classifying the link. In this work we are interested only in binary relations, i.e. relations between two entities, and we want both to detect and to classify them.

The first methods developed for relation extraction where based on patterns, manually defined or automatically extracted [14]. In the medical domain we can cite the Sem-Rep system [22] which used the UMLS in order to define semantic patterns, [1] used lexicalized patterns and [7] resorted to multi-layer patterns (word forms, lemmas and parts-of-speech). The RelEx system [11], designed to extract protein-protein interaction relations, made use of rules based on dependency trees. First, the dependency path that linked two entities of interest (proteins) was extracted, then the dedicated rules identified relations of three types: effector-relation-effectee, relation-of-effectee-by-effector, relation-between-effector-and-effectee.

Supervised machine learning methods are used when the size of the annotated training data is big enough. In the biomedical domain, the most popular techniques are SVM (e.g. [18,23]) and CRF, with a large variety of features (surface, semantic, syntactic, etc.). [6] hypothesized that "instances containing similar relations will share similar substructures in their dependency trees". Therefore they developed a system based on SVM and augmented dependency trees (addition of features on the nodes, such as part-of-speech, chunk tag, etc.) in order to extract relations from newswire documents (located, citizen-of, part-of, etc.). Their experiments show an improvement of performances using dependency tree kernels instead of bag-of-words kernels. More recently, a lot of methods have been proposed based on deep learning techniques (for example [16]). As in our task we do not have any training data, we cannot use these supervised methods.

In order to counter the issue of the availability of annotated data, we had a look at unsupervised methods. The Open Information Extraction task enables to extract relations from frequent syntactic patterns, in particular using the subject-verb-object structure (see ReVerb [8]). It assumes that the relations of interest are always expressed by similar simple syntactic structures or by verbs. We have observed complex syntactic structures in our corpus, and openIE systems did not bring usable results. Finally, [20] take advantage of an existing database in order to automatically build an annotated corpus from a projection of the relations saved in it. This method is called distance supervision and requires a database instead of a manually annotated corpus. In our project, we do not have a database on input, and the goal is precisely to automatically build one, so we cannot use distance supervision.

3   Corpus and Entity Extraction

This work is based on the CAS corpus - French Corpus with Clinical Cases [12] - used in multiple editions of DEFT (*Défi Fouille de Textes*) [5, 13], an evaluation campaign of systems dedicated to French medical text processing. The corpus is composed of 167 clinical cases in French (100 cases for training and 67 for testing) and it contains 13548 entities. The clinical cases came from publications in scientific literature and teaching samples for medical students. The cases are anonymized and describe real or fictitious situations from different medical domains (cardiology, urology, oncology, etc.). Some parts of the corpus were manually annotated from scratch (by two annotators), while others were automatically pre-annotated (by CRF-based methods) and then manually corrected.

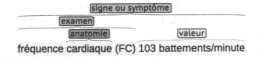

Fig. 2. Nested entities in the CAS corpus. Trans.: *'heart rate (HR) 103 beats/minute'*

The corpus contains annotation for 13 types of entities: 2 for temporal expressions (DATE, MOMENT), which will be ignored in this paper, 4 for medical objects and facts (ANATOMIE 'ANATOMY', EXAMEN 'EXAM', PATHOLOGIE 'PATHOLOGY', and SIGN OU SYMPTÔME - SOSY 'SIGN OR SYMPTOM - SOSY'), and 7 for patient treatment (DOSE, DURÉE 'DURATION', FRÉQUENCE 'FREQUENCY', MODE, SUBSTANCE, TRAITEMENT 'TREATMENT', and VALEUR 'VALUE').[4] Shorter entities can be nested in larger ones, for example an exam entity can contain the body part (anatomy) on which the exam is carried out (Fig. 2). The statistics of the entity types and their nesting are shown in Table 1 for the subset of CAS corpus used in this paper.

---

[4] Entity types are listed in French and with the English translation, if it differs from French. In the rest of the paper, the English names are used in the core of the texts, while the French ones appear in figures. The entity types SUBSTANCE, ANATOMY and TREATMENT will sometimes be abbreviated by SUB, ANAT and TREAT, respectively.

In previous work [19], we used the CAS corpus for automatic entity extraction with a CRF cascade. This approach was motivated by the frequent nesting of entities (e.g. 1831 SOSY entities contain 1909 nested entities). The idea was to extract entities of each nesting level with a different CRF model, so that the output of earlier CRF layers is used on input of further layers. The method showed an overall precision of 0.839, a recall of 0.613 and an F1 score of 0.708. These performances allow us to expect reasonable quality of automatic entity extraction in new in-domain texts.

The situation is, however, more difficult as far as relation extraction is concerned, since the CAS corpus contains no annotations for relations. Therefore, we constructed our own small development corpus. A typology of relations to annotate was first defined (cf. Sect. 4). Then, the 11 longest documents in CAS were selected and annotated by 4 annotators (each text was manually annotated by a single annotator, and some texts were double-checked by another one). The resulting corpus is composed of 6289 words, and contains annotations for 1421 entities and 742 relations. It was split into a development and a test subcorpus of respectively 6 and 5 files.

Table 1. Entities in the subset of the CAS corpus used in this paper

Entity	Number	Nested entities
SOSY	277	SUB, ANAT, EXAM, VALUE
SUBSTANCE	26	
ANATOMY	212	
EXAM	146	SUB, ANATOMY
DOSE	125	
MODE	96	
FREQUENCY	88	
VALUE	85	
PATHOLOGY	55	ANAT, VALUE
TREATMENT	44	ANAT
DURATION	25	
Total	1421	

Finally, since our relation extraction methods rely notably on syntactic patterns (cf. Sect. 5), the CAS corpus was pre-processed with the SpaCy[5] parser using the fr_core_news_md model[6] trained on the UD French Sequoia v2.8 corpus[7]. This corpus contains, in particular, a number of texts from the European Medicines Agency.

4 Typology of Relations

To define a typology of relations, we started from existing annotation schemas for medical texts in French, in particular [4] which is mainly based on UMLS [3] and composed of 37 relations classified into 5 types: aspect relations, assertion relations, drug-attribute relations, temporal relations and event-related relations, and [21] which uses 10 relations, defined jointly with experts in radiology: localisation, target, sign, cause...

Among these relations we selected those which: (i) are binary, (ii) occur frequently in texts, (iii) are specific enough to apply to a low number of entity types, (iv) have stable behavior (to be able to extract them with rules). We decided not to work on temporal and causal relations for the moment. They are generic (not domain-specific), more

---

[5] https://spacy.io/.

[6] https://spacy.io/models/fr.

[7] https://github.com/UniversalDependencies/UD_French-Sequoia.

Table 2. Relations treated in this work and types of entities to which they apply.

Relation's name	Entity type pairs	Examples
MEANS	MODE-SUBSTANCE	*une [crme]*[MODE] *de [nistatin]*[SUB] *a t prscrite* 'an [ointment][MODE] of [nistatin][SUB] was prescribed'
MEASURE_OF	DOSE-SUBSTANCE VALUE-SUBSTANCE	*la [doxorubicine]*[SUB] *raison de [37,5 mg/m2/dose]*[DOSE] '[37,5 mg/m2/dose][DOSE] of [doxorubicine][SUB]'
ACCOMPAGNIES	PATHOLOGY-PATHOLOGY SOSY-SOSY PATHOLOGY-SOSY SUBSTANCE-SUBSTANCE	*le patient prsentait une [fatigue importante]*[SOSY] *de mtme* *qu'une [hyperthermie]*[SOSY] 'the patient displayed [great tiredness][SOSY] as well as [hyperthermia][SOSY]'
REVEALS/ SEARCHES/ TESTS	EXAM-VALUE EXAM-SOSY EXAM-PATHOLOGY	*l'[chographie abdominale et pelvienne]*[EXAM] *rvle la* *prsence d'une [masse surrnalienne droite]*[SOSY] 'the [abdominal and pelvic ultrasonography][EXAM] reveals a [growth in the right adrenal gland][SOSY]'
LOCATION	ANATOMY-PATHOLOGY ANATOMY-SOSY ANATOMY-EXAM ANATOMY-TREATMENT	*elle a subi une [rsection au niveau* *du [lobe suprieur droit]*[ANAT]*]*[TREAT] 'she underwent a [resection of the [upper right lobe][ANAT]][TREAT]'

complex to extract and covered by a separate task in information extraction. The resulting set contains 5 relations: MOYEN 'MEANS', MESURE_DE 'MEASURE_OF', ACCOMPAGNE 'ACCOMPANIES', RÉLÈVE/RECHERCHE/TESTE 'REVEALS/SEARCHES/TESTS' and LOCALISATION 'LOCATION'. Table 2 shows the 5 relations as well as types of entities to which they apply.

5　Rule-Based Relation Extraction

The aim of this work is to automatically extract relations between entities in a medical text, to feed a knowledge graph. Most of the state-of-the-art methods in relation extraction (Sect. 2) are hardly applicable in our case due to the lack of data. Medical texts are most often concerned by severe privacy constraints and are rarely available outside of the strictly clinical context. The CAS corpus, the one we use, is one of the rare (if not the only) French dataset of clinical cases available for NLP research. It is not annotated for relations, thus, supervised relation extraction methods are currently excluded. Non-supervised methods, such as OpenIE, are not appropriate either, since they do not apply to a pre-defined set of relations and have low precision. Finally, distant supervision requires a large pre-existing knowledge base of relevant relations, which is not available for the medical domain in French. Under these strong data constraints, we resorted to rule-based methods relying on syntactic, lexical and surface clues.

Syntactic Rules. The hypothesis behind the syntactic rules is that syntactic dependencies between words signal semantic relations between entities containing these words. For instance, Fig. 3 shows a one-word entity [*dyphenhydramine*][SUB] and a 2-word entity [*voie IV*][MODE] '[intraveneous route][MODE]' connected by the MOYEN 'MEANS' relation. The corresponding dependency graph reveals that a syntactic dependency exists between the headwords (*dyphenhydramine* and *voie* 'route') of these two entities.

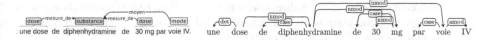

Fig. 3. The syntactic dependency of type NMOD (nominal modifier) between *diphenhydramine* and *voie* 'route' signals the semantic relation of type MEANS between a mode and a substance. Trans.: *a dose of dyphenhydramine of 30 mg via intravenous route*

Fig. 4. The common head *reçoit* 'receives' signals the relation of type MEASURE_OF between a dose and a substance. Also, the occurrence of *300 mg* between *acétaminophène* 'acetaminophen' and *voie orale* 'oral route', as well as a dependency between *mg* and *voie* 'route', signal the relation of type MEANS between a mode and a substance. Trans.: *receives acetaminophen 300 mg by oral route.*

Note that (in Table 2) the type of the relation between two entities (if any) is fully determined by the types of these entities. Entities of types MODE and SUBSTANCE can only occur in a relation of type MEANS, those of types DOSE and SUBSTANCE in a relation MEASURE_OF, etc. Given two entities E_1 and E_2 of types T_1 and T_2, respectively, as well as a relation type R, we will say that E_1 and E_2 are R-compatible if R can occur between entities of types T_1 and T_2 according to Table 2. For instance, [*dyphenhydramine*]^{SUB} and [*30 mg*]^{DOSE} in Fig. 3 are MEASURE_OF-compatible. The above observations and definitions allow us to formulate three generic syntactic rules :

Syn1 If two entities E_1 and E_2 are R-compatible and a direct syntactic link occurs between any two of their components, then a relation of type R should be inserted between E_1 and E_2. This rule is illustrated by Fig. 3, with E_1, E_2 and R equal to [*dyphenhydramine*]^{SUB} and [*voie IV*]^{MODE} '[*intraveneous route*]^{MODE}' and MOYEN 'MEANS', respectively.

Syn2 If two entities E_1 and E_2 are R-compatible and (any two of their components) have the same head[8], then a relation of type R should be inserted between E_1 and E_2. For instance, in Fig. 4, the entities [*acétaminophène*]^{SUB} and [*300 mg*]^{DOSE} are MEASURE_OF-compatible. They have incoming dependencies OBL:ARG (oblique nominal) and OBJ (object) outgoing from the same word *reçoit* 'receives', and they are, indeed, connected by a relation of type MEASURE_OF.

Syn3 If two entities E_1 and E_2 are R-compatible, if a third entity E_3 occurs between E_1 and E_2, and if (any component of) either E_1 or E_2 has a direct syntactic link with (any component of) E_3, then a relation of type R should be inserted between E_1 and E_2. For instance, in Fig. 4, the entities [*acétaminophène*]^{SUB} '[acetaminophen]^{SUB}' and [*voie orale*]^{MODE} '[oral route]^{MODE}' are separated by a

---

[8] Word w_1 is a syntactic head of word w_2 if there is a syntactic dependency link outgoing from w_1 and incoming in w_2. Most dependency parsing models ensure that each word (except the root of the sentence) has exactly one head, i.e. the dependency graph is a tree.

third entity $[300\,mg]^{\text{DOSE}}$. A direct dependency (of type NMOD) connects *mg* with *voie* 'route', and there is, indeed, a relation (of type MOYEN 'MEANS') between $[acétaminophène]^{\text{SUB}}$ and $[voie\ orale]^{\text{MODE}}$.

While the syntactic rules have a relatively good coverage, they suffer from at least two weaknesses. Firstly, the parsing results may be erroneous: some dependencies may be missing or spurious. Secondly, the large variety of possible syntactic structures in which entities occur would require a large number of specific rules whose precision might be low. So, we also resorted to lexical rules which abstract away from syntax and look at the relative position and the context in which two entities occur in a sentence.

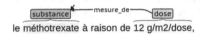

le méthotrexate à raison de 12 g/m2/dose,

Fig. 5. Relation of type MEASURE_OF between a substance and a dose signaled by the trigger *à raison de* 'at the rate of'. Trans.: *methotrexate at the rate of 12 g/m2/dose.*

Lexical Rules. A precise relation of type R is sometimes signaled by precise trigger words which occur between two entities. We call such words *R-compatible triggers.* For instance, in Fig. 5 the relation of type MEASURE_OF between the non-adjacent entities $[méthotrexate]^{\text{SUB}}$ and $[12\ g/m2/dose]^{\text{DOSE}}$ is signaled by the trigger sequence *à raison de* 'at the rate of'. We experimentally established short lists of triggers for two relations: MEASURE_OF and ACCOMPANIES. They are listed in Table 3. This, allows us to formulate the following lexical extraction rule:

Lex1 If entities E_1 and E_2 are R-compatible and an R-compatible trigger occurs between them, then a relation of type R should be inserted between E_1 and E_2.

Table 3. Relations detectable by trigger words

Relation	Trigger words	Example
MEASURE_OF	*à raison de* 'at the rate of',	$[cidofovir]^{\text{SUB}}$ $[intraveneux]^{\text{MODE}}$ ***à raison de*** $[375\ mg]^{\text{DOSE}}$
	dosé 'measured out', concentration	'$[intraveneous\ route]^{\text{MODE}}$ $[cidofovir]^{\text{SUB}}$ **at the rate of** $[375\ mg]^{\text{DOSE}}$'
ACCOMPANIES	*ainsi que* 'as well as', *sans* 'without'	*un traitement intraveineux de* $[métoclopramide]^{\text{SUB}}$ ***associé à*** *de la* $[diphénhydramine]^{\text{SUB}}$
	associé à 'associated with', *avec* 'with'	'an intraveneous treatment of $[metoclopramide]^{\text{SUB}}$ **associated with** $[diphenhydramine]^{\text{SEB}}$'

The observation of the entities and relations in the CAS corpus led us to formulate and implement other lexical rules, which we finally did not retain due to their weak performances. In particular, we extracted a list of predicates such as *révèle* 'reveals', *confirme* 'confirms', *demeure* 'remains', etc. signaling the REVEALS/SEARCHES/TESTS relation as shown in the examples below. Notice that the variety of such predicates is huge and hinders the reliability of the corresponding lexical patterns.

– L'[*échographie abdominale*]^{EXAMEN} ne **révèle** [*aucune anomalie*]$^{\text{SOSY}*9}$
The [abdominal ultrasonogram]^{DOSE} reveals [no anomaly]$^{\text{SOSY}*}$
– l'[*imagerie par résonance magnétique abdominale*]^{EXAMEN} **confirme** la présence d'une [*masse surrénalienne à droite*]^{SOSY}
the [abdominal magnetic resonance imaging]^{EXAM} confirms the presence of a [growth in the right adrenal gland]^{SOSY}'

Sequence Rules. Sometimes a relation R can be detected by the sheer proximity of entities of the relevant types. For instance, Fig. 6 shows two entities [*échographie de l'appareil urinaire*]^{EXAMEN} '[ultrasonogram of the urinary system]^{EXAM}' and [*reins de taille normale*]^{SOSY} '[kidneys of regular size]^{SOSY}'. Their sequence signals the relation of type REVEALS/SEARCHES/TESTS. We experimentally checked that this relation type can be reliably detected by this principle.

: une échographie de l'appareil urinaire montrant des reins de taille normale,

Fig. 6. Entities of type EXAM and SOSY occuring in a sequence and connected by a relation of type REVEALS/SEARCHES/TESTS. Trans.: *ultrasonogram of the urinary system showing the kidneys of regular size.*

A sequence of two entities, connected by a particular relation, can also include a third entity of another specific type. For instance, in Fig. 4, the entities [*acétaminophène*]^{SUB} '[acetaminophen]^{SUB}' and [*voie orale*]^{MODE} '[oral route]^{MODE}' are separated by entity [*300 mg*]^{DOSE}, and a relation of type MEANS occurs between the first two. While rule 5 allows us to extract the MEANS relation in this examples, it misses other cases with an erroneous syntactic analysis. We experimentally checked that the principle of spotting sequences of 3 entities of specific types is especially reliable for the MEANS relation type. Thus, we formulated the two following sequence-based rules:

Seq1 If two entities E_1 and E_2 are REVEALS/SEARCHES/TESTS-compatible and occur one after another with no other intervening entity, then a relation of this type should be inserted between E_1 and E_2. Figure 6 illustrates this rule.

Seq2 If two entities E_1 and E_2 are MEANS-compatible and a third entity of type DOSE occurs between them, then a relation of type MEANS should be inserted between E_1 and E_2. This rule is illustrated by Fig. 4.

Nesting Rules. Recall (from Table 1) that the CAS corpus has a high rate of nested entities. In some cases, the precise types of the nesting and nested entity are a sufficient evidence of a relation existing between the two. For instance, in Fig. 6 the entity

---

^{9} An asterisk following an entity type signals a negated entity occurrence.

[*échographie de l'appareil urinaire*]^{EXAMEN} '[ultrasonogram of the urinary system]^{EXAM}' includes [*appareil urinaire*]^{ANATOMIE} '[urinary system]^{ANATOMY}' and both are connected by the LOCATION relation. We experimentally found that this principle is quite accurate for LOCATION-compatible entities, which yields the following nesting-based rule:

Nest1 If two entities E_1 and E_2 are LOCATION-compatible, E_2 is of type ANATOMY and is included in E_1, then a relation of this type should be inserted between E_1 and E_2. Figure 6 illustrates this rule.

The final relation extraction system consists in applying all the rules formalized above (Syn1, Syn2, Syn3, Lex1, Seq1, Seq2 and Nest1) and retaining all the relations inserted by them. The order of the rules does not matter for the final outcome. If the same relation was inserted by more than one rule, only one of its occurrences is retained. The following section describes the evaluation of this system on the test corpus.

6   Experimental Results

As discussed in Sect. 3, we evaluate our approach on a test corpus composed of 5 files with 548 entities and 230 relations. The corpus contains not only sentences, but also text representation of tables. The latter was removed from the corpus beforehand, because our rules are not designed for it and, thus, false negatives can easily be generated.

The system was evaluated using a confusion matrix of the *true positives*, TP (i.e. relations correctly extracted), the *false positives* FP (i.e. relations extracted in wrong places) and the *false negatives*, FN (i.e. missed relations). The results are given in Table 4 along with precision, recall and F1-score. We notice an overall relatively high F1 score of

Table 4. Relation extraction results

Relation	TP	FN	FP	Precision	Recall	F1
MEASURE_OF	38	8	4	0,90	0,83	0.86
MEANS	23	7	2	0.92	0.76	0.84
ACCOMPANIES	5	2	1	0.83	0.71	0.77
REVEALS	52	3	3	0.95	0.95	0.95
LOCATION	65	12	0	1.00	0.84	0.92
Total	183	32	10	0.95	0.84	0.89

0.89. The number of false positives is low compared to the false negatives, indicating that our rules are specific and fail to cover many potential relations. This is mainly due to the selection of unambiguous relations. For example, the precision of 1 for LOCATION is due to nested entities in which this relation is sure to occur.

An analysis of the texts annotated by the system, reveals that some errors came from wrong segmentation of text into sentences and words. However, most errors are due to incorrect dependency trees. As the majority of the relations are extracted using syntactic rules, the system is sensible to the syntactic variability. For example, in *La dose totale reçue lors de cette* [*perfusion*]^{MODE} *a été égale à* [*30 mg*]^{DOSE} " 'The total dose received during this [perfusion]^{MODE} was equal to [30 mg]^{DOSE}' there is no direct dependency between [*perfusion*]^{MODE} and [*30 mg*]^{DOSE}, while there is one in the synonymous phrase [*perfusion*]^{MODE} *de* [*30 mg*]^{DOSE} " '[perfusion]^{MODE} *of* [30 mg]^{DOSE}'.

As we are exploiting syntactic rules, we cannot cover all patterns and may also need a larger corpus to identify other instances of the relations. After analysing the test corpus, we discover new relation patterns, only present in the test corpus. Nevertheless, the medical field remaining very specific, it limits this variability and makes it possible to obtain good results with few rules.

7 Conclusions

We have shown initial experiments and results in extracting semantic relations between entities in a medical corpus. A generic syntactic parser applied to a specialized text proved accurate enough to approximate semantic relations via syntactic dependencies. We also strongly exploited the fact that the relation between two entities is fully determined by the types of these entities. This allowed the rules to cover many relations at the same time. Another opportunity is the specificity of the text genre under study (clinical cases), in which dedicated lexico-syntactic constructions and entity sequences often repeat and can yield targeted rules.

Future work includes extending the sets of rules and experiments to larger corpora (e.g. by annotating relations in the whole CAS corpus). The resulting annotations might then be used to train a model in a supervised setting. We will also further investigate the interface between information extraction from text on the one hand, and designing the schema of a knowledge graph and populating it from text on the other hand.

Acknowledgements. Work partly supported by the ICVL federation and RTR-DIAMS. It is done in the context of DOING action of the GDR-MADICS.

References

1. Abacha, A.B., Zweigenbaum, P.: Automatic extraction of semantic relations between medical entities: a rule based approach. J. Biomed. Semant. **2**(Suppl 5), S4+ (2011)
2. Amavi, J., Halfeld Ferrari, M., Hiot, N.: Natural language querying system through entity enrichment. In: Bellatreche, L., et al. (eds.) TPDL/ADBIS -2020. CCIS, vol. 1260, pp. 36–48. Springer, Cham (2020). https://doi.org/10.1007/978-3-030-55814-7_3
3. Bodenreider, O.: The unified medical language system (UMLS): integrating biomedical terminology. Nucleic Acids Res. **32**(Database-Issue), 267–270 (2004)
4. Campillos, L., Deléger, L., Grouin, C., Hamon, T., Ligozat, A.L., Névéol, A.: A French clinical corpus with comprehensive semantic annotations: development of the medical entity and relation LIMSI annotated text corpus (MERLOT). Lang. Resour. Eval. **52**(2), 571–601 (2017)
5. Cardon, R., Grabar, N., Grouin, C., Hamon, T.: Présentation de la campagne d'évaluation DEFT 2020 : similarité textuelle en domaine ouvert et extraction d'information précise dans des cas cliniques. In: Cardon, R., Grabar, N., Grouin, C., Hamon, T. (eds.) 6e conférence conjointe Journées d'Études sur la Parole (JEP, 33e édition), Traitement Automatique des Langues Naturelles (TALN, 27e édition), Rencontre des Étudiants Chercheurs en Informatique pour le Traitement Automatique des Langues (RÉCITAL, 22e édition). Atelier DÉfi Fouille de Textes, pp. 1–13. ATALA, Nancy, France (2020). https://hal.archives-ouvertes.fr/hal-02784737

6. Culotta, A., Sorensen, J.: Dependency tree kernels for relation extraction. In: Proceedings of the 42nd Annual Meeting of the Association for Computational Linguistics (ACL-2004), pp. 423–429 (2004)
7. Embarek, M., Ferret, O.: Learning patterns for building resources about semantic relations in the medical domain. In: Proceedings of the International Conference on Language Resources and Evaluation, LREC 2008, 26 May–1 June 2008, Marrakech, Morocco (2008)
8. Fader, A., Soderland, S., Etzioni, O.: Identifying relations for open information extraction. In: Proceedings of the 2011 Conference on Empirical Methods in Natural Language Processing, pp. 1535–1545. Association for Computational Linguistics, Edinburgh, Scotland, UK, July 2011. https://aclanthology.org/D11-1142
9. Francis, N., et al.: Cypher: an evolving query language for property graphs. In: Das, G., Jermaine, C.M., Bernstein, P.A. (eds.) Proceedings of the 2018 International Conference on Management of Data, SIGMOD Conference 2018, Houston, TX, USA, 10–15 June 2018, pp. 1433–1445. ACM (2018)
10. Franciscus, N., Ren, X., Stantic, B.: Dependency graph for short text extraction and summarization. J. Inf. Telecommun. 3(4), 413–429 (2019)
11. Fundel, K., Küffner, R., Zimmer, R.: RelEx-relation extraction using dependency parse trees. Bioinformatics 23, 365–371 (2007)
12. Grabar, N., Grouin, C., Hamon, T., Claveau, V.: Corpus annoté de cas cliniques en français. In: TALN 2019–26e Conference on Traitement Automatique des Langues Naturelles, pp. 1–14. Toulouse, France, July 2019. https://hal.archives-ouvertes.fr/hal-02391878
13. Grouin, C., Grabar, N., Illouz, G.: Classification de cas cliniques et évaluation automatique de réponses d'étudiants : présentation de la campagne DEFT 2021. In: Denis, P., et al. (eds.) Traitement Automatique des Langues Naturelles, pp. 1–13. ATALA, Lille, France (2021). https://hal.archives-ouvertes.fr/hal-03265926
14. Hearst, M.A.: Automatic acquisition of hyponyms from large text corpora. In: COLING 1992 Volume 2: The 14th International Conference on Computational Linguistics (1992). https://aclanthology.org/C92-2082
15. Lafferty, J.D., McCallum, A., Pereira, F.C.N.: Conditional random fields: probabilistic models for segmenting and labeling sequence data. In: Proceedings of the Eighteenth International Conference on Machine Learning, pp. 282–289. ICML 2001, Morgan Kaufmann Publishers Inc., San Francisco, CA, USA (2001)
16. Li, Z., Yang, Z., Shen, C., Xu, J., Zhang, Y., Xu, H.: Integrating shortest dependency path and sentence sequence into a deep learning framework for relation extraction in clinical text. BMC Med. Inform. Decis. Mak. 19, 22 (2019)
17. Mihalcea, R., Tarau, P.: Textrank: bringing order into text. In: Proceedings of the 2004 Conference on Empirical Methods in Natural Language Processing, EMNLP 2004, A meeting of SIGDAT, a Special Interest Group of the ACL, held in conjunction with ACL 2004, 25–26 July 2004, Barcelona, Spain, pp. 404–411. ACL (2004)
18. Minard, A.L., Ligozat, A.L., Grau, B.: Multi-class SVM for relation extraction from clinical reports. In: Recent Advances in Natural Language Processing, RANLP 2011, 12–14 September, 2011, Hissar, Bulgaria, pp. 604–609 (2011)
19. Minard, A.L., Roques, A., Hiot, N., Halfeld Ferrari, M., Savary, A.: DOING@DEFT: cascade de CRF pour l'annotation d'entités cliniques imbriquées. In: Cardon, R., Grabar, N., Grouin, C., Hamon, T. (eds.) 6e conférence conjointe Journées d'Études sur la Parole (JEP, 33e édition), Traitement Automatique des Langues Naturelles (TALN, 27e édition), Rencontre des Étudiants Chercheurs en Informatique pour le Traitement Automatique des Langues (RÉCITAL, 22e édition). Atelier DÉfi Fouille de Textes, pp. 66–78. ATALA, Nancy, France (2020). https://hal.archives-ouvertes.fr/hal-02784743

20. Mintz, M., Bills, S., Snow, R., Jurafsky, D.: Distant supervision for relation extraction without labeled data. In: Proceedings of the Joint Conference of the 47th Annual Meeting of the ACL and the 4th International Joint Conference on Natural Language Processing of the AFNLP, pp. 1003–1011. Association for Computational Linguistics, Suntec, Singapore, August 2009. https://aclanthology.org/P09-1113
21. Ramadier, L., Lafourcade, M.: Patrons sémantiques pour l'extraction de relations entre termes - Application aux comptes rendus radiologiques. In: TALN: Traitement Automatique des Langues Naturelles. jep-taln2016, Paris, France, July 2016. https://hal.archives-ouvertes.fr/hal-01382323
22. Rindflesch, T.C., Bean, C.A., Sneiderman, C.A.: Argument identification for arterial branching predications asserted in cardiac catheterization reports. In: AMIA Annual Symposium Proceedings, pp. 704–708 (2000)
23. Uzuner, O., Mailoa, J., Ryan, R., Sibanda, T.: Semantic relations for problem-oriented medical records. Artif. Intell. Med. **50**, 63–73 (2010)

Privacy Operators for Semantic Graph Databases as Graph Rewriting

Adrien Boiret[✉], Cédric Eichler, and Benjamin Nguyen

INSA Centre Val de Loire, Laboratoire d'Informatique Fondamentale d'Orléans, Bourges, France
{adrien.boiret,cedric.eichler,benjamin.nguyen}@insa-cvl.fr

Abstract. Database sanitization allows to share and publish open (linked) data without jeopardizing privacy. During their sanitization, graph databases are transformed following graph transformations that are usually described informally or through ad-hoc processes.

This paper is a first effort toward bridging the gap between the rigorous graph rewriting approaches and graph sanitization by providing basic generic graph rewriting operators to serve as a basis for the construction of sanitization mechanisms. As a proof of concept, we formalize two operators, blank node creation and weighted relation randomization, using an algebraic graph rewriting approach that takes into account semantic through the equivalent of WHERE and EXCEPT clauses. We show that these operators can be used to achieve pseudonymity and local differential privacy. Both operators and all related rewriting rules are implemented using the Attributed Graph Grammar System (AGG), providing a concrete tool implementing formal graph rewriting mechanisms to sanitize semantic graph databases.

1 Introduction and Related Work

Database Sanitization. In many collaborative data centric applications that collect personal data, such as car pooling, or smart metering, the data collectors (i.e. the entity managing the application) need to publish and share raw data with various parties, which can range from internal developers who need test data, to data analysts in charge of producing predictive or explicative models, or simply publish their results to the scientific community. Since this database is composed of personal data, the European General Data Protection Regulation (GDPR) rules apply. In this paper, we consider the case where the database is sanitized (anonymized) prior to its release, which is compatible with the GDPR. Sanitization of a database means that it is no longer possible (or at least very difficult, costly and time consuming) to reidentify individuals in the dataset.

Private Publication of Semantic Graph. A massive amount of work has focused on privacy in data presented as tables. They have resulted in multiple well-established models, such as k-anonymity [19] and differential privacy [7]. In this article, we consider *semantic databases*, such as RDF, which are *graph databases.* Such databases present the advantage of managing the semantics of

© Springer Nature Switzerland AG 2022
S. Chiusano et al. (Eds.): ADBIS 2022, CCIS 1652, pp. 366–377, 2022.
https://doi.org/10.1007/978-3-031-15743-1_34

the data, which we view as an advantage in the context of anonymization, since it helps the database designer better understand which data is sensitive, and how it can be modified. Indeed, the aforementioned general anonymization concepts have been translated and applied to graph representations, but mainly in the context of social networks or computer networks [21,22]. Only a small batch of work has tackled the publication of semantic data-graphs: Delanaux et al. [5] developed a declarative framework for anonymizing RDF graphs by replacing sensitive nodes by blank nodes. Anonymisation of RDF data was also studied in [10,15] where the anonymisation model is inspired by the k-anonymity model. These methods rely on graph modifications and, more often than not, include an ad-hoc transformation process.

Graph Rewriting for Database Modification. To generalize and abstract consistent updating methods, different works have used formalisms such as tree automata or grammars for XML (see [16] for a survey) or first order logic for graph databases (*e.g.* [3,8]). Despite of the importance of graphs in databases and ontology representations, the use of formal graph rewriting techniques to model database evolutions is seldom studied. Formal graph rewriting techniques are usually based on *category theory*, an abstract way to deal with different algebraic mathematical structures and the relationships between them. Algebraic approaches of graph rewriting allow a formal yet visual specification of rule-based systems characterizing both the effect of transformations and the contexts in which they may be applied.

Few approaches relying on graph rewriting to formalize ontology evolutions have already been proposed [4,14,18]. They usually focus on formalization but do not provide an implementation. Recently, Chabin et al. [1,2] proposed SETUP, a tool for the management of RDF/S updates. In SETUP, graph rewriting rules formalize atomic updates and guarantee the preservation of RDF/S intrinsic constraints.

Positioning. In this paper, we aim at bridging the gap between rigorous graph rewriting and graph sanitization. More precisely, we aim at formalizing basic operators using attributed graph rewriting rules to serve as a basis for building graph anonymisation mechanisms. Such mechanisms could include adaptations of aforementioned techniques.

Contributions. In this paper, we formalize and implement two basic operators using AGG – The Attributed Graph Grammar System [20], one of the most mature development environments supporting the definition and application of typed graph rewriting systems [17]. These two operators demonstrate the feasibility of the approach and should be enriched with other operators to build a library of operators supporting various anonymisation schemes. The basic operators we choose to model are:

- BLANK, that replaces a set of nodes with a blank copy of themselves while preserving all or a subset of incoming and outgoing edges. This can be used *e.g.* to enforce pseudonymisation or the approach proposed in [5]. Note that even though pseudonymisation is not sufficient to ensure anonymisation, it is the basis of multiple anonymisation schemes (e.g. directly identifying data are usually suppressed as a first step of k-anonymity mechanisms) and is generally encouraged by the GDPR (see e.g. Recital 28).

- RANDOMIZE, that adds noise to a relation R by randomizing its target with a bias in favour of the initial value. We show that it can be used to achieve Local Differential Privacy (LDP).

We show that, given our formalisation and implementation it is easy to describe sets of nodes or edges as the scope of our operators. This can be intuitively seen as clauses:

- WHERE allows restriction of the operator to a subset of nodes defined by a relation (e.g. nodes that share a type or specific attribute).
- EXCEPT allows exclusion by the operator of a subset of nodes or relations defined by a relation.

In Sect. 2, we present the formalism of graph databases and graph rewriting we base our approach on, and the visual conventions used in the AGG rewriting tool. We also provide a running example for the rest of the paper. Section 3 describes our formalization and implementation of the operators BLANK and RANDOM, as well as how to translate and implement the keywords WHERE and EXCEPT in the chosen algebraic graph rewriting approach for restriction and exclusion in their scope.

2   Background and Setting

2.1   Attributed Multigraphs

We consider databases to be modelled as attributed oriented multigraphs. In such models, it is customary for nodes and edges to have properties (among a finite set) and attributes (as words on a signature). In [1], RDF/S databases are modelled as a typed graph with 4 node types and 6 edge types. These types are inherent to RDF and thus the model can not be applied natively to arbitrary graph databases (*e.g.* neo4j).

We argue that considering a single node type and a single edge type having a single attribute (named *att* and *prop*, respectively) is in fact at least as expressive. Indeed, typing and additional properties can be encoded via special kinds of relation (see Sect. 2.2). We believe this model to be able to capture most –if not all- graph database representations. In addition to facts, represented in black in the figures, the proposed rewriting system considers *artifact types* represented in blue. Artifacts are used in intermediary steps of the rewriting procedures to provide information related to the transformations (see Sect. 3).

2.2   Running Example and Motivation

As a running example, we consider a graph database that contains information on travels, both professional and personal.

It has nodes for relevant entities, people and travels, whose attributes are an identifier. It also has nodes for every literal describing informations on those entities, e.g. last name, first name and address for people, date and destination for travels. We do not differentiate nodes representing entities or literals.

Its edges describe both relations between entities, e.g. "this person participated in this travel", represented by an edge of attribute ``attends``, but also relations between entities and their information, e.g. "this person's name is in this literal", represented by an edge of attribute ``name``. Typing falls within this second case e.g. "this node is a person" or "this literal is a city", represented by an edge of attribute ``type``.

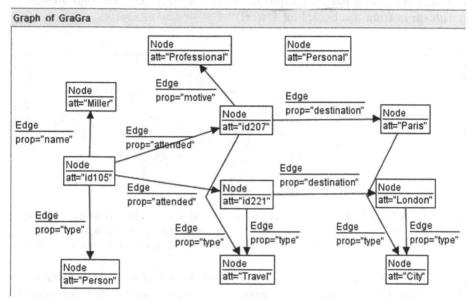

Graph of GraGra

In this instance, id105 (named Miller) attended travel id207 to Paris for professional reasons.

We give two examples of applications of our operators in this database, that we write in an as-of-yet unspecified toy-language for illustrative purposes:

– We want to create a blank node to replace non-professional travel. This blank would inherit the **destination** attribute from its original, but nothing else. We can write this as wanting a blank for nodes where the relation **type** has value travel, that is to say that have an edge labelled ``type`` towards a node labelled ``travel``. However we would like to exclude professional travels, i.e. when the relation **motive** has value professional. On these we would only redirect edges of label ``destination``. We can write this as: BLANK WHERE type = travel EXCEPT motive = professional REDIRECT destination
– We want to randomize the destination of all travel to any city. For this we must specify the relation (edges labelled ``destination``), and the target (**type** has value city).
We can write this as: RANDOM destination TO WHERE type = city

2.3   Graph Rewriting Rules

We adopt the Single Push Out (SPO) formalism ([13]) to specify rewriting rules and *Negative Application Conditions* (NACs) [9], one of its extensions to specify additional application conditions and restrict their applicability. These rules may be *fully specified graphically*, enabling an easy-to-understand yet formal graphical view of the graph transformation.

The SPO approach is an algebraic approach based on category theory. A rule is defined by two graphs – the Left and Right Hand Side of the rule, denoted by L and R – and a partial morphism m from L to R (*i.e.*, an edge-preserving morphism m from a subgraph of L to R).

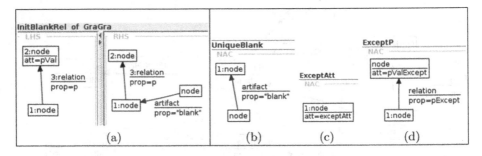

Fig. 1. Identifying nodes having some relation with some value to be replaced by a blank: (a) SPO core and its NACs (b) uniqueness of the blank replacing a node, (c) EXCEPT a specific node, (c) EXCEPT nodes with some relation having some value.

Example 1. Figure 1a formalizes the SPO core of the first step of the *blank* operator. It identifies the nodes to be replaced and creates blanks. Its L is composed of a relation p from some node (1) to a node (2) with an attribute pVal. R has the same pattern plus a new node that is source of an artifact edge labeled ''blank'' whose target is node (1). Note that the attributes of two nodes of R are not represented. In general, this stems from three possibilities: (i) it does not matter, e.g. an unattributed node in L or NAC will match any node; (ii) it can be inferred, e.g. a node in R or NAC has the same attributes as the node it is matched with in R. This is the case for node (2) in the example; (iii) a node in R is created without an attribute value. This is the case here for the new node.

   By convention, an attribute value within quotation mark (e.g. ''blank'') is a fixed constant, while *a value noted without quotation mark (e.g. p) is a variable that is either a wildcard matching any value or whose value is given as input.*

   The partial morphism from L to R is specified in the figure by tagging graph elements - nodes or edges - in its domain and range with a numerical value. An element with value i in L is part of the domain of m and its image by m is the graph element in R with the same value i. For instance, in Fig. 1a, the notation *1:* for the nodes with an unspecified attribute in L and R indicates that they are mapped through m. In the following, we refer to such nodes as 1:node.

A graph rewriting rule $r = (\mathsf{L}, \mathsf{R}, m)$ is applicable to a graph G iff there exists a total morphism $\tilde{m} : \mathsf{L} \rightarrow G$. The result of the application of r to G w.r.t. $\tilde{m}$ is the object of the push-out of the diagram composed by L, R, G, m, and $\tilde{m}$. Informally, the application of r to G with regard to $\tilde{m}$ consists in modifying G by (1) removing the image by $\tilde{m}$ of all elements of L that are not in the domain of m (*i.e.*, removing $\tilde{m}(\mathsf{L} \backslash Dom(m))$); (2) removing all dangling edges (*i.e.*, deleting all edges that were incident to a node suppressed in step (1)); (3) adding an isomorphic copy of all elements of R that are not mapped through m.

Example 2. In Fig. 1a, the rule is applicable to any non-artifact edge with the possibility to specify the attribute of the edge (p) and/or of its target (pVal) as input of the rule. Indeed, such attributes are either wildcard that match any value or their value may be given as input. For example, with the input (''type'', ''travel'') the rule only maps 1:node of *type travel*. The application of the rule consists in adding an unattributed (blank) node and a ''blank'' artifact edge from the new node to 1:node, representing the fact that the latter will be replaced by the former. In our example, this matches id207 and id221, but none of the people on the left of cities on the right.

NACs are well-studied extensions that restrict rule application by forbidding certain patterns in the graph. EXCEPT clauses will mostly be encoded through NACs. A NAC for a rule r is defined as a constraint graph which is a super-graph of its left-hand side. An SPO rule $r = (L, R, m)$ with NACs is applicable to a graph iff: (*i*) there exists a total morphism $\tilde{m} : \mathsf{L} \rightarrow G$ (this is the classical SPO application condition); (*ii*) for all NAC N associated with r, there exists *no* total morphism $\tilde{m} : N \rightarrow G$ whose restriction to L is $\tilde{m}$.

By convention, since NACs are super-graphs of L, unnecessary parts of L are not depicted when illustrating a NAC. Graph elements common to L and NAC are identified by a numerical value similarly to elements mapped by m.

Example 3. Figures 1b, 1c, and 1d represent NACs associated to the SPO core of Fig. 1a. The first specifies that 1:node must not be the target of a ''blank'' artifact edge, ensuring that we will not create several blank nodes linking to the same existing node. The second maps 1:node with a node whose attribute exceptAtt is given as input. This forbids the application of the rule to any node with such attribute, akin to an EXCEPT clause excluding a particular node (e.g. exclude specifically the travel id207). The third NACs forbids the application of the rule to any 1:node source of a relation pExcept and/or with value pValExcept. For instance, to exclude travels made for professional reasons, we give the input ''motive'' for *pExcept* and ''professional'' for pValExcept. This excludes all nodes with an edge ''motive'' linking to a node of attribute ''professional'', which in this case is exclusively the travel id207. With *pExcept* given as ''motive'' the rule excludes all travels with a declared motive (this would still exclude id221). With pValExcept given as ''professional'', the rule will exclude all 1:node that are linked to the a node with attribute ''professional'' (regardless of the relation). Note that here, at least one of the two should be given an input value, or the NAC would match L and the rule would never be applicable.

A rewriting procedure –or rule sequence– as we consider it here is a succession of steps. Each step is the application of a rewriting rule as long as the rule applies or a specified number of times. We consider that *when a rule is applicable w.r.t. several morphism, one is chosen uniformly at random.*

3   Operators

This section introduces the BLANK and RANDOM operators. The implementation in AGG [20] of the rewriting rules involved in these operators is available online.[1] For those operators, we focus on an initial set S of sources, a set T of targets, and a set R of edges that can be defined with WHERE or EXCEPT and identified through pattern matching as described in Sect. 2.3. They are built as a set of transformations that can be easily composed and used to construct a privacy mechanism.

3.1   Blank Node Creation

The goal of this operator is to replace nodes of S by blanks. It creates a blank copy of each $v \in S$ and then reroutes the in- and out-edges of v to its copy. To illustrate the expressiveness of this operator, we recall the use of BLANK proposed as an example in Sect. 2.2 and show how the clauses can be parametrized.

Initialisation: The first step of the transformation is the creation of the blank node (see Fig. 1). To ensure that we only create one blank per node, we use the NAC of Fig. 1b. The sets T and R can be defined positively (WHERE) by p and/or pVal (as input parameters), while nodes to exclude (EXCEPT) are described in the NAC of Fig. 1c and Fig. 1d. To obtain "BLANK WHERE type = travel EXCEPT motive = professional" , one should therefore provide p = "type", pVal = "travel", pExcept = "motive", pValExcept = "professional".

Rerouting: The second step reroutes all incoming and outgoing edges for nodes with a blank. *e.g.* for incoming edges, if an edge was of source 1:node, it is now of source 2:node, where 2:node is identified as the unique blank of 1:node thanks to the artifact edge. We then do the same for outgoing edges (see Fig. 2). We note that we can limit positively (WHERE) this rerouting to edges of a certain relation prop=p, or add a NAC to exclude (EXCEPT) the redirection of edges of a given relation prop=pExcept. The clause "REDIRECT destination" is constructed by providing p = "destination" and discarding the NAC.

---

[1] univ-orleans.fr/lifo/evenements/sendup-project/index.php/privacy-operators/.

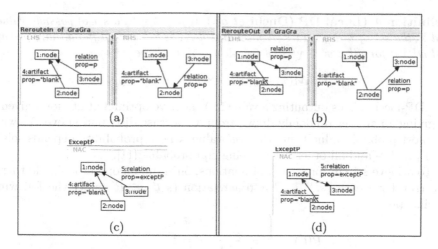

Fig. 2. Rerouting of the incoming (a) and outgoing (b) edges from 1:node to its blank 2:node. The NAC (c) and (d) correspond to EXCEPT exclusions.

Cleanup: The operators proposed create nodes and edges. Every artifact node or edge should be deleted at the end of the rewriting. However in the case of blank node creation, there is an additional step to consider: the actual deletion of the original node. Indeed, if the rerouting is complete (all edges have been rerouted to and from 1:node to its blank 2:node), we may wish to delete 1:node altogether using the rule formalized in Fig. 3. The two NACs (noIn and noOut) ensure that the deleted node has no incident edge left.

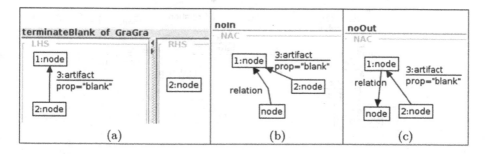

Fig. 3. If 1.node has no incoming (b) or outgoing (c) edge, we delete it (a).

3.2 Randomized Edge Selection for Local Differential Privacy

In this section we propose a randomizing operator that can be used for example to achieve local differential privacy (LDP) as defined and used in [6, 12].

Definition 4 (Local DP (Duchi *et al.*) [6]). *Let χ be a set of possible values and Y the set of noisy values. A mechanism $\mathcal{M}$ is ε-locally differentially private (ε-LDP) if for all $x, x' \in \chi^2$ and for all $y \in Y$ we have*

$$Pr[\mathcal{M}(x) = y] \leq e^{\varepsilon} \times Pr[\mathcal{M}(x') = y]$$

LDP-mechanisms outputing a value in Y achieve optimal utility for a given ε by giving an answer randomly drawn from a staircase distribution over Y, with the most probable value being the *real* value -whose probability depends solely on $|Y|$ and ε- and all other values being equiprobable [11].

To achieve ε-LDP for a set R of relations, our random operator should therefore transform each $(s, t) \in R$ into a relation $(s, t') \in R'$ under the following specification:

$$P(t') = \begin{cases} 0 & t' \notin T \\ \frac{K}{|T|-1+K} & t' = t \\ \frac{1}{|T|-1+K} & t' \in T \wedge t' \neq t \end{cases}$$

with K an integer approximation of e^{ε}, $K = \lfloor e^{\varepsilon} \rfloor$.

To do so, we pick a new target at random, but, to obtain a staircase distribution from a uniform distribution –used to chose the morphism with regard to which the transformation rule is applied–, we skew the odds by creating clones of the true answer to recreate a staircase distribution. This procedure is to be repeated for every edge to randomize, rather than once for the whole set.

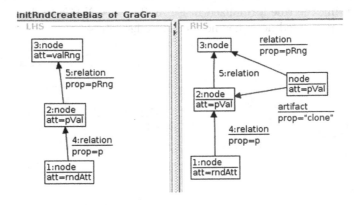

Fig. 4. Creating a clone of the real target, to bias the process towards truth.

Initialisation: We start by creating a bias for the correct value, by creating a clone node (see Fig. 4). This rule is to be repeated as many times as required to obtain ε-LDP. For instance, creating one clone makes it twice as likely to choose the right answer rather than a false one, hence it suits $\varepsilon \geq \ln(2)$. Creating two clones makes the truth three times likelier, and suits $\varepsilon \geq \ln(3)$. If we create 0 clones, then we are 0-LDP as the edges' targets will be uniformly randomized. In general we need to apply this rule $K - 1$ times.

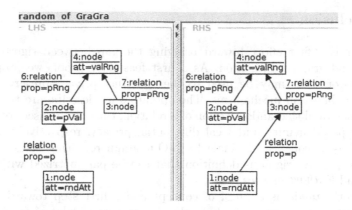

Fig. 5. Redirect the edge to any target, potentially clones.

Rerouting: We then redirect the edge we consider towards any target (see Fig. 5). The target is picked randomly among nodes of T and the clones. We have one chance to pick the original node, and $e^\varepsilon - 1$ chances to pick a clone. This means we are e^ε times more likely to pick one of these nodes than any other. This rule only has to be applied once. To construct "RANDOM destination TO WHERE type = city" one would need to provide as input p = ''destination'', pRng = ''type'', and valRng = ''city''.

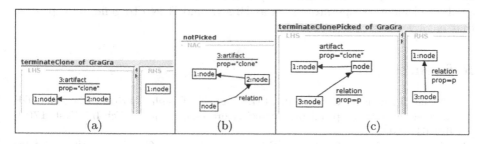

Fig. 6. Erasing clones (a) that have not been randomly selected (b) or redirect the relation and erase if they have been picked (3).

Termination: Since clones are additional chances to tell the truth, they have to be ultimately deleted (see Fig. 6). If no non-artifact edge is incident to the clone 2.node (Fig. 6b), then the clone can be safely deleted without loss of information (Fig. 6a). If there is such an edge, then this means that it was the new target chosen during the rerouting step. We will redirect this edge to the original 1.node, before deleting the clone (Fig. 6c).

In the end, the rerouting has a probability $\frac{K}{|T|-1+K}$ to have been redirected to the original target and $\frac{1}{|T|-1+K}$ to have landed on another node/value. Therefore the mechanism grants $\ln(K)$-LDP and thus ε-LDP since $\ln(K) \le \varepsilon$.

4   Conclusion

This paper is a first effort toward bridging the gap between rigorous graph rewriting and graph sanitization. As a first feasibility proof, we propose two basic operators that may be used to the construct privacy-preserving graph databases publication mechanisms. The operators we describe are the creation of blank nodes and the randomization of a relation, that can be used for example to achieve pseudonymity and local differential privacy, respectively. They are formalized as a sequence of Single Push-Out graph rewriting rules, providing a generic expressive rigorous definition that can be parametrized with a set of WHERE and EXCEPT clauses.

This work stands as a proof of concept and a first step towards a graph rewriting based approach to graph database sanitization. A lot of considerations and work remain open on the topic. Notably, both operators we present here are designed to work uniformly on a single relation. In real world usecases, most queries are joins of several relations (or to speak in graph terms, path queries on more than one edge). For a sanitization mechanism to preserve good qualities on such requests while still providing differential privacy, it is likely that we will need operators adapted to the preservation of invariants on composite paths. It is possible, but not yet shown, that some such operators can be built as compositions of one-relation operators. Another guarantee that we aim to establish in future work is the preservation of the database's schema.

Finally, as a more general goal, we aim to create a descriptive tool that would allow a user to specify semantic and privacy constraints. Such a tool would then combine elementary operators such as BLANK and RAND to generate a process that enforce those constraints.

References

1. Chabin, J., Eichler, C., Ferrari, M.H., Hiot, N.: Graph rewriting rules for RDF database evolution: optimizing side-effect processing. Int. J. Web Inf. Syst. **17**(6), 622–644 (2021)
2. Chabin, J., Eichler, C., Halfeld-Ferrari, M., Hiot, N.: Graph rewriting rules for RDF database evolution management. In: Proceedings of the 22nd International Conference on Information Integration and Web-Based Applications & Services, pp. 134–143. ACM (2020)
3. Chabin, J., Halfeld-Ferrari, M., Laurent, D.: Consistent updating of databases with marked nulls. Knowl. Inf. Syst. **62**(4), 1571–1609 (2019). https://doi.org/10.1007/s10115-019-01402-w
4. De Leenheer, P., Mens, T.: Using graph transformation to support collaborative ontology evolution. In: Schürr, A., Nagl, M., Zündorf, A. (eds.) AGTIVE 2007. LNCS, vol. 5088, pp. 44–58. Springer, Heidelberg (2008). https://doi.org/10.1007/978-3-540-89020-1_4
5. Delanaux, R., Bonifati, A., Rousset, M.-C., Thion, R.: Query-based linked data anonymization. In: Vrandečić, D., et al. (eds.) ISWC 2018. LNCS, vol. 11136, pp. 530–546. Springer, Cham (2018). https://doi.org/10.1007/978-3-030-00671-6_31

6. Duchi, J.C., Jordan, M.I., Wainwright, M.J.: Local privacy and statistical minimax rates. In: 54th Annual IEEE Symposium on Foundations of Computer Science, pp. 429–438. IEEE Computer Society (2013)
7. Dwork, C.: Differential privacy. In: Bugliesi, M., Preneel, B., Sassone, V., Wegener, I. (eds.) ICALP 2006. LNCS, vol. 4052, pp. 1–12. Springer, Heidelberg (2006). https://doi.org/10.1007/11787006_1
8. Flouris, G., Konstantinidis, G., Antoniou, G., Christophides, V.: Formal foundations for RDF/S KB evolution. Knowl. Inf. Syst. **35**(1), 153–191 (2013)
9. Habel, A., Heckel, R., Taentzer, G.: Graph grammars with negative application conditions. Fundam. Inf. **26**(3,4), 287–313 (1996)
10. Heitmann, B., Hermsen, F., Decker, S.: k-RDF-neighbourhood anonymity: combining structural and attribute-based anonymisation for linked data. PrivOn@ ISWC 1951 (2017)
11. Kairouz, P., Oh, S., Viswanath, P.: Extremal mechanisms for local differential privacy. J. Mach. Learn. Res. **17**(17) (2016). http://jmlr.org/papers/v17/15-135.html
12. Kasiviswanathan, S.P., Lee, H.K., Nissim, K., Raskhodnikova, S., Smith, A.D.: What can we learn privately? SIAM J. Comput. **40**(3), 793–826 (2011)
13. Löwe, M.: Algebraic approach to single-pushout graph transformation. Theoret. Comput. Sci. **109**(1–2), 181–224 (1993)
14. Mahfoudh, M., Forestier, G., Thiry, L., Hassenforder, M.: Algebraic graph transformations for formalizing ontology changes and evolving ontologies. Knowl.-Based Syst. **73**, 212–226 (2015)
15. Radulovic, F., García-Castro, R., Gómez-Pérez, A.: Towards the anonymisation of RDF data. In: SEKE, pp. 646–651 (2015)
16. Schwentick, T.: Automata for XML - a survey. J. Comput. Syst. Sci. **73**(3), 289–315 (2007)
17. Segura, S., Benavides, D., Ruiz-Cortés, A., Trinidad, P.: Automated merging of feature models using graph transformations. In: Lämmel, R., Visser, J., Saraiva, J. (eds.) GTTSE 2007. LNCS, vol. 5235, pp. 489–505. Springer, Heidelberg (2008). https://doi.org/10.1007/978-3-540-88643-3_15
18. Shaban-Nejad, A., Haarslev, V.: Managing changes in distributed biomedical ontologies using hierarchical distributed graph transformation. Int. J. Data Min. Bioinform. **11**(1), 53–83 (2015)
19. Sweeney, L.: k-anonymity: a model for protecting privacy. Internat. J. Uncertain. Fuzziness Knowl.-Based Syst. **10**(5), 557–570 (2002)
20. Taentzer, G.: AGG: a graph transformation environment for modeling and validation of software. In: Pfaltz, J.L., Nagl, M., Böhlen, B. (eds.) AGTIVE 2003. LNCS, vol. 3062, pp. 446–453. Springer, Heidelberg (2004). https://doi.org/10.1007/978-3-540-25959-6_35
21. Wu, X., Ying, X., Liu, K., Chen, L.: A survey of privacy-preservation of graphs and social networks, pp. 421–453. Springer, Cham (2010). https://doi.org/10.1007/978-1-4419-6045-0_14
22. Zheleva, E., Getoor, L.: Privacy in social networks: a survey. In: Social Network Data Analytics, pp. 277–306. Springer, Cham (2011). https://doi.org/10.1007/978-1-4419-8462-3_10

K-GALS: 1st Workshop on Knowledge Graphs Analysis on a Large Scale

K-GALS: 1st Workshop on
Knowledge Graphs Analysis on a Large Scale

Workshop Chairs

Mariella Bonomo   University of Palermo, Italy
Simona E. Rombo   University of Palermo, Italy

Program Committee Members

Lorenzo Di Rocco   University of Rome "La Sapienza", Italy
Valeria Fionda   University of Calabria, Italy
Umberto Ferraro Petrillo   University of Rome "La Sapienza", Italy
Antonia Russo   University "Mediterranea" of Reggio Calabria,
                           Italy
Edoardo Serra   Boise University, USA
Cristina Serrao   University of Calabria, Italy
Blerina Sinaimeri   LUISS University, Italy
Francesca Spezzano   Boise University, USA
Filippo Utro   IBM T. J. Watson Research, USA
Leonardo Alexandre   INESC-ID, Instituto Superior Técnico, Portugal
Mónica de Mendonça Silva   INESC-ID, Portugal
João Tiago Aparicio   Tècnico LOSBOA, Portugal

A Collaborative Filtering Approach for Drug Repurposing

Simone Contini[1] and Simona E. Rombo[1,2(✉)]

[1] Department of Mathematics and Computer Science, University of Palermo, Palermo, Italy
simone.contini@community.unipa.it, simona.rombo@unipa.it
[2] Kazaam Lab s.r.l., Palermo, Italy

Abstract. A recommendation system is proposed based on the construction of Knowledge Graphs, where physical interaction between proteins and associations between drugs and targets are taken into account. The system suggests new targets for a given drug depending on how proteins are linked each other in the graph. The framework adopted for the implementation of the proposed approach is Apache Spark, useful for loading, managing and manipulating data by means of appropriate Resilient Distributed Datasets (RDD). Moreover, the Alternating Least Square (ALS) machine learning algorithm, a Matrix Factorization algorithm for distributed and parallel computing, is applied. Preliminary obtained results seem to be promising.

Keywords: Drugs · Machine learning algorithms · Big Data technologies · Latent factors

1 Introduction

One of the most important challenges in the field of drug discovery is the prediction of new targets for existing drugs [4]. Indeed, drug discovery is a costly and time-consuming task, and new drugs have to pass through several steps before being approved. The repurposal of old drugs in the context of new therapies has shown to be a successful approach, allowing conspicuous money and time saving. In the last few years computational techniques have started to be proposed to this aim (see [9] for a good review on the topic and more recent approaches such as [1,2,10,12]). However, often the proposed approaches fail in properly combining large amounts of data which may be extracted from heterogeneous sources. A pipeline taking into account also a suitable integration strategy would provide a valuable help to physicians by supporting their decision making through novel knowledge that can be discovered this way.

On the other hand, many problems of the real life can be modeled as graphs, able to take into account important relationships between interacting "actors". In particular, physical interaction between proteins has received a large attention in the literature (e.g., [3,5,6,11,15,16]), showing that the study of complex

© Springer Nature Switzerland AG 2022
S. Chiusano et al. (Eds.): ADBIS 2022, CCIS 1652, pp. 381–387, 2022.
https://doi.org/10.1007/978-3-031-15743-1_35

graphs modelling this process globally for a given organisms may provide important insights for several applications. This direction may be considered interesting to be investigated also in the context of drug repurposal, due to the fact that the way proteins interact each other often influences the effect of drugs on specific targets.

Here we propose a recommendation system relying on knowledge graphs built upon protein-protein interactions (PPI) and associations between drugs and targets (DTA). The main assumption is that, depending on how proteins are linked in the graph, given an input drug new targets can be suggested for further investigation.

As the involved data is notably increasing, Big Data technologies started to be applied for the implementation of bioinformatics tools on a large scale (e.g., [14,18]). Here a Big Data framework has been applied, that is, Apache Spark [17], useful for loading, managing, and manipulating data by means of appropriate Resilient Distributed Datasets (RDD), and for providing suitable libraries for the implementation of the Alternating Least Square (ALS) machine learning algorithm, a matrix factorization algorithm for distributed and parallel computing [8].

The obtained results on data coming from real laboratory experiments show that the proposed approach is successfull in correctly identifying new target for existing drugs.

2   Methods

We consider a graph G where nodes represent human proteins, and there is an edge between two nodes if a physical interaction between the corresponding two proteins has been discovered and experimentally demonstrated in the laboratory [13]. Given a node p of G, its neighborhood is the set of nodes linked to p by an edge in G. Each node in G may also be associated to a drug if the corresponding protein is target for that drug.

The main idea at the basis of the proposed approach is that novel targets for existing drugs may be searched for among "neighbors" of their known targets in G.

For the purposes considered here, it is reasonable to observe that not every protein that physically interacts with one of the targets of a drug has to be a possible target in its turn. However, if a set of existing drug-target associations is used to train the recommendation system, then only those proteins which effectively present a behavior similar to that of the real targets can be identified. To this aim, a parallel version of the ALS machine learning algorithm [8] is applied. It is worth to pint out that the system also allow a visual inspection of the predicted new targets in the graph.

2.1   Pipeline

The recommendation system presented here is based on the following steps, graphically illustrated in Fig. 1:

1. Creation, cleaning, filtering and analysis of the model integrating protein-protein interactions and known drug-target associations.
2. Creation of the ALS model and data fitting.
3. Implementation and evaluation of the main core recommendation system.
4. Graphs generation and visual overview.

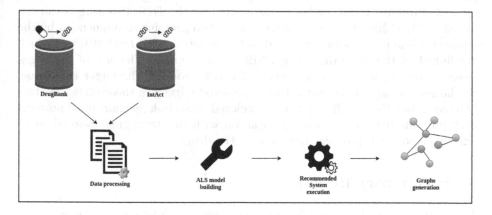

Fig. 1. The pipeline of the proposed approach.

2.2   Implementation of the Recommendation System

The implementation is provided in PySpark. In particular, a DataFrame *indexedDF* is built in which the features of drugs and interactors are added. For the training step, first indexedDF is randomly split (80% for training and 20% for testing), then the ALS function is used and the model created with a combination of some of them. After that the model has been generated, the *recommendForAllUsers* function from PySpark is used which takes as a parameter the number n of proteins that the system will recommend for each drug, and returns, for each drug, n recommended $< protein, rating >$ pairs. Then Knowledge Graphs are generated and filtered such that they contain only nodes linked with a specific protein (recommended by the system) at a certain distance, that is, linked to that protein by a pathway of at most a chosen number of edges. To this aim, the *ego_graph* function from NetworkX library is used, which takes as parameters the reference graph, the central node, and the radius. Then a new list is created, *nodes_list*, containing these nodes with their relative weight.

2.3   Visualization

For the representation of interactive graphs, appropriate functions from the PyVis library have been used. In particular, for a given recommended protein, two types of graphs can be generated: a complete one, containing all the interactions of this protein, and a partial one, containing only the interactions of this protein with proteins that are directly associated with the reference drug, and any interactions with other recommended proteins. This is due to the fact that in the Knowledge Graph each protein could have many interactions, and often a complete representation of the graph, albeit interactive, due to such a high number of interactions would not be very intuitive.

Figure 2 shows an example, with a partial graph having as central node the protein $P61981$ recommended by the system for the drug $DB12010$ and radius set to 1. Proteins recommended by the system for the $DB12010$ drug are shown in red ($P61981$ has an interaction with other two proteins recommended by the system), while proteins that have a direct association with $DB12010$ and which are linked at the same time with $P61981$ are in green. The size of the edges refers to the confidence value between the two proteins: the larger this value, the larger the size of the edge. The red-colored edges are those ones that link two recommended proteins; the orange-colored edges link recommended proteins with proteins that have a direct association with the drug; green-colored edges link two associated proteins directly with the drug.

3   Preliminary Results

Two datasets have been used for the experimental validation as follows:

- Intact database (https://www.ebi.ac.uk/intact/) for the human PPI network, containing 15 features [ID(s) interactor A, ID(s) interactor B, Alt. ID(s) interactor A, Alt. ID(s) interactor B, Alias(es) interactor A, Alias(es) interactor B, Interaction detection method(s), Publication 1st author(s), Publication Identifier(s), Taxid interactor A, Taxid interactor B, Interaction type(s), Source database(s), Interaction identifier(s), and Confidence value(s)] and $1,054,920$ records.
- DrugBank database (https://go.drugbank.com/) for direct associations between drug and target, containing five features (DrugBank ID, Name, Type, UniProt ID, and UniProt Name) and $20,941$ records.

As usual in the performance evaluation of collaborative filtering approaches, we have used Root Mean Square Error (RMSE) to test the accuracy of our system on the considered datasets. RMSE computes the mean value of all the differences squared between the true and the predicted ratings, and then proceeds to calculate the square root out of the result [7].

In particular, with references to the DrugBank known associations and according to the procedure explained above, the **RMSE** for the recommendation system proposed here is equal to **0.17**, showing that it is able to correctly predict novel drug-target associations successfully and with a large accuracy.

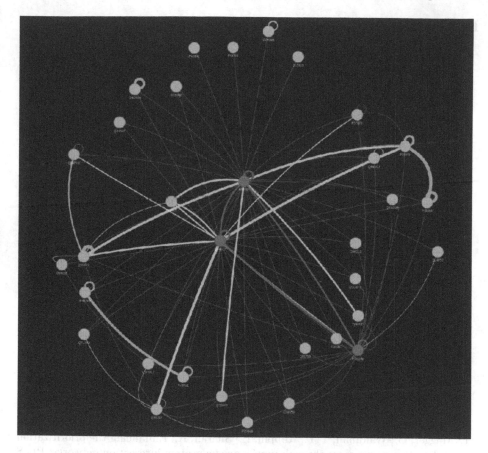

Fig. 2. Visualization example. (Color figure online)

4   Conclusion

Identifying new targets for existing drugs (drug repurposing) is an important task in drug discovery. Indeed, its purpose is the detection of new clinical uses for existing drugs. Since in vitro experiments are extremely expensive and time-consuming, and considering that known drug-target interactions based on wet-lab experiments are limited to a very small number, computational prediction has become relevant in the last few years.

We have proposed here a recommendation system relying on Knowledge Graphs implemented on Big Data technologies to predict new drug-target associations. In carrying out this work we have used human PPI and known drug-target associations, in order to train the system. Preliminary results are promising, showing a good accuracy of the presented approach.

References

1. Che, J., Chen, Z., Guo, L., Wang, S., Aorigele: Drug target group prediction with multiple drug networks. Comb. Chem. High Throughput Screening **23**(4), 274–284 (2020)
2. Cheng, F., et al.: Network-based approach to prediction and population-based validation of in silico drug repurposing. Nat. Commun. **9**(1), 2691 (2018)
3. Chindelevitch, L., Ma, C.-Y., Liao, C. S., Berger, B,: Optimizing a global alignment of protein interaction networks. Bioinformatics **29**(21), 2765–2773 (2013)
4. Chong, C.R., Sullivan Jr., D.J.: New uses for old drugs. Nature **448**(7154), 645–646 (2007)
5. Fionda, V., Palopoli, L., Panni, S., Rombo, S.E.: Protein-Protein interaction network querying by a "Focus and Zoom" approach. In: Elloumi, M., Küng, J., Linial, M., Murphy, R.F., Schneider, K., Toma, C. (eds.) BIRD 2008. CCIS, vol. 13, pp. 331–346. Springer, Heidelberg (2008). https://doi.org/10.1007/978-3-540-70600-7_25
6. Fionda, V., Palopoli, L., Panni, S., Rombo, S.E.: A technique to search for functional similarities in protein-protein interaction networks. Int. J. Data Mining Bioinform. **3**(4), 431–453 (2009)
7. Karunasingha, D.S.K.: Root mean square error or mean absolute error? Use their ratio as well. Inf. Sci. **585**, 609–629 (2022)
8. Koren, Y., Bell, R.M., Volinsky, C.: Matrix factorization techniques for recommender systems. Computer **42**(8), 30–37 (2009)
9. Li, J., Zheng, S., Chen, B., Butte, A.J., Swamidass, S.J., Lu, Z.: A survey of current trends in computational drug repositioning. Brief Bioinform. **17**(1), 2–12 (2016)
10. Luo, H., Li, M., Wang, S., Liu, Q., Li, Y., Wang, J.: Computational drug repositioning using low-rank matrix approximation and randomized algorithms. Bioinformatics **34**(11), 1904–1912 (2018)
11. Magger, O., Waldman, Y.Y., Ruppin, E., Sharan, R.: Enhancing the prioritization of disease-causing genes through tissue specific protein interaction networks. PLoS Comput. Biol. **8**(9), e1002690 (2012)
12. Olayan, R.S., Ashoor, H., Bajic, V.B.: DDR: efficient computational method to predict drug-target interactions using graph mining and machine learning approaches. Bioinformatics **34**(7), 1164–1173 (2017)
13. Panni, S., Rombo, S.E.: Searching for repetitions in biological networks: methods, resources and tools. Briefings Bioinform. **16**(1), 118–136 (2015)
14. Ferraro Petrillo, U., Sorella, M., Cattaneo, G., Giancarlo, R., Rombo, S.E.: Analyzing big datasets of genomic sequences: fast and scalable collection of K-MER statistics. BMC Bioinform. **20-S**(4), 138:1–138:14 (2019)
15. Pizzuti, C., Rombo, S.E.: *PINCoC*: a co-clustering based approach to analyze protein-protein interaction networks. In: Yin, H., Tino, P., Corchado, E., Byrne, W., Yao, X. (eds.) IDEAL 2007. LNCS, vol. 4881, pp. 821–830. Springer, Heidelberg (2007). https://doi.org/10.1007/978-3-540-77226-2_82
16. Pizzuti, C., Rombo, S.E.: An evolutionary restricted neighborhood search clustering approach for PPI networks. Neurocomputing **145**, 53–61 (2014)

17. Zaharia, M., Chowdhury, M., Das, T., et al.: Resilient distributed datasets: a fault-tolerant abstraction for in-memory cluster computing. In: Gribble, S.D., Katabi, D. (eds.) Proceedings of the 9th USENIX Symposium on Networked Systems Design and Implementation, NSDI 2012, San Jose, CA, USA, 25–27 April 2012, pp. 15–28. USENIX Association (2012)
18. Zhou, W., Li, R., Yuan, S., Liu, C., Yao, S.: MetaSpark: a spark-based distributed processing tool to recruit metagenomic reads to reference genomes. Bioinformatics **33**(7), 1090–1092 (2017)

A Review: Biological Insights
on Knowledge Graphs

Ylenia Galluzzo[(⊠)]

Department of Mathematics and Computer Science,
University of Palerme, Palerme, Italy
`ylenia.galluzzo01@community.unipa.it`

Abstract. Knowledge graphs in the biomedical context are spreading rapidly attracting the strong interest of the research due to their natural way of representing biomedical knowledge by integrating heterogeneous domains (genomic, pharmaceutical, clinical etc.). In this paper we will give an overview of the application of knowledge graphs from the biological to the clinical context and show the most recent ways of representing biomedical knowledge with embeddings (KGE). Finally, we present challenges, such as the integration of different knowledge graphs and the interpretability of predictions of new relations, that recent improvements in this field face. Furthermore, we introduce promising future avenues of research (e.g. the use of multimodal approaches and Simplicial neural networks) in the biomedical field and precision medicine.

Keywords: Biomedical knowledge graph · Knowledge graph embeddings · Text mining · Graph Neural Network

1 Introduction

Knowledge graphs are an area of research of great interest in both academia and industry, since they facilitate the extraction of information (facts and hypotheses) thanks to well-defined interconnections between relevant entities (abstract and concrete) to the case study domain. In addiction, they are equally interesting for trying to understand how to form new relationships through the use of data semantics and linkages. Knowledge graphs are represented in their original form by the knowledge base graphs in the Resource Description Framework (RDF). Information (Resource, a Property and a Property value) is represented with assertions that constitute the SPO triples (subject, predicate, object) to express direct complex relationships between different resources [1]. Knowledge graphs can be described as an ontology. An ontology is a data model that represents knowledge about a topic as set relations of concepts of a domain and instances of an object representing the topic. The Web Ontology Language (OWL) is a markup language used to express ontologies. RDF and OWL became important

---

Y. Galluzzo—Now working at Expleo Italia S.p.a.

S. Chiusano et al. (Eds.): ADBIS 2022, CCIS 1652, pp. 388–399, 2022.
https://doi.org/10.1007/978-3-031-15743-1_36

standards of the Semantic Web. Google in 2012 made knowledge graphs gain popularity by introducing its "Google Knowledge Graph" [2]. Its uses Knowledge Vault which combines probabilistic knowledge extracted from Web content with prior knowledge derived from existing knowledge repositories and allows users to receive relevant information based on the search queries they use. Currently, many open knowledge bases or ontologies have been published, such as WordNet [3], DBpedia [4], Wikidata [5] etc. and industry Knowledge Graph (e.g. Google, Microsoft's Bing, Facebook, eBay, IBM etc.) [6].

There are many papers summarising the state of research on knowledge graphs. Among the most recent, Hogan et al. (2021) [7] presents a comprehensive introduction to knowledge graphs in which existing data query models and languages are compared and methods for creating, evaluating and publishing knowledge graphs are summarised.

Knowledge plays an important role in reasoning-driven Natural Language Processing (NLP) tasks. In fact, knowledge graphs turn out to be an important method of solving various NLP problems, such as Question Answering (KGQA) [8–10]. Semantics in information can help to extrapolate information more *semantically close* to the query. Structured knowledge is also a key element in conversational AI in which virtual assistants (e.g. Alexa, Siri or Cortana) answer questions in an advanced way (open questions), as opposed to common chatbots programmed only to be able to answer strictly controlled questions (closed questions).

Recently, research works have been collected different Knowledge Graph (KG) construction techniques and their application [11]. In particular, KGs have various application perspectives on different domains: medical, financial, cybersecurity, news and education, social network de-anonymization, classification, geoscience.

Although several surveys on knowledge graph embeddings in general [12–14] and on the biological topic [15] have been published over the past few years, this paper aims to explore and summarise the most recent advances in the application of KGs, giving a quick overview of the topic. The aim is to distinguish between biological and clinical domains, and the issues that may arise in the careless construction of KGs, as well as to provide information on the ability of KGs to support semantic knowledge. Current uses of the latest NLP models for the creation of clinical KGs are showed in this regard. We introduce the most recent and promising future research paths (e.g., the use of multimodal approaches and Simplicial neural networks) in the biomedical and precision medicine fields. In addition, the study of usable resources for the construction of a biomedical KG is expanded in this paper.

2 Knowledge Graphs in Bioinformatics

The application of KGs in the field of biomedical data for decision support ranges from application in the clinical to the biological field. One of the first and most famous rule-based systems for medical diagnosis is MYCIN [16] with a knowledge base of 600 rules. There is a close connection between KG and biomedical NLP, which on one hand makes it possible to improve the amount and representation of data in KG and on the other hand KG makes it possible to improve predictions for solving NLP tasks (e.g. named entity recognition (NER) [17] and relation extraction [18]).

Relationship extraction systems are required to find the interconnection between a wide variety of topics: to assess the relationship between non-pharmacological variables to COVID-19 pandemic and to support policy-making on COVID-19 in public health [19].

A knowledge graph in the biomedical field is used to relate a considerable amount of interrelated information: genes with biological processes, molecular functions, and cellular components in which they are located; genes with their phenotype or interaction with other genes; drugs with the diseases they are able to cure; genes that are responsible for diseases; generic symptoms related to diseases, etc. Using graphs as a representation of biomedical data appears to be the best solution for modelling objects of this type in a natural way.

In Fassetti et al. [20], graphs are used for the identification of features to characterise and at the same time discriminate gene expression of sets of healthy/diseased samples through the identification of patterns between the graphs belonging to the sets of samples with a complementary health status.

In the following Table 1, most of the data resources that are used for the construction and integration of knowledge in a KG.

Table 1. Biomedical data source.

Database	Description
Hetionet [21]	Biomedical knowledge assembled from 29 different databases of genes, compounds, diseases, etc.
Drug repurposing knowledge graph [22]	biological knowledge graph relating genes, compounds, diseases, biological processes, side effects and symptoms
iDISK [23]	Contains a variety of attributes and relationships describing information about each dietary supplement
KEGG [24]	Integrated database resource that contains genomic information, chemical information and health information
PharmGKB [25]	Knowledge about clinically actionable gene-drug associations and genotype-phenotype relationships
Bgee [26]	Comparison of gene expression patterns
Comparative toxicogenomics database [27]	Manually curated information about chemical-gene/protein interactions, chemical-disease and gene-disease relationships
Reactome [28]	Metabolic pathways, each annotated as a set of biological events, dealing with genes and proteins
DrugBank [58]	Global provider of structured drug information

(*continued*)

Table 1. (*continued*)

Database	Description
Supertarget [59]	Integrates drug-related info associated with adverse drug effects, drug metabolism, pathways and Gene Ontology terms for target proteins
HGNC [29]	Contains relation between gene symbol and a list of corresponding entry in other database
UniprotKB [30]	Collection of annotated functional information on proteins
SIDER [31]	Collects information on drug classification and side effects and links to further information, e.g. drug-target relations
TISSUE [32]	Tissue expression from manually curated literature, proteomics and transcriptomics screens, and automatic text mining
SIGNOR [33]	Manually annotated causal relationships between human proteins, chemicals of biological relevance, stimuli and phenotypes
STRING [34]	Database of known and predicted protein-protein interactions
STITCH [35]	Database of known and predicted interactions between chemicals and proteins
SMPDB [36]	Database containing pathways found in model organisms such as humans, mice, E. coli etc.
OncoKB [37]	Oncology knowledge, contains biological and clinical information about genomic alterations in cancer
Brenda tissue ontology [38]	Collection of enzyme functional data
Disease ontology [39]	Ontology for human disease
Cell ontology [40]	Repository for biomedical ontologies
Gene ontology (GO) [41]	Contains annotations for gene products in biological processes, cellular components and molecular functions
miRBase [42]	miRNAs sequences and annotations, associated with names, keywords, genomic locations, and references
mirCancer [43]	Contains associations between miRNAs and human cancers.
miRanda [44]	miRNA-target interactions
miRNASNP [45]	Contains miRNA-related mutations
ChEMBL [61]	Manually curated database of bioactive molecules with drug-like properties
ChEBI [46]	Ontology containing information about chemical entities
PubChem [60]	Chemistry database: information about chemical structures, identifiers, properties, biological activities

In literature, although using different methods and algorithms, graphs are mostly used to solve common problems: making inferences about biomedicine, creating alternative ways to represent graphs on the same knowledge domain and extending information extraction.

A large part of research is currently devoted to the identification of similar entities within a KG. Embeddings generated using neural networks are used to calculate knowledge-based similarities between, for example, drugs, proteins and diseases [53].

Many ways are used to extend knowledge in the biomedical domain to discover latent information or missing information in KGs.

Completion of the knowledge graph (KGC) aims to complete the structure of the knowledge graph by predicting the missing entities or relationships in the knowledge graph and extracting unknown facts. KGC technologies may involve the use of traditional methods, such as rule-based reasoning and the probability graph model (Markov logic network). Recently, KGC techniques use methods of learning through embeddings representation: methods based on semantic correspondence models, based on learning of representation and other methods based on neural network models.

The use of models based on a generative approach to learn the embeddings of entities and relationships allows to generate hypotheses regarding the relationships associated with a connection score between graph embeds through multiple techniques: tensor factorisation (DistMult model [47]) and latent distance similarity (TransE model [48]). This type of techniques are used in polypharmacy, to evaluate the side effects that are caused by the interaction of drug combinations [49,50].

2.1   Methods of Knowledge Graph Embedding

A very common application grown recently is to create entity embeddings or assertions on KGs by training deep learning networks such as autoencoders from inputs constructed by KG nodes [51,52]. The purpose of representing graphs in general to high-dimensional in a low-dimensional space, is to capture the essence of a graph while preserving the intrinsic (global and/or local) structure of the graph in the form of a dense vector representation, both of arcs [47] and of single nodes [54]. This type of approach has been used to analyze knowledge graphs of different domains and allows, starting from compressed and *meaningful* information, to apply classification techniques and build predictors, which are able to help researchers to face and solve problems related to identifying associations between diseases and biomolecules [55], finding new treatments for existing drugs [21] etc.

In the following section, we will examine how the most recent techniques of *representative learning* are used in the biochemical and clinical context.

Application on Biochemical Data. One of the studies that enabled a comprehensive KG in the biomedical context is PharmKG [56]. Multi-omics data are aggregated from disease-related words, gene expression and chemical structure, maintaining biological and semantic features using the latest KGE approaches. A major study was directed at the use of drugs, mainly dealing with their reuse and adverse reactions, which is necessary to prevent significant harm to patients, but also at potential drug-drug interactions (DDIs), drug-protein (DPIs), drug-disease, drug-target interactions by modelling the problem of predicting links between graph nodes with KGs. In the study by Zhu et al. [57], the authors provided a detailed overview of the existing drug knowledge bases and applications. This type of work used datasets containing the main properties of drugs (DrugBank [58] and SuperTarget [59]) and datasets including the main information about chemical compounds (PubChem [60] and ChEMBL [61]). In Lin et al. [62], a Bio2RDF created from DrugBank is used to assess the relationships between a potential drug and its neighbours by basing the prediction on GNN models applied to the biological KG. Innovative work involving drugs is the creation of a BioDKG-DDI model [63] that aims to identify DDI relationships to support experimental work in the laboratory for drug construction. BioDKG-DDI uses an innovative self-attention mechanism on DNNs (deep neural networks) that allows the attenuation of embeddings multi-features: molecular structures, drug structures, and drug similarity matrix. Furthermore, in this type of work, the latest DNNs are used to predict the search for similar drugs, e.g. to find molecules with antibiotic properties by approximating the retrosynthesis from the graph [64]. Drug-protein interactions are another type of prediction that researchers are concerned about. In BridgeDPI [65], convolutional CNN and feed-forward network layers are used to encode SMILES drug and protein sequence information. For predicting DPI, as in other similar works in literature, GNN is used to build *bridge* nodes between interactions and predict new connections between nodes. In order to build a complete picture of the current state of research, it is worth mentioning that Neural networks called Hyperbolic Graph Neural Networks (HGNN) have been applied to the DISEASE dataset, based on the SIR disease spreading model [66], proving that they achieve excellent results in link prediction. Recently, however, these new sophisticated neural networks have been outperformed by Simplicial neural networks (SNNs) for link prediction [67], which achieve better results in terms of ROC AUC on the same dataset.

Application on Clinical Data. Studies of KG-based recommendation systems built from electronic medical records (EMRs) are attempting to support medical decision-making for the benefit of proper patient care. Creating KGs from medical record texts that contain a patient's treatment history (medical diagnoses, therapies, etc.) is certainly a way to create a knowledge base with lower costs than building KGs based on the deeper biological aspects (relationships between genes, diseases, chemical composition of drugs, etc.) that require more attention. The research work on this type of application is still at an early stage and thanks to the use of the most recent models for extracting information from texts (use of LSTM, BERT and NER models) can provide information extracted from unstructured data that enables the biomedical knowledge bases of KGs to be enriched with non-trivial connections [68–70]. Recently, thanks to the work of Zhang et al. [71] using attention mechanisms and convolutional graphs, embedded KG features were created to improve the classification and generation of radiological reports in order to improve diagnosis and support the work of physicians.

3 Open Research Problems

3.1 Construction and Integration of Knowledge Graphs

Biomedical knowledge graphs are usually manually curated by expert researchers. One database that has been constructed by a group of domain experts is COSMIC [72], which was built from the literature by associating genes with the related cancer type. However, biological knowledge is constantly evolving and it becomes extremely necessary to cope with changes through scalable intelligent systems that integrate real-time updates. In a way, the problem of updating knowledge bases is also strongly related to the challenge of aligning knowledge representation to make KGs and their relationships increasingly reliable. The alignment of entities of different KGs, based on their similarity, is one of the widely studied techniques, but only recently in Xiang et al. [73] take into account the ontology by including hierarchies and disjunctions of classes to avoid mismapping the entities to be aligned. The quality of the available knowledge graphs has been analysed and it has been shown that low data quality propagates into the embedding models and thus the accuracy of KGE predictions decreases in inverse proportion to the scarcity and unreliability of the data [15]. Missing knowledge and errors in KG integration can quickly propagate into the prediction tasks, continuing to proliferate incorrect and misleading domain knowledge, which becomes particularly relevant and problematic if we are working on the biomedical domain.

3.2 Performance

Complex biological systems are modelled as graphs, but graph exploration, training and prediction techniques require a considerable amount of resources and time and therefore this implies limited scalability. This problem of exploratory models is partly addressed by KGEs that operate with linear time and space complexity, but the problem of dynamically encoding new entities that integrate into the graph is currently unresolved. KGEs are highly dependent on prior knowledge of embeddings for each type of information in the knowledge base, capable of maintaining local and global information and this issue propagates into the scalability of predictions.

3.3 Explainable Predictions

Lack of interpretability is a recurring problem of deep learning models [74]. This can be problematic because of the increasing use of neural networks in decision-making in biomedical applications. Efforts have been made in this regard. CrossE [75] explores the process of explaining graph search paths using embeddings to interpret link prediction. In KGE, learning meaningful embeddings using specific optimisation techniques is a process that leads to predictions that are difficult to interpret. In data analysis, GNN models, which are often used in the biomedical context, generate relevant information for each data node, which makes the model somewhat more interpretive. Recently, efforts have been made to constrain the training of information, so that KGE models can be partially interpretable thanks to a set of applied constraints (e.g. type constraints and basic relation axioms) [76,77]. Methods of extracting information from NLP also pose further problems in the construction of a reliable KG in the health domain. There are still so many problems with the use of complex models for understanding natural language [78]. Indeed, the *bias* that can be derived from the extraction of information in EMHs should not be underestimated. Inevitably these biases in the data will propagate to some extent in the results of the predictions. This is the reason why it is even more important for research to focus on creating increasingly reliable and explainable models for the health sector.

4   Discussion and Conclusion

This survey aims to present the latest models and strategies to use knowledge graphs in the biomedical context. Their use in the past years has become more and more widespread and research is currently focused on growing the results obtained from their application in the biomedical context. As presented in this survey, many knowledge graphs are generally constructed from data sources, which have been created either manually by experienced researchers or by sophisticated NLP techniques (NER, relation extraction). We subsequently pointed out the potential errors in the biomedical context, which this type of approach can generate in the construction of knowledge graphs.

The knowledge extension process in KGEs can certainly be addressed by the low-dimensional representation of the characteristics of each entity and/or relation of which the graph is made up. This type of compressed and representative representation of the knowledge graph can make it possible to find potential inconsistencies during the process of integrating the graphs and partially solve some of the problems associated with knowledge graph construction errors caused by misaligned entities. Indeed, KGEs at present turn out to be a very active field of research, due to their ability to provide a generalisable context on the KG and to probabilistically deduce new relations missing in the existing graph structure. This characteristic allowed in many works to speed up the discovery of new drugs by evaluating the interaction between the properties of molecules that are present in the KG. The importance of KG feature representation, expressed in this discussion, tends to emphasise its effectiveness in the construction of increasingly complete KGs.

A further step forward is the introduction in recent research work of the construction of a multimodal knowledge network. This means that additional information is fed into the KG to enhance reasoning. The construction of a multimodal network tends to utilise the combination of several features of interactions between KG entities with the aim of improving predictions (e.g. on drug repositioning) [79]. The multimodal approach is also recently used in precision medicine where more detailed knowledge and a specific focus is needed to create KGs that represent and generate *reliable* knowledge [80].

In conclusion, this discussion presents the current open challenges in the use of KGs in the biomedical field, focusing on the necessary improvement of the interpretability and quality of biomedical KG data in order to increase the community's confidence in the predictions and thus support specialised medicine.

References

1. Lassila, O., Swick, R.R.: Resource description framework (RDF) model and syntax specification (1998)
2. Dong, X., et al.: Knowledge vault: a web-scale approach to probabilistic knowledge fusion. In: Proceedings of the 20th ACM SIGKDD International Conference on Knowledge Discovery and Data Mining (2014)
3. Miller, G.A.: WordNet: a lexical database for English. Commun. ACM **38**(11), 39–41 (1995)
4. Auer, S., Bizer, C., Kobilarov, G., Lehmann, J., Cyganiak, R., Ives, Z.: DBpedia: a nucleus for a web of open data. In: Aberer, K., et al. (eds.) ASWC/ISWC -2007. LNCS, vol. 4825, pp. 722–735. Springer, Heidelberg (2007). https://doi.org/10.1007/978-3-540-76298-0_52
5. Erxleben, F., Günther, M., Krötzsch, M., Mendez, J., Vrandečić, D.: Introducing Wikidata to the linked data web. In: Mika, P., et al. (eds.) ISWC 2014. LNCS, vol. 8796, pp. 50–65. Springer, Cham (2014). https://doi.org/10.1007/978-3-319-11964-9_4
6. Noy, N., et al.: Industry-scale knowledge graphs: lessons and challenges. Commun. ACM **62**(8), 36–43 (2019)
7. Hogan, A., et al.: Knowledge graphs. Synth. Lect. Data Semant. Knowl. **12**(2), 1–257 (2021)
8. Lukovnikov, D., et al.: Neural network-based question answering over knowledge graphs on word and character level. In: Proceedings of the 26th International Conference on World Wide Web (2017)
9. Huang, X., et al.: Knowledge graph embedding based question answering. In: Proceedings of the Twelfth ACM International Conference on Web Search and Data Mining (2019)
10. Purkayastha, S., et al.: Knowledge graph question answering via SPARQL silhouette generation. arXiv preprint arXiv:2109.09475 (2021)
11. Zou, X.: A survey on application of knowledge graph. J. Phys. Conf. Ser. **1487**(1), 012016 (2020)
12. Choudhary, S., et al.: A survey of knowledge graph embedding and their applications. arXiv preprint arXiv:2107.07842 (2021)
13. Dai, Y., et al.: A survey on knowledge graph embedding: approaches, applications and benchmarks. Electronics **9**(5), 750 (2020)
14. Wang, Q., et al.: Knowledge graph embedding: a survey of approaches and applications. IEEE Trans. Knowl. Data Eng. **29**(12), 2724–2743 (2017)
15. Mohamed, S.K., Nounu, A., Nováček, V.: Biological applications of knowledge graph embedding models. Brief. Bioinform. **22**(2), 1679–1693 (2021)
16. Van Melle, W.: MYCIN: a knowledge-based consultation program for infectious disease diagnosis. Int. J. Man-Mach. Stud. **10**(3), 313–322 (1978)

17. Karampatakis, S., Dimitriadis, A., Revenko, A., Blaschke, C.: Training NER models: knowledge graphs in the loop. In: Harth, A., et al. (eds.) ESWC 2020. LNCS, vol. 12124, pp. 135–139. Springer, Cham (2020). https://doi.org/10.1007/978-3-030-62327-2_23
18. Hoffmann, R., et al.: Knowledge-based weak supervision for information extraction of overlapping relations. In: Proceedings of the 49th Annual Meeting of the Association for Computational Linguistics: Human Language Technologies (2011)
19. Yang, Y., et al.: Constructing public health evidence knowledge graph for decision-making support from COVID-19 literature of modelling study. J. Saf. Sci. Resilience **2**(3), 146–156 (2021)
20. Fassetti, F., Rombo, S.E., Serrao, C.: Discovering discriminative graph patterns from gene expression data. In: Proceedings of the 31st Annual ACM Symposium on Applied Computing (2016)
21. Himmelstein, D.S., et al.: Systematic integration of biomedical knowledge prioritizes drugs for repurposing. Elife **6**, e26726 (2017)
22. Ioannidis, V.N., et al.: DRKG-drug repurposing knowledge graph for COVID-19. arXiv preprint arXiv: 2010.09600 (2020)
23. Rizvi, R.F., et al.: iDISK: the integrated DIetary supplements knowledge base. J. Am. Med. Inform. Assoc. **27**(4), 539–548 (2020)
24. Ogata, H., et al.: KEGG: Kyoto encyclopedia of genes and genomes. Nucleic Acids Res. **27**(1), 29–34 (1999)
25. Whirl-Carrillo, M., et al.: Pharmacogenomics knowledge for personalized medicine. Clin. Pharmacol. Ther. **92**(4), 414–417 (2012)
26. Bastian, F.B., et al.: The Bgee suite: integrated curated expression atlas and comparative transcriptomics in animals. Nucleic Acids Res. **49**(D1), D831–D847 (2021)
27. Davis, A.P., et al.: The comparative toxicogenomics database: update 2019. Nucleic Acids Res. **47**(D1), D948–D954 (2019)
28. Gillespie, M., et al.: The reactome pathway knowledgebase 2022. Nucleic Acids Res. **50**(D1), D687–D692 (2022)
29. Tweedie, S., et al.: Genenames.org: the HGNC and VGNC resources in 2021. Nucleic Acids Res. **49**(D1), D939–D946 (2021)
30. The UniProt Consortium: UniProt: the universal protein knowledgebase in 2021. Nucleic Acids Res. **49**(D1), D480–D489 (2021)
31. Kuhn, M., et al.: The SIDER database of drugs and side effects. Nucleic Acids Res. **44**(D1), D1075–D1079 (2016)
32. Santos, A., et al.: Comprehensive comparison of large-scale tissue expression datasets. PeerJ **3**, e1054 (2015)
33. Licata, L., et al.: SIGNOR 2.0, the SIGnaling network open resource 2.0: 2019 update. Nucleic Acids Res. **48**(D1), D504–D510 (2020)
34. Szklarczyk, D., et al.: The STRING database in 2021: customizable protein-protein networks, and functional characterization of user-uploaded gene/measurement sets. Nucleic Acids Res. **49**(D1), D605–D612 (2021)
35. Szklarczyk, D., et al.: STITCH 5: augmenting protein-chemical interaction networks with tissue and affinity data. Nucleic Acids Res. **44**(D1), D380–D384 (2016)
36. Jewison, T., et al.: SMPDB 2.0: big improvements to the small molecule pathway database. Nucleic Acids Res. **42**(D1), D478–D484 (2014)
37. Chakravarty, D., et al.: OncoKB: a precision oncology knowledge base. JCO Precis. Oncol. **1**, 1–16 (2017)
38. Chang, A., et al.: BRENDA, the ELIXIR core data resource in 2021: new developments and updates. Nucleic Acids Res. **49**(D1), D498–D508 (2021)

39. Schriml, L.M., et al.: The human disease ontology 2022 update. Nucleic Acids Res. **50**(D1), D1255–D1261 (2022)
40. Jupp, S., et al.: A new ontology lookup service at EMBL-EBI. In: Malone, J. et al. (eds.) Proceedings of SWAT4LS International Conference (2015)
41. Gene Ontology Consortium: The gene ontology resource: enriching a GOld mine. Nucleic Acids Res. **49**(D1), D325–D334 (2021)
42. Kozomara, A., Griffiths-Jones, S.: miRBase: integrating microRNA annotation and deep-sequencing data. Nucleic Acids Res. **30**(suppl 1), D152–D157 (2010)
43. Xie, B., et al.: miRCancer: a microRNA-cancer association database constructed by text mining on literature. Bioinformatics **29**(5), 638–644 (2013)
44. John, B., et al.: Human microRNA targets. PLoS Biol. **2**(11), e363 (2004)
45. Gong, J., et al.: Genome-wide identification of SNPs in microRNA genes and the SNP effects on microRNA target binding and biogenesis. Hum. Mutat. **33**(1), 254–263 (2012)
46. Hastings, J., et al.: ChEBI in 2016: improved services and an expanding collection of metabolites. Nucleic Acids Res. **44**(D1), D1214–D1219 (2016)
47. Bordes, A., et al.: Translating embeddings for modeling multi-relational data. In: Advances in Neural Information Processing Systems, vol. 26 (2013)
48. Yang, B., et al.: Embedding entities and relations for learning and inference in knowledge bases. arXiv preprint arXiv:1412.6575 (2014)
49. Malone, B., García-Durán, A., Niepert, M.: Knowledge graph completion to predict polypharmacy side effects. In: Auer, S., Vidal, M.-E. (eds.) DILS 2018. LNCS, vol. 11371, pp. 144–149. Springer, Cham (2019). https://doi.org/10.1007/978-3-030-06016-9_14
50. Nováček, V., Mohamed, S.K.: Predicting polypharmacy side-effects using knowledge graph embeddings. AMIA Summits Transl. Sci. Proc. **2020**, 449 (2020)
51. Zitnik, M., Agrawal, M., Leskovec, J.: Modeling polypharmacy side effects with graph convolutional networks. Bioinformatics **34**(13), i457–i466 (2018)
52. Ma, T., et al.: Drug similarity integration through attentive multi-view graph auto-encoders. arXiv preprint arXiv:1804.10850 (2018)
53. Zhang, X.-M., et al.: Graph neural networks and their current applications in bioinformatics. Front. Genetics **12**, 690049 (2021)
54. Grover, A., Leskovec, J.: node2vec: scalable feature learning for networks. In: Proceedings of the 22nd ACM SIGKDD International Conference on Knowledge Discovery and Data Mining (2016)
55. Shen, Z., et al.: miRNA-disease association prediction with collaborative matrix factorization. Complexity **2017**, 2498957 (2017)
56. Zheng, S., et al.: PharmKG: a dedicated knowledge graph benchmark for biomedical data mining. Briefings Bioinform. **22**(4), bbaa344 (2021)
57. Zhu, Y., et al.: Drug knowledge bases and their applications in biomedical informatics research. Briefings Bioinform. **20**(4), 1308–1321 (2019)
58. Wishart, D.S., et al.: DrugBank: a knowledgebase for drugs, drug actions and drug targets. Nucleic Acids Res. **36**(suppl_1), D901–D906 (2008)
59. Hecker, N., et al.: SuperTarget goes quantitative: update on drug-target interactions. Nucleic Acids Res. **40**(D1), D1113–D1117 (2012)
60. Kim, S., et al.: PubChem substance and compound databases. Nucleic Acids Res. **44**(D1), D1202–D1213 (2016)
61. Gaulton, A., et al.: ChEMBL: a large-scale bioactivity database for drug discovery. Nucleic Acids Res. **40**(D1), D1100–D1107 (2012)
62. Lin, X., et al.: KGNN: knowledge graph neural network for drug-drug interaction prediction. In: IJCAI, vol. 380 (2020)

63. Ren, Z.-H., et al.: BioDKG-DDI: predicting drug-drug interactions based on drug knowledge graph fusing biochemical information. Briefings Funct. Genomics **21**(3), 216–229 (2022)
64. Liu, C.-H., et al.: RetroGNN: approximating retrosynthesis by graph neural networks for de novo drug design. arXiv preprint arXiv:2011.13042 (2020)
65. Wu, Y., et al.: BridgeDPI: a novel Graph Neural Network for predicting drug-protein interactions. Bioinformatics **38**(9), 2571–2578 (2022)
66. Chami, I., et al.: Hyperbolic graph convolutional neural networks. In: Advances in Neural Information Processing Systems, vol. 32 (2019)
67. Chen, Y., Gel, Y.R., Poor, H.V.: BScNets: block simplicial complex neural networks. In: Proceedings of the AAAI Conference on Artificial Intelligence, vol. 36(6) (2022)
68. Harnoune, A., et al.: BERT based clinical knowledge extraction for biomedical knowledge graph construction and analysis. Comput. Methods Prog. Biomed. Update **1**, 100042 (2021)
69. Li, L., et al.: Real-world data medical knowledge graph: construction and applications. Artif. Intell. Med. **103**, 101817 (2020)
70. Gong, F., et al.: SMR: medical knowledge graph embedding for safe medicine recommendation. Big Data Res. **23**, 100174 (2021)
71. Zhang, Y., et al.: When radiology report generation meets knowledge graph. In: Proceedings of the AAAI Conference on Artificial Intelligence, vol. 34(07) (2020)
72. Forbes, S.A., et al.: COSMIC: somatic cancer genetics at high-resolution. Nucleic Acids Res. **45**(D1), D777–D783 (2017)
73. Xiang, Y., et al.: OntoEA: ontology-guided entity alignment via joint knowledge graph embedding. arXiv preprint arXiv:2105.07688 (2021)
74. Guidotti, R., et al.: A survey of methods for explaining black box models. ACM Comput. Surv. (CSUR) **51**(5), 1–42 (2018)
75. Zhang, W., et al.: Interaction embeddings for prediction and explanation in knowledge graphs. In: Proceedings of the Twelfth ACM International Conference on Web Search and Data Mining (2019)
76. Krompaß, D., Baier, S., Tresp, V.: Type-constrained representation learning in knowledge graphs. In: Arenas, M., et al. (eds.) ISWC 2015. LNCS, vol. 9366, pp. 640–655. Springer, Cham (2015). https://doi.org/10.1007/978-3-319-25007-6_37
77. Minervini, P., Costabello, L., Muñoz, E., Nováček, V., Vandenbussche, P.-Y.: Regularizing knowledge graph embeddings via equivalence and inversion axioms. In: Ceci, M., Hollmén, J., Todorovski, L., Vens, C., Džeroski, S. (eds.) ECML PKDD 2017. LNCS (LNAI), vol. 10534, pp. 668–683. Springer, Cham (2017). https://doi.org/10.1007/978-3-319-71249-9_40
78. Helwe, C., Clavel, C., Suchanek, F.M.: Reasoning with transformer-based models: deep learning, but shallow reasoning. In: 3rd Conference on Automated Knowledge Base Construction (2021)
79. Xiong, Z., Huang, F., Wang, Z., Liu, S., Zhang, W.: A multimodal framework for improving in silico drug repositioning with the prior knowledge from knowledge graphs, p. 1. IEEE/ACM Trans. Comput. Biol, Bioinform (2021). https://doi.org/10.1109/TCBB.2021.3103595
80. Zhu, C., et al.: Multimodal reasoning based on knowledge graph embedding for specific diseases. Bioinformatics **38**(8), 2235–2245 (2022)

CiteRank: A Method to Evaluate Researchers Influence Based on Citation and Collaboration Networks

Fabrizio Angiulli, Fabio Fassetti, and Cristina Serrao[⊠]

University of Calabria, Via Pietro Bucci, 41C, Rende, CS, Italy
{f.angiulli,f.fassetti,c.serrao}@dimes.unical.it

Abstract. The scientific impact of researchers is often evaluated based on the citations they receive from the others, thus the definition of citation metrics has long been analysed to find criteria able at jointly consider the quantity and the quality of the exchanged citations. Here, we propose a network based approach aimed at estimating the researchers influence through the analysis of the citations they receive and the authors from which they come. Particularly, we introduce a concept of fairness based on the idea of properly weighting the citations among each pair of researchers starting from the assumption that the ability of attracting citations from beyond the group with which a researcher collaborates normally provides an evidence of the broader impact of his research.

Provided with this revised citation network, we use the well known PageRank algorithm to score the authors and get a reliable measure of their influence.

Keywords: Citation networks · Collaboration networks · PageRank

1 Introduction

The evaluation of the impact of researchers activities is a key issue in the academic field and there has always been the need to define quantitative measures to evaluate and possibly make comparable their scientific production [1]. Such evaluation metrics mainly consider the publications [2] and, although both their quantity and their quality should be assessed, the definition of quality measures is much more complex.

Papers perceived to be interesting or relevant in a certain research field are usually highly cited by other researchers, therefore, to provide a measure of scientific impact, it is reasonable to take into account the amount of citations a certain researcher gets from the others. To this aim, the definition of *citation networks* is a subtle way to model the relations, in terms of exchanged citations, among researches; their analysis through the strategies which are commonly used in network science can provide a valuable contribution toward the definition of new scientific impact measures.

© Springer Nature Switzerland AG 2022
S. Chiusano et al. (Eds.): ADBIS 2022, CCIS 1652, pp. 400–410, 2022.
https://doi.org/10.1007/978-3-031-15743-1_37

However, when evaluating citations, one should consider that some unfair behaviours, such as extreme self-citations or relatively small clusters of authors massively citing each other's papers, make citation metrics spurious and meaningless. In this paper, we propose a strategy to analyse *citations networks* which attempt to introduce a concept of *fairness* based on the idea of properly weighting the citations among each pair of researchers on the basis of the collaboration relations inferred from the available data. Particularly, we want to reward the ability of a researcher to attract citations from people which are far from the group with which he collaborates normally, as we argue that this provide an evidence of the broader impact of his research.

To support this kind of analysis, we consider the idea of the PageRank algorithm [3] and specialize it to the domain of citation network analysis with the aim of providing a score function involving the number of citations an author receives and also the authors from which they comes. Our algorithm, namely CiteRank, makes the role of collaboration relationships more explicit and joins this knowledge with the one coming from the citation network to provide a more fair criteria to evaluate the scientific impact of researchers.

The rest of the paper is organized as follows. In Sect. 2 we discuss some of the approaches available in literature to deal with the citation networks analysis and the researchers impact evaluation in general; in Sect. 3 we present our method by first providing a definition of citation network and then by discussing the main features of the PageRank algorithm and our contribution toward the definition of the CiteRank strategy; in Sect. 4 the experimental analysis is reported; finally Sect. 5 draws conclusions.

2   Related Works

The use of citation metrics has become widespread to asses the quality and impact of researchers but is fraught with difficulties. Some challenges relate to what citations and related metrics fundamentally mean and how they can be interpreted or misinterpreted as a measure of impact or excellence [1].

Many metrics are available [4], mainly relying on the exchanged citations among authors. One of the pioneer in this filed was probably Hirsch who proposed the h-index [5] which has inspired also the definition of further indices which try to solve some of the h-index issues [6–8]; moreover, there are also some attempts to define composite indicators [9].

The possibility of evaluating authors of scientific publications based on citation networks has also been explored as a way to rank or compare authors in a given research area [10]. Some tools are available to visualize and analyse citation networks [11] and measures which are traditionally used in the field of network analysis have been exploited to rank authors within citations and co-authorsip networks [12].

Hence, the possibility of using the PageRank algorithm to measure the popularity and prestige of researchers from a citation perspective seems a promising approach [13,14].

In [15] a co-authorship network has been used to model the relationships among the authors and infer an indicator of their impact. The frequency of co-authorship and the total number of co-authored papers have been taken into account and several metrics have been used to analyse the network. Among them, the PageRank looks particularly suitable for this kind of analysis and provides some interesting knowledge. Similarly, in [16], a weighted version of the PageRank algorithm is proposed to consider citation and co-authorship network topologies. According to both these approaches, a higher score is expected for those authors who collaborate with different authors and also to authors who collaborate with a few highly coauthored ones. This idea differs from our perspective as we attempt to recognize and reward the citations coming from outside the group of collaborators.

In [17] a couple of PageRank modifications are proposed that weigh citations between authors on the basis of their co-authorship relations; additionally, they emphasise the time when articles are published and citations are made.

Recently, the PageRank algorithm has been used also to study citation patterns among universities [18] with the underlining assumption that scientific citations are driven by the reputation of the reference; thus, the PageRank algorithm is expected to yield a rank which reflects the reputation of an academic institution in a specific field.

Finally, an interesting PageRank version has been proposed in [19] based on the no-linear combination of the scores distributed to a node from its downstream neighbors. The method has been applied to publications networks and indirectly provides a measure of impact of their authors. These last two contributions consider some scenarios which are a bit different from ours, however their insights could be used to enrich our technique.

3   Method

In this section we first discuss how we discover and model the relations among the researchers based on their publications and collaborations, then we present the algorithmic strategy we propose to provide a quantitative measure of the impact of their research. Particularly, we exploit a network representation derived from the citations that the researchers exchange among themselves, we opportunely include the knowledge inferred about their collaborations and then we run the well known PageRank algorithm [3] to get a quantitative score of the impact of their research.

3.1   Citation Network

Suppose you are provided with a list of scientific papers, then you can derive some information about the citations each author receives from the others. This kind of knowledge can be effectively modeled through a *citation network*.

A *Citation Network* is an oriented weighted graph $\mathbf{G_{cit}} = (V, E, W)$, such that each node $v \in V$ is associated to an author and an edge between a certain pair (u, v) of authors state that u cites v in one of his papers. A weight $w_{i,j}$ is associated with each edge stating in how many of his papers u cites v. The network can be described also through its *adjacency matrix* A, such that $A_{i,j} = 1$ if i cites j in at least one of his papers and 0 otherwise.

We argue that the greater the ability of an author to gain citations from research groups with which he does not cooperate normally and the greater the impact of his scientific production, therefore it would be interesting to recognise this kind of citations and modify the weight matrix W to reward them.

Such a new matrix should be able to properly adjust the weights according to the origin of the citations each author receives, thus we define $\widetilde{W} = W \odot S$, i.e. we multiply each edge weight $w_{i,j}$ for a correction factor $s_{i,j}$ which rewards or penalises the citations that the author j receives from i.

The values in S have to comply with the following requirements:

1. $0 \leq s_{ij} \leq 1 \quad \forall i, j = 1 \ldots n$.
2. $s_{ij} \approx 0$ when the relationship between author i and author j is perceived to be strong, thus the citations they exchange between one-another could be penalised.
3. $s_{ij} \approx 1$ when the relationship between author i and author j is perceived to be weak, thus it is fair to consider the citations they exchange without any penalty.

Note that, the more the statistics associated to an author i tend to a value close to 1, the more the research activity of i has a wide impact and involves researches who are completely far from his collaboration group.

It is possible to establish different reward/penalty criteria, thus we take the prospective of defining the elements in S as the linear combination of m different statistics as follows:

$$s_{ij} = \sum_{h=1}^{m} \alpha_h \frac{1}{1 + x_h^{i,j}} \quad \text{with } \alpha_h \geq 0 \wedge \sum_{h=1}^{m} \alpha_h = 1 \tag{1}$$

where α_h are the coefficients of the linear combination that allow to tune the impact of each statistic. The values $x_h^{i,j}$, which are dependent from i and j, model and quantify the relationship among the authors. Thus, providing a statistic means to specify how x_h is defined for a given pair of researchers according to the type of relation we want to manage. Here, two different statistics have been proposed.

Citation Statistic. If the number of papers of an author a_j cited by an author a_i is significantly higher than the average number of citations made by a_i to the papers of all the other authors he cites, then it is worth penalising such an outlying behaviour. This idea is modeled by defining $x_1^{i,j}$ as follows:

$$x_1^{i,j} = w_{ij}/\mu_{i/j}^{cit}$$

where $\mu_{i/j}^{cit}$ is the arithmetic mean of the non-zero values on the ith row of matrix W, except for the value w_{ij}.

Collaboration Statistic. Citations coming from researchers which do not collaborate with a given author can be perceived as more interesting as they denote an impact of his research also outside his own collaboration group. Therefore, the second statistic intends to take into account the collaboration among the authors and penalises the citations coming from those with whom a collaboration relationship exists. We infer such a relation by deriving a co-authorship network. an author i and an author j are connected if they have co-authored at least one paper, and the weight $co(i,j)$ of their connection denotes the number of papers they have written together. Here, we want to award the citations of an author i to researchers with whom he does not regularly collaborate; to this aim we define:

$$x_2^{i,j} = co(i,j)/\mu_{i/j}^{co}$$

where $\mu_{i/j}^{co}$ is the average number of papers that i has written with other co-authors (except for j). Note that this statistic impacts only the edges of the citation network connecting two authors such that they have co-authored at least a paper and one cites the other.

These two statistics catch two different aspects of the problem we are dealing with and their combination through formula (1) can properly adjust the edge weights.

3.2   The PageRank Algorithm

To analyse the *citation network* presented so far, we consider the well know PageRank algorithm [3]. The main idea from Sergey Brin and Larry Page was that of providing a ranking of the information available on the Web by taking into account the "link popularity", thus a web page gets a high rank if it is referenced by many other web pages which are highly ranked, too. More in details, a link to a web page B *supports* it, but its contribution to the PageRank score of B depends on the score of the pages which contain a link to it: a web page referenced by other *important* web pages, i.e. with a high PageRank score, should receive a high score as well.

Let $\{B_1, B_2, \cdots, B_n\}$ be a set of n web pages and consider the adjacency matrix A modelling the connections among these pages, i.e.

$$A_{i,j} = \begin{cases} 1, & \text{if there is a link between } B_i \text{ and } B_j \\ 0 & \text{otherwise} \end{cases}$$

Note that the sum of all the values on a certain row i provides the number of links outgoing from a page B_i. Let call r_i such a value.

A person surfing the Web and visiting B_i can either follow a link available on page B_i or move to a completely new page. Suppose that the first event occurs with probability δ and the other with probability $1 - \delta$, then the probability $\mathbf{p}_{i,j}$ of a certain page B_j to be reached starting from B_i is

$$\mathbf{P}_{i,j} = \begin{cases} (1-\delta)/n + \delta A_{i,j}/r_i, & \text{if } r_i \neq 0 \\ 1/n & \text{if } r_i = 0 \end{cases} \tag{2}$$

We will refer to the matrix $\mathbf{P}$, whose elements are the probabilities $\mathbf{p}_{i,j}$, as *transaction probability matrix*. Note that each node has a probability greater than zero of being reached by any other node, thus $\mathbf{P}$ results to be *irreducible*; moreover, as all its element are non-negative and the sum of the values on each row is equal to 1 it is also *stochastic*. Due to this properties, the Fobeniuous-Perron Theorem holds and the largest eigenvalue of $\mathbf{P}$ is $\lambda_0 = 1$.

At a given time k, the web-user will be on a page $B(k)$ among all the possible ones. Let call $\pi_i(k)$ the probability of having $B(k) = B_i$. The probability of visiting B_i at time $k + 1$ is the sum of the probabilities of reaching B_i from any page B_j, weighted by the probability of being on page B_j at time k. In formula:

$$\pi_i(k+1) = \mathbf{p}_{1i}\pi_1(k) + \mathbf{p}_{2i}\pi_2(k) + \ldots + \mathbf{p}_{ni}\pi_n(k) \tag{3}$$

In matrix form, you can define the vector $\pi(k) = [\pi_1(k), \pi_2(k), \ldots \pi_n(k)]$ which contains, for each web-page, the probability for a user to visit it at time k; then, Eq. (3) can be rewritten as:

$$\pi(k+1) = \pi(k)\mathbf{P} \tag{4}$$

If you can define the vector $\pi(0)$ of probabilities of starting the navigation from each web page, then the probability of reaching a certain page in k time steps is:

$$\pi(k) = \pi(0)\mathbf{P}^k \tag{5}$$

Due to the properties of $\mathbf{P}$, it is possible to prove that $lim_{k\to+\infty}\mathbf{P}^k = \mathbf{P}^*$, where $\mathbf{P}^*$ is a matrix whose rows are all the same and equal to the left eigenvector of $\mathbf{P}$ associated with its largest eigenvalue. Thus, such an eigenvector provide a stationary probability, i.e. if a person surfed the web for an infinite amount of time, then he would visit each web page with a probability provided by the left eigenvector of $\mathbf{P}$, regardless of the starting page. These probabilities define the PageRank scores.

3.3   From PageRank to CiteRank

The PageRank strategy described so far provides a quantitative measure of the importance of a web pages based on its connections with the other pages; we argue that the same idea can be effectively used to analyse a citation network like the one introduced in Sect. 3.1. To this aim, you need to define the *transaction probability matrix* associated with the citation network, taking into account also the weights associated to the edges.

Let $n = |V|$ and rw_i be the sum of the values on the ith row of W, then an element $\mathbf{pw}_{i,j}$ of the *transaction probability matrix* $\mathbf{PW}$ is defined as:

$$\mathbf{pw}_{i,j} = \begin{cases} (1-\delta)/n + \delta W_{i,j}/rw_i, & \text{if } rw_i \neq 0 \\ 1/n & \text{if } rw_i = 0 \end{cases}$$

Similarly, you can use $\widetilde{W}$ to derive a *transaction probability matrix* $\mathbf{PW}$ which considers the corrections provided by the statistics discussed so far. Starting from these transaction matrix, it is possible to evaluate the PageRank score of each researcher and quantify the influence of his scientific work.

4   Experimental Analysis

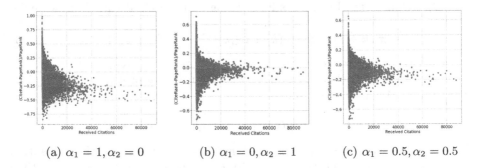

(a) $\alpha_1 = 1, \alpha_2 = 0$ (b) $\alpha_1 = 0, \alpha_2 = 1$ (c) $\alpha_1 = 0.5, \alpha_2 = 0.5$

Fig. 1. Gain or loss in the score due to the correction on the edge weight matrix W by the statistics.

The method has been tested on a citation network we build from data available in the well known repository DBLP and made available by *aminer.org* [20]. We consider version 11 updated at 2019–05–05. The dataset contains a row for each scientific paper, thus we move from the papers to the authors and derive the citation network whose nodes are the researchers.

Due to a limited availability of memory, for the purpose of our experiments, we perform also a sampling procedure by randomly selecting some rows, taking care to include all the rows connected to authors cited by the randomly selected ones. The resulting *citation network* includes 1,142,786 researchers.

Provided with this network, we first study the influence of each statistic in the evaluation of the CiteRank score of each researcher in order to opportunely tune the coefficient of the linear combination. To this aim, we first run the PageRank algorithm on the *transaction probability matrix* derived from W; we will refer to the scores derived in this scenario as PageRank scores. On the contrary, the CiteRank scores have been calculated using as input the *transaction probability matrix* derived from $\widetilde{W}$. We first run the algorithm with $\alpha_1 = 1$ and $\alpha_2 = 0$ to consider the *citation statistic* only; then with set $\alpha_1 = 0$ and $\alpha_2 = 1$ to take into account the *collaboration statistic* only.

Figures 1a and 1b show the impact of each of the two statistics. Particularly, we have reported, for each researcher, the difference between his CiteRank score and his PageRank score, divided by his PageRank. In this way, we are able to highlight the gain or loss in the score due to the correction introduced by each statistic. Such a change has been reported w.r.t. the number of citations a researcher has received throughout his career.

It is worth noting that the impact of the statistics seems to be stronger for researchers who still have few citations: their CiteRank can even double when the citation statistic is considered only (Fig. 1a) or can be penalised to become a quarter of the original PageRank score; these researchers, who are probably at the beginning of their career, are more sensitive to the perturbations introduced by the statistics due to their few incoming edges, however, the ones who increase their score are worth of attention because most of the citations they get actually comply with our notion of *fairness*, thus are not affected by strong penalization.

On the contrary researchers having a lot of citations have a widespread network of people citing them, thus the penalties that may occur on some of their incoming edges in the citation network weakly impact their score.

As for the collaboration statistic, it is interesting to point out that it corrects only the citations from i to j if i and j are also co-authors and i has written with j many more papers than the ones he has written with the other co-authors. Therefore, such a statistic involves only a subset of the edges outgoing from i, namely the ones to his co-authors. Despite this, Fig. 1b shows a trend similar to the one described for Fig. 1a, with young researchers more affected by the effect of the corrections on the weight matrix of the citation graph and older researcher almost maintaining their score.

Based on these evidence, we have decided to equally weight the two statistics in the linear combination of formula (1) and their joint contribution is reported in Fig. 1c.

Provided with the scores assigned to the researchers, it is possible to define two rankings, namely $rank_p$, based on the PageRank scores, and $rank_c$, based on the CiteRank scores. Then, you can evaluate for each researcher the change in position when moving from the first ranking to the second one as

$$\Delta_{pos}(i) = \frac{pos(rank_c, i) - pos(rank_p, i)}{max\{pos(rank_c, i), pos(rank_p, i)\}}$$

where $pos(\cdot)$ provides the position of the researcher i in the given ranking. In this way, we get the number of positions the researcher has gained w.r.t. the ranking in which he occupies the worst position, i.e. the positions gained compared to the maximum he could gain. If such a gain occurs within the PageRank-based ranking, $\Delta_{pos}(i)$ is negative, thus the corrections provided by the statistics have negatively influenced the score; on the contrary, a positive value shows that the author is robust against the statistics due to the *fairness* of his incoming edges, so that, although, some of his citations are eventually penalised, he complexly maintains his score or even improves it. Note that, for any value of the CiteRank score, it is possible to gain or lose positions in the ranking (see Fig. 2a), but a

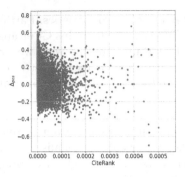
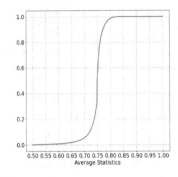

(a) Δ_{pos} w.r.t. CiteRank scores.

(b) Cumulative number of researchers gaining positions in the ranking w.r.t to their average statistics.

Fig. 2. Change of position within the ranking when moving from PageRank scores to CiteRank scores.

gain of positions in the higher part of the ranking is much more significant, thus values of $\Delta_{pos}(\cdot)$ are computed so that they take into account the original position of the author.

Furthermore, we have analysed the gain of positions also w.r.t. the values of the statistics in S. We argue that the researchers whose statistics tend to assume values close to 1 have an impact that go much beyond their connections. Such an attitude should be rewarded by our strategy by making these researchers gaining positions within the ranking calculated on the basis of the CiteRank scores.

To prove this claim, we have considered, for each researcher i, the mean of the statistics assigned to the edges outgoing from i, i.e. we have evaluated the mean of non-zero elements over the rows of S. In Fig. 2b we have reported, for each value of average statistic μ_s, the number of researchers who gain positions in the ranking and are associated with an average statistic lower than μ_s. Note that the counting has been normalized dividing by the whole number of researchers gaining positions. It is interesting to note that when the average statistic is about 0.75 there is a significant increase in the number of researchers gaining positions. This proves that the method is able at recognising and reward virtuous behaviours.

5   Conclusion

The method proposed through this work provides an alternative perspective to measure researchers influence based on the analysis of their citation patterns joined with knowledge about their collaborations.

We have introduced a concept of *fairness* based on the idea of properly weighting the citations in a way that rewards virtuous attitudes, such as the ability of a researcher to obtain citations from other researchers that are beyond the group of people with which he normally collaborates. To support this analysis, we have exploited the well known PageRank algorithm in a version able to deal with weighted graphs and we have derived a score function involving the number of citations an author receives and the authors from which they come.

The preliminary experiments discussed here show that our strategy is promising to appropriately score the researchers according to their impact.

Up to this point, the analysis based on DBLP dataset is a bit skewed to computer science data, thus it would be interesting in the future to extend the experiments to other scientific communities to highlight common attitudes and different behaviours.

Some drawbacks still exist with the technique as we have highlighted that young researchers may be penalised due to the few citations they still make and receive. Therefore, we plan to perform a deeper analysis to shed light on the impact of the statistics we have proposed and we would consider also the possibility of designing further statistics or evaluate different strategies to combine them, even with already existing metrics. These ideas would be the main topics of future research.

References

1. Hicks, D., Wouters, P., Waltman, L., De Rijcke, S., Rafols, I.: Bibliometrics: the Leiden manifesto for research metrics. Nature **520**(7548), 429–431 (2015)
2. Ioannidis, J.P.A., Baas, J., Klavans, R., Boyack, K.W.: A standardized citation metrics author database annotated for scientific field. PLoS Biol. **17**(8), e3000384 (2019)
3. Brin, S., Page, L.: The anatomy of a large-scale hypertextual web search engine. Comput. Netw. ISDN Syst. **30**(1–7), 107–117 (1998)
4. Roldan-Valadez, E., Salazar-Ruiz, S.Y., Ibarra-Contreras, R., Rios, C.: Current concepts on bibliometrics: a brief review about impact factor, eigenfactor score, citescore, scimago journal rank, source-normalised impact per paper, h-index, and alternative metrics. Irish J. Med. Sci. (1971-) **188**(3), 939–951 (2019)
5. Hirsch, J.E.: An index to quantify an individual's scientific research output. Proc. Natl. Acad. Sci. **102**(46), 16569–16572 (2005)
6. Jin, B., Liang, L.M., Rousseau, R., Egghe, L.: The r-and ar-indices: complementing the h-index. Chin. Sci. Bull. **52**(6), 855–863 (2007)
7. Burrell, Q.L.: Hirsch's h-index: a stochastic model. J. Informetr. **1**(1), 16–25 (2007)
8. Farooq, M., Khan, H.U., Iqbal, S., Munir, E.U., Shahzad, A.: Ds-index: ranking authors distinctively in an academic network. IEEE Access **5**, 19588–19596 (2017)
9. Ioannidis, J.P.A., Klavans, R., Boyack, K.W.: Multiple citation indicators and their composite across scientific disciplines. PLoS Biol. **14**(7), e1002501 (2016)
10. Nykl, M., Campr, M., Ježek, K.: Author ranking based on personalized pagerank. J. Informetr. **9**(4), 777–799 (2015)
11. Eck, N.J.V., Waltman, L.: Citnetexplorer: a new software tool for analyzing and visualizing citation networks. J. Informetr. **8**(4), 802–823 (2014)
12. Amjad, T., Daud, A., Aljohani, N.R.: Ranking authors in academic social networks: a survey. Library Hi Tech (2018)
13. Ding, Y., Yan, E., Frazho, A., Caverlee, J.: Pagerank for ranking authors in co-citation networks. J. Am. Soc. Inf. Sci. Technol. **60**(11), 2229–2243 (2009)

14. Fiala, D., Tutoky, G.: Pagerank-based prediction of award-winning researchers and the impact of citations. J. Informetr. **11**(4), 1044–1068 (2017)
15. Liu, X., Bollen, J., Nelson, M.L., Van de Sompel, H.: Co-authorship networks in the digital library research community. Inf. Process. Manag. **41**(6), 1462–1480 (2005)
16. Yan, E., Ding, Y.: Discovering author impact: a pagerank perspective. Inf. Process. Manag. **47**(1), 125–134 (2011)
17. Fiala, D.: Time-aware pagerank for bibliographic networks. J. Informetr. **6**(3), 370–388 (2012)
18. Massucci, F.A., Docampo, D.: Measuring the academic reputation through citation networks via pagerank. J. Informetr. **13**(1), 185–201 (2019)
19. Yao, L., Wei, T., Zeng, A., Fan, Y., Di, Z.: Ranking scientific publications: the effect of nonlinearity. Sci. Rep. **4**(1), 1–6 (2014)
20. Tang, J., Zhang, J., Yao, L., Li, J., Zhang, L., Su, Z.: Arnetminer: extraction and mining of academic social networks. In: Proceedings of the 14th ACM SIGKDD International Conference on Knowledge Discovery and Data Mining, pp. 990–998 (2008)

MADEISD: 4th Workshop on Modern Approaches in Data Engineering and Information System Design

MADEISD: 4th Workshop on Modern Approaches in Data Engineering and Information System Design

Workshop Chairs

Ivan Luković   University of Belgrade, Serbia
Sonja Ristić   University of Novi Sad, Serbia
Slavica Kordić   University of Novi Sad, Serbia

Program Committee Members

Paulo Alves   Polytechnic Institute of Bragança, Portugal
Moharram Challenger   University of Antwerp, Belgium
Boris Delibašić   University of Belgrade, Serbia
Dražen Drašković   University of Belgrade, Serbia
João M. Fernandes   University of Minho, Portugal
Krešimir Fertalj   University of Zagreb, Croatia
Krzysztof Goczyła   Gdańsk University of Technology, Poland
Ralf-Christian Härting   Aalen University, Germany
Dušan Jakovetić   University of Novi Sad, Serbia
Miklós Krész   InnoRenew CoE and University of Primorska,
                            Slovenia

Dragan Maćoš   Beuth University of Applied Sciences Berlin,
                            Germany

Zoran Marjanović   University of Belgrade, Serbia
Sanda Martinčić-Ipšić   University of Rijeka, Croatia
Cristian Mihaescu   University of Craiova, Romania
Nikola Obrenović   University of Novi Sad, Serbia
Maxim Panov   Skolkovo Institute of Science and Technology,
                            Russia

Rui Humberto Pereira   Polytechnic Institute of Porto, Portugal
Aleksandar Popović   University of Montenegro, Montenegro
Patrizia Počšić   University of Rijeka, Croatia
Adam Przybyłek   Gdansk University of Technology, Poland
Kornelije Rabuzin   University of Zagreb, Croatia
Igor Rožanc   University of Ljubljana, Slovenia
Nikolay Skvortsov   Russian Academy of Sciences, Russia
William Steingartner   Technical University of Košice, Slovakia
Vjeran Strahonja   University of Zagreb, Croatia
Slavko Žitnik   University of Ljubljana, Slovenia

Abstract Machine for Operational Semantics of Domain-Specific Language

William Steingartner[✉][iD], Róbert Baraník, and Valerie Novitzká[iD]

Faculty of Electrical Engineering and Informatics, Technical University of Košice,
Košice, Slovakia
{william.steingartner,valerie.novitzka}@tuke.sk,
robert.baranik@student.tuke.sk

Abstract. In this paper, we focus on some aspects of structural operational semantics for a selected domain-specific language for robot control, similar to the approach for Karel the Robot. For a given language, we formulate and develop a method of an abstract implementation on an abstract machine for structural operational semantics. The achieved results as well as the mentioned research are a part of the research in the field of semantic methods, where we focus on the formalization of semantic methods for software engineering. This area is also very important in the training of young IT experts, as semantic methods can help to understand program behavior and detect errors in program design. To make the teaching of formal semantics in the field of domain-specific languages more attractive, we have also prepared an application that serves to visualize the individual steps of the program on an abstract machine – simulation of translated code with visualization of a robot's movement.

Keywords: Abstract machine · Containerization · Domain-specific language · Formal semantics · Micro-service · Online teaching · Teaching software · University didactic

1 Introduction

The formal semantics offers students studying computer science and IT experts the opportunity to better understand how code execution works on the machine using abstraction, thus removing some unnecessary details which can be ignored. Knowledge of structural operational semantics and the abstract machine is an important foundation in the teaching of programming languages. However, the related abstract machine theory can be unfriendly or difficult for students to learn. To overcome this problem, we follow the idea of making formal methods visual in particular steps of calculations. The current situation requires that students also acquire skills in working with this type of educational software, as

This work was supported by the project "A development of the new semantic technologies in educating of young IT experts", project no. KEGA 011TUKE4/2020, by the Ministry of Education, Science, Research and Sport of the Slovak Republic.

S. Chiusano et al. (Eds.): ADBIS 2022, CCIS 1652, pp. 413–424, 2022.
https://doi.org/10.1007/978-3-031-15743-1_38

stated by the author in [15]. In educating of young IT experts, there is a need for tools which can illustrate the semantic methods to students and make it clearer how for example an abstract machine operates in an interactive way for not only self-study for students but also for teachers to present abstract machine in their lectures. Many times, the introduction to programming is explained in the languages similar to the original idea of Karel the Robot (or simply *Karel*). Therefore, we were also inspired by this approach and expanded our research in the field of domain-specific languages, where we work with a language for robot control. The domain-specific language we work with is the language of Karel which is an entity moving in two-dimensional space by executing given commands.

An abstract implementation of the robot language is a great motivation since the other semantic approaches are defined (denotational and operational semantics are defined in [6], natural semantics in [19]). The soundness of our approach is verified and proven by the proof of equivalence.

The aim of this paper is to define an abstract machine for structural operational semantics of the robot language and for its abstract implementation. After that, we designed an application that serves as an emulation tool of abstract machine defined. For its design and implementation, a modular approach was taken. The application is divided into 4 micro-services: translation logic, translation graphical user interface, simulation logic, and simulation graphical user interface, with a database in the fifth container.

This paper is a part of our project to prepare young engineers and IT experts to deal with semantic methods. We have prepared the applications for teaching and learning denotational semantics [18], operational semantics [17] and abstract machine [16]. Our needs are to specific that we had problems to find similar applications. We followed the methods published in [14] with some modifications. The aim of our project is to provide an integrated system for learning and teaching the semantic methods. This paper serves for illustrating how domain-specific languages differ in using the operational semantics.

The purpose of the logic components is to parse the given input source code and to provide an output form (bytecode) for visual processing and calculation. The processed input for the translation logic is afterwards translated to abstract machine code and for the simulation logic, the data is used for simulating the execution on the abstract machine, returning simulation data. The web is used as a graphical user interface for both components using Spring with Thymeleaf template engine for displaying variable data on the HTML page which uses Javascript scripts for additional calls. Web pages had both English and Slovak localization with the link from translator to simulator and back.

The paper is structured as follows: in Sect. 2, we present a syntax of the language and preliminaries for defining the semantics. Section 3 contains a definition of the abstract machine(s) for the robot language(s). Section 4 focuses on the design of an emulation of abstract machine, and Sect. 5 presents its specification and functionality. Finally, the Sect. 6 concludes our paper.

2   Domain-Specific Languages and Language for Robot

A domain-specific language (DSL) is a programming language or executable specification language that, through appropriate notation and abstraction, offers expressive power focused on, and usually limited to, a specific domain of the problem [4]. Domain-specific languages are languages tailored to a specific application domain. They offer substantial gains in expressiveness and ease of use compared with general-purpose programming languages in their domain of application [9]. By the creation of a DSL, a user can focus on the problem and its solution in a specific area to further processing. One of the possible benefits of using DSL is lower implementation time, reduced maintenance costs, as well as better portability, reliability, optimizability and testability [3]. The language of the robot is an external DSL which means it is a fully independent language with its syntax and semantics [7]. Language workbench is a great tool for creating such a language – it is easy to define not the only parser, but also a custom editing environment [5].

The operational semantics of the language of a robot is a base for abstract machine definition. The robot is an entity in two-dimensional space with specified x and y coordinates which are typically changed during the execution of a program. The robot can move horizontally and vertically, diagonal movement is not possible in this simplified version. Language has two syntactic areas [6]:

- $n \in$ **Num** – numerals specifying number of steps which robot has to move;
- $C \in$ **Comm** – commands.

The syntactic category is specified by an abstract syntax giving the basis elements and the composite elements. The syntactic category of well-structured commands **Comm** is defined as follows:

$$C:: = \textbf{left} \mid \textbf{right} \mid \textbf{up} \mid \textbf{down} \mid \textbf{left } n \mid \textbf{right } n \mid \textbf{up } n \mid \textbf{down } n \mid$$
$$\textbf{skip} \mid \textbf{reset} \mid C; C.$$

Semantic domain **Point** $= \mathbb{Z} \times \mathbb{Z}$ represents the robot's position (denoted e.g. p). A position is considered as a state. Then change of the position of robot is considered as a change of a state. Semantics of the commands is given by evaluation of the semantic function

$$\mathscr{C} : \textbf{Comm} \rightarrow (\textbf{Point} \rightarrow \textbf{Point}).$$

Structural operational semantics is known as a small-step semantics. The transitions in this method are of the form

$$\langle C, p \rangle \Rightarrow \alpha,$$

where α stands for:

- a new state p', if the command C is performed in one step (typically movement in some direction in one step or the teleportation to the initial position) and the execution has terminated, or

– an intermediate configuration (a tuple) of a form $\langle C', p' \rangle$ if the execution of command C is not completed and remaining computation continues.

For example, the transition of the command that is performed in more steps:

$$\langle \mathbf{up}\ n, p \rangle \Rightarrow \langle \mathbf{up}; \mathbf{up}\ m, p \rangle$$

where $[\![\,n\,]\!] = [\![\,m\,]\!] \oplus 1$, for $[\![\,m\,]\!], [\![\,n\,]\!] \in \mathbb{N}_0$ and $[\![\,\cdot\,]\!]$ is a semantic function that sends numerals to naturals (or zero) [6]:

$$[\![\,\cdot\,]\!] : \mathbf{Num} \to \mathbb{N}_0.$$

For making this language more general, we change its abstract syntax by adding commands for turning into the given direction (according to the angle, typically 90 degrees) and moving in the current heading. Abstract syntax is then:

$$C:: = \mathbf{forward} \mid \mathbf{forward}\ n \mid \mathbf{turn\ left} \mid \mathbf{turn\ right} \mid$$
$$\mathbf{skip} \mid \mathbf{reset} \mid C; C.$$

These commands enable the robot to rotate around its axis and move forward in the direction of current heading. Robot can rotate only 90 degrees in single step. Values of angles come from the semantic domain $\mathbf{Angle} = \mathbb{Z}$ and its elements $\alpha \in \mathbf{Angle}$. We consider the following convention: if the robot is facing north, then current angle is $\mathbf{0}$, for east it is $\mathbf{90}$, for south $\mathbf{180}$ and for west $\mathbf{270}$. The **forward** command calculates the next position by trigonometric functions sine and cosine. For the possible values of angle, these functions return values of either $-\mathbf{1}$, $\mathbf{0}$ or $\mathbf{1}$.

Because angle is an element that provides the direction of robot, a new semantic domain must be defined for this form of language:

$$\mathbf{Config} = \mathbf{Point} \times \mathbf{Angle}.$$

Execution of former commands won't be changed after latter commands addition, only **reset** command will also reset robot's rotation. Since the commands **up**, **down**, **left**, **right** and their variants with the number of steps do not change the robot's rotation, they can be separated from the commands **turn left**, **turn right**, **forward** and **forward** n by an imaginary line for commands that use robot rotation and which do not. Commands of different groups should not be combined, although they can be executed on the same abstract machine – for this reason two ANTLR grammars were created during implementation. The **reset** command, **skip** command and the concatenation command then belong to both groups.

Further extension of the language, which were not proposed in abstract machine proposal, are $\mathbf{init}(x, y)$ command, which can change the starting position and position to reset to by **reset** command as the first command in the code. Conditional and loop commands can be added for shortening the code and opening a lot of options. Another extension can be addition of flags on the environment which the robot can place and lift from. Furthermore, battery or energy management can be added, so each robot's movements deplete its energy which needs to be recharged later on [6].

3 Abstract Machine for DSL – Theory and Development of Method

In the first phase of development in language-oriented programming, it is necessary to design a formal syntax and semantics of the language [8]. Generally, syntax is the set of rules that defines the combinations of symbols that are considered to be correctly structured statements or expressions in a computer language, as defined by the syntax of the language itself. Language surface forms are determined by their syntax. Under semantics we understand the meaning of such structures – the interpretation of the code or its associated meaning, such as symbols, characters, or any other part of the code. Several semantic methods are widely used in practice. We focus on structural operational semantics which, in contrast to natural semantics (known as *big-step semantics*), is a semantic method that describes a meaning of programs in particular steps, and it is also called *small-step semantics* [10].

Abstract implementation of a program is used to verify the correctness of implementation of programming languages and consists of the definition of the abstract machine, and the translation of the program into a sequence of the instructions, which are then executed on the abstract machine [10]. Definition of the abstract machine consists of *configuration* and *instructions* of the abstract machine. Abstract machine configuration can be written as $\langle c, v, s \rangle$, where c is the instruction sequence, v is an evaluation stack of values and s is the state of the abstract machine. As *state* we generally understand some mathematical (formal) abstraction of computer memory, here state is an actual position of robot (as defined in Sect. 2).

Our definition of abstract machine for structural operational semantics is defined for domain-specific language of robot, namely for both versions of the language. The state of the abstract machine determines the position of the robot in the environment

$$p \in \textbf{Point} = \mathbb{Z} \times \mathbb{Z}$$

and the angle of its rotation $\alpha \in \textbf{Angle} = \mathbb{Z}$. Hence, for the first version of the language containing the directional commands, the state is an element of semantics domain **Point**. For the second version of the language, comprising the rotations, a state is an element of semantic domain

$$s \in \textbf{Config} = (\textbf{Point} \times \textbf{Angle}).$$

Evaluation stack is not used in this abstract machine(s), but will be needed for possible extension of robot language. Code $c \in \textbf{Code}$ consists of instructions by production rules:

$$ins:: = \textbf{LEFT} \mid \textbf{LEFT-}n \mid \textbf{RIGHT} \mid \textbf{RIGHT-}n \mid$$
$$\textbf{UP} \mid \textbf{UP-}n \mid \textbf{DOWN} \mid \textbf{DOWN-}n \mid$$
$$\textbf{FORWARD} \mid \textbf{FORWARD-}n \mid \textbf{TLEFT} \mid \textbf{TRIGHT} \mid$$
$$\textbf{SKIP} \mid \textbf{RESET},$$

$$c:: = \quad \varepsilon \mid ins : c$$

Some of the instructions contain numeric parameters which represent the number of steps in given direction (it simply means that command is performed n-times, where $\mathbf{n} = [\![\, n \,]\!]$). For situation, when numeral n is converted to boundary values ($\mathbf{0}$) or not supported values (negative numbers), an abstract machine (and in simulation program, as well) does not do anything. Hence, compilation of such commands provides an error.

From the semantic equivalence (proved for natural semantics in [19]) it follows that (for any parametric instruction, denoted **INST-**n):

$$\langle \mathbf{INST}\text{-}n, v, p \rangle =\!\!\gg \langle \mathbf{INST}; \mathbf{INST}\text{-}m, v, p \rangle$$
$$\langle \mathbf{INST}\text{-}1, v, p \rangle =\!\!\gg \langle \mathbf{INST}, v, p \rangle$$

for $[\![\, n \,]\!] = [\![\, m \,]\!] \oplus 1$, $[\![\, n \,]\!], [\![\, m \,]\!] \in \mathbb{N}_0$. We note that all instructions with the value $\mathbf{0}$ of the parameter don't do anything and they are (semantically) equivalent to the instruction **SKIP**.

Generating of abstract machine code from the input language is done by the translation function

$$\mathscr{TC} : \mathbf{Comm} \to \mathbf{Code}$$

which sends an input source code written in the robot language into a sequence of instructions of abstract machine. Although we defined two forms of robot language, only one translation function for our purposes is defined:

$\mathscr{TC}[\![\, \text{left} \,]\!] = \mathbf{LEFT}$ $\qquad\qquad$ $\mathscr{TC}[\![\, \text{left } n \,]\!] = \mathbf{LEFT}\text{-}n$

$\mathscr{TC}[\![\, \text{right} \,]\!] = \mathbf{RIGHT}$ $\qquad$ $\mathscr{TC}[\![\, \text{right } n \,]\!] = \mathbf{RIGHT}\text{-}n$

$\mathscr{TC}[\![\, \text{up} \,]\!] = \mathbf{UP}$ $\qquad\qquad\quad$ $\mathscr{TC}[\![\, \text{up } n \,]\!] = \mathbf{UP}\text{-}n$

$\mathscr{TC}[\![\, \text{down} \,]\!] = \mathbf{DOWN}$ $\qquad$ $\mathscr{TC}[\![\, \text{down } n \,]\!] = \mathbf{DOWN}\text{-}n$

$\mathscr{TC}[\![\, \text{turn left} \,]\!] = \mathbf{TLEFT}$ $\qquad$ $\mathscr{TC}[\![\, \text{turn right} \,]\!] = \mathbf{TRIGHT}$

$\mathscr{TC}[\![\, \text{forward} \,]\!] = \mathbf{FORWARD}$ $\quad$ $\mathscr{TC}[\![\, \text{forward } n \,]\!] = \mathbf{FORWARD}\text{-}n$

$\mathscr{TC}[\![\, \text{reset} \,]\!] = \mathbf{RESET}$ $\qquad\quad$ $\mathscr{TC}[\![\, \text{skip} \,]\!] = \mathbf{SKIP}$

$\mathscr{TC}[\![\, C_1 : C_2 \,]\!] = \mathscr{TC}[\![\, C_1 \,]\!] : \mathscr{TC}[\![\, C_2 \,]\!]$

The colon symbol ":" serves as an instruction separator in the abstract machine instruction sequence listing.

For example, let

$$\text{forward } 2; \text{turn right}; \text{forward}; \text{turn right};$$
$$\text{forward } 1; \text{turn left}; \text{forward } 4;$$

be a correct program in robot language. A code of abstract machine for this program is the following:

FORWARD-2 : **TRIGHT** : **FORWARD** : **TRIGHT** :
FORWARD-1 : **TLEFT** : **FORWARD-**4

As the next step, we define the meaning of the abstract machine code using a partially defined *execution function*:

$$\mathscr{M} : \mathbf{Code} \to \mathbf{Point} \rightharpoonup \mathbf{Point},$$

defined as follows:

$$\mathscr{M}[\![\,c\,]\!]p = \begin{cases} p', & \text{if } \langle c, \varepsilon, p \rangle =\!\gg^* \langle \varepsilon, \varepsilon, p' \rangle, \\ \bot, & \text{otherwise.} \end{cases}$$

By composing the translation and execution functions, we get the semantic function for the abstract machine:

$$\mathscr{S} : \mathbf{Comm} \rightarrow \mathbf{Point} \rightharpoonup \mathbf{Point},$$
$$\mathscr{S}[\![\,C\,]\!] = (\mathscr{M} \circ \mathscr{TC})[\![\,C\,]\!].$$

This definition applies to the language version with directional commands. We note that the definition of the execution ($\mathscr{M}'$) and semantic function ($\mathscr{S}'$) would be analogous for the second types of language variants (with rotations), taking into account the semantic area.

Semantics of abstract machine instructions (for the language with directional commands) is defined as follows:

$$\langle \mathbf{LEFT} : c, \varepsilon, (x, y) \rangle =\!\gg \langle c, \varepsilon, (x \ominus 1, y) \rangle$$
$$\langle \mathbf{LEFT}\text{-}n : c, \varepsilon, (x, y) \rangle =\!\gg \langle \mathbf{LEFT} : \mathbf{LEFT}\text{-}m : c, \varepsilon, (x, y) \rangle$$
$$\langle \mathbf{RIGHT} : c, \varepsilon, (x, y) \rangle =\!\gg \langle c, \varepsilon, (x \oplus 1, y) \rangle$$
$$\langle \mathbf{RIGHT}\text{-}n : c, \varepsilon, (x, y) \rangle =\!\gg \langle \mathbf{RIGHT} : \mathbf{RIGHT}\text{-}m : c, \varepsilon, (x, y) \rangle$$
$$\langle \mathbf{UP} : c, \varepsilon, (x, y) \rangle =\!\gg \langle c, \varepsilon, (x, y \oplus 1) \rangle$$
$$\langle \mathbf{UP}\text{-}n : c, \varepsilon, (x, y) \rangle =\!\gg \langle \mathbf{UP} : \mathbf{UP}\text{-}m : c, \varepsilon, (x, y) \rangle$$
$$\langle \mathbf{DOWN} : c, \varepsilon, (x, y) \rangle =\!\gg \langle c, \varepsilon, (x, y \oplus 1) \rangle$$
$$\langle \mathbf{DOWN}\text{-}n : c, \varepsilon, (x, y) \rangle =\!\gg \langle \mathbf{DOWN} : \mathbf{DOWN}\text{-}m : c, \varepsilon, (x, y) \rangle$$
$$\langle \mathbf{RESET} : c, \varepsilon, (x, y) \rangle =\!\gg \langle c, \varepsilon, p^* \rangle$$
$$\langle \mathbf{SKIP} : c, \varepsilon, (x, y) \rangle =\!\gg \langle c, \varepsilon, (x, y) \rangle$$

for $[\![\,n\,]\!] = [\![\,m\,]\!] \oplus 1$, $[\![\,n\,]\!], [\![\,m\,]\!] \in \mathbb{N}_0$ and p^* stands for an initial (starting) position defined by user specification.

Similarly, semantics of abstract machine instructions (for the language with rotations) is defined as follows:

$$\langle \mathbf{FORWARD} : c, \varepsilon, ((x, y), \alpha) \rangle =\!\gg \langle c, \varepsilon, ((x \oplus \sin \alpha, y \oplus \cos \alpha), \alpha) \rangle$$
$$\langle \mathbf{FORWARD}\text{-}n : c, \varepsilon, (p, \alpha) \rangle =\!\gg \langle \mathbf{FORWARD} : \mathbf{FORWARD}\text{-}m : c, \varepsilon, (p, \alpha) \rangle$$
$$\langle \mathbf{TLEFT} : c, \varepsilon, ((x, y), \alpha) \rangle =\!\gg \langle c, \varepsilon, (p, (\alpha \oplus 270) \bmod 360) \rangle$$
$$\langle \mathbf{TRIGHT} : c, \varepsilon, ((x, y), \alpha) \rangle =\!\gg \langle c, \varepsilon, ((\alpha \oplus 90) \bmod 360) \rangle$$
$$\langle \mathbf{RESET} : c, \varepsilon, ((x, y), \alpha) \rangle =\!\gg \langle c, \varepsilon, (p^*, \mathbf{0}) \rangle$$
$$\langle \mathbf{SKIP} : c, \varepsilon, (p, \alpha) \rangle =\!\gg \langle c, \varepsilon, (p, \alpha) \rangle$$

We note, that in this specification, an initial configuration (state) is a tuple consisting of the position p^* and the angle $\mathbf{0}$. The initial value for the angle can be changed according to the user specification.

4   Application Design

We followed the idea of containerizing the application. We based on the methodological design, where we separated the individual layers (front-end, back-end) and implemented each functional unit as a separate Docker module. Docker is an open-source containerization platform which enables developers to package applications into containers and offers isolation of applications into containers which run safely [2] and independently thanks to Docker Engine directly on the host computer's operating system [13]. Docker containers also offer scalability and portability [12], so the system is easy to extend or change, and reusability of containers in other systems thereby achieving reproducible research [1]. For our application system, we designed five containers using Docker Compose for better configuration:

- *translator logic* – accepts translation request, performs translation and returns abstract machine code with possible errors in source code in robot's language which is also stored to the database,
- *translator graphical user interface* – application displaying a web page that the user can interact with by writing source code in the robot's language, requesting a translation to translation logic and displaying results;
- *simulator logic* – searches for translation request in the database, parses the abstract machine code, performs simulation and provides the results;
- *simulator graphical user interface* – the second web page which can be accessed through the translator page which sends the latest translation request id to be sent from this component to simulator logic, displays a visualization of a robot on the canvas, currently executed command and transitions of configurations, which all can be controlled by simulation controller offering control buttons and simulation speed slider;
- *database* – PostgreSQL database where translation requests with their errors are saved by translator logic and searched by simulator logic.

All components of applications are running on Spring Boot which is being used in a big amount of projects in areas like cloud computing, big data, reactive programming and client applications development [20]. Spring Boot data can be displayed on web pages using a template engine like Thymeleaf which is used for our GUI components. Visualization, HTTP requests and other logic are implemented using Javascript, jQuery and other libraries.

For reading and tranasforming the user input, a technology of ANTLR has been used [11]. Three ANTLR grammar files are designed for our application for compiler and simulator logic components:

- simple directional commands (without rotations) for translator – translates input source code to abstract machine language where only **up**, **down**, **left**, **right** are present, then commands with parameter n as number of steps and **reset** commands are valid;
- commands with rotations for translator – translates input source code to abstract machine language where only **forward**, **forward** n, **turn left**, **turn right** and **reset** commands are valid;
- parser for simulator – parses the abstract machine code into data structures (bytecode) on which simulation can be performed.

5   Functionality of Application

An application can be accessed through a browser. This allows to users to work with the application from anywhere. An application allows to insert the user's input source code or to load a code into the application from external storage. The idea is that user input is sent to the back-end layer and compiled. If no error occurs, the source is correct and compiled bytecode is sent back for performing the visualization.

The user interface helps the user to enter the input source by providing also hints for writing the code in the robot language. The hint disappears when the user starts to write the code. Two main functionalities are available – simulation (visual computation) of abstract machine and compilation (translation) of input source to internal bytecode (a data structure according to the specification).

For the implementation of the compilation module, the ANTLR tool was used. Compilator reads an input source and identifies in which variant of language the program is written:

- if code contains rotation commands, appropriate grammar is applied;
- if code does not contain rotation commands, only directional-stepping commands, then grammar for this language is applied;
- otherwise, no code is generated and the program provides error information (since the languages cannot be mixed).

If the commands are not mixed, but during the compilation, the lexical or syntactic error occurs, the compilation does not provide a bytecode, only information about the compilation error. Then, abstract machine code is annotated by adding HTML tags and sent back to the front-end where all errors are highlighted.

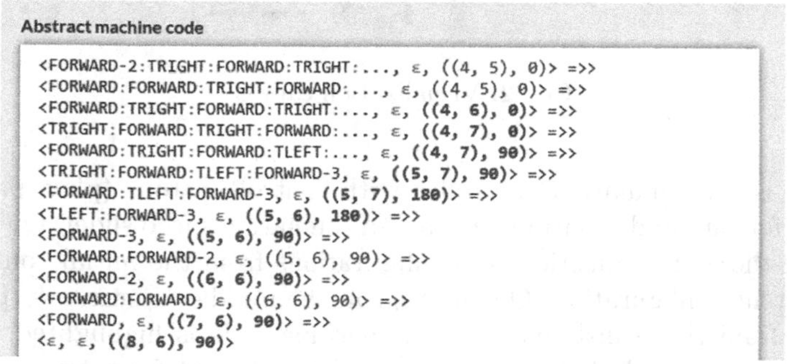

```
Abstract machine code

<FORWARD-2:TRIGHT:FORWARD:TRIGHT:..., ε, ((4, 5), 0)> =>>
<FORWARD:FORWARD:TRIGHT:FORWARD:..., ε, ((4, 5), 0)> =>>
<FORWARD:TRIGHT:FORWARD:TRIGHT:..., ε, ((4, 6), 0)> =>>
<TRIGHT:FORWARD:TRIGHT:FORWARD:..., ε, ((4, 7), 0)> =>>
<FORWARD:TRIGHT:FORWARD:TLEFT:..., ε, ((4, 7), 90)> =>>
<TRIGHT:FORWARD:TLEFT:FORWARD-3, ε, ((5, 7), 90)> =>>
<FORWARD:TLEFT:FORWARD-3, ε, ((5, 7), 180)> =>>
<TLEFT:FORWARD-3, ε, ((5, 6), 180)> =>>
<FORWARD-3, ε, ((5, 6), 90)> =>>
<FORWARD:FORWARD-2, ε, ((5, 6), 90)> =>>
<FORWARD-2, ε, ((6, 6), 90)> =>>
<FORWARD:FORWARD, ε, ((6, 6), 90)> =>>
<FORWARD, ε, ((7, 6), 90)> =>>
<ε, ε, ((8, 6), 90)>
```

Fig. 1. Computational sequence of an abstract machine code (cutted screenshot)

The successful compilation, if no error appears, provides bytecode for abstract machine computation steps and visualization. The result of compilation is then stored in an internal database and sent back to the GUI component.

The main screen of the application contains two visual areas. An abstract machine code is displayed on the right-hand side of an application window (Fig. 1). The application allows also to save the source code. The user interface supports to be displayed in English and Slovak languages, a color theme can be changed from light to dark, as well.

By clicking on the execution button, the user is redirected to the simulator page with the id of the latest translation request. On the simulator web page, the given id is sent to the back-end and then to simulator logic. There the id is searched in the database for a given request. After finding the request, the abstract machine code is parsed according to the third grammar to a list of commands from which the initial configuration is created and execution of these commands is started. The result of an execution is a list of states or rather configurations of the abstract machine and this list is added to the execution request which is sent back to the user and front-end.

On the left-hand side, there is the visualization canvas showing the robot's position and rotation in an orthogonal grid (Fig. 2).

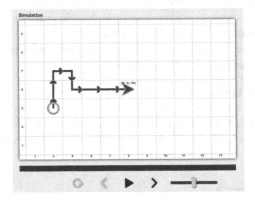

Fig. 2. Visualization canvas

Bellow is the simulation controller with control buttons (play, pause, step back, step forward and reset buttons) and a simulation speed slider. On the right-hand side, there are transitions of configurations from the initial configuration to the current configuration. On the top, in the middle, a currently performed command from the translated abstract machine code is highlighted. The user can download a record of the simulation in four different formats:

- standard CSV format with position, rotation, value and code stacks for each configuration in time,
- XML format with structured data similar to CSV data,
- PDF file with a title, current time, a snapshot of the canvas and transitions of configurations which are generated using jsPDF library by MrRio[1],

---

[1] https://mrrio.github.io/jsPDF/.

– a TEXsource format and picture with a snapshot of the canvas which looks similar to the PDF and is stored in a ZIP archive using JSZip[2].

Visualization is performed using Javascript following states or configurations from simulator logic to move the robot in its environment and draw its path.

6 Conclusion

In this paper, we presented our approach to abstract implementation for selected domain-specific language describing the controlling of robot. We developed and defined our semantic approach for both versions for natural semantics in [19]. For the existing approach in structural operational semantics (defined in [6]), we defined our approach of abstract implementation and we defined two kinds of appropriate abstract machines. For this semantic approach, we also developed a visualizing software that provides compilation of input source code written in robot language and emulates the calculation of abstract machine for both versions of the language, and moreover, it accepts direct input in abstract machine code and provides also the calculations and emulations. Design of micro-service system is created by using containerization, so each micro-service can be reused in other systems later (if we assume the future work oriented to a complex software visualizing environment). Created application offers many possible extensions, also thanks to the modular approach applied by containerization, and improvements. List of commands can be extended in future based on a new specification. Such developed application is ready to be integrated into teaching process for the courses oriented to formal semantics, and (possibly) formal languages. Its added value is a significant degree of interactivity, clarity and illustrativeness and, above all, the possibility to use the application in the process of present and distance teaching as well as during the independent preparation of students.

Because we see the potential for expanding and consolidating formal methods for software engineering, mainly due to the fact that all formal methods are based on formal semantics, we will focus our research on semantic methods with the possibility of integrating the results into practice and into teaching young IT specialists and experts.

Acknowledgement. The authors express their gratitude to prof. Marjan Mernik for the original idea of how to formulate a domain-specific language for a given application domain. The authors also would like to thank to Dániel Horpácsi and Judit Horpácsiné Kőszegi for their approach to DSL semantics for the robot that motivated us in our research.

References

1. Chamberlain, R., Schommer, J.: Using docker to support reproducible research. Figshare **1101910**, 44 (2014). https://doi.org/10.6084/m9.figshare.1101910

---

[2] https://github.com/Stuk/jszip.

2. Combe, T., Martin, A., Di Pietro, R.: To docker or not to docker: a security perspective. IEEE Cloud Comput. **3**(5), 54–62 (2016)
3. Deursen, A., Klint, P.: Domain-specific language design requires feature descriptions. J. Comput. Inf. Technol. **10**, 1–17 (2002)
4. Deursen, A., Klint, P., Visser, J.: Domain-specific languages: an annotated bibliography. SIGPLAN Not. **35**, 26–36 (2000)
5. Fowler, M.: Domain-Specific Languages. Pearson Education, Boston (2010)
6. Horpácsi, D., Kőszegi, J.: Formal semantics (2014). regi.tankonyvtar.hu/en/tartalom/tamop412A/2011-0052_05_formal_semantics/index.html. Accessed 14 Dec 2020
7. Johanson, A.N., Hasselbring, W.: Hierarchical combination of internal and external domain-specific languages for scientific computing. In: Proceedings of the 2014 European Conference on Software Architecture Workshops, pp. 1–8 (2014)
8. Kollár, J., Porubän, J., Chodarev, S.: Modelovanie a generovanie softvérových architektúr. elfa s.r.o (2012)
9. Mernik, M., Heering, J., Sloane, A.: When and how to develop domain-specific languages. ACM Comput. Surv. **37**, 316–344 (2005). https://doi.org/10.1145/1118890.1118892
10. Nielson, R.H., Nielson, F.: Semantics with Applications: An Appetizer. Springer, London (2007). https://doi.org/10.1007/978-1-84628-692-6
11. Parr, T., et al.: What's ANTLR (2004). www.antlr3.org/share/1084743321127/ANTLR_Reference_Manual.pdf
12. Patil, S.: Study of container technology with docker. Int. J. Adv. Res. Sci. Commun. Technol. **5**, 504–509 (2021)
13. Rad, B.B., Bhatti, H.J., Ahmadi, M.: An introduction to docker and analysis of its performance. Int. J. Comput. Sci. Netw. Secur. (IJCSNS) **17**(3), 228 (2017)
14. Schreiner, W.: Theorem and algorithm checking for courses on logic and formal methods. In: Quaresma, P., Neuper, W. (eds.) Proceedings 7th International Workshop on Theorem Proving Components for Educational Software, ThEdu@FLoC 2018, Oxford, United Kingdom, 18 July 2018. EPTCS, vol. 290, pp. 56–75 (2018). https://doi.org/10.4204/EPTCS.290.5
15. Seidametova, Z.: Some methods for improving data structure teaching efficiency. Educ. Dimension **58**, 164–175 (2022). https://journal.kdpu.edu.ua/ped/article/view/4509
16. Steingartner, W.: Compiler module of abstract machine code for formal semantics course. In: 2021 IEEE 19th World Symposium on Applied Machine Intelligence and Informatics (SAMI), pp. 000193–000200 (2021). https://doi.org/10.1109/SAMI50585.2021.9378696
17. Steingartner, W.: On some innovations in teaching the formal semantics using software tools. Open Comput. Sci. **11**(1), 2–11 (2021). https://doi.org/10.1515/comp-2020-0130
18. Steingartner, W., Gajdoš, E.: The visualization of a graph semantics of imperative language. Polytechnica J. Technol. Educ. **2**(5), 7–14 (2021). https://doi.org/10.36978/cte.5.2.1
19. Steingartner, W., Novitzká, V.: Natural semantics for domain-specific language. In: Bellatreche, L., et al. (eds.) ADBIS 2021. CCIS, vol. 1450, pp. 181–192. Springer, Cham (2021). https://doi.org/10.1007/978-3-030-85082-1_17
20. Walls, C.: Spring Boot in Action. Manning Publications, New York (2016)

Machine Learning-Based Model Categorization Using Textual and Structural Features

Alireza Khalilipour[1] ⓘ, Fatma Bozyigit[2](✉) ⓘ, Can Utku[3],
and Moharram Challenger[2] ⓘ

[1] Computer Engineering Department, Islamic Azad University, Qazvin Branch, Qazvin, Iran
[2] Department of Computer Science, University of Antwerp, AnSyMo/CoSys Corelab, Flanders Make Strategic Research Center, Antwerp, Belgium
`{fatma.bozyigit,moharram.challenger}@uantwerpen.be`
[3] Department of Computer Engineering, Işık University, Istanbul, Turkey

Abstract. Model Driven Engineering (MDE), where models are the core elements in the entire life cycle from the specification to maintenance phases, is one of the promising techniques to provide abstraction and automation. However, model management is another challenging issue due to the increasing number of models, their size, and their structural complexity. So that the available models should be organized by modelers to be reused and overcome the development of the new and more complex models with less cost and effort. In this direction, many studies are conducted to categorize models automatically. However, most of the studies focus either on the textual data or structural information in the intelligent model management, leading to less precision in the model management activities. Therefore, we utilized a model classification using baseline machine learning approaches on a dataset including 555 Ecore metamodels through hybrid feature vectors including both textual and structural information. In the proposed approach, first, the textual information of each model has been summarized in its elements through text processing as well as the ontology of synonyms within a specific domain. Then, the performances of machine learning classifiers were observed on two different variants of the datasets. The first variant includes only textual features (represented both in TF-IDF and word2vec representations), whereas the second variant consists of the determined structural features and textual features. It was finally concluded that each experimented machine learning algorithm gave more successful prediction performance on the variant containing structural features. The presented model yields promising results for the model classification task with a classification accuracy of 89.16%.

Keywords: Model Driven Engineering · Model management · Metamodel · Text mining · Machine learning

1 Introduction

In order to address the challenges and reduce the complexity of software development, one of the key approaches is Model-driven Engineering (MDE), which is a modern

© Springer Nature Switzerland AG 2022
S. Chiusano et al. (Eds.): ADBIS 2022, CCIS 1652, pp. 425–436, 2022.
https://doi.org/10.1007/978-3-031-15743-1_39

software engineering paradigm, focuses on using models as first-class entities. It has gained popularity with academic and industrial communities in software engineering, leading to a plethora of models. Accordingly, intelligent model management is essential for different purposes such as clustering and classification with this number of models. Models are utilized in different engineering and science disciplines, and it is necessary to manage and analyze them using data science techniques to find hidden patterns [1]. Text, audio, image, and video are the forms of data that have drawn a great deal of attention in data science so far. However, model data structure has received little emphasis from the data science.

Due to the specific structure of models, it is not possible to directly use machine learning and data mining algorithms on them. Since many models have textual components, text mining become necessary to handle the texts' implicit structure to perform machine learning algorithms correctly. Text mining includes two basic steps: pre-processing and creating feature sets for data representation. There are two sorts of research commonly applied for text representation, indexing, and term weighting [2]. This paper employed Term Frequency-Inverted Document Frequency (TF-IDF) unsupervised term weighting and word2vec neural language representation methods after the pre-processing for model vectorization. Then, Logistic Regression (LR), Naïve Bayes (NB), k Nearest Neighbors (kNN), Support Vector Machine (SVM), Random Forest (RF), and Artificial Neural Network (ANN) classifiers have been experimented on classification standalone and also together with structural features (number of classes, methods, attributes, and association links, weighted methods and attributes per class, depth of inheritance tree, number of children).

This study contributes to the literature by experimenting with different classifiers with different feature representation strategies. It applies Term Frequency-Inverted Document Frequency (TF-IDF) unsupervised term weighting and word2vec neural language representation methods. Moreover, it is the first attempt to use hybrid feature vector including both textual and structural features for metamodel classification task to the best of our knowledge. As a result, it was concluded that using utilizing machine learning algorithms on hybrid feature vectors significantly improve the classification performance of textual features confirming the motivation point of the paper.

The paper consists of the following sections. Section 2 reviews the literature, whereas Sect. 3 gives information about the dataset, data pre-processing steps, feature engineering, and methods used in the study. Section 4 presents the details of the experimental study with metrics and results. Finally, Sect. 5 draws a conclusion.

2   Literature Review

There are many studies implementing model categorization in the literature. It has been observed what kind of information (textual or structural) does the existing methods employ for learning operations. In this respect, previous studies can be classified into two categories, i.e., analysis of textual information and analysis of structural information.

Basciani et al. [3] proposed a hierarchical clustering method for metamodels. Although this paper benefits from a vector-based learning method, the description of a metamodel is considered an ordinary text, and models are compared based on completely textual information for clustering. More advanced techniques inspired by information retrieval and Natural Language Processing (NLP) were employed by Babur [4] to extract features, develop the vector space, and finally evaluate the proposed method through clustering. This study also failed to put sufficient emphasis on the structure of the models. In another study, Babur and Cleophas [5] experimented neural network classifier on a dataset of 555 metamodels.. The experiments were conducted in n-gram states for feature extraction.

Addressing clone detection in metamodels, the paper presented by Babur [6] retrieves the input metamodels for clustering. In every cluster, similar fragments were then sought separately. In other words, an n-gram (n = 2) was extracted from the corresponding graph of the metamodel and then stored in vectors. After that, clones were detected through comparisons drawn between those bigrams. Similarly, metamodels were first clustered in the study of Babur et al. [7]; however, subtrees of depth one were used for clone detection. The extracted features contained only the textual information of metamodels and included no graph structures.

Literature search shows that the current studies on model categorization mainly use textual or structural contents separately [23]. Therefore, we used text embedding techniques in combination with kernel-based approaches covering both textual and structural information in the model management to obtain high precision. Overall, the determined features were used to improve machine learning algorithms.

3 Proposed Approach

To address the challenge of using textual information of the models in learning, the proposed approach aims to vectorize to embed the textual information of each model in the resultant vector. The ultimate goal is to enhance accuracy in learning the models developed through machine learning algorithms. This section discusses establishing vectors through models based on textual information in two (related) phases: pre-processing and feature engineering. The Fig. 1 illustrates the proposed framework.

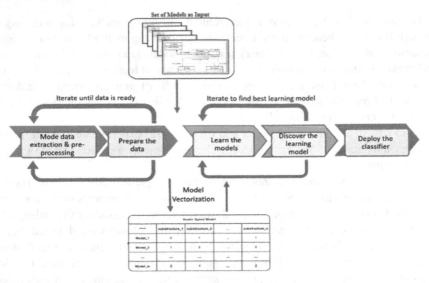

Fig. 1. The proposed approach processes.

3.1 Pre-processing

Step 1: In this step, we formed the raw data in the experimental dataset using PyEcore [8], a MDE framework written for Python. It allows to handle metamodels, the information of classes, references, attributes, and methods were easily obtained. The second step aims to find domain-specific and context-sensitive similarities and semantic relations in the best way possible. Therefore, after lexical refinement through NLP techniques (e.g., tokenization, stemming), the proposed method considers semantic similarity rather than apparent similarity of words. Therefore, WordNet [9], which is an extensive lexical database of English, is used to assess the similarity of words and elements within the related domain. We used WordNet synset instances which are the groupings of synonymous lemmas expressing the same concept, to expand certain lemmas.

By conducting step 1 on the running example (see Fig. 2.), metamodels 1, 2, and 3 of the same domains (the education domain) are detected, whereas Student, Pupil, and Educate elements refer to a common element (Student) in that domain.

Step 2: After detecting semantic similarities, similar elements of the same domain are detected among all models. The common element has a name resembling all similar elements and includes all of the fields existing in common elements (common and uncommon fields). Hence, there are many similar elements in the same-domain models with respect to the common element. They are similar to the common element in different but close ratios. At the end of this step, all common elements of all similar elements are recognized.

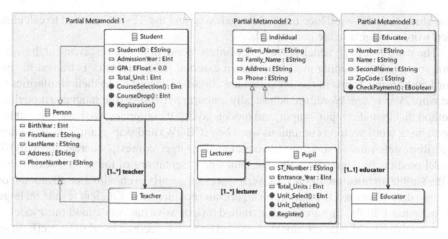

Fig. 2. The partial metamodel examples

In the running example, this step resulted in extracting a common element from three similar elements (Student, Pupil, and Educatee). Figure 3 shows the common element.

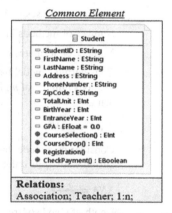

Fig. 3. The common element of Student, Pupil, and Educatee concepts

3.2 Feature Vectorization

Unstructured textual data in models is challenging to process and needs to be described by term sets to represent their contents. The vector space method [10] is one of the most used text representation models for a host of information retrieval operations. This method also appeals to the underlying metaphor of practicing spatial proximity for semantic proximity [11]. There are two sorts of research commonly applied for text representation: indexing and term weighting. Indexing assigns indexing terms for documents, whereas term weighting assigns each term's weight to show its importance.

This study uses the word2vec method for indexing and the TF-IDF method to calculate each word's weights in the metamodels.

The word2vec [12], which represents words as continuous vectors, is one of the most common word embedding models. After an external neural network is trained for the word embedding, terms in the document are classified according to their similarities in the word2Vec space. Word2vec is basically a neural network-based language modeling method that includes input, output, and hidden layers. It comprises two basic algorithms for training word vectors: continuous word bag (CBOW) and skip-gram. The skip-gram algorithm determines the terms surrounding the target context, whereas the CBOW model predicts by aggregating the distributed representations of the main context. Due to the simple architecture of the CBOW, it works efficiently even with a small amount of training data. However, the skip-gram algorithm provides more efficient results on large datasets than CBOW. The ability to be trained on extensive datasets allows this model to learn complex word relationships such as vec(Ecore) + vec(metamodel) $\cong$ vec(Eclipse Modeling Framework).

Table 1. Structural features of the models

Structural feature	Description
Number of classes	Number of classes in the metamodel
Number of methods	Number of methods in the metamodel
Number of attributes	Number of attributes in the metamodel
Average number of methods	Considering a model having i classes and c_i with methods number of m_i *then* the metric is calculated using: $$ANM = \frac{\sum_0^i m_i}{i}$$
Average number of attributes	Considering a model having i classes and c_i with attributes of a_i *then* the metric is calculated using: $$ANA = \frac{\sum_0^i a_i}{i}$$
Depth of inheritance tree	Depth of a node of tree refers to the length of the maximal path from the node to the root of the tree
Number of children	Number of immediate descendants of the class
Number of association links	Number of association relationships of the class
Average number of association links	Considering a model having i classes and c_i with association relationships of r_i *then* the metric is calculated using: $$ANR = \frac{\sum_0^i ar_i}{i}$$

TF-IDF is obtained by multiplying the term frequency (TF) and inverse document frequency (IDF) for a term in the text [11]. While TF gives the occurrence frequency of a word in the document, the value IDF (inverse document frequency) indicates this term's occurrence frequency in other documents. The main idea in TF-IDF is to classify terms as much as possible into the same category considering their high appearance in one document and high absence in other documents. When a term appears with a high TF frequency in a text document and rarely appears with low IDF frequency in other documents, it is accepted that the term has a good classification accuracy.

In this paper, we also included the structural metrics during classification process of Ecore diagrams presented in [13] number of classes, number of methods, number of attributes, average number of methods, average number of attributes, depth of inheritance tree, number of children, number of association links, and average number of association links.

3.3 Baseline Machine Learning Algorithms

After the pre-processing and feature selection steps were utilized, some baseline algorithms, commonly used to classify the textual data, were implemented in the first part of this section.

Logistic Regression (LR): LR is used to analyze a data set within one or more independent features [14]. This supervised learning method assigns a new sample to one of the specified discrete classes by employing a logistic function. Logistic regression is a statistical method used to analyze a data set within one or more independent features determining a result.

Naive Bayes (NB): NB depends on the common principle of Bayes Theorem, i.e., a distinct feature in a class is independent of any other feature's presence [15]. It describes the probability of an event based on prior knowledge of conditions. The two main advantages of NB are not requiring a large amount of training data and being able to train comparatively fast than sophisticated models.

k-Nearest Neighbor (kNN): kNN is a supervised learning method that classifies unlabeled observations by assigning them to the label of the most similar k neighbors. The distance metric to find the nearest neighbors of the active instance can be calculated by different methods such as Euclidean, Manhattan, Minkowski, and Hamming.

Support Vector Machine (SVM): SVM trademarked by Vapnik [16] is one of the widely used learning-based pattern classification techniques for classification, regression, and outlier detection. It depends on a solid theoretical background built on statistical learning theory and structural risk minimization techniques. The primary purpose of support vector machines is to find a function in a multidimensional space that can separate the data by a maximal margin.

Random Forest (RF): RF is a commonly used machine learning algorithm merging the output of many classification trees to achieve a single result [17]. The input from samples in the initial dataset is loaded into each decision tree. The prediction of a tree having the most votes is chosen as the outcome. It enables any classifiers with weak correlations to create a robust classifier.

Artificial Neural Network (ANN): ANN is one of the commonly used machine learning algorithms which adopts brain-style information processing consisting of neurons [18]. It has multiple connected layers of nodes with weights and activation functions. The network's processing unit is divided into input, output, and hidden layers. The input layer accepts input data, hidden layers process this instance, and the output layer assembles the result of the system processing result. The power to manage noisy and missing data makes the ANN preferable in data science research.

4 Experimental Study

In this study, we demonstrated our approach on a labeled Ecore metamodel dataset [19] for domain clustering, including 555 models from 9 different categories (bibliography, conference management, bug/issue tracker, build systems, document/office products, requirement/use case, database/sql, state machines, and petri nets). First, data preparation was utilized to transform the raw data (texts in the models) in a useful and efficient format. After pre-processing, two feature representation models (TF-IDF and word2vec) were proposed to observe their effects on categorization accuracy. For the second evaluation, structural features included in textual features and the performance of the machine learning algorithms were observed on the hybrid feature representation. The evaluation results of each method were obtained by dividing the data set into ten pieces by cross-validation.

4.1 Evaluation Metrics

The precision (pr) is obtained by the ratio of the correct data to the total data, and it is calculated according to Eq. 1. TP (true positive) denotes the number of objects that are correctly extracted by the system. FP (false positive) refers to the number of objects that the system confirms as true when it is not.

$$precision(pr) = \frac{TP}{TP + FP} \tag{1}$$

The recall (re)metric is calculated by the ratio of the correct data to the expected accurate data given in Eq. 2. FN (false negatives) in the equation refers to the number of correct data which could not be found.

$$recall(re) = \frac{TP}{TP + FP} \tag{2}$$

F1-measure of the proposed approach is defined as the harmonic means of the precision and recall values as demonstrated in Eq. 3.

$$F1 - measure = \frac{2 \times pr \times re}{pr + re} \tag{3}$$

4.2 Implementation Details

Firstly, we formed the raw data in the experimental dataset by using PyEcore, an MDE framework written for Python. The machine learning algorithms were designed using the Python Scikit-learn library [20], which provides a high-level construct for classifiers efficiently. It is built on NumPy, SciPy, and Matplotlib. Other Python libraries used in the study are Pandas [21] and Gensim [22]. NumPy is a widely used library for its fast-mathematical computation on arrays and matrices. Pandas library provides rapid, flexible, and expressive data structures. It is an ideal tool for cleaning, modeling, and organizing the analysis results appropriately. Gensim is a Python library generally utilized on large datasets for topic modeling, document indexing, and similarity detection.

4.3 Experimental Results

Table 2 compares the F1-Measure scores obtained using algorithms with TF-IDF and word2vec (CBOW and skip-gram) representations. Considering the experimental results, the highest F1-Measure values were achieved by the ANN classifier in all feature representations. On the other hand, the NB classifier with two representations had the lowest F1-Measure values among all classifiers.

Table 2. Evaluation of different algorithms and feature representations (non-floating)

Algorithm	Feature representation model		Precision	Recall	F1-measure
LR	TF-IDF		0.73	0.76	0.74
	word2vec	skip-gram	0.50	0.66	0.57
		CBOW	0.34	0.58	0.43
NB	TF-IDF		0.80	0.83	0.74
	word2vec	skip-gram	0.53	0.54	0.54
		CBOW	0.59	0.60	0.60
kNN	TF-IDF		0.81	0.78	0.79
	word2vec	skip-gram	0.56	0.58	0.57
		CBOW	0.53	0.54	0.52
SVM	TF-IDF		0.83	0.81	0.82
	word2vec	skip-gram	0.50	0.67	0.57
		CBOW	0.34	0.58	0.43
RF	TF-IDF		0.80	0.82	0.81
	word2vec	skip-gram	0.69	0.74	0.71
		CBOW	0.59	0.66	0.62
ANN	**TF-IDF**		**0.86**	**0.86**	**0.86**
	word2vec	**skip-gram**	**0.70**	**0.75**	**0.72**
		CBOW	0.66	0.69	0.67

Another point to be noticed is that the algorithms with the TF-IDF encoding method performed better than ones with the word2vec method (see Fig. 4). According to our assumption, the poor performances of word2vec feature representations (both cbow and skip-gram) are based on the limited training data. ANN with TF-IDF was evaluated as the most accurate classifier with an 86.45% F1-Measure value. The closest criteria result to ANN were achieved by SVM with TF-IDF with an 82.28% F1-Measure value. On the other hand, TF IDF-based NB was the worst-performing algorithm with a 74.23% F1-Measure score.

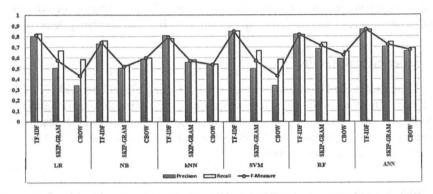

Fig. 4. Experimental results of machine learning classifiers based on TF-IDF and word2vec

The critical point of the experimental study results is that each experimented machine learning algorithms including structural features in addition to the textual features achieved better results on the dataset. For example, using the structural features expressed in Table 1 enhanced the ANN performance by 2.8% (F1-Measure) compared to inputting only textual features. As a result, the study's motivation, "creating hybrid vectors including both textual and structural features could increase the accuracy score of model management." is validated in the second experimented study in the paper (see Table 3).

Table 3. Comparison of textual and hybrid feature representations (non-floating)

Algorithm	Textual features (TF-IDF)			Textual + Structural features		
	Pre	Re	F1-measure	Pre	Re	F1-measure
LR	0.73	0.76	0.74	0.76	0.78	0.77
NB	0.80	0.83	0.81	0.82	0.84	0.83
kNN	0.81	0.78	0.79	0.84	0.80	0.82
SVM	0.83	0.81	0.82	0.84	0.85	0.85
RF	0.80	0.82	0.81	0.83	0.84	0.84
ANN	**0.86**	**0.86**	**0.86**	**0.88**	**0.90**	**0.89**

5   Conclusion

Due to a large number of models available nowadays, it is now more necessary than ever to employ intelligent methods to manage these models and their repositories. Using machine learning techniques, intelligent methods can properly meet this need. In this study, the textual information of each model is first summarized in its elements through text processing and NLP techniques, as well as the ontology of synonyms within a specific domain. The final results were formed as vectors (TF-IDF and word2vec), in which the columns and rows refer to features and models, respectively. Finally, six machine learning classifiers (LR, NB, kNN, SVM, RF, and ANN) with TF-IDF and word2vec feature representations were experimented with to detect models' categories. Experimental results showed that the best-performing method is ANN with TF-IDF weighting scheme, and it achieved 86.45% F1-Measure score. In second experimental study, the performances of machine learning classifiers were observed on different on two different variants of the datasets. The first variant includes only textual features (represented both in TF-IDF and word2vec representations), whereas the second variant consists of the determined structural features and textual features. ANN algorithm was again the most achieved classifier on hybrid vectors with an 89.15% F1-Measure score. It was proved that each machine learning algorithm showed a more successful performance scores in a hybrid feature vector than a pure textual one. The other considerable point is that word2vec based machine learning classifiers showed poor performance in terms of F1-Measure compared to the TF-IDF term weighting scheme. Considering the results, it can be concluded that this study appears promising for future studies on model categorization. As future work, we plan to apply these techniques on industrial models in Model-Driven Engineering [24, 25].

References

1. Tekinerdogan, B., Babur, Ö., Cleophas, L., van den Brand, M., Akşit, M.: Introduction to model management and analytics. In: Model Management and Analytics for Large Scale Systems, pp. 3–11. Elsevier (2020)
2. Harish, B.S., Guru, D.S., Manjunath, S.: Representation and classification of text documents: a brief review. Int. J. Comput. Appl. **2**, 110–119 (2010)
3. Basciani, F., Rocco, J., Ruscio, D., Iovino, L., Pierantonio, A.: Automated clustering of metamodel repositories. In: Nurcan, S., Soffer, P., Bajec, M., Eder, J. (eds.) CAiSE 2016. LNCS, vol. 9694, pp. 342–358. Springer, Cham (2016). https://doi.org/10.1007/978-3-319-39696-5_21
4. Babur, O.: Statistical analysis of large sets of models. In: 2016 31st IEEE/ACM International Conference on Automated Software Engineering (ASE), pp. 888–891. IEEE, Singapore (2016)
5. Babur, Ö., Cleophas, L.: Using n-grams for the automated clustering of structural models. In: Steffen, B., Baier, C., van den Brand, M., Eder, J., Hinchey, M., Margaria, T. (eds.) SOFSEM 2017. LNCS, vol. 10139, pp. 510–524. Springer, Cham (2017). https://doi.org/10.1007/978-3-319-51963-0_40
6. Babur, O.: Clone detection for Ecore metamodels using n-grams. In: Proceedings of the 6th International Conference on Model-Driven Engineering and Software Development, MODELSWARD 2018, pp. 411–219. SciTePress, Portugal (2018)

7. Babur, O., Cleophas, L., Brand, M.: Metamodel clone detection with Samos. J. Comput. Lang. **51**, 57–74 (2019)
8. Steinberg, D., Budinsky, F., Merks, E., Paternostro, M.: EMF: Eclipse Modeling Framework. Pearson Education (2008)
9. Fellbaum, C.: WordNet. In: Poli, R., Healy, M., Kameas, A. (eds.) Theory and Applications of Ontology: Computer Applications, pp. 231–243. Springer, Dordrecht (2010). https://doi.org/10.1007/978-90-481-8847-5_10
10. Salton, G., Wong, A., Yang, C.S.: A vector space model for automatic indexing. Commun. ACM **18**(11), 613–620 (1975)
11. Zhang, W., Yoshida, T., Tang, X.: A comparative study of TF*IDF, LSI, and multi-words for text classification. Exp. Syst. Appl. **38**(3), 2758–2765 (2011)
12. Church, K.W.: Word2vec. Nat. Lang. Eng. **23**(1), 155–162 (2017)
13. Chidamber, S.R., Kemerer, C.F.: A metrics suite for object-oriented design. IEEE Trans. Softw. Eng. **20**(6), 293–318 (1994)
14. Bozyiğit, A., Utku, S., Nasibov, E.: Cyberbullying detection: utilizing social media features. Exp. Syst. Appl. **179**, 115001 (2021)
15. Bozyiğit, A., Utku, S., and Nasibov, E.: Cyberbullying detection by using artificial neural network models. In: 2019 4th International Conference on Computer Science and Engineering, pp. 520–524 (2019)
16. Cortes, C., Vapnik, V.: Support-vector networks. Mach. Learn. **20**(3), 273–297 (1995)
17. Basaran, K., Bozyiğit, F., Siano, P., Taser, P., Kilinc, D.: Systematic literature review of photovoltaic output power forecasting. IET Renew. Power Gener. **14**(19), 3961–3973 (2020)
18. Mishra, M., Srivastava, M.: A view of artificial neural network. In: 2014 International Conference on Advances in Engineering & Technology Research, pp. 1–3 (2014)
19. Babur, O.: A labeled ecore metamodel dataset for domain clustering (2019)
20. Pedregosa, F., et al.: Scikit-learn: machine learning in Python. J. Mach. Learn. Res. **12**, 2825–2830 (2011)
21. McKinney, W.: Pandas: a foundational Python library for data analysis and statistics. Python High Perform. Sci. Comput. **14**(9), 1–9 (2011)
22. Srinivasa-Desikan, B.: Natural Language Processing and Computational Linguistics: A Practical Guide to Text Analysis with Python, Gensim, spaCy, and Keras. Packt Publishing Ltd., Birmingham (2018)
23. Khalilipour, A., Bozyigit, F., Utku, C., Challenger, M.: Categorization of the models based on structural information extraction and machine learning. In: Cengiz Kahraman, A., Tolga, C., Onar, S.C., Cebi, S., Oztaysi, B., Sari, I.U. (eds.) Intelligent and Fuzzy Systems: Digital Acceleration and The New Normal - Proceedings of the INFUS 2022 Conference, Volume 2, pp. 173–181. Springer, Cham (2022). https://doi.org/10.1007/978-3-031-09176-6_21
24. Challenger, M., Erata, F., Onat, M., Gezgen, H., Kardas, G.: A model-driven engineering technique for developing composite content applications. In: 5th Symposium on Languages, Applications and Technologies, SLATE 2016, pp. 11:1–11:10 (2016)
25. Asici, TZ., Karaduman, B., Eslampanah, R., Challenger, M., Denil, J., Vangheluwe, H.: Applying model driven engineering techniques to the development of Contiki-based IoT systems. In: IEEE/ACM 1st International Workshop on Software Engineering Research & Practices for the Internet of Things (SERP4IoT), pp. 25–32 (2019)

Architecture Design of a Networked Music Performance Platform for a Chamber Choir

Jan Cychnerski$^{(\boxtimes)}$ (iD) and Bartłomiej Mróz (iD)

Faculty of Electronics, Telecommunications and Informatics,
Gdańsk University of Technology, Gdańsk, Poland
jan.cychnerski@eti.pg.edu.pl, bartlomiej.mroz@pg.edu.pl

Abstract. This paper describes an architecture design process for Networked Music Performance (NMP) platform for medium-sized conducted music ensembles, based on remote rehearsals of Academic Choir of Gdańsk University of Technology. The issues of real-time remote communication, in-person music performance, and NMP are described. Three iterative steps defining and extending the architecture of the NMP platform with additional features to enhance its utility in remote rehearsals are presented. The first iteration uses a regular video conferencing platform, the second iteration uses dedicated NMP devices and tools, and the third iteration adds video transmission and utilizes professional low-latency audio and video workstations. For each iteration, the platform architecture is defined and deployed with simultaneous usability tests. Its strengths and weaknesses are identified through qualitative and quantitative measurements – statistical analysis shows a significant improvement in rehearsal quality after each iteration. The final optimal architecture is described and concluded with guidelines for creating NMP systems for said music ensembles.

Keywords: Real-time system · Networked music performance · Platform design

1 Introduction and Related Work

In recent years, advances in technology have resulted in the increasing use of real-time remote working devices. Also, the COVID-19 pandemic made many aspects of everyday life transfer to the virtual world. For some fields of work this change does not affect the efficiency and quality of work. Nevertheless, remote creative and artistic work, particularly music performance, remains a challenge.

1.1 Real-Time Communication

Recently, various real-time conferencing tools (e.g. Zoom, MS Teams, Google Meet) rapidly expanded and amply gained new users [16]. The success of conferencing platforms was largely caused by the general features that make typical Internet communication simple and effective, which include [8,10]: ease of

© Springer Nature Switzerland AG 2022
S. Chiusano et al. (Eds.): ADBIS 2022, CCIS 1652, pp. 437–449, 2022.
https://doi.org/10.1007/978-3-031-15743-1_40

installation, convenience of use, ability too operate under varying network conditions, resistance to audio noise (low quality microphone, background noises), video and screen sharing, chat, file transfer, shared whiteboard, integration with office/school software.

In order to achieve the aforementioned goals, these applications must exhibit very high automation and compatibility, which is achieved at a cost of [13]:

1. long buffers, high transmission delays reaching several hundred milliseconds
2. very aggressive noise reduction, cutting out a large part of the bandwidth, and even changing the audio signal content using AI
3. strong, automated emphasis on the main speaker while muting the others

1.2   In-Person Music Performance

A different set of requirements is posed by music performance in typical medium-sized ensembles (e.g. chamber choirs), especially those with a conductor. During normal stationary in-person rehearsals, communication is characterized by [14, 18, 20]:

1. no audiovisual delays of any kind – the performers are in the same room
2. the performers hear each other and adjust their performance in real time to the close performers and the overall sound of the ensemble (self-feedback)
3. a conductor in front of the performers ensures overall cohesion, interpretation, volume and tempo, conducting visually in real time
4. the physical location of the performers matters – people standing closer (so performing similar musical parts) are heard louder (intra-section feedback)

1.3   Networked Music Performance

The features of typical conferencing tools are highly inadequate for a music performance. For this reason, the topic of remote music performance is a separate branch in research and industry, called Networked Music Performance [9]. In recent years, advances in technology have resulted in the increasing use of real-time remote working devices. Nevertheless, remote creative and artistic work, particularly music performance, remains a challenge. One-way connections, through which artistic events are transmitted in real-time, have become available using networks capable of supporting high bandwidth, low-latency packet routing, and guaranteed quality of service (QoS) [5, 7]. These can be described as unidirectional, allowing artists to remotely connect only between the performance venue and the audience, not each other. It is much more difficult when performers in two or more remote locations attempt to perform an established composition or improvisation together in real-time. Weinberg [21] calls this way of performing music the "bridge approach". In such cases, the inevitable delay caused by the physical transit time of network packets is known to affect performance [1, 3, 6, 12]. Therefore, attempts have been made to account for these delays by design or composition [2, 4].

A number of NMP experiments and frameworks are described in [15,17, 19]. Some platforms, like MusiNet or Diamouses, are large-scale projects run by universities; other, like LoLa or JackTrip require a subscription or licensing plan. The Jamulus platform stands out as an open-source, community-driven and easy to use NMP tool [11]. It allows joint performance via freely available servers, or via dedicated server configured with Jamulus software on a local machine. These aspects are especially appealing to smaller, non-professional ensembles (e.g. students, local communities, etc.) which are not capable of huge financial investments or involvement in research projects regarding remote rehearsing.

1.4 Contribution

This paper discusses the process of defining and practically realizing a platform architecture that meets the requirements for networked music performance of medium-sized conducted amateur music ensembles through the example of remote rehearsals of a chamber choir. The proposed architecture in contrast to existing publications and commercial solutions takes into account: (1) feasibility study in a highly ecologically valid, choral environment (vs artificial environments in NMP papers [19]), (2) use of available, open-source platforms for low-latency transmission of both audio and video (vs difficult to obtain, scientific projects like Internet2, Musinet, Diamouses, GigaPoP, etc.), (3) consideration for the lack of technical capabilities of performers, (4) consideration for a low-cost hardware options, (5) proposal for an architecture for the NMP and a set of good practices.

2 Experimental Setup

In the experiment, 3 NMP architectures were proposed and investigated for a ~20-member choral ensemble: Zoom@Home, Jamulus@Home and Jamulus@Univ. Each architecture was implemented in practice as means for the rehearsals of the Academic Choir of the Gdansk University of Technology, during the 2021 COVID-19 outbreak in Poland. After deployment, the architectures were used without major modifications for nearly 2 months each. Architectures were evaluated qualitatively by the choristers through a survey and by the authors responsible for the implementation. Quantitative assessment was performed through a round trip time (RTT) audio and video latency measurements. After each deployment, various aspects affecting rehearsal quality were determined, and based on them, decisions were made to implement modifications in the next iteration. Comparative features of all architectures are shown in Table 1, and described in detail in the following subsections.

2.1 Zoom@Home Architecture: A Simple Conferencing Tool

The first proposed architecture for remote choir rehearsals was a standard web-based communication platform using popular tools. Therefore, Skype, Jitsi, MS

Table 1. Comparison of proposed architecture features

Architecture	Zoom@Home	Jamulus@Home	Jamulus@Univ
Connection	Any	Broadband	LAN
Choir audio	Any	Semi-professional; ASIO drivers	Professional; ASIO interface
Conductor audio	Any	Professional, using ASIO interface	
Audio RTT	300–1000 ms	63–135 ms	40–85 ms
Choir video	Any	–	–
Conductor video	Any	–	Low latency camera
Video latency	500–1000 ms	–	25–100 ms
Feedback	–	Self & intra-section feedback, choir mix	
Setup	Automatic, self-supervised	Manual	Manual, semi-supervised
Assistance	–	Remote	Remote + local
Rehearsal scope	Individual singing with muted mic, solo singing	singing section parts or tutti with piano	Jamulus@Home + singing tutti a'cappella

Teams and Zoom platforms were initially tested. Eventually Zoom was selected due to its popularity, ease of use, compatibility and very low entry threshold. No changes were made to the typical conferencing architecture. Simultaneous singing was effectively impossible due to significant delays and the automated speaker highlight feature that amplifies one person and mutes the others. The choir could rehearse only in the following capacities: (1) casual conversation as in a typical conference, where everyone has microphone on and everyone can talk, (2) singing together with microphones turned off; the only active microphone is the conductor's, who leads the rehearsal by playing choral parts on the piano, and (3) solo singing with occasional help and commentary from the conductor. From a technical standpoint, the Zoom@Home architecture posed no problems. There were no requirements for hardware, Internet connection, or location. Installation and configuration were possible on any device; each chorister configured the Zoom application without technical assistance; most choristers used a webcam to enhance the feeling of presence at a rehearsal. RTT latencies ranged from 300–1000 ms for the audio channel and 500–1000 ms for the video channel. Such high RTTs prevented more than one person from singing at a time, making it impossible to measure relative latency between choristers.

2.2   Jamulus@Home Architecture: NMP with Choristers at Home

The second iteration of the architecture – presented in Fig. 1 – introduced several improvements. Jamulus software was chosen due to its main features: opensource, cross-platform, built by the NMP community, focused on low latency audio connection, no limitations on the number of participants, and no requirements for specialized hardware. Several inexpensive headsets were reviewed for latency, compatibility, sound quality, and comfort; then one optimal model (Logitech PC 960 USB) was selected. Additionally, semi-professional low-latency compatible audio equipment was allowed. To ensure minimal delays and high quality of the conductor's audio connection, he was provided with professional audio equipment. The video connection got discarded, further lowering the latency; a dedicated Jamulus server was deployed at the University, ensuring broadband connection in the nearest area. A requirement was imposed for a minimum internet connection type (wired broadband) and a maximum geographical distance from the Jamulus server (100 km). A correctly configured low-latency ASIO driver was required. That resulted in RTT in the range of 63–135 ms, which allowed for real-time collaborative singing.

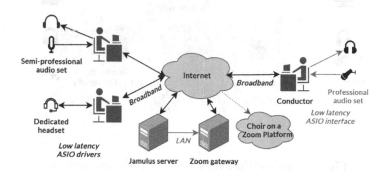

Fig. 1. Jamulus@Home NMP platform architecture

Despite attempts to unify the hardware and imposed requirements, the hardware and software configuration had to be done manually by users, with active remote support (via Windows Remote Assistance or TeamViewer) provided by two designated technicians. The real-time NMP solution required a complete revision of remote workflow. Simultaneous singing became viable – it allowed the whole ensemble to sing together (tutti) for the first time under pandemic restrictions. Mechanisms included in the Jamulus software enabled the choristers to adjust the volume of particular persons, allowing them for intra-section feedback (strengthening their sections and weakening the rest); similar to traditional rehearsals. The conductor could communicate with the choristers on a real-time basis and lead through accompaniment on a piano. It became possible to rehearse in sections and hear the overall sound of the pieces.

Not all choristers could meet the requirements of NMP (equipment, connectivity, etc.). A solution to this problem was an additional gateway server to the Zoom platform. As a countermeasure for high delays in the audio signal of the Zoom platform, the communication with the gateway server was one-way. Thus, it allowed choristers not capable of NMP for passive participation in rehearsals.

2.3   Jamulus@University Architecture: Low-Latency Audio + Video, Choristers and Conductor at the University

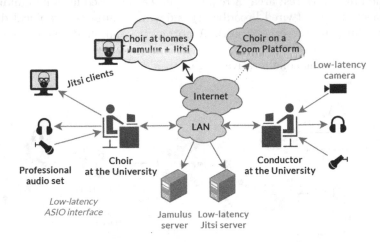

Fig. 2. Jamulus@University NMP platform architecture

Despite the ability to sing together in the previous iteration, persisting transmission delays caused frequent loss of synchronization between choristers, requiring the conductor to enforce the tempo through either accompaniment or counting aloud. Still, maintaining a steady performance was very difficult and unattainable without a conductor. Moreover, due to the lack of video input, it was impossible for the conductor to lead visually, making it impossible for him to control the tempo variability of the pieces (accelerations, decelerations, fermatas, tempo rubato).

In the third iteration of the architecture design, namely Jamulus@University, choristers got invited to a dedicated classroom at the university, equipped with pre-configured professional audio hardware, providing high quality audio with very low latency. The conductor's workstation was supplied with low-latency audio-video equipment and moved to another classroom with a piano. In order to ensure a low latency video connection, a dedicated Jitsi server was deployed, providing the conductor's video with a latency of 25–100 ms. Such a delay allowed the conductor to feasibly lead the choir in a real-time manner visually, which was previously inaccessible. All communication was done over a LAN, reducing the audio latency to 40–85 ms RTT. Controlled environment allowed creation of semi-automated scripts facilitating deployment of necessary software for workstations (2-person assistance was still in place, but in smaller capacity).

The Jamulus@University was the most extensive of the architectures tested, as presented in Fig. 2. It was used for two months of remote rehearsals and was crucial in sustaining choral activities during the COVID-19 lockdown.

3   Results

Each iteration contained a questionnaire assessing NMP architecture's usability. Since the remote rehearsals lasted several months, some choristers participated in several iterations. In total, 23 choir members (9 male, 14 female, aged 18–30) participated in the experiments. The Zoom@Home architecture assessment consists of 8 responses; the Jamulus@Home and the Jamulus@University assessments comprise 16 and 13 answers, respectively.

3.1   Assessment of the Zoom@Home Architecture

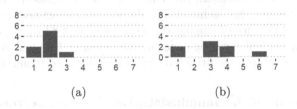

(a)   (b)

Fig. 3. Exercising difficulty and rehearsing comfort of the Zoom@Home NMP, respectively. Likert scale: 1–3 = more difficult/less comfortable, 4 = about the same, 5–7 = easier/more comfortable

The first examined was the Zoom@Home NMP architecture. As it did not differ much from typical conferencing platforms, only two rehearsal quality aspects were measured – exercising difficulty and rehearsing comfort, referring to traditional, non-remote rehearsals – as shown in Fig. 3. Most of the participants rated the difficulty as 2: more difficult than on traditional rehearsal. Some participants reported shyness with solo singing; the rating might reflect that intricacy. Interestingly, the rehearsing comfort was rated mainly as 3: a bit less comfortably, than on traditional rehearsal, and 4: about the same, as on traditional rehearsal. Many choristers reported high comfort in rehearsing at home, especially those with conducive conditions. However, some participants reported difficulties when singing at home; the most highlighted were other household members, neighbors getting easily irritated, or a noisy environment with construction or traffic noise from nearby places.

3.2   Assessment of the Jamulus@Home Architecture

The Jamulus@Home architecture was evaluated by the following quantitative factors: exercising difficulty and rehearsing comfort as compared to traditional, non-remote rehearsals, inter-chorister latency, and setup difficulty. Most participants rated the difficulty as 2: more difficult, and 3: a bit more difficult, than on traditional rehearsals. This rating appears slightly higher than for the previous platform. The ratings for rehearsing comfort are somewhat higher than for the

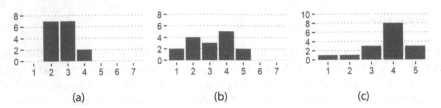

Fig. 4. Exercising difficulty, rehearsing comfort, and setup difficulty of Jamulus@Home NMP. Likert scale for (c): 1 = very difficult, a lot of help needed; 5 = very easy, everything highly understandable.

Zoom platform; most choristers rated the comfort as 4: about the same, as on traditional rehearsal. This architecture allowed inter-chorister latency measurement: the time difference of singing the same note was 83 ms ±57 ms. Furthermore, the setup difficulty rating concentrated around 4 – as shown in Fig. 4c – indicating minor setup difficulties with remote assistance required.

3.3   Assessment of the Jamulus@University Architecture

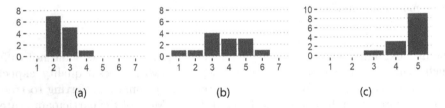

Fig. 5. Exercising difficulty, rehearsing comfort and safety of Jamulus@University. Likert scale for (3): 1 = strong fear of infection, 5 = no fear of infection

The Jamulus@University assessment comprised exercising difficulty and rehearsing comfort ratings in relation to traditional, non-remote rehearsals (Fig. 5), as well as inter-chorister latency. Most participants rated the difficulty as 2: more difficult, and 3: a bit more difficult, than on traditional rehearsals. The ratings for rehearsing comfort seem to be more spread but mainly concentrated around score 4: about the same, as on traditional rehearsal. Higher comfort resulted also from a reduced sense of shyness in these conditions. Additionally, the choristers were asked for a rating of perceived safety, since it involved spending a considerable amount of time in one classroom. The majority of the choristers rated the safety as 5: no fear of infection (Fig. 5c). Such a result might reflect the convenient placement of dedicated workstations in separate compartments made with plywood walls. This setup gave the choristers a level of separation, especially in terms of loudness and personal distance. It is also worth noting that no COVID-19 outbreak occurred within the choir at that time.

With reduced latency and the addition of video, it became possible to perform tutti a'cappella pieces without accompaniment, metronome, or other enforced tempo control. Overall, keeping the tempo became easier, even though some of the choisters were still using the Jamulus@Home platform. The time difference between different choristers singing the same note dropped to 47 ± 46 ms.

3.4 Statistical Analysis

All statistical calculations were performed with R software (version 4.1.2). The most notable packages were MASS (7.3–54), car (3.0–12), emmeans (1.7.1–1), ordinal (2019.12–10) and RVAideMemoire (0.9–80). The assessment of exercising difficulty and rehearsing comfort was repeated for each architecture's iteration, allowing direct comparison. Some participants answered more than one question; therefore, a generalized linear mixed model approach was selected. Since the answers were on a 7-point Likert scale, the Cumulative Link Mixed Model (CLMM) was chosen for the analysis. The dependent variable, Platform, being a categorical variable, was encoded using Helmert contrast coding. Also, the analysis was performed on both unweighted and weighted response data. The weights were derived from the number of participants and were assigned as follows: 1–2 rehearsals – 1; 3–6 rehearsals – 2; 7 and more rehearsals – 3. Results are shown in Fig. 6.

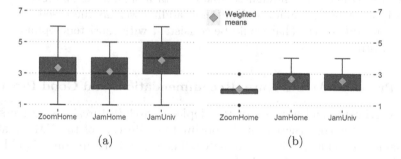

Fig. 6. Rehearsing comfort and exercising difficulty statistics

An analysis of variance for unweighted data based on ordinal logistic regression indicated no statistical effect on rehearsing comfort ($\chi^2(2,N=37)$ = 1.93, n.s.) However, testing on weighted data indicated a statistical effect ($\chi^2(2,N=86)$ = 10.84, $p < 0.01$). Since the statistical significance was detected on weighted data, post-hoc analysis was performed for this scenario. Pairwise comparisons using Z-tests, corrected with Holm's sequential Bonferroni procedure, indicated that Likert scores for Jamulus@Home vs. Jamulus@Univ were statistically significantly different (Z = −3.1, $p < 0.01$), but Likert scores for Jamulus@Home vs. Zoom@Home and for Jamulus@Univ vs. Zoom@Home were not significantly different (Z = −1.4, n.s., and Z = 1.7, n.s.).

The statistical analysis is consistent with what is shown in Fig. 6a - the rehearsing comfort is similar for participants in all three platforms. However, weighted Likert scores induced a statistically significant difference between Jamulus@Home platform and Jamulus@Univ platform, suggesting higher comfort of rehearsing for the latter.

An analysis of variance for unweighted data based on ordinal logistic regression indicated a statistical effect on exercising difficulty ($\chi^2(2$, N = 37) = 29.5, $p < 0.001$). For weighted data, this test also indicated a statistical effect ($\chi^2(2$, N = 86) = 46.4, $p < 0.001$). Thus, post-hoc analysis was performed for both cases. For unweighted Likert scores, pairwise comparisons using Z-tests, corrected with Holm's sequential Bonferroni procedure, indicated that Likert scores for Jamulus@Home vs. Jamulus@Univ were statistically significantly different (Z = −1910, $p < 0.001$), also Likert scores for Jamulus@Home vs. Zoom@Home and for Jamulus@Univ vs. Zoom@Home were statistically significantly different (Z = 24025, $p < 0.001$, and Z = 29152, $p < 0.001$). The comparisons for weighted Likert scores indicated the higher magnitude of statistically significant differences: Jamulus@Home vs. Jamulus@Univ (Z = −6494, $p < 0.001$), Jamulus@Home vs. Zoom@Home Z = 31852, $p < 0.001$), and Jamulus@Univ vs. Zoom@Home (Z = 42875, $p < 0.001$). The statistical analysis is somewhat inconsistent with what is in Fig. 6b – the exercising difficulty on the Zoom@Home platform is clearly higher than on any of the Jamulus platforms, but both Jamulus platforms seem to have the same exercising difficulty. However, the statistical analysis showed for both weighted and unweighted Likert data a significant difference between them, in favor of the Jamulus@Univ one; adding weights increased the magnitude of this difference. That would be consistent with choristers' opinions about this platform.

3.5 Proposed Architecture, Recommendations and Good Practices

The experiments and analysis of the deployed architectures' assessments concluded with the construction of the optimal architecture of the NMP platform for chamber ensembles with a conductor. The general architecture should fulfill the following requirements and good practices (also shown in Fig. 7):

1. affordable, cross-platform, low-latency (RTT not exceeding 100 ms) NMP software, along with a dedicated server located near the performers;
2. dedicated, low-latency (not exceeding 100 ms) video streaming server;
3. conductor's station with professional equipment, fast network, close to the servers, allowing unrestricted spoken, accompanied and visual conducting;
4. performer stations with professional audio equipment, located close to the servers, allowing for adjustable self-feedback, the ability to hear the overall ensemble's mix, with additional amplification of their own section;
5. performers who cannot use (4) should be equipped with dedicated audio equipment and a broadband Internet connection; unification of hardware reduces installation problems and allows for the creation of config scripts;

6. performers who cannot use (5) may participate in rehearsals passively (watch/listen only) through a one-way gateway to popular communicators;
7. remote assistance must be available for all performers at all times, according to the users technical skills; it is worthwhile to test the connection in groups to make sure that the configuration is correct and to fix problems in advance

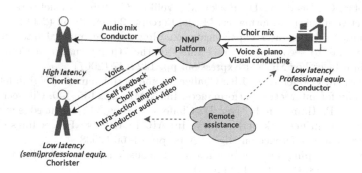

Fig. 7. Proposed generalized NMP platform architecture

4 Summary

This paper outlines three proposed Networked Music Performance architectures dedicated to chamber musical ensembles with a conductor, which were defined, deployed and examined in practice. In contrast to other NMP systems and publications, proposed architectures combine uncommon and conflicting requirements of medium-sized choirs, like intra-section feedback, conductor's real-time video, ease of deploy and more. Subsequent architecture defisn iterations were assessed against their usefulness. The conclusive architecture is described along with listed guidelines for its implementation in Sect. 3.5.

The Zoom@Home architecture was easy to use; however, it did not allow for rehearsals similar in quality to in-person rehearsals. The Jamulus@Home architecture, implementing software and hardware dedicated to NMP, allowed for actual remote real-time music performance. However, it required much effort from performers, the conductor, and extensive technical assistance. The final Jamulus@University architecture enabled visual input from the conductor and reduced delays between performers, effectively increasing the overall quality of rehearsals, and allowing the choir to effectively work on the new repertoire.

This work proves that it is possible and useful to implement NMP platforms for said musical ensembles using existing affordable tools and equipment. However, overall quality of rehearsals needs to be improved. Work needs to be done to decrease network and audio latencies by incorporating low-latency drivers into operating systems. A lot of effort must be put to lower the entry threshold for NMP tools: automatic configuration and volume control, intuitive user interfaces, integration of video and audio signals, remote assistance helpers. Enabling technologies like peer-to-peer communication, VST plugins, ambisonics and virtual reality techniques would improve musicians' experience even more, allowing to create better NMP platforms in the future.

References

1. Bartlette, C., Headlam, D., Bocko, M., Vellklc, G.: Effect of network latency on interactive musical performance. Music. Percept. **24**(1), 49–62 (2006)
2. Bouillot, N.: nJam user experiments: enabling remote musical interaction from milliseconds to seconds. In: Proceedings of the 7th international Conference on New interfaces For Musical Expression, pp. 142–147 (2007)
3. Bouillot, N., Cooperstock, J.R.: Challenges and performance of high-fidelity audio streaming for interactive performances. In: NIME, pp. 135–140. Citeseer (2009)
4. Cáceres, J.P., Hamilton, R., Iyer, D., Chafe, C., Wang, G.: To the edge with China: explorations in network performance. In: ARTECH 2008: Proceedings of the 4th International Conference on Digital Arts, pp. 61–66 (2008)
5. Chafe, C.: Tapping into the Internet as an acoustical/musical medium. Contemp. Music. Rev. **28**(4–5), 413–420 (2009)
6. Chafe, C., Caceres, J.P., Gurevich, M.: Effect of temporal separation on synchronization in rhythmic performance. Perception **39**(7), 982–992 (2010)
7. Chafe, C., Wilson, S., Leistikow, R., Chisholm, D., Scavone, G.: A simplified approach to high quality music and sound over IP. In: Proceedings of the COST G-6 Conference on Digital Audio Effects (DAFX-00), pp. 159–164. Citeseer (2000)
8. Correia, A.P., Liu, C., Xu, F.: Evaluating videoconferencing systems for the quality of the educational experience. Distance Educ. **41**(4), 429–452 (2020). https://doi.org/10.1080/01587919.2020.1821607
9. Dessen, M.: Networked music performance: an introduction for musicians and educators (2020). ujeb.se/mrEpS5. Accessed 15 May 2022
10. Faadhilah, A.F., Elfitri, I.: Comparison of audio quality of teleconferencing applications using subjective test. In: Audio Engineering Society Convention 152. Audio Engineering Society (2022)
11. Fischer, V.: Case study: Performing band rehearsals on the Internet with Jamulus (2015)
12. Gu, X., Dick, M., Kurtisi, Z., Noyer, U., Wolf, L.: Network-centric music performance: practice and experiments. IEEE Commun. **43**(6), 86–93 (2005)
13. Hoßfeld, T., Binzenhöfer, A.: Analysis of Skype VoIP traffic in UMTS: end-to-end QoS and QoE measurements. Comput. Netw. **52**(3), 650–666 (2008)
14. King, E.C.: Collaboration and the study of ensemble rehearsal. In: Eighth International Conference on Music Perception and Cognition (ICMPC8) (2004)
15. Mall, P., Kilian, J.: Low-latency online music tool in practice (2021)
16. Matulin, M., Mrvelj, Š, Abramović, B., Šoštarić, T., Čejvan, M.: User quality of experience comparison between Skype, Microsoft Teams and Zoom videoconferencing tools. In: Perakovic, D., Knapcikova, L. (eds.) FABULOUS 2021. LNICST, vol. 382, pp. 299–307. Springer, Cham (2021). https://doi.org/10.1007/978-3-030-78459-1_22
17. Onderdijk, K.E., Acar, F., Van Dyck, E.: Impact of lockdown measures on joint music making: playing online and physically together. Front. Psychol. **12**, 1364 (2021)

18. Pennill, N.: Ensembles working towards performance: emerging coordination and interactions in self-organised groups. Ph.D. thesis, University of Sheffield (2019)
19. Rottondi, C., Chafe, C., Allocchio, C., Sarti, A.: An overview on networked music performance technologies. IEEE Access **4**, 8823–8843 (2016)
20. Volpe, G., D'Ausilio, A., Badino, L., Camurri, A., Fadiga, L.: Measuring social interaction in music ensembles. Philos. Trans. R. Soc. B Biol. Sci. **371**(1693), 20150377 (2016)
21. Weinberg, G.: Interconnected musical networks: toward a theoretical framework. Comput. Music. J. **29**(2), 23–39 (2005)

Defining Software Architecture Modalities Based on Event Sourcing Architecture Pattern

Olga Jejić[(✉)] [iD], Milica Škembarević[iD], and Slađan Babarogić

Faculty of Organizational Sciences, University of Belgrade, Belgrade, Serbia
{olga.jejic,milica.skembarevic,sladjan.babarogic}@fon.bg.ac.rs

Abstract. The main focus of this paper is the development of data-intensive systems. One of the key issues is maintaining consistency while being able to promptly process frequent data requests. The central premise is that CQRS (*Command Query Responsibility Segregation*) and *Event sourcing* concepts can be utilized for addressing these challenges. To ensure the development of a system that would be capable of executing the immense number of required operations, yet at the same time provide the desired reliability, a software architecture (combining CQRS, *Event sourcing,* and the *Service Fabric*) is proposed based on a real-life project. Furthermore, several modalities of the architecture are defined to be used in different scenarios, depending on the volume of data that is to be processed. The presented modalities of the described software architecture were then implemented as a part of the information system that supports the organization and grading of exams for an immense number of candidates. Consequently, a large volume of data is generated, and the proposed architecture has proven best suited for reporting purposes which will be described in this paper.

Keywords: Architecture · CQRS · Event sourcing · Event store · Service fabric

1 Introduction

Well-designed software architecture is a prerequisite for the successful development of an application that satisfies both market needs and client requirements. This is the main rationale for adopting different architectural styles for different circumstances. Designing a system becomes even more challenging when it is expected that numerous clients will frequently query and update data during a relatively short period of time. To be able to track the state of the system at any point in time, it is necessary to establish mechanisms for storing all the data entering the system or originating within it. However, as complex systems do not usually rely on a single database it may be difficult to determine the source of truth. Moreover, to achieve its full potential, the system must have dedicated mechanisms for maintaining the consistency and accuracy of the large volume of data received through different system components (so that all segments of the system work with the same data). Conversely, the more functionality the system provides, the more complex and expensive it becomes for maintenance, management, and expansion. The main goal of this paper is to present a software architecture that can

© Springer Nature Switzerland AG 2022
S. Chiusano et al. (Eds.): ADBIS 2022, CCIS 1652, pp. 450–458, 2022.
https://doi.org/10.1007/978-3-031-15743-1_41

be used for developing modern complex systems which are based on events as the only source of truth.

The second part of this paper is dedicated to the proposed software architecture which is intended for systems that are focused on reporting. Several modalities of this architecture are provided to allow adequately handling different requirements related to the amount of data that needs to be processed. This software architecture has been implemented as a part of a data-intensive information system that is used for testing around 65.000 candidates. Each question on the test represents a task for the evaluators. In this case, tasks are graded by approximately 1.800 evaluators. The tests need to be scanned, and since they are organized into packages, packages first need to be imported. On average, there are around 4000 packages imported to the system in total, resulting in circa 200.000 tasks processed by the system and graded by the evaluators by the hour. The rest of this section gives an overview of the theoretical background related to the existing architectural styles and patterns that are incorporated into the proposed software architecture. Although they originated a long time ago, they have only recently started to be used more intensively given that they lead to better and more competitive systems [1]. Furthermore, they allow for processing large amounts of data to tackle the issues related to maintaining consistency [2] and resolving inevitable failures.

1.1 Command Query Responsibility Segregation - CQRS

The CQRS pattern evolved from the CQS (*Command-Query Separation*) pattern which was introduced in [3]. The main idea behind the CQS pattern is to detach commands from queries. Queries are actions that just provide pieces of information about the stored objects, while commands represent actions that create or change objects in the database. When the CQS pattern is utilized, the whole system becomes more reliable and the modularity and reusability of systems' components increase. CQRS represented a big step on the macro scale (compared to CQS which is considered to be a micro-scale pattern) towards the separation of the querying and the data managing responsibilities thereby allowing for the system to function properly even with larger payloads. According to [4], the adoption of CQRS can entail the following stages:

1. Assessing the typical application data access (Fig. 1a)
2. Separating the read and write APIs
3. Separating the read and write models
4. Separating the read and write database (Fig. 1b)

One of the main recommendations is to segregate the database i.e., maintain separate databases for read and write operations. This requires a mediator for the synchronization of those databases. It is important to highlight that CQRS does not support the *Read your own writes* principle. All commands in CQRS follow the *fire-and-forget* principle.

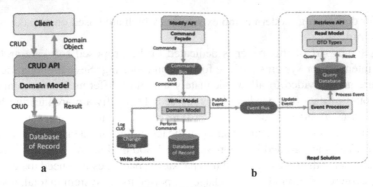

Fig. 1. CQRS stages

1.2 Event Sourcing

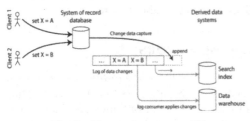

Fig. 2. Change data capture [6]

The notion of *Event sourcing* originated within the Domain-Driven Design
(DDD) community and is used alongside CQRS [5]. The main idea behind
this pattern is that object states should
be persisted as a sequence of immutable
state-changing events which are considered to be the only source of truth. Every
event that occurs in the system is stored
in a log. Each event is stored in the log with a unique identifier that is based on the
time of its occurrence. The log is hidden from the users, but they can use it indirectly
to restore the database to a previous state and detect the exact place and time where
an error occurred. Another relevant concept is known as *change data capture* (depicted
in Fig. 2) which is the process of observing all data changes written to a database and
extracting them in a form in which they can be replicated to other systems [6]. The *Event
sourcing* pattern is tightly coupled with CQRS (the *Event sourcing pattern* must be used
with CQRS, but CQRS can be used without the *Event sourcing* pattern).

1.2.1 Event Store

An *Event store* is a database that is used to store events. Relational databases are generally
not appropriate for storing events because they only adequately store the latest state
of an object. It is possible to improve the relational database by implementing some
mechanisms that are used as a substitute for databases that store events. Each extension
on the relational database is very expensive and requires a lot of developer time. In
the *Event sourcing* pattern, all the events related to an object need to be stored. The
mechanism that is used for this purpose is a *stream* – an ordered sequence of events
related to a single object in the system. Events are written to a stream in the same order
in which they occur, and it is not possible to edit or delete records in the stream. A stream

has its own unique identifier, and each event is marked with a timestamp and (optionally) metadata that carries information about the action that triggered the event. The three most important pieces of information that need to be stored in the *Event store* are [7]: Event type, Sequential event number, and Serialized data (e.g., in a JSON format).To obtain the current state of an object, it is necessary to go through each event in the stream that is associated with that object. This might be acceptable for streams with a small number of events, which is rarely the case. For a stream with numerous events, a new solution emerged: a *snapshot*. A *snapshot* represents the state of an object at a certain point in time in which all the previous events from the stream have been read [1]. A *snapshot* can be taken every time an event is written, after a certain number of events, or after a certain period. The *Event store* also provides projections [8]. Projections refer to a different representation of data than its original version. They do not manipulate the original data (events). What is obtained as a projection, from the angle of relational databases, is a view (queried data from database) that has no state, does not change the data, and thus has no impact on the system.

1.3 Mediator

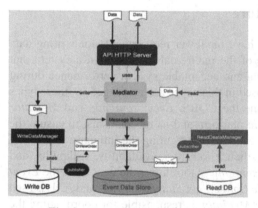

Fig. 3. The Mediator pattern [9]

The purpose of the *Mediator* [9] is to route and synchronize write and read requests. When the data enters the system, it is processed as a *write command* and sent to the *Write Data Manager* component. The *Write Data Manager* stores the data in the *write database* and publishes an *event*. The *Message Broker* is responsible for adding the event to the *Event Data Store* and sending a message to a dedicated *message queue*. As it is guaranteed the event reaches its destination, issues with writings (e.g., an operational failure, network problems) are resolved using idempotency [10]. Even if the system tries to rewrite an event, each event has to be recorded only once. All interested parties (e.g., the *Read Data Manager* as depicted in Fig. 3) subscribed to that particular message queue will receive the message and process it in accordance with the business requirements. The *Read Data Manager* imports the processed data into its database, transforms it into the required format, and returns it to the client.

1.4 Service Fabric

The *Microsoft Azure Service Fabric* platform facilitates the development and implementation of scalable, distributed applications without requiring the development of a complex infrastructure. Its main components are services, clusters, and nodes. A *cluster* is a set of virtual or physical machines, connected via a network, on which the services reside. Each machine represents a *node* in the system [11]. The platform allows for the creation and management of easily accessible services (both stateful and stateless) that process large amounts of information by exposing a *Reliable Services* APIs framework for application development. Stateful services can store states by utilizing built-in persisted and replicated state providers, i.e., *Reliable Collections* (*Reliable Dictionaries* and *Reliable Queues*). Commands are stored in a *Reliable Dictionary,* key/value pair collection while messages are stored in a *Reliable Queue.* Writing to both collections is done under a single transaction i.e., either both operations are executed, or neither one is. Stateful service is convenient to survive node crashes as Service Fabric is able to restore the service state by playing back the logs. Stateful service possesses append-only files that track and record all service changes to local disks. *Stateful services* and *Reliable Collections* are of key importance when it comes to defining modalities.

2 Proposed Software Architecture

Given that nowadays most applications have far fewer requirements for writing data into the system than for reading it, one of the main goals is to design a system that provides strong consistency when writing and acceptable system performance during reading. The software architecture proposed in this paper was envisaged for these types of systems. The main idea is to combine the CQRS, *Event sourcing*, and *Mediator* patterns, together with the *Service Fabric* platform and its mechanisms, to ensure the development of a system that would be capable of executing the immense number of required operations, and with greater resilience to errors, availability, scalability, and better manageability. The role of *Event sourcing* is to retrieve the data from the *Event store*. The *Service Fabric* (i.e., its *Reliable Collections, Stateful service,* and API) is used for processing and storing the data. The *Mediator* is responsible for coordinating the entire process. Finally, CQRS enables the separation of concerns, so that the software architects and developers can focus solely on one segment of the system. The architecture of the reporting segment of the system is given in Fig. 4a. The API is defined on one of the nodes of the *Service Fabric* cluster, with the various services residing on the other nodes. *Stateful services* and their *replicas* are a significant part of the architecture. Services have one or more partitions for load distribution. Each partition has one primary and multiple secondary replicas, and the replica states are automatically synchronized. Primary and secondary replicas are used as a mechanism for ensuring system availability and they are allocated to different nodes. To ensure the adequate operation of the system, when the primary replica fails the secondary replica automatically takes its place. All writes are performed on the primary, and each reading is executed from the secondary replicas, hence there is no lock. Given that the focus of this paper is on reporting capabilities, writing to the database, as a different segment of the system, will not be regarded in the rest of the section. It should be mentioned that client applications and databases are also crucial to this architecture. Even though they do not belong to the cluster, they are included to fully represent the overall picture of the system. Depending on the amount of data that needs to be processed, different read mechanisms are used to ensure adequate system operation, without downtime. Consequently, different modalities of the proposed software architecture are also offered.

2.1 Modality 1

This architecture modality in Fig. 4b was envisaged for situations in which thousands of messages are generated in a short period of time. Since the focus of the proposed architecture is on querying and reporting, writing data to the database is based on the aforementioned *Event sourcing* pattern and the rest is abstracted. At the beginning of the scanning and evaluation process, each package is imported into the system. This modality is used for generating the number of imported and scanned packages. When a client (e.g., mobile or desktop application), at a certain point in time, needs information about the number of (created, deleted, or uploaded packages) it sends a request to the API which forwards it to the *Stateful service*. The *Stateful service* is subscribed to the relevant stream from which it collects events (i.e., the change of a state of a package). However, it is not necessary for the *Stateful service* to go through all the events in a stream, every time a new request arrives, because it is possible to record the last accessed position and continue processing events from there. The most recent state of the object is obtained from previous cumulative states. Storing the cumulative state of the object in memory is not an expensive operation if it handles events for only a few thousand entities that change their state several times with longer intervals in between the changes. When an event appears in the *Event store*, the *Stateful service* retrieves it, processes it, and immediately saves it in the *Reliable Dictionary*. The state of the *Reliable Dictionary* is replicated automatically. The *Stateful service* and the *Event store* are designed to withstand and process vast amounts of data, where thousands of events take place every second, while not overly burdening the system. However, if the system needs to process thousands and millions of events that are constantly generated in a short time interval, it would take a toll on the performance and resources of the system. Therefore, if we need a system that can process a larger amount of data and return accurate information it would be better to opt for a different modality of the presented architecture.

2.2 Modality 2

In systems in which tens of thousands of entities (i.e., students' tests for assessment) change their state several times per second (i.e., created, uploaded, processed, evaluated, validated, canceled), it is necessary to define the way such a large number of events will be efficiently processed. Writing one event at a time into a *Reliable Dictionary* and then replicating its state across all the replicas can be a very demanding operation. Consequently, another modality is defined for scenarios that require processing each event while avoiding writing to *Reliable Dictionary* and constantly replicating the state. The *Stateful service* is subscribed to the *Event store* as was the case in the previous modality. In other words, it takes events and processes them every X seconds. Within this time frame, events are collected, processed, and stored in an in-memory, key/value collection (*Dictionary*). In the test scenario, every 30 s (e.g., empirically proved to be the best practice) all processed events are relocated from the *in-memory Dictionary* into the *Reliable Dictionary*, and, subsequently, the state is synchronized with the other replicas. After replication, the contents of the *in-memory Dictionary* are deleted and event processing resumes. When a client requests specific data, the request is sent to the API, the API then forwards it to the appropriate service, and the service retrieves the data from the *Stateful service*. The tradeoff of this modality is responsiveness versus *eventual consistency* (i.e., processing events and appending them to the database may result in an "outdated" image of the data). In other words, if new entries were made in the meantime, the client will not be shown the latest cumulative state of the required object. However, even with systems that are not *eventually consistent*, there is a certain type of delay. For example, an application that is based on a three-tier architecture and uses the MVC pattern also has a slight delay. The time it takes to retrieve the data from the database, pass it through all the application tiers, and respond to the client's request is never equal to zero. Even though these are negligibly small-time intervals, technically speaking, an outdated image could be delivered. If the user does not require the cumulative state of the entity (e.g., when diagrams and charts are to be used for reporting), it is preferable to take *snapshots* every X seconds (e.g., 30 s) and plot them instead. Furthermore, if it is necessary to process a large amount of data and queries it can be beneficial to use *partitions* so that every request is routed to the relevant partitions, which process it in parallel, thereby reducing the load. The only issue here is whether the predetermined number of partitions is sufficient, as it is not possible to dynamically increase the number of partitions.

2.3   Modality 3

When a large set of data is to be loaded from the *Event store* and processed to reduce it to a smaller set of aggregated data (events) on which the reporting will be based, the smaller set can first be stored in a *Reliable Queue*. Subsequently, the events from *Reliable Queue* will be processed and stored in the *Reliable Dictionary*. Consequently, the relevant data, in its final format, can be found in the *Reliable Dictionary*. This segment of the system has two roles: the handling of the numerous raw events and the handling of the smaller aggregated datasets. In the case scenario, this modality is used for the assessment of the performance of test evaluators based on a defined business rule. As was the case in the previous modality, the *Stateful service* is subscribed to the *Event store* stream and collects the events every 180 s. Each event is then processed in accordance with the business requirements. In this example, the Stateful service takes tests every 180 s and assigns them to evaluators. *Reliable Queue* contains the tests for each evaluator. The second part of the process takes events from *Reliable Queue* and applies the set business rule for the specific evaluator.

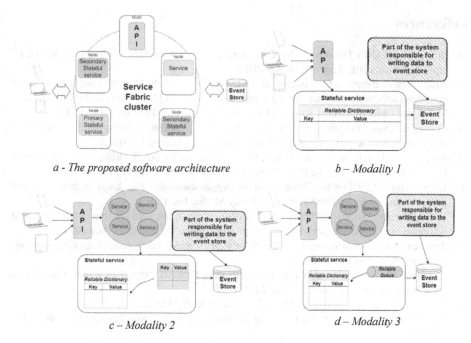

Fig. 4. The proposed software architecture and its modalities

3 Conclusion

If a software system that processes large amounts of data is to adequately fulfill client requirements, while maintaining performances at the desired level, the software architect must make various trade-offs. For systems that use multiple databases, it is difficult to establish a single source of truth for the entire system. Therefore, it is important to ensure that all events that occur within the system are stored and that the data is properly replicated. Consequently, after a system failure, the system can be restored to its previous state. A software architecture is proposed, which combines the CQRS, *Event sourcing*, and *Mediator* patterns, together with the *Service Fabric* platform and its mechanisms. Regarding the number of events that need to be processed, three variations (modalities) of the software architecture based on the *Event sourcing* architectural pattern are presented. Although in terms of responsiveness, scalability, and system availability, this architecture provides significant improvement in system performance, it is always necessary to consider if the trade-off is worth it. If the system handles less than a few thousand events a day, it is better to turn to traditional approaches. The application of *Event sourcing* patterns can sometimes be detrimental to system performance. Future work could be aimed at expanding the presented modalities in accordance with the needs of the system and changes in client requirements.

References

1. Young, G.: CQRS Documents by Greg Young. https://cqrs.files.wordpress.com/2010/11/cqrs_documents.pdf. Accessed 08 Apr 2022
2. Nlogn Team. Eventual Consistency and Strong Consistency. https://levelup.gitconnected.com/eventual-consistency-and-strong-consistency-d0b882133ca5. Accessed 08 Apr 2022
3. Meyer, B.: Object-Oriented Software Construction, 2nd edn. Prentice Hall (2000)
4. Command Query Responsibility Segregation (CQRS) Pattern. https://www.ibm.com/cloud/architecture/architectures/event-driven-cqrs-pattern/. Accessed 15 Apr 2022
5. Zimarev, A.: Hands-on Domain-Driven Design with .NET Core: Tackling Complexity in the Heart of Software by Putting DDD Principles into Practice. Packt Publishing (2019)
6. Kleppmann, M.: Designing Data-Intensive Applications the Big Ideas Behind Reliable, Scalable, and Maintainable Systems. O'Reilly Media, Inc. (2017)
7. Event Sourcing Pattern – IBM Cloud Architecture Center. https://www.ibm.com/cloud/architecture/architectures/event-driven-event-sourcing-pattern. Accessed 22 Aug 2021
8. Zimarev, A.: Event Sourcing and CQRS. https://www.eventstore.com/blog/event-sourcing-and-cqrs. Accessed 16 Apr 2021
9. Reselman, B.: An Illustrated Guide to CQRS Data Patterns. https://www.redhat.com/architect/illustrated-cqrs. Accessed 22 Aug 2021
10. Stopford, B.: Designing Event-Driven Systems. O'Reilly Media, Inc. (2018)
11. Bai, H.: Programming Microsoft Azure Service Fabric, 2nd edn. Pearson Education, Inc. (2018)

Supporting Data Types in Neo4j

Kornelije Rabuzin, Maja Cerjan[✉], and Snježana Križanić

Faculty of Organization and Informatics, University of Zagreb, Varaždin, Croatia
`macerjan@foi.hr`

Abstract. The use of graph databases has grown over the years, creating the
need to use many of the features available in relational databases. Several of these
features are already in use, but since graph databases are still in a development
phase, many concepts supported in relational databases are still not supported in
graph databases. Neo4j supports data entry of various types that can be associated
with nodes. However, in the node creation phase it is not possible to specify the
property data type. The aim of this paper is to demonstrate how assigning data
types to each property can be implemented. Also, it is demonstrated how to check
whether a particular data is of the correct type. For that purpose, the change of
the syntax used for creating nodes is proposed, and a trigger that would check the
correctness of the entered data types for certain attributes is defined.

Keywords: Graph databases · Data types · Neo4j · Cypher · NoSQL

1 Introduction

Nowadays, the growth of data is an increasingly common phenomenon in all industries
and in all spheres of business. Over the years, alternative ways of displaying and processing data have been sought. This has the ability of improving data management, but
also the presentation of data itself, turning it much more efficient and enabling a simpler and faster manipulation. Relational databases are satisfactory in most areas, and are
therefore considered the most popular type of database. However, they sometimes do not
have satisfactory characteristics that allow a proper, efficient, and effective operation,
especially given the large connected systems that manage millions of records. In such
cases, non-relational databases are needed, which facilitate and speed up the work of
large data. In NoSQL, this is done in a more specific way. There are different approaches
to store data that can be applied, depending on the need but also on the application. For
example, data can be stored as a key-value pair, document, etc. Another possibility is
to display the data in a graph form, using nodes and connections between nodes, which
will be the focus of this paper.

In this paper we discuss how to execute the command to create nodes, while immediately defining data types or performing a type check, to ensure that each property is
entered correctly. Likewise, if there was to be a large database that had nodes with similar or the same properties, it would be important to have the ability to create a specific
data type, with the definition of name and constraints. That was the idea of this paper,
i.e., the implementation of data types that would have a wide application over the entire
database, but also check whether the data type is the correct one.

S. Chiusano et al. (Eds.): ADBIS 2022, CCIS 1652, pp. 459–466, 2022.
https://doi.org/10.1007/978-3-031-15743-1_42

2 Literature Overview

This article is a pioneering one because there are no publications directly comparable to ours on the subject, and we anticipate that the improvements it suggests will be implemented in some future Neo4j releases. In such way, the emphasis was on generic graph database concerns, which are crucial for comprehending the advancements in this study. The presentation of relevant papers will be divided into sections so that the reader understands the issues addressed in the papers and why they are essential.

The first section of related papers discusses research on generic facts about graph databases. The research is offered to give the reader an understanding of what graph databases are in terms of component analysis, advantages and disadvantages, methods of creation, comparisons with other databases, comparisons of different database management systems, and so on. In this paper (Guia et al. 2017), an analysis of the main components of graph databases was performed, so it was possible for readers to understand what would be the main advantages of this type of database, but also what characteristics are fundamental to get a real graph database. The most well-known system for graph databases (Neo4j) was also analyzed, while some examples are given using Cypher. Constraints as well as the languages used to create queries in such databases have been elaborated. In Maleković and Rabuzin (2016), detailed explanations of Indexed Sequential Access Model (ISAM) databases, Hierarchical and Network databases are given, as well as graphic representations of some examples of the application of these databases are also given.

The second section discusses non-relational database data type issues. In Engle (2018) it is found that graph databases, as well as column databases, do not allow to define data types at the very beginning when defining the schema itself, which is contrary to the case in relational databases when new queries or new tables are created. This work also puts a focus in what and how to deal with data types in other types of non-relational databases. Also in this paper, 12 evaluation criteria are proposed in order to compare relational and non-relational databases. In Buneman et al. (1996), authors show that several decades ago an attempt was made to develop a new scheme that would be slightly different from the traditional one that is used in structured data, with a special emphasis on unstructured data. This work achieved a new way to present the data and the scheme itself, in the form of graphs, which were named as boundary graphs. It was also examined the closeness of schemas under queries, as well as possibilities of query optimization.

The third section contains a proposal for introducing a data type constraint as one of the possible solutions to the challenge of defining data types in Neo4j. Looking for possible ways to assign data types to certain node properties, an interesting concept suggested by Tony Chiboucas was found on the Neo4j forum (Chiboucas 2018). Namely, as Tony himself points out, the syntax for creating constraints was originally developed so it could be expanded in the future. Since this was not fully integrated, in this paper we suggest upgrading the existing Cypher constraint parse and adding the ability to check data type. The current syntax points out that ASSERT is a somewhat parsed result that says something must be Boolean. If this thesis is correct, then a constraint could be written as follows:

```
CREATE CONSTRAINT ON (course: Course) ASSERT
apoc.meta.type(course.id) = 'INTEGER';
```

The proposal for program changes in Neo4j itself was proposed at the Github commit (Finné 2013).

3 Data Types

Defining data types is fundamental when it comes to databases, regardless of the type. In order to properly manage the data, all the attributes that make up a relation are assigned a type, according to which it will be known how its manipulation should be done. The type itself needs to be defined at the very beginning in the planning phase, even before creating the scheme.

In Neo4j a database containing nodes, properties, and connections can be created. Before a property of a node is created, it must be known what that property represents and how the user will ultimately enter the data. For this reason, it is extremely important that before creating the database, a scheme is made according to which the implementation of data in Neo4j will be performed. As stated on the official website of the system, data types that are categorized into several groups can be used, which are the following (Neo4j 2021): Property types, Structural types and Composite types.

The first one can be considered as the most important one - property types. The data types that belong to this group are: Integer, Float, String, Boolean, Point, Date, Time, LocalTime, DateTime, LocalDateTime, and Duration. The specificity of this group lies in the fact that these data types have the ability to return directly from the query and save that as a property.

The second group cited by Neo4j is structural types. The structure in graph databases is composed of the three most important components: nodes, relations and paths that make up the whole set of nodes and relations. After creating nodes and connections between them, it is possible to perform a query that returns the desired data and display the nodes and connections that are requested.

The third group of data types are composite types, which include maps and lists. What is specific about maps is that they contain two components (key and value). The key needs to be defined first, and each key contains a name that must be a string in each case. These keys are assigned different values.

Another component of this group are lists. Similar to other areas, the list is a method by which some values are obtained in sequence. Each value is defined in square brackets and separated by a comma, which is similar to the way of defining arrays in several programming languages.

The main characteristic of this group is that these types cannot be stored as properties, while other possibilities that are allowed in the first group can be used. The next section will show how to implement data types in Neo4j. Currently, the data type cannot be specified when creating nodes and/or connections, so we'll offer a suggestion in the following paragraphs.

4 Implementation of Custom Data Types in Neo4j

Before making changes to the database itself, it is necessary to make certain changes in the configuration and install certain plugins. The config file from the selected database must be edited first, with the addition of the following commands:

- dbms.unmanaged_extension_classes = n10s.endpoint = / rdf
- dbms.security.procedures.unrestricted=gds.*,apoc.*
- apoc.trigger.enabled = true

```
CREATE(
    John: Person{
        id: 0, type: Integer,
        name:'John', type: String,
        lastname:'Doe', type: String,
        city:'Zagreb', type: String,
        birth_date:localdatetime('1978-07-21T21:40:32.142'), type: dateTime
    }
);
```

Fig. 1. Example of new syntax for creating node with data types

First line in config enables HTTP export endpoint. Second line in config allows all users to use all APOC procedures. Third line in config enables APOC triggers.

Afterwards it is necessary to install the following plugins: Awesome Procedures On Cypher (APOC) and Neosemantics (n10s). APOC is a Neo4j add-on library with numerous of procedures and methods that extend the capabilities of the database. Neosemantics is an RDF plugin for Neo4j. The RDF data transfer paradigm is a W3C standard. Neosemantics essentially makes it feasible to store RDF data in Neo4j without loss (imported RDF can subsequently be exported without losing a single triple in the process) (Neo4j 2021). After the installation of the plugins is completed, it is necessary to restart the server. The syntax for creating new nodes looks like this:

```
CREATE (Andy: Person {name: 'Andy', lastname: 'Developer', city:
'Tokio', birth_date: localdatetime('2015202T21')})
```

To implement custom data types for each attribute in the node, we suggest the following syntax. As can be seen from the example (Fig. 1), the data type would be placed behind the attribute value. After that, it is important to create a trigger that will validate a certain attribute and check if the correct data type was entered. The trigger itself validates the meta type of the value assigned to a property and returns an error if the value does not correspond to the respective data type. In the rest of the section, three ways of checking for a data type will be presented.

4.1 Implementation Using SHACL Validation

Using the Neosemantics plugin, data validation in graph databases can be performed using Shapes Constraint Language (SHACL), which belongs to the World Wide Web Consortium (W3C) standard. The goal of this check is to create certain constraints that will be applied to data types. The SHACL constraints can be described as Resource Description Framework (RDF). For this reason, a new so-called shape that will be valid for Person node is created.

```
call n10s.validation.shacl.im-
port.inline('
   @prefix neo4j:
<neo4j://graph.schema#>.
   @prefix sh:
<http://www.w3.org/ns/shacl#>.
   neo4j:PersonShape a
sh:NodeShape ;
   sh:targetClass neo4j:Person
;
   sh:property [
      sh:path neo4j:id;
      sh:maxCount 1;
      sh:datatype xsd:integer;
   ];
   sh:property [
      sh:path neo4j:name;
      sh:maxCount 1;
      sh:datatype xsd:string;
   ];
   sh:property [
      sh:path neo4j:lastname;
      sh:maxCount 1;
      sh:datatype xsd:string;
   ];
   sh:property [
      sh:path neo4j:city;
      sh:maxCount 1;
      sh:datatype xsd:string;
   ];
   sh:property [
      sh:path neo4j:birth_date;
      sh:maxCount 1;
      sh:datatype xsd:local-
datetime;
   ];
.
','Turtle');
```

Five constraints have been placed on data types, which will be explained afterwards. The first constraint was placed on property "id", obliging this data to be of type integer, and to have a unique value in the database. The second constraint was on the property "name", which must be of type string, and must also have a unique value (maxCount property). Then there are constraints on "lastname" and "city" properties which must be of type string, and also have a unique value, exactly like name property. Last constraint was on property called "birth_date". This property must be of type localdatetime and must have a unique value. Once the shape and constraints have been created, it is necessary to create an APOC trigger that will validate each transaction in the phase before saving to the database, and will check the shape and constraints shown above.

```
CALL apoc.trigger.add('shacl-validate','call n10s.valida-
tion.shacl.validateTransaction($createdNodes,$createdRelation-
ships, $assignedLabels, $removedLabels, $assignedNodeProperties,
$removedNodeProperties, $deletedRelationships, $deletedNodes)',
{phase:'before'})
```

If the user tried to enter a new Person in the wrong format, i.e., using the wrong data type, we would get an error (Fig. 2).

```
1  CREATE(
2      John:Person{
3          id:1,
4          name: 3,
5          lastname: 4.2,
6          city: 'Zagreb',
7          birth_date: localdatetime('2015-07-21T21:40:32.142')
8      }
9  )
```

ERROR Neo.ClientError.Transaction.TransactionHookFailed

Error executing triggers {shacl-validate=Failed to invoke procedure `n10s.validation.shacl.validateTransaction`: Caused by: org.neo4j.exceptions.CypherTypeException: Don't know how to treat that as a boolean: 2015-07-21T21:40:32.142}

Fig. 2. Error triggering trigger 'shacl-validate'

4.2 Implementation Using APOC Trigger

Another way to create a check on the proposed data type is to use the APOC trigger. APOC trigger contains statements when user tries to change data. By changing data we mean on creating, updating or deleting data. Triggers can be executed in two phases: before or after commit. The trigger itself validates the meta type of the value assigned to a property and returns an error if the value is the wrong data type. The trigger is created as follows:

```
CALL apoc.trigger.add("checkEmployeeId", "UNWIND apoc.trig-
ger.propertiesByKey($assignedNodeProperties, 'id') AS prop
CALL apoc.util.validate(apoc.meta.type(prop) <> 'INTEGER', 'ex-
pected integer property type, got %s', [apoc.meta.type(prop)])
RETURN null", {phase:'before'})
```

Once all triggers are created, the user can no longer enter the wrong data type for the specified properties.

4.3 Comparison of Two Suggested Implementations

A brief comparison of the two methods described above will be presented below. Both proposed methods are very similar in their functionality, but there are differences between them. When discussing SHACL verification, it's important to note that there are three fundamental modes for performing the verification (Neo4j 2021). The first mode validates the entire graph; more specifically, all constraints are checked throughout the graph, and only wrongly entered data is returned as a consequence. The second mode involves validating a specific section of the graph, such as a node. This mode, like the first, merely returns wrongly entered data as a result. Transaction validation is the third mode, which is initiated by a trigger. The method of operation is that if the validation yields a result that is not empty, the transaction has failed and a rollback is required. In our example, the third mode was used, i.e. transaction validation, using a trigger that was triggered in the pre-commit phase of the transaction itself. The APOC trigger on the property value, on the other hand, was employed, which used a different sort of validation. The trigger validates the meta type of value associated with the specified property. An error will be returned if the value is supplied in the improper data type, and a rollback transaction will be conducted. The outcome is a Boolean value in this scenario. When we examine the two ways, we can observe that the key difference is in the data validation method itself. The first approach, SHACL validation, does validation directly through the defined shape and the constraints entered in it. Validation directly via the APOC trigger, on the other hand, validates the meta values assigned to the property being checked. Another distinction is the value returned by the validator. SHACL validation yields an empty or non-empty result, but trigger-based validation yields a Boolean value.

5 Discussion

When being unable to forecast the structure of data, NoSQL offers greater flexibility for data insertion. But subsequently running optimized queries becomes more challenging. This means that while NoSQL can be designed to be optimized for a specific form of query, it becomes less optimum for all other sorts of queries. The more flexible data format also forces an end user to create a lot of code in applications to support business rules. This means that the format of every single piece of data must be verified (String, Integer, etc.).

With the practical examples presented in this paper, it can be seen how restrictions on data types can be made in the current version of Neo4j in several ways. The very proposal presented for the new syntax when creating nodes is technically feasible, and could be accepted in some future versions of Neo4j. Tony Chiboucas provided a new syntax for defining data type restrictions for the Cypher language, which was also an innovative solution to the data type problem. Because Cypher is Neo4j's official query language, the proposed improvements would be beneficial to other related languages like Apache TinkerPop, Gremlin, and SPARQL. The proposed changes can be implemented in other prominent graphics databases, such as Amazon Neptune, JanusGraph, OrientDB, CayleyGraph, DGraph, and others, in addition to Neo4j.

Like any work, this one has the possibility of improvement that could be achieved in the form of creating a visual element. For example, an application could be developed that would merge with an existing database created for the purpose of this paper, where the real benefits of creating such a constraint could be better demonstrated. It is considered that the way of creating a new type of data is well designed and implemented, but it would certainly be easier if the proposed syntax was accepted and included in some new versions of Neo4. Also, it is planned to investigate domain support.

6 Conclusion

When observing everything that has been done so far around the topic of graph database, it can be concluded that these solutions are being upgraded day by day, with even more possibilities to overcome the existing limitations, so they can be used at the same level as relational databases. This paper is a pioneering effort because there are currently no similar works in this field of study.

The importance of properly assigning data types is of utmost importance in order to be able to properly manage the data contained in the database. In other words, a database's functionality is significantly impacted by the wise selection of the data type that is kept within. With proposed changes, it is possible to significantly improve the speed of the database itself and decrease the time needed to conduct both basic and complicated queries by making an informed choice about the kind of data to be kept in the database. The time required for the programmer to process the data is greatly decreased by knowing the type of data that was fetched from the database.

There are different types available that can be used applying the proposed implementation. By accepting the syntax for the data type, it can be much easier to work with data in terms of use and restrictions. There should be no additional triggers to check the data type, which would speed up the operation of the database itself. Checking the effectiveness of the database with and without the suggested improvements is one of the next steps in this planned implementation. Additionally, non-relational databases need the exploration of new implementation strategies and other limitations that are present in relational databases.

References

Buneman, P., Davidson, B.S., Fernandez, M., Suciu, D.: Adding structure to unstructured dana. In: ICDT 1997: Proceedings of the 6th International Conference on Database Theory, pp. 336–350 (1996). https://repository.upenn.edu/cgi/viewcontent.cgi?referer=&httpsredir=1&article=1038&context=db_research

Chiboucas, T. [tony.chiboucas]: The syntax for creating constraints was designed with future extensibility in mind. [Comment on the online forum post Data type of a property]. Neo4j forum, August 2018. https://community.neo4j.com/t/data-type-of-a-property/1309/9

Engle, R.D.: A methodology for evaluating relational and NoSQL databases for small-scale storage and retrieval (2018)

Finné, M.: Added CREATE/DROP CONSTRAINT syntax in Cypher [Github commit] (2013). https://github.com/neo4j/neo4j/commit/49ae8fcb64a3470421736609d54cbf0104a66a8a

Guia, J., Gonçalves Soares, V., Bernardino, J.: Graph databases: Neo4j analysis. In: Proceedings of the 19th International Conference on Enterprise Information Systems - Volume 3: ICEIS, pp. 351–356 (2017). ISBN 978-989-758-247-9; ISSN 2184-4992. https://doi.org/10.5220/0006356003510356

Maleković, M., Rabuzin, K.: Uvod u baze podataka, Udžbenik Sveučilišta u Zagrebu, Varaždin (2016)

Maleković, M., Rabuzin, K., Šestak, M.: Graph Databases - are they really so new. Int. J. Adv. Sci. Eng. Technol. 4(4), 8–12 (2016). https://urn.nsk.hr/urn:nbn:hr:211:997990

Neo4j: Expressions (2021). https://neo4j.com/docs/cypher-manual/current/syntax/expressions/. Accessed 19 Feb 2022

Systematic Literature Review for Emotion Recognition from EEG Signals

Paulina Leszczełowska[(✉)] and Natalia Dawidowska[(✉)]

Faculty of Electronics, Telecommunications and Informatics, Gdańsk University of Technology, Gdańsk, Poland
paulinaleszczelowska@gmail.com, dawidowska.natalia.99@gmail.com

Abstract. Researchers have recently become increasingly interested in recognizing emotions from electroencephalogram (EEG) signals and many studies utilizing different approaches have been conducted in this field. For the purposes of this work, we performed a systematic literature review including over 40 articles in order to identify the best set of methods for the emotion recognition problem. Our work collects information about the most commonly used datasets, electrodes, algorithms and EEG features, as well as methods of their extraction and selection. The number of recognized emotions was also extracted from each paper. In the analyzed articles, the SEED dataset turned out to be the most frequently used. The two most prevalent groups of electrodes were frontal and parietal. Evaluated papers suggest that alpha wavelets are the most beneficial band for feature extraction in emotion recognition. FFT, STFT, and DE appear to be the most popular feature extraction methods. The most prominent algorithms for feature selection among analyzed studies were classifier-dependent wrappers, such as the GA or SVM wrapper. In terms of predicted emotions, developed models in more than half of the papers were designed to predict three emotions. The predictive algorithms that were mostly used by researchers are neural networks or vector machine-based models.

Keywords: Emotion recognition · Electroencephalogram (EEG) · Affect · Signal processing · Emotion classification · Machine learning

1 Introduction

Emotions have been investigated in various fields providing the foundation of Affective Neuroscience [2]. To better comprehend particular emotions, researchers began to link them to physiological responses such as facial expressions, heart rate, skin conductance response, blood pressure, and respiration rate. According to many neuroscience sources, human emotions are closely tied to activity in a number of brain subregions [5].

Given these assumptions, one of the most common physiological signals used to recognize emotions is an electroencephalogram (EEG). These signals have been shown to obtain high classification accuracy for emotion recognition in laboratory settings, and have emerged as a viable tool for describing how cognition and emotional behavior are

© Springer Nature Switzerland AG 2022
S. Chiusano et al. (Eds.): ADBIS 2022, CCIS 1652, pp. 467–475, 2022.
https://doi.org/10.1007/978-3-031-15743-1_43

associated on a physiological level [6]. The goal of this paper is to perform the systematic literature review to discover different approaches to the subject of emotion recognition using electroencephalogram (EEG) signals.

2   Literature Review Process Design and Execution

We put together the 4 research questions that compactly presented our interests and directed our further steps during the review (Table 1).

Table 1. Selected research questions.

ID	Question
RQ1	Which emotions are recognizable from EEG in adults?
RQ2	Which methods are used in emotion recognition from EEG signals?
RQ3	Which methods are used in processing EEG signals?
RQ4	Which of the gathered EEG signals are most significant in emotion recognition?

We established criteria (only English articles published within 10 years, focusing on emotion recognition from EEG, necessarily with DOI, abstract and full text available) and searched only by title through the IEEE Xplore, Web of Science, Scopus and Pubmed databases. After removing the duplicates, we had 519 articles (out of 875 found) that were qualified for the next step, i.e. tagging by title, to narrow the range of papers to those most related to our needs.

Seeing the results of title tagging and a large number (122) of articles that received the maximum score from all 3 annotators, we decided to qualify only these for the next step - tagging by abstract. During the title tagging process, we managed to isolate 7 articles, which seemed to be valuable reviews that could contribute a lot to our research. Hence our decision to directly qualify these articles for reading and analysis.

Reading 3 out of 7 papers marked as a review provided us with vital information concerning the use of music for emotion stimulation while acquiring EEG signals for datasets. In the review [12] author noticed that in a study regarding music video excerpts, it was observed that higher frequency bands such as gamma were detected more prominently when subjects were listening to unfamiliar songs. Therefore, for further analysis, we decided to classify only those articles that used a dataset not based on a musical stimulus (i.e. exclude mainly the DEAP dataset). After this exclusion, we obtained 66 papers for further analysis. After combining the two reviews: [3] and [4] that were to form the basis of our analysis, we obtained 34 papers published between 2004 and 2019. We chose 9 other articles published between 2020 and 2021 to have a complete set including also the latest papers on the subject. After analyzing them fully on our own, we ended up with a sum of 43 articles. Our key findings included information about: the most popular EEG signals datasets and stimuli used to evoke emotion, analyzed electrodes, used EEG features, methods of their extraction and selection, recognized emotions, and utilized classifiers.

3 Results

3.1 Datasets

The information about the dataset was found in 20 out of 43 articles. The most commonly used dataset in the articles was SJTU Emotion EEG Dataset (SEED). This dataset was used in twelve out of 20 articles. Other common types of datasets were those created by the authors (8 articles). Only in 1 article, the MAHNOB-HCI database was used, and only in 1 paper, the AMIGOS database was utilized.

3.2 Electrodes

28 out of 43 articles included information about which electrodes were measured and/or used in the model. The largest set of electrodes included in a single study was 62 electrodes in articles [5, 9, 10] and [11] and all of them used the SEED dataset. Figure 1 shows that the most frequently used electrode was F4 (21 articles), followed closely by F3 (20 articles).

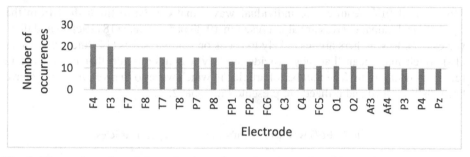

Fig. 1. The total number of times the given electrode was mentioned in articles. Only electrodes used 10 or more times are shown.

Different electrodes can be grouped based on the international 10–20 system and regions of the brain. For the sake of aggregation, we have distinguished 5 groups – Frontal (F), Central (C), Parietal (P), Temporal (T), and Occipital (O). Electrodes are named based on their placement, some electrodes are placed over the border of 2 regions of the brain, which results in electrodes named with two capital letters e.g. FP1, PO3, etc. In the case of these electrodes, we decided to include them in both groups i.e. FP1 belongs to groups Frontal and Parietal. Some electrodes have been labeled differently i.e. CMS and DRL. These two electrodes belong to the Temporal group.

In total there were 522 mentions of different electrodes in 28 papers. From Fig. 2 one can clearly see that most frequently used electrodes were these placed over the frontal lobe (219 mentions). Other quite popular groups were Parietal and Central, 171 and 153 mentions respectively. The least often used groups: Temporal and Occipital were mentioned 66 and 65 times. This disproportion, however, could be possibly caused by the disproportion of area size. There are 24 different electrodes belonging to the Frontal group, 25 electrodes belonging to Parietal group, 23 electrodes in Central group, Temporal group consisting of 12 electrodes, and Occipital of 10 electrodes.

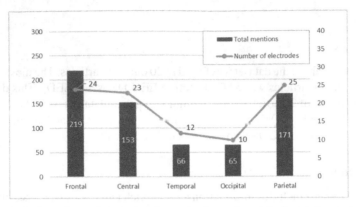

Fig. 2. Number of mentions of particular electrode groups in articles in relation to total number of electrodes per group.

3.3 EEG Features and Features Extraction Methods

Mostly used EEG features are individual waves that can be distinguished from the signal. The dominant waves are alpha (used in 19/43 papers) beta (18 articles) and theta waves (17). Researchers are less likely to focus on gamma waves (14) and delta waves (13) which may mean that they provide slightly less information that is valuable in recognizing emotions. The number of articles in which waves of a given frequency have been distinguished and utilized are presented in Table 2.

Table 2. EEG wavelets - number of their usages in articles.

Waves	Number of occurrences
Alpha	19
Beta	18
Theta	17
Gamma	14
Delta	13

The most frequently used feature extraction algorithms were FFT (Fast Fourier Transform) or STFT (Short-Time Fourier Transform) - analysis methods in the frequency domain which were utilized in 15 out of 43 articles. In 25% of the analyzed works (11/44 articles) DE (Differential Entropy) was used for feature extraction. Another method was the calculation of statistical measures which enable extraction of signal statistics such as standard deviation (9 papers), mean (7 papers), etc. which were used a bit less frequently. Other algorithms used by the researchers were: PSD (Power Spectral Density) - 7 articles, FD (Fractal Dimension) - 7 articles, PSD Welch's method - 5 articles, DASM (Differential Asymmetry) - 5 articles, Hjorth features - 5 articles, DWT

(Discrete Wavelet Transform) - 4 articles, RASM (Rational Asymmetry) - 4 articles, WT (Wavelet Transform) - 4 articles.

In the analyzed works, other features and features extraction algorithms were also used, but due to their appearance in less than 4 articles and the clarity of our analysis, we decided to omit them in this report (Table 3).

Table 3. Identified feature extraction methods and number of their occurrences in articles.

Feature extraction method	Number of occurrences
FFT or STFT	15
DE	11
Statistical measures - std	9
Statistical measures - mean	7
PSD	7
FD	7
PSD (Welch's method)	5
DASM	5
Hjorth Features	5
DWT	5
RASM	4
WT	4

3.4 Feature Selection Methods

The most frequently used method for feature selection was classifier-dependent wrapper. In addition, researchers willingly used methods such as mRMR (min-Redundancy Max-Relevance), PCA (Principal Component Analysis), and Correlation coefficient analysis - each of them was used in 3 articles. Max Pooling algorithm was used in 2 articles and other methods that were used only in individual articles and were therefore omitted in this report.

3.5 Emotions

21 out of 43 articles included information about which emotions were recognized in a model. There were 16 different types of emotions identified. The maximal number of emotions recognized in a model was 8 (neutral, disgust, anger, joy, amusement, tenderness, fear, sadness) in the article [7]. The minimal found number of emotions recognized was 1 - disgust in the article [8]. Mode and median amount of emotions recognized are equal to 3 and mean amount of emotions recognized equals 3.28.

The most frequently included emotion was neutral emotion (in 13 articles out of 21), followed by positive and negative emotions (each 11 out of 21 articles). Three out of 21 articles used models which recognized emotions based on valence-arousal planes. One article differentiated 2 classes per valence and arousal - high and low valence, high and low arousal. Figure 3 contains a chart showing the number of articles recognizing particular emotion in the model.

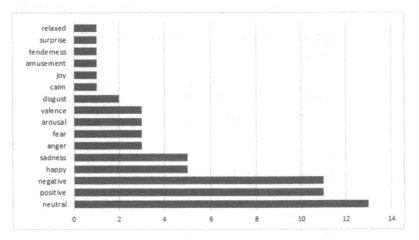

Fig. 3. Number of articles recognizing particular emotion in the model.

3.6 Algorithms

21 of the 43 articles mentioned 18 types of prediction algorithms used to recognize emotions. The most frequently used algorithm was Support Vector Machine (SVM), applied in 11 articles. Other types of algorithms that were mentioned multiple times were k Nearest Neighbors (kNN) (2 articles), Library for Support Vector Machines (LIBSVM) (2 articles), Linear Discriminant Analysis (LDA) (2 articles), Convolutional Neural Network (CNN) (3 articles), Artificial Neural Network (ANN) (3 articles). The rest of the algorithms (Linear Regression (LR), Relevance Vector Machine (RVM), Deep Belief Network (DBN), Extreme Machine Learning (ELM), Own Neural Network, Bi-hemispheres Domain Adversarial Neural Network (BiDANN), Regularized Graph Neural Network (RGNN), Deep Neural Network (DNN), Graph regularized Extreme Learning Machine (GELM), Graph Regularized Sparse Linear Regression (GRSLR), Naive Bayes, Takagi–Sugeno fuzzy model) were mentioned only in one paper each.

Given this disproportion of usages between SVM and other algorithms, we have decided to divide all of them into five groups based on how they work. The identified groups are Clustering, Regression, Vector Machine, Neural Network, and Other (Table 4 and Fig. 4).

Table 4. Categorization of predictive algorithms based on their behavior.

Group	Algorithms
Clustering	Knn
Regression	LR, GRSLR
Vector Machine	SVM, RVM, LIBSVM
Neural Network	DBN, ELM, Own NN, BiDANN, RGNN, CNN, DNN, GELM, ANN
Other	LDA, Naive Bayes, Takagi–Sugeno fuzzy model

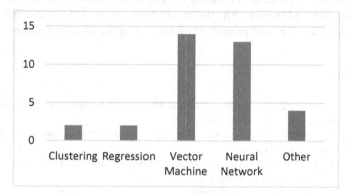

Fig. 4. Number of mentions of particular groups of algorithms in articles.

Two most often used types of algorithms were Vector Machines and Neural Network algorithms (13 articles each). In the Other group, individual algorithms were placed that could not be classified to any of previously mentioned groups (5 articles). The least mentioned groups were Clustering and Regression (2 articles each).

4   Summary

When considering datasets, there were found only 4 datasets that were using visual stimuli. The most commonly used dataset was the SEED dataset, followed by datasets developed by the authors of the articles. Other datasets were used only 1 time each.

Two most prevalent groups of electrodes were Frontal and Parietal. Our findings support the widely held assumption that the frontal lobe stores more emotional activation than other areas of the brain [12]. Considering filtering electrodes to those from the Frontal group could possibly have positive influence on the accuracy of the model.

Evaluated papers suggest that alpha wavelets are the most beneficial band for emotion recognition. In terms of the number of mentions, beta and theta waves were only marginally behind alpha. As we found out in one of the reviews, high-frequency bands such as alpha, beta, and gamma are more effective at distinguishing emotions [12]. FFT, STFT, and DE appear to be the most popular methods of feature extraction. Many researchers were also using simple statistical measures as extracted features.

The most prominent feature selection algorithms were classifier-dependent wrappers, such as the GA or SVM wrapper, which were used to successfully limit the number of features without losing the signal properties that best describe the EEG, boost accuracy and lower the likelihood of overfitting.

In more than half of the papers, developed models predicted three emotions, which is consistent with the finding that more than half of the articles predicted positive, neutral, and negative emotions. These findings are useful in terms of model quality, because model accuracy often decreases as the number of classes predicted increases.

Finally, after conducting an analysis of predictive algorithms, we discovered that most researchers used either a neural network or a vector machine-based models. The Neural Network group included a variety of neural network derivatives, each of which was only mentioned a few times. The Vector Machine group, on the other hand, contained only three elements, the most prominent of which was the SVM algorithm.

References

1. Alarcao, S., Fonseca, M.: Emotions recognition using EEG signals: a survey. IEEE Trans. Affect. Comput. **10**(3), 374–393 (2019)
2. Panksepp, J.: Affective Neuroscience: The Foundations of Human and Animal Emotions. Oxford University Press (2004)
3. Jenke, R., Peer, A., Buss, M.: Feature extraction and selection for emotion recognition from EEG. IEEE Trans. Affect. Comput. **5**(3), 327–339 (2014)
4. Torres, E., Torres, E., Hernández-Álvarez, M., Yoo, S.: EEG-based BCI emotion recognition: a survey. Sensors **20**(18), 5083 (2020)
5. Li, Y., Zheng, W., Cui, Z., Zhang, T., Zong, Y.: A Novel Neural Network Model Based on Cerebral Hemispheric Asymmetry for EEG Emotion Recognition. In: 27th International Joint Conf. on Artificial Intelligence (2018)
6. Martinez-Tejada, L., Yoshimura, N., Koike, Y.: Classifier comparison using EEG features for emotion recognition process. In: 2020 IEEE 18th World Symposium on Applied Machine Intelligence and Informatics (SAMI) (2020)
7. Liu, Y., Yu, M., Zhao, G., Song, J., Ge, Y., Shi, Y.: Real-time movie-induced discrete emotion recognition from EEG signals. IEEE Trans. Affect. Comput. **9**(4), 550–562 (2018)
8. Iacoviello, D., Petracca, A., Spezialetti, M., Placidi, G.: A real-time classification algorithm for EEG-based BCI driven by self-induced emotions. Comput. Methods Programs Biomed. **122**(3), 293–303 (2015)
9. Zhang, W., Wang, F., Jiang, Y., Xu, Z., Wu, S., Zhang, Y.: Cross-subject EEG-based emotion recognition with deep domain confusion. In: Yu, H., Liu, J., Liu, L., Ju, Z., Liu, Y., Zhou, D. (eds.) ICIRA 2019. LNCS (LNAI), vol. 11740, pp. 558–570. Springer, Cham (2019). https://doi.org/10.1007/978-3-030-27526-6_49
10. Li, Y., Zheng, W., Cui, Z., Zong, Y., Ge, S.: EEG emotion recognition based on graph regularized sparse linear regression. Neural Process. Lett. **49**(2), 555–571 (2018). https://doi.org/10.1007/s11063-018-9829-1
11. Hwang, S., Ki, M., Hong, K., Byun, H.: Subject-independent EEG-based emotion recognition using adversarial learning. In: 2020 8th Intern. Winter Conference on Brain-Computer Interface (BCI) (2020)
12. Suhaimi, N., Mountstephens, J., Teo, J.: EEG-based emotion recognition: a state-of-the-art review of current trends and opportunities. Comput. Intell. Neurosci. **2020**, 1–19 (2020)

13. Ascertain-dataset.github.io. ASCERTAIN dataset. [online] (2021). https://ascertain-dataset.github.io/. Accessed 6 Aug 2021
14. Bcmi.sjtu.edu.cn. SEED Dataset. [online] (2021). https://bcmi.sjtu.edu.cn/home/seed/index.html. Accessed 13 Aug 2021
15. Correa, J.: AMIGOS: A Dataset for Affect, Personality and Mood Research on Individuals and Groups. [online] Eecs.qmul.ac.uk. (2021). http://www.eecs.qmul.ac.uk/mmv/datasets/amigos/index.html. Accessed 3 Aug 2021
16. Koelstra, S.: DEAP: A Dataset for Emotion Analysis using Physiological and Audiovisual Signals. [online] Eecs.qmul.ac.uk (2021). https://www.eecs.qmul.ac.uk/mmv/datasets/deap/. Accessed 26 July 2021

Semantically-Driven Secure Task Execution over Wireless Sensor Networks

Niki Hrovatin[1,2](✉) ⓘ, Aleksandar Tošić[1,2] ⓘ, and Michael Mrissa[1,2] ⓘ

[1] Faculty of Mathematics, Natural Sciences and Information Technologies, University of Primorska, 6000 Koper, Slovenia
niki.hrovatin@famnit.upr.si
[2] InnoRenew CoE, Livade 6, 6310 Izola, Slovenia

Abstract. The growing adoption of low-cost sensors is raising valid concerns with regards to data, user, computation, and network privacy. In this paper we propose a semantically-driven secure task execution on wireless sensor networks. We rely on blockchain smart contracts and onion routed task execution driven by semantic descriptions to respectively provide role-based access control (RBAC) for query, and support local privacy-preserving task execution. We validate the feasibility of our approach in terms of query time, relative to size of the payload, and number of sensor in the network through NS3 simulations.

Keywords: Blockchain · Privacy · Semantic web · WSN

1 Introduction

Nowadays, Wireless Sensor Networks (WSNs) nodes are powerful enough to support edge computing, thus allowing to execute data processing tasks on site, and to avoid cloud-related drawbacks (high latency, security, privacy...) [6]. However, edge computing raises security concerns (outside access to code), and semantic heterogeneity concerns (different nodes provide different functionalities and data) due to lack of explicit semantic description.

In this paper, we combine lightweight semantic reasoning with onion routing to enable semantic-driven decentralized execution of user tasks. Our solution uses semantic annotations to describe sensor capabilities and semantic reasoning to match them to tasks from user queries. We show how nodes contribute to distributed execution of semantically matched obfuscated tasks with onion routing. We remove any single point of failure, and decouple users from direct access to the WSN by using a permissioned blockchain network running a proof of authority consensus. We deploy a set of smart contracts implementing decentralized RBAC, which limits access to publicly exposed functions.

© Springer Nature Switzerland AG 2022
S. Chiusano et al. (Eds.): ADBIS 2022, CCIS 1652, pp. 476–483, 2022.
https://doi.org/10.1007/978-3-031-15743-1_44

2   Related Work

Firstly, we identify related work on semantic annotation of sensors. The review in [8] references 30 ontologies from 2004 to 2018 to semantically describe sensors or measurements. Most ontologies have a general purpose, so they can be utilized in any application field, and a few of them are specialized to a specific field, such as weather forecast or manufacturing. The most widely adopted proposal nowadays is the W3C Semantic Sensor Network (SSN)[1]. In a similar fashion, the Sensor Web Enablement initiative from the Open Geospatial Consortium [1] provides data models and service interfaces to facilitate access to sensor data. In our work, we rely on the SSN ontology and combine it with well-known ontologies such as QUDT[2] to explicitly describe data concepts and context.

Secondly, our work relates to Onion Routing (OR) [7] as the most used systems for enabling anonymous communication over the Internet. Notable mentions are The Onion Router (TOR network)[3], Invisible Internet Project (I2P)[4], and Lokinet[5]. The original technique described in [7] makes use of a particular message named the onion. The onion is used to establish a bi-directional communication channel for data interchange and guarantee anonymity since nodes involved in the onion relaying do not know the entire path of the onion.

The technique of encoding the message path information in the message itself is known as source routing [15], and it was proposed in several privacy-preserving schemes for WSNs. The source routing technique was applied in [2] to route a declarative query privately to one aggregator node of a WSN. The aggregator node is then executing the query sourcing data from its owned region. Moreover, due to the broadcasting nature of the wireless communication that could disclose the aggregator node, the described technique hides the identity of the aggregator node by issuing multiple bogus queries.

Even though OR is computationally demanding, many researchers are proposing its application to WSNs [5,13]; however, these techniques use OR to establish an anonymous communication channel. In WSNs, this could lead to data origin deanonymization due to the open communication medium. The technique developed in [9] establishes an onion route that leads the message through a circular path and allows only specific nodes in the path to access the message content, thus preserving privacy both from internal and external threats. In this work, we propose an extension with semantic matching to provide distributed task execution and preserve privacy.

Thirdly, our work makes use of blockchain as a means to substitute the need for two or more interacting systems/parties to trust each other or a third party. In recent years, researchers experimented with replacing central trusted parties in many areas such as medical records [3], privacy preservation, telecommunication [11], wireless sensor networks [16]. Of particular relevance in this paper

---

[1] https://www.w3.org/TR/vocab-ssn/.
[2] http://www.qudt.org/.
[3] https://www.torproject.org/.
[4] https://geti2p.net/en/.
[5] https://lokinet.org/.

is the application of blockchain for secure, and privacy preserving access control as described in [4] where the authors propose a RBAC system based on Ethereum smart contracts. The solution implements a challenge-response protocol to realize endorsement relationships between users and their roles. A later simplified implementation was provided by OpenZeppelin[6], which we followed in this paper. Commonly, solutions are implemented in permissioned environments such as private Ethereum [12] in order to avoid the cost of smart contract executions in permission-less environments.

3   Contribution

We propose a semantically-driven solution to execute tasks in a decentralized fashion. We rely on semantic matching to compare the data that nodes provide and the data required in a semantically described query. Typically, a data collection task consists in querying data that is relevant to the data mining process. Therefore, for each sensor, there is a need to describe the collected data (concept) and the conditions of its collection (context). Based on the semantic description of each sensor, it is then possible, using reasoning techniques such as subsumption, to align sensor data for the purpose of a query. For example, in a IAQ monitoring scenario, a query might need only the average of temperatures that are related to a window of 1 day before a certain date. The precision of the temperature value should also be accurate enough to participate in the collection. Then, the unit must also match the units of other values that collect similar data, especially in the case where a building is monitored using heterogeneous equipment (which is mostly the case in shared housing). Indeed, the location of the sensor is relevant for data interpretation. Figure 1 gives a simplified overview (namespaces are not included) of a temperature concept and its context.

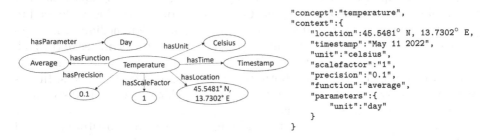

Fig. 1. Graphical and JSON description of the temperature concept and context

Therefore, we propose to semantically describe the data concepts that each node can provide, together with its context. Being given a task on one side and a node description on the other, we are then able to apply semantic matching techniques to evaluate if the data matches any task from the query.

---

[6] https://docs.openzeppelin.com/contracts/2.x/access-control.

To do so, we need to describe as a list of subtasks that can be executed with some order dependency. A task should be described as a couple (`concept`, `context`) where context is a set of descriptions about the conditions of the data to be collected. For example, a node could receive the task described in Fig. 1.

Through an onion route, each node on the network can look at the tasks and identify the parts it can realize. Data can be selected according to the semantic matching of the concepts described in the query with the concepts that describe the sensor and its data. Semantic matching offers the opportunity to not exactly match data, but to adapt to semantically equivalent or replaceable terms. In order to enable semantic matching, we rely on the work described in [14] where concepts are matched to describe functionalities of services. Considering that our solution follows the REST architectural style, our interfaces are generic. Therefore, instead of matching functionality, we use the semantic matching technique on data. That means that the concept described in the query must be equivalent to, or subsume, the one of the sensor description.

Concerning context, we match context similarly. Additional SWRL rules and builtin operators for comparison allow to describe more advanced matching[7]. For example, a data value that describes a precision of 0.01 matches a query that expects a precision of 0.1. Similarly, for the location, distance can be calculated so that the data from close sensors might be acceptable to fulfill the query.

3.1 Onion Routing for Secure Task Execution over WSN

The system described in this paper uses the alternative OR technique that was first proposed in [9]. The proposed scheme does not anonymize the sender and receiver, but it uses an onion message to create an anonymity set[8]. The anonymity set consists of nodes that perform an operation and nodes that only route the message. Therefore, the identity of nodes performing the operation remains hidden even to external actors eavesdropping on the wireless broadcasting. In [9], the anonymity set is established by delivering encryption keys in onion layers only to specific nodes in the onion path.

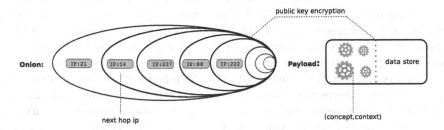

Fig. 2. Graphical representation of a message in the proposed system.

---

[7] https://www.w3.org/Submission/SWRL/#6.1.
[8] https://datatracker.ietf.org/doc/html/draft-hansen-privacy-terminology-02.

Here, we propose to establish the anonymity set via semantic matching and use the onion to provide privacy for task execution. Therefore, differently than in [9], each layer of the onion includes only path details. As can be seen from Fig. 2, the onion is accompanied by the payload, which consists of a set of tasks as described in Sect. 3 and a data store. The payload is protected using public-key cryptography applied on a hop-by-hop basis.

In the following, we resume the operations performed by a node of our system at message receipt. The message payload is deciphered using the node's private key. The node performs semantic matching completing the supported tasks and storing eventual results in the data store of the payload. The onion is deciphered using the node's private key, revealing the next-hop IP address and the inner onion layer. The next-hop IP address is used to determine the encryption key for payload encryption. The payload is encrypted, and the message consisting of the inner onion layer and the payload is forwarded to the node at the next-hop IP.

3.2 Query Construction and Execution

A set of Ethereum smart contracts were deployed on a private instance of the Ethereum network running a proof of authority PoA consensus mechanism. The solution features three modules, the RBAC module for protecting the query capabilities of the underlying WSN, the registry contract that stores a list of pairs (public key, address) of each sensor, and a union of all tasks the underlying WSN is supporting, and the query execution contract responsible executing queries, storing the onion messages and their corresponding results.

In its simplest form, the RBAC contract derives the public key from the calling wallet and maps it to a role. All exposed public functions are secured by the RBAC, which limits access to specific roles. An admin role is responsible for curating the list of users and their role assignments can be modified. The sensor role is given access to insert query results, and the query role is given to users who can query the WSN.

The registry contract exposes the register function limited to the admin role, which is responsible for adding a deployed sensor to the network by supplying the pair (public key, address). The contract simply stores a map of all pairs (public key, address), and the map of template pairs (concept, context) supported by the network. Respectively, the execution contract is responsible for storing the user created onion message, and the corresponding result.

To query the WSN, the user performs the following operation as shown in Fig. 3: 1) Query the registry contract to obtain the list of pairs (public key, address) of all sensors, and available template pairs (concept, context). 2) Shuffle the pairs (public key, address), and select the query tasks. 3) Perform multi layer encryption using the pairs (public key, address) to create an onion and binds it with the selected tasks. The last layer of the onion includes the caller's public key for result submission. 4) Call the execution contract by submitting the onion message. 5) Monitor incoming blocks for the result submitted by the last sensor in the path, encrypted with the caller's public key. 6) Decrypt the result using the corresponding private key.

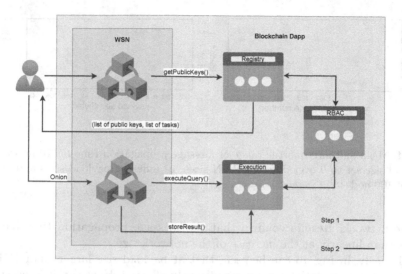

Fig. 3. Workflow diagram of user interactions with the smart contracts

4 Evaluation and Results

To evaluate the presented scheme for distributed task execution, we examine the propagation time of messages at varying onion and payload sizes. Our publicly released simulator [10] relies on the well-known network simulator NS3[9]. We ran two experiments: **a)** Examine the propagation time in networks of different sizes. **b)** Examine the propagation time at different payload sizes. The propagation time is metered starting from the message emanation by the origin node to the message's return to that same node. Each message is constructed to follow a randomized circuit-like path that leads the message through all the network nodes, with the last encryption layer of the onion containing the address of the origin node. Message size is kept fixed through padding.

We setup the simulator described in [9] to simulate the emission of messages in WSNs of various sizes. The WSN is constructed by deploying nodes following a grid structure. The distance between two neighbouring nodes is 60m, the wireless communication is based on the IEEE 802.11n standard at 2.4 GHz and a data rate of 13 Mbps. Messages are transmitted over the TCP protocol, multi-hop routing relies on the Optimized Link State Routing Protocol (OLSR)[10], and cryptography is handled with the Libsodium library[11].

In Fig. 4a we present the propagation time of messages in simulated networks of different sizes. For each network size, 20 onion messages were routed in the network. Each onion message was constructed to be routed through all the nodes

---

[9] https://www.nsnam.org/.
[10] https://www.ietf.org/rfc/rfc3626.txt.
[11] https://doc.libsodium.org/.

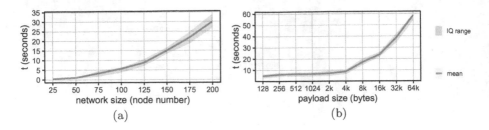

Fig. 4. Mean and interquartile range of message propagation time for 20 messages. a) payload size set to 0 over different WSN sizes. b) different payload sizes and network size of 100 nodes.

in the network. Results confirm that the message propagation time increases faster than linearly at the increase of the network size.

Similarly, in Fig. 4b can be seen, that at payload size larger than $4k$ bytes, the message propagation time is significantly affected. However, it is possible to notice that the propagation time does not double at a fold increase in payload size. Therefore, the increase in payload size has a smaller effect on the message propagation time than increasing the network size.

5   Conclusion

In this work, we presented a scheme for the semantically-driven secure task execution over WSNs. The scheme allows authorized parties to obtain semantic descriptions of services provided by the WSN and construct a layered object for retrieving required results. The scheme employs semantic matching in onion routed messages to establish an anonymity set and avoid disclosing details about services provided by individual sensors to the service consumer or eventual malicious actors.

The scheme was evaluated using the simulation tool presented in [10], results show that applying the proposed technique introduces a substantial latency before obtaining the result. However, the introduced delay was smaller than 35s at a network size of 200 nodes; therefore, we consider such delay acceptable due to the added preservation of privacy.

Acknowledgments. The authors gratefully acknowledge the European Commission for funding the InnoRenew project (Grant Agreement #739574) under the Horizon2020 Widespread-Teaming program, the Republic of Slovenia (Investment funding of the Republic of Slovenia and the European Regional Development Fund) and the Slovenian Research Agency ARRS for funding the project J2-2504.

References

1. Botts, M., Percivall, G., Reed, C., Davidson, J.: OGC® sensor web enablement: overview and high level architecture. In: Nittel, S., Labrinidis, A., Stefanidis, A. (eds.) GSN 2006. LNCS, vol. 4540, pp. 175–190. Springer, Heidelberg (2008). https://doi.org/10.1007/978-3-540-79996-2_10
2. Carbunar, B., Yu, Y., Shi, W., Pearce, M., Vasudevan, V.: Query privacy in wireless sensor networks. ACM Trans. Sensor Netw. (TOSN) **6**(2), 1–34 (2010)
3. Chen, Y., Ding, S., Xu, Z., Zheng, H., Yang, S.: Blockchain-based medical records secure storage and medical service framework. J. Med. Syst. **43**(1), 1–9 (2019)
4. Cruz, J.P., Kaji, Y., Yanai, N.: Rbac-sc: role-based access control using smart contract. IEEE Access **6**, 12240–12251 (2018)
5. El Mougy, A., Sameh, S.: Preserving privacy in wireless sensor networks using onion routing. In: 2018 International Symposium on Networks, Computers and Communications (ISNCC), pp. 1–6. IEEE (2018)
6. Garcia Lopez, P., et al.: Edge-centric computing: vision and challenges. ACM SIG-COMM Comput. Commun. Rev. **45**(5), 37–42 (2015)
7. Goldschlag, D.M., Reed, M.G., Syverson, P.F.: Hiding routing information. In: Anderson, R. (ed.) IH 1996. LNCS, vol. 1174, pp. 137–150. Springer, Heidelberg (1996). https://doi.org/10.1007/3-540-61996-8_37
8. Honti, G.M., Abonyi, J., Natella, R.: A review of semantic sensor technologies in internet of things architectures. Complexity **2019** (2019). https://doi.org/10.1155/2019/6473160
9. Hrovatin, N., Tošić, A., Mrissa, M., Vičič, J.: A general purpose data and query privacy preserving protocol for wireless sensor networks. arXiv preprint arXiv:2111.14994 (2021)
10. Hrovatin, N., Tošić, A., Vičič, J.: Ppwsim: privacy preserving wireless sensor network simulator. SoftwareX **18**, 101067 (2022)
11. Khalaf, O.I., Abdulsahib, G.M., Kasmaei, H.D., Ogudo, K.A.: A new algorithm on application of blockchain technology in live stream video transmissions and telecommunications. Int. J. e-Collabor. (IJeC) **16**(1), 16–32 (2020)
12. Leal, F., Chis, A.E., González-Vélez, H.: Performance evaluation of private ethereum networks. SN Comput. Sci. **1**(5), 1–17 (2020)
13. Palmieri, P.: Preserving context privacy in distributed hash table wireless sensor networks. In: Qing, S., Okamoto, E., Kim, K., Liu, D. (eds.) ICICS 2015. LNCS, vol. 9543, pp. 436–444. Springer, Cham (2016). https://doi.org/10.1007/978-3-319-29814-6_37
14. Paolucci, M., Kawamura, T., Payne, T.R., Sycara, K.: Semantic matching of web services capabilities. In: Horrocks, I., Hendler, J. (eds.) ISWC 2002. LNCS, vol. 2342, pp. 333–347. Springer, Heidelberg (2002). https://doi.org/10.1007/3-540-48005-6_26
15. Sunshine, C.A.: Source routing in computer networks. ACM SIGCOMM Comput. Commun. Rev. **7**(1), 29–33 (1977)
16. Tošić, A., Hrovatin, N., Vičič, J.: A WSN framework for privacy aware indoor location. Appl. Sci. **12**(6), 3204 (2022)

MegaData: 2nd Workshop on Advanced Data Systems Management, Engineering, and Analytics

MegaData: 2nd Workshop on Advanced Data Systems Management, Engineering, and Analytics

Workshop Chairs

Feras M. Awaysheh   University of Tartu, Estonia
Fahed Alkhabbas   Malmö University, Sweden
Said Alawadi   Uppsala University, Sweden

Program Committee Members

Pablo Rodríguez-Mier   INRAE, France
Victor M. Muñoz   Open University of Catalonia, Spain
Manisha Sirsat   INESC, Portugal
Arturo Gonzalez-Escribano   Universiadad de Valladolid, Spain
James Benson   University of Texas at San Antonio, USA
Rosa Filgueira   The University of Edinburgh, UK
Imed Romdhani   Edinburgh Napier University, UK
Sattam Almatarneh   Middle East University, Jordan
Said Alawadi   Uppsala University, Sweden
Syed Attique Shah   University of Tartu, Estonia
Pablo Caderno   University of Santiago de Compostela, Spain
Ahmad Aburomman   University of A Coruña, Spain
Maanak Gupta   Tennessee Technological University, USA
Mehdi Gheisari Guangzhou   University, China
Mohamed Ragab   Tartu University, Estonia
Houshyar Honar Pajooh   Masey University, New Zealand
Xoan C. Pardo   University of A Coruña, Spain
Jose R. R. Viqueira   University of Santiago de Compostela, Spain
Fahed Alkhabbas   Malmö University, Sweden

Stored and Inherited Relations with PKN Foreign Keys

Witold Litwin[✉]

Université Paris Dauphine, Paris, France
Witold.litwin@dauphine.fr

Abstract. A stored and inherited relation (SIR) is 1NF stored relation enlarged with inherited attributes (IAs). The latter make SIRs as base tables the only known view-savers for logical navigation free (LNF) or calculated attribute free (CAF) queries, without any denormalization. Recall that LN means joins among base tables, while calculated attributes serve as the virtual ones do at some popular DBSs, but can be more general, e.g., with aggregate functions or sub-queries. The overall advantage of SIRs is substantially less procedural queries and DB schemes. We now show that usual schemes of stored relations with foreign keys implicitly define IAs forming so-called natural SIRs. The exclusive advantage of the latter is the LNF queries with zero procedurality to define the IAs. We then discuss SIRs with FKs and explicit IAs, the calculated ones especially. We show that generalizing a typical relational DBS to SIR-enabled one should be simple. Preexisting applications could remain not affected, while new ones could profit from LNF and CAF queries. We conclude that major relational DBS should evolve to SIR-enabled "better sooner than later". To make LNF and CAF queries standard, at last, for the benefit of, likely, millions of SQL users.

Keywords: Relational model · Foreign key · Inheritance · Logical navigation · SQL · Stored and inherited relation

1 Introduction

The relational model as defined by Codd has two 1NF constructs (abstractions), [6, 7]. A *stored* relation (SR), also called *base relation or table*, consists of stored attributes, (SAs), only. Values of these attributes are not calculable from other attributes in the DB (that is why they have to be stored). An *inherited* relation, more commonly called *view* or *view table*, consists of (relationally) *inherited* attributes, (IAs), only. One calculates every IA from SAs or other IAs, through a stored (relational) query called *view scheme*. Originally, one supposed every IA calculable only. Later, it appeared practical sometimes to maintain a (stored) snapshot of selected IAs, refreshed whenever needed. Such views and IAs were termed *materialized*, [13–16]. Although stored, a materialized view is not an SR. It is indeed entirely calculable through its (view) scheme.

Recently, we proposed to add the stored and inherited relation (SIR) construct to this model, [1–4]. The construct roots in [17], part of the popular in nineties trend to harness

© Springer Nature Switzerland AG 2022
S. Chiusano et al. (Eds.): ADBIS 2022, CCIS 1652, pp. 487–499, 2022.
https://doi.org/10.1007/978-3-031-15743-1_45

inheritance in the relational DBs. E.g., see [20] or Postgres, [19], or later proposals, [11]. A SIR, say R, is a 1NF relation with both SAs and IAs, the primary key being SAs only. We refer by default to the projection of R on its all and only SAs as to R_ and call it *base* of R. We also say that the IAs *enlarge* R_ and we refer to the IAs scheme as to *Inheritance Expression* (IA). The crucial advantage of SIRs as base tables over the logically the same base tables, but SRs only, as required by the present model, is that no IAs may create normalization anomalies. Unlike it would often happen if the same attributes as IAs were SAs instead. Two important advantages for queries to a DB with SIRs without any normalization anomaly follow, with respect to the equivalent queries to the DB with normalized SRs only, i.e., the queries providing for the same output, [1–4]:

(1) A query Q addressing any SAs or IAs of SIR R can be *Logical Navigation Free* (LNF), while an equivalent query Q', addressing normalized R_ as stand-alone SR named R, would typically require some LN. Recall that the concept designates joins among base tables, typically equijoins on foreign and referenced keys, [18]. Recall also that the normalized SRs as base tables of an SQL DB suffice for every SQL query to the DB. If Q' is such a select-project-join query, Q consists typically from the select-project part of Q' addressing SIR R only. Q is then in practice always less procedural, i.e., requires fewer characters, than Q'. In addition, joins are often felt dreadful, the outer ones especially, [9, 15]. Not surprisingly, clients typically at least dislike the LN. We designate any SIRs free of it for some queries as SIRs *for LNF queries.*

(2) IAs can be *calculated* attributes (CAs). These are defined through relational and value expressions or sub-queries, perhaps with scalar or aggregate functions. A query Q to SIR R with CAs may then be free of defining any of these, selecting CAs by names only. I.e., Q can be a *CAF* query, avoiding the procedurality of the CAs specification within the equivalent Q' to R_ alone. SAs with the same names and values as CAs would do in theory as well, but most often would denormalize the base table to 2NF at best (as we recall by examples later on). We designate any SIRs with CAs as SIRs *for CAF queries.*

A SIR R can provide for both LNF and CAF queries. At present, the only practical way to provide for the capabilities of any SIR R is view R that we call *conceptually equivalent* to or the *canonical* view of SIR R, *C-view* R in short. Every C-view R is logically, i.e., mathematically, equal to SIR R. I.e., the attribute names and order are the same, as well as every tuple. The only difference is physical: every SA A in the base R_ of SIR R is IA A in view R, inherited from R_ as the stand-alone base SR. The latter may actually be a pre-existing base table R we referred to in (1) that one had to rename somehow to create view R. Recall that SQL prohibits any same name relations in a DB.

The "price" to pay for (1) or (2) for SIR R with respect to R_ alone as a base table is the procedurality of the IE, i.e., the minimal number of characters or keystrokes to define it. For SQL, it is some additional procedurality for Create Table R for SIR R, [1]. The similar price for C-view R in SQL is the procedurality of Create View R. The general advantage of every SIR R is that the IE can be less procedural than the Create View R, [1]. The rationale is that the latter has to redefine as an IA every SA of R_.

This obviously must cost some procedurality. By the same token, to create SIR R is always less procedural than to create R_ and C-view R. In popular terms, every SIR R is a *view-saver* for C-view R. Actually, SIRs are also less procedural to maintain, [1].

All this is Our Rationale for SIRs. We follow the general trend in DB-science and in entire CS in fact. Recall that this is why the relational model took over the Codasyl one, although the latter was already in use, e.g., in Oracle Codasyl DBS. Likewise, it is why it took also over every other earlier DB model. The assertional (declarative...) relational algebra queries or, better, the predicative ones, were indeed in general considerably less procedural than any equivalent navigational ones in any of these models. Also, it is lower procedurality of the higher-level programming languages for general programming that wiped out the use of assemblers for it.... See oldies on the subject, e.g., early editions of [8].

In [1], we proposed Create Table extended to SIRs, providing thus also for the IE. One specifies there every SA as one would do within Create Table at present. However the former may interlace with the IAs. Create Table of a SIR includes similarly any table options one may define at present. Recall that these specify the primary key, foreign keys (FKs), etc. The IE may define every IA as C-view R would do. We term every such IE *explicit*. Accordingly, we call *explicit* every Create Table R for SIR R with an explicit IE.

An explicit Create Table R for SIR R first enumerates thus all the SAs and IAs, perhaps interlacing. The clauses From with, perhaps, its sub-clauses follows, as well as any table options. We qualified of *SIR-enabled* every DBS (or DBMS, as some prefer) supporting any Create Table with an explicit IE. We showed that making popular DBSs SIR-enabled should be simple.

Additional rules for SIRs in [1], including SQL '*', providing for implicit IAs for queries and views at present, may provide analogously for an *implicit* IE in Create Table for SIRs. In practice, an implicit IE may have implicit IAs or an implicit From clause. The latter may lack of sub-clauses clauses necessary for the C-View or may even be entirely omitted (an empty IE). We call *implicit* every Create Table R with an implicit IE. We also talk about the *implicit* scheme of R. An implicit IE is always less procedural than the explicit one.

We supposed further that every SIR-enabled DBS typically transparently preprocesses an implicit IE into the explicit one for any further processing, e.g., as for '*' at present. An exception is every (implicit) CA declared as if it was, on some popular DBSs at present, a so-called *virtual* attribute (column), (VA), also called computed, generated, dynamic.... We deal more with the VAs later.

In [2], we conjectured from a motivating example that a usual scheme of an SR R with FKs could also be an implicit scheme of a specific SIR R. That one would have (i) an empty implicit IE, (ii) SAs and table options of SR R as R_ and (iii) would provide for (1). I.e., it would be the least procedural SIR scheme for LNF queries. Accordingly, clients could profit from LNF queries without any additional work for the DBA to define the IE, not to mention C-view R. In the same time, no normalization anomaly with respect to R_ could ever follow. Furthermore, the rules for the explicit IE given such an implicit one seemed simple. We conjectured in [2] that every relational DBS should therefore support such SIRs. This would make typical LNF queries the *de facto* standard, simplifying the life of SQL clients, likely in millions today. We studied the conjecture further in [3]. We have shown that it holds provided that (a) SIR R is a so-called in [3] *natural* SIR and (b) one basically considers FKs as E. Codd originally did in [7], thus more generally than at present.

Below, we continue analyzing SIRs. We first illustrate the above overview of SIRs with motivating examples. We also recall our definition of FKs, based upon Codd's one [7, 12]. Next, we discuss natural SIRs more in depth than in [3]. We then discuss implicit schemes for SIRs with FKs other than the natural ones, with explicit CAs in particular. We show that such schemes should usually be substantially less procedural that the explicit ones.

Afterwards, we discuss Alter Table for SIRs, possibly enlarging an SR R with FKs into the natural SIR R. One may provide accordingly LNF queries to a DB that preexisted the upgrading of its DBS to a SIR-enabled one as discussed here, while basically keeping the legacy applications running. We conclude that to provide for the latter should be easy, hence we maintain our postulate of every DBS evolving towards SIR-enabled "better sooner than later".

2   Natural SIRs

2.1   SIRs by Example

Our framework for motivating examples is the "biblical" S-P DB, Fig. 1. S-P seems the first DB illustrating the relational model, [8]. It is also the mold for about every present DB. Hence, properties of S-P generalize accordingly.

Example 1. Suppose that DBA actually declares the base table S-P.SP as follows:

(1)   Create Table SP (S# Char 5, P# Char 5, QTY INT Primary Key (S#, P#));

Here, the Primary Key clause is a table option. Suppose further that DBA declares the following view, after renaming SP to SP_:

(2)   Create View SP AS Select SP_.S#, SNAME, STATUS, S.CITY, SP_.P#,PNAME, COLOR,WEIGHT,P.CITY,QTY From SP_ Left Join S On (SP_.S# = S.S#) Left Join P On (SP_.P# = P.P#);

Recall that no two relations in an SQL DB can share a name. To rename S-P.SP somehow is thus necessary for (2). Next, observe that (i) every SP tuple (S#, P#, QTY) at Fig. 1 is also logically the proper sub-tuple (S#, P#, QTY) of a tuple of view SP defined by (2) with the same (S#, P#), (ii) that view SP can contain only one such tuple since S.S# and P.P# are keys, and that, finally, (iii) (2) does not define any other tuples. These properties make view SP (2) the C-view R of SIR SP, declared as follows through Create Table SP in SQL extended to SIRs:

(3)   Create Table SP (S# Char 5 {SNAME, STATUS, S.CITY} P# Char 5 {PNAME, COLOR, WEIGHT, P.CITY} QTY INT {From SP_ Left Join S On (SP_.S# = S.S#) LEFT JOIN P On (SP_.P# = P.P#)} Primary Key (S#, P#));

Here, the brackets { } contain successive parts of the IE. They are not a necessity, but simplify the implementation of a SIR-enabled DBS reusing a present DBS that we designate usually as the *kernel* DBS. Observe that the IE is the part of (2) defining the SQL projection on every IA in (3) and only on such IAs. But, SP_ referred to in (3) is the base of SIR SP, that is S-P.SP (1), preserved in (3), although implicitly renamed for the referencing in From clause, the same besides as in (2). Below, we refer to DB S-P with SIR SP (3) instead of SP (2) as to S-P1. Figure 2, on the last page, shows the scheme of S-P1. Figure 3 shows S-P1 content, given that of S-P at Fig. 1. For convenience, every IA at Fig. 3 is *Italics*. The SP content at Fig. 3 could actually be also the one of C-view SP (2). Provided that, with our notation, every column of view SP is *Italic.@*.

Example 2. In S-P there is no referential integrity between SP and S. Hence SP could have tuple t with S# not in S, e.g., $t = $ (S6, P1, 100). Suppose now that instead of From clause in view SP, one declares: From SP Inner Join S On.... View SP resulting from could not be C-view SP. Indeed, the former would not have any tuples with the sub-tuple t. In contrast, such a view could be C-view SP if the referential integrity between S and SP was required.@

Example 3. Recall that for a DML or a DDL statement S, the procedurality, say $p (S)$, is the minimal number of characters (keystrokes) to express S, without convenience spacing especially. In Introduction we recalled that for every SIR R and C-view R, p (IE) in Create Table R is always smaller than p (Create View R). In our example, p (Create View SP) (2) is $p_1 = 156$. For the IE in (3), $p_2 = 112$, with the character count including '}' as part of the IE, but excluding '{ ', as replacing the usual SQL separator ',' that would be there without the IE. So C-view SP is at least $(p_1 - p_2)/p_2*100 = 39\%$ more procedural than the IE. In other words, the IE saves $(p_1 - p_2)/p_1*100 = 28\%$ of p (C-view SP). All these savings for the DBA work provide for the same service for the client, i.e., for the same queries. SIR SP (3) is thus a view-saver for view SP (2). Simply put, on a SIR-enabled DBS, adding view SP (3) to S-P, instead of creating SIR SP (3), would be just a waste of time.@

Example 4. Consider the need for every PNAME supplied by Smith. The SQL query to S-P, say Q_1, requires the LN through the same equijoins between SP and S, and P in From clause as in (3) or equivalent joins. Hence, for the same need expressed as query, say Q_2, to S-P1, the From clause in (3) would do, while the selection on SNAME and projection on PNAME in Q_2 would be the same as in Q_1. Q_2 would be thus an LNF query. Being free of any LN, Q_2 would be always substantially less procedural than Q_1, regardless of the actual LN in the latter. View SP (2) could provide for Q_2 as well, although through substantially more procedural scheme, as we've seen. The view would be the only possibility at present.

The possibility of an equivalent LNF Q_2 to S-P1 instead of Q_1 to S-P with LN like in (3) or equivalent clearly extends to any select-project part of Q_1. The extension implies only that, sometimes, some IAs in Q_2 may need qualified names. E.g., consider any query retrieving supplier's and part's CITY.@

Example 5. Suppose that P.WEIGHT is in pounds, while queries often need it in KGs. Suppose the latter provided by the attribute named, say, WEIGHT_KG, calculated as INT(WEIGHT * 0.454) and preceding WEIGHT. Adding WEIGHT_KG as SA would create the normalization anomaly in P. Making it CA in every query in need of, would increase the procedurality of those. The classical solution is to rather create the convenient view P. A query to P searching WEIGHT_KG could then invoke it by name only, becoming a CAF query for it, accordingly. It is easy to see however that view P could then be also C-view P for SIR P with explicit Create Table P as follows:

(4) Create Table P As (P# Char 4, PNAME Char 20, COLOR Char 10 {INT(WEIGHT * 0.454) As WEIGHT_KG} WEIGHT Int, CITY Char (30) {From P_} Primary Key (P#));

Observe that, again, we have p (IE) $< p$ (Create View P). Observe also that for popular DBSs providing for VAs we spoke about in Introduction, WEIGHT_KG could alternatively enlarge SR P accordingly through an implicit IE. E.g., at yet hypothetical SIR-enabled MySQL one could declare P as:

(5) Create Table P As (P# Char 4, PNAME Char 20, COLOR Char 10, WEIGHT_KG As INT(WEIGHT * 0.454), WEIGHT Int, CITY Char (30), Primary Key (P#));

The obvious benefit is further reduction of procedurality with respect to (4), thus with respect to view P as well therefore. See more on all this in Example 5 in [21]. @

Example 6. Suppose that, in addition to LNF queries, SP clients wish for queries selecting QTY for some supplies that this attribute is always followed by the one named PERCENTAGE. The latter should provide for every supply selected, the percentage that the QTY there constitutes with respect to the entire supply of the part supplied. For the reasons detailed in Example 6 in [21], the most practical SIR SP should be then the one with the explicit scheme:

(6) Create Table SP (S# Char 5 {SNAME, S.CITY, STATUS} P# Char 5 {PNAME, COLOR, WEIGHT, P.CITY} QTY Int {Round(100*Qty/(select sum(X.qty) from SP_ X where X.p# = SP_.p#), 3) as PERCENTAGE From SP_ Left Join S On (SP_.S# = S.S#) LEFT JOIN P On (SP_.P# = P.P#)} Primary Key (S#, P#));

2.2 Foreign Keys for SIRs

Despite being fundamental to the relational model, the concept of the foreign key appears still surprisingly imprecise. The original definition is in [7]. Codd amended it later several times, [12]. The present definitions in textbooks or for popular DBSs differ from the original and are not all equivalent. For SIRs, we basically stick to the original. We thus call *foreign key* (FK) an SA and an SA only, perhaps composite, that (i) is usually not a (stored) primary key, (PK), but every of its values could be that of some uniquely chosen PK. Then, (ii), FK "cross-references" its (own) relation and the one with PK. Accordingly, one qualifies the latter usually also of the *referenced* key (RK). Likewise, one qualifies so the relation with the RK, say R'. In turn, the FK may be qualified of the *referencing* one, as well as its relation, say R.

As [7] details, (i) implies that FK domain is a sub-domain of that of RK. Originally, it meant in particular that FK and RK share also the proper name. The distinction between domains and attributes came indeed later. Then (ii) means that every FK-value v, references tuple t with RK value v, provided that t exists. FK idea realizes thus the "cross-referencing" through logical pointers, unlike through the physical one, the basic mode for referencing by the times of [7]. The benefit claimed is the *logical/physical data independence*. In particular, as known, if a query needs some values in R together with some referenced ones in R', then the *(left) FK-join*: R left outer join R' on FK = RK in the query expresses the referencing, regardless of underlying physical data structures and changes to these. Likewise does the equivalent right FK-join, or, sometimes, the equivalent inner FK-join if the referential integrity is enforced. Notice that FK-joins in queries constitute the already mentioned LN. See oldies for more on the theme.

Below, we consider that for any SIR R an SA (named) F, perhaps composite, is an FK for either of the two reasons:

a. F is declared so through the familiar FOREIGN KEY clause in Create Table R or Alter Table R. Every such F is a *declared* FK. For a declared F, RK may be a candidate key on some popular DBSs. This is nevertheless at best, a debatable choice. Unlike for the PK, the absence of the table option for candidate keys in SQL makes it indeed error-prone. In every case, a declared F is subject to the usual referential integrity. As at present besides, since according to any definitions of the concept we are aware of, especially in SQL, an attribute F can be an FK iff one declares F so.

b. F is atomic and is neither PK of R nor a declared FK or within the latter. Also, prior to the processing of Create Table R, there is in the DB one and only one R' with PK R'.F (notice that this makes R' $\neq$ R necessarily). We call any such F a *natural FK*. R' is then the referenced relation and R'.F is the RK. A natural FK is not subject to the referential integrity. We follow the original, claiming the referential integrity definable only, [7, 12]. That is why apparently Codd constrained the referencing values to the referenced domain only, not to the actually stored RKs.

As already stated, natural FKs seem the original meaning of the concept and are also the most popular FKs, perhaps surprisingly for some. The rationale is the least procedural FK-joins in queries. Atomic declared FKs do the same, but require the declarations, while the referential integrity is not always the must. Observe also that our definition of an FK implies that every composite FK must be declared. The rationale is that the referencing FK -> RK is by attribute position at present, not by the name sharing. It may happen however that the ordered set of the proper names forming a composite FK is the same as the ordered set of the referenced proper names. One may consider then that FK and RK share a (proper) name. Also, even the composite RK is usually PK of R'. We call accordingly *PK-Named* (PKN) every FK, atomic or composite, that shares the name of an RK being a PK. A PKN FK can thus be natural or declared, while every natural FK is PKN by definition.

Furthermore, we suppose that, in every Create Table R submitted for processing by a SIR-enabled DBS, every PKN FK F and only such F implies specific IAs in the explicit Create Table R. The latter is the final R scheme, we recall. In the nutshell, for each such F, the IAs added to the implicit R scheme, mirror all the non-PK attributes of R'. Namely, they have the names and values defined by the FK-join over R, for an SR R or over R_ for a SIR R, and over R'. We call these IAs the *natural inheritance*, (NI), *from* R' or *through* F in R...). We define them formally in the next section. In contrast, a non-PKN FK in R scheme implies the referential integrity only.

It results from the above that for every PKN FK, declaring it implies both the referential integrity and NI. If one does not want the former, one should not declare the FK. For every atomic PKN FK, a natural FK will result and fit the goal. This cannot work for any composite PKN FK. One solution is (i) to add to R' a *surrogate* SA, say C that is an atomic PK and (ii) add an SA C to R as well, while dropping the composed FK. C will be a natural FK then, providing for NI only, as wished. Example 10 later on illustrates the point.

Finally, we suppose that, as at present, no Create Table R" or Alter Table R" propagates to any existing table R ≠ R". In particular no SA F of some R can become implicitly a natural FK, because one issued Create Table R" with PK F or issued Alter Table R that ended up with attribute F named upon PK of some R". In practice, it means that no such statement can enlarge R with NI from R". A dedicated Alter Table R we discuss in Section_ is necessary.

Example 7. Natural FKs are present in S-P, assuming tacitly that DBA creates S and P first. SP.S# and SP.P# are natural FKs then, with PKs S.S# and P.P# being the respective RKs. The original verbal description of S-P scheme indicates indeed that each pair has a common domain. Finally, as the natural FK, SP.S# in SP scheme (1), will imply NI from S. Likewise, SP.P# will imply NI from P. As it will appear formally in next section, SP scheme (1) will lead then to SP scheme (3) as the explicit scheme, with (1) as the implicit one.

The original description of S-P also does not mention any referential integrity. Nevertheless, at Fig. 1, every SP tuple respects that constraint for each FK. Regardless, one may insert, e.g., P7 into SP, without the presence of P7 in P. The feature can be useful, e.g., if DBA allows for the data for P7 in P to be inserted later.

S	S#	SNAME	STATUS	CITY		SP	S#	P#	QTY
	S1	Smith	20	London			S1	P1	300
	S2	Jones	10	Paris			S1	P2	200
	S3	Blake	30	Paris			S1	P3	400
	S4	Clark	20	London			S1	P4	200
	S5	Adams	30	Athens			S1	P5	100
							S1	P6	100
P	P#	PNAME	COLOR	WEIGHT	CITY		S2	P1	300
	P1	Nut	Red	12	London		S2	P2	400
	P2	Bolt	Green	17	Paris		S3	P2	200
	P3	Screw	Blue	17	Rome		S4	P2	200
	P4	Screw	Red	14	London		S4	P4	300
	P5	Cam	Blue	12	Paris		S4	P5	400
	P6	Cog	Red	19	London					

Fig. 1. S-P database

If the referential integrity was required for a pair, e.g., (SP.S#, S.S#), one should declare it in Create Table SP or Alter Table SP. This, using the usual: Foreign Key (S#) References S(S#)… SP.S# would be a declared (atomic) PKN FK then. On the other hand, if in S-P as on Fig. 1, S would have been created after SP, then SP.S# would not be a natural FK anymore. Consequently, there would not be NI through it in the explicit SIR SP scheme resulting from Create Table SP (1).@

2.3 Basic Natural SIRs

We will now show that Create SP (1), can be an implicit scheme for SP (3). Recall that (1) defines all and only SAs of SIR SP (3), hence, we have $p(IE) = 0$ there. The property frees thus DBA in need to create (3), from any additional procedurality otherwise required. We will show that the property generalizes in fact to any SIR R qualified of *natural* in [2].

Definition 1. Let R be a SIR with FKs, atomic or composite, denoted for the PKN ones only in the left to right order as F1,F2.... Let us denote respectively (i) the relations referenced by these FKs as R'1..., (ii) for each R', the ordered set of all the non-PK attributes in R', say A1,A2... whether SAs or IAs, as $\underline{A}'$. Next, suppose that for a composite PKN FK, the notation R_.F = R'.F means R_.f1 = R'.f1 And R_.f2 = R'.f2 And.... Finally, for every $\underline{A}'$, consider that $\underline{A}$ denotes the set of IAs in R defined by the pseudo-SQL query: Select $\underline{A}'$ From R_ Left Join R' On R_.F = R'.F. Then, we say that R is *natural* if (a) in the attribute list in the explicit Create Table R, every $\underline{A}_i$ immediately follows (the last SA of) F_i: $i = 1,2..$; (b) R has no other IAs.@

The following easy proposition follows.

Proposition 1. If R is natural, then the (explicit) From clause is as follows or equivalent:

From R_ Left Join R'$_1$ On R_.F$_1$ = R'.F$_1$ Left Join R'$_2$ On R_.F$_2$ = R'$_2$.F$_2$ Left...

Indeed, the first join defines $\underline{A}_1$. Likewise, the 2nd joins defines $\underline{A}_2$, without affecting $\underline{A}_1$ names and values, by properties of left outer joins. Etc.@

Definition 2. For every natural SIR R, we qualify every IA A within of *naturally inherited (in R or from R' or through F*, for some FK F) or simply of a *natural* IA. Alternatively we say that R *naturally inherits* every A and $\underline{A}$ (*from* R', etc.). Also, each $\underline{A}$ constitutes in R the NA *through* F or *from* R'. Likewise, $\underline{A}_1$, $\underline{A}_2$... together form the NI *through* all PKN FKs and they also *naturally* enlarge R_. Next, we say for every R.R'.A in some R.R'.$\underline{A}$, that R'.A' is the *source* of A. Finally, we say that R is a *basic* natural SIR iff every R' is an SR.

Example 8. On a SIR-enabled DBS, S-P1.SP illustrated at Fig. 2 is a natural SIR. We leave the proof as the exercise or see Example 8 in [21]. In contrast, S-P1.SP enlarged further with PERCENTAGE would not be a natural SIR, (why ?).

Accordingly, in our terminology, every IA following S# till SA P# is sourced in S. Next, SP naturally inherits each and all of them. Respectively, same is true for every IA following P# till QTY. All these IAs together constitute for SP its NI through the foreign keys and they naturally enlarge SP_. Finally, SP is a basic natural SIR.@

Observe the following easy properties of natural SIRs:

Proposition 2. Suppose that Create Table R in some DB defines at present an SR R with FKs defined in Definition 1. Let $\underline{S}_0$ denote all the SAs preceding the last SA of F_1, the only one for an atomic F_1 of course, S_1 – all these preceding in this sense F_2 etc. Accordingly, suppose the following generic formula for Create Table R:

(7) Create Table R (S_0, F_1, S_1, F_2.... <table options>);

Then, (a) the following formula defines the explicit Create Table R for the natural SIR R, with From clause including perhaps the Where one:

(8) Create Table R (S_0, F_1, A_1, S_1, F_2, A_2,... <From clause>, <table options>);

Also, (b) one may consider Create Table R (7) as the implicit one of the natural SIR R. Moreover, the implicit IE is then empty, i.e., $p(\text{IE}) = 0$, we recall. Next, (c) the SAs in R (7) and the table options there, define also the base R_ of the natural R (8). Finally,

(d) for every SR R as above and every base R_ thus, there is only one natural R in the DB and vice versa.@

Because of space limitations here, the (easy) proof is in [21]. Besides, given (d), for every natural SIR R, we say sometimes that R is natural *for* R_ or *for* SR R. Example 9 and 10 in [21] detail all the above. Furthermore, as we already hinted for S-P1, the clients would gain LNF queries to SP, regardless of whether the DBA uses (3) or, better, (1) for S-P1.3P. We have hinted also that this may apply more generally to natural SIRs. It applies in fact to every basic natural SIR. See Proposition 3 in [21].

Example 9. Consider the need for SP.S#, SNAME, CITY for every supply. Every Q_1 to S-P must have then the LN through the FK-join SP Left Join S. E.g., one may issue Q_1 as:

(9) Select SP.S#, SNAME, CITY From SP Left Join S On SP# = S.S#;

Given transformations (10) – (11) in [21], for S-P1, Q_1 boils down to Q_2:

(10) Select S#, SNAME, S.CITY From SP;

Observe that $p(Q_1) = 56$ and $p(Q_2) = 31$. Thus, LN alone in Q_1 is almost as procedural as Q_2. Hard to see why an S-P client having choice could ever prefer Q_1 to Q_2. See Example 10 in [21] for more on all this, as well as Proposition 3bis there. @

Recall also that every IA A of a natural SIR R, is a natural one itself. By definition, it thus has the same name as an attribute of some base table R' that R references, called also *source* of R.A. Thus one may consider that for every query Q to R only that we qualified of LNF, for every IA A that Q perhaps addresses, Q addresses then in fact some R'.A. One may say then that Q is an LNF query not only to R, but also, indirectly through every IA A, to every base table R' that A is sourced in. For some, that meaning of an LNF query is perhaps the primary one even, [17].

Observe finally that if Create Table R defining at present an SR R only, may define the natural SIR R instead, then it provides for the discussed attractive LNF queries, at no additional data definition cost for DBA. We now describe the algorithm effectively inferring the explicit natural SIR R scheme from the one of the SR R, on any popular DBS.

2.4   Compound Natural SIRs

A *compound* natural SIR R inherits through some FKs from SIRs. These can be natural perhaps compound themselves, or others. In other words, a non-PK attribute of an R' can now be an SA or an IA. Operationally, as usual today, we suppose again every R' being created before R. By the same token, we suppose that no later alterations of any of R' schemes cascade to R. Here are motivating examples of compound natural SIRs. They seem framework for frequent future practical cases.

Example 11. Suppose one alters S-P scheme as follows. An additional relation CG (CITY, GPS) stores uniquely for each city the GPS location. Suppose further that on a SIR-enabled DBS, one creates CG first, then S and P with their S-P schemes as at Fig. 1

and SP through its scheme (1), at last. Then, SP is a compound natural SIR. See why in Example 13 in [21], as well as for more examples there.

2.5 Inferring Explicit Schemes of Natural SIRs

See [21] for the discussion of the algorithms termed Alg. 1 and 2 there. These find every PKN FK from SYSTABLES, present at any popular relational DBSs.

3   SIRs with PKN FKs and Explicit IAs

SIR R with some PKN FKs may have also an implicit scheme with some explicit IAs. The explicit R scheme contains then in addition every NI. The FK-joins defining the latter enlarge the implicit From clause declared for the explicit IAs. The latter is the explicit From clause without however every FK-join defining NI in R through some PKN FK. An empty implicit From clause may ultimately result. For every implicit IE of R with explicit IAs, we must obviously have p (IE) > 0. Still, the latter may provide for substantial procedurality savings with respect to the explicit IE and even more as the view-saver.

The typical needs for the explicit IAs seem as follows. (i) R has one or more CAs, defined through a value expression inheriting from SAs or other IAs in R or from some R', or defined by a sub-query, or defined as if A was a VA for the kernel DBS supporting such attributes. Then, (ii) for privacy, for some FK F, R may have only some or even none of the natural IAs through F. Or, (iii), for some F, R may have the same IAs as in NI through F, but, for query convenience, displaced within R or renamed. Finally, (iv) F_1 and F_2; $F_1 \neq F_2$; may share R'; unlike for the assumptions for NI, we recall. The motivating examples for all these needs are in [21], as well as deeper discussion skipped here, given the space limit.

4   Altering SIRs with FKs

One can alter every SIR R through the explicit Alter Table R for SIRs, [1]. For even lesser procedurality, we now consider that for any R with PKN FKs, Alter Table R can have also an implicit clause denoted IE (). The clause recalculates the IE. Specifically, it may add the IE even if R preexists the upgrade of a DBS to a SIR-enabled one. Recall that such upgrade would enlarge R to a natural SIR R, without affecting any data in R. It would bring the free bonus of the LNF queries to any SA of R and to any of the non-PK attributes of any R'. Given space limits here, see [21] for more details and the motivating Example 16 there.

5   Implementing SIRs

The *canonical* implementation of SIR-enabled DBS proposed in [1], extended to implicit schemes above discussed, seems easy [5, 10, 14, 21]. See the rationale for this claim in [21]. Recall simply that every such implementation would provide for the additional *SIR-layer* above the kernel DBS. That one would offer SIRs and all the usual data management services provided by the kernel internally.

Create Table S (	Create Table P (	Create Table SP (
S# Char 5,	P# Char 5,	S# Char 5 {SNAME, STATUS, S.CITY}
SNAME Char 30,	PNAME Char 30,	P# Char 5 {PNAME, COLOR, WEIGHT,P.CITY}
STATUS Int,	COLOR Char 30,	QTY Int
CITY Char 30,	CITY Char 30,	{From (SP_ Left Join S On SP_.S#=S.S#) Left Join
Primary Key (S#));	WEIGHT Int,	P On SP.P#=P.P#)}
	Primary Key (P#));	Primary Key (S#, P#));

Fig. 2. Explicit S-P1.SP scheme. Implicit one is that of S-P.SP, outside { } brackets here.

Table S

S#	SNAME	STATUS	CITY
S1	Smith	20	London
S2	Jones	10	Paris
S3	Blake	30	Paris
S4	Clark	20	London
S5	Adams	30	Athens

Table P

P#	PNAME	COLOR	WEIGHT	CITY
P1	Nut	Red	12	London
P2	Bolt	Green	17	Paris
P3	Screw	Blue	17	Rome
P4	Screw	Red	14	London
P5	Cam	Blue	12	Paris
P6	Cog	Red	19	London

Table SP

S#	SNAME	STATUS	S.CITY	P#	PNAME	COLOR	WEIGHT	P.CITY	QTY
S1	Smith	20	London	P1	Nut	Red	12	London	100
S1	Smith	20	London	P2	Bolt	Green	17	Paris	200
S1	Smith	20	London	P3	Screw	Blue	17	Oslo	400
S1	Smith	20	London	P4	Screw	Red	14	London	200
S1	Smith	20	London	P5	Cam	Blue	12	Paris	100
S1	Smith	20	London	P6	Cog	Red	19	London	100
S2	Jones	10	Paris	P1	Nut	Red	12	London	300
S2	Jones	10	Paris	P2	Bolt	Green	17	Paris	400
S3	Blake	30	Paris	P2	Bolt	Green	17	Paris	200
S4	Clark	20	London	P2	Bolt	Green	17	Paris	200
S4	Clark	20	London	P4	Screw	Red	14	London	300
S4	Clark	20	London	P5	Cam	Blue	12	Paris	400

Fig. 3. S-P1 data. IAs are *Italic*. S-P1.SP is the natural SIR SP for S-P.SP at Fig. 1.

6 Conclusion

Typical present DB schemes managed by a SIR-enabled DBS provide for LNF queries to base tables at no data definition cost whatever to DBA. Also, SIRs with PKN FKs and with CAs provide for LNF or CAF queries, through Create Table substantially less procedural than known till now. This would be a bonus for the DBA as well. Future work should therefore start with prototype implementations of SIR-level for popular DBSs.

Altogether, it appears that the usual relational DB schemes were not read as they should be. The often felt dreadful LN in otherwise simple base table queries uselessly bothered generations. Likewise did the need for the views, the only tool to offset this shortcoming in practice till now. Same for the present limitations of CAs in base tables either restrained to VAs or requiring views till now as well. Major DBs should become SIR-enabled "better sooner than later". Making LNF & CAF queries to the base tables the standard, at last. It will be a long overdue service to, likely, millions of SQL clients at present.

Acknowledgments. We are grateful to Ron Fagin for invitation to present this material at IBM Almaden Research Cntr., March 2020. Likewise, we thank Darrell Long for his March 2020 invitation to talk about at UCSC Eng. as well.

References

1. Litwin, W.: SQL for stored and inherited relations. In: 21st International Conference on Enterprise Information Systems, (ICEIS 2019) (2019). 12 p. http://www.iceis.org/?y=2019
2. Litwin, W.: Manifesto for improved foundations of relational model (EICN-2019). Proc. Comput. Sci. **160**, 624–628 (2019)
3. Litwin, W.: Natural stored and inherited relations. In: (EUSPN-ICTH 2021), November 1-4, Procedia Computer Science, vol. 198. Elsevier, pp. 171–178 (2022).
4. Litwin, W.: Stored and Inherited Relations. arXiv preprint arXiv:1703.09574 (2017)
5. Litwin, W.: Supplier-Part Databases with Stored and Inherited Relations Simulated on MS Access. Lamsade Tech. E-Note (2016)
6. Codd, E.F.: Derivability, Redundancy and Consistency of Relations Stored in Large Data Banks. IBM Res. Rep. RJ 599 #12343 (1969)
7. Codd, E.F.: A relational model of data for large shared data banks. CACM **13**, 6 (1970)
8. Date, C.J.: An Introduction to Database Systems. Pearson Education Inc. (2004). ISBN 0-321-18956-6
9. Date, C.J., Darwen, H.: Watch out for outer join. In: Date and Darwen Relational Database Writings (1991)
10. Date, C.J.: View Updating and Relational Theory. O'Reilly (2012)
11. Date, C.J.: Type Inheritance & Relational Theory. O'Reilly (2016)
12. Date, C.J.: E.F. Codd and Relational Theory. Lulu (2019)
13. Goldstein, J., Larson, P.: Optimizing Queries Using Materialized Views: A Practical, Scalable Solution. ACM SIGMOD (2001)
14. Halevy, A.Y.: Answering queries using views: a survey. VLDB J. **10**, 270–294 (2001)
15. Jajodia, S., Springsteel, F.N.: Lossless outer joins with incomplete information. BIT **30**(1), 34–41 (1990)
16. Larson, P., Zhou J.: Efficient Maintenance of Materialized Outer-Join Views. ICDE (2007)
17. Litwin, W., Ketabchi, M., Risch, T.: Relations with Inherited Attributes. HPL. Palo Alto, CA. Tech. Rep. HPL-DTD-92-45, 30 (1992)
18. Maier, D., Ullman, J.D., Vardi, M.Y.: On the foundations of the universal relation model. ACM-TODS **9**(2), 283–308 (1984)
19. Postgres SQL. https://www.postgresql.org/
20. Stonebraker, M., Moore, D.: Object-Relational DBMSs: The Next Great Wave, 2nd edn. Morgan Kaufmann (1998)
21. Litwin, W.: Stored and Inherited Relations with Natural or Declared Foreign Keys. Lamsade Res. Rep. Report June 16, 2020. Latest update: 11 August 2022. 12p. https://www.lamsade.dauphine.fr/~litwin/Designing%20an%20RDB%20with%20SIRs.pdf

Implementing the Comparison-Based External Sort

Michael Polyntsov[1,2], Valentin Grigorev[2], Kirill Smirnov[1,2],
and George Chernishev[1,2(✉)]

[1] Saint-Petersburg State University, Saint Petersburg, Russia
polyntsov.m@gmail.com, kirill.k.smirnov@gmail.com, chernishev@gmail.com
[2] PosDB Team, Saint-Petersburg, Russia
valentin.d.grigorev@gmail.com

Abstract. In the age of big data, sorting is an indispensable operation for DBMSes and similar systems. Having data sorted can help produce query plans with significantly lower run times. It also can provide other benefits like having non-blocking operators which will produce data steadily (without bursts), or operators with reduced memory footprint.

Sorting may be required on any step of query processing, i.e., be it source data or intermediate results. At the same time, the data to be sorted may not fit into main memory. In this case, an external sort operator, which writes intermediate results to disk, should be used.

In this paper we consider an external sort operator of the comparison-based sort type. We discuss its implementation and describe related design decisions. Our aim is to study the impact on performance of a data structure used on the merge step. For this, we have experimentally evaluated three data structures implemented inside a DBMS.

Results have shown that it is worthwhile to make an effort to implement an efficient data structure for run merging, even on modern commodity computers which are usually disk-bound. Moreover, we demonstrated that using a loser tree is a more efficient approach than both the naive approach and the heap-based one.

Keywords: Query engines · Query processing · External sort

1 Introduction

Sorting is a very important operation in any data processing system. For example, in DBMSes leveraging sorted data may allow efficient implementation of operators [27] such as sort-merge join, which is one of the most popular approaches to join two tables larger than available memory in a reasonable time. Another example is performing aggregation over data sorted on aggregation attributes. In this case, it is possible to implement a non-blocking aggregation operator, i.e., an operator which can start producing results without having to read all input records.

There are two classes of sorting operators—internal (in-memory) and external (disk-based) sorts. The former assumes that all data fits into main memory and the latter has to write intermediates to disk. Recent research mostly focuses on an

© Springer Nature Switzerland AG 2022
S. Chiusano et al. (Eds.): ADBIS 2022, CCIS 1652, pp. 500–511, 2022.
https://doi.org/10.1007/978-3-031-15743-1_46

in-memory type because of a surge of interest that in-memory processing currently experiences [12, 22, 25]. Such systems try to avoid external sorting, since it signifi-cantly degrades performance. Despite all this, disk-based systems are still in use[1] and research in this area is relevant.

Thus, efficient implementation of the sorting operator is crucial for the perfor-mance of a DBMS as a whole. Such implementation heavily depends on the amount of an engineering effort put into the source code, which is specific for each particular system. For example, it may include adjusting implementation to various system parameters (e.g. disk block size, cache line size, etc.) and writing efficient code in general. However, some high-level techniques can be reused between systems, and thus they are of scientific interest. For example, such techniques as key normaliza-tion or record surrogate usage are discussed in Graefe's survey [9].

In this paper we study the implementation of efficient external sorting in a disk-based column-store. Aside from the aforementioned application, in column-stores, external sort has another important one: it addresses the out-of-order prob-ing problem [1]. The particular research focus of this paper is the impact of data structure choice on the performance of the comparison-based external sort. This data structure is used during run merging and allows to reduce the number of com-parisons. More specifically we pose two following research questions: 1) is it ben-eficial to use a specialized data structure on the merge step in a contemporary environment (commodity PCs), and 2) which kind is the best.

To answer them, we have implemented an external sort operator inside PosDB [3, 4], a distributed column-store engine. In the operator's core, we have devised three approaches to implementing the merge: naive (comparing all to all), a loser tree, and a binary heap.

To validate the quality of our implementation we first compare it with Post-greSQL's external sort. Then we evaluate these three data structures by comparing them with each other using a synthetic and a real dataset.

The paper is organized as follows. In Sect. 2 we provide a concise introduction into external sorting. Then, in Sect. 3 we discuss the related work and provide a motivation for our study. Next, in Sect. 4 we describe the proposed approaches and their implementation. In Sect. 5 we explain conducted experiments and their results. Finally, we conclude this paper with Sect. 6.

2   Background

Sorting, because of its importance, has been studied very extensively. Sorting algo-rithms fall into two categories: external and in-memory (or internal [13]) sorting. In this paper we will focus on external sorting. An external sorting algorithm can be either comparison-based or partition-based. Algorithms of the partition-based class distribute input tuples into buckets using several pivot values (selected by the algorithm) while ensuring the following properties:

---

[1] As of July 2022, disk-based systems PostgreSQL and MySQL were in the top-5 most used according to DB-Engines Ranking (https://db-engines.com/en/ranking).

- each bucket covers some data range,
- all data ranges do not overlap,
- a union of all data ranges covers the whole range.

Then the algorithm sorts the contents of individual buckets and in doing so it is ensured that each bucket fits into memory. The result is an ordered sequence of buckets, each containing a sorted range.

The partition-based sorting algorithm has two degrees of freedom: the algorithm that sorts the buckets in memory and pivot selection.

Comparison-based algorithms generate sorted runs (i.e., in this case, data ranges intersect) and then merge them into one long run. They consist of two stages: run generation and run merging. These stages can be either completely independent and consecutive, or alternate (known as oscillating sort [13]). Consequently, there are two approaches to run generation:

1. Read tuples from disk for as long as there is enough available memory; sort these tuples using some internal sorting algorithm;
2. The algorithm maintains a data structure in memory with a priority queue interface. On each iteration the lowest value from the data structure is retrieved and moved to a buffer or written to the disk. Then algorithm inserts the next value from the input data into the data structure. If the next value is smaller than the last retrieved, it should be marked as belonging to the next run. When the data structure contains only such marked values, the current run is treated as completed. After this, the new one is started. This method is known as the replacement selection [13]. The tournament tree [13] can be used to implement this method.

The second approach was more popular in the past. On average, it generates runs that are twice the size of available memory, which can be useful in a memory-constrained environment. Nowadays, available RAM sizes have grown significantly and the Quicksort became more popular.

In the past, tapes were used instead of disks for persistent storage. Of course, there can be more runs than tapes. Different algorithms exist to merge runs and distribute them over tapes, e.g. multiway merge, polyphase merge, cascade merge, etc. These algorithms are called merge patterns. Disks are the ubiquitous storage type nowadays, but the concept of tape as an abstraction can still be useful in order to save disk space by reusing input tapes as output tapes. This technique is still used in industrial systems[2].

In addition to the merge pattern, comparison-based sorting algorithms differ in the algorithm for selecting the smallest value among the first elements of the merged runs. The naive approach, and therefore the asymptotically slowest, is to check all candidates. A more efficient approach here is to use a data structure that speeds up finding the smallest element. For this purpose, a heap or a tournament tree (either loser or winner) are the most common choices.

---

[2] https://github.com/postgres/postgres/blob/REL_14_STABLE/src/backend/utils/sort/tuplesort.c.

3   Related Work and Motivation

3.1   Related Work

Over the years, there has been a lot of research on external sorting. The algorithmic theoretical aspects of external comparison-based sorting are very well understood. That is, in-memory sorting algorithms, run merge algorithms and the appropriate data structures for single-processor environment are known and have not largely changed over time since Knuth's survey [13]. Most of the current research in non-distributed external sorting is focused on efficient implementation of an algorithm on modern hardware, i.e., implementing sorting that fully utilizes all available hardware capabilities. The two most popular concerns are how to utilize caches efficiently and how to mitigate disk bandwidth limitations. More generally, one can classify the most of the current research papers on external sorting as follows:

- techniques to improve sorting performance by using hardware resources more efficiently, namely: CPU and caches, HDD/SSD, GPU;
- novel external sorting algorithms that make heavy use of parallelism;
- distributed external sorting algorithms.

In the paper [16] authors describe the AlphaSort algorithm. It was designed with memory hierarchy in mind and therefore exhibits cache-friendly behavior. The Quicksort is used for in-memory sorting. Runs are merged using a tournament tree. A parallel version of the algorithm for multiprocessor systems is also presented.

The authors of the study [19] focus on external sort operator in memory-inconsistent environment. They consider real-time or goal-oriented DBMSes where the available memory for a query (or transaction) may change on-the-fly due to other concurrent transactions. The authors conclude that using replacement selection with block writes at the run generation stage and their approach called dynamic splitting at the run merge stage is the most efficient way to handle memory fluctuations.

In the paper [20] the authors present the Leyenda algorithm. It is a novel parallel external sorting algorithm with state-of-the-art performance. Hybrid approach is proposed to generate runs, namely an adaptive parallel most-significant-digit Radix sort. K-way-merge algorithm is used to merge runs. In order to achieve efficient I/O Leyenda leverages parallelism and memory mapping. mmap() is used to map disk memory to OS pages, allowing parallel I/O operations to be performed on it.

Paper [5] is focused on the splitting stage of a parallel multiprocessor external sort in shared-nothing environment. It is assumed that the sorting algorithm works in a Samplesort [6] fashion. The study compares two approaches to splitting input tuples at the distribution stage: exact splitting [11] and probabilistic splitting. The authors conclude that the latter is more efficient and, with the growth of the input file size, the performance gap only increases.

The authors of the study [7] are focused on sorting on GPUs: they propose a novel external sorting algorithm GPUTeraSort. The idea is that all

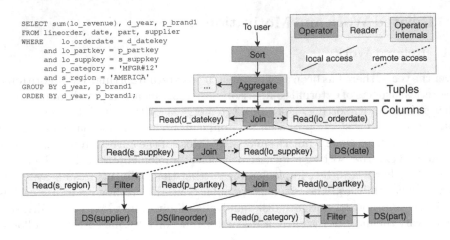

Fig. 1. Ultra-late materialization plan example

computationally-intensive and memory-intensive operations are handled by the GPU while the CPU manages resources and I/O operations. The algorithm sorts key-pointer pairs using an improved bitonic sort on the GPU. Disk striping is used to achieve peak I/O performance.

The authors of the paper [26] describe various techniques for speeding up I/O operations of a comparison-based external sort on HDD. They review and compare two layout strategies: contiguous layout and interleaved layout. The authors compare three reading strategies: forecasting, double buffering and planning strategy. They conclude that a contiguous layout and the planning read strategy are the most efficient.

Study [24] is focused on external sorting on a system with NAND flash memory as a secondary storage (SSD). Unclustered sorting and the concept of reused pages are proposed. Authors state that clustered sorting can be more efficient if input tuples are small enough. A decision rule to select a more suitable algorithm at run time is presented.

In the paper [2] authors propose external sorting algorithms on Network of Workstations, measure their performance and compare it to performance of algorithms on shared-memory computers. Comparison-based algorithms are used for external sorting. Also, disk striping and overlapping of computation and I/O are discussed in-depth.

In the study [14], the authors are focused on external sorting on the shared-nothing architecture. They propose a novel parallel load-balanced multiple-input multiple-output algorithm. The authors conclude that their approach is more efficient than exact splitting and comparable to probabilistic splitting.

3.2   Motivation

Thus, to the best of our knowledge, there are no studies that consider the design of a comparison-based sort and evaluate the choice of the data structure used on the merge step for implementing the external sort operator in DBMSes.

At the same time, it is of interest to industry and leads to a number of important questions. It is a well-known fact that commodity computers with a single disk are I/O-bound and thus, there may be no point to invest effort into perfecting the in-memory part of the algorithm. Next, if it is nevertheless beneficial, further questions arise—which data structure is the best and how does varying parameters affect it.

4   Proposed Solution

4.1   PosDB Basics

To experimentally evaluate the proposed solution, we used PosDB [3,4], a column-oriented parallel distributed disk-based query engine. PosDB uses the Volcano model [8] to represent query execution plans. That is, each plan is an operator tree with the edges describing data flows. Each operator has an iterator interface. PosDB implements block-oriented processing (operators exchange blocks of tuples), which is more efficient [18] than tuple oriented processing. Since PosDB is column-oriented, i.e., stores tables in a columnar format, at some point during the query execution, it is necessary to convert the column representation to a tuple-based one. This moment is called the materialization point, and the stage of query execution where this conversion takes place is determined by the materialization strategy. In the query plan presented on Fig. 1 materialization point is indicated by a brown dotted line. PosDB supports several materialization strategies, namely: early, late and ultra-late materialization, and partially supports hybrid materialization. Early materialization strategy is the most common one in contemporary column-stores, its idea is to perform materialization before (or sometimes in) filter operators. Late materialization, by contrast, tries to postpone materialization for as long as possible (until first or second join, depending on system). Ultra-late materialization is a PosDB refinement for late materialization described in [4]. Finally, hybrid materialization is a strategy where positions are processed simultaneously with the tuple values.

To handle all of these strategies, different types of operators are implemented in PosDB: position-based, tuple-based, and hybrid operators. Position-based operators use the generalized join index [21] to effectively represent positional data. The join index only contains information about the positions, not the values themselves, which is not sufficient to perform some operations, such as joining or filtering. The functionality to retrieve values via positions in PosDB is provided by auxiliary entities called readers [3]. There are several types of readers, for example, ColumnReader, PartitionReader, SyncReader. ColumnReader retrieves values of a specified attribute locally, PartitionReader retrieves values of a specified partition either locally or remotely. SyncReader is a composite reader that manages several simpler ones to extract the values of several attributes synchronously. Readers, in turn, use access methods [10] to provide their functionality.

Thus, PosDB is able to emulate row-based systems by exploiting early materialization. Such emulation was used to experimentally evaluate the implemented external sorting algorithms.

4.2   Operator Implementation

External sort is implemented as a separate operator in the Volcano model. There are several external sort operators in PosDB.

1. **Position-based value sort**. In this case, the operator accepts join index as input and has to obtain values since it needs them for sorting.
2. **Position-based position sort**. Input is the same, but in this case the operator does not need values, since it sorts positions.
3. **Tuple-based**. This is the classic sort, where operator accepts blocks of tuple data and sorts them. Since in PosDB tuple-based representation lacks positions, it is a value sort.

Position-based value sort operator may be beneficial for late materialization (and therefore interesting for further research), since it sorts key-position pairs similar to unclustered (tag) sorting, which is stated [24] to be more efficient on SSD if tuples are big enough. However, it is beyond the scope of this paper. The same is true regarding the position-based position sort. Sorting positions is essential for addressing the out-of-order probing problem inherent for column-stores [1]. The focus of this paper is the latter, classic variant of external sorting—the tuple-based operator. It implements a comparison-based external sorting algorithm on tapes with the Introsort [15] for run generation and polyphase merge [13] for run merging. The algorithm works as follows:

1. Calls GetNext() of child operator until all available memory is filled;
2. Sorts collected tuples in-memory using std::sort implemented as Introsort;
3. Using the polyphase merge pattern, selects the abstract tape to which the current run will be flushed;
4. Flushes current run to the selected tape;
5. Repeats steps 1–4 until input is fully processed;
6. Merges runs from the tapes according to the polyphase merge pattern until there is only one run left.

The implemented external sort tuple-based operator sorts pointers to tuples, not tuples themselves. Even though run merging can be efficiently performed in a single pass in many simple queries, multilevel merge still may be necessary to efficiently execute a complex query plan, e.g. a plan with several sort and hash operations [9]. Taking this into account, it was decided to implement abstract tapes to store runs. Run merging (and distributing over tapes) is performed using polyphase merge, which, however, falls back to ordinary multiway merge when there are more tapes than runs. Abstract tapes can also be used to efficiently reuse disk space during the merge stage, as is done in PostgreSQL (See footnote 2). External sort operator completely bypasses PosDB buffer manager and writes generated runs directly to the disk, it is essential [9] for optimized I/O. Introsort was chosen for the in-memory sorting algorithm for the same reasons that make Quicksort superior [16] to the replacement selection, namely:

1. Quicksort results in far fewer cache misses, which is very important for the performance of modern CPUs;
2. Quicksort is less CPU intensive;
3. With today's main memory sizes, there is no need for runs to be twice the amount of available memory (this is what made the replacement selection very attractive in the past);
4. With Quicksort (or any other in-place sorting algorithm) it is possible to copy each record only once during run generation stage, while replacement selection needs at least two copy operations [9].

For all of these reasons, replacement selection is widely considered obsolete, and PostgreSQL abandoned it in favor of the Quicksort algorithm a few years ago. Also, none of top five solutions in the ACM SIGMOD 2019[3] Programming Contest (in which external sorting was one of the challenges) used it. Among them, two solutions implemented some kind of radix sort for in-memory sorting, other two implemented Quicksort with various optimizations and Leyenda implemented a hybrid approach. Despite the fact that the Radix sort is the most efficient sorting algorithm for some specific data types (e.g. integers), it is difficult to implement it as a general-purpose algorithm for arbitrary data.

During the run merging stage, it is necessary to select the smallest element among the frontmost tuples of runs being merged. It can be done using various approaches, and we implemented and experimentally evaluated three, namely: 1) naive, 2) using the tree of losers data structure, and 3) using a binary heap. All disk operations are buffered via basic `std::fstream` buffers.

Naive. The naive approach is straightforward and does not use an auxiliary data structure. It simply iterates over first tuples of input runs and selects the smallest one. The selected tuple is then flushed to the output tape and replaced with the next tuple from its corresponding run.

Binary Heap. This approach maintains a binary heap in memory to speed up finding the smallest element. First, it constructs the binary heap and fills it with pointers to the first tuples of runs. Second, it removes the root from the heap and flushes tuple pointed to by the root to the output tape. It then reads the next tuple from the run of the previously removed tuple and inserts a pointer to it to the heap.

---

[3] http://sigmod19contest.itu.dk/.

Loser Tree. This is the same approach as with a binary heap, but instead of a heap, a loser tree [13] data structure is used.

5   Experiments

To evaluate and compare the performance of the implemented algorithms in a DBMS environment, we have performed the following experiments: 1) Comparison with PostgreSQL in order to "validate" the quality and correctness of our implementation; 2) Evaluation of the three approaches on the synthetic Star Schema Benchmark [17]; 3) Evaluation of the three approaches on a real dataset, namely, TripData [23].

Experiments were performed using the following hardware and software configuration. Hardware: Intel Core i5-7200U CPU @ 2.50 GHz × 4, 8 GiB RAM, 240 GB KINGSTON SA400S3. Software: Ubuntu 20.04.4 LTS ×86_64, Kernel 5.13.0-40-generic, gcc 9.4.0, PostgreSQL 12.10.

For the first two experiments, we decided to use a simple query with an ORDER BY clause over the lineorder table from the Star Schema Benchmark. SSB was used with a scale factor of 35 (over 10 GBs). In this experiment the table was modified by excluding columns that do not participate in any query that comes with this benchmark. The query under consideration over lineorder was as follows:

```
1  SELECT
2    LO_CUSTKEY, LO_DISCOUNT, LO_EXTENDEDPRICE, LO_ORDERDATE,
3    LO_ORDERPRIORITY, LO_ORDTOTALPRICE, LO_PARTKEY,
4    LO_QUANTITY, LO_REVENUE, LO_SUPPKEY, LO_SUPPLYCOST
5  FROM LINEORDER ORDER BY LO_ORDERDATE;
```

The PosDB query plan of the considered query is shown in Fig. 2. Essentially, it is equivalent to the PostgreSQL one. Despite the fact that PostgreSQL is a row-based system, whereas PosDB is column-based, the comparison is still valid since the query projects all attributes. As the result, PosDB behaves identically to the row-based system: readers of the Materialize operator access the same amount of data from disk, albeit from several different files. It is also worth mentioning that PostgreSQL uses a heap to merge runs. In all conducted experiments, the number of tapes was greater than the number of runs. This means that the polyphase merge algorithm behaved like a multiway merge both in PosDB and PostgreSQL.

In the first two experiments, the total number of runs was chosen to be 84. PosDB generates 84 runs with 150 megabytes of memory available for the sort operator for all three considered algorithms. PostgreSQL at the same time needs 413 megabytes of work_mem[4] for this, probably because of internal buffers for optimized I/O. To make this comparison more accurate, we measured the performance of PostgreSQL with a total of 84 runs (work_mem=413MB) and with 231 runs (work_mem=150MB, the same amount of memory was given to the PosDB sort operator). In these experiments, the average query execution time for 40 iterations was taken with a confidence interval of 95%. The results of the first experiment are

---

[4] https://www.postgresql.org/docs/current/runtime-config-resource.html.

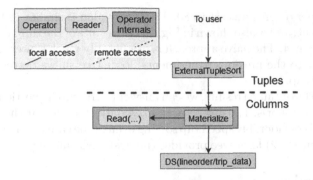

Fig. 2. PosDB plans for queries used in experiments.

shown in Fig. 3. PostgreSQL sorts 231 runs with 150MB of `work_mem` a bit slower than the naive approach, but it sorts 84 runs considerably faster. There could be several reasons for this: PostgreSQL implements prereading for a better disk access pattern, plus it sorts ⟨key prefix, pointer⟩ pairs, which is more efficient [9,16] than pointer sorting.

Due to the space constraints, the results of the second experiment are shown in Fig. 3 as well. It is easy to see that the naive approach is the worst, binary heap is the second, and loser tree is the best one out of the considered three.

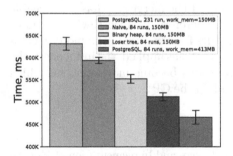
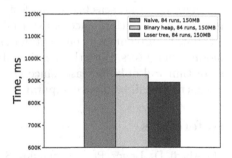

Fig. 3. PostgreSQL comparison (exp 1), evaluation on synthetic data (exp 2).

Fig. 4. evaluation on real data (exp 3), TripData dataset.

In the third experiment we compared the performance of the implemented algorithms using the `trip_data` table of the TripData dataset. Unlike the previous experiment, this one uses real data. The table has approximately 128 million rows (over 20 GBs). We used a simple query of the same kind as the one in the first experiment:

```
1  SELECT * FROM trip_data ORDER BY trip_time_in_secs
```

The number of runs was set to 80. The query plan is similar to the one in the first experiment, and is also shown in Fig. 2. The results of the third experiment are presented in Fig. 4. The naive approach is predictably much slower than the other two. Similarly to the previous experiments, loser tree shows better performance than binary heap.

Based on the obtained results, we can answer the research questions given in the introduction as follows: 1) using a specialized data structure at the merge step is beneficial and considerably speeds up sorting even in the case when disk is expected to be a bottleneck, 2) loser tree provides the best performance.

6　Conclusion

In this paper, we have considered the design of a comparison-based sort and evaluated the choice of the data structure used on the merge step for implementing the external sort operator in DBMSes. First, in our review of the related work, we demonstrated that no study has yet touched on the problem of selecting a data structure for run merging in external sort for DBMSes. After that, we posed several related research questions. To answer them, we have implemented three different approaches and experimentally evaluated them using the PosDB query engine. As the result, we have shown that a loser tree is the most efficient method for run merging.

Thus, we have demonstrated that even without complex I/O optimizations, such as disk striping, parallel I/O using `mmap`, elegant read ahead techniques, etc., the algorithm that selects the smallest element at the merge step has a noticeable impact on the performance of external sort. Therefore, it is worthwhile to make an effort to implement an efficient data structure for run merging on modern commodity computers. With the mentioned optimizations applied, the influence of the algorithm will only increase, since I/O operations become more efficient, which means there will be more computational load on the CPU.

References

1. Abadi, D., Boncz, P., Harizopoulos, S.: The Design and Implementation of Modern Column-Oriented Database Systems. Now Publishers Inc., Delft (2013)
2. Arpaci-Dusseau, A.C., Arpaci-Dusseau, R.H., Culler, D.E., Hellerstein, J.M., Patterson, D.A.: High-performance sorting on networks of workstations. SIGMOD Rec. **26**(2), 243–254 (1997)
3. Chernishev, G.A., Galaktionov, V.A., Grigorev, V.D., Klyuchikov, E.S., Smirnov, K.K.: PosDB: an architecture overview. Program. Comput. Softw. **44**(1), 62–74 (2018). https://doi.org/10.1134/S0361768818010024
4. Chernishev, G.A., Galaktionov, V., Grigorev, V.V., Klyuchikov, E., Smirnov, K.: A comprehensive study of late materialization strategies for a disk-based column-store. In: Proceedings of DOLAP@EDBT/ICDT 2022. CEUR Workshop Proceedings, vol. 3130, pp. 21–30 (2022). http://ceur-ws.org/Vol-3130/paper3.pdf
5. DeWitt, D.J., Naughton, J.F., Schneider, D.A.: Parallel sorting on a shared-nothing architecture using probabilistic splitting. In: Proceedings of PDIS 1991, pp. 280–291 (1991)

6. Frazer, W.D., McKellar, A.C.: Samplesort: a sampling approach to minimal storage tree sorting. J. ACM **17**(3), 496–507 (1970)
7. Govindaraju, N., Gray, J., Kumar, R., Manocha, D.: Gputerasort: High performance graphics co-processor sorting for large database management. In: Proceedings of SIGMOD 2006, pp. 325–336 (2006)
8. Graefe, G.: Query evaluation techniques for large databases. ACM Comput. Surv. **25**(2), 73–169 (1993)
9. Graefe, G.: Implementing sorting in database systems. ACM Comput. Surv. **38**(3), 10-es (2006)
10. Hellerstein, J.M., Stonebraker, M., Hamilton, J.: Architecture of a database system. Found. Trends Databases **1**(2), 141–259 (2007)
11. Iyer, B.R., Ricard, G.R., Varman, P.J.: Percentile finding algorithm for multiple sorted runs. In: Proceedings of the VLDB 1989, pp. 135–144 (1989)
12. Jin, W., Qian, W., Zhou, A.: Efficient string sort with multi-character encoding and adaptive sampling. In: Proceedings of SIGMOD 2021, pp. 872–884 (2021)
13. Knuth, D.E.: The Art of Computer Programming, vol. 3, 2nd edn. Sorting and Searching. Addison Wesley Longman Publishing Co., Inc., USA (1998)
14. Kumar, A., Lee, T.T., Tsotras, V.J.: A load-balanced parallel sorting algorithm for shared-nothing architectures. Distrib. Parallel Databases **3**(1), 37–68 (1995)
15. Musser, D.R.: Introspective sorting and selection algorithms. Softw. Pract. Exper. **27**(8), 983–993 (1997)
16. Nyberg, C., Barclay, T., Cvetanovic, Z., Gray, J., Lomet, D.: Alphasort: a cache-sensitive parallel external sort. VLDB J. **4**(4), 603–628 (1995)
17. O'Neil, P., O'Neil, E., Chen, X., Revilak, S.: The star schema benchmark and augmented fact table indexing. In: Nambiar, R., Poess, M. (eds.) TPCTC 2009. LNCS, vol. 5895, pp. 237–252. Springer, Heidelberg (2009). https://doi.org/10.1007/978-3-642-10424-4_17
18. Padmanabhan, S., Malkemus, T., Jhingran, A., Agarwal, R.: Block oriented processing of relational database operations in modern computer architectures. In: Proceedings of ICDE 2001, pp. 567–574 (2001)
19. Pang, H., Carey, M.J., Livny, M.: Memory-adaptive external sorting. In: Proceedings of VLDB 1993, pp. 618–629 (1993)
20. Shi, Y., Li, Z.: Leyenda: an adaptive, hybrid sorting algorithm for large scale data with limited memory. CoRR abs/1909.08006 (2019)
21. Valduriez, P.: Join indices. ACM Trans. Database Syst. **12**(2), 218–246 (1987)
22. Watkins, A., Green, O.: A fast and simple approach to merge and merge sort using wide vector instructions. In: Proceedings of IA3, pp. 37–44 (2018)
23. Whong, C.: New York City taxi fare data 2013 (2013). http://chriswhong.com/open-data/foil_nyc_taxi/
24. Wu, C.H., Huang, K.Y.: Data sorting in flash memory. ACM Trans. Storage **11**(2), 1–25 (2015)
25. Yin, Z., et al.: Efficient parallel sort on avx-512-based multi-core and many-core architectures. In: Proceedings of HPCC/SmartCity/DSS, pp. 168–176 (2019)
26. Zheng, L., Larson, P.R.: Speeding up external mergesort. IEEE Trans. Knowl. Data Eng. **8**(2), 322–332 (1996)
27. Özsu, M.T., Valduriez, P.: Principles of Distributed Database Systems, 3rd edn. Springer, Heidelberg (2011). https://doi.org/10.1007/978-3-030-26253-2

Document Versioning for MongoDB

Lucia de Espona Pernas(✉) and Ela Pustulka

FHNW University of Applied Sciences and Arts Northwestern Switzerland,
4600 Olten, Switzerland
{lucia.espona,elzbieta.pustulka}@fhnw.ch
http://www.fhnw.ch

Abstract. Data versioning is required in various business and science contexts, including governance, risk and compliance (GRC) and is essential for security audits, legal compliance and business strategy development. We present a data versioning library for MongoDB to support an innovative enterprise resource planning (ERP) system for small and medium enterprises (SMEs) which aims to be flexible and adapt to changing business needs. We exploit the fact that the volume of archival data is orders of magnitude larger than of the currently valid documents and that historic data is rarely accessed. Experiments with eight sets of 1 million mutations/queries on 100K of valid documents (average size 2.3 kB), carried out over a period of 60 h on a local PC show stable average versioning write/read operation performance per document in the range of 12.3/1.2 ms which proves that the solution is viable in an SME scenario.

Keywords: Database · NoSQL · Document versioning · CRUD · ERP · MongoDB · Performance

1 Introduction

Early work on data versioning for relations was done by Bernstein and Goodman [4] who presented complex multi-versioning scenarios with algorithms guaranteeing versioning correctness. Stonebraker et al. [33] outlined three types of versioning requirements for a relational database: *no archive* where no historical access to a relation is needed, *light archive* where archival is needed but will not be accessed frequently, and *heavy archive* when the system needs to look up and update timestamps of previous transactions. In an ERP system we expect to see *no archive* in business objects which do not change and *light archive* to be used for compliance, security and strategy queries, with no *heavy archive*.

In business, governance, risk and compliance (GRC) are of primary importance. Burns and Peterson [18] suggested a number of mechanisms supporting compliance. In this context, data versioning can be used as a foundation of system security or information movement tracking [6]. Versioning systems retain earlier versions of modified documents, allowing recovery from user mistakes or system corruption [31]. In a business system, versioning is important in three

S. Chiusano et al. (Eds.): ADBIS 2022, CCIS 1652, pp. 512–524, 2022.
https://doi.org/10.1007/978-3-031-15743-1_47

settings: legal compliance which requires that we can access the details of financial transactions for the last ten years, support of security audits for a few years back, and the use of old data for business strategy development. In this last case we may ask how often suppliers replace products with new ones, which customers cancel or modify orders, or perform other business intelligence queries. Versioning queries will not be carried out on all data and will not be frequent.

Some DBMSs have built in data versioning, with or without redundancy, with automatic or manual backup controlled by an administrator and even an advanced support as a combination of all of those, but in most cases this is not enough [7] to satisfy compliance requirements. Mainstream NoSQL DBMSs including MongoDB [20] only provide limited support for data versioning [12]. However, S3 [2,32] has introduced versioning in 2016.

Our ERP targets SMEs and the business system has to comply with the law and provide access to older data versions for financial transactions up to 10 years into the past. Some data has to be versioned, as it is expected to change, and some data will not change or change rarely.

For example, in a purchase order scenario which is part of two benchmarks, TPC-R and TPC-H [25], with relations PART, PARTSUPPLIER, LINEITEM, SUPPLIER, CUSTOMER, ORDERS, NATION and REGION, see [37], we expect that the following data will be versioned: PART may change its price from time to time, a PARTSUPPLIER tuple may be deleted as a part becomes obsolete or the supplier is no longer active, a LINEITEM can be deleted once from an existing order but not changed or a new LINEITEM may be added, SUPPLIER or CUSTOMER may change the details from time to time, and an ORDER will be the most often updated relation, as sometimes the SME cannot source an order item and may delete it or replace items, based on availability or customer decision.

Here, we focus on adding versioning support to an ERP based on MongoDB and measuring performance. Earlier steps in the ERP development were presented in [26] where we show that the ERP fits the business use case as defined by Enablerr [1]. We offer three contributions. First, our versioning algorithm for MongoDB ensures that the query time on the currently valid document versions is kept constant as we store historical data in a separate collection. Second, we guarantee ACID versioning consistency by using a transaction spanning two collections (the current data and the historical data). Third, we demonstrate good versioning performance on a dataset of 100K documents undergoing 1 million mutations, with CRUD operations needing between 1.2 and 12.3 ms, which satisfies the needs of an SME. We show that the requirements of compliance, security audit and business strategy reporting can be implemented without sacrificing performance.

The paper is organized as follows. Section 2 presents related work, Sect. 3 outlines the versioning solution, and Sect. 4 presents the experiments and results. In Sect. 5 we discuss and conclude.

2   Related Work

The need to support historical queries, auditing, provenance, and reproducibility leads to an increase in data volumes. This has led to many efforts at building efficient data management systems that support versioning as a first class construct. We increasingly see version control embedded in many collaborative tools such as word processors, spreadsheets and wikis [3].

In databases, schema evolution and versioning have been researched extensively [4,27,33]. Versioning for relational databases includes TardisDB [28] and Decibel which is a relational dataset branching system [17]. Versioning for XML was studied by Chien et al. [5]. Here, we focus only on data versioning and not schema versioning, with application to an ERP scenario on top of MongoDB. Despite advances in data versioning, none of the prior systems fit NoSQL databases. The existing NoSQL work focuses on schema changes necessitated by the flexibility of NoSQL data models [29,34].

The only document versioning library for MongoDB, to the best of our knowledge, is Vermongo [24,35], but it does not support the range of operations we require. The core idea is to store the current and past document versions in separate collections. One can efficiently access the most heavily requested data, i.e. the latest versions. Implementations are available for Couchbase [36] and even for SQL in the context of temporal changes, keeping the historical data in a separate table [23]. However, Vermongo has several deficiencies which make it unsuitable for an ERP system as it lacks ACID guarantees, does not support all data types (particularly DBRefs which are the equivalent of foreign keys in a relational database) and all the CRUD operations defined by Mongoose, such as *updateOne* or *updateMany*. In Sect. 3 we discuss our solution to versioning which provides full versioning support as needed in an ERP.

3   The Versioning Solution

We now describe the new versioning solution starting with the versioning dimensions discussed by Van de Sompel et al. [30].

1. **Identification:** *how are different versions identified?* Every document has an unique id and a version number.
2. **Strategy:** *how do we assign identifiers to versions?* New versions inherit the id and get a new version number that increases sequentially.
3. **Relationships:** *How are version relationships expressed?* Relationships are expressed by sharing the object id. Higher version numbers correspond to more recent versions.
4. **Timestamping:** *How are the date and time associated with versions?* The date and time are stored as the start (inclusive) and end timestamp (exclusive). Current valid versions have only a start timestamp.

3.1   The Document Versioning Pattern

Our solution follows the MongoDB document versioning pattern [8] which aims to keep the document version history available and usable. To accomplish this, a version number is added to each document and the database has to contain two collections for each original collection under version control: one holding the latest (and most queried) version and another that holds all previous versions. Figure 1 shows the two collections defined by this pattern, the main collection and the shadow collection.

   The pattern assumes that most of the queries refer to the current document version and the historical data are only accessed offline for batch reporting or exceptionally in case of failure, security breach or legal inquiry. The second assumption is that the frequency of document reads (current version) is much higher than the frequency of editing. This way the overhead of writing (insert, update and delete) is compensated by the reduction of query time for current versions.

   These assumptions hold for most SMEs which archive many versions of documents during long periods (10–15 years) since they are critical for internal investigation, regulatory compliance, and electronic discovery [13]. A long period of retention implies a large growth of historic versions [15] while the much smaller current data is the one used on a daily basis. For this reason, the *Document Versioning Pattern* is a suitable option for the ERP scenario for which our versioning solution has been developed.

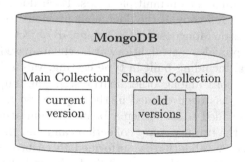

Fig. 1. Versioning uses a main and a shadow collection.

3.2   Implementation

Following the pattern described above, when a new business object is added to the system, a main collection for this object is created, and our library automatically creates an additional collection to hold old document versions. The originally defined collection becomes the *Main Collection* and stores only the current versions. Documents have only a validity start date. Object ids and other user defined unique fields are unique.

The *Shadow Collection* stores the historical versions. Documents have both a start and an end date and validity periods do not overlap. The *(objectId, version)* combination is unique but the original object ids are not unique and any other uniqueness constraints are automatically removed, to support the storage of multiple old versions of the same document sharing some field values. The documents in the shadow collection are stored for compliance and by default only two indexes are defined. The first index is on the unique object id in the shadow collection (combination of the main collection object id and version number) and it is created automatically by MongoDB as in every collection. The second index is on the original identifier field from the main collection and the validity timestamps, which allows us to retrieve efficiently the version of a document valid at a particular time point. If the shadow collection is to be used for reporting, additional indexes can be defined. The efficiency impact of creating such indexes should be kept in mind since the archival collections grow over time at a higher pace than the main collection. The versioning algorithm works as follows.

1. On *create*, a new document is written to the main collection.
2. On *update*, a new version is written to the main collection and the previous version is added to the shadow collection, within a transaction.
3. On *delete*, the no longer valid version is moved to the shadow collection, within a transaction.
4. On *read*, both main and shadow collections can be queried.

The whole document is moved to the shadow collection on any update, no matter if it is on a single or on multiple fields. Each document version in the shadow collection is a full copy of the document previous to the modification.

Our solution relies on Mongoose [16], a MongoDB object modelling library for NodeJS. Mongoose provides a straightforward, schema-based solution to model the application data, including built-in type casting, validation, query building and business logic hooks among other features. In our implementation, once the document model has been defined in the application using Mongoose, the versioning plugin is added to the schema before instantiating the model. This plugin will generate the two necessary collections and add the versioning related fields to the schema. A validation is performed to ensure the original schema definition does not use any of the reserved fields shown below. The fields *editor* and *deleter* have been added to comply with the ISO 15926-2 [14]. Further details of our solution are available as open-source [11] and ready to use as a npm library [10].

1. Version: integer, starting at 1 (required)
2. Validity timestamps: old versions have a mandatory start and end, and the current version has a mandatory start and an empty end.
3. Editor: User that performs creation or update (required)
4. Deleter: User that performs a deletion
5. Other reserved fields for auxiliary usage are: *session*, to handle transactions, *edition*, to define custom edition values, and *deletion*, to define custom deletion values.

The versioning plugin offers new versioning specific query methods.

1. *findVersion (id, version)* finds a document (in the main or shadow collection) matching the given id and version number
2. *findValidVersion (id, date)* finds a document (in the main or shadow collection) matching the given id and being valid on the specified date timestamp.
3. *findAll(offset, limit)* retrieves all the documents in the main collection. It accepts pagination parameters offset and limit.

In the shadow collection two indexes are defined on the versioned documents: the default MongoDB index on the _id field (composed by the _id of the main collection document and the version number) and an index on the _id of the main document plus the validity period(_id._id, validityStart, validityEnd). This index supports the *findValidVersion* query method.

The Mongoose middleware pre-hooks implement the versioning on save (insert, update) and delete, performing the necessary actions on the main and shadow collection. We support all Mongoose model and query API operations defined at [21, 22]. Operations involving multiple document modifications such as *deleteMany, findAndDelete* and *updateMany* are not typical in our event-driven ERP system and were implemented for the sake of completeness (see [11]).

The current document version has to be provided on every update or delete and has to match the latest existing version to prevent concurrency issues. The application engine refreshes the data if a write attempt fails and provides the latest version for editing.

In our approach, transactions are required in updates and deletes to guarantee operation atomicity, as those operations affect both the main and shadow collections. The Mongoose Vermongo package [35] does not provide any assurance regarding data consistency during versioning, so it is not suited for an ERP. To remedy this, our solution makes use of MongoDB transactions to secure the update and delete. Transactions are not needed for inserts as those only affect the main collection.

Transactions in MongoDB are implemented as follows. MongoDB uses multi-granularity locking that allows operations to lock at the global, database or collection level, and allows for individual storage engines to implement their own concurrency control below the collection level (e.g., at the document-level in the WiredTiger engine we use).

MongoDB uses reader-writer locks that allow concurrent readers shared access to a resource, such as a database or collection. In addition to a shared (S) locking mode for reads and an exclusive (X) locking mode for writes, intent shared (IS) and intent exclusive (IX) modes indicate an intent to read or write a resource using a finer granularity lock. When locking at a certain granularity, all higher levels are locked using an intent lock. The intent locks are high level locks that act as a traffic signal. They are added to a conflict FIFO queue while other existing locks are not released on the finer granularity object. Once the intent locks are in place (e.g. at collection level), the lower level locks are set (e.g. at document level). This design reduces the processing needed for managing locks as the concurrency control system only has to check the intent locks and not all the lower level locks.

The transactions are handled by the user and passed to the versioning library for each enveloped operation. It is therefore possible to perform multi-document updates inside a single transaction but this is rather uncommon on our business scenario as the event driven paradigm of the ERP involves mainly single document atomic operations. In addition, transactions spanning multiple documents or collections are not recommended for efficiency reasons and they lock too much data.

Beside using ACID transactions, we have solved the management of DBRefs [19] that were not supported by Vermongo. DBRefs are references from one document to another which combine the referenced document's _ id, collection name, and, optionally, the database name. This data type is fully supported in our library and correctly stored in the shadow collection. No specific handling is needed for the versioning of the referenced documents since the document _ id should never be modified on update. It is straightforward to find the referenced version for each document version in the shadow collection by accessing the referenced collection using the DBRef _id and the validity timestamp as parameters for the *findValidVersion* query method described above.

4   Evaluation

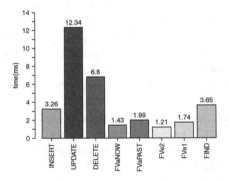

Fig. 2. Our versioning solution performance. Operation time (ms) for the insert, update, delete and various find operations, averaged over 1M executions. Update is the slowest, followed by delete, while insert and read operations perform similarly. Both update and delete use a transaction spanning two collections.

We performed an experiment simulating ERP database use to measure the versioning overhead. We assume that archival data grows over time while the current data stays at a similar size. We compare the performance of three approaches: no versioning, versioning without using a shadow collection (all data current and old in one collection), and our versioning which uses two collections, as outlined previously. We use 100K valid documents at any time point where each document contains employee data of average size 2.3 KB. Each employee document holds complex data including large nested documents and document arrays storing personal data, skills, projects and other information. One million operations of each type were executed in groups of 100K. The experiments were executed on a MacBook Pro (13-inch, M1, 2020) with 16 GB RAM, 256 GB SSD, running macOSX Big Sur 11.1. A local MongoDB (5.0.3 Community) Replica Set with three nodes was used, with the default database configuration. Some system background tasks may affect results but should be similar for all tests. Performance was measured at the application level, via a Mongoose API call, including a transaction where needed, since the software introduces an overhead inherent to the solution design that needs to be reported, as reporting the times from MongoDB directly does not correspond to our use scenario. Each of the eight operations was tested with 1M repetitions, and the entire experiment took approximately 60 h. Experiments can be easily reproduced, see [9].

- INSERT: insert a new document
- UPDATE: update an existing document
- DELETE: delete an existing document
- FVaNOW: query the current valid version by object id
- FVaPAST: retrieve a past valid version by id and date
- FVe2: query current version by id and version number
- FVe1: query past version by id and version number
- FIND: find by non-indexed field on currently valid documents.

Figure 2 gives a first impression of performance. To gain a deeper understanding, we compare three approaches: VERSIONING which is our solution, PLAIN VERSIONING which uses the same versioning method but does not use a shadow collection, so all data are in one main collection, and NO VERSIONING which is a baseline, see Table 1.

We now discuss the results. Figure 2 shows an overview of our solution and shows that updates and deletes are the slowest, as they involve writes in both the main and the shadow collection, which involves a transaction spanning both collections. The times we report are adequate for an SME ERP and even the slowest operation, the UPDATE, needs 12–13 ms, which is acceptable for a business user. Table 1 shows the comparison of three versioning approaches. The UPDATE, at 12–13 ms in our approach is faster than an update using plain versioning with just one main collection, at 19 ms. The overhead of versioning in comparison to no versioning during an update is around 7.5 ms. Plain versioning using a single collection gives a faster DELETE than our versioning, by about 4 ms. Other operations perform similarly to plain versioning, with the exception of FIND which searches for an unindexed field and probably causes a collection scan. FIND in our versioning needs as long as a FIND with no versioning, i.e. versioning produces no overhead. However, in plain versioning a FIND is extremely slow, 594 ms, as all versions are kept together and the collection is very large. This confirms our expectation that our solution has similar performance in most operations and performs better for current documents.

Table 1. Performance of the three versioning solutions for eight operations executed each 1M times in batches of 100K, average time in ms. The slowest operation, update, is in bold.

Operation	VERSIONING	PLAIN VERSIONING	NO VERSIONING
INSERT	3.26	3.87	3.89
UPDATE	**12.34**	**19.11**	**4.81**
DELETE	6.80	2.72	1.25
FVaNOW	1.43	1.80	0.81
FVaPAST	1.99	1.71	–
FVe2	1.21	1.45	0.79
FVe1	1.74	1.45	–
FIND	3.65	594.04	3.02

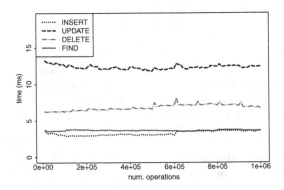

Fig. 3. VERSIONING. Average operation time (ms) of our solution, executed over 1M repetitions in batches of 100K.

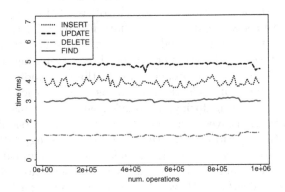

Fig. 4. NO VERSIONING. Average operation time (ms) without versioning over 1M operations.

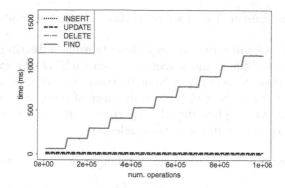

Fig. 5. PLAIN VERSIONING (no shadow collection). Average operation time (ms) over 1M operations.

Figures 3, 4, and 5 show the insert, update, delete and find in all three approaches, during a run of 1 million versioning events. Figure 3 which shows our versioning and Fig. 4, no versioning, show almost flat lines as the experiment progresses, so we can expect good performance as time progresses and objects are versioned, where the total pool of valid documents has the same size. This corresponds to a company which is in business and whose number of employees stays constant, as we are using employee data. We observe a different behaviour in PLAIN VERSIONING, Fig. 5, where the FIND looks like a staircase. As the collection grows, a linear scan to find an unindexed value requires more and more time, as the collection needs to be scanned.

As the queries on the valid documents show similar execution times with and without versioning and such queries are going to be the most heavily performed operations in an ERP, we confirm that performance is satisfactory for this use case.

5   Conclusion

We presented a new solution for data versioning. We observe good performance which can support a flexible and adaptable ERP system. Our work is motivated by the popularity of MongoDB and the trend towards maintaining data evolution history.

We designed and implemented a NodeJS library to manage many document versions of a MongoDB collection by splitting those into two collections. Data consistency is maintained by the use of transactions when required. We evaluated the performance of our solution and compared it to two alternatives, plain versioning and no versioning. Our experiments show clearly that our versioning approach performs well enough to be used in the planned production environment. Beside showing a small performance overhead of keeping historical data, we show performance gains on querying current document versions even in non indexed fields, as compared to a solution that does not separate the data into two collections.

In the future, we will rerun our experiments using a distributed production environment in a real-life business scenario from enablerr. We want to evaluate the solution further, by writing a NodeJS connector on Chronos [38]. Later, we will investigate the impact of the maintenance of versioned collections in our ERP environment. Other plans include experiments in using automated indexing and machine learning to automate index selection.

Acknowledgements. We acknowledge funding from www.innosuisse.ch, grant 44824.1 IP-ICT.

References

1. pier4all AG (Ltd.): enablerr - the revolutionary business solution (2021). https://www.enablerr.ch/en/
2. Amazon Web Services, I.: Amazon simple storage service user guide, API v. 2006–03-01 (2006). https://docs.aws.amazon.com/AmazonS3/latest/userguide/s3-userguide.pdf
3. Bayoudhi, L., Sassi, N., Jaziri, W.: A survey on versioning approaches and tools. In: Abraham, A., Piuri, V., Gandhi, N., Siarry, P., Kaklauskas, A., Madureira, A. (eds.) ISDA 2020. AISC, vol. 1351, pp. 1155–1164. Springer, Cham (2021). https://doi.org/10.1007/978-3-030-71187-0_107
4. Bernstein, P.A., Goodman, N.: Multiversion concurrency control-theory and algorithms. ACM Trans. Database Syst. **8**(4), 465–483 (1983)
5. Chien, S.Y., Tsotras, V.J., Zaniolo, C.: XML document versioning. . SIGMOD Rec. **30**(3), 46–53 (2001)
6. Cioranu, C., Cioca, M.: Database versioning, a transparent SQL approach. J. Mobile Embedded Dist. Syst. **5**, 1–5 (2013)
7. Cioranu, C., Cioca, M., Novac, C.: Database versioning 2.0, a transparent SQL approach used in quantitative anagemment and decision making. Proc. Comput. Sci. **55**, 523–528 (2015)
8. Coupal, D., Alger, K.W.: Building with Patterns: The Document Versioning Pattern (2019). https://www.mongodb.com/blog/post/building-with-patterns-the-document-versioning-pattern
9. De Espona, L.C.: Data Versioning Experiment Repository (2021). https://github.com/pier4all/data-versioning
10. De Espona, L.C.: Versioning Module MongoDB (2021). https://www.npmjs.com/package/mongoose-versioned
11. De Espona, L.C.: Versioning MongoDB Repository (2021). https://github.com/pier4all/mongoose-versioned
12. Felber, P., et al.: On the Support of Versioning in Distributed Key-Value Stores. In: IEEE International Symposium on Reliable Distributed Systems- SRDS, pp. 95–104 (2014)
13. Ferrada, H., Navarro, G.: A lempel-Ziv compressed structure for document listing. In: Kurland, O., Lewenstein, M., Porat, E. (eds.) SPIRE 2013. LNCS, vol. 8214, pp. 116–128. Springer, Cham (2013). https://doi.org/10.1007/978-3-319-02432-5_16

14. ISO: Industrial automation systems and integration - Part 2: Data model, ISO 15926-2:2003. ISO (2003)
15. Jin, X., Agun, D., Yang, T., Wu, Q., Shen, Y., Zhao, S.: Hybrid Indexing for Versioned Document Search with Cluster-Based Retrieval. In: CIKM 2016, pp. 377–386. ACM (2016)
16. LearnBoost: Mongoose (2010). https://www.npmjs.com/package/mongoose
17. Maddox, M., Goehring, D., Elmore, A.J., Madden, S., Parameswaran, A., Deshpande, A.: Decibel: the relational dataset branching system. Proc. VLDB Endow. **9**(9), 624–635 (2016)
18. Mitchell, S.L.: GRC360: a framework to help organisations drive principled performance. Int. J. Discl. Gov. **4**(4), 279–296 (2007)
19. MongoDB I: BSON Types (2021). https://docs.mongodb.com/manual/reference/bson-types
20. MongoDB I: MongoDB (2021). https://www.mongodb.com
21. OpenCollective: Mongoose API model (2022). https://mongoosejs.com/docs/api/model.html
22. OpenCollective: Mongoose API Query (2022). https://mongoosejs.com/docs/api/query.html
23. Phungtua-Eng, T., Chittayasothorn, S.: Slowly changing dimension handling in data warehouses using temporal database features. In: Nguyen, N.T., Gaol, F.L., Hong, T.-P., Trawiński, B. (eds.) ACIIDS 2019. LNCS (LNAI), vol. 11431, pp. 675–687. Springer, Cham (2019). https://doi.org/10.1007/978-3-030-14799-0_58
24. Planz, T.: Vermongo: Simple Document Versioning with MongoDB (2012). https://github.com/thiloplanz/v7files/wiki/Vermongo
25. Poess, M., Floyd, C.: New TPC benchmarks for decision support and web commerce. SIGMOD Rec. **29**(4), 64–71 (2000)
26. Pustulka, E., von Arx, S., Espona, L.: Building a NoSQL ERP. In: ICICT 2022. Springer (2023). https://doi.org/10.1007/978-981-19-1610-6_59
27. Roddick, J.F.: Schema versioning. In: Liu, L., Özsu, M.T. (eds.) Encyclopedia of Database Systems, 2nd edn. Springer, Cham (2018). https://doi.org/10.1007/978-1-4614-8265-9_323
28. Schüle, M.E., Schmeißer, J., Blum, T., Kemper, A., Neumann, T.: TardisDB: extending SQL to support versioning. In: SIGMOD/PODS 2021, pp. 2775–2778. ACM (2021)
29. Sevilla Ruiz, D., Morales, S.F., García Molina, J.: Inferring versioned schemas from NoSQL databases and its applications. In: Johannesson, P., Lee, M.L., Liddle, S.W., Opdahl, A.L., López, Ó.P. (eds.) ER 2015. LNCS, vol. 9381, pp. 467–480. Springer, Cham (2015). https://doi.org/10.1007/978-3-319-25264-3_35
30. de Sompel, H.V., Sanderson, R., Nelson, M.L., Balakireva, L., Shankar, H., Ainsworth, S.: An HTTP-based versioning mechanism for linked data. In: Workshop on Linked Data on the Web, LDOW, vol. 628. CEUR-WS.org (2010)
31. Soules, C.A.N., Goodson, G.R., Strunk, J.D., Ganger, G.R.: Metadata efficiency in versioning file systems. In: FAST 2003, pp. 43–58 (2003)
32. Staff, C.: A Second Conversation with Werner Vogels. Commun. ACM **64**(3), 50–57 (2021)
33. Stonebraker, M., Rowe, L.A.: The design of POSTGRES. In: SIGMOD 1986, pp. 340–355. ACM (1986)
34. Störl, U., Klettke, M., Scherzinger, S.: NoSQL schema evolution and data migration: state-of-the-art and opportunities. In: EDBT 2020, pp. 655–658 (2020)
35. Sutunc, M.: Vermongo Mongoose Plugin (2016). https://www.npmjs.com/package/mongoose-vermongo

36. Team, C.: How to : Implement Document Versioning with Couchbase (2013). https://blog.couchbase.com/how-implement-document-versioning-couchbase/
37. TPC: TPC BenchmarkTM H Standard Specification Revision 3.0.0. TPC (2022). https://www.tpc.org/tpc_documents_current_versions/pdf/tpc-h_v3.0.0.pdf
38. Vogt, M., Stiemer, A., Coray, S., Schuldt, H.: Chronos: the swiss army knife for database evaluations. In: EDBT 2022, pp. 583–586 (2020)

SWODCH: 2nd Workshop on Semantic Web and Ontology Design for Cultural Heritage

SWODCH: 2nd Workshop on Semantic Web and Ontology Design for Cultural Heritage

Workshop Chairs

Béatrice Markhoff	University of Tours, France
Roberta Ferrario	ISTC-CNR, Trento, Italy
Antonis Bikakis	University College London, UK
Alessandro Mosca	Free University of Bozen-Bolzano, Italy
Marianna Nicolosi Asmundo	University of Catania, Italy
Stéphane Jean	University of Poitiers – ENSMA, France

Program Committee Members

Khalid Belhajjame	Paris-Dauphine University, France
Emilio M. Sanfilippo	ISTC-CNR Laboratory for Applied Ontology, Italy
Géraud Fokou Pelap	Think Semantic Web & IA, France
Daniele Metilli	University College London, UK
Kalliopi Kontiza	University College London, UK
Douglas Tudhope	University of South Wales, UK
Genoveva Vargas-Solar	CNRS-LIRIS, France
Laura Pandolfo	University of Sassari, Italy
Michalis Sfakakis	Ionian University, Greece
Gayo Diallo	University of Bordeaux, France
Dusko Vitas	University of Belgrade, Serbia
Jouni Tuominen	University of Helsinki, Finland
Sofia Stamou	Ionian University, Greece
Trond Aalberg	Norwegian University of Science and Technology, Norway
Amedeo Napoli	CNRS, Inria, Université de Lorraine, France
Prince Sales Tiago	Free University of Bozen-Bolzano, Italie
Maria Rosaria Stufano Melone	Politecnico Di Bari, Italie
Enrico Daga	The Open University, UK
Fayçal Hamdi	CEDRIC - CNAM, France
Ludger Jansen	University of Rostock
Ranka Stankovic	University of Belgrade, Serbia
Daria Spampinato	CNR, Italy
Carmen Brando	EHES, France
Efstratios Kontopoulos	Catalink Ltd

Eero Hyvönen	Aalto University and University of Helsinki, Finland
Cédric Pruski	Luxembourg Institute of Science and Technology, Luxembourg
Martin Lopez-Nores	University of Vigo, Spain
Valentina Bartalesi	ISTI-CNR, France
Catherine Roussey	INRAE, France
Andreas Vlachidis	University College London, UK
Carlo Meghini	CNR ISTI, Italy
Daniele Francesco Santamaria	University of Catania, Italy
Sotirios Batsakis	University of Huddersfield, UK
Maria Theodoridou	ICS, FORTH, Greece

Developing and Aligning a Detailed Controlled Vocabulary for Artwork

Luana Bulla[1(✉)], Maria Chiara Frangipane[1], Maria Letizia Mancinelli[3],
Ludovica Marinucci[1], Misael Mongiovì[1], Margherita Porena[1],
Valentina Presutti[2], and Chiara Veninata[3]

[1] ISTC - National Research Council, Rome and Catania, Italy
{luana.bulla,chiara.frangipane,ludovica.marinucci,
margherita.porena}@istc.cnr.it, misael.mongiovi@cnr.it
[2] LILEC - University of Bologna, Bologna, Italy
valentina.presutti@unibo.it
[3] ICCD, Ministry of Culture, Rome, Italy
{marialetizia.mancinelli,chiara.veninata}@beniculturali.it

Abstract. Controlled vocabularies have proved to be critical for data interoperability and accessibility. In the cultural heritage (CH) domain, description of artworks are often given as free text, thus making filtering and searching burdensome (e.g. listing all artworks of a specific type). Despite being multi-language and quite detailed, the Getty's Art & Architecture Thesaurus –a *de facto* standard for describing artworks– has a low coverage for languages different than English and sometimes does not reach the required degree of granularity to describe specific niche artworks. We build upon the Italian Vocabulary of Artworks, developed by the Italian Ministry of Cultural Heritage (MIC) and a set of free text descriptions from ArCO, the knowledge graph of the Italian CH, to propose an extension of the Vocabulary of Artworks and align it to the Getty's thesaurus. Our framework relies on text matching and natural language processing tools for suggesting candidate alignments between free text and terms and between cross-vocabulary terms, with a human in the loop for validation and refinement. We produce 1.166 new terms (31% more w.r.t. the original vocabulary) and 1.330 links to the Getty's thesaurus, with estimated coverage of 21%.

Keywords: Controlled vocabularies · Cultural heritage · String-matching · Semantic similarity

1 Introduction

Maintaining updated catalogs of artwork objects is an arduous yet fundamental task in managing, enhancing and preserving cultural heritage. Most catalogs to

---

This work was supported by the project POR FESR Lazio 2014–2020: "ReAD - Representation of Architectural Data".

S. Chiusano et al. (Eds.): ADBIS 2022, CCIS 1652, pp. 529–541, 2022.
https://doi.org/10.1007/978-3-031-15743-1_48

date rely on free text type descriptions, which makes classification arbitrary and integration between different sources difficult. For instance, in ArCo [3] –the Italian knowledge graph of cultural heritage– data are not aligned with a vocabulary, but types of artworks are specified by free text descriptions. Although guidelines and cataloging standards[1] have been provided, a perfectly uniform cataloging is impossible to obtain, due to subjectivity, human error, and the necessity of handling specific cases.

Controlled vocabularies enable overcoming the semantic heterogeneity commonly encountered in artwork catalogs and provide a common terminology for describing and cataloging artwork objects. A well-known example is the *Getty's Art & Architecture Thesaurus (AAT)*[2], which defines a detailed thesaurus with descriptions in multiple languages. However, the AAT has a limited number of descriptions in languages different than English, thus making it barely usable with resources in other languages. To fill this gap and to provide a finer-grain type hierarchy, the *Institute of the General Catalog and Documentation* (ICCD) has developed a standard Italian vocabulary for cataloging artworks, namely ICCD *Vocabulary of Artworks*[3].

Although detailed, the ICCD Vocabulary of Artworks is still incomplete for two reasons. First because of an uncontrolled multiplication and specialization of descriptions that define cultural assets, which are difficult to validate by domain experts. Secondly because of missing terms in the vocabulary. For example, in ArCo, out of 9.531 artwork type descriptions only 2.515 have correspondent terms. Another important limit of this vocabulary is that it is not aligned with Getty's AAT, thus making it not interoperable with resources and systems in English and other languages.

To overcome the limits of the state of the art discussed above, we apply a bottom-up approach for proposing new terms for the ICCD Vocabulary of Artwork and a semi-automatic procedure for aligning the vocabulary with Getty's AAT. Our method performs automatic alignment of Italian free text descriptions of artworks to terms of the vocabulary and considers remaining non-aligned descriptions as candidate terms. Next we adopt a threefold approach to align terms of the ICCD vocabulary with Getty's AAT. First, we connect terms that match with Italian labels of Getty's AAT. This procedure achieves low coverage since few AAT terms contain Italian labels. Second, we employ Wikidata[4] to find a correspondence between Getty's AAT terms and WordNet, and use Open Multilingual WordNet[5] to connect terms of our Italian vocabulary with WordNet. Last, we employ a language agnostic sentence embedding model to map terms of both vocabularies to a Euclidean space and match semantically similar terms by cosine similarity. This step achieves a significant increase in recall and

---

[1] http://www.iccd.beniculturali.it/it/normative.

[2] http://www.getty.edu/research/tools/vocabularies/aat/aat_faq.html#number.

[3] http://www.iccd.beniculturali.it/it/ricercanormative/139/thesaurus-per-la-definizione-dei-beni-storici-artistici.

[4] https://www.wikidata.org/.

[5] http://compling.hss.ntu.edu.sg/omw/.

strongly reduces the manual effort since the associations it proposes need just to be classified as correct or incorrect.

The paper is organized as follows. In Sect. 2, we present an overview of the state of the art, including initiatives for creating and aligning Italian and English controlled vocabularies. Section 3 describes the resources we started with, including the ICCD Vocabulary of Artworks and the ArCo knowledge graph. In Sect. 4, we detail our bottom-up approach to automatically match and align free text descriptions to controlled vocabularies and our threefold approach for aligning terms across two vocabularies of different languages. In Sect. 5, we present our results, including the evaluation of our tools in terms of precision, recall, F1 and accuracy. Section 6 concludes the paper and outlines future work.

2 Related Work

Linking a resource to a controlled vocabulary means to organize information and promote consistency in the usage of terms. A well-known resource in the Cultural Heritage domain is the *The Art & Architecture Thesaurus* (AAT)[6] created by the Getty Vocabulary Program [4,12]. This multi-language vocabulary aims to help describe and categorize cultural properties. It is the most used resource for controlled terms in CH and currently stores more than 400,400 terms. The vocabulary is linked by Wikidata [19] –a free collaborative knowledge base linked to various resources, including Wikipedia and WordNet– through the property P1014[7]. This direct link allows us to associate AAT terms to corresponding synsets in WordNet [7], which is one of the best known lexical resources in English, representing words with their meanings and their relations (e.g., synonymy, hyponymy, meronymy). Although it supports several languages, its coverage for languages other than English is limited. For example only 8% of concepts have associated Italian terms. Moreover, despite its complex and comprehensive content, the AAT cannot cover all the nuances and specialized terms required for cataloging large data sources (e.g., ArCo).

Several processes and methodologies for semantic enrichment have been developed, soliciting experiments and research in the European scene[8]. In the context of ARIADNE research infrastructure [1], some attempts (e.g., [2]) were made for improving interoperability of the AAT Vocabulary by means of an interactive and manual mapping tools using vocabulary linked data. Our work follows this direction by leveraging on Italian resources and an automatic alignment process.

As detailed in Sect. 3, the ICCD *Vocabulary of Artworks*[9] and ICCD *Vocabulary of Archaeological finds*[10] are the most important resources for describing

---

[6] http://www.getty.edu/research/tools/vocabularies/aat/aat_faq.html#number.

[7] https://www.wikidata.org/wiki/Property:P1014.

[8] https://pro.europeana.eu/project/evaluation-and-enrichments.

[9] http://www.iccd.beniculturali.it/it/ricercanormative/139/thesaurus-per-la-definizione-dei-beni-storici-artistici.

[10] http://www.iccd.beniculturali.it/it/ricercanormative/108/thesaurus-per-la-definizione-dei-reperti-archeologici.

types of objects in the Italian cultural heritage domain. Their creation, maintenance and updating involve a bottom-up approach for constantly updating and integrating the existing resources with new items.

Furthermore, our vocabulary alignment process relies on ontology matching techniques and can be addressed by means of diverse approaches [6]. In [16], the matching of two controlled vocabularies is carried out using Concept Facets (CF), which contain distinct features (semantic relations) for each concept and are able to represent them in a comprehensive way. Then the combination of the two vocabularies' CF is performed with lexical comparison algorithms. This approach merges multiple methods, but do not address translation concerns. Simpler methodologies, such as string matching techniques, seem to work relatively well given the differences of the considered vocabularies [18]. In particular, we relied on previous experiments on string-matching algorithms in English texts. The work of [11] present limits in the management of small variations of the text (e.g., due to typos), which are disregarded. This problem can be solved by the approximate string matching algorithms that are intended to find a sub-string close to a given pattern string. As in our work, Hakak and colleagues describe a similarity match approach by applying the *Levenshtein distance* function [13]. Another significant work is represented by an automatically textual classification method based on a controlled vocabulary developed in the field of engineering [10]. The authors propose a string-matching algorithm based on the Information Engineering thesaurus and classification scheme. This approach was implemented with different methods such as weighting schemes of terms, cut-offs and exclusion of some terms. In our work, we also rely on semantic text similarity to match terms from different vocabularies [20]. Specifically we employ a language agnostic sentence embedding model [8] to associate terms of different languages. Modern approaches similar to ours are based on neural language models [15,17], some of them specifically aimed at handling multiple languages [5,14].

3   Materials: ICCD Controlled Vocabularies and ArCo KG Textual Descriptions

The *Institute of the General Catalog and Documentation* (ICCD) of the Italian *Ministry of Cultural Heritage* (MIC) is in charge of maintaining the *General Catalog of Italian Cultural Heritage*[11] (GC) that is the result of a collaborative effort involving many and diverse organizations that administer cultural properties all over the Italian territory. They submit their catalog records through a collaborative platform, named *SIGECweb*[12]. To date, ICCD has collected and stored ~2.5M records, which are part of ArCo *knowledge graph* (KG) [3] of Italian Cultural Heritage (CH).

---

[11] http://www.catalogo.beniculturali.it/sigecSSU_FE/Home.action?
timestamp=1521647516354.

[12] http://www.iccd.beniculturali.it/it/sigecweb.

As detailed in Sect. 4.1, ArCo textual descriptions were used to integrate the ICCD Vocabulary of Artworks in our semiautomatic enrichment and alignment work. It is composed of 4.225 Italian terms, organized into seven general categories, , with their sub-categories. All terms refer to physical objects.

Furthermore, we used the Vocabulary of Archaeological finds (VAF) from ICCD as gold standard for testing the automatic alignment tools presented in Sect. 4. This ICCD vocabulary aiming at cataloging the types of finds in the Archaeological domain, is composed of 1.784 Italian terms, referred to physical objects, hierarchically organized into nine general categories, with their sub-categories. It has been manually aligned to the *The Art & Architecture Thesaurus* (AAT) by domain experts, using three link types: broad match, close match, exact match. In total 1.191 terms were linked: 455 are broad match, 104 are close match and 632 are exact match.

4 Methodology

In the following we describe each step of our methodology as depicted in Fig. 1. In Sect. 4.1, we detail the extraction process from ArCo textual descriptions and the string matching and alignment process on both ArCo descriptions and ICCD Vocabulary of Artworks; in Sect. 4.2, we describe our tree-fold approach for linking our extended version of the Vocabulary of Artworks with the Getty's AAT Vocabulary.

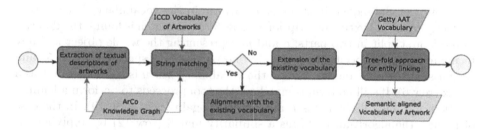

Fig. 1. Flowchart of our methodology process for the creation of a detailed and semantic aligned Vocabulary of Artworks. Red rectangles connected with black arrows represent the sequence of each action performed. Yellow parallelograms denote the input data, while the blue parallelogram represents the final output. (Color figure online)

4.1 Matching and Alignment Algorithm

Starting from the analysis of the ICCD Vocabulary of Artworks, we have identified a number of (mostly morphological and functional) relationships between specific and generic terms. For instance, in the ICCD Vocabulary of Artworks, the term "pugnale" (dagger) is linked to the term "pugnale ad anello" (ring dagger), that is a new term related to the shape of the dagger; it also has a relation with the term "pugnale da caccia" (hunting dagger), which identify a new term based on the function of the dagger; finally, it is also linked to the term "lama di pugnale" (dagger blade), which is a part of the object.

After the analysis of the vocabulary, we extracted textual descriptions of artworks from ArCo KG, to compare them with the analyzed Vocabulary. We found 8.443 different descriptions entered by catalogers. We identified the most frequent cases of misalignment, which can be classified into seven categories: (i) New terms; (ii) Typos; (iii) Terms containing other information (materials, techniques, repertories); (iv) Terms with special characters (*, -); (v) Terms referring to numismatic properties; (vi) Terms with the indication "frammenti" (fragments) or "forma ricostruibile" (rebuildable shape); (vii) Plural/singular mismatches. From this analysis we also noticed that sometimes parts of objects are described by the object and the part separated by the symbol /. For instance, "ciotola/ ansa" (bowl/ handle) in place of "ansa di ciotola" (handle of bowl).

The analysis of the ArCo descriptions led us to the creation of transformation rules aimed at the optimization of the extracted data. These rules are part of the normalization process included in the first step (row 2) of the overall matching procedure described in Algorithm 1. It is possible to have a complete overview of the rules directly by consulting the available code on GitHub[13].

We designed a simple algorithm that takes as input the list of vocabulary terms T and a generic free text description d of a cultural asset, and returns the corresponding term of the Vocabulary of Artworks. The algorithm first normalizes both the list of terms and the textual description (rows 1, 2 in Algorithm 1) by general rules (i.e., normalization of capital letters) and specific rules (See Footnote 13). Subsequently, it looks for a term in the vocabulary that matches perfectly with the textual description d (row 3). If no match is found, the description d is first split in two parts: d and p, representing the whole object and any terms matching a specification of only one part of it, respectively (row 5) and then assigned to the longest term of the vocabulary which is contained in it as a prefix (row 6). If still no term is matched, the tool proceeds to perform a lemmatization of d and all terms $t \in T$ and attempt again (rows 8, 9, 10). In the case of failure, the algorithm performs a similarity match (row 12) by applying the Levenshtein Distance [13] with a threshold of 3 and returning the less distant term or a failure. The tool proceeds by classifying all descriptions by processing them through the algorithm described above. The choice of matching as a prefix depends on the annotation policies, which can be observed in the data, where the object is described at the beginning with more general words followed by more specific words. The tool offers the possibility of tracing the description to the longest term of the vocabulary that contains it as a suffix (in place of prefix). This makes it adaptable also to controlled vocabularies that are based on different hierarchical orderings or different languages.

---

[13] https://github.com/LuanaBulla/Controlled-Vocabularies-for-Cultural-Heritage/tree/main/code.

At the end of the process, the description could be classified through the identification of its equivalent within the controlled vocabulary, through the entry 'Not present' or through the label 'Untraceable'. If during the execution more than a term is matched, the algorithm returns a failure (not specified in Algorithm 1). This case denotes the ambiguity of the description provided, which, associated with several items within the vocabulary, cannot be traced back uniquely.

Algorithm 1: Overall procedure for matching a textual description with its corresponding vocabulary term

Data: T = list of vocabulary terms; d = textual description
Result: term t that matches d
1   $T \leftarrow \{normalize(t) \text{ for all } t \in T\}$;
2   $d \leftarrow normalize(d)$;
3   $t \leftarrow exact\_match(d, T)$;
4   **if** $t = fail$ **then**
5      $d, p \leftarrow split(d)$;
6      $t \leftarrow longest\_prefix\_match(d, T)$;
7   **if** $t = fail$ **then**
8         $T \leftarrow \{lemmatize(t) \text{ for all } t \in T\}$;
9         $d \leftarrow lemmatize(d)$;
10        $t \leftarrow longest\_prefix\_match(d, T)$;
11   **if** $t = fail$ **then**
12          $t \leftarrow similarity\_match(d, T)$;

13 return t;

4.2 Linking with Getty's AAT

To achieve state-of-the-art alignment, we link the Vocabulary of Artworks (VA) and its extension deriving from the alignment of the new terms given by the ArCo's catalogers to the Getty Art & Architecture Thesaurus (AAT).

On the international stage, the AAT vocabulary offers a standard for the domains of art, architecture and decorative arts by embracing a perspective of continuous evolution and growth. In this regard, the Getty's AAT also promotes multilingual inclusion, which entails the translation of some of its elements into a variety of languages. Despite the effectiveness of this initiative, the entries aligned with terms in Italian represent only the 8% of the total items. To ensure that all words in our controlled vocabularies are aligned with the target resource (AAT), we followed two steps, one automatic and one manual. To align terms of our Italian vocabulary with the Getty's AAT, we adopted a threefold automatic approach followed by manual validation.

The first approach consists in matching the terms in our controlled vocabulary to the Italian translations already contained in the Getty's AAT. We compare all the Italian concepts in the Getty vocabulary to those in the VA and in its extension to determine if an exact match occurs. If so, the terms are aligned.

The second approach involves WikiData[14], a free and open knowledge base that serves as a central repository for structured data from Wikimedia projects, such as Wikipedia and Wiktionary. This huge database includes an alignment between some AAT vocabulary entries and the reference WordNet synsets. Since the latter are in English, we translate the synsets into Italian by means of Open Multilingual Wordnet[15], a database that aims at facilitating multilingual navigation. The first and second approaches were successful in linking 14% and 9% of the entries in the Vocabulary of Artworks and our extension, respectively.

The third approach aims at mapping all terms into our controlled vocabularies by applying a language-agnostic BERT sentence embedding model (LaBSE) [8], which supports 109 languages. Trained and optimized to produce similar representations exclusively for bilingual sentence pairs that are translations of each other, the model promotes tasks of classification, translation, and similarity from a multilingual perspective. Following an initial vectorization through embedding of all terms to be compared, we verify the degree of similarity between each item of our controlled vocabularies and the single items contained in the AAT corpus using cosine similarity. For every term of the Vocabulary of Artwork we choose the AAT term with the higher degree of similarity. This method allow us to match our controlled vocabularies with the Getty AAT lexicon in a fully automatic way. This series of operations was followed by a manual validation of the results which led to a complete and detailed assessment described in Sect. 5.

5   Results and Evaluation

In this section, we report the results of our methodology and the evaluation of our methods in terms of precision, recall and coverage. The final version of the extended Vocabulary of Artworks[16] consists of 6.091 terms. 3.819 terms were derived from the clean-up process of the original ICCD vocabulary and 2.272 from the integration of ArCo free text descriptions. As a result, the size of the new vocabulary version grew by 37% over the prior version. The Vocabulary of Artworks' terms were semantically aligned with the Getty's AAT vocabulary. As we show below, the alignment tool has an expected 24% recall, which was evaluated on a different vocabulary for which we have the links with Getty's AAT: the manually annotated Vocabulary of Archaeological Finds (VFA), which represents our gold standard. Detailed results of the automatic alignment modules are given below.

5.1   Evaluation of the Text Matching and Alignment Algorithm

First, we evaluated the text matching tool and its performance on the whole set of artwork free text descriptions (8.443 descriptions) in the ArCo KG. The

---

[14] https://www.wikidata.org/.

[15] http://compling.hss.ntu.edu.sg/omw/.

[16] https://github.com/LuanaBulla/Controlled-Vocabularies-for-Cultural-Heritage.

alignment process between textual descriptions and terms from the Vocabulary of Artwork was validated manually by checking the whole output.

We computed precision and recall for all the steps performed by the tool. In order to obtain detailed information, we evaluated the impact of each step of the algorithm (exact match, prefix match, lemmatized prefix match and similarity match) and we computed precision and recall adapted for the multi-class setting [9] for each step. The validation process was conducted as follows: we first checked the alignments when an exact match was found, to verify if the normalization process established in the algorithm (rows 1-2 of Algorithm 1) was valid for all occurrences. For matches based on lemmatization and prefix match, we checked the controlled vocabulary to find out if it contains a specific term. The alignment based on similarity was validated by the analysis of the strings, to verify if the change of a letter produced a different word. Finally, if a match was not found we checked the controlled vocabulary to find out generic terms to which the description could be traced back. For all the cases of doubt, we submitted the validation of the alignments to ICCD domain experts. We also created a list of annotation to better identify the types of error.

The results of the application of all the individual steps of the algorithm, as well as the whole process, are reported in Table 1. The exact match and prefix match (rows 1-6) produce the highest precision, while the similarity phase (row 12) produces the best recall. In general, the whole Algorithm 1 performs 98% precision and 99% recall.

Table 1. Performance of the Matching and Alignment algorithm in terms of precision and recall.

Step	Precision	Recall
Exact match and prefix	**0,994**	0,780
Lemmatization	0,993	0,807
Similarity	0,989	0,962
Complete	0,982	**0,993**

Further analysis of the errors has shown that incorrect matches fall into 3 main error types. The first error cause concerns the application of the similarity score due to the overuse of the similarity metrics mismatches (e.g., the description "blocco" (block) paired with the term "bricco" (jug). The second error cause is to be attributed to incorrect descriptions made by users. For instance, the improper use of prepositions, such as "di" (of) instead of "per" (for) is the main example of this error type. The third error cause consists of the misalignment with terms that already exist in the controlled vocabularies.

5.2 Evaluation of the Linking with the Getty's AAT Vocabulary

Evaluation of the Vocabulary of Artworks Alignment. The second result of our work is the alignment between the new Vocabulary of Arworks (VA) and Getty's AAT entries. We present (Tables 2 and 3) both the precision of our technique and the proportion of vocabulary terms covered by it for both the initial VA and the proposed extension. Due to a lack of prior manual alignment between our controlled vocabularies and the AAT vocabulary, we cannot compute recall and hence we report coverage, as the percentage of linked terms over all terms.

Table 2. Performance of the Getty's AAT vocabulary alignment with the VA.

Step	Precision	Coverage
Getty@it	**81%**	8%
WordNet	66%	2%
SBert	12%	10%
Total	21%	**21%**

Table 3. Performance of the AAT vocabulary alignment on the VA's extension.

Step	Precision	Coverage
Getty@it	**84%**	3%
WordNet	67%	3%
SBert	17%	15%
Total	23%	**23%**

As expected, the combination of the terms of our controlled vocabularies with the Italian translations already contained in the AAT (indicated in both tables as "Getty@it" Step) achieved higher precision in the alignment process both in the VA and in its extension. However, the application of this first approach achieved only limited coverage of the controlled vocabularies' terms. A similar trend is shown by the second approach, which involves linking the WordNet Synsets extracted from WikiData (indicated in the tables as "WordNet" Step). Higher coverage is instead achieved by the third method, which takes into consideration the semantic similarity between the items of the controlled vocabularies and the AAT entries (indicated in the tables as "SBert" Step). In this case, the precision is considerably lower (12% and 17%, respectively). However, note that high precision is not essential for the aim of our research because the terms processed have been subjected to a thorough manual validation that has highlighted and filtered the incorrect alignments.

Alignment Evaluation with Vocabulary of Archaeological Finds as Gold Standard. To further test the validity of our methodology, we evaluated the alignment tool on the Vocabulary of Archaeological finds (VAF), which was already manually aligned to the Getty's AAT vocabulary by domain experts. In Table 4 we report the results in terms of recall and precision calculated by taking into consideration the total number of items manually validated by domain experts (indicated in table as "Precision w/o") and the totality of the vocabulary entries, including the ones that don't have a correspondence with Getty AAT (indicated in the table as "Precision w").

Table 4. Performance of the AAT vocabulary alignment on 67% of the VAF vocabulary (Precision w/o) and the total number of entries in it (Precision w).

Step	Precision w/o	Precision w	Recall
Getty@it	**72%**	**63%**	6%
Getty@it + WordNet	51%	48%	10%
Getty@it + WordNet + SBert	10%	4%	**24%**
Total	20%	10%	**24%**

In the first scenario, we evaluate the 67% of the entire lexicon, which was aligned with the AAT vocabulary and correspond to a total of 1.191 items. Due to the non-classified items by the domain experts of the remaining 33% of the VAF, all the 1.784 terms that compose it are contemplated in the second case for the purpose of completeness. In this instance, we consider the added entries as not attributable to the Getty vocabularies and therefore we evaluate the link made by our alignment tool as incorrect. The total recall is 24%, while the accuracy of the first and second scenarios is 20% and 10%, respectively. Precision is lower than on VA, probably due to the limited coverage of alignment techniques via Getty and WordNet, and the difficulty of dealing with more specific entries inscribed in the archaeological heritage field. This area comprises items that are more terminologically sought after and less common than the Vocabulary of Artwork entries. The lower performances on VAF w.r.t. VA suggest that we can expect higher recall on VA and hence 24% recall on VAF might be an underestimation for VA.

6   Conclusion and Future Work

Our work proposed an agile methodology for the automatic alignment of textual descriptions to controlled vocabularies in the CH domain. Its application led to the creation of a Vocabulary of Artworks (VA) extension starting from free text descriptions in the KG ArCo. We also link the terms in the VA and in its extension to the Getty's AAT vocabulary, testing our methods on the VAF vocabulary as a golden standard.

Following the results achieved, we plan to improve both of our tool. First, given the limits of the *Levenshtein Distance* (cf. Sect. 4.1), we intend to test different semantic similarity metrics for matching free text descriptions with vocabulary terms. We also plan to enhance the alignment approaches between our controlled vocabularies and the Getty's AAT Vocabulary by taking into consideration their hierarchy structure. This would enable better filtering of information and more accurate matching of entries.

References

1. Aloia, N., et al.: Enabling european archaeological research: the ariadne e-infrastructure. Internet Archaeol. **43** (2017)
2. Binding, C., Tudhope, D.: Improving interoperability using vocabulary linked data. Int. J. Digit. Libr. **17**(1), 5–21 (2015). https://doi.org/10.1007/s00799-015-0166-y
3. Carriero, V.A., et al.: ArCo: the italian cultural heritage knowledge graph. In: Ghidini, C., et al. (eds.) ISWC 2019. LNCS, vol. 11779, pp. 36–52. Springer, Cham (2019). https://doi.org/10.1007/978-3-030-30796-7_3
4. Cobb, J.: The journey to linked open data: the getty vocabularies. J. Libr. Metadata **15**(3–4), 142–156 (2015)
5. Conneau, A., et al.: Unsupervised cross-lingual representation learning at scale. arXiv preprint arXiv:1911.02116 (2019)
6. Euzenat, J., Shvaiko, P.: Ontology matching. Springer, Heidelberg (2013). https://doi.org/10.1007/978-3-642-38721-0
7. Fellbaum, C.: Wordnet: An electronic lexical database: Bradford book. MIT Press, Cambridge (1998)
8. Feng, F.a.o.: Language-agnostic bert sentence embedding. arXiv preprint arXiv:2007.01852 (2020)
9. Godbole, S., Sarawagi, S.: Discriminative methods for multi-labeled classification. In: Dai, H., Srikant, R., Zhang, C. (eds.) PAKDD 2004. LNCS (LNAI), vol. 3056, pp. 22–30. Springer, Heidelberg (2004). https://doi.org/10.1007/978-3-540-24775-3_5
10. Golub, K., et al.: Automated classification of textual documents based on a controlled vocabulary in engineering. KO **34**(4), 247–263 (2007)
11. Hakak, S., et al.: Exact string matching algorithms: survey, issues, and future research directions. IEEE Access **7**, 69614–69637 (2019)
12. Harpring, P.: Introduction to controlled vocabularies: terminology for art, architecture, and other cultural works. Getty Publications (2010)
13. Levenshtein, V.I., et al.: Binary codes capable of correcting deletions, insertions, and reversals. In: Soviet Physics Doklady, vol. 10, pp. 707–710. Soviet Union (1966)
14. Liu, Y., et al.: Multilingual denoising pre-training for neural machine translation. Trans. Assoc. Comput. Linguist. **8**, 726–742 (2020)
15. Luan, Y., et al.: Sparse, dense, and attentional representations for text retrieval. Trans. Assoc. Comput. Linguist. **9**, 329–345 (2021)
16. Morshed, A.u., Sini, M.: Creating and aligning controlled vocabularies. In: Workshop on AT4DL 2009, p. 50 (2009)
17. Reimers, N., Gurevych, I.: Sentence-bert: sentence embeddings using siamese bert-networks. arXiv preprint arXiv:1908.10084 (2019)
18. Tordai, A., et al.: Aligning large skos-like vocabularies: Two case studies. In: ESWC (2010)

19. Vrandečić, D.: Wikidata: A new platform for collaborative data collection. In: Proceedings of the 21st International Conference on World Wide Web, pp. 1063–1064 (2012)
20. Zad, S., et al.: A survey of deep learning methods on semantic similarity and sentence modeling. In: 12th IEMCON, pp. 0466–0472. IEEE (2021)

Using Wikibase for Managing Cultural Heritage Linked Open Data Based on CIDOC CRM

Joonas Kesäniemi[1]([✉])(iD), Mikko Koho[1,2](iD), and Eero Hyvönen[1,2](iD)

[1] Semantic Computing Research Group (SeCo), Aalto University, Espoo, Finland
joonas.Kesaniemi@aalto.fi
[2] HELDIG – Helsinki Centre for Digital Humanities, University of Helsinki,
Helsinki, Finland

Abstract. This paper addresses the problem of maintaining CIDOC CRM-based knowledge graph (KG) by non-expert users. We present a practical method using Wikibase and specific data input conventions for creating and editing linked data that can be exported as CIDOC CRM compliant RDF. Wikibase is a proven and maintained software for generic KG maintenance with a fixed but flexible data model and easy-to-use user interface. It runs the collaboratively edited Wikidata KG, as well as increasing amount of domain specific services. The proposed solution introduces a set of data input conventions for Wikibase that can be used to generate CIDOC CRM compliant RDF without programming. The process relies on the aforementioned data input rules combined with generic mapping implementations and metadata stored as part of the KG. We argue that this convention over coding makes the system more easily approachable and maintainable for users that want to adhere to the CIDOC CRM principles, but are not ontology experts. As part of the preliminary evaluation of the proposed solution, an example on managing Cultural Heritage data in the military history domain with discussion on the limitations of the approach is presented.

Keywords: Wikibase · CIDOC CRM · Knowledge graph · Cultural heritage · Linked data

1 Introduction

While thousands of Linked Data datasets of Cultural Heritage (CH) have become openly available in the Linked Open Data (LOD) cloud[1] and elsewhere, an ever-more serious challenge is how to manage the knowledge graphs (KG) when the underlying ontologies and metadata evolve over time [4]. This is especially true for datasets that are products of unique research projects with minimal resources available for maintenance in the long run. When the graph is created using data

---

[1] https://lod-cloud.net/.

© Springer Nature Switzerland AG 2022
S. Chiusano et al. (Eds.): ADBIS 2022, CCIS 1652, pp. 542–549, 2022.
https://doi.org/10.1007/978-3-031-15743-1_49

exported from existing systems or from manually created sets of heterogeneous source files, it might be easy to make changes to the source data, but the transformation pipelines still require maintenance in order keep the graph version of the data relevant. In order to minimize the need for maintaining transformations, one can also choose to maintain the data graph directly. This is the approach taken, for example, in [9] in the context of culinary tradition and in [7] in archaeology domain. Both examples are using the CIDOC Conceptual Reference Model (CRM)[2] as their ontological foundation. CRM is a standard for interoperable information in the CH field [2]. It is an event-based top-level ontology with high expressive power to address the intricacies of describing CH data. CRM is also a complex ontology and its implementations produce equally complex networks of data, which makes creating easy-to-use tooling a challenge. The CRM events can make references to places where the event occurred, to the people and physical objects involved, and to the time-span of the occurrence. As the information of individual CH objects and actors relating to them is split into multiple events and a large amount of references between entities, making changes to the data is not as straight-forward as with more simple document or object-based data models. For example, adding a previously unknown birth date for a person would involve either creating a new birth event linked to the person, or modifying a birth event linked to the person, if one exists already.

This paper presents ongoing work related to the maintenance of CRM-based linked data using the Wikibase software[3]. Wikibase is developed to serve as the platform for Wikidata[4], which is a massive open and free knowledge base maintained by both humans and machines and hosted by the Wikimedia Foundation. Wikidata currently contains structured information about more than 90 million things and maintain an edit history of more than 1.6 billion actions, so it has been proven to be able to handle massive datasets if necessary. Wikibase is available as an open-source software suite and comes out-of-box with generic and simple user interface, search functionality and support for complete change history. It is designed for open and collaborative management of KGs and has been adopted by libraries, archives, and research groups for their data management needs [1]. We present a practical framework called WB-CIDOC for creating and editing CRM based data with Wikibase instance with an option to publish it as a custom CRM compliant RDF[5] dataset. Our approach tries to strike a balance between the structures required by the reference ontology and data input conventions needed to generate them by re-purposing features of the Wikibase data model. In Sect. 2 related works are first presented and the proposed method is introduced in Sect. 3. Experiences of using the method in our case study are then presented (Sect. 4). Finally, in conclusion, contributions of this paper and current limitations are summarized.

---

[2] https://www.cidoc-crm.org/.
[3] https://wikiba.se/.
[4] https://www.wikidata.org/.
[5] https://www.w3.org/RDF/.

2   Related Work

Existing works that combine technical solutions for manipulating KGs directly and use the CRM ontology are few, but there are some notable related works to be found. WissKI [10] is an open source Virtual Research Environment and content management system for Cultural Heritage data built on top of CIDOC CRM ontology. WissKI uses so-called ontology paths to map simple data inputs such as "date created" into more complete CRM data structures. Ontology paths are similar to the data input conventions presented in this paper. The main difference is how these shortcuts or mappings are used in the user interface. WissKI uses ontology paths to compile the input forms, whereas our approach relies on Wikibase's generic form interface. Also, in WissKI the ontology paths are added manually as opposed to a set of predefined input rules of WB-CIDOC.

Another more recent solution that combines KGs and CIDOC CRM is the ResearchSpace platform[6]. ResearchSpace advertises itself as a collaborative and contextualizing knowledge system that allows users to connect qualitative and quantitative research. It provides integrated tools for building dataset as well as semantic web applications. Many features of the ResearchSpace rely heavily on the CIDOC CRM ontology. It provides read and write access to the data through semantic forms, which in contrast to Wikibase's built-in generic implementation, must be implemented on a case-by-case basis. Examples [7] and [9] mentioned in the Sect. 1 are implemented with ResearchSpace. More detailed description of the system and more examples of research projects utilizing ResearchSpace to collect, enrich, and analyze data can be found in [8].

Gayo et al. [5] use Wikibase for representing and maintaining Shared Authority Files (SAF) in CIDOC-CRM in a project by Luxembourg Competency Network on Digital Cultural Heritage. SAFs are used by memory organizations to maintain authoritative records of identities of important entities, such as persons, places, and organizations. The approach is identical to ours, as it uses Wikibase qualifiers to provide data for linked resources. They even provide example SPARQL queries for mapping between the Wikibase data model and CIDOC CRM. However, these queries seems to be manually created examples, where as we use the metadata stored in the KG to automatically generate necessary transformations. The work also has a much narrower scope than our work, focusing solely of managing person related SAF data, which can be modelled with only a small subset of CIDOC CRM, namely the Appelations and their related Symbolic objects. In fact, their work is complementary to ours, since our WB-CIDOC approach does not consider CIDOC CRM Appelations yet.

3   WB-CIDOC – Wikibase for CIDOC CRM

This section describes the WB-CIDOC approach for using Wikibase as the source for CIDOC CRM-compliant RDF datasets. The potential user of the WB-CIDOC is someone who needs to be able to create CRM-based data and

---

[6] https://github.com/researchspace/researchspace

might even understand why, but who does not need nor want to understand the ontological nuances of the CRM. Our potential user is also more inclined to implement data processing automations through graphical user interface instead of scripting.

WB-CIDOC consists of two parts: Wikibase items and properties that make up the schema for of the data to be inputted with metadata mapping to CRM and a set of simple data input conventions that take advantage of Wikibase's data model and user interface to denote shortcuts for resources required by the CRM. The former provides the model and mapping of the domain specific data to the CRM ontology and the latter gives the rules on how to use to the model in different situations. Finally, an external software component is used to implement the mapping from Wikibase's data model to the CRM RDF output with the help of the aforementioned schema and data adhering to the rules as described below. WB-CIDOC is a generic approach and should be suitable for any CRM project that can work with currently supported CRM features.

Stemming from its collaborative nature, Wikibase's data model is designed to represent statements about items of interest and their references. It also allows for every statement to be qualified with one or more additional statements, similarly to RDF reification [6]. For example, the Wikidata item for Douglas Adams[7] contains a statement with property occupation and value novelist with a qualifying information start time 1979, i.e., a qualifier provides additional information about its linked statement. A more detailed description of the Wikibase's data model can be found in [11].

Items and properties are the basic entity types in Wikibase. In order to be able to denote that some item "X" represents a person, one could, for example, first create an item for Person and a property instance of, and then add a statement instance of Person to the item "X". WB-CIDOC requires at least three properties to function: instance of, URI mapping and event type. The property instance of is used as explained above to link an item to a specific type or class of entities. Properties event type and URI mapping are used in the RDF export process to identify and generate CRM events, and to translate Wikibase generated URIs with Q and P numbers to external schemas, respectively. Other items for types and properties are determined by the domain model of the data to be inputted. See Sect. 4 for an example from the military history domain. When the required and domain specific properties and items are in place, it is possible to start inputting the actual data. It should be noted that the domain model can be modified at any point if and when the need arises.

Data input rules of the current alpha version of the WB-CIDOC are focused on creating CRM events and time-spans. We are reporting on ongoing work and more rules can be added in the future.

A so called *implicit event* is created by adding a statement to an item with a property that has an aforementioned event type statement. For example, we can have a property called was born with identifier P2 with following statements: event type crm:E67_Birth and URI mapping crm:P98i_was_born. The data

---

[7] https://www.wikidata.org/wiki/Q42.

type of the property was born must be monolingual text[8]. Finally, the implicit event denoting the birth for an item describing John Doe could be added with the following statement: John Doe was born "Birth of John Doe". The value of the was born property would become the preferred label of the generated crm:E67_Birth event and it would be linked to the person with the property crm:P98i_was_born as metadata recorded as part of the was born (P2) property. An implicit event can be further described by adding qualifiers to the statement. Continuing with the previous example, we might want to record the birth place and mother of John Doe. This can be achieved by adding qualifiers to the "was born" statement. Again URI mapping values of both properties took place at and by mother would be used as part of the RDF generation to map Wikibase URIs to something that is valid CRM. It should be noted here, that crm:Event instances are only generated for properties that are associated with event type metadata and they is used to describe items. Therefore, using the same property as a qualifier would not trigger an event generation.

If the event is linked to multiple actors or things, the user must decide on which related item to add the implicit event statement. For example, E67_Birth event involving a child and a mother, can be added either to the mother's (Q1) side as Q1 gave birth "Birth of Y" with the qualifier brought to life Q2 or to the item representing the child (Q2) as Q2 has birth event "Birth of Y" with the qualifier by mother Q1. The problem with this approach is that the event is not directly visible from the item page of the other items involved. For example, in the latter case there would not be a link from mother to the child. These connections can be discovered on-demand by using Wikibase's "What links here" feature. Another option is to make the participation in an event explicit by adding inverse links to other event participants. For example from mother to child with a statement brought to life John Doe. This kind of property pairs are handled by the mapping implementation in a way that both persons refer to the same event even if both was born and brought to life properties contain event type metadata.

Another basic data input convention is used to handle time-spans. Time-span related properties, such as crm:P82a_begin_of_the_begin and crm:P79_beginning_is_qualified_by, can be added directly to implicit events as qualifiers. During export a new resource with the type crm:E52 Time-Span is created from all the properties from the qualifiers where their URI mapping metadata refers to a CRM property with the domain crm:E52_Time-Span.

4   Case WarSampo – Finnish World War II Data

WarSampo[9] [3] is a semantic portal and a linked data service about Finland in the Second World War. It consists of a harmonized KG in a SPARQL endpoint assembled from several heterogeneous sources, as well as a web portal with nine

---

[8] https://www.mediawiki.org/wiki/Wikibase/DataModel.

[9] Project: https://seco.cs.aalto.fi/projects/sotasampo/en; portal: https://www. sotasampo.fi/en.

application perspectives to the underlying KG. The WarSampo KG makes extensive use of CRM and its event-based model. The National Archives of Finland has received significant amount of correction suggestions to the data through the portal, and WB-CIDOC is one of the approaches we are currently experimenting with for incorporating changes back to the KG.

WB-CIDOC's implicit events are a good match for events that are not significant enough to warrant their own item and involve limited amount of participants. Let's take the aerial victory <http://ldf.fi/warsa/events/event_lv32101> as an example. The event has the pilot and his squadron as participants with details about the location, time period and aircraft involved. Figure 1 shows a partial description of the data in Wikibase using WB-CIDOC. In addition to the statements visible in the figure, the item John Doe is defined as instance of Person and Squadron 32 as instance of MilitaryUnit. Also, Person and MilitaryUnit are defined with metadata URI mapping crm:E21_Person and URI mapping crm:E74_Group respectively.

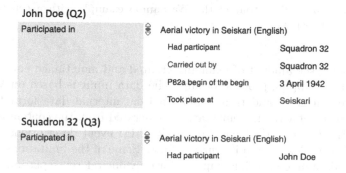

Fig. 1. Partial description of the event http://ldf.fi/warsa/events/event_lv3288 in Wikibase user interface.

The event described in Fig. 1 includes two participants which both use the property **participated in** that has event type crm:E5_Event. However, only one instance of an event is generated since mapping implementation can detect through qualifiers that they should refer to the same event. Figure 2 shows a snippet from the generated RDF export, which includes the processing of an implicit event with time-span.

5   Discussion

This paper presented an approach called WB-CIDOC for maintaining cultural heritage data that is compiliant with CIDOC CMR using the Wikibase platform. The presented approach simplifies data input task by introducing certain input conventions that can be used to infer some of the required CRM resources,

```
warsa:Q2 skos:prefLabel "John Doe" ;
    a crm:E21_Person ;
    crm:P11i_participated_in warsa-events:Q2-3690b3d7 .
warsa:Q3 skos:prefLabel "Squadron 32" ;
    a crm:E74_Group ;
    crm:P11i_participated_in warsa-events:Q2-3690b3d7 .
warsa-event:Q2-3690b3d7
    a crm:E5_Event ;
    skos:prefLabel "Aerial victory in Seiskari"@en ;
    crm:P11_had_participant warsa:Q3, warsa:Q2 ;
    crm:P14_carried_out_by warsa:Q3 ;
    crm:P7_took_place_at warsa:Q4 ;
    crm:P4_has_time-span warsa:Q2-3690b3d7-ts-1 ;
warsa-event:Q2-3690b3d7-ts-1
    a crm:E52_Time-Span ;
    crm:P82a_begin_of_the_begin "1942-04-03"^xsd:date ;
    crm:P82b_end_of_the_begin "1942-04-03"^xsd:date ;
```

Fig. 2. Subset of RDF output of the WarSampo example with a generated event instance and time-span.

hence lowering the amount of manually created and maintained entities. It also simplifies the mapping process, because the data input is based on Wikibase's generic user interface and transformation from internal data format to CRM RDF is done by generic implementations generated from the content of the KG.

WB-CIDOC currently focuses on basic CRM event descriptions and is not suitable for complex data in its current form. Some of the limitations stem from Wikibase's data model, such as support for only one reference per implicit event. For example, it is not possible to denote that time-span information came from "source X" and participant information from "source Y". The use of monolingual text as the data type of properties with event type metadata means that event labels are currently generated in only one language. One possible workaround could be to define special label property for adding other languages as qualifiers. Also, RDF export related event generation cannot currently properly link entities to shared events if there are more than one implicit event with same property and pair of participants. This kind of situation will lead to duplicate events instead of one shared event.

Implementation of the proposed solution is currently in an experimental stage, as we are working with the National Archives of Finland to develop different ways of maintaining the WarSampo data. Applying WB-CIDOC approach to other published datasets with more elaborate use of CRM would be an interesting area of further research as well as a step towards validation of the proposed solution.

Acknowledgements. Our work is funded by the Memory Foundation for the Fallen, National Archives of Finland, and FIN-CLARIAH project of the Academy of Finland. Computing resources of the CSC – IT Center for Science are used.

References

1. Diefenbach, D., Wilde, M.D., Alipio, S.: Wikibase as an infrastructure for knowledge graphs: the EU knowledge graph. In: Hotho, A., et al. (eds.) ISWC 2021. LNCS, vol. 12922, pp. 631–647. Springer, Cham (2021). https://doi.org/10.1007/978-3-030-88361-4_37
2. Doerr, M.: The CIDOC conceptual reference module: an ontological approach to semantic interoperability of metadata. AI Mag. **24**(3), 75–75 (2003)
3. Hyvönen, E., et al.: WarSampo data service and semantic portal for publishing linked open data about the second world war history. In: Sack, H., Blomqvist, E., d'Aquin, M., Ghidini, C., Ponzetto, S.P., Lange, C. (eds.) ESWC 2016. LNCS, vol. 9678, pp. 758–773. Springer, Cham (2016). https://doi.org/10.1007/978-3-319-34129-3_46
4. Koho, M., Ikkala, E., Heino, E., Hyvönen, E.: Maintaining a linked data cloud and data service for second world war history. In: Ioannides, M., et al. (eds.) EuroMed 2018. LNCS, vol. 11196, pp. 138–149. Springer, Cham (2018). https://doi.org/10.1007/978-3-030-01762-0_12
5. Gayo, L., Michelle, P.: Representing the luxembourg shared authority file based on cidoc-crm in wikibase (2021). https://www.youtube.com/watch?v=MDjyiYrOWJQ (Accessed 20 Jun 2022)
6. Manola, F., Miller, E., McBride, B., et al.: Rdf primer. W3C Recommendation **10**(1–107), 6 (2004)
7. Marlet, O., Francart, T., Markhoff, B., Rodier, X.: Openarchaeo for usable semantic interoperability. In: ODOCH 2019@ CAiSE 2019 (2019)
8. Oldman, D., Tanase, D.: Reshaping the knowledge graph by connecting researchers, data and practices in researchSpace. In: Vrandečić, D., et al. (eds.) ISWC 2018. LNCS, vol. 11137, pp. 325–340. Springer, Cham (2018). https://doi.org/10.1007/978-3-030-00668-6_20
9. Partarakis, N., et al.: Representation and presentation of culinary tradition as cultural heritage. Heritage **4**(2), 612–640 (2021). https://doi.org/10.3390/heritage4020036
10. Scholz, M., Goerz, G.: Wisski: a virtual research environment for cultural heritage. In: Raedt, L.D., et al. (eds.) ECAI 2012–20th European Conference on Artificial Intelligence. Including Prestigious Applications of Artificial Intelligence (PAIS-2012) System Demonstrations Track, Montpellier, France, 27–31 August 2012. Frontiers in Artificial Intelligence and Applications, vol. 242, pp. 1017–1018. IOS Press(2012). https://doi.org/10.3233/978-1-61499-098-7-1017, https://doi.org/10.3233/978-1-61499-098-7-1017
11. Vrandečić, D., Krötzsch, M.: Wikidata: a free collaborative knowledgebase. Commun. ACM **57**(10), 78–85 (2014)

Megalithism Representation in CIDOC-CRM

Ivo Santos[1]([📧]) [ID], Renata Vieira[1] [ID], Cassia Trojahn[2] [ID], Leonor Rocha[3] [ID], and Enrique Cerrillo Cuenca[4] [ID]

[1] CIDEHUS, University of Évora, Évora, Portugal
{ifs,renatav}@uevora.pt
[2] Institut de Recherche en Informatique de Toulouse, Toulouse, France
cassia.trojahn@irit.fr
[3] CEAACP, University of Algarve and University of Évora, Évora, Portugal
lrocha@uevora.pt
[4] Department of Prehistory, Ancient History and Archaeology,
Complutense University of Madrid, Madrid, Spain
enriqcer@ucm.es

Abstract. The definition of Megalithism is found in specialized and academic works, on Archaeology and Prehistory, as well as in generalist dictionaries, glossaries and encyclopedias. However we are not aware of formal definitions specifically to this particular domain. This paper presents an under development proposal of Megalithism Knowledge Representation that relies on CIDOC CRM to represent the monuments and concepts of European Megalithism. It includes the granularity required to represent this form of architecture: from composite parts to single elements, such as standing stones. A structured way for representing such definitions is required in order to guide future knowledge extraction from Megalithism reports.

Keywords: Megalithism · Knowledge representation · CIDOC CRM

1 Introduction

The epistemological root of the concept of Megalithism (from the Greek *megas*, large, and *lithos*, stone) is based on the dimensions and nature of the monuments. Although its apparently simple definition, the heterogeneity and intrinsic polymorphism of Megalithism stands out, emerging a heterogeneous and polymorphic reading of this cultural manifestation. This emerging reading reflects the multiple actors that shaped the phenomenon through construction, reconstruction or even the interaction between monuments and regions, or even through the multiplicity of archaeological interpretations.

The information treated by History and Archaeology is made up of fragments that allow the portrayal of a past reality and, as such, allow the construction of

This work is funded by national funds through the Foundation for Science and Technology, under the project UIDB/00057/2020.

S. Chiusano et al. (Eds.): ADBIS 2022, CCIS 1652, pp. 550–558, 2022.
https://doi.org/10.1007/978-3-031-15743-1_50

multiple narratives [3]. Thus, archaeological information systems must be able to deal with subjectivity, multivocality, temporality and uncertainty. As stated in [3], "without a data model capable of adequately describing not only the archaeological data contained within records, but the archaeological processes used to generate records, any database is limited in terms of its capabilities and suitability for reporting, assessment and analysis and ultimately its suitability as an archive" [3, p. 489].

Under the principle of protection by archaeological record, there is an evolution in the *outputs* of archaeological works that have become increasingly larger, more complex and mostly digital. In this context of proliferation of the digitization of Cultural Heritage, new challenges have risen, the range of applications has become more comprehensive and the number of actors interested in obtaining and sharing knowledge related to heritage has increased. In this context, digital sharing of Cultural Heritage information implies challenges such as: interoperability, structuring and transforming data into information, information into knowledge and knowledge into application, etc. [12], and ideally, to adapt to FAIR Principles. Knowledge-based approaches, structured with Linked Open Data (LOD) concepts and ontologies, are recognized as a basis for efficient solutions [12]. However, many institutions and/or research groups continue to independently develop their databases and management application, isolated from the others, already existing, producing a fragmented accumulation of data [8], leaning to create "data silos".

In this paper, we propose a CIDOC CRM extension, regarding the representation of megalithic monuments, having in mind the usual aspects included in archaeology grey literature. There are other proposals of CIDOC CRM extensions for archaeological works. In [6], the authors present the components of archaeological buildings, with a focus on construction techniques and materials, describing the visual representations, the buildings' chronology and provenance of buildings. In [9] the authors propose extensions, referring to archaeological experiments mainly concerning digital analysis of virtual samples. We are not aware of an extension specific to the definition of megalithic monuments with the objective of organize usual official reports content. The goal is to set up the basis for sharing the rich existing data about Megalithism in a principled way, including the granularity needed to represent this form of architecture: from composite parts to single elements, such as standing stones.

In Europe megalithism appears in prehistoric and protohistoric chronology, but similar symbolic expressions, funerary and non-funeral, appear in non-coeval periods in other regions of the globe. In archaeological literature, the origins of Megalithism is a long-term debate, associated with "oriental missionaries", "segmentary societies", "environmental changes" and, in some regions, mimetical behavior [4]. "This vast array of explanations reflect how difficult is to identify the historical reasons laying beyond such a wide phenomenon were global and local features are combine in a multitude of ways" [4].

The rest of the paper is organised as follows. Section 2 introduces the CIDOC CRM model and Sect. 3 presents our proposal, followed of a discussion in Sect. 4.

2   CIDOC CRM

The CIDOC CRM (*standard* ISO21127) is a continuously developing *Conceptual Reference Model* (CRM) that provides formal frameworks for describing implicit and explicit concepts and relationships used in cultural heritage documentation [8]. This initiative is managed by the *International Committee for Documentation of the International Council of Museums* and has the participation of an extensive community of volunteers who meet annually with the aim of integrating and linking data on Cultural Heritage from the basic relationships between objects and events.

According to [5], CIDOC CRM encourages information sharing and is based on the epistemological idea of phases of the academic/scientific process:

- collect and organize evidence (observation and primary sources);
- link facts through relationships;
- interpretation of evidence—contextualize and build hypotheses;
- display results—publication;

As a base, CRM describes general characteristics of objects (identifiers, typology, title, material, dimension, notes), but also the history of the object through events and activities (transfer of custody, location – former and current, origin, discovery, attribution of classifications, measurements) as well as relationships between objects and parts of objects (bibliography, composition, likeness) and their representations (drawings, paintings, inscriptions) [11].

This high-level *framework* is seen as a common, modular and extensible semantic standard, providing the chance to develop extensions for various purposes.

3   Defining Megalithism

Our proposal is an extension to concepts related to Megalithism. Figure 1 illustrates the concept of Megalithic monument (MM) defined on the basis of CIDOC CRM elements. We refer to Version 7.1.1 of CIDOC CRM. Due to space limitations, we present only the description of some classes and relations. The concept MM is a subclass of (*E24 Physical Human Made Thing* and also of its subclass *E22 Human Made Object*), which is defined along with other specific relations. We pictured relations regarding its identification (appellation), location, legal protection (rights), type, and composition. There are several other relevant relations such as association with artifacts and others.

In [6], a CIDOC CRM extension is proposed specifically to incorporate components of archaeological buildings, they define (B1 Built Work), our concept for MM may be integrated as its subclass. Other extensions for archaeology will not be discussed in this paper, but on later future work.

The idea is that these concepts, presented in more principled ways, may help to find and compare monuments. This structure is also intended to guide information extraction from the megalithic sites reports that allow description of the monuments.

Next we detail some of these concepts and relations. We will use as an example one specific monument, "Anta Grande da Comenda da Igreja" (Fig. 2), a well-preserved dolmen (*anta*) located in Alentejo (Portugal) known by its dimensions relative to its neighbours, with a long history of archaeological works and other publications [7].

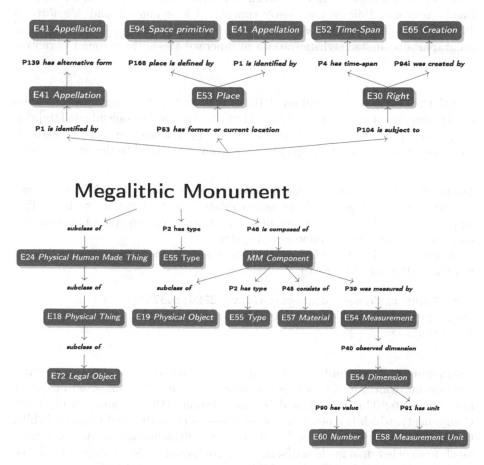

Fig. 1. *Megalithic monument* description elements

Toponymy. An MM is identified by its appellation, which may have alternative forms. It might be a code assigned to the monument by certain institutions, such as the CNS (Portuguese national archaeological site code) code. The monument we are using as an example is identified by appellation "Anta Grande da Comenda da Igreja". It has the CNS code 616 and alternative names "Anta Grande da Herdade da Igreja" and "Anta Grande da Herdade da Comenda".

Type. This class comprises concepts denoted by terms from thesauri and controlled vocabularies used to characterize and classify instances of CRM classes. Instances of (E55 Type) represent concepts in contrast to instances of (E41 Appellation), which are used to name instances of CRM classes. Examples of the architecture type are "Dólmen", "Menir" and "Recinto Megalítico", but others can be added.

The MM "Anta Grande da Comenda da Igreja", for example, has the types "funerary megalithism", "Anta/Dólmen" and "Megalitismo Eborense". In Portugal, *Anta* and *Dólmen* are mostly considered as synonyms and *Megalitismo Eborense* is a concept used only among Portuguese archaeologists to describe a geographically limited stylistic variant of funerary Megalithism found in central Alentejo.

Legal Protection. The relation (P104 is subject to) (E30 Right) specifies the legal protection of a monument. The MM "Anta Grande da Comenda da Igreja" is subject to a protection (right) that was created by a protection event (Creation). This protection has a a time span of time from 20-01-1936 to the present day.

Location. The relation (P56 has former or current location) (E53 Place) allows the introduction of general information related to the location, such as (E41 Appelations) and (E94 Space primitive) that may contain spacial coordinates and may be linked to Geoinformation systems.

In our example, "Anta Grande da Comenda da Igreja" (P168 place is defined by) the (E94 Space Primitive):

```
<gml:Point srsName=''urn:ogc:def:crs:EPSG::4326''>
 <gml:pos>38.757998,-8.203181</gml:pos>
</gml:Point>
```

Component. The record of an MM may include one or more MM architectural components which, as far as it is concerned, may also be composed of one or more elements. This is expressed by the relation (P46 is composed of) (MM Component), which is a new concept, subclass of (E19 Physical Object). While, from the CRM perspective, the definition of megalithic monument does not differ much from other man made artifacts, being composed by MM components is the key to its differentiation. The basic MM component is denominated megalith, and it is usually understood as a single piece of stone roughly shaped or in definite shapes for symbolic purposes. Different arrangements of megaliths (including single stones) are taken here as MM Components.

Types of MM components are chambers, corridors, orthostats, etc. In this way, the configuration designed allows the registration of MM components with their respective descriptions, but also, if necessary, the registration of each standing stone. Regarding [6] , mentioned earlier, this would be related to (B2 Morphological Building Section).

In our example, "Anta Grande da Comenda da Igreja" (P46 is composed of) a part that (P2 has type) "Chamber", that (P45 consists of) (E57 Material) Granite and has (P40 observed dimension) diameter which (P90 has value) of 4,5 (E60 Number) in meters (Measurement Unit), as in Fig. 2.

Besides the concepts described here, there are some other relevant classes and relations associated to the extension for MM being developed, such as the relation with archaeological works undertaken on the monuments and the elements found, such as artifacts (E22 Human Made Objects). Description of artifacts are important to allow users to filter the monuments by the typology of artifacts associated with them. For example, filtering all monuments with geometrics or a certain ceramic typology, trace of reuse.

Subclasses of Megalithic Monument. Another key aspect in the definition of megalithism are its specializations or subclasses. There are defined types of monuments which are related to a unique or isolated standing stone or megalith, such as the concept of a Menhir. Other sub concepts or type of monuments include multiple stones, which may be aligned, or placed in circles (Cromlechs, see Fig. 3). A Dolmen (Fig. 2) is an example of a more complex megalithic monument in which stones create a chamber, often considered to be used as a tomb or burial chamber. With the proposed model we intend to formalize all these concepts and their specific compositions, trying to push forward supra-regional studies.

4   Discussion

This extension proposal allows for the identification of the main elements and relations that define and characterize an MM. It is intended to better describe monuments in order to help finding and comparing them. It will also serve as a basis for the organization of current reported knowledge constituted of non-structured data.

As introduced early in the paper, in the literature, we find other proposals of CIDOC CRM extensions in the area of archaeology. In [9] the authors also propose CIDOC CRM extensions, referring to archaeological experiments mainly concerning digital analysis of virtual samples. In [6], a CIDOC CRM extension is proposed specifically to incorporate components of archaeological buildings, which in some aspects are related to our proposal. However, they deal mainly with construction techniques, materials, visual representations, chronology and provenance of the buildings. For megalithic site description, construction techniques were very basic and are mostly unknown.

Differently from these other works, our main focus here is to describe Megalithism concepts. Our approach focuses on relevant elements for allowing better search and comparison, and considers the usual components of official reports of megalithic sites studies (prospections, excavations, environmental assessments) and their particular relations with subclasses of monuments.

Fig. 2. Anta Grande da Comenda da Igreja components: chamber (composition) and megaliths (units). Image taken from Montemorbase 3D model.

We plan to make the ontology available in OWL format so that we can rely on community reuse, refinment and extensions (https://github.com/IvoSantos/MegalithismCIDOC).

Following [1], we intend to develop information extraction (IE) mechanisms for this type of specialized textual corpora. For advancing in that direction, in a previous work [10], we presented the corpus on Portuguese Megalithism. Such domain specialized corpus will be manually annotated for the development of IE tools.

Fig. 3. Almendres cromlech, Évora, Portugal. Adapted from [2]. Photo by R. de Balbín

Knowledge organization and representation is an important step in order to make it possible to preserve the context of the megalithic monuments in its various facets: intrinsic context (archaeological), archaeological interventions and intervenients. In other words, the use of Semantic *Web* premises allow the implementation of an archaeological information structure in which all data are related in a known way, searchable and reusable, without exception. Simultaneously, by combining semantic capability and 3D modeling, this implementation creates the basis for further investigations. We plan to develop this interdependence between 3D modeling and monument information in future works.

The context in Archaeology is so important that it can determine the meaning and (archaeological) value of monuments and artefacts. However, the databases used in the discipline, as well as the academic projects and works themselves, usually do not include all the existing technological capabilities to safeguard this same context.

As archaeological practice incessantly seeks to complement the record with more and more digital techniques, archaeological information systems must follow this trend, including these digital data and safeguarding its context. The importance of context validates the demonstrated need for clarity, which only the principles of *Web* Semantics can grant to archaeological information. If context is everything for Archaeology and, technologically, the best way to describe a context is through semantically rich attributes, then, for this discipline, and in accordance with good management and data sharing practices currently in force, the use of computer application that values semantics, becomes essential.

References

1. Brandsen, A., Verberne, S., Lambers, K., Wansleeben, M.: Can bert dig it? - named entity recognition for information retrieval in the archaeology domain. J. Comput. Cult. Herit. (2021). https://doi.org/10.1145/3497842
2. Cerrillo-Cuenca, E., Bueno-Ramírez, P., de Balbín-Behrmann, R.: "3dmeshtracings": A protocol for the digital recording of prehistoric art. its application at almendres cromlech (Évora, portugal). J. Archaeol. Sci. Rep. **25**, 171–183 (2019). https://doi.org/10.1016/j.jasrep.2019.03.010, https://www.sciencedirect.com/science/article/pii/S2352409X19300379
3. Cripps, P.J.: Places, people, events and stuff; building blocks for archaeological information systems. In: CAA 2012 (2011)
4. Diniz, M.: The origins of megalithism in western iberia: resilient signs of a symbolic revolution? De Gibraltar aos Pirenéus-Megalitismo, Vida e Morte na Fachada Atlântica Peninsular, pp. 303–320 (2018)
5. Dörr, M.: Harmonized models for the digital world cidoc crm and extensions (video of part of the special interest group of cidoc crm meeting). https://www.youtube.com/watch?v=AILB8eUNYgA, (Accessed 1-8 2018)
6. Gergatsoulis, M., Papaioannou, G., Kalogeros, E., Mpismpikopoulos, I., Tsiouprou, K., Carter, R.: Modelling archaeological buildings using CIDOC-CRM and its extensions: the case of Fuwairit, Qatar. In: Ke, H.-R., Lee, C.S., Sugiyama, K. (eds.) ICADL 2021. LNCS, vol. 13133, pp. 357–372. Springer, Cham (2021). https://doi.org/10.1007/978-3-030-91669-5_28
7. Leisner, G.: Antas do concelho de Reguengos de Monsaraz: Instituto para a alta cultura. Materiais para o estudo da cultura megalítica em Portugal, Editora gráfica portuguesa (1951)
8. Migliorini, S., Grossi, P., Belussi, A.: An interoperable spatio-temporal model for archaeological data based on ISO standard 19100. J. Comput. Cult. Heritage **11**(1), 1–28 (2017). https://doi.org/10.1145/3057929, https://doi.org/10.1145/F3057929
9. Niccolucci, F.: Documenting archaeological science with cidoc crm. Int. J. Digit. Libr. **18**(3), 223–231 (2017)
10. Santos, I., Vieira, R.: Semantic information extraction in archaeology: challenges in the construction of a portuguese of megalithism. In: Research Conference on Metadata and Semantics Research. pp. 236–242. Springer (2022). https://doi.org/10.1007/978-3-030-98876-0_21
11. Tchienehom, P.: Modref project: from creation to exploitation of cidoc-crm triplestores. In: The Fifth International Conference on Building and Exploring Web Based Environments (WEB 2017) (2017)
12. Tibaut, A., Kaučič, B., Dvornik Perhavec, D., Tiano, P., Martins, J.: Ontologizing the heritage building domain. In: Ioannides, M., Martins, J., Žarnić, R., Lim, V. (eds.) Advances in Digital Cultural Heritage. LNCS, vol. 10754, pp. 141–161. Springer, Cham (2018). https://doi.org/10.1007/978-3-319-75789-6_11

Ontology for Analytic Claims in Music

Emilio M. Sanfilippo[1]($\boxtimes$) and Richard Freedman[2]

[1] ISTC-CNR Laboratory for Applied Ontology,
Via alla Cascata 56/c, 38123 Trento, Italy
emilio.sanfilippo@cnr.it
[2] Haverford College, Haverford, PA 19041, USA
rfreedma@haverford.edu

Abstract. The Semantic Web is increasingly used in research about music. A challenge is how to conceive musical works considering the hot debate about their nature in areas like musicology, philosophy, and library science. In addition, scholars ask for approaches representing research *claims* since these can be useful to document the scholarly debate. Building on a research project in musicology, we present an ontology for musical works and claims about them. The development of the ontology is work in progress.

Keywords: Ontology · Claim · Musicology · Early Music

1 Introduction

The concept of *musical work* is so basic to composers, performers, and listeners that it seems almost self-evident. And yet the entity we know, e.g., as Beethoven's *Symphony No. 9* is to a considerable extent not analogous to other kinds of artistic productions. Like paintings or novels, e.g., musical works are generally presumed to have an author and some structure. But musical works, like theatrical plays, are in some other ways different from either of those types. The physical substance of Leonardo's *La Gioconda*, e.g., is clearly the paint and wood of which it is made. We can hardly be so definitive about the substances of Beethoven's *Symphony No. 9*. Is it the movement of the air during a particular performance? The original manuscript? The printed orchestral score? In some sense, one may argue that the *Ninth* is a surprisingly unstable abstraction related to all of these. This is the reason why many scholars nowadays refer to the concept of work as a philosophical construction that is particularly complicated in the case of music [7]. We should add that this abstraction is also a cultural construction, since it can be differently interpreted across epochs [19]. Also, considering the variety of opinions with respect to musical works, one may even doubt whether scholars refer to the same thing even when they use the same terms [13].

These kinds of considerations are not only a concern for musicologists or philosophers. When dealing with music data in computer science, we need to understand how to draw subtle distinctions between different musical pieces.

© Springer Nature Switzerland AG 2022
S. Chiusano et al. (Eds.): ADBIS 2022, CCIS 1652, pp. 559–571, 2022.
https://doi.org/10.1007/978-3-031-15743-1_51

Ontologies are used nowadays to, e.g., store, publish or give access to data [5].
A clear conceptualization of musical works is therefore needed [13]. The purpose
of the paper, in brief, is to present some insights on musical works building on
an approach at the intersection between musicology, philosophy, and knowledge
representation. In addition, we present a manner to represent analytic claims to
explicitly document scholarly opinions about musical phenomena. This discussion will lead us to introduce some aspects of the *Ontology for Analytic Claims
in Music* (OMAC). To exemplify the discussion, we will borrow examples from
the research project CRIM – *Citations: The Renaissance Imitation Mass.*[1] Along
the way, we will review the existing literature about the notion of musical work
from different research perspectives, introduce the basic structure of OMAC, and
conclude with some suggestions for future work.

2   Musical Works: A (Partial) Review of the State of the Art

The investigation on musical works is a challenging research topic in disciplines
like philosophy, musicology, and library science (e.g., [19,20,22]). Before com-
menting on how digital resources treat musical works, we must acknowledge that
ours is hardly a discourse about all music (the consideration of which would be
too wide a mandate). Our scope is comparatively modest, focussing on the art
music tradition of Western Europe, and, more specifically, to music of the years
before 1600. Music from this tradition is in turn certainly not the same as that
of, e.g., Beethoven's day. Pieces by Josquin des Prez (active around 1500) and
Beethoven (active around 1800) nevertheless share some basic concepts about
how and why they exist, and what scholars might say about them.

The first assumption is to conceive a musical work as being intentionally
created at a certain time within a socio-cultural context. Hence, *contra* philosophers embracing Platonic attitudes (see the review by [22]), a musical work is
something human-made which could not exist if not composed on purpose.

The second assumption is that a musical work is not identical with its score.
The same work can be indeed represented through different scores in different
notations. We can do a similar consideration for performances, since there can be
different executions for the same work happening at different times and places.
In principle, a musical work can exist without being ever played. From these
perspectives, musical works enjoy a more abstract nature in comparison to both
scores and performances, and cannot be reduced to *sets* of such entities [22].

The third assumption is that musical works have various features, some of
which are, e.g., based on historical sources, whereas others result from scholarly
analyses. For instance, it is a matter of historical evidence that the motet *Tota
pulchra es*[2] was composed by Claudin de Sermisy (a French composer active in
the middle years of the sixteenth century), whereas it is a matter of analysis
that the motet includes a *fuga* (see Sect. 3.2).

---

[1] https://crimproject.org/.

[2] https://crimproject.org/pieces/CRIM_Model_0011/.

The fourth assumption is that it could be misleading to force all different types of music produced over the centuries within a single concept of musical work [7,13]. Talbot [20], for example, claims that "within the tradition of what we call [...] Western art music, it has seemed axiomatic until quite recently that the basic unit of artistic production and consumption is the 'work' - a hard-edged artefact with a clear identity." Talbot concludes arguing that "this common-sense or perhaps naive view is increasingly coming under fire from several sides." Indeed, scholars of Renaissance music often confront rival versions of a musical text that strain our very notion of the stable work in the first place. And students of Chopin's piano music (to take an example from the less distant past) often encounter the unsettling fact that the great composer frequently crafted and authorized seemingly contradictory versions of what would otherwise seem to be the same composition, and even published editions that contain mutually incompatible performance indications, repeats and even pitches.

Turning our attention to digital projects, the Functional Requirements for Bibliographic Records (FRBR) [16] was introduced in 1998 in the domain of library science to represent bibliographic data. FRBR has been reused for different ontologies (e.g., [6,9,15,23]) and frameworks, including the Music Encoding Initiative.[3] It has been also aligned to CIDOC-CRM, leading to FRBRoo [1]. Hence, considering its relevance, we will comment on the manner in which especially this latter version conceptualizes musical entities.

The documentation on FRBRoo recognizes that the notion of work is intentionally only vaguely characterized within the model [1, p.26]. Quoting from [1], "[a]n instance of F1 Work begins to exist from the very moment an individual has the initial idea that triggers a creative process in his or her mind. [...] Unless a creator leaves at least one physical sketch for [the] work, the very existence of that instance of F1 Work goes unnoticed [...]" [1, p.27]. This consideration can be interpreted in at least two ways: work as 1) the creator's idea or 2) the result of the creator's idea. In both cases, the quotation suggests that a work can exist without being represented in any text or score. The first interpretation, which is documented in the literature on FRBRoo (e.g., [8,9]), recalls intentionalist perspectives in philosophy [20]. As an objection to this philosophical view, Levinson argues that "if compositions are private intuitive experiences in the mind of composers, [...] they become inaccessible and unshareable" (quoted in [3]). In further excerpts, FRBRoo suggests that a work is the *abstract content* of expressions [1, p.54]. There is however something confusing: in the first two interpretations, a work is only possibly related to expressions, whereas it is now conceived as an abstraction from expressions. Also, reference to abstract content leaves open space for questioning about the relation between content and interpretation [18]. As a consequence of these ambiguities, we refrain from the adoption of FRBRoo, as well as from the use of existing resources based on it.

Keeping on the context of library science, Smiraglia [19] considers musical works as sorts of abstract entities that he calls *documentary entities* and that are neither coincident with expressions, nor with mental ideas, and that are useful

---

[3] https://music-encoding.org/guidelines/v4/content/metadata.html.

for classification purposes; e.g., to classify multiple editions of a novel. In our experience, practitioners working with archives and digital libraries sometimes speak of works as *identifiers* or *entry points* for resembling expressions (see [13]).

From our perspective, we mitigate domain experts' views (as discussed above) with a pragmatic attitude to manage music data. In particular, we rely on the proposal by Masolo et al. [11] on the notion of literary work and grounded on an approach at the intersection between AI, philosophy, and human sciences. The idea discussed by the authors is to conceive works as *socio-cultural entities* resulting from agreements among the members of expert communities about the manner in which texts are interpreted. The assumption is that agents interpret texts and engage in debates to check how similar their interpretations are. The situation in which they agree on the similarity of their interpretations leads to a literary work as a common shared interpretation. Hence, in this perspective, a literary work is something that exists only with respect to a group of (agreeing) agents but nothing prevents the existence of multiple, perhaps even incompatible, works related to the same text. Although Masolo et al.'s proposal is centered on literature, we think that their analysis can be applied to music, too.

In our understanding, indeed, this approach matches well with domain experts' considerations, as well as with the analyses of music proposed by scholars like Goher [7] and Talbot [20], among others. It seems also compatible with cataloging approaches [19]. Collecting, indeed, multiple entities under a documentary work means to make a choice about what counts as an "exemplar" of, e.g., Beethoven's *Ninth* and what not. From this perspective, a work is the result of an agreement process among the members of a community about how to consider the relations between the entities possibly grouped together.

An evident consequence of this approach is that musical works tend to proliferate, e.g., if interpretations are incompatible or non-overlapping. This could be a disadvantage if one wishes to classify, e.g., multiple scores under a *single* work. At the same time, however, from a scholarly criticism perspective, it could be an interesting position to stress the difference between multiple interpretations. To trade-off this pluralistic view with a pragmatic attitude for data classification, one can assume that scholars (within specific contexts) *do* share a common "minimal" interpretation of multiple expressions. For instance, if we say that the *Symphony No. 9 in D-minor* (as a musical work) was composed by Beethoven, is a symphony, and consists of four movements, we talk of the *Ninth* as the interpretation shared by all scholars in our domain. This is a simplifying assumption compared to Masolo et al.'s proposal [11] but it proves useful for our purposes. We shall see in the next sections how this view leads to a computational ontology.

3   Ontology for Analytic Claims in Music (OMAC)

We introduce here some aspects of the *Ontology for Analytic Claims in Music* (OMAC) and present some examples based on CRIM. The ontology is under development; our purpose is not to present its full axiomatization but the main ideas behind it. To reflect the distinction between musical works and claims, we develop OMAC in a modular architecture consisting of two interrelated modules, i.e., the Musical Work module (Sect. 3.1), and the Analytic Claim module (Sect. 3.2). The ontology can be extended with further elements. OMAC is specified in the Web Ontology Language (OWL);[4] we have maximized the reuse of existing resources like DBpedia (see [12] for an example of use of DBpedia in music). For the sake of shortness, we present the ontology only partially via some Description Logic formulas; the reader can refer to the available OWL files.[5]

3.1   OMAC – Musical Work Module

The Musical Work module allows representing musical works and their attributes. From a high-level perspective, it focuses on the distinction between musical works and musical works' parts. The distinction is based on authorial structure, namely, the division of a work into sections and subsections as given by composers and identified in scores through, e.g., formal divisions [17]. Formulas (a1)–(a6) present the taxonomy of musical entities, whereas (a7)–(a13) present some parthood-like relations (see the ontology files for more information).[6]

a1 $MusicalWork \sqsubseteq MusicalEntity$

a2 $MusicalWorkPart \sqsubseteq MusicalEntity$

a3 $MusicalWork \sqcap MusicalWorkPart \sqsubseteq \bot$

a4 $MusicalSection \sqsubseteq MusicalWorkPart$

a5 $MusicalSubsection \sqsubseteq MusicalWorkPart$

a6 $MusicalSection \sqcap MusicalSubsection \sqsubseteq \bot$

a7 $authorialPartOf \sqsubseteq partOf$

a8 $sectionOf \sqsubseteq authorialPartOf$

a9 $subsectionOf \sqsubseteq authorialPartOf$

a10 $MusicalEntity \sqsubseteq \forall hasPart.MusicalEntity$

a11 $MusicalWorkPart \sqsubseteq \exists authorialPartOf.MusicalWork$

a12 $MusicalWorkSection \sqsubseteq \exists sectionOf.MusicalWork$

a13 $MusicalWorkSubsection \sqsubseteq \exists subsectionOf.MusicalSection$

To comment on the formulas, we use the general notion of musical entity to cover both musical works and their parts. It is therefore at this level that musical entities correspond to shared interpretations (see previous section). Also, the formal representation does not explicitly bind musical entities to interpreting agents or agreeing mechanisms between the agents. This is because the ontology is a computational artifact to support data management, hence we aim at a simple formal representation. The introduction of interpreters would inevitably

---

[4] https://www.w3.org/TR/owl-ref/.

[5] https://github.com/HCDigitalScholarship/OMAC.

[6] The mereology of musical entities might be further refined; e.g., to understand whether mereological sums of sections still count as works' sections. One needs however to pay attention in representing models that are significant from a musicological perspective.

lead to a more complex formal characterization (see [11]). The ontology is however coherent with what previously said. Also, by axiom (a10), instances of the
class *MusicalWork* (a subclass of *MusicalEntity*) do not necessarily have parts.
An example in Early Music is *Tota pulchra es* (by Claudin de Sermisy), a Latin
motet in one continuous section for four voices. An Italian madrigal by Cipriano
de Rore, *Ite rime, dolenti,* in contrast is divided into two sections separated by
a pause. A cyclic Mass brings more complexity to the fore. Jean Guyon's *Missa
je suis desheritèe,*[7] for instance, consists of the customary five liturgical sections
of the Ordinary of the Catholic Mass (i.e., Kyrie, Gloria, Credo, Sanctus, and
Agnus Dei). These movements represent a standard setting used in almost every
Mass from this period. Yet some of these movements are further divided into
subsections; e.g., the Kyrie is normally divided into three distinct subsections,
each of which closes with a definitive musical cadence.

Figure 1 shows the (partial) RDF graph for the authorial structure of the
Missa je suis desheritèe according to (some of) the elements introduced above.[8]

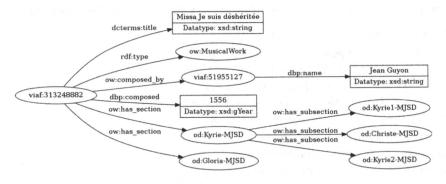

Fig. 1. RDF graph for the authorial structure of the mass (partial view)

One can reason over data through the axioms of the ontology deducing the
structure of the mass. For instance, since od:Kyrie-MJSD is section of the work
viaf:313248882, and od:Christe-MJSD is subsection of od:Kyrie-MJSD, then
od:Christe-MJSD is subsection of the whole work. Also, since the mass was
composed by Jean Guyon, it is possible to infer that all mass' authorial parts
have the same composer.[9] These inferences can be useful to retrieve information
which can be derived from the data even though it is not explicitly stated.

By axiom (a11), a musical work part can be included in different musical
works; e.g., one and the same section could be the authorial part of multiple
works. As a thought experiment, if, e.g., the Benedictus of a Sanctus movement,
enjoyed separate authorized circulation as a contrapuntal duo, is it still the

---

[7] https://crimproject.org/masses/CRIM_Mass_0006.

[8] RDF models for all examples presented in the paper are available at https://github.
com/HCDigitalScholarship/OMAC (see the files for namespaces).

[9] For these inferences, readers can browse the object property chains in the ontology.

same part of the original Mass, or is it a new composition? In our experience, such instances are controversial among music scholars since they depend on the manner in which the identity of works' (authorial) parts is conceived. For instance, one may claim that since a section has been intentionally conceived as part of a specific work, its identity strictly depends on that; hence, it cannot be detached from the work while preserving its identity. At the current state, we do not take a specific commitment on this matter since further research is needed.

Besides the representation of musical works structure, the Musical Work module allows characterizing musical entities under different aspects, e.g., by making explicit genre, performing forces, composition date, etc. For limits of space, we cannot dig into the presentations of these aspects (see ontology files). To further comment on Fig. 1, RDF triples like viaf:313248882 ow:composed_by viaf:51955127 and viaf:313248882 dbp:composed "1556", representing who composed the musical work and when, respectively, are shortcuts to ease data management. Musical works (and their authorial parts) are indeed the outputs of events carried out by composers that took place in certain contexts. In a first-order logic setting, the relations could be defined on the lines of (f1) where the event is now explicit.

f1 $composedBy(x, p) \equiv (MusicalWork(x) \lor MusicalWorkPart(x)) \land$
$Person(p) \land \exists e(CompositionEvent(e) \land carryOut(p, e) \land output(e, x))$

Because of the restricted expressivity of OWL (f1) cannot be expressed. Also, at the current state, we do not include events in OMAC. Should application needs require their treatment, they can be represented by extending the ontology.

3.2 OMAC – Analytic Claim Module

The Analytic Claim module is devoted to the representation of scholarly claims. The intuition is to conceive them as statements through which experts characterize and analyze musical entities [4]. A scholar, for instance, may make a claim about the composition date of a musical work for which such information is not available in the sources. The claim could be however wrong with respect to reality. In addition, nothing prevents multiple scholars from expressing the same claim, as well as the formulation of incompatible claims about the same entities, or claims expressing different granular information (e.g., a composition date in terms of a time interval or time point within that interval).

We find it useful to build our approach on the framework proposed by Masolo et al. [10]. In their proposal, the authors introduce the notion of *observation* to represent the classification of entities under properties as result of certain procedures. For instance, that John is observed as satisfying the property of *being 1,70 mt* tall according to a measurement procedure. Hence, an observation does not express something that necessarily holds in reality. Masolo et al.'s [10] approach is more complex than what just said, although, for the sake of the presentation, this short introduction suffices. A similar strategy has been proposed by Carriero et al. [2] in the context of cultural heritage. Also, Weigl et al. [23] use the Web Annotation Data Model to represent music experts' analyses, among others. Although some of the authors' ideas are close to ours, we think that the approach in [10] is better suited for our purposes; e.g., differently from the latter, it makes sense that claims do not necessarily express true facts.

In the case of musicology, debates about date, or even authorship of works from the years before 1600 are common, and entirely understandable given the sparse documentary record about the specific occasions that led to the creation of a particular work. Surprisingly few works from the period survive in copies associated directly with their composers. Thus even basic statements of fact about a piece might better be modeled as contingent observations rather than given facts. The situation is even more complex (and hotly debated) in the case of analytic, structural, or stylistic assertions about a musical work. Scholars might well want to claim that a particular work is beautiful, or that a particular part in the work represents something remarkable. Such claims are typical in the world of analysis and criticism, yet they lack any formal vocabulary by which they can be made discoverable to any beyond those who take the time to read complex narrative arguments or analytic charts.

A vast analytic database of thousands of analytic observations compiled in the course of the CRIM project nevertheless provides a good testing ground for ways to model such critical claims. They are formulated according to regular data structures, and contain highly formalized observations about particular passages in the works under consideration. To show some examples, Fig. 2 represents a portion, called *analytic segment*, of *Tota Pulchra es.*[10] An analytic segment is part of either a musical work or a work part that is not authorial; hence, it is not identified in the corresponding scores via formal divisions. Rather, it is only singled out by analysts for musicological purposes. We might call this claim an assertion about the structure of the piece. Considering again Fig. 2, the analyst who identified the segment characterizes it with the attributes shown on the right side of the figure. The list of attributes has been defined by domain experts in the scope of the project.[11]

Another example of claim relevant for CRIM is about the similarities identified by analysts between two musical entities, one considered as *model* and the other one as *derivative*. For example, Fig. 3 (partially) shows the relation of non-mechanical transformation holding between *Tota pulchra es* and the *Credo* of the *Missa Tota pulchra es.*[12] The details of the claims do not need to delay us here, since they are part of the specialized vocabulary developed for the project.[13]

To give a first representation of claims in the scope of CRIM, we need to introduce further modeling elements. By (d1), an analytic segment is a musical entity analyzed by a person (i.e., the analyst) that is related – via the relation

---

[10] The data in Fig. 2 is available at: http://crimproject.org/observations/1/.

[11] CRIM attributes to characterize analytic segments: https://sites.google.com/haverford.edu/crim-project/vocabularies/musical-types?authuser=0.

[12] The data in Fig. 3 is available at: https://crimproject.org/relationships/2/.

[13] CRIM about similarity relations: https://sites.google.com/haverford.edu/crim-project/vocabularies/relationship-types.

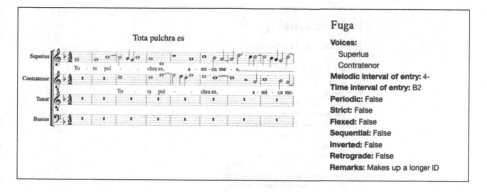

Fig. 2. Example of CRIM structural claim

segmentOf (a subproperty of *partOf*) – to either a musical work or a musical work part. Axioms (a14)–(a16) introduce only some axioms for the primitive classes of claims. By (a15), *StructureClaim* models claims of the sorts represented in Fig. 2. Similarly (a16) introduces the class *SimilarityClaim* between two musical entities (as in Fig. 3), i.e., the model and the derivative, represented through the relations *hasModel* and *hasDerivative* (subproperties of *refersTo*). OMAC covers a taxonomy of similarity claims according to CRIM. Other predicates are used to capture the intended meaning of claims in the project (see diagrams below).

d1 $AnalyticSegment \equiv MusicalEntity \sqcap \exists analyzedBy.Person \sqcap$
$$\exists segment.Of(MusicalWork \sqcup MusicalWorkPart)$$
a14 $Claim \sqsubseteq \exists refersTo.MusicalEntity \sqcap \exists statedBy.Person$
a15 $StructureClaim \sqsubseteq Claim \sqcap \exists refersTo.AnalyticSegment$
a16 $SimilarityClaim \sqsubseteq Claim \sqcap \geq 1hasModel.MusicalEntity \sqcap$
$$\leq 1hasModel.MusicalEntity \sqcap \geq 1hasDerivative.MusicalEntity \sqcap$$
$$\leq 1hasDerivative.MusicalEntity$$

Figure 4 shows the RDF graph for the (partial) representation of the claim in Fig. 2. In particular, od:claim_P1 is a *StructureClaim* attributing the schema *fuga* to od:segment_P1, which is segment of od:Tota_pulchra_es. Relations like *periodic* and *strict* etc. are specific to the CRIM vocabulary.

Figure 5 shows the RDF graph for the (partial) representation of the claim about the similarity shown in Fig. 3. In particular, od:claim_R2 is the similarity claim instantiating *NonMechanicalTransformation* (a subclass of *SimilarityClaim*), referring to the model od:Tota_pulchra_es and derivative od:Credo_MTPE (a section of *Missa Tota Pulchra es*). According to the analyst, the similarity is characterized through the relations *new counter subject* and *old counter subject shifted metrically* (among others). By reasoning over the data

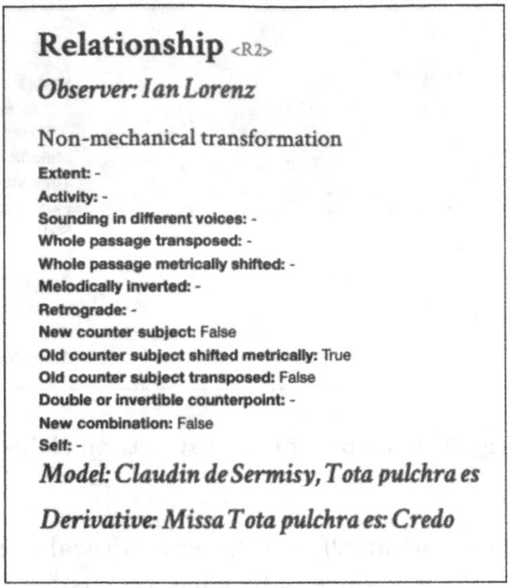

Relationship <R2>
Observer: Ian Lorenz
Non-mechanical transformation
Extent: -
Activity: -
Sounding in different voices: -
Whole passage transposed: -
Whole passage metrically shifted: -
Melodically inverted: -
Retrograde: -
New counter subject: False
Old counter subject shifted metrically: True
Old counter subject transposed: False
Double or invertible counterpoint: -
New combination: False
Self: -
Model: Claudin de Sermisy, Tota pulchra es
Derivative: Missa Tota pulchra es: Credo

Fig. 3. Example of CRIM similarity claim

rdf:type
oc:StructureClaim
oc:stated_by
od:Ian_Lorenz
oc:refers_to
od:segment_P1
oc:segment_of
od:Tota_pulchra_es
od:claim_P1
oc:has_schema
oc:Fuga_schema
oc:periodic
oc:strict
oc:flexed
false
Datatype: xsd:boolean

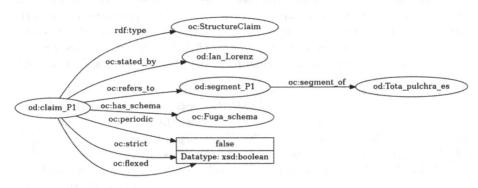

Fig. 4. RDF graph for the structural claim in Fig. 2 (partial view)

through the axioms of the ontology, including object property chains, one can derive, among other things, a derivation relation holding between *Tota pulchra es* and the *Credo*. As said in the previous section, this kind of inference can be useful for data retrieval e.g., to retrieve which musical entities derive from a certain work or work part, hence, to explore the data in the knowledge base.

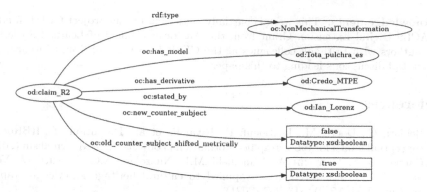

Fig. 5. RDF graph for the similarity claim shown in Fig. 3 (partial representation)

4 Conclusions

We presented (some portions of) the OMAC ontology and showed how it allows representing some aspects relative to musical works and claims. The ontology is grounded on an approach leveraging theoretical insights with a practical attitude for digital applications. With respect to the state of the art, we employ a notion of musical work that reflects the current debate across different disciplines. Also, the explicit treatment of claims opens the way to a formal approach for documenting, sharing, linking, and comparing scholarly observations on the same phenomena.

Future work is needed to strengthen the proposal. This includes the further development of the ontology to meet domain experts' knowledge and needs, e.g., relative to events relevant from a musicological view. The representation of claims needs improvements, too, from both a theoretical and modeling perspective; e.g., various classes of claims need to be introduced and formally characterized. One could even think to attach sorts of "trust values" to claims; e.g., if there are incompatible claims about the same entity, scholars may wish to trust one more than the others. It could be also explored how claims relate to works; the latter could be indeed conceived as complex scholarly statements on resembling expressions. We also plan to interact with scholars involved in digital projects about music(-ology) to check how OMAC can interact with other models, especially if one wishes to compare and possibly merge or link multiple datasets. In addition, we plan to plug OMAC in the information system for the management of CRIM data to make them FAIR and usable through Semantic Web technologies. This could allow users to navigate through the data through RDF graphs and, by exploiting automatic reasoning procedures, to discover data that are only implicitly stated. In this context, we will investigate the feasibility of an ontology-based data access (OBDA) architecture [14] to connect the ontology to the CRIM database.

Acknowledgements. This work is partially supported by the project CRIM, funded by ACLS Digital Extension Grant from the American Council of Learned Societies. The authors wish to thank colleagues at the CESR University of Tours (France) for the fruitful discussions leading to this paper.

References

1. Bekiari, C., Doerr, M., Le Boeuf, P., Riva, P., et al.: Definition of FRBRoo: a conceptual model for bibliographic information in object-oriented formalism (2017)
2. Carriero, V.A., Gangemi, A., Mancinelli, M.L., Nuzzolese, A.G., Presutti, V., Veninata, C.: Pattern-based design applied to cultural heritage knowledge graphs. Semantic Web **12**(2), 313–357 (2021)
3. Cray, W.D., Matheson, C.: A return to musical idealism. Australas. J. Philos. **95**(4), 702–715 (2017)
4. Cristofaro, S., Sanfilippo, E.M., Sichera, P., Spampinato, D.: Towards the representation of claims in ontologies for the digital humanities. In: Proceedings of the International Joint Workshop on Semantic Web and Ontology Design for Cultural Heritage (SWODCH), CEUR vol. 2949 (2021)
5. Daquino, M., et al.: Characterizing the landscape of musical data on the web: State of the art and challenges. In: Second Workshop on Humanities in the Semantic Web - WHiSe II, 21–25 Oct 2017, Vienna, Austria (2017)
6. Fields, B., Page, K., De Roure, D., Crawfordz, T.: The segment ontology: bridging music-generic and domain-specific. In: 2011 IEEE International Conference on Multimedia and Expo, pp. 1–6. IEEE (2011)
7. Goehr, L.: The Imaginary Museum of Musical Works: An Essay in the Philosophy of Music: An Essay in the Philosophy of Music. Clarendon Press (1992)
8. Le Boeuf, P.: A strange model named FRBRoo. Cataloging Classif. Q. **50**(5–7), 422–438 (2012)
9. Lisena, P., Troncy, R.: Representing complex knowledge for exploration and recommendation: the case of classical music information. In: Applications and Practices in Ontology Design, Extraction, and Reasoning, pp. 107–123. IOS Press (2020)
10. Masolo, C., Botti Benevides, A., Porello, D.: The interplay between models and observations. Appl. Ontol. **13**(1), 41–71 (2018)
11. Masolo, C., Sanfilippo, E.M., Ferrario, R., Pierazzo, E.: Texts, compositions, and works: A socio-cultural perspective on information entities. In: Proceedings of the 5th International Workshop on Foundational Ontologies (FOUST), CEUR vol. 2969 (2021)
12. Passant, Alexandre: dbrec — music recommendations using DBpedia. In: Patel-Schneider, P.F., et al. (eds.) ISWC 2010. LNCS, vol. 6497, pp. 209–224. Springer, Heidelberg (2010). https://doi.org/10.1007/978-3-642-17749-1_14
13. Pietras, M., Robinson, L.: Three views of the "musical work": bibliographical control in the music domain. Libr. Rev. **61**, 551–560 (2012)
14. Poggi, A., Lembo, D., Calvanese, D., De Giacomo, G., Lenzerini, M., Rosati, R.: Linking data to ontologies. In: Spaccapietra, S. (ed.) Journal on Data Semantics X. LNCS, vol. 4900, pp. 133–173. Springer, Heidelberg (2008). https://doi.org/10.1007/978-3-540-77688-8_5
15. Raimond, Y., Abdallah, S.A., Sandler, M.B., Giasson, F.: The music ontology. In: ISMIR, vol. 2007, p. 8th. Citeseer (2007)
16. Riva, P., Le Boeuf, P., Zumer, M.: IFLA library reference model. a conceptual model for bibliographic information (2017)

17. Sadie, S.: Movement (2001). https://www.oxfordmusiconline.com/grovemusic/view/-10.1093/gmo/9781561592630.001.0001/omo-9781561592630-e-0000019258
18. Sanfilippo, E.M.: Ontologies for information entities: state of the art and open challenges. Appl. Ontol. **16**(2), 1–25 (2021)
19. Smiraglia, R.P.: Musical works as information retrieval entities: epistemological perspectives. In: ISMIR, pp. 85–91 (2001)
20. Talbot, M.: Introduction. In: The Musical Work: Reality or Invention?, pp. 1–13. Oxford University Press (2000)
21. Talbot, M.: The musical work: reality or invention? Oxford University Press (2000)
22. Thomasson, A.L.: The ontology of art. In: The Blackwell guide to aesthetics, Malden, MA, pp. 78–92. Blackwell (2004)
23. Weigl, D.M., et al.: Read/write digital libraries for musicology. In: DLfM 2020: 7th International Conference on Digital Libraries for Musicology, pp. 48–52. ACM Digital Library (2020)

Testing the Word-Based Model in the Ontological Analysis of Modern Greek Derivational Morphology

Nikos Vasilogamvrakis[1]([✉]) [iD], Maria Koliopoulou[2], Michalis Sfakakis[1] [iD], and Giannoula Giannoulopoulou[3]

[1] Department of Archives, Library Science and Museology, Ionian University, Corfu, Greece
{120vasi,sfakakis}@ionio.gr
[2] Department of German Language and Literature, National and Kapodistrian University of Athens, Athens, Greece
mkoliopoulou@gs.uoa.gr
[3] Department of Italian Language and Literature, National and Kapodistrian University of Athens, Athens, Greece
giannoulop@ill.uoa.gr

Abstract. In the present article we explore the Item and Process (IP) approach - frequently known as Word-Based (WB) - as a theoretical model to ontologically represent the interconnection between derivatives of Modern Greek (MG). The model puts emphasis on the word as an indivisible base unit, the template rules to which words are subsumed to form new ones and the kind of relationships they establish. After a brief MG morphological analysis and the representation of various WB formation rules we proceed to test those on the MMoOn model in order to check its ontological expressiveness. In doing so we adopt an as possible top-down approach so that templates dynamically link to their respective lexical instances. Although the model generally satisfies the IP paradigm specifications it seems deficient in dealing with MG language-specific derivational rules or directional peculiarities and not very persuasive in terms of input-output categorial change representation as well as in dealing with derivatives at lexical-to-hyper-lexical level. To tackle these issues, we propose possible solutions and present their advantages in each case.

Keywords: IP morphology · Word-based morphology · MMoOn · Ontologies · Linguistic linked data · Modern Greek derivation

1 Introduction

In respect of derivational theory in morphology, the debate approach has been mainly expressed by two major word-formation paradigms: the morpheme-based or Item and Arrangement (IA) emphasizing on the lexical presence of morphemes and their participation in word formation by concatenative arrangement and the Item and Process (IP)

S. Chiusano et al. (Eds.): ADBIS 2022, CCIS 1652, pp. 572–584, 2022.
https://doi.org/10.1007/978-3-031-15743-1_52

or Word-based[1] which regards the word as an indivisible base unit and derivation as the application of template rules to it. It is generally assumed that in languages of synthetic morphology such as Modern Greek (MG) morpheme-based processes are indispensable, if nothing else, to discover step-by-step word structure synthesis or to deconstruct lexemes into their parts [11]. On the other hand, there are cases that strict concatenative affixation procedures do not suffice for explaining derivational formations when certain diachronic or morpho-phonological phenomena occur which seem more suitable to be dealt with a series of descriptive rules and standardized templates. In the present study we will focus on the extent that an ontological model (MMoOn) expresses the IP paradigm rules and the variosity of derivational relations between words.

In Sect. 2 we elaborate on IP templates of MG derivation covering representative lexical cases while in Sect. 3 we explore how these templates may be applied to or extended in the MMoOn ontological model. In Sect. 4 we discuss the results and implications of our testing and in Sect. 5 we conclude on the topic defining areas for future research.

2 Morphological Analysis

It is most common that the IP approach is elaborated in contrast to morpheme-based or concatenative processes. As stated by Haspelmath and Sims [2], in a word-based model "the fundamental significance of the word is emphasized and the relationship between complex words is captured not by splitting them up into parts and positioning a rule of concatenation, but by formulating word-schemas that represent the features common to morphologically related words". This means that the word-based approach gives precedence to the derivational relationship between an input and an output word, i.e. the specific rule that outputs a derivative and not to the microanalysis of its constituent units. A major difference between the two theories is that added morphemes are handled as parts of a word-schema [1] and not as separate lemmatic entries in the lexicon.

In support of the IP model it is assumed that just by putting morphemes together in a row cannot by itself give ground to specific phenomena or to the rules occurring in derivation. Based on standardized IP abstract templates [1, 2, 6] in Table 1 we originally analyze various cases of MG that are subject to derivational rules such as affixation, conversion, reanalysis, back-formation, synizesis, subtraction, reduplication and cross-formation. All these MG processes are represented as template rules with respective lexical instances[2]. In the IP template column the binary relational rule is given as a $X_A \rightarrow X_{A/B}$ or $X_A \leftrightarrow X_{A/B}$ template (X comprises the string that remains unchanged, be it a stem or a word or an unspecified lexical part, whereas subscripts A and B the part-of-speech categorial change). It must be noted that template formulation should define a unique template of a given rule, reaching a pre-lexical level, usually through a characteristic affix (e.g. $Xía_N \rightarrow Xiá_N$), and being as inclusive as possible. All template examples represent a substantial account of MG derivational rules coming from relevant theoretical analyses [7–9], however, they do not comprise indisputable - but rather suggested - formulas of

---

[1] The two terms are used interchangeably.

[2] Lexical instances in the morphological analysis are phonetically transcribed according to the International Phonetic Alphabet (IRA) (cf. https://www.internationalphoneticassociation.org/content/ipa-chart).

linguistic representation. They all point to the binary process of derivation of a word from a previous simple lexeme or a derived word including the individual attributes triggered (additional morphs and category change). The process can be either one-way (→) referring to a defined source-target relationship or bidirectional (↔) when it needs to be seen bilaterally [2].

A most common non-concatenative formation rule is Conversion, i.e. a derivation of a lexeme just by category change *(nomikós* 'juristic' > *nomikós* 'lawyer')* but without always being clear which word is derived from the other *(oδiγ-ó* 'to drive' <> *oδiγ-ós* 'driver')*. Reduplication, which is the repetition of a lexical unit, can be either morpheme- or word-based. It is a process not very productive in MG and usually takes the form of lexical emphasis *(pro-páppos* 'great-grandfather' > *pro-propápos* 'great-great-grandfather', *líγo* 'a little' > *líγo-líγo* 'little by little')* and often involves onomatopoeia *(psi-psi* 'sound of calling a cat' > *psi-psí-na* 'cat', *tsi-tsi* 'sound of sizzling > *tsitsirízo* 'to sizzle')*. Most notably, although such a lexical unit repetition parallels an affixational process (as in *pro-pro-páppos*) there is a clear difference in the formation rule (affixation vs. reduplication). Contrastive to concatenation, a Subtraction rule creates new lexemes by deleting material [5, 10]. This often happens with lived-on Ancient Greek (AG) forms in MG that have triggered more simplified or popularized (+popular) equivalents (e.g. *erotó* > *rotó* 'to ask or query'). The economy of language has also led to other morpho-phonological simplification modes as e.g. lexemes derived from a Synizesis rule (a double syllable turned into one), which often involves stress metathesis *(poniría* > *poniriá* 'cunning' or *vasiléas* > *vasiliás* 'king')*. In different cases a diachronic Reanalysis process can reinstate forms and moreover establish them in everyday use (e.g. ex-contracted AG verbs like *aγapáo* > *aγapó* > *aγapáo* 'to love') [9]. Diachrony can also in some cases bring up "back-formed" lexemes (produced by a Back-formation rule) with the help of analogy e.g. nouns from verbs *(anaséno* 'to breathe' > *anása* 'breath')* [7].

Table 1. Indicative MG template rules representation of the IP paradigm

IP template	Lexical instance	Specific rule type	Broader rule category
X_V → xeX_V	léo 'say' > xeléo 'to unsay'	Prefixation	Affixation
$Xti-s_N$ → $Xik-os_{ADJ}$	stratiótis 'soldier' > stratiotikós 'military'	Suffixation	
X_{ADJ} → X_N	nomikós 'juristic' > nomikós 'laywer'		Conversion
Xos_N ↔ Xo_V	oδiγós 'driver' <> oδiγó 'to drive'		
$Xó_V$ → $Xáo_V$	αγαpó > αγαpáo 'to love'		Reanalysis
$Xeno_V$ → $Xα_N$	αnaséno 'to breethe' > αnása 'breath'		Back-formation

(*continued*)

Table 1. (*continued*)

IP template	Lexical instance	Specific rule type	Broader rule category
Xíα_N → Xiá$_N$ Xéαs$_N$ → Xiás$_N$	poniría > poniriá 'cunning' vαsiléas > vasiliás 'king'		Synizesis
eX$_V$ → X$_V$	erotó > rotó 'to ask'		Subtraction
proX$_N$ → proproX$_N$	propápos 'great-grandfather' > pro-propápos 'great-great-grandfather'		Reduplication
X$_{ADV}$ → XX$_{ADV}$	liγo 'a little' > liγo-liγo 'little by little'		
Xmα_N ↔ Xsi$_N$	θé-ma 'topic' <> θé -si 'position'		Cross-formation
xeX$_V$ ↔ pαrαX$_V$	xeléo 'to unsay' <> paraléo 'to exaggerate'		

One interesting word formation model is the so-called Cross-formation where two or more output words can be simultaneously derived from one - synchronically justified or not - input word that provides a common base, usually by a simple paradigmatic replacement of an affix or by analogy (*tíθimi* (AG) > *θe-* > *θé-ma* 'topic' and *θé-si* 'position', *θéma* <> *θési*). Finally, Affixation can also be depicted as a templatic rule, whether morpheme-based (*stratióti-s* 'army' > *stratiot-ik-ós* 'military') or word-based. In the latter case, it takes mostly the form of a prefixed or re-prefixed lexeme, where a word - and not a stem - appears as the structural base unit for the formation of successive lexemes (*léo* 'to say' > *xeléo* 'to unsay' or *dia-γráf-o* 'to cross off' > *pro-dia-γráfo* 'to specify'), with the derivational process moving from lexical to hyper-lexical[3] level.

3 Ontological Analysis

From the previous analysis it has been clear that the IP paradigm focuses on the relationship between input and output words that are derivatives of the first as well as the representation of the specific derivational rules involved. As we have done previously [11] we will test MG word-based templates onto the MMoOn[4] model, focusing on the respective ontological section. We have chosen MMoOn [4] because it "is currently the only existing comprehensive domain ontology for the linguistic area of morphological language data" [3] and because of its use as a modeling template for the development of the Ontolex Morphology Module[5] [3]. In what follows we make some purposeful remarks on the

---

[3] By the term hyper-lexical we refer to a lexeme that at least one of its morphological constituents is another lexeme (cf. a derived prefixed word, e.g. *pros-δiorízo* 'to define', a compound word, e.g. *thallas-o-taraxi* 'sea storm' or an analytical word form, e.g. *θa trekso* 'I will run').

[4] https://aksw.org/Projects/MMoOn.html and https://github.com/MMoOn-Project/MMoOn.

[5] https://www.w3.org/community/ontolex/wiki/Morphology.

MMoOn representation efficiency and present respective onto-morphological sugges-
tions so that MG lexical data are represented in the best possible way. Accordingly, in
Fig. 1 we extend the ontology with appropriate *ell_schema*[6] classes starting from the
MorphologicalRelationship class downwards:

- The model differentiates only between two main derivational rules: conversion or
zero derivation and stem-based affixation (though not overtly stated) in the form
of *DerivedAdjective, DerivedAdverb, DerivedNoun, DerivedVerb* classes for every
derived grammatical category. Furthermore, it seems that derivational rules (conversion, affixation) are subdivided into categorial change subclasses incorporating
semantically identical instances (e.g. *VerbalNoun has type VerbalNoun* which is a
subclass of *Conversion* in the *deu_inventory*[7]). In this point, instead of adding rules
as an one-to-one class-instance representation, it would be more conceptually right to
define them as hierarchically specialized entities, i.e. as subclasses belonging to the
general *Derivation* class that are further specialized into rule template instances. More
specifically, in accordance to our analysis of Table 1, we would rather specify a rule
not by its input-output category but by its specific formation template (i.e. *Xti-s_n-Xik-
os_adj*[8] *has type ell_schema:Suffixation* that is a subclass of *ell_schema:Derivation*
(see Fig. 1[9])). We, therefore, introduce MG various derivational template instances that
have the respective abstract rule class as their type (Affixation, Conversion, Synizesis,
Subtraction, Back-formation, Reanalysis, Reduplication). With regard to affixation in
particular, despite the fact that it is sufficiently explicated by part-whole relationships
as shown in the morpheme-based approach [11], the type of rule and template is still
missing and thus is given by an IP template.
- The fact that derivational rules (i.e. conversion and affixation) in the MMoOn core
are further subdivided into categorial input-output subclasses would inevitably result
in repeated categorial combinations. For example, there is the *Conversion* subclass:
AdjectiveNoun and the *DerivedNoun* subclass: *DeadjectivalNoun* that both - independently of the rule they belong to - bear exactly the same categorial change meaning.
This could prove verbose, impractical and bring up inconsistencies when more MG
language rules are added as subclasses of the *Derivation* superclass, because input-
output category combinations would have been repeated for each rule and with a
unique IRI to prevent a derivative from belonging to more than one rule. In dealing
with this problem we could for one thing transfer categorial change information to the
more specialized rule template instance as represented in Table 1 but with precaution
taken so that all formulated template instances are unique and do not conflict with

---

[6] The current version files of the ell_schema and its ell_inventory can be found in https://github.
com/nvasilogamvrakis/mmoon_project/blob/main/ell_schema/ell_schema_02.owl and https://
github.com/nvasilogamvrakis/mmoon_project/blob/main/inventory/ell_inventory_02.owl
respectively and can be used as proof of concept.

[7] https://github.com/MMoOn-Project/OpenGerman/blob/master/deu/inventory/og.ttl.

[8] Capital X represents the unchanged lexical part (stem, word or unspecified lexical string) while
the final lowercase letters the input-output categorial change.

[9] All instances in the ontology, be it rule templates in the schema or morphological entities in
the inventory, are directly given in MG morphologically transcribed IRIs.

each other. For example as shown in Fig. 1 a *Conversion* formation template would be as *X-os_n-X-o_v* for the derived sequence οδιγός > οδιγό.

- The representation of the derivational relationship between a target (derived) and a source word is given only with the target as domain in the *is derived from* object property (OP). Regarding this, it would be more convenient that the relationship be also represented inversely so that one can express the active derivational relation from a source to an output word. We would therefore propose an *ell_schema:generates* OP as *Inverse Of is derived from* (Table 2).

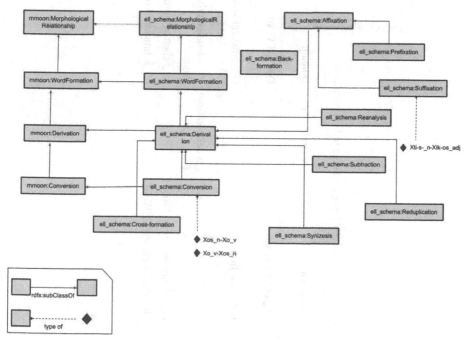

Fig. 1. The WB paradigm representing derivational rule type

- Contrary to derivation with directionality is the fact that in certain cases we cannot tell the direction of the process as this has become obscure in the course of time (e.g. conversion: οδιγός <> οδιγό) or has been subsumed to a formation process by analogy or replacement (e.g. cross-formation: θέma <> θέsi). Therefore, we need the relation to be characterized by symmetry and at the same time to be consistent with the MMoOn *is derived from* OP. We accordingly propose a symmetric *ell_schema:bidirectionalDerivation* OP (Table 2) as subproperty of *is derived from* which in turn is *Inverse Of ell_schema:generates*. By those two we achieve to have asserted statements in Turtle syntax such as:

    ell_schema:"οδηγώ" a ell_schema:DerivedWord.
    ell_schema:"οδηγός" a ell_schema:DerivedWord;
    ell_schema:bidirectionalDerivation ell_schema:"οδηγώ".

Table 2. OPs attributes and restrictions relating to WB paradigm

OP	Symmetric	Inverse Of	Subproperty Of	Domain	Range
is derived from		ell_schema:generates		Derived Word	Grammatical Word OR Lexical Entry
ell_schema_generates		is derived from		Grammatical Word OR Lexical Entry	Derived Word
ell_schema:bidirectionalDerivation	yes		is derived from	ell_schema:Derived Word	ell_schema:Derived Word
ell_schema:consistsOfWord		ell_schema:belongsToWord		ell_schema:Word	ell_schema:Word
ell_schema:belongsToWord		ell_schema:consistsOfWord		ell_schmema:Word	ell_schmema:Word

and the following inferred statements by the Reasoner[10]:

  ell_schema:"οδηγός":isDerivedFrom ell_schema:"οδηγώ".
  ell_schema:"οδηγώ" ell_schema:generates ell_schema:"οδηγός".
  ell_schema:"οδηγώ" ell_schema:bidirectionalDerivation
  ell_schema:"οδηγός".
  ell_schema:"οδηγώ":isDerivedFrom ell_schema:"οδηγός".
  ell_schema:"οδηγός" ell_schema:generates ell_schema:"οδηγώ".

- For the affixational word-based template $X\_v \rightarrow xe\text{-}X\_v$ we have to notice that: a) There is no defined relationship in the model between a *DerivedWord* and a *Word* class (*DerivedWord* is not defined as domain in the *consistsOfWord* OP) since derived words in general - besides compounds or analytic word forms - may too consist of previously formed simple lexemes or other derived words and affixes (e.g. *armóz-o* 'to be suitable' > *pros-armózo* 'to adapt' > *ana-prosarmózo* 'to readapt') b) Regarding the *isComposedOf* OP for decomposing a compound to its constituent words we have to note that an inverse relation is absent and the *belongsTo* OP cannot be used since it has *Morph* as domain pairing with the inverse *consistsOfMorph* OP to cover the sublexical-to-lexical level and c) It is also unclear why there are two similar OPs i.e. *isComposedOf* and *consistsOfWord* (*CompoundWord* is domain at both OPs), which as it seems can be used indiscriminately for compounds (*consistsOfWord* for analytic word forms as well) but not for derived word-based words. To overcome these limitations we would propose a new *ell_schema:consistsOfWord* and an inverse *ell_schema:belongsToWord* OP setting their domain and range to the general *ell_schema:Word* (Table 2) since we do not want to alter *consistsOfWord* OP legacy semantics. In doing so we would succeed in covering equally composition and decomposition of compounds, derived words (as we have seen in word-based concatenation e.g. *léo* > *xe-léo*) or analytic word form cases at hyper-lexical-to-lexical level.

In Fig. 2, we ontologically analyze four MG derivational formations: (a) *Synizesis* in πονηρία *(poniría)* > πονηριά *(poniriá)* has an one way derivational relation and a *Xía_n-Xiá_n* template, (b), (c) *Conversion* in οδηγός *(odiγós)* < > οδηγώ *(odiγó)* has a symmetric *bidirectionalDerivation* relation and a *Χoς_n-Χω_v* or a *Χω_v-Χoς_n* template while νομικός *(nomikós)* > νομικός *(nomikós)* has a clear derivational direction with a *X_adj-X_n* template (d) Prefixation with ξελέω 'to retract' and παραλέω 'to exaggerate' have a word-based relation with the simple lexeme λέω 'to say' (ξελέω *(xeléo)* < λέω *(léo)* > παραλέω *(paraléo)*) each belonging to *X_v-ξεX_v* and *X_v-παραX_v* templates respectively. It is worth noting that both word-based derived words ξελέω and παραλέω are further decomposed into their prefixes ξε- *(xe-)* and παρα- *(para-)* respectively and the simple lexeme λέω, actually recalling morpheme-based decomposition but with word as a base. Finally, in e) we have a Back-formation rule i.e. ανασαίνω *(anaséno)* > ανάσα *(anása)* with a *Χαινω_v-Xa_n* template.

---

[10] To test our ontological MG instance, we have used the ontology editor Protégé v.5.5.0.

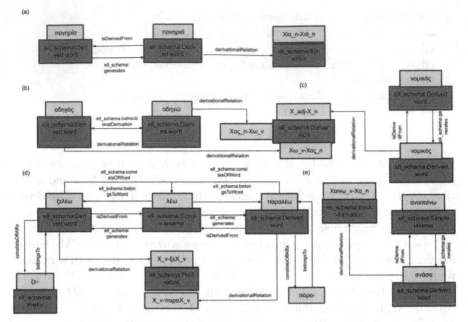

Fig. 2. Onto-lexical representation of MG derivational templates of Synizesis, Conversion Prefixation and Back-formation in MMoOn

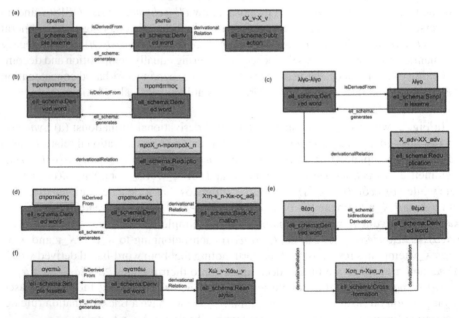

Fig. 3. Onto-lexical representation of MG derivational templates of Subtraction, Reduplication, Suffixation, Reanalysis and Cross-formation in MMoOn

In Fig. 3, we further analyze the derivational templates for (a) Subtraction (*ερωτώ (erotó)* > *ρωτώ (rotó)* *εX_v-X_v*), (b), (c) Reduplication (*προπάππος (propáppos)* > *προ-προπάππος (pro-propáppos)*, *προX_n-προπροX_n* and *λίγο (lígo)* > *λίγο-λίγο (lígo-lígo)*, *X_adv-XX_adv* respectively), (d) morpheme-based Suffixation (*στρατιώτης (stratiótis)* > *στρατιωτικός (stratiotikós)*, *Xτη-ς_n-Xικ-ος_adj* (e) Cross-formation (*θέμα (théma)* < > *θέση (thési)*, *Xμα_n-Xση_n*) and (f) Reanalysis (*αγαπάω (agapáo)* > *αγαπώ (agapó)*, *Xώ_v-Xάω_v*).

In all formation cases, derivational analysis goes along with the respective belonging rule and its lexical template by using the MG-specific subclasses of the core *Derivation* class. Input and output word instances are related to each other either via the *isDerived-From* Inverse of *ell_schema:generates* or the *ell_schema:bidirectionalDerivation* OPs and the output word in turn to the specific derivational rule via the *derivationalRelation* OP. As it was previously mentioned, the specific rules presented comprise substantial MG derivational templates reflecting most of MG morphological analyses texts [7–9].

4   Discussion

In the previous analysis we focused on derivational cases of MG where mostly a strict concatenative approach by itself cannot justify the resulting lexical structures because a formation process is also dependent on some intra-linguistic transformational processes. Next to traditional cases of conversion, MG morphology creates new words via morpho-phonological rules (e.g. synizesis, subtraction), diachronic changes (e.g. reanalysis, back-formation), reduplication and of course affixation, be it either morpheme- or word-based, all of which can be refined to specialized templates. These specifications have been tested on their representation in the MMoOn model resulting in proposed MG schema classes and formulated MG word formation templates as instances of rules.

However, it is important to point out that rules would rather reach a pre-lexical linguistic level and be as inclusive as possible linking economically to lexical instances instead of being lexical instances themselves. The choice of how downward a template should be would actually depend on the specific descriptive needs of an implementation and the uniqueness of the template. Most likely it would follow a generic formula such as $X\_a \rightarrow X\_a/b$ or $X\_a \leftrightarrow X\_a/b$ (accompanied with the appropriate lexical string, which is often a converting affix) relating to a group of lexical instances[11]. Based on that, one would easily make up theoretical assumptions on the productiveness of a rule after enough lexical data are populated in the ontology.

In this respect, what has come to be problematic in the MMoOn is the predefined specification of rules (e.g. conversion, affixation) as input-output category combinatorial classes ($X \rightarrow Y$) which would practically result in their repetition for each rule with superfluous entity definitions or in possible ontological inconsistencies with a derivative being subsumed to more than one formation rule. Given a language that would specify all four part-of-speech categories (V, N, Adj, Adv) such as MG there could practically

---

[11] As long as templates between different rules do not overlap, a more abstract template for a less productive rule could also be possible.

be nine input-output category combinations excluding the ones with no category change referred to as *Expansion*, a fact that would rather make the ontology verbose and complex. Representing categorial change at a pre-lexical rule instance level as presented in Sect. 3 would seem an appropriate solution if it was assured that all formulated templates were unique and did not conflict with each other.

From another point of view (Fig. 4), categorial change could be transferred to the *DerivationalMeaning* class (of the *Meaning* superclass) of the core MMoOn and specifically to the more precise *Expansion* and *FunctionalDerivation* subclasses, further subcategorized by each one of the input-output category instances with X and Y denoting the different grammatical categories. Based on that, it might be better to follow a rather decentralized approach by detaching categorial change from the *MorphologicalRelationship* to the *Meaning* class so that it would link to each distinct formation rule via an OP (*derivationalMeaning*) and not a subclass (as it is now in the core MMoOn). This would allow us to represent the change uniformly and separately from rules.

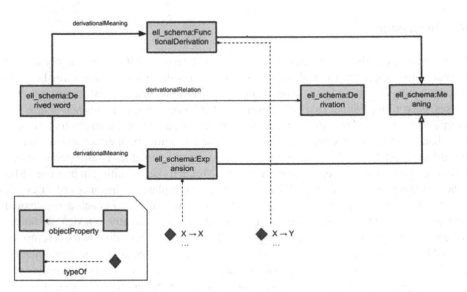

Fig. 4. Derivational relation of a derived word to the specific *Meaning* subclass

Another problematic point has been the lack of expressing derivational directionality. More specifically, the *is derived from* OP alone cannot account for the diversity of MG derivational cases as with certain rules (conversion, cross-formation) the source-target relation moves symmetrically to both directions. Additionally, the *is derived from* OP only depicts a target-from-source relation while the analysis would also benefit from a source-to-target one. That is to say, that a starting lexeme such as *stratós* 'army' would generate a series of lexical derivatives i.e. *stratiá* 'troops', *stratiótis* 'soldier' and *stratiotikós* 'military' if we would exploit the transitive closure of *stratós* using the *ell_schema_generates* OP [11]. Therefore, in either case symmetric and inverse properties would seem a reasonable choice.

Finally, MMoOn does not seem to cover sufficiently constituent relationships at lexical-to-hyper-lexical level and in a holistic way as it does with lexical-to-sub-lexical level relationships. It seems that its focus lies in decomposing a compound or an analytic word form and not in the analytics of a derived word (based on word) nor in an inverse word-to-word *belongsTo* relationship, triggering thus the proposition of adequate *ell_schema* OPs.

5 Conclusion and Future Research

In the present article we analyzed MG IP lexical cases and afterwards tested them ontologically in the MMoOn model. The most representative MG derivational rules i.e. conversion, morpho-phonological (synizesis etc.), diachronic rules (reanalysis etc.), cross-formation and affixation were chosen for testing. To host our MG lexical templates, we appropriately extend and modify the ontology so that groups of lexical instances can refer to their belonging rules. Although the ontology generally complies with the IP approach either in terms of derivational relations or of the rules involved it has proven deficient in hosting specialized MG language templates, cases with alternative directionality or derivatives at lexical-to-hyper-lexical level. Moreover, the combination of expressing rules with input-output categorial change initiated by the model may seem impractical as it could lead to verbosity and semantic inconsistencies. To each of these problems and in accordance with our MG morphological analysis we have proposed alternative solutions presenting their advantages each time. Our future plans include a contrastive analysis of the word-based to the morpheme-based approach as well as exploring the sequential relations between morphemes. As soon as a more finalized version of the Ontolex Morphology module is available a comparative study with the latter is also within our purposes.

Acknowledgements. This research was supported by the project: "Activities of the Laboratory on Digital Libraries and Electronic Publishing of the Department of Archives, Library Science and Museology".

References

1. Booij, G.: The Grammar of Words: An Introduction to Linguistic Morphology. Oxford University Press (2012)
2. Haspelmath, M., Sims, A.: Understanding Morphology. Routledge (2013)
3. Klimek, B., et al.: Challenges for the representation of morphology in ontology Lexicons. In: Kosem, I., et al. (eds.) Electronic Lexicography in the 21st Century. Proceedings of the eLex 2019 Conference: Smart Lexicography, pp. 570–591 (2019). https://elex.link/elex2019/wp-content/uploads/2019/09/eLex_2019_33.pdf
4. Klimek, B., et al.: MMoOn core - the multilingual morpheme ontology. Semant. Web **4**, 1–30 (2020). http://www.semantic-web-journal.net/system/files/swj2549.pdf
5. Lieber, R.: Derivational morphology. In: Oxford Research Encyclopedia of Linguistics. Oxford University Press (2017). https://doi.org/10.1093/acrefore/9780199384655.013.248

6. Lieber, R.: Theoretical issues in word formation. In: Audring, J., Masini, F. (eds.) The Oxford Handbook of Morphological Theory, pp. 33–55 Oxford University Press (2018). https://doi.org/10.1093/oxfordhb/9780199668984.013.3
7. Manolessou, I., Ralli, A.: From ancient Greek to modern Greek. In: Müller, P., et al. (eds.) Word-Formation. An International Handbook of the Languages of Europe, pp. 2041–2061. De Gruyter (2015). https://doi.org/10.1515/9783110375732-027
8. Ralli, A.: 172. Greek. In: Word-Formation, pp. 3138–3156. De Gruyter Mouton (2016). https://doi.org/10.1515/9783110424942-004
9. Ralli, A.: Morfologia (in Greek). Patakis, Athens (2005)
10. Spencer, A.: Morphophonological operations. In: Spencer, A., Zwicky, A.M. (eds.) The Handbook of Morphology, pp. 123–143. Blackwell Publishing Ltd, Oxford (2017). https://doi.org/10.1002/9781405166348.ch6
11. Vasilogamvrakis, N., Sfakakis, M.: A morpheme-based paradigm for the ontological analysis of modern Greek derivational morphology. In: Garoufallou, E., et al. (eds.) Metadata and Semantic Research, pp. 389–400. Springer, Cham (2022). https://doi.org/10.1007/978-3-030-98876-0_34

Analyses of Networks of Politicians Based on Linked Data: Case ParliamentSampo – Parliament of Finland on the Semantic Web

Henna Poikkimäki[1,2](✉)[ID], Petri Leskinen[1,2][ID], Minna Tamper[1,2][ID], and Eero Hyvönen[1,2][ID]

[1] Semantic Computing Research Group (SeCo), Aalto University, Espoo, Finland
henna.poikkimaki@aalto.fi
[2] HELDIG – Helsinki Centre for Digital Humanities, University of Helsinki, Helsinki, Finland

Abstract. In parliamentary debates the speakers make reference to each other. By extracting and linking named entities from the speeches it is possible to construct reference networks and use them for analysing networks of politicians and parties and their debates. This paper presents how such a network can be constructed automatically, based on a speech corpus 2015–2022 of the Parliament of Finland, and be used as a basis for network analysis.

Keywords: Parliamentary studies · Network analysis · Linked data · Digital humanities

1 Introduction

Parliamentary speech data of plenary sessions provide a wealth of information about the state and functioning of democratic systems, political life, language, and culture [2]. Arguably, using Linked Data (LD) for representing metadata about the speeches and the politicians is useful: LD provides well-defined semantics for representing and enriching knowledge[1] aggregated from heterogeneous data sources [4,26], as well as methods for publishing the data for data-analyses and for developing applications, such as semantic portals [9].

This paper argues for using textual speech data for analyzing networks of politicians and parties. It is shown, how Named Entity Recognition (NER) and Linking (NEL) [18] can be used for extracting reference networks from parliamentary speeches, and be used for network analyses. As a case study, speeches and politicians of the Parliament of Finland are considered. We first present a data model and data service for representing reference networks of parliamentary speeches, and how such a network was created based on the ParliamentSampo data [9,15,22]. After this, a system for making network analysis is presented with examples from a Finnish dataset 2015–2019 of 65 000 speeches. Finally, contributions of the work are discussed and directions for further research outlined.

---

[1] https://www.w3.org/standards/semanticweb/.

© Springer Nature Switzerland AG 2022
S. Chiusano et al. (Eds.): ADBIS 2022, CCIS 1652, pp. 585–592, 2022.
https://doi.org/10.1007/978-3-031-15743-1_53

2   Related Works

Several parliamentary corpora have been formed from the minutes of the plenary debates, see, e.g., [14] and the CLARIN list of parliamentary corpora[2]. The related ParlaMint project[3] [6] brings together Parla-CLARIN-based national corpora. Parliamentary materials have also been transformed into the form of LD when creating the LinkedEP [26] system on the European Parliament's data, the Italian Parliament[4], the LinkedSaeima for the Latvian parliament [4], and the Finnish ParliamentSampo [9,15,22] whose data is used in this paper. The corpora have been used mostly for linguistic analyses, not for network science as suggested in our paper. For example, in [3] the content of women's parliamentary speeches in the British Parliament are analyzed, and in [1,7,10,12,20] thematic and conceptual analyses of language and the opinions were made.

Network science [21,27,29], a field revolutionized around 20 years ago, has been successful in explaining phenomena and fundamental concepts in a wide array of systems from societies to brain and cellular biology. The tools and ideas developed for this range from ideas characterising the whole network to diagnostics computed for individual nodes, such as centrality measures, node roles, and local clustering coefficients. Given the unexploited potential, there is a need to provide networks hidden in parliamentary data for researchers to study in Digital Humanities (DH) using methods of network science.

3   A Data Model and Service for Speech Networks

The ParliamentSampo data publication consists of two parts: 1) a knowledge graph (KG) of all parliamentary debate speeches in Finland from 1907 [22] and 2) a KG of the Members of Parliament (MP) and the parliament organizations [15]. In our earlier research, sociocentric and egocentric networks connecting the actors could be constructed from texts based on, e.g., mentioned names, hypertext links, genealogical relations, or similarities in characteristics such as lifetime events [5,16,24]. In this paper, the same idea is applied to parliamentary speeches that make mutual references to each other through mentioned MPs.

In order to extract such reference networks, the original RDF speech graph was enriched with Natural Language Processing (NLP) methods, such as named entity recognition (NER) and linking (NEL) [23]. The extraction was done using the upgraded Nelli tool [25] by querying the textual speeches from the ParliamentSampo SPARQL endpoint[5]. The speeches were first cleaned from the interruptions and then lemmatized using the Turku Neural Parser[6] [11]. The FinBERT-NER model [28] was used on lemmatized texts for NER. After NER,

---

[2] https://www.clarin.eu/resource-families/parliamentary-corpora.

[3] https://www.clarin.eu/content/parlamint-towards-comparable-parliamentary-corpora.

[4] http://data.camera.it.

[5] See the datamodel for speeches [22], MPs [15], and their enrichments respectively.

[6] http://turkunlp.org/Turku-neural-parser-pipeline.

the mentioned people, places, groups, organizations, and their related information were then linked internally to the ParliamentSampo knowledge graph of MPs using the ARPA tool [17] on the extracted named entities. For a broader data enrichment, linkings to external data sources were created, including the Kanto[7] vocabulary for Finnish actors provided by the National Library, the YSO Places ontology[8], PNR[9] gazetteer of Finnish place names by the National Survey, and the Semantic Finlex[10] [19] legal KG of the Ministry of Justice.

The NEL results were transformed into RDF as instances of the class Named-Entity described in Table 1. Namespace *provo* refers to the PROV-O ontology[11]. The instances link the resources behind the mentions to speakers.

Table 1. Metadata schema for the class for NamedEntity.

Element URI	C	Range	Meaning of the value
:surfaceForm	1	xsd:string	Original surface form in text
:count	1	xsd:integer	Number of entity mentions in a speech
:category	1	xsd:string	Type of the named entity
skos:relatedMatch	0..*	rdfs:Resource	Links to ontologies for named entities
provo:wasAssociatedWith	1..*	:NamedEntityMethod, provo:SoftwareAgent	Provenance information about the method used to extract the named entity

The accuracy of the NER was estimated for 100 randomly selected mentions of people, places, organizations, and expressions of time. The precision was 97%, recall 77%, and F1-score 86%. Based on our initial evaluation, 88% of the speeches contained a named entity of which 30% contained a person name. From this evaluation set, roughly 20% contained only person names and for them the F1-score was 100%. The linking of person names was more tricky. Currently, the person names were linked internally to ParliamentSampo successfully only if a person was mentioned using the full name. The results for linking people were calculated for 50 randomly selected speeches (containing 105 person mentions) where precision was 95%, recall 80%, and F1-score 87%. The family name references were not linked to MPs and it remains as a future work to connect the family names and the full names properly.

4   Analyzing Networks of Politicians

This section gives examples of analysing networks of MPs using the Python package NetworkX [8]. Two different reference networks were constructed based on speeches given during the electoral term 2015–2019 and so far linked person

---

[7] https://finto.fi/finaf/en/.

[8] https://finto.fi/yso-paikat/en/?clang=en.

[9] http://www.ldf.fi/dataset/pnr.

[10] https://data.finlex.fi.

[11] https://www.w3.org/TR/prov-o/.

names. Speeches that do not contain any linked person names were excluded from the analysis. In addition, administrative speeches of the Speaker of the Parliament of Finland were not taken into account. Analyses of reference networks can, e.g., reveal MPs who are most active in parliamentary debates and help to recognize possible disputes between MPs or parties.

The first reference network has MPs as nodes and the other one parties. In the former case, links point from the speaker to the mentioned person, and the weight of the link corresponds to the total number of speeches with at least one mention. The network has in total 209 MPs that have been mentioned or have mentioned someone. The total number of mentions to other MPs extracted from the speeches is 2108. Mentions of people who were not MPs or ministers at the chosen electoral term were filtered out of the result set.

To study and visualize the network, hub and authority values were calculated using the HITS algorithm [13]. Ten MPs with highest authority values and ten nodes with highest hub values are shown in Fig. 1. From the MPs with the highest authority values, Juha Sipilä, Timo Soini, and Petteri Orpo were ministers and leaders of their parties during the 2015–2019 term. During the same years, Antti Rinne, Ville Niinistö, and his successor Touko Aalto from the opposition served also as leaders of their parties. Three MPs, Touko Aalto, Pia Viitanen, and Ben Zyskowicz, are both top hubs that make references as well as top authorities often mentioned by other MPs. None of the MPs with highest hub values were ministers.

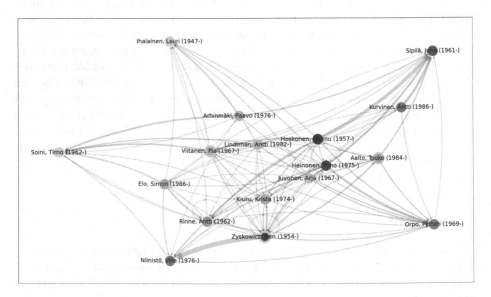

Fig. 1. Ten MPs with highest hub and authority values based on the HITS algorithm. The darker red, the larger authority value, and the darker blue, the larger hub value. (Color figure online)

The second reference network uses parties as nodes instead of MPs. Mentions between MPs were obtained in a similar manner as for the first reference network. MPs were then grouped into 11 parties by their memberships. The Sankey diagram in the Fig. 2 depicts how MPs in government parties refer to MPs in other parties and Fig. 3 the other way around. For example, MPs in the Green League refer mostly to MPs in the Centre Party and National Coalition Party, probably due to disputes related to, e.g., environmental issues. The Sankey diagrams were rendered using the Python module pySankey[12].

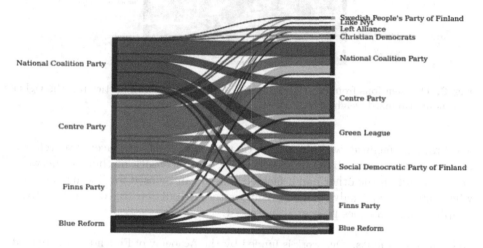

Fig. 2. The mentions from government parties (on the left) to all parties (on the right) with nonnormalized weights.

Analyses like these can be used to point out phenomena of potential interest in parliamentary discussion, but in order to really interpret the results traditional close reading is needed. The faceted search engines and other tools of the ParliamentSampo portal make it easy to filter out, e.g., those speeches of an individual MP or all members of a party that mention particular MPs [9].

5   Discussion

This paper presented the idea of creating reference networks of MPs based on references in their speeches. As a case study, reference networks based on 65 000 parliamentary speeches were created using methods of NER and NEL, and resulting examples of network analysis were presented. Our first experiments suggest that network analysis can be used to detect possibly interesting phenomena in the discussions but interpreting the results require close reading the related texts. In order to support both distant and close reading tasks, the ParliamentSampo system under development aims to integrate seamlessly semantic faceted search

---

[12] https://github.com/anazalea/pySankey.

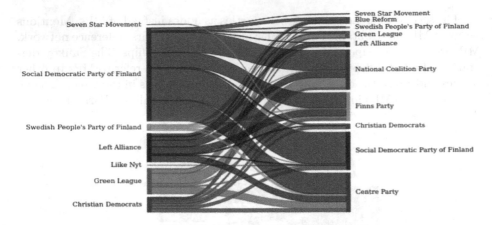

Fig. 3. The mentions from opposition parties (on the left) to all parties (on the right) with nonnormalized weights.

and browsing facilities with data-analytic tools. On one hand, faceted search can be used to filter out subsets of speeches to be studied with methods of network analysis, and on the other hand, faceted search can be used for finding speeches when interpreting analyses made using external tools on top of the SPARQL endpoint, like the ones presented in Sect. 4.

Acknowledgements. Our work is funded by the Academy of Finland and is related to the EU project InTaVia (https://intavia.eu) and the EU COST action Nexus Linguarum (https://nexuslinguarum.eu). Computing resources of the CSC – IT Center for Science were used.

References

1. Beelen, K., et al.: Digitization of the Canadian parliamentary debates. In: Säily, T., Nurmi, A., Palander-Collin, M., Auer, A. (eds.) Exploring Future Paths for Historical Sociolinguistics, pp. 83–107. John Benjamins, Amsterdam (2017). https://doi.org/10.1017/S0008423916001165
2. Benoît, C., Rozenberg, O. (eds.): Handbook of Parliamentary Studies: Interdisciplinary Approaches to Legislatures. Edward Elgar Publishing (2020). https://doi.org/10.4337/9781789906516
3. Blaxill, L., Beelen, K.: A feminized language of democracy? The representation of women at Westminster since 1945. Twentieth Century Br. Hist. **27**(3), 412–449 (2016). https://doi.org/10.1093/tcbh/hww028
4. Bojārs, U., Daris, R., Lavrinovičs, U., Paikens, P.: LinkedSaeima: a linked open dataset of Latvia's parliamentary debates. In: Acosta, M., Cudré-Mauroux, P., Maleshkova, M., Pellegrini, T., Sack, H., Sure-Vetter, Y. (eds.) SEMANTiCS 2019. LNCS, vol. 11702, pp. 50–56. Springer, Cham (2019). https://doi.org/10.1007/978-3-030-33220-4_4
5. Elson, D.K., McKeown, K., Dames, N.J.: Extracting social networks from literary fiction. aclweb.org https://www.aclweb.org/anthology/P10-1015.pdf

6. Erjavec, T., et al.: The ParlaMint corpora of parliamentary proceedings. Lang. Resour. Eval. (2022). https://doi.org/10.1007/s10579-021-09574-0
7. Guldi, J.: Parliament's debates about infrastructure: An exercise in using dynamic topic models to synthesize historical change. Technol. Cult. **60**(1), 1–33 (2019). https://doi.org/10.1353/tech.2019.0000
8. Hagberg, A.A., Schult, D.A., Swart, P.J.: Exploring network structure, dynamics, and function using networkx. In: Varoquaux, G., Vaught, T., Millman, J. (eds.) Proceedings of the 7th Python in Science Conference, Pasadena, CA, USA, pp. 11–15 (2008)
9. Hyvönen, E., et al.: Finnish parliament on the semantic web: using ParliamentSampo data service and semantic portal for studying political culture and language. In: DIPADA, Digital Parliamentary data in Action. CEUR Workshop Proceedings, vol. 3133 (2022). https://ceur-ws.org/Vol-3133/paper05.pdf
10. Ihalainen, P., Sahala, A.: Evolving conceptualisations of internationalism in the UK parliament: collocation analyses from the league to Brexit. In: Fridlund, M., Oiva, M., Paju, P. (eds.) Digital histories: Emergent Approaches Within the New Digital History, pp. 199–219. Helsinki University Press (2020). https://doi.org/10.33134/HUP-5-12
11. Kanerva, J., Ginter, F., Miekka, N., Leino, A., Salakoski, T.: Turku neural parser pipeline: an end-to-end system for the CoNLL 2018 shared task. In: Proceedings of the CoNLL 2018 Shared Task: Multilingual Parsing from Raw Text to Universal Dependencies, 31 October–1 November 2018, Brussels, Belgium, pp. 133–142. Association for Computational Linguistics, Brussels, Belgium (2018)
12. Kettunen, K., La Mela, M.: Semantic tagging and the Nordic tradition of everyman's rights. Digit. Sch. Humanit. (2021). https://doi.org/10.1093/llc/fqab052
13. Kleinberg, J.M.: Authoritative sources in a hyperlinked environment. J. Assoc. Comput. Mach. **46**, 604–632 (1999). https://doi.org/10.1145/324133.324140
14. Lapponi, E., Søyland, M.G., Velldal, E., Oepen, S.: The Talk of Norway: a richly annotated corpus of the Norwegian parliament, 1998–2016. Lang. Resour. Eval. **52**(3), 873–893 (2018). https://doi.org/10.1007/s10579-018-9411-5
15. Leskinen, P., Hyvönen, E., Tuominen, J.: Members of Parliament in Finland knowledge graph and its linked open data service. In: Graphs. Proceedings of the 17th International Conference on Semantic Systems, 6–9 September 2021, Amsterdam, The Netherlands, pp. 255–269 (2021). https://doi.org/10.3233/SSW210049, https://ebooks.iospress.nl/volumearticle/57420
16. Leskinen, P., Rantala, H., Hyvönen, E.: Analyzing the lives of Finnish academic people 1640–1899 in Nordic and Baltic countries: AcademySampo data service and portal. In: 6th Digital Humanities in Nordic and Baltic Countries Conference, Long Paper. CEUR Workshop Proceedings (March 2022, forth-coming)
17. Mäkelä, E.: Combining a REST lexical analysis web service with SPARQL for mashup semantic annotation from text. In: Presutti, V., Blomqvist, E., Troncy, R., Sack, H., Papadakis, I., Tordai, A. (eds.) ESWC 2014. LNCS, vol. 8798, pp. 424–428. Springer, Cham (2014). https://doi.org/10.1007/978-3-319-11955-7_60
18. Martinez-Rodriguez, J.L., Hogan, A., Lopez-Arevalo, I.: Information extraction meets the semantic web: a survey. Semant. Web Interoperability Usability Applicability **11**(2), 255–335 (2020)
19. Oksanen, A., Tuominen, J., Mäkelä, E., Tamper, M., Hietanen, A., Hyvönen, E.: Semantic Finlex: transforming, publishing, and using Finnish legislation and case law as linked open data on the web. In: Knowledge of the Law in the Big Data Age. Frontiers in Artificial Intelligence and Applications, vol. 317, pp. 212–228. IOS Press (2019)

20. Quinn, K., Monroe, B., Colaresi, M., Crespin, M.H., Radev, D.R.: How to analyze political attention with minimal assumptions and costs. Am. J. Polit. Sci. **54**, 209–228 (2010). https://doi.org/10.1111/j.1540-5907.2009.00427.x
21. Saramäki, J., Moro, E.: From seconds to months: an overview of multi-scale dynamics of mobile telephone calls. Eur. Phys. J. B **88**(6), 1–10 (2015). https://doi.org/ 10.1140/epjb/e2015-60106-6
22. Sinikallio, L., et al.: Plenary debates of the parliament of Finland as linked open data and in Parla-CLARIN markup. In: 3rd Conference on Language, Data and Knowledge, LDK 2021, pp. 1–17. Schloss Dagstuhl- Leibniz-Zentrum fur Informatik GmbH, Dagstuhl Publishing, August 2021. https://drops.dagstuhl.de/opus/ volltexte/2021/14544/pdf/OASIcs-LDK-2021-8.pdf
23. Tamper, M., Leal, R., Sinikallio, L., Leskinen, P., Tuominen, J., Hyvönen, E.: Extracting knowledge from parliamentary debates for studying political culture and language, May 2022, forth-coming
24. Tamper, M., Leskinen, P., Hyvönen, E., Valjus, R., Keravuori, K.: Analyzing biography collection historiographically as linked data: case national biography of Finland. Semant. Web Interoperability Usability Applicability (2021, accepted). https://seco.cs.aalto.fi/publications/2021/tamper-et-al-bs-2021.pdf
25. Tamper, M., Oksanen, A., Tuominen, J., Hietanen, A., Hyvönen, E.: Automatic annotation service APPI: named entity linking in legal domain. In: Harth, A., et al. (eds.) ESWC 2020. LNCS, vol. 12124, pp. 208–213. Springer, Cham (2020). https://doi.org/10.1007/978-3-030-62327-2_36
26. Abercrombie, G., Batista-Navarro, R.: Sentiment and position-taking analysis of parliamentary debates: a systematic literature review. J. Comput. Soc. Sci. **3**(1), 245–270 (2020). https://doi.org/10.1007/s42001-019-00060-w
27. Vespignani, A.: Twenty years of network science. nature **558**(7711), 528–529 (2018). https://doi.org/10.1038/d41586-018-05444-y
28. Virtanen, A., Kanerva, J., Ilo, R., Luoma, J., Luotolahti, J., Salakoski, T., Ginter, F., Pyysalo, S.: Multilingual is not enough: BERT for Finnish (2019)
29. Watts, D.J., Strogatz, S.H.: Collective dynamics of 'small-world' networks. nature **393**, 440–442 (1998). https://doi.org/10.1038/30918

Neural Word Sense Disambiguation to Prune a Large Knowledge Graph of the Italian Cultural Heritage

Erica Faggiani, Stefano Faralli(✉)(iD), and Paola Velardi(iD)

Sapienza University of Rome, Rome, Italy
faggiani.1212510@studenti.uniroma1.it, {faralli,velardi}@di.uniroma1.it

Abstract. In this paper, we describe our recent findings in interlinking the ArCo Italian cultural heritage entities to the well known Getty Art and Architecture (GVP) Thesaurus through the automated extraction of candidate entities from textual descriptions and the subsequent pruning of ambiguous out-of-domain entities using Neural Word Sense Disambiguation. The disambiguation task is particularly complex since, as detailed in this paper, we map Italian entities in the Arco cultural heritage onto lexical concepts in English (such as those in the GVP Thesaurus). To date, the majority of entity linking and word sense disambiguation systems are designed to work with English and to operate with general purpose sense inventories and knowledge bases, such as DBpedia, BabelNet and WordNet. To address this challenging entity linking and disambiguation task, we adapted a state-of-the-art Neural Word Sense Disambiguation to work in this multi-language setting. We here describe our adaptation process and discuss preliminary experimental results.

Keywords: Cultural heritage · Entity linking · Neural word sense Disambiguation · ArCo · GVP

1 Introduction

In the past decades, there has been a growing number of experimental investigations on knowledge-based applications [2]. Specifically, knowledge-based systems have become more and more popular and effective for a variety of reasons, e.g., the increased availability of large datasets [15,28], the recent advancements in the field of deep learning [13], and the capability of knowledge-based systems to provide explanations of systems outcomes [23]. As a result, knowledge-based applications begin to meet the requirements for enabling novel industrial and commercial automated applications [18,20,26,27].

© Springer Nature Switzerland AG 2022
S. Chiusano et al. (Eds.): ADBIS 2022, CCIS 1652, pp. 593–604, 2022.
https://doi.org/10.1007/978-3-031-15743-1_54

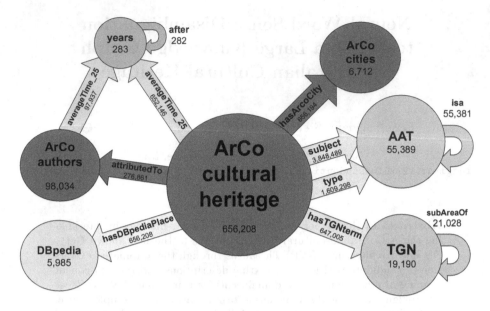

Fig. 1. A high-level representation of the current noisy interlinked graph. In this paper, we experiment with a state-of-the-art NWSD algorithm to mitigate the noisy interlinks we generated due to polysemy.

Unfortunately the state of the art is not as advanced in more specialized domains, thus limiting the performance of automated systems - a well known problem described, e.g., in [8]. In this paper, we address the problem of creating large knowledge graphs in the specific domain of the Italian Cultural Heritage and the smart tourism. Within the activities of the "SMARTOUR: intelligent platform for tourism" project, our research [7] is aimed at automating the process of: i) interlinking the ArCo [5] cultural heritage entities with the concepts belonging to external knowledge bases, such as DBpedia [1] and the GVP [12] ontologies, and of, ii) enriching the concepts with geographic, temporal and authorship annotations.

The resulting high-coverage interlinked knowledge base (see Fig. 1 for a high-level representation of the current graph) will be leveraged for a variety of tourism and cultural related applications, such as cataloging activities in museums and recommending personalized tours based on the artistic and historical interests of the users [4].

As a first result, we obtained high quality interlinks for ArCo cultural heritage properties (such as locations to DBpedia places)[1]. However, our resource includes a number of linking errors due to polysemy, a problem that we address in this paper.

---

[1] https://sites.google.com/uniroma1.it/agdli/.

For instance, noisy interlinks have been generated among the 3,848,489 and 1,609,298 links generated for the *subject* and *type* textual properties (see Fig. 1).

To prune out-of-domain concepts originated by polysemy, we adapted to our task recent advancements in the field of link prediction [21] and *Word Sense Disambiguation* (WSD) [3] algorithms. The contribution of this paper is threefold:

- we describe and discuss our approach for adapting a state-of-the-art neural word sense disambiguation system (NWSD) to work with a multilingual setting and to operate with an art and architecture-specific sense inventory;
- we discuss the preliminary experimental results;
- we release the source code and the related resources at: https://github.com/arumdauo/BERT-WSD-Adapted.

The remaining of this paper is structured as follows: i) in Sect. 2 we describe the motivations and the literature related to our investigation; ii) in Sect. 3, we describe the adopted word sense disambiguation method for pruning out-of-domain nodes in the graph; iii) in Sect. 4, we describe the experimental setting and discuss the evaluation outcomes; iv) finally, in Sect. 5, we discuss the limitations of our approach and the future research directions.

2   Motivations and Related Work

In our research activity, we are aimed at linking entities occurring in textual descriptive properties of cultural heritage entities belonging to the ArCo ontology [5]. The ArCo project started in 2017 with the aim of providing a knowledge base of the information contained in a variety of catalogs. Nowadays, ArCo is considered the most prominent knowledge graph of Italian cultural heritage. The outcomes of the project include the definition of a set of 7 ontologies as a ground for the formal representation of Italian cultural heritage entities. Thus, in this specific domain, ArCo is a fundamental (and Linked Open Data compliant) ground for the publication of data.
Overall, ArCo provides ~172.5 million RDF triples representing the data from the General Catalogue of the Italian Ministry of Cultural Heritage and Activities (MiBAC).[2]

While in our previous work [7], we described the overall interlinking initiative, aimed at connecting the ArCo entities to the *concepts* of external knowledge bases, in this paper we focus on a graph pruning task, to reduce linking errors caused by polysemy when connecting the entities of ArCo with one of the tree ontologies of the GVP, the Getty Art and Architecture Thesaurus [12].

The ontology of the GVP includes three controlled ontologies, i.e., the above mentioned AAT, the *Union List of Artist Names* (ULAN) e the *Getty Thesaurus*

---

[2] ArCo "indirectly reuses: DOLCE-Zero, DOLCE+DnS, CIDOC-CRM, EDM, BIBFRAME, FRBR, FaBiO, FEntry, OAEntry" ARCO - PRIMER GUIDE V1.0 http://wit.istc.cnr.it/arco/primer-guide-v1.0-en.html.

of Geographic Names (TGN). GVP is considered the main reference for the categorization of art and architecture artifacts. AAT, ULAN e TGN are the most complete formal representation of concepts in the art and culture domains, and are of tremendous help in applications devoted to the cataloging, documenting and retrieving. Note that ArCo and AAT are in a sense complementary, since, as discussed in [7], ArCo includes mostly entities, with just a few high-level concepts, while AAT is a taxonomy of concepts. Integrating these two resources is particularly challenging also because AAT is a lexical ontology of English concepts, while ArCo entity names and descriptions are in Italian.

In our research activity, entity linking [22] is performed in two main steps. First, we mine for the occurrences of domain specific words in textual descriptive properties of cultural heritage entities in ArCo. We then look for the candidate word senses defined in the AAT lexicalized ontology and try to identify the most suitable word meaning (concepts) by means of word sense disambiguation (WSD) [19]. In Fig. 2, we show an example of the noisy links we generate, due to polysemy.

To mitigate this problem[3], we exploit neural word sense disambiguation, as described in Sect. 3.

WSD is the ability to identify the meaning of words in context in a computational manner. WSD is considered as an "AI-Complete" problem in Natural Language Processing (NLP) domain.[4]

Nowadays, semi-supervised methodologies with neural network architectures, such as Bidirectional Encoder Representations from Transformers (BERT) [6], represent the state of the art for the WSD problem, as well as other NLP problems. BERT is a pre-trained language model which has been proven to be effective in extracting features from texts, being able to pre-train models on large corpora through self-supervised learning, obtaining vectorial word representations (embeddings), subsequently used in downstream tasks such as WSD. In the past few years, BERT has been integrated in several WSD approaches, such as (among others) [11,16,24]. Specifically, for our objectives, we are interested in Neural WSD (NWSD) systems which exploit word senses definitions (glosses), represented in our case by the AAT concepts' scope notes. We report in Fig. 3, an example scope note for the *motif* AAT concept.

Previous works performing this task using WordNet definitions are [11,17]. Among the NWSD approaches based on BERT and using glosses, we mention [14] and [25], that frame the WSD task as a sentence-gloss pair classification problem. Specifically, the sentence-gloss pairs used during training are composed by the context in which a target word occurs and the gloss for the corresponding word meaning concept in WordNet [9]. In this work, we decided to adapt the system [25] to perform word sense disambiguation on Italian terms occurring

---

[3] Further worsened, as previously mentioned, by the automated translation in English of the concepts extracted from the textual field in ArCo, in Italian.

[4] "The idea of an *AI-complete* problem has been around since at least the late 1970s, and refers to the more formal idea of the technique used to confirm the computational complexity of NP-complete problems." [10].

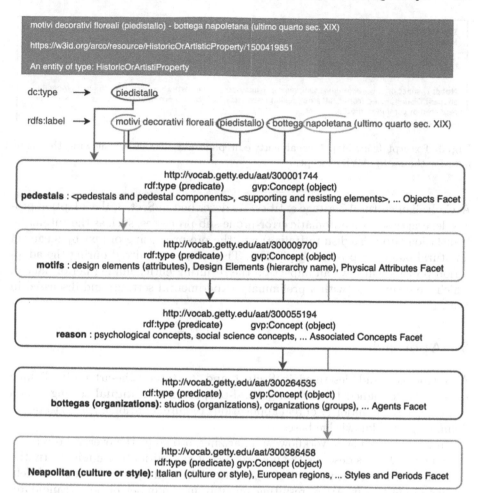

Fig. 2. Example of the noisy entity links we generated while mining candidates concepts for the Italian word *motivi*, whose best suitable word meaning, in the descriptive property context of the ArCo *HistoricOrArtisticProperty* entity with id 1500419851, corresponds to the concept *motif* ("design elements") of the AAT.

on ArCo entity descriptions, thus replacing WordNet with the GVP AAT, and BERT with Italian BERT[5]. The decision was based on both the availability of the source code and on the reported state-of-the-art performances among various challenging benchmarking datasets.

In summary, in this study, we are aimed at investigating about the feasibility of an adaptation approach to automatic domain-specific entity linking for the

---

[5] MDZ Digital Library team (dbmdz) at the Bavarian State Library https://huggingface.co/dbmdz/bert-base-italian-xxl-cased.

```
ID: 300009700                                           Record Type: concept
Page Link: http://vocab.getty.edu/page/aat/300009700

⚘ motifs (design elements (attributes), Design Elements (hierarchy name))

Note: Distinct or separable design elements, usually decorative, whether occurring singly as individual
shapes. Distinguished from "patterns (design elements)," which are ornamental designs composed of
repeated or combined motifs.
```

Fig. 3. Excerpt from http://vocab.getty.edu/page/aat/300009700 showing the scope note for the *motif* AAT concept.

ArCo Italian cultural heritage entities. As described in Sect. 3, the adaption process leverages several automatic error-prone sub-processes, such as the automatic translation (from English to Italian) and the pre-processing of text by means of a natural language processing pipeline. The errors committed during the adaptation process will have an impact on the resulting entity linking performances, which we estimated (with a preliminary experimental setting) and discussed in Sect. 4.

3   Approach

As introduced and described in Sects. 1 and 2, state-of-the-art entity linking systems are designed to work in general-purpose **monolingual** settings. As a consequence, the majority of real applications require additional efforts to target domain-specific knowledge bases.

In Fig. 4, we show a workflow summarizing in 6 steps the overall process.

The first 4 steps describe the pre-processing and initial text mining activities devoted at finding occurrences of candidate AAT concepts in the textual definitions of entities in ArCo, resulting possibly in a number of noisy candidates caused by polysemy and translation errors.

The subsequent steps 5 and 6 describe the creation of a ground truth and the adaptation of an existing NWSD algorithm to reduce these linking errors.

Pre-processing and Text Mining

Step 1: In this preliminary step we queried the beniculturali.it SPARQL endpoint[6] and collected the RDF triples describing the $\sim656K$ ArCo cultural entities' properties;

Step 2: To address the multilinguality issue, we performed an automatic translation[7], from English to Italian for all the terms and the scope notes of the 55K AAT concepts. We report in Table 1 an example excerpt of the translated terms.

---

[6] https://dati.cultura.gov.it/sparql.
[7] Google Translation AI, https://cloud.google.com/translate.

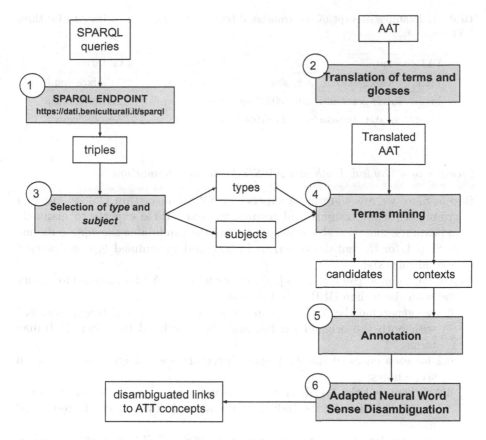

Fig. 4. Workflow summarizing the steps of our adaptation task.

Step 3: From the resulting RDF triples retrieved after Step 1, for each ArCo entity, we extracted the *type* and *subject* textual properties. Examples of such descriptive properties are reported in Fig. 2, where we extracted "piedistallo" (pedestal) and "motivi decorativi floreali (piedistallo) - bottega napolatana (ultimo quarto sec. XIX)"[8] as the *type* and *subject*, for the ArCo HistoricOrArtisticProperty entity with id 1500419851;

Step 4: To determine AAT concepts occurrences, we applied standard text mining techniques and identified all possible candidates entity links to AAT concepts. For example (see Fig. 2), we mine the term *motivi* and we identify *motifs*(design elements) and *reasons*(psychological concepts) as candidate entity links;

---

[8] floral decorative motifs (pedestal) - Neapolitan workshop (last quarter of the 19th century).

Table 1. Example excerpt of the translated terms (from English to Italian) for three AAT concepts.

AAT concept	EN terms	IT terms
http://vocab.getty.edu/aat/300379603	Salinization	Salinizzazione
http://vocab.getty.edu/aat/300379600	Thermochemistry	Termochimica
http://vocab.getty.edu/aat/300379601	aerobiology	Aerobiologia

Creation of a Ground Truth and NWSD Algorithm Adaptation

Step 5: Since we are adapting a supervised NWSD algorithm [25], to create a ground truth for training and testing, we asked three experts to manually indicate the most suitable concepts among the candidates concepts obtained at Step 4, for the entities occurring in 600 random sampled *type* and *subject* textual contexts;

Step 6: We adapted the system in [25] to work with the AAT as a sense inventory and with the Italian BERT. To this end:

6.1 we generated the AAT sense inventory resource, as a dictionary indexed with both the original English and the translated (see Step 2) Italian terms;

6.2 for each entry of the AAT sense inventory we collected the translated scope notes;

6.3 we replaced the WordNet sense inventory, with our Italian AAT dictionary and to handle the Italian BERT instead of the original pre-trained model;

6.4 we developed a script to generate a 2009 CONLL format[9] compliant ground truth dataset for the NWSD system's training and evaluation (an example excerpt is reported in Fig. 5). The dataset generation leverages a standard preprocessing NLP pipeline, based on SpaCy (Industrial-Strength Natural Language Processing[10]), for part of speech tagging.

We remind that, to enable replicability and reprodicibility of our study, both source code and data are available at: https://github.com/arumdauo/BERT-WSD-Adapted.

4   Evaluation

To estimate the effectiveness of the resulting adapted NWSD we performed a preliminary 6-fold cross-validation experiment with the 600 annotated samples, and calculated the resulting F1 performances. To this end, we performed the fine tuning of the pretrained Italian BERT (dbmdz/bert-base-italian-cased) with a learning rate of $2e-5$ and 15 epochs of training.

---

[9] https://www.signll.org/conll/.
[10] https://spacy.io/.

```
<arco id="https://w3id.org/arco/resource/HistoricOrArtisticProperty/0800430453"
    ↪ >
<sentence id="https://w3id.org/arco/resource/HistoricOrArtisticProperty
    ↪ /0800430453">
<instance id="https://w3id.org/arco/resource/HistoricOrArtisticProperty
    ↪ /0800430453.0" lemma="cofanetto" pos="NOUN">cofanetto</instance>
<wf lemma="-" pos="PUNCT">-</wf>
<wf lemma="ambito" pos="NOUN">ambito</wf>
<instance id="https://w3id.org/arco/resource/HistoricOrArtisticProperty
    ↪ /0800430453.1" lemma="russo" pos="ADJ">russo</instance>
<wf lemma="(" pos="PUNCT">(</wf>
<wf lemma="sec" pos="NOUN">sec</wf>
<wf lemma="." pos="PUNCT">.</wf>
<wf lemma="xx" pos="ADJ">XX</wf>
<wf lemma=")" pos="PUNCT">)</wf>
</sentence>
</arco>
```

Fig. 5. Example excerpt of the resulting 2009 CONLL format compliant ground truth dataset, where the noun *cofanetto* is a target word for the WSD task.

We repeated the same experimental setting to assess the performances of a random baseline, which has been designed to pick a random word meaning among the candidates.

As a result, the adapted NSWD system obtained a F1 score of ∼0.65 (standard deviation ∼0.15) while the random baseline a F1 score of ∼0.21 (standard deviation ∼0.03). During this preliminary evaluation, we obtained a significant improvement over the random baseline (+0.44), but due to the small number of samples in our manually annotated dataset, we obtained F1 score below the performance of the original system, tested in a monolingual setting and with WordNet sense inventory (∼0.79) [25]. Upon analyzing the different errors made by the system, we categorized errors into the following groups of causes:

- erroneous translations of AAT terms and scope notes. For instance the term *manufactories* (obsolete word for "factory") has been automatically translated into the Italian *manifatture* ("manufacture", "craftsmanship") instead of *fabbriche*.
- missing scope notes for many AAT concepts (only 10,262 out of the 55,375 AAT concepts supply scope notes).
- erroneous text prepossessing, more specifically, errors caused by part of speech tagging of Italian sentences. For instance, the part of speech for the Italian word *campana* is often erroneously determined as *Adjective* rather than *Noun* when the meaning in the context is related to the sense of the English word *bell* and vice-versa, as *Noun* rather than *Adjective*, when in context is describing the provenance from the Italian Region Campania.

5 Conclusions

With the aim of investigating the feasibility of an adaptation approach to automatic domain-specific entity linking for the ArCo Italian cultural heritage entities, in this paper, we described and experimented with the adaptation process of a state-of-the-art NWSD system to operate in a multilingual setting and to target a domain-specific sense inventory (i.e., the GVP AAT). The adaption process involved several automatic error-prone sub-processes, such as the automatic translation (from English to Italian) and the pre-processing of text by means of a natural language processing pipeline.

The errors committed during the adaptation process have an impact on the resulting entity linking performances. We discussed our preliminary experimental results and identified the main causes of errors.

We are currently working to improve the system's performance by addressing: i) the small dimension of our ground-truth dataset and ii) the quality of automated translations.

Specifically, we believe that the creation of a larger ground truth dataset will both provide more robust performance estimation, and overall better performance in the task of entity linking. Finally, we are also experimenting with other families of algorithms, such as state-of-the-art graph completion and pruning algorithms applied to noisy knowledge graphs.

Acknowledgements. This work was carried out within the research project "SMAR-TOUR: intelligent platform for tourism" funded by the Italian Ministry of University and Research with the Regional Development Fund of European Union (PON Research and Competitiveness 2007-2013).

References

1. Auer, S., Bizer, C., Kobilarov, G., Lehmann, J., Cyganiak, R., Ives, Z.: DBpedia: a nucleus for a web of open data. In: Aberer, K., et al. (eds.) ASWC/ISWC -2007. LNCS, vol. 4825, pp. 722–735. Springer, Heidelberg (2007). https://doi.org/10.1007/978-3-540-76298-0_52

2. de Bem Machado, A., Secinaro, S., Calandra, D., Lanzalonga, F.: Knowledge management and digital transformation for Industry 4.0: a structured literature review. Knowl. Manage. Res. Pract. **20**(2), 320–338 (2022). https://doi.org/10.1080/14778238.2021.2015261

3. Bevilacqua, M., Pasini, T., Raganato, A., Navigli, R.: Recent trends in word sense disambiguation: a survey. In: Zhou, Z.H. (ed.) Proceedings of the Thirtieth International Joint Conference on Artificial Intelligence, IJCAI-21, Survey Track, pp. 4330–4338. International Joint Conferences on Artificial Intelligence Organization, August 2021. https://doi.org/10.24963/ijcai.2021/593

4. Binucci, C., De Luca, F., Di Giacomo, E., Liotta, G., Montecchiani, F.: Designing the content analyzer of a travel recommender system. Exp. Syst. Appl. **87**, 199–208 (2017). https://doi.org/10.1016/j.eswa.2017.06.028

5. Carriero, V.A., et al.: ArCo: the Italian cultural heritage knowledge graph. In: Ghidini, C., et al. (eds.) ISWC 2019. LNCS, vol. 11779, pp. 36–52. Springer, Cham (2019). https://doi.org/10.1007/978-3-030-30796-7_3

6. Devlin, J., Chang, M.W., Lee, K., Toutanova, K.: BERT: pre-training of deep bidirectional transformers for language understanding. In: Proceedings of the 2019 Conference of the North American Chapter of the Association for Computational Linguistics: Human Language Technologies, Volume 1 (Long and Short Papers), Minneapolis, Minnesota, pp. 4171–4186. Association for Computational Linguistics, June 2019. https://doi.org/10.18653/v1/N19-1423

7. Faralli, S., Lenzi, A., Velardi, P.: AGDLI: ArCo, GVP and DBpedia linking initiative. In: Seneviratne, O., Pesquita, C., Sequeda, J., Etcheverry, L. (eds.) Proceedings of the ISWC 2021 Posters, Demos and Industry Tracks: From Novel Ideas to Industrial Practice co-located with 20th International Semantic Web Conference, ISWC 2021, Virtual Conference, CEUR Workshop Proceedings, 24–28 October 2021, vol. 2980. CEUR-WS.org (2021). https://ceur-ws.org/Vol-2980/paper304.pdf

8. Feigenbaum, E.A.: Knowledge engineering. Ann. NY Acad. Sci. **426**(1), 91–107 (1984). https://doi.org/10.1111/j.1749-6632.1984.tb16513.x

9. Fellbaum, C. (ed.): WordNet: An Electronic Lexical Database. Language, Speech, and Communication. MIT Press, Cambridge, MA (1998)

10. Goebel, R.: Folk reducibility and AI-complete problems. In: Dengel, A.R., Berns, K., Breuel, T.M., Bomarius, F., Roth-Berghofer, T.R. (eds.) KI 2008. LNCS (LNAI), vol. 5243, pp. 1–1. Springer, Heidelberg (2008). https://doi.org/10.1007/978-3-540-85845-4_1

11. Hadiwinoto, C., Ng, H.T., Gan, W.C.: Improved word sense disambiguation using pre-trained contextualized word representations. In: Inui, K., Jiang, J., Ng, V., Wan, X. (eds.) Proceedings of the 2019 Conference on Empirical Methods in Natural Language Processing and the 9th International Joint Conference on Natural Language Processing, EMNLP-IJCNLP 2019, Hong Kong, China, 3–7 November 2019, pp. 5296–5305. Association for Computational Linguistics (2019). https://doi.org/10.18653/v1/D19-1533

12. Harpring, P.: Development of the getty vocabularies: AAT, TGN, ULAN, and CONA. Art Documentation J. Art Libr. Soc. North Am. **29**(1), 67–72 (2010). https://www.jstor.org/stable/27949541

13. Hoppe, F., Dessì, D., Sack, H.: Deep learning meets knowledge graphs for scholarly data classification. In: Companion Proceedings of the Web Conference 2021, WWW 2021, pp. 417–421. Association for Computing Machinery, New York, NY, USA (2021). https://doi.org/10.1145/3442442.3451361

14. Huang, L., Sun, C., Qiu, X., Huang, X.: GlossBERT: BERT for word sense disambiguation with gloss knowledge. In: Inui, K., Jiang, J., Ng, V., Wan, X. (eds.) Proceedings of the 2019 Conference on Empirical Methods in Natural Language Processing and the 9th International Joint Conference on Natural Language Processing, EMNLP-IJCNLP 2019, Hong Kong, China, 3–7 November 2019, pp. 3507–3512. Association for Computational Linguistics (2019). https://doi.org/10.18653/v1/D19-1355

15. Janev, V., Graux, D., Jabeen, H., Sallinger, E. (eds.): Knowledge Graphs and Big Data Processing. LNCS, vol. 12072. Springer, Cham (2020). https://doi.org/10.1007/978-3-030-53199-7

16. Jiaju, D., Fanchao, Q., Maosong, S.: Using BERT for Word Sense Disambiguation, September 2019. https://doi.org/10.48550/arXiv.1909.08358

17. Luo, F., Liu, T., He, Z., Xia, Q., Sui, Z., Chang, B.: Leveraging gloss knowledge in neural word sense disambiguation by hierarchical co-attention. In: Proceedings of the 2018 Conference on Empirical Methods in Natural Language Processing, Brussels, Belgium, October–November 2018, pp. 1402–1411. Association for Computational Linguistics (2018). https://doi.org/10.18653/v1/D18-1170

18. Mladineo, M., Crnjac Zizic, M., Aljinovic, A., Gjeldum, N.: Towards a knowledge-based cognitive system for industrial application: case of personalized products. J. Ind. Inf. Integr. **27**, 100284 (2022). https://doi.org/10.1016/j.jii.2021.100284

19. Navigli, R.: Word sense disambiguation: a survey. ACM Comput. Surv. **41**(2), 1–69 (2009). https://doi.org/10.1145/1459352.1459355

20. Pokojski, J., Szustakiewicz, K., Woźnicki, Ł., Oleksiński, K., Pruszyński, J.: Industrial application of knowledge-based engineering in commercial CAD/CAE systems. J. Industr. Inf. Integr. **25**, 100255 (2022). https://doi.org/10.1016/j.jii.2021.100255, https://www.sciencedirect.com/science/article/pii/S2452414X21000546

21. Rossi, A., Barbosa, D., Firmani, D., Matinata, A., Merialdo, P.: Knowledge graph embedding for link prediction: a comparative analysis. ACM Trans. Knowl. Discov. Data **15**(2), 1–49 (2021). https://doi.org/10.1145/3424672

22. Shen, W., Wang, J., Han, J.: Entity linking with a knowledge base: issues, techniques, and solutions. IEEE Trans. Knowl. Data Eng. **27**(2), 443–460 (2015). https://doi.org/10.1109/TKDE.2014.2327028

23. Tiddi, I., Schlobach, S.: Knowledge graphs as tools for explainable machine learning: A survey. Artif. Intell. **302**, 103627 (2022). https://doi.org/10.1016/j.artint.2021.103627

24. Wiedemann, G., Remus, S., Chawla, A., Biemann, C.: Does BERT make any sense? Interpretable word sense disambiguation with contextualized embeddings. In: Proceedings of the 15th Conference on Natural Language Processing, KONVENS 2019, Erlangen, Germany, 9–11 October 2019 (2019). https://corpora.linguistik.uni-erlangen.de/data/konvens/proceedings/papers/KONVENS2019_paper_43.pdf

25. Yap, B.P., Koh, A., Chng, E.S.: Adapting BERT for word sense disambiguation with gloss selection objective and example sentences. In: Findings of the Association for Computational Linguistics, EMNLP 2020, Online, pp. 41–46. Association for Computational Linguistics, November 2020. https://www.aclweb.org/anthology/2020.findings-emnlp.4

26. Zheng, C., et al.: Knowledge-based program generation approach for robotic manufacturing systems. Robot. Comput. Integr. Manuf. **73**, 102242 (2022). https://doi.org/10.1016/j.rcim.2021.102242

27. Zheng, W., Cheng, J., Wu, X., Sun, R., Wang, X., Sun, X.: Domain knowledge-based security bug reports prediction. Knowl. Based Syst. **241**, 108293 (2022). https://doi.org/10.1016/j.knosys.2022.108293

28. Zhuang, Y., Wu, F., Chen, C., Pan, Y.: Challenges and opportunities: from big data to knowledge in AI 2.0. Front. Inf. Technol. Electron. Eng. **18**(1), 3–14 (2017). https://doi.org/10.1631/FITEE.1601883

Doctoral Consortium

Doctoral Consortium

Workshop Chairs

Genoveva Vargas Solar   CNRS, LIRIS, France
Ester Zumpano   University of Calabria, Italy

Program Committee Members

Judith Awiti   Université libre de Bruxelles, Belgium
Andrea Calì   Birkbeck University of London, UK
Barbara Catania   University of Genoa, Italy
Jérôme Darmont   University Lumiére Lyon 2, France
Marlon Dumas   University of Tartu, Estonia
Javier-Alfonso Espinosa-Oviedo   University of Lyon, France
Abir Farouzi   LIAS/ISAE-ENSMA, France
Sergio Flesca   University of Calabria, Italy
Cristian Molinaro   University of Calabria, Italy
Laura Po   University of Modena and Reggio Emilia, Italy
Nicolas Travers   University Eiffel Paris, France
Raquel Trillo   University of Zaragoza, Spain
Domenico Ursino   Marche Polytechnic University, Italy
José-Luis Zechinelli-Martini   University of the Americas, Mexico

Querying Sensor Networks Using Temporal Property Graphs

Erik Bollen[1,2]([✉]) [iD]

[1] Hasselt University and Transnational University Limburg, Data Science Institute,
Diepenbeek, Belgium
erik.bollen@uhasselt.be
[2] Flemish Institute for Technological Research, Data Science Hub, Mol, Belgium

Abstract. In this paper, *sensor networks* are considered, which are (stable) transportation networks equipped with sensors. We abstract transportation networks as graphs with sensor measurements that can be attributed to the nodes and edges. A sensor network is thus modelled as a graph with properties that can change throughout time, and which can be viewed as *time series*. To the best of our knowledge, a model, in which graph databases and time series analysis are combined to create a temporal property graph model, is new. This work also describes a language to query such temporal property graphs and discusses how both can be implemented and realised by using Neo4j and its query language Cypher. In short, this paper presents a database system for storing and querying sensor networks that enables future projects to reduce set-up times and prevent use-case specific implementation of database and querying applications.

Keywords: Graphs · Time series · Sensor networks · Query languages

1 Introduction

Many real life problems are studied by gathering data and then analysing them. Often these data can be modelled as graphs. In this setting, examples include social networks, traffic data, routing problems and transportation data. All these examples can be considered as graph problems. This paper focuses on a specific subset of graph problems, namely sensor networks. *Sensor networks* are (stable) transportation networks, equipped with sensors. When modelling transportation networks as graphs, sensor measurements in the network can be attributed to the nodes and edges of this graph and they can be viewed as *time series*. We assume that transportation networks are *stable*, meaning that the topology or structure of the physical network hardly ever changes. The temporal aspect of our applications resides in the aspect of the sensors that produce series of measurements over time. Examples of sensor networks include road networks equipped with sensors that measure the density or speed of the traffic, but also river networks on which sensors are placed to measure the water height

© Springer Nature Switzerland AG 2022
S. Chiusano et al. (Eds.): ADBIS 2022, CCIS 1652, pp. 607–614, 2022.
https://doi.org/10.1007/978-3-031-15743-1_55

or pollution. In order to facilitate research and projects on sensor networks, the data of these networks need to be stored, managed and need to be queryable and analysable. For example, for a river system, in which sensors measure the salinity of the water at a regular basis, both the network and results of the measurement need to be stored for further analysis (used, for example, in a support system for a water management agency). This implies that we need a suitable system to store a sensor network and to query it.

Many existing database systems can deal with sensor networks. The relational database model is well established, but because we work with graph-based data, there is a mismatch between the relational model and the real life situation [3]. A more graph-oriented approach is needed, and graph databases seem the suitable solution for these problems. Their history starts before the year 2000, but only recently, implementations have pushed this field to a higher level of practical usability.

The *property graph model* is often used in this context (see, for example, Angles [2]). Newly developed query languages such as Cypher and G-Core also have a great impact. For work on this topic, we refer to Bonifati et al. [4], Francis et al. [7] and Libkin et al. [9]. Especially the Cypher implementation in Neo4j[1] provides a big step towards more feasible graph database backed projects. This language treats nodes and edges as first class citizens and often provides a plethora of possibilities to work with properties of nodes. But in sensor networks, time series are an essential part of the nodes and edges, and until now, there is little support to structurally handle time series together with graphs.

On the other hand, time and time series have their own history of research. For example, the famous interval algebra of Allen explains how to deal with relationships between time intervals [1]. Earlier, Velain [13] defined relationships between time points, which are better suited in the context of time series data. On top of this foundational work, a huge amount of research exists with respect to the analysis of times series, interpolation and prediction, to name a few categories [10]. Much of this research is geared to specific applications, like the financial market and its stock market series. But in many cases, the time series are the only focus and their connection to a graph-based network is not considered.

Temporal graphs do have a vast literature but, in most cases, these graphs refer to networks that change over time (such as social networks). For example, Rost et al. [12] describe a technique to add temporal properties to nodes and edges. They build on earlier work of Junghanns et al. [11] in which Gradoop, a graph database system built on Hadoop, is described. The analysis in these cases, often focuses on paths, communities and other metrics or algorithms about the structure of the graph and how they evolve over time. Two examples are Erlebach et al. [6], who describe the temporal graph exploration problem and its complexity on dynamics graphs, and Froese et al. [8], where the main topic is a distance measure to determine how similar two temporal graphs are. However, this type of graphs is not applicable to our sensor network setting, since the

---

[1] https://neo4j.com/product/neo4j-graph-database/.

underlying transportation network is stable (edges and nodes are not added or deleted). The work of Debrouvier et al. [5] is more related to our work, since they consider graphs in which the property values can change throughout time. However, a difference is that they use time intervals connected to the values of the properties where our approach uses values linked with a timestamp.

This implies that there is a gap between the application of sensor networks, on the one hand, and storing, managing and analysing sensor networks in graph-based data management systems, on the other hand. This paper tries to bridge this gap by providing (1) a theory to model sensor networks; and (2) an approach to store and manage these data, supported by a query language to retrieve information of the network-based sensor data and to analyse it.

2   The Proposed Approach

The underlying model for our approach is based on the property graph model (see, for example, Bonifati et al. [4]). Let $\mathcal{O}$ be a set of objects with labels and properties appropriate in the application context, let $\mathcal{L}$ be a finite set of labels, let $\mathcal{K}$ be a set of property keys and let $\mathcal{V}$ be a set of values. These sets are pairwise disjoint and assumed to be finite. In order to accommodate time series in graphs, the definitions from Bonifati et al. [4] is modified by differentiating between temporal property keys $\mathcal{K}_{\mathcal{T}} \subseteq \mathcal{K}$ and static property keys $\mathcal{K}_{\mathcal{S}} \subseteq \mathcal{K}$ (with $\mathcal{K}_{\mathcal{T}} \cap \mathcal{K}_{\mathcal{S}} = \emptyset$). In addition, let $\mathcal{T}$ be the set of all possible timestamps, which we assume to be a discrete set that is ordered by the relation $\leq$. A *temporal property graph* G, is then defined to be a structure $G = (N, E, \eta, \lambda, v_S, v_T)$ where:

- $N \subseteq \mathcal{O}$ is a finite set of objects, which we call nodes;
- $E \subseteq \mathcal{O}$ is a finite set of objects (with $N \cap E = \emptyset$), which we call edges;
- $\eta : E \to N \times N$ is a function assigning an ordered pair of nodes to each edge;
- $\lambda : N \cup E \to \mathcal{P}(\mathcal{L})$ is a function assigning to each node and each edge a finite set of labels (here, $\mathcal{P}(\mathcal{L})$ denotes the power set of $\mathcal{L}$);
- $v_S : (N \cup E) \times \mathcal{K}_{\mathcal{S}} \to \mathcal{V}$ is a partial function assigning values to static properties of nodes and edges; and
- $v_T : (N \cup E) \times \mathcal{K}_{\mathcal{T}} \to \mathcal{P}((\mathcal{T} \times \mathcal{V}))$ is a partial function assigning a set of (timestamp, value)-pairs (that is, a set of measurements) to temporal properties of nodes and edges. We require that in such a set no timestamp appears twice and we can therefore call them *time series*.

The first three items of the above definition allow us to think of (N, E, η) as a directed graph. Within a time series there is total ordering on the measurements (induced by the natural order on timestamps), which in turn induces a "next"-relationship. Let m_1 and m_2 be two measurements (that is, elements of $\mathcal{T} \times \mathcal{V}$). The temporal order $\leq$ on $\mathcal{T}$ then induces an order (also denoted by $\leq$) on $\mathcal{T} \times \mathcal{V}$, as follows: $m_1 \leq m_2$ if and only if $\tau(m_1) \leq \tau(m_2)$, with $\tau : \mathcal{T} \times \mathcal{V} \to \mathcal{T}$ being the projection on the time component (that is, τ returns the timestamp of a measurement). Similarly, we define $\nu(m)$ to be the value of the measurement m (that is, its second component). In Vilain [13], there are three relations available for time points that can be used for measurements: "before", "after", and "equals". Our work adds four predicates: "previous", "next", "first", and "last". Previous and next refer to the measurements within one time series directly before and after another (considering the timestamp). The first and last predicates are true for measurements within a time series where there is no previous or next measurement, respectively.

Based on this model, the existing Cypher query language can be extended with the same pattern matching idea. In the `MATCH` part of the query, patterns of nodes and edges are described. In addition, properties can be described as `propertyKey: propertyValue` and it is assumed, for the sake of consistency, that properties values can be assigned to variables by placing the variable in place of the value: `propertyKey: variable`. For example, the Cypher query `MATCH (n name:"Hasselt") RETURN n;` which selects a node `n` with as value `"Hasselt"` in the property `name` can also be written as `MATCH (n name:v) WHERE v = "Hasselt" RETURN n;`. The new elements in the proposed query language are *time series patterns* and *measurement patterns*. A *time series pattern* consists of one or multiple measurements that describe together a situation in the series that then can be matched. A *measurement pattern* is denoted as `<timestamp, val>` where `timestamp` can be value of $\mathcal{T}$ or a variable and `val` can be a value of $\mathcal{V}$ or a variable. For example, the pattern `<2022-01-01T13:00, 10>` matches measurements m where $\tau(m) = $ 2022-01-01T13:00 and $\nu(m) = 10$. In the case of `<t, 10>` all measurements m are matched where the $\nu(m) = 10$, with, for each of them, `t` $= \tau(m)$. The patterns `<t, v>` matches all measurements and `v` $= \nu(m)$ and `t` $= \tau(m)$. These patterns can be combined to a time series pattern by concatenating them to a sequence of measurements. Implicitly, two measurements concatenated will describe two measurement for which the next or previous predicate holds. Thus, the concatenation `<2022-01-01T13:000, v><u,w>` matches two measurements m_1 and m_2 in a time series s where $\tau(m_1) = $ 2022-01-01T13:00 and $next(m_1, m_2, s)$ holds (meaning that m_2 is the next measurement after m_1 in time series s). If these measurements exists then `v` $= \nu(m_1)$, `u` $= \tau(m_2)$ and `w` $= \nu(m_2)$. Time series patterns can be placed within the existing Cypher structure on value places of properties. Thus, `MATCH (n {name:"Hasselt", temperature:<t, v>}) RETURN t, v;` corresponds to a node that has a property `name` with value `"Hasselt"` and that has a temporal property or time series, called `temperature`. All measurements are matched in this case and returned as a result of the query.

To store the graph and the time series data, different options are available each with specific advantages and disadvantages, but an in-depth study of the approaches is still needed. We consider three possible options, being the *External Model*, the *Object Model*, and the *Full Graph Model*. The *External Model* assumes that the graph and the time series are stored in separate locations. The time series are stored in an external database next to the graph database. This approach is interesting, because it allows to optimise each storage. For example, Neo4j can be used for the graph part and InfluxDB for the time series. But for this approach, querying and adding or deleting data is more complicated, since two different systems need to be managed. The *Object Model* is the one that is closest to the actual conceptual model. Here, the time series are stored as properties in the nodes and edges. In Neo4j, for example, it is possible to store lists and JSON objects as properties. However, these structures and the related functionalities are often not designed to handle the amount of data a time series has. In addition, Neo4j currently only supports string based JSON which requires constant JSON (de-)serialisation. The *Full Graph Model* uses the nodes and edges in the graph to represent the time series. In this case, a series is a linked list of nodes, where each node is a measurement with two properties for storing the timestamp and the value. These nodes are linked in the chronological order by edges. This linked list has a head node which contains the information about the time series and it connects to the most recent measurement. An example of such construction is given in Fig. 1, where a node E has a time series S with three measurements. The full graph model embeds all data in one graph and enables easier analysis on the data. For example, an aggregation tree can be build on top of it.

Fig. 1. An example of how a time series S, that is a property of node E, would be stored in the full graph model.

As first step, we choose the full graph model, as it leaves many possibilities for the analysis and querying. In addition, the limits of the graph databases used are not yet encountered in terms of number of nodes and edges in the use-cases and applications tested. However, this model has a considerable impact on the storage, because for each measurement a node and an edge is used. For example, an one year time series with a 15 min resolution, results in 35041 nodes and edges on top of nodes and edges needed for the network itself. Since in this model each time series is a graph in itself, the new proposed query language can be implemented by a translation that converts time series patterns and measurement patterns to the corresponding node and edge patterns. Next, the conversion of the time series and measurement patterns is described. It is assumed that the full graph model uses edges with label :Has to link time series to the object nodes in the graph and edges with the label :Previous to link measurements within a time series. These labels should not be used for other edges in the graph. The conversion will, for each measurement, create a node pattern, that are linked together by edge patterns when there are multiple measurements, and create a series node pattern which is linked to the measurements with a variable length edge pattern, as well as linked to the original node by a single edge pattern.

The complete conversion algorithm is:

1. Take the name of the time series, S, (the property key of the time series) and create a pattern for the series node: (:Series {name: S}).

2. Take each measurement pattern, `<t, v>` and convert it to a node pattern by describing the timestamp and/or value in the pattern together with a label for the measurement node: `(:Measurement {timestamp: t, value: v})`.
3. Connect the measurement nodes with edge pattern of length one: `-[:Previous]->`.
4. Connect the most recent measurement (in the timestamp order) to the series node with the edge pattern `-[:Previous*1..]->`.[2]
5. Finally, connect the series node pattern to the original node pattern by one edge directed from the original node to the series node: `-[:Has]->`.

3   Results

To demonstrate the new query language and the usability of the model, we introduce a small example, which is shown in Fig. 2 with the topological part as the graph and the time series as properties on the graph. This corresponds to the temporal property graph model, but in reality, it is stored with the full graph model in the Neo4j database. Figure 2 depicts a river network consisting of segments with, on most locations, a sensor that measures water temperature. For the sake of this example, only a small subset of the time series is shown, that is, the measurements on a certain day taken hourly from 10:00 until 15:00. On the graph, the time series is shown as a list of pairs where each pair is a measurement. In reality, the timestamp would also include the full date, but this is removed for clarity.

A basic query on this network would be

```
MATCH (:segment {id:6, temperature:<14:00, v>}) RETURN v;
```

which returns as result 19. That is the value that was measured on segment with id 6 at 14:00. A more advanced query that also explores the network is

```
MATCH (:segment {id:1, temperature:<t, v>})-[:flows*1..]->
      (n:segment {temperature:<t, v>}) RETURN n.id, t, v;.
```

This query looks for all locations downstream of the segment with id 1 and returns measurements of those locations that have the same value at the same time as location 1. This means the result contains records: `[(3, 10:00, 14), (3, 11:00, 14), (5, 10:00, 14), (5, 11:00, 14)]`. The last example shows how the query language can be used to ask for aggregation results on the network. We could, for example, be interested what the average measured temperature was of the measurements after 12:00 for the segment with id 7. The result is `17.667` and the query:

```
MATCH (:segment {id:7, temperature:<t, v>})
      WHERE t > 12:00 RETURN avg(v);
```

---

[2] It is important to use a variable length pattern, because the place of the measurements in the time series is unknown.

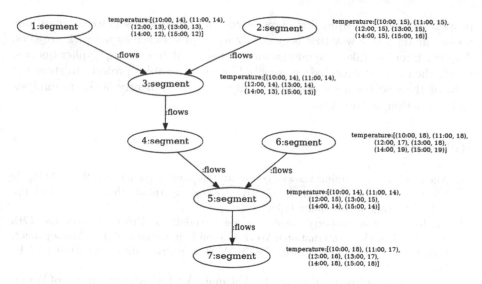

Fig. 2. An example river with segments that are connected if water flows from on to another segment, and where on certain segments water temperature is measured.

4 Future Work and Conclusion

There are limitations that have not yet been resolved. One of which is that there are no time series on edges when using the full graph model. This requires a more complicated model and it might be interesting to investigate if the full graph model is still the best approach. Preliminary research has shown that not all interesting queries can be described with the query language introduced here. For example, the query, that tries to find a sequence of (an unknown number of) measurements in a time series where the values of the measurements are strictly increasing, is not expressible in the proposed language. This issue can be studied by looking at the Regular Property Graph Logic (see, for example, Bonifati et al. [4]). Therefore, future research should aim at determining which queries can be expressed and which cannot. Finally, it should be mentioned that there is still work to be done in optimising the system in terms of storage and elaborating on the possibilities to aggregate and analyse data.

In conclusion, the contributions of this work can be summarised as follows. A sensor network is considered to be stable transportation network in which nodes and edges represent objects that interact or are related to each other, but these relations do not change. At these nodes and edges there are properties that change and are modelled as time series. By creating a temporal property graph model, this work defines a general approach to deal with these types of data. In addition, some storage models are given and one of them is used as an example. Based on the temporal property graph model, a query language is created and demonstrated, that extends the idea of pattern matching of the graph to the time series. This model and query language treats nodes, edges, and temporal properties (time series) as first class citizens and enables users to ask questions about the data in a way that is generally applicable in sensor network projects. A query interface takes the queries and translates it to existing Cypher queries, which then can be executed on Neo4j. The above enables projects to store the data of their sensor networks more easily, and subsequently facilitate analysis and prediction on top of it.

References

1. Allen, J.F.: Maintaining knowledge about temporal intervals. In: Weld, D.S., de Kleer, J. (eds.) Readings in Qualitative Reasoning About Physical Systems, pp. 361–372. Morgan Kaufmann (1990)
2. Angles, R.: The property graph database model. In: Proceedings of the 12th Alberto Mendelzon International Workshop on Foundations of Data Management, Cali, Colombia, 21–25 May 2018. CEUR Workshop Proceedings, vol. 2100. CEUR-WS.org (2018)
3. Bollen, E., Hendrix, R., Kuijpers, B., Vaisman, A.: Towards the Internet of Water: Using graph databases for hydrological analysis on the Flemish river system. Trans. GIS **25**(6), 2907–2938 (2021)
4. Bonifati, A., Fletcher, G., Voigt, H., Yakovets, N.: Querying Graphs. Morgan & Claypool Publishers (2018)
5. Debrouvier, A., Parodi, E., Perazzo, M., Soliani, V., Vaisman, A.: A model and query language for temporal graph databases. VLDB J. **30**(5), 825–858 (2021). https://doi.org/10.1007/s00778-021-00675-4
6. Erlebach, T., Hoffmann, M., Kammer, F.: On temporal graph exploration. In: Halldórsson, M.M., Iwama, K., Kobayashi, N., Speckmann, B. (eds.) ICALP 2015. LNCS, vol. 9134, pp. 444–455. Springer, Heidelberg (2015). https://doi.org/10.1007/978-3-662-47672-7_36
7. Francis, N., et al.: Cypher: an evolving query language for property graphs. In: Proceedings of the 2018 International Conference on Management of Data, SIGMOD 2018, pp. 1433–1445. Association for Computing Machinery (2018)
8. Froese, V., Jain, B., Niedermeier, R., Renken, M.: Comparing temporal graphs using dynamic time warping. Soc. Netw. Anal. Mining **10**(1), 1–16 (2020). https://doi.org/10.1007/s13278-020-00664-5
9. Green, A., et al.: Updating graph databases with cypher. Proc. VLDB Endow. **12**(12), 2242–2253 (2019)
10. Hamilton, J., Press, P.U.: Time Series Analysis. No. 10 in Book collections on Project MUSE, Princeton University Press, Princeton (1994)
11. Junghanns, M., Petermann, A., Gómez, K., Rahm, E.: GRADOOP: scalable graph data management and analytics with Hadoop. CoRR abs/1506.00548 (2015)
12. Rost, C., Thor, A., Rahm, E.: Analyzing temporal graphs with gradoop. Datenbank-Spektrum **19**(3), 199–208 (2019)
13. Vilain, M.B.: A system for reasoning about time. In: Proceedings of the Second AAAI Conference on Artificial Intelligence, AAAI 1982, pp. 197–201. AAAI Press (1982)

Formalization of Data Integration Transformations

Théo Abgrall(✉)

Free University of Bozen-Bolzano, Bolzano, Italy
tabgrall@unibz.it

Abstract. In our research work, we attempt to bridge two fields once closely interconnected: Information Preservation and Data Integration. This relative split is due both part to the growth of the web and its associated challenges and the lack of proper formalism and tools for lossless schema transformation. To answer that concern, our approach is to propose a catalogue of lossless schema transformation templates, called transformation patterns, based on operations commonly used in data integration. As some of these operations include the promotion and/or demotion of metadata in a relational schema, we introduce an extension of transformation patterns by incorporating a dynamic aspect. We also want to deal with the leftover operations, those that lose information, and turn them into lossy transformation patterns. Finally, we present the early implementation of a schema transformation framework using transformation patterns.

Keywords: Schema transformation · Information preservation · Data integration

1 Introduction

The problem of information preservation is almost as old as the history of relational databases itself. Two relational schemas could hold the same capacity of information despite being written differently and made to contain dissimilar knowledge. Coincidentally, this problem is akin to another quite preeminent one in Data Integration, which is Data Heterogeneity. In data integration [1] the main ambition is to overview and manages multiple data sources through the use of a global schema. The problem of **Data Heterogeneity** is thus to overcome the disparate nature of these data sources to find matches between them. Despite their seemingly shared purpose, the two fields progressively grew apart since the popularization of the Web. The reasons are twofold: On one hand, the massive growth of accessible information has, somehow, made it less valuable than it was previously. On the other hand, there has been a lack of proper tools or formalisms to prove schema dominance and equivalence.

Under the supervision of Enrico Franconi.

© Springer Nature Switzerland AG 2022
S. Chiusano et al. (Eds.): ADBIS 2022, CCIS 1652, pp. 615–622, 2022.
https://doi.org/10.1007/978-3-031-15743-1_56

Our motivation is to focus on the second concern by reviewing the current formalisms for information preservation before expanding one of them into a data integration context. The formalism we use is a template-based formalism describing the structure and constraints of a database on both sides of a transformation expressed via mappings. In this paper, we begin by presenting a brief history of information capacity preservation. Following that, we review the literature at the crossroad between information preservation and data integration. Then, we describe the transformation pattern formalism, the foundations upon which our research is based on. In section five We detail our ambitions to develop that formalism even further, notably by representing more complex transformation and introducing a new notation for flexibility. We finally conclude by exhibiting our current proof-of-concept schema transformation framework and by discussing our intentions to insert it in a data integration context, as well as the potential consequences this choice carries.

2   Information Preservation

At its core, most of the literature regarding the preservation of information during a schema transformation emerged from Hull's [2] definition of four decreasingly strict notions of schema dominance. The main idea behind schema dominance is that for two relational schemas, any valid database instance in one of the schemas can be found in the other.

Definition 1 (Schema Transformation). *Let there be two schema S and T. A **schema transformation** from S to T is a mapping function $f_{S \to T} : S \to T$ such that $f_{S \to T} I(S) \to I(T)$ where I correspond to all valid database instances of S and T, respectively. Furthermore, a mapping function $f_{S \to T} : S \to T$ is **total** if $\forall x \; \exists y \; f_{S \to T}(x) = y$, **injective** if $\forall x, x' \; f_{S \to T}(x) = f_{S \to T}(x')$, **surjective** if $\forall y \; \exists x \; f_{S \to T}(x) = y$ and **bijective** if $f_{S \to T}$ is total, injective and surjective.*

Definition 2 (Schema Dominance). *Let there be two schema S and T and the two mapping functions $f_{S \to T}$ and $f_{T \to S}$. We say that S **dominates** T if for any instance $I(T)$ of T: $f_{S \to T} \circ f_{T \to S} I(T) = I(T)$. If S dominates T, then the mapping function $f_{T \to S}$ is injective and total.*

Definition 3 (Schema Equivalence). *Let there be two schema S and T. We say that S and T are **equivalent** if and only if S dominates T and T dominates S. If S and T are equivalent, then the mapping function $f_{S \to T}$ is bijective.*

An earlier version of these definitions was presented in [3] under the name of query-dominance where a single query replaces the mapping function. Later, these definitions were expanded [4] by answering the conjuncture that schema transformations could preserve primary keys, thus implying the preservation of not only instances but also relational constraints. Once more, the notions of schema dominance and equivalence were extended in [5] which defines a correct schema transformation as one which preserves both constraints and instances, thus differentiating between instance preservation, constraint preservation and their combination, information capacity preservation.

3   Related Work

On study that encapsulate very well the purpose of bijective functions in a relational database context as a way of guaranteeing updates through transformations is [6]. Furthermore, bijectives functions are also useful at the query level when encoding or decoding values or attributes name, thus allowing merging, splitting and value modification under constraints. The problem of dependency preservation is discussed in [7] where information capacity is tested with the chase procedure. A stepping stone in including information capacity in a schema integration context is [8]. With a new formalism, they were able to assess schema dominance and specify transformations by adding constraints. As the field of data integration and conceptual models are closely related, an approach defines by [9] is to focus on the equivalence of conceptual models. In a framework for data integration based on a both-as-view (BAV), [10] propose to use a sequence of basic operation, sometimes compiled into templates, to dismantle and recompose data sources. However, This approach does not preserve embedded dependencies. Around this point in time, the quality of information preservation became less prevalent in data integration as the number of accessible data sources skyrocketed. The field of information preservation thus become more closely related to the one of database reverse engineering, where the goal is to generate a conceptual model from an usually old and decrepit database, or, more properly, legacy databases. Yet, another field at the crossroad of data integration and schema transformation is the one of data exchange [11] grew in that period. While the main focus of data exchange is the generation of a proper database instance from constraints, the problem of schema exchange introduced in [12] is closer to our ambitions. Mostly with regards to the grouping of similar schemas into templates making an automatic schema transformation possible.

4   Transformation Patterns

4.1   Definition

Some operations seemed capable of preserving information at first glance, such as the horizontal decomposition illustrated in [3]. However, a proper formalism that clearly defines the requirements necessary for such transformation to be lossless was still lacking. It was in the context of Database Reverse Engineering, a set of techniques generating a conceptual model from excerpts of a relational database, that the following approach originated. Transformation patterns [13] are crafted templates describing a particular schema transformation and the constraints necessary to ensure its lossless quality. They applied to first-order logic schemas defined as the couple $[\mathbb{A}, \mathbb{C}]$ with $\mathbb{A}$ the alphabet, a set of predicates, and $\mathbb{C}$ a set of constraints. Transformation patterns are thus composed of:

Database Pattern: Representing a relational schema and the constraints applied over it, a database pattern contains both a set of relation patterns and a set of constraint patterns. A **Relation Pattern** is defined as $R(ATT_1, ..., ATT_n)$ with R the name of the relation and ATT_j corresponding to a set of distinct variables. A **Variable** is a set of attributes grouped via the constraints applied to them. For example, a relation $Person(id, name, dob)$ with a sole primary key constraint on id fit the following relation pattern $R(PK, REST)$. A **Constraint Pattern** represents relational constraints over the set of relation patterns. Of which, we can mention primary keys, functional dependency and, more generally, most embedded dependencies. To pursue our former example, a constraint pattern setting id as a primary key would be written as $pk[R](PK)$. Naturally, every variable written in each constraints patterns must also exist in their respective relation patterns. A transformation pattern contains two database patterns, thus describing the relation schema and its constraints on either side of a transformation. We abusively tend to call these source and target database patterns (or input and output) but as each transformation is bidirectional and reversible there is no real direction.

Mapping Pattern: To map each variable from one schema to another, we define a mapping pattern as a couple of query sets written in relational algebra. Each relation in a source schema has to be expressed as a set of queries over the target schema, and conversely.

The consequence of working with first-order logic schema is that a proof of equivalence between two database schema can be given their mutual entail through some logic theorem solver, here Prover9 [19] in our case.

4.2   Example

We present a short, slightly simplified, example of a transformation defining the horizontal decomposition of a table based on a specific condition:

DB Pattern A

$R : (ATT)$
$\sigma_{cond}(R) \neq \emptyset$
$const^k[R](ATT^k)$
$const^l[R](ATT^l)$

Mappings

$R1 = \sigma_{cond}(R)$
$R2 = \sigma_{\neg cond}(R)$

$R = R1 \bigcup R2$

DB Pattern B

$R1 : (ATT)$
$R2 : (ATT)$
$\sigma_{cond}(R1) = R1$
$\sigma_{cond}(R2) = \emptyset$
$const^k[R1](ATT^k)$
$const^l[R2](ATT^l)$

Fig. 1. Simplified transformation pattern for horizontal decomposition

This transformation pattern split in two a relation based on a condition applied over at least one attribute contains within the variable ATT. Let say we have $Worker$: $(id, name, salary)$ and the condition $\sigma_{salary>2000}(Worker) \neq \emptyset$, then we can pretty easily arrive to the two new relations $Junior$: $(id, name, salary)$ and $Senior$: $(id, name, salary)$ and their subsequent constraints. On that topic, the constraint $const$ corresponds to any additional, residual constraints. In this case, any primary key within ATT would be conserved.

At this point, transformation patterns serve as templates properly defining the constraints necessary to ensure lossless transformation. As they stand, the main purpose of transformation patterns is to serve as guidance for database engineers whenever schema transformation is wanted. As the applications are limited to the available number of written templates, a solution is to add new transformation patterns to the list of existing ones. Thus started our research work, expending this list, or, more appropriately, the *catalogue* of available transformation patterns.

5   Expanding the Catalogue

We began expanding the catalogue browsing schema transformation commonly used in data integration, such as those defined in [14]. While diving through the many possible schema transformations to turn into templates, one stood out, provokingly: the pivot operation [15]. The pivot transformation takes one column of distinct values and turns it into that many new attributes with associated values taken from another column. At first glance, this should be impossible to write as a template since no mappings made in relational algebra can express the promotion (and demotions) of values into attributes (and the inverse for the unpivot). However, we can already circumnavigate the limits of relational algebra in transformation patterns. Thus, it should be possible for us to write transformation patterns representing more complex operations, such as some introduced in [16] and more notably, operations encompassing all metadata of a relational schema. Opening the floodgates that way however caused a new concern to surfaces: dynamicity.

So far, transformation patterns were tailor-made to a specific operation where the number of relations, variables and instances are clearly described before and after. It, unfortunately, implies that there are no generic patterns available. This applies to complex operations such as the pivot, but let us illustrate the issue with a simple example.

Example 1. Going further than Fig. 1, what if we wanted to generate a full partition of a relational schema $Worker$: $(id, name, salary)$ based on every distinct salary values? We could already do with regular transformation pattern by applying the same ones sequentially, but is there a way to execute this transformation in a single step? In the first pattern we want the following constraints $\sigma_{cond_1}(R) \neq \emptyset$, ..., $\sigma_{cond_n}(R) \neq \emptyset$ for any number $n \geq 1$ of distinct values. In the second, we need to write $R1$: (ATT), ..., $Rn(ATT)$ plus every related constraints. Similarly for the mappings where we have $R1 = \sigma_{cond_1}(R)$, ..., $Rn = \sigma_{cond_n}(R)$ and $R = R1 \cup ... \cup Rn$.

Our proposal is an extension of the current transformation pattern formalism focused on the addition of two annotations: a dynamic one ^{d} and their associated indexes $_i$. The **Dynamic Annotation** ^{d} added to a relation or a constraint the repetition of that line several times, based on the indexes found in the body. Added to a variable it generates a new set of variables going from 1 to n. The indexes $_i$ denotes the relation, variable or expression, targeted. We call a transformation pattern containing these annotations **dynamic**. Applied to our previous example, we can now write a dynamic version of the pattern described in Fig. 1 in Fig. 2

DB Pattern A

$R : (ATT)$
$^d\sigma_{cond_i}(R) \neq \emptyset$
$^dconst^{k_i}[R](ATT^{k_i})$

Mappings
$^dRi = \sigma_{cond_i}(R)$

$R = \bigcup_1^n(Ri)$

DB Pattern B

$^dRi : (ATT)$
$^d\sigma_{cond_i}(Ri)$
$^dconst^{k_i}[Ri](ATT^{k_i})$

Fig. 2. Simplified dynamic transformation pattern for horizontal decomposition

In this example, relation and constraints preceded by a dynamic ^{d} indicate its appurtenance to a set of an arbitrary size n. Once n is fixed, the value of the index i contained in each dynamic term will range from 1 to n. From this new pattern, it is very simple to generate the one described in Fig. 1 by setting $n = 2$ and defining $cond_1 = salary > 2000$ and $cond_2 = \neg cond_1$.

Dynamic patterns not only provide a generic version of regular transformation patterns, but they also allow for more intuitive solutions to complex transformation than otherwise. It is possible to write the pivot operation as a sequence of transformation patterns transposing a single distinct value at each iteration. Besides the elegance of resolving a complex operation in a single step, it avoids using other patterns at some point in the sequence, and thus simplify the process. So far, every dynamic pattern we wrote can to be instantiated into lossless patterns quite simply through the use of a formal language parser [17]. But is the opposite so? Is there a function or algorithm that automatically gives the generic, dynamic version of any transformation patterns? This issue is currently worked on as we believe so. Finally, while transformation patterns where originally designed with relational database in mind, it seem plausible to apply them over different data structure, such as XML (Fig. 3).

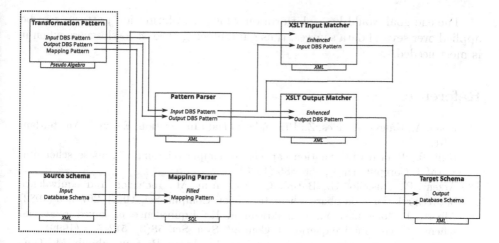

Fig. 3. Schema transformation framework using transformation patterns. Each step outside of the border is generated.

6 Conclusion and Future Work

In this paper, we presented a recent formalism for lossless schema transformation and introduced an extension to that formalism to represent more complex operations. Introducing dynamic annotations solves an issue of flexibility as many, if not all, transformation patterns should have their own generic version, ready to be rewritten for any arbitrary values. In their current forms, transformation patterns can simply serve as guidance for database engineers, but we can do more than that. Our main motivation is the creation of a schema transformation framework based on transformation patterns and thus, allowing for the sequential, automatic transformation of relational databases. As all lossless patterns act like bijective functions, such a framework would allow updates on the sources databases from any generated views as prescribed in [8]. We have already written some parts of this transformation framework and implemented as part of the KPrime Project [20], which current process is illustrated in Fig 1.

Here, a source database written in XML is matched with a corresponding transformation pattern and consequently, automatically transformed into the expected new form. From this point, our motivation is to adapt this current framework into a data integration context. Naturally, this comes with new difficulties, starting with a current, obvious, blank spot in our approach: not all transformations are lossless. Many are instead what we call *lossy*, meaning that either values or constraints are lost in the process. Thus, the formalization of lossy transformation, as well as potential ways to mitigate them is intuitively the next ambition of our research work. Following that, further development in the schema transformation framework is another aspiration of ours. Notably, the incorporation database identity resolution [18] data provenance or data evolution seems to be relevant notions to take into account.

The end goal would be a platform on which transformation patterns can be applied over several data sources, thus guaranteeing at least losslessness when it is most needed.

References

1. Doan, A., Halevy, A., Ives, Z.: Principles of Data Integration. Elsevier, Amsterdam (2012)
2. Hull, R.: Relative information capacity of simple relational database schemata. SIAM J. Comput. **15**(3), 856–886 (1986)
3. Atzeni, P., Ausiello, G., Batini, C., Moscarini, M.: Inclusion and equivalence between relational database schemata. Theor. Comput. Sci. **19**(3), 267–285 (1982)
4. Albert, J., Ioannidis, Y., Ramakrishnan, R.: Equivalence of keyed relational schemas by conjunctive queries. J. Comput. Syst. Sci. **58**(3), 512–534 (1999)
5. Qian, X.: Correct schema transformations. In: Apers, P., Bouzeghoub, M., Gardarin, G. (eds.) EDBT 1996. LNCS, vol. 1057, pp. 114–128. Springer, Heidelberg (1996). https://doi.org/10.1007/BFb0014146
6. Kobayashi, I.: Losslessness and semantic correctness of database schema transformation: another look of schema equivalence. Inf. Syst. **11**(1), 41–59 (1986)
7. Abiteboul, S., Hull, R., Vianu, V.: Foundations of Databases, vol. 8. Addison-Wesley, Reading (1995)
8. Miller, R.J., Ioannidis, Y.E., Ramakrishnan, R.: The use of information capacity in schema integration and translation (1993)
9. McBrien, P., Poulovassilis, A.: A formalisation of semantic schema integration. Inf. Syst. **23**(5), 307–334 (1998)
10. McBrien, P., Poulovassilis, A.: Data integration by bi-directional schema transformation rules. In: Proceedings 19th International Conference on Data Engineering (Cat. No. 03CH37405), pp. 227–238. IEEE (2003)
11. Fagin, R., Kolaitis, P.G., Miller, R.J., Popa, L.: Data exchange: semantics and query answering. Theor. Comput. Sci. **336**(1), 89–124 (2005)
12. Papotti, P., Torlone, R.: Schema exchange: generic mappings for transforming data and metadata. Data Knowl. Eng. **68**(7), 665–682 (2009)
13. Ndefo, N., Franconi, E.: A study on information-preserving schema transformations. Int. J. Semant. Comput. **14**(01), 27–53 (2020)
14. Raman, V., Hellerstein, J.: Potters wheel: an interactive framework for data cleaning and transformation. Working draft (2001)
15. Wyss, C.M., Robertson, E.L.: A formal characterization of PIVOT/UNPIVOT. In: Proceedings of the 14th ACM International Conference on Information and Knowledge Management, pp. 602–608 (2005)
16. Wyss, C.M., Robertson, E.L.: Relational languages for metadata integration. ACM Trans. Database Syst. (TODS) **30**(2), 624–660 (2005)
17. Parr, T.: The Definitive ANTLR 4 Reference. Pragmatic Bookshelf, Raleigh (2013)
18. Borgida, A., Toman, D., Weddell, G.: On referring expressions in information systems derived from conceptual modelling. In: Comyn-Wattiau, I., Tanaka, K., Song, I.-Y., Yamamoto, S., Saeki, M. (eds.) ER 2016. LNCS, vol. 9974, pp. 183–197. Springer, Cham (2016). https://doi.org/10.1007/978-3-319-46397-1_14
19. Prover9 and Mace4. https://www.cs.unm.edu/mccune/prover9/
20. KPrime. https://www.kprime.it/index.html

Models and Query Languages
for Temporal Property Graph Databases

Valeria Soliani[1,2]([⊠]) (iD)

[1] Hasselt University, Hasselt, Belgium
valeria.soliani@uhasselt.be
[2] Instituto Tecnológico de Buenos Aires, Buenos Aires, Argentina
vsoliani@itba.edu.ar

Abstract. Although property graphs are increasingly being studied by the research community, most authors do not consider the evolution of such graphs over time. However, this is needed to capture a wide range of real-world situations, where changes normally occur. In this work, we propose a temporal model and a high level query language for property graphs and analyse the real-world cases where they can be useful, with focus on transportation networks (like road and river networks) equipped with sensors that measure different variables over time. Many kinds of interesting paths arise in this scenario. To efficiently compute these paths, also path indexing techniques must be studied.

Keywords: Property graphs · Temporal graphs · Sensor networks

1 Problem Statement and Motivation

Property graphs are graphs whose nodes an edges are annotated with (property, value) pairs [3]. They are widely used for modeling and analyzing different kinds of networks. The property graph data model underlies most graph databases in the marketplace.[1] Typically, the graphs used in practice do not change over time. However, in real-world problems, time is present in most applications and graphs are not the exception. Many changes may occur in a property graph as the world they represent evolves over time: edges, nodes and properties can be added and/or deleted, property values can be updated, to mention the most relevant ones. Social networks are clear examples of this statement: If u and v are vertices modeling persons, and an edge represents a relationship between u and v telling that u follows v or u isFriendOf v, these relationships may change over time or even the persons may unregister from the network. Accounting for these changes would allow queries like "Who were friends of Mary while she was working at Hasselt University", that otherwise could not be answered. Another example are graphs representing road networks, where vertices model locations and edges represent roads or highways that exist at different periods of time and

---

[1] For example, http://www.neo4j.com, http://janusgraph.org/.

© Springer Nature Switzerland AG 2022
S. Chiusano et al. (Eds.): ADBIS 2022, CCIS 1652, pp. 623–630, 2022.
https://doi.org/10.1007/978-3-031-15743-1_57

whose properties (like road condition) also vary over time. This allows queries like "Compute the time that we saved going from Buenos Aires to Pinamar after the construction of Highway Number 11."

Our work builds on the hypothesis that keeping the history of changes in graphs is relevant in many interesting real-world applications and that this has been largely overlooked in property graph database modeling. We study how temporal databases concepts can be applied to graph databases in order to model, store, and query temporal property graphs.

In Sect. 2, we review related work. Section 3 presents our approach to the problem and the main results obtained so far. Section 4 discusses ongoing work and open problems. We conclude in Sect. 5.

2   Related Work

Literature on temporal graphs is mostly oriented to address certain path problems in homogeneous graphs (graphs whose nodes are all of the same kind), not tackling property graphs. This is the case of [13,14], where only edges are timestamped with the initial validity time of the relationship and the duration of the relationship, and there is only one kind of relationship in the graph. Wu et al. [13] study temporal paths and introduce the notions of earliest-arrival path, latest-departure path, fastest path, and shortest path. Along the same lines, Chronograph [5] is a temporal model for graphs that enables temporal traversals such as: temporal breadth-first search, temporal depth-first search, and temporal single source shortest path.

A first approach to the problem of temporal property graphs modeling was presented by Campos et al. [6]. Over this work we build the model we introduce in Sect. 3. Further, the notion of Continuous path used in our work was initially introduced in [12], where a model and index for temporal XML documents was proposed. Pokorny et al. [11] index graph patterns in Neo4j, using a structure stored in the same database as the graph, an approach we follow in our work for indexing temporal paths. Another approach we build on is the temporal database index proposed by Elmasri et.al. [8] to process temporal selections and temporal join operations. In [10], an index structure is designed for temporal attributes with various frequency and rates of changes.

3   Our Approach and First Results

The first result of our work is a temporal graph data model, called *TGraph*, that accounts for changes in nodes and relationships in property graphs [7]. Together with this model, we proposed and implemented a high-level query language, T-GQL, that not only considers temporal operators to query a TGraph, but also deals with temporal path structures, the actual first-class citizens in our model. We introduce the model and query language next.

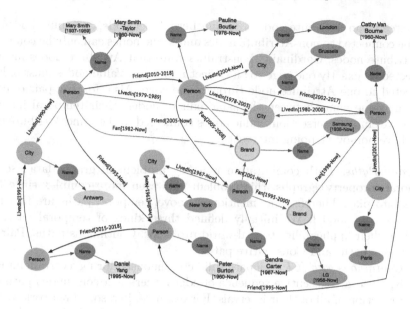

Fig. 1. A temporal property graph.

Temporal Model. We mentioned that in property graphs, nodes and edges are labeled with a sequence of (property,value) pairs. All of them can evolve over time. Thus, to keep the history of the graph we need a data model that can deal with all of these kinds of changes. In our model, entities are represented as *Object* nodes. These nodes may have attributes that may be either static or which can change over time. The former are represented as classic properties of the Object node. The latter are represented as a different kind of node, called *Attribute* node, connected to the Object node. For example, Object nodes may represent a person as an entity, a static attribute of such person may be her date of birth and a temporal attribute may be her name, which can change when a person gets married. The values that this attribute can take are represented as *Value* nodes connected to the corresponding Attribute node. Figure 1 depicts an example of the TGraph model representing a social network. There are three kinds of Object nodes: Person, Brand and City. There are also three types of temporal relationships: LivesIn, Fan, and Friend. The first one is labeled with the periods when someone lived somewhere, the second one with the periods when a person was fan of a brand, and the last one, with the period when two people were friends, for example, Mary Smith-Taylor has been a friend of Peter Burton since 1993. The temporal Attribute node Name represents the name associated with a Person node. We see for example that "Mary Smith" became "Mary Smith-Taylor" in 1960. For clarity, if a node is valid throughout the complete history of the graph, the temporal labels are omitted.

Nodes and edges in TGraph must satisfy a set of temporal constraints that state how the nodes in the graph must be connected. An Object node can only be connected to an Attribute node or to another Object node, Attribute nodes can only be connected to non-Attribute nodes and Value nodes can only be connected to Attribute nodes. Cardinality constraints state that Attribute nodes must be connected by exactly one edge to an Object node, and Value nodes must only be connected to one Attribute node through one edge. Also, for any pair of edges with the same name between the same pair of nodes, their temporal intervals must have empty intersection. Value nodes connected to the same Attribute node must have non-overlapping intervals.

Temporal Paths. Path computation is a key problem in graph databases. In temporal (property) graphs, this problem gets even more complex since time comes into play. Therefore, as mentioned above, temporal paths are first-class citizens of our model. We initially defined three kinds of temporal paths [7], based on their applicability to real-world problems: Continuous paths, Pairwise Continuous paths and Consecutive paths.

A *Continuous path (CP)* is a path valid continuously during a certain interval. That is, given a consecutive sequence of edges, there is a continuous path over the intersection of all of their intervals. For example, in a social network we may be interested in finding out chains of people that where friends during the same period. In many cases, a weaker condition over temporal paths suffices. The user may be interested in paths such that there is an intersection in the intervals of consecutive edges pairwise. These paths are called *Pairwise Continuous paths (PCP)*. *Consecutive paths (CSP)*, are useful for scheduling. In these cases, we require that consecutive edges do not overlap, for example, to leave a certain time for a train or flight connection. Thus, CSPs are sequences of edges such that the pairwise intersection between consecutive edges is empty. Different kinds of Consecutive paths can be defined, according to the use that is given to them. For example, in scheduling problems, the *earliest-arrival path (EAP)* is the path that can be completed in a given interval such that the ending time of the path is minimum; the *latest-departure path (LDP)* is the path that can be completed in a given interval such that the starting time of the path is maximum; the *fastest path (FTP)* is the path that can be completed in a given interval such that its duration is minimum; finally, the *shortest path (STP)* is the path that can be completed in a given interval such that its number of edges is minimal.

Query Language. TGraph comes equipped with a high-level SQL-like query language called T-GQL. The T-GQL language has a mixed flavour between SQL and Cypher [9], Neo4j's high-level query language. The syntax of the language has the typical SELECT-MATCH-WHERE form. The SELECT clause performs a selection over variables defined in the MATCH clause. The MATCH clause may contain one or more path patterns and function calls. The result of the query is always a temporal graph (analogous to relational temporal databases theory), although the query may not mention temporal attributes. This can be modified by the SNAPSHOT operator, which allows retrieving the state of the graph at a certain point in time. T-GQL supports the three path semantics above, namely Continuous, Pairwise Continuous, and Consecutive paths, implemented as functions included in a library of Neo4j plugins. Consider the query "Compute the friends of the friends of each person and the period when the relationship occurred through all the path." For example, in Fig. 1, Pauline Boutler was a friend of Cathy Van Bourne between 2002 and 2017. Also, Mary Smith-Taylor was a friend of Pauline between 2010 and 2018. Thus, the path $(MarySmith - Taylor \xrightarrow{\text{Friend}} Pauline \xrightarrow{\text{Friend}} Cathy, [2010, 2017])$ will be in the answer since the whole path was valid in this interval. The query reads in T-GQL (cPath computes the CPs of length two for every node):

```
SELECT path
MATCH (n:Person), path = cPath((n)-[:Friend*2]->(:Person))
```

Analogously, PCPs can be also computed using the pairCPath function, and four functions are supported for CSPs: fastestPath, earliestPath, shortestPath, and latestDeparturePath.

The intermediate results of a query can be filtered by an interval I, provided by the user, that filters out the paths whose interval does not intersect I. The temporal granularity of the starting and the ending instants of the interval must be the same. Also, the BETWEEN operator receives a temporal interval and performs a temporal filter. The WHEN clause has the form MATCH-WHERE-WHEN and can reference variables in an outer query. This clause is used to compute events occurring during concurrent intervals.

4   Current Work and Open Problems

We are currently working on different topics that extend the work presented in the previous section. We comment on this work next.

Indexing Continuous Paths. Computing temporal paths over the TGraph model, particularly CPs, turns out to be costly. We studied how to improve the computation of CPs by indexing them. For this, two index structures are proposed and implemented, one that indexes all the paths and another one that indexes all paths of length two. In the latter case, computing the paths of length higher than two requires additional processing. We implemented additional commands in T-GQL to create and make use of indices. We also consider reducing the search space by limiting the time window to consider the one in which queries will most likely fit. Although we have already implemented this proposal, this is a fertile field to work in and improve current results.

Temporal Modeling of Sensor Networks. The model in the previous section can represent graphs whose nodes and edges change over time, like in the case of social networks. There are other kinds of interesting networks in real-world scenarios where the model can be applied. This is the case of transportation networks, like roads or river networks, which differ from social networks in that there is an element that flows through the network (e.g., cars, water). These networks are rather *stable*, in the sense that changes occur occasionally. For example, the

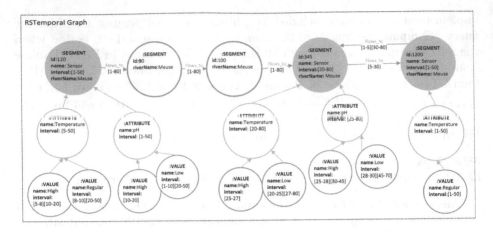

Fig. 2. A temporal graph for a river network equipped with sensors.

direction of the water flow in a river may change due to a flood or a branch may disappear during long dry weather periods. More interesting is the case where sensors are attached to segments in transportation networks, producing time-series data. These are called *sensor-equipped transportation networks* [1], sensor networks for short. The data captured by sensors can be used in various application areas, like traffic control and river monitoring. For example, the Internet of Water project of the Flemish government in Belgium will deploy a network with 2,500 wireless water quality sensors. Bollen et al. [4] proposed a formal model and a calculus to query sensor networks, where the network topology and the time-series data are stored in a graph database for querying and analysis. However, this model does not account for changes over the network. Thus, we propose to model sensor networks using the TGraph model, where Object nodes represent network segments, Attribute nodes represent the temporal properties of the segments, and Value nodes represent the time-series data captured by the sensors.

Modeling sensor networks requires extending TGraph in many ways and opens up a wide variety of paths that are worth studying using Allen's Algebra [2]. We call this model *SN-TGraph*. In this model we distinguish Object nodes that hold a sensor from the ones which do not, and call them *Sensor* and *Segment* nodes, respectively. Also, a list of time intervals indicates the periods of time where a segment had a working sensor on it. Properties that do not change across time are represented as usual in property graphs. Figure 2 shows a scheme of the SN-TGraph model for a river network. Shaded nodes are Sensor nodes. All nodes are of type :**Segment**, and sensor nodes have the value *Sensor* for the property name. There are two kinds of Attribute nodes, for representing series of temperature and pH values. Also, in the Value nodes there is an ordered sequence of time intervals for each value of the variables.

The temporal paths presented in Sect. 3 are more involved in SN-TGraphs, since paths must now be defined based on a function over the sensors measurements. This way, CPs are now defined as paths where the *value* of some function is the same throughout the path during a certain interval. At this time, we only consider functions over categorical rather than continuous variables. This is the case, for example, where there is a sequence of consecutive sensors such that each sensor has measured the value of a variable Temperature categorized as *High* throughout a certain interval. We denote this as an SN-CP. PCPs over a sensor network are defined as paths such that for every pair of consecutive sensors there is an overlapping interval where the value of the variable is the same. We denote these paths SN-PCP. In the case of Consecutive paths, there is a sequence of consecutive sensors where the value of the variable is the same and the time intervals do not overlap. We denote these paths as SN-CSP.

We note that the paths above cannot completely capture the flow in a transportation (sensor) network, since an event captured by a sensor (e.g., a traffic jam in a road network or the presence of a pollutant in a river) will probably be detected by a sensor located after the previous one in the sense of the movement. Further, if the sensors were placed close to each other, it is possible that a value measured by one sensor would still be valid while it is valid in the next one. If this condition is fulfilled for each pair of sensors, the path can be considered a PCP, while in case the sensors were placed far apart, it can be a CSP. To combine both situations in one path in which the valid time of a sensor measurement just starts before the next one we introduce the notion of *Flow path (SN-FP)*.

Temporal Relations Between Paths. The difference between the four kinds of paths introduced so far lies in the way in which the temporal interval of each node (or edge) is related to the next one. Temporal relations between every pair of consecutive intervals in those paths can be described in terms of Allen's Algebra. There are thirteen possible Allen's relations covered by CPs, PCPs, CSPs, and SN-FPs. Even though not all possible combinations of Allen's intervals are interesting in real-world situations, we want to study if there are interesting paths that can be identified in addition to the ones already defined. In order to determine if our paths are not covering some important combination, we need to define a path taxonomy, and map paths to real-world cases. We are currently carrying out this study.

5   Conclusion

In this Ph.D. project, we study the evolution of different kinds of property graphs over time. As our first result, we proposed a temporal model (called TGraph) and a high-level query language (called T-GQL) to account for changes on nodes and edges in property graphs. Temporal paths are first-class citizens in this model. Although TGraph is applicable to, for example, social networks, it cannot capture transportation networks (i.e., networks were some element flows from one node to another) equipped with sensors which produce time-series data measuring the evolution of different variables over time. Thus, we proposed SN-TGraph, a model that extends TGraph. In SN-TGraph, new kinds of temporal paths can be defined in terms of the variables measured by the sensors. We are currently studying the impact and applicability of these paths to real-world cases, like river systems and road networks, two typical cases of transportation networks. In addition to the above, since computing temporal paths is costly, we are studying different ways of indexing such paths.

Acknowledgements. Valeria Soliani was partially supported by Project PICT 2017-1054, from the Argentinian Scientific Agency.

References

1. Akyildiz, I.F., Su, W., Sankarasubramaniam, Y., Cayirci, E.: A survey on sensor networks. IEEE Commun. Mag. **40**(8), 102–114 (2002)
2. Allen, J.F.: Maintaining knowledge about temporal intervals. Commun. ACM **26**(11), 832–843 (1983)
3. Angles, R., Thakkar, H., Tomaszuk, D.: RDF and property graphs interoperability: status and issues. In: Hogan, A., Milo, T. (eds.) Proceedings of AMW, Asunción, Paraguay, 3–7 June 2019. CEUR Workshop Proceedings, vol. 2369. CEUR-WS.org (2019)
4. Bollen, E., Hendrix, R., Kuijpers, B., Vaisman, A.A.: Time-series-based queries on stable transportation networks equipped with sensors. ISPRS Int. J. Geo Inf. **10**(8), 531 (2021)
5. Byun, J., Woo, S., Kim, D.: Chronograph: enabling temporal graph traversals for efficient information diffusion analysis over time. IEEE Trans. Knowl. Data Eng. **32**(3), 424–437 (2020)
6. Campos, A., Mozzino, J., Vaisman, A.A.: Towards temporal graph databases. In: Pichler, R., da Silva, A.S. (eds.) Proceedings of AMW, Panama City, Panama, 8–10 May 2016. CEUR Workshop Proceedings, vol. 1644. CEUR-WS.org (2016). http://ceur-ws.org/Vol-1644/paper40.pdf
7. Debrouvier, A., Parodi, E., Perazzo, M., Soliani, V., Vaisman, A.: A model and query language for temporal graph databases. VLDB J. **30**(5), 825–858 (2021). https://doi.org/10.1007/s00778-021-00675-4
8. Elmasri, R., Wuu, G.T.J., Kim, Y.: The time index: an access structure for temporal data. In: Proceedings of VLDB 1990, August 13–16, 1990, Brisbane, Queensland, Australia, pp. 1–12. Morgan Kaufmann (1990)
9. Francis, N., et al.: Formal semantics of the language cypher. CoRR abs/1802.09984 (2018)
10. Kvet, M., Matiasko, K.: Impact of index structures on temporal database performance. In: EMS 2016, Pisa, Italy, 28–30 November 2016, pp. 3–9. IEEE (2016)
11. Pokorný, J., Valenta, M., Troup, M.: Indexing patterns in graph databases. In: DATA 2018, Porto, Portugal, July 26–28, 2018, pp. 313–321. SciTePress (2018)
12. Rizzolo, F., Vaisman, A.A.: Temporal XML: modeling, indexing, and query processing. VLDB J. **17**(5), 1179–1212 (2008)
13. Wu, H., Cheng, J., Huang, S., Ke, Y., Lu, Y., Xu, Y.: Path problems in temporal graphs. Proc. VLDB Endow. **7**(9), 721–732 (2014)
14. Wu, H., et al.: Core decomposition in large temporal graphs. In: 2015 IEEE International Conference on Big Data, Big Data 2015, Santa Clara, CA, USA, October 29–November 1 2015, pp. 649–658 (2015)

Termination and Confluence
of an Extended CHASE Algorithm

Andreas Görres[✉]

Computer Science Department, University of Rostock, 18051 Rostock, Germany
`andreas.goerres@uni-rostock.de`

Abstract. Many requirements of systems managing and analyzing large volumes of data are interconnected and should therefore be realized together. In this research project, we use a fundamental algorithm of database theory – the CHASE – to address those requirements in a unified manner. While highly expressive, the language of the CHASE is still inadequate to formulate many problems of practical importance. Extending the CHASE with additional features would increase its range of applications, but might jeopardize its key features confluence, safe termination and efficiency.

In this work, we demonstrate that calculating basic linear algebra operations with the CHASE is feasible after extending the algorithm with negation and a restricted set of scalar functions. We discuss how confluence and termination of the CHASE are influenced by these extensions.

Keywords: CHASE · Data science pipeline · Termination · Confluence · Efficieny · Relational algebra · Linear algebra · Negation

1 Introduction and Motivation

While processing large amounts of data is a traditional part of database management, big data analytics becomes more and more ubiquitous. Especially applications like the Internet of Things generate large amounts of information, opening up new areas of application for classic database algorithms, e.g., analyzing big data by extended database languages, such as linear algebra extensions to relational algebra.

Our research is mainly based on two application areas: Research data management and smart assistive systems. Use cases for research data management result from our long-time cooperation with the Institute for Baltic Sea Research (IOW). Resulting from this research, we gain insights into the (e.g. statistical) data analysis operations used in real world applications. Furthermore, the historical development of the used database is base for our research on schema evolution and on database transformation operations. Basically, published research

Supervised by Prof. Andreas Heuer.

data referring to a historical database schema should be reproducible even in the future. To allow the reproduction of results, the involved tuples need to be saved, not the entire database of raw measurement data. Identifying those tuples motivates our research on provenance. However, simply publishing the entire used data set is not always possible. Especially medical data might contain personal information, resulting in the problem to guarantee privacy while still allowing the reproduction of results. This problem is even more pronounced in our second application area, smart assistive systems. The behavior of a user, e.g. a patient, is observed by a large amount of sensor data, so there current activities and intentions can be predicted. If parts of the analysis can be done on an early level of computation, in a completely distributed manner near the sensors or even by the sensors themselves, collecting personal information on a central server is no longer needed. This not only results in more privacy (privacy by design), but even in a more efficient, parallel evaluation of data. As before, our need for privacy counteracts our interest in reproducibility. However, tracing back results to the original data allows us to focus on a small amount of sensors having an actually significant influence on the calculation. This way, we could not only increase efficiency by disregarding unneeded data, but even realize privacy. Ideally, our three main objectives provenance, privacy and efficiency exhibit synergistic effects if we include them into a unified framework.

In our research, we try to realize those features using a fundamental algorithm of database theory: the CHASE.

2   Formalizing Problems in Terms of the CHASE

More than forty years ago, the CHASE algorithm was introduced in two seminal works [3,10]. Right from the start, seemingly unrelated use cases – query optimization and schema construction – were examined in a unified way . Later on, the number of application areas (for slightly adapted variants of the CHASE) became even broader and the concept of "universal solution" was introduced to describe the connection between the different problem cases [8]. In this tradition, we understand the CHASE as a universal algorithm that is able to process a variety of parameters (e.g. integrity constraints, privacy constraints, and view definitions) into a variety of database objects (e.g. database instances or database queries).

Formally, let P be a parameter and O be an object, then the $\text{CHASE}_P(O)$ applies the parameter P to the database object O. If P is a query and O a relational database, the result of the CHASE is the result relation of the query P. If P are integrity constraints and O is a database query, the result of the CHASE is a query implicitly satisfying all the constraints, which is needed, e.g., to apply semantic query optimization techniques. We refer to the result of the CHASE as target $T = \text{CHASE}_P(O)$.

For our research in research data management and smart assistive systems, we need to combine the following steps of a data science pipeline: (1) Designing and evolving database schemas; (2) Analyzing data by means of a relational database language, extended by linear algebra operations; (3) Supporting reproducibility of the data analysis by data provenance techniques such as why-provenance; and (4) Guaranteeing privacy (as well as integrity) constraints within the analysis and reproducibility steps. To be able to unify and combine these four tasks, we use the CHASE algorithm in the following ways.

- To describe the evolution of database schemes, we interpret P as a schema mapping and O as the source database schema, resulting in the target database schema T after the evolution.
- To perform an extended query as a representation of a data analysis, we interpret P as a query and O as the relational database, resulting in the target relation, i.e. the query result T.
- To calculate the why-provenance of the query used for the data analysis, we interpret the query result T as the new database object O and *invert* the query operations to represent the provenance query, i.e. the new parameter P. The new result T of this provenance query is a sub-database of the original database, the set of witnesses for the data analysis.
- To satisfy all the constraints, such as integrity or privacy constraints, while performing the data analysis by an extended relational query, we use the extended query as the database object O and the constraints as the CHASE parameter P. The result of the CHASE is a query respecting all the constraints in P.

The CHASE can be interpreted as a rule system, applying different rules in P to the logical representation of O. Since P is a set of rules, there is no fixed order in which the rules are applied to the object O. So we have the basic problems of **termination** (does the application of rules stop?) and **confluence** (if we apply the rules in a different order, is the algorithm always calculating the same result?). As a third problem, we have to check the **efficiency** of the CHASE, since in general the CHASE is often of higher complexity than specialized solutions for the above problems considered in isolation.

Even though, the classical CHASE is too restricted for most practical applications and therefore needs to be extended. In the classical CHASE, P and O are simple formulas of first-order predicate logic. For example, if P or O are used as database queries, the queries are restricted to conjunctions of positive select-project-join queries, with equality tests being the only permitted select operation.

3   Main Tasks of the PhD Thesis

For our PhD project, we define two main tasks: Unification and extension of the CHASE. As we described before, parameter and object of the CHASE depend on the application area. Even though the same parameters (e.g. integrity constraints) could be used for different CHASE objects, the actual behavior of the algorithm differs. For example, the same constraint might lead to the creation of an existentially quantified variables in a query and to the introduction of a new null value in a database instance. Most implementations of the CHASE algorithm are tailored towards a certain use case, for instance query optimization or database repair. Even in this paper, we focus on the CHASE on database instances. Ultimately, the different variants of the CHASE need to be unified, resulting in a truly universal algorithm.

To address real life use cases in an effective manner, the CHASE needs to be extended, for example with negation, functions and arithmetic comparisons. However, extending the CHASE might influence its termination behavior, confluence and efficiency. Concerns regarding those properties are not restricted to the extended CHASE. Originally, CHASE parameters were restricted to functional dependencies and full inclusion dependencies, which ensured a finite and unique CHASE result. Once embedded dependencies were introduced to the CHASE, termination and confluence were no longer guaranteed. Still, those problems of the original algorithm can be addressed, for example by testing for termination beforehand [5]. There were attempts to extend the CHASE in the past (Sect. 4). However, those extensions are usually tailored towards a specific use case and not the universal CHASE. The semantics of most CHASE extensions depends on the CHASE object: Negation in a CHASE parameter refers to tuples absent from a CHASE instance, but to explicitly negated atoms in a query. Furthermore, it is unclear if the results of a specialized extended CHASE (e.g. annotations encoding conditions for certain tuples) can be interpreted by the next step of data processing.

As described in Sect. 2, the second step of the data science pipeline considers linear algebra operations as an extension of relational database languages. At the end of this paper, we present a simple CHASE program calculating matrix addition. Using this simple example, we explain the procedure of the CHASE and demonstrate the necessity for two CHASE extensions, negation and scalar functions. Furthermore, we examine why the presented program is confluent and terminates. In future works, we plan to evaluate practical effectiveness of the described extended universal CHASE based on a prototypical implementation.

4   Related Work

Previous efforts to extend the CHASE focused on arithmetic comparisons and negation. Along the way, complexity and termination behavior of these CHASE variants were examined. In contrast, arithmetic functions seem to be of lesser interest for current CHASE research. In [2], a CHASE variant extended with arithmetic comparisons is used to describe data exchange. This AC-CHASE tree considers all possible orders of null values, therefore generating an exponential number of possible results. In a similar manner, [5] uses the C-CHASE tree to answer queries with negation on ontologies. Recently, [7] simulated the disjunctive CHASE with the non-disjunctive variant of the algorithm. This way, disjunctive properties of arithmetic comparisons and negation can be described using the standard CHASE. However, this solution is not guaranteed to terminate in polynomial time and is therefore not necessarily more efficient than the tree-like CHASE variants. In fact, if we restrict the AC-CHASE to certain combinations of arithmetic comparisons, we preserve the so called *homomorphism property*, thereby achieving polynomial data complexity of the terminating CHASE. This result extends to other application areas of the CHASE, like query optimization [1].

5   The CHASE Algorithm

The CHASE is a fixpoint algorithm incorporating CHASE parameters into objects, so that the resulting objects implicitly contain the parameters. CHASE parameters are expressed as logical implications. For equality generating dependencies (EGDs) and tuple generating dependencies (TGDs), a general algorithm can be found in [6]. In this work, we focus on TGDs:

$$TGD : \phi(\boldsymbol{x}, \boldsymbol{y}) \rightarrow \exists \boldsymbol{Z} : \psi(\boldsymbol{x}, \boldsymbol{Z})$$
$$EGD : \phi(\boldsymbol{x}) \rightarrow x_1 = x_2; x_1, x_2 \in \boldsymbol{x}.$$

In the next subsections, we extend the general schema of a TGD with negation and simple scalar functions. Similar to [5] and unlike [8], we do not restrict negation to single atoms. CHASE objects, which include database instances and queries, are encoded as sets of relational atoms.

The CHASE consists of a sequence of CHASE steps. In each step, a homomorphism (the trigger) between all atoms of a (nondeterministically chosen) CHASE parameter's body and some atoms of the CHASE object is defined. If the image of the TGD head atoms (possibly extended for existentially quantified variables) is not present in the CHASE object, an image of the head atoms is materialized in the CHASE object. For each existentially quantified variable, a fresh marked null value (or existentially quantified variable) is generated. The CHASE continues until a fixpoint is reached. If the TGDs are cyclic and existentially quantified variables are present in any head atom, the CHASE might not terminate. The problem of CHASE termination is, in general, undecidable, but well researched [5].

5.1   Introduction to CHASE Extensions Using Matrix Addition as an Example

Basic linear algebra operations are fundamental for machine learning algorithms used in big data analytics. In the following, we show how matrix addition can be defined using extended TGDs, a necessary requirement for reasoning about the algorithm with the CHASE. Let us consider matrices A and B, encoded in relations $A(I, J, V)$ and $B(I, J, V)$. Coordinates of a matrix field are encoded in attributes I and J and the field value in attribute V. Matrix AB encodes the sum of the matrices A and B.

$$\begin{array}{ccc} A & B & AB \\ \begin{pmatrix} 1 & 0 \\ 0 & 1 \end{pmatrix} & \begin{pmatrix} 1 & 0 \\ 1 & 0 \end{pmatrix} & \begin{pmatrix} 2 & 0 \\ 1 & 1 \end{pmatrix} \end{array}$$

To illustrate the basic working of an extended CHASE algorithm, we define a simple program calculating matrix addition:

$$r_1 : A(i_1, j_1, v_1), B(i_1, j_1, v_2) \rightarrow AB(i_1, j_1, sum(v_1, v_2))$$
$$r_2 : A(i_1, j_1, v_1), \neg(\exists V_2 : B(i_1, j_1, V_2)) \rightarrow AB(i_1, j_1, v_1)$$
$$r_3 : \neg(\exists V_1 : A(i_1, j_1, V_1)), B(i_1, j_1, v_2) \rightarrow AB(i_1, j_1, v_2).$$

The only new values created by the standard CHASE are marked null values (and variables). In addition to this, we can generate new constants if we allow scalar function terms, like $sum()$, in the TGD head.

We define a homomorphism from the body of r_1 to the matrix fields with the coordinates $(1, 1)$ in both matrices. Consequently, the CHASE generates a tuple $AB(1, 1, 2)$, the image of r_1's head. Notice that we immediately calculated the result of the scalar function instead of generating a nested term in AB. Of course, by saving the nested function term in relation AB, we might be able to optimize the term later on using arithmetic transformations, but the CHASE is ill-equipped for this kind of optimization.

While r_1 is sufficient to calculate the result of matrix addition if all matrix fields are represented by tuples in the database, this is not the case if we use the compressed database representation of matrices defined in [11], which allows a more efficient treatment of sparsely populated matrices. Here, fields with value zero are represented by missing relational tuples. Therefore, the tuple $B(2, 2, 0)$ is absent from this representation and we are unable to find a homomorphism for the coordinates $(2, 2)$. Consequently, no tuple with those coordinates is generated in AB, even though we expect this field to have a value of one. In [11], this problem is addressed using an outer join. We adapt this approach with additional TGDs r_2 and r_3 containing negation.

After defining the mapping $\{i_1 \mapsto 2, j_1 \mapsto 2, V_1 \mapsto 1\}$ for the positive atoms of r_2, we rewrite the negative atom into the following boolean subquery $\exists V_2 : B(2, 2, V_2) \rightarrow ()$ by substituting the variables i_1 and j_1 with their respective mappings. A boolean query is basically a TGD whose head consists of an empty tuple. Since we are unable to define a consistent homomorphism for the boolean query, we are unable to generate this tuple, which is interpreted as "false". Consequently, we proceed by materializing the image of the head atom, $AB(2, 2, 1)$. Notice that the body of a boolean query might have multiple atoms and can contain existentially quantified variables (in this example V_2). Variables in negated atoms not present in any positive atoms are known as *unsafe*. In this example, we could avoid unsafe negation by defining a view that projects over attributes containing the safe variables. However, by interpreting negation as a negated subquery (with explicit quantification of unsafe variables) we show that multiatomic unsafe negation does not pose a challenge to the CHASE on database instances.

5.2   Confluence and Termination of the Example

While negation is often (even in the previous example) restricted to single atoms (negative subgoals) and safe variables, exceeding those limitations is surprisingly natural for the extended CHASE. In fact, the CHASE algorithm already utilizes this kind of generalized negation when testing trigger activity. After finding a valid homomorphism for the body atoms of a TGD (the trigger), we test if the image of the head atoms is already present in the CHASE object. This active trigger test can be interpreted as an implicit multiatomic negation, treating the TGD head as a negated conjunction of body atoms. For this, existentially quantified variables from the TGD head are implicitly renamed and act as unsafe variables. While the CHASE on full TGDs is confluent, the general CHASE algorithm is not. One main reason for this behavior is the previously described test for trigger activity. Since this test can be expressed using negation, it is unsurprising that general negation leads to additional cases of non-confluence. If a TGD is blocked by tuples a second TGD generates, the order in which both TGDs are applied directly influences the result. This remains valid even if we restrict ourselves to stratified negation (that is, there are no circular dependencies between TGDs containing negation). Stratified negation simply guarantees there is an order of rule application in which blocking TGDs are applied before the respective blocked TGDs. The presented CHASE program, however, is confluent since it is "semi-positive": Only tuples are negated whose relations never appear in any TGD head.

The main purpose of scalar functions in the given example is the generation of new constants. In this regard, they have similar effects as existentially quantified variables in head atoms, which also contribute to the creation of new values (null values or variables). Similar to existentially quantified variables, scalar functions might lead to a non-terminating CHASE sequence. However, adjustment of classical termination tests, like *Weak Acyclicity* [9], would still guarantee CHASE termination. Being non-recursive, the given CHASE program is weakly-acyclic and guaranteed to terminate. However, there is a major difference between null values and the constants generated by a scalar function: Only by applying an explicitly defined EGD, we can unify two different null values. This way, the application of an EGD can terminate a CHASE sequence that might otherwise be infinite. It might also unify the results of two alternative CHASE sequences, thereby guaranteeing confluence. For scalar functions, this unification is not defined explicitly by the CHASE parameter, but by arithmetic rules instead (e.g. the commutativity of addition). Furthermore, two attribute values might not even be identical, but converge to the same constant in the progress of an infinite CHASE sequence.

6 Conclusions and Future Work

In this work, we extended the CHASE on database instances with negation and scalar functions, exemplified by a simple CHASE program. In a similar manner, we have defined more complex, but still confluent and terminating programs. Since these programs are unions of (extended) TGDs, CHASE techniques used for optimizing unions of conjunctive queries could be used to optimize them. However, while unsafe negation was smoothly integrated into the CHASE on instances, it poses a serious challenge to the CHASE on queries. In future works, we will describe our solution to these problems. Furthermore, we intend to evaluate this solution by implementing it into our prototypical CHASE software ChaTEAU [4].

The CHASE algorithm of classic database theory can be applied to a multitude of problem cases, solving them in a unified manner. In this regard, interactions between the requirements provenance, privacy and efficiency are of particular interest to us. However, for practice-oriented use cases, extensions of the algorithm are needed. These extensions might affect efficiency, confluence and termination of the algorithm in a negative way.

In this work, we illustrated the prospects of an extended CHASE algorithm with a simple framework calculating linear algebra operations. As a next step of our research, we will examine how CHASE programs can be modified using the CHASE algorithm, thereby contributing to the solution of the initially formulated requirements.

Acknowledgements. This work was supported by a scholarship of the Landesgraduiertenförderung Mecklenburg-Vorpommern.

References

1. Afrati, F.N.: The homomorphism property in query containment and data integration. In: IDEAS, pp. 2:1–2:12. ACM (2019)
2. Afrati, F.N., Li, C., Pavlaki, V.: Data exchange in the presence of arithmetic comparisons. In: EDBT. In: ACM International Conference Proceeding Series, vol. 261, pp. 487–498. ACM (2008)
3. Aho, A.V., Sagiv, Y., Ullman, J.D.: Efficient optimization of a class of relational expressions. ACM Trans. Database Syst. 4(4), 435–454 (1979)
4. Auge, T., Heuer, A.: ProSA—using the CHASE for provenance management. In: Welzer, T., Eder, J., Podgorelec, V., Kamišalić Latifić, A. (eds.) ADBIS 2019. LNCS, vol. 11695, pp. 357–372. Springer, Cham (2019). https://doi.org/10.1007/978-3-030-28730-6_22
5. Baget, J., Garreau, F., Mugnier, M., Rocher, S.: Revisiting chase termination for existential rules and their extension to nonmonotonic negation. CoRR abs/1405.1071 (2014)
6. Benedikt, M., et al.: Benchmarking the chase. In: PODS, pp. 37–52. ACM (2017)
7. Bourgaux, C., Carral, D., Krötzsch, M., Rudolph, S., Thomazo, M.: Capturing homomorphism-closed decidable queries with existential rules. In: KR, pp. 141–150 (2021)
8. Deutsch, A., Nash, A., Remmel, J.B.: The chase revisited. In: PODS, pp. 149–158. ACM (2008)
9. Fagin, R., Kolaitis, P.G., Miller, R.J., Popa, L.: Data exchange: semantics and query answering. Theor. Comput. Sci. 336(1), 89–124 (2005)
10. Maier, D., Mendelzon, A.O., Sagiv, Y.: Testing implications of data dependencies. ACM Trans. Database Syst. 4(4), 455–469 (1979)
11. Marten, D., Meyer, H., Dietrich, D., Heuer, A.: Sparse and dense linear algebra for machine learning on parallel-RDBMS using SQL. OJBD 5(1), 1–34 (2019)

Towards Prescriptive Analyses
of Querying Large Knowledge Graphs

Mohamed Ragab[(✉)]

Institute of Computer Science, Tartu University, Tartu, Estonia
mohamed.ragab@ut.ee

Abstract. Leveraging relational Big Data (BD) processing frameworks
to process large-scale (RDF) graphs yields a great interest in optimizing
query performance. Modern BD systems are yet complicated data systems, where the configurations notably affect the performance. Benchmarking different frameworks and configurations provides the community with best practices for better performance. However, most of these
benchmarking efforts are classified as *descriptive* and *diagnostic* analytics. Moreover, there is no standard for comparing these benchmarks
based on *quantitative* ranking techniques. In this paper, we discuss
how our work fills this timely research gap. Particularly, we investigate how to enable *prescriptive* analytics via ranking functions (called
"*BenchRank*"). We present a research plan that builds on the state-of-
the-art benchmarking efforts in the area of querying large RDF graphs.
Finally, we present our research results of the proposed plan.

Keywords: Graphs querying · RDF · Benchmarking · Big data

1 Introduction

The increasing adoption of Knowledge Graphs (KGs) in industry and academia
requires scalable systems for taming linked data large volumes and velocity [6,15]. In absence of a scalable native system for *querying* large (knowledge)
graphs [12], most approaches fall back to using relational storage engines and Big
Data (BD) frameworks (e.g., *Apache Spark* or *Impala*) for handling large graph
query workloads [13,14]. Despite its flexibility, the relational model require several additional *design decisions* when used for representing graphs, which cannot
be decided automatically, e.g., the choice of the *schema*, the *partitioning technique*, and the *storage formats*.

BD System performance analysis is already *time-consuming* even for simple
workloads because those systems expose hundreds of configuration parameters
for tuning. The problem aggravates when the processing tasks become complex,
like in the case of graphs [12], because the additional design decisions entail performance *trade-offs*. Thus, the results are often *situational*, i.e., there is rarely
an absolute winner [8]. The analysis limits emerge from existing works that

M. Ragab—Supervised by Riccardo Tommasini, LIRIS Lab, INSA Lyon, France.

S. Chiusano et al. (Eds.): ADBIS 2022, CCIS 1652, pp. 639–647, 2022.
https://doi.org/10.1007/978-3-031-15743-1_59

query large RDF graphs using BD frameworks. Indeed, the result discussions are merely *descriptive* or at most *diagnostic*, i.e., they answer the question *what happened?* or at most *why did it happen?*. Such analyses require additional work from the data engineers to make a decision, especially with a complex experimental solution space that hides *unknown trade-offs* [9].

Conversely, we advocate that the expected answer from performance experimentation is *what should we do?* [7]. Thus, *prescriptive* performance analysis (PPA) is essential for complex analytical task like querying large graphs. PPA reduces the need for human intervention even further by making the insights actionable by relying on *statistical* and *mathematical* models.

In this paper, we investigate the problem of enabling prescriptive analytics in the context of BD systems that query large RDF graphs. This kind of analysis aims to guide the practitioner directly to actionable decisions, navigating complex solution spaces without ignoring the underlying experimental dimensions.

2   State of the Art

This section presents the state of the art on querying large (RDF) graphs, highlighting the *(C)*hallenges related to our problem statement.

The literature focuses on optimizing the performance of the systems [1,13,14]. However, different experimental dimensions that affect their performance are not systematically studied (*C1*) nor compared (*C2*), as each work focuses only on *one* dimension at a time, e.g., schema [3], partitioning [2], or storage [5]. Focusing *solely* on one experimental dimension neglects the presence of other dimensions' trade-offs. Thus, the proposed optimizations cannot be generalized in presence of new experimental dimensions (*C3*). For instance, changing data partitioning or storage formats affects the performance of schema advancements [8,11].

Finally, there is a gap in the performance analysis for querying RDF graphs (*C4*). Prior works stop their analyses at describing the performance results of engines around a specific experimental dimension. For instance, *Shätzle et al.* [13,14] described the performance of big relational systems. They explain *why* their systems outperform others RDF systems in terms of schema optimizations. Similarly *Abdelaziz et al.* [1] conducted a comprehensive *descriptive* and *diagnostic* survey of several RDF processing systems, describing their performance in terms of *scalability*, *query efficiency*, and *workload adaptability* metrics.

However, these efforts cannot guide the practitioner to make informed decisions. They lack the prescriptions on what should be selected from a variety of experimental dimensions' options.

3   Methodology

This work uses the *Macro, Mezzo, Micro* framework for framing the research problem, which requires to formulate the research questions at *three* levels of analysis: the *Macro* question is broad and helps capturing the research context; the *Mezzo* question restricts the scope posing some requirements. Finally, the *Micro* question poses a problem that can be evaluated.

"Can we guarantee fair benchmarking assessment of relational BD systems performance while querying big graphs?" is the *Macro* question that frames our research. The question allows multiple answers and thus, we shall reformulate it into more specific ones. Thus, we list some *requirements* which help us to narrow down the problem and validate our approach.

R1 **Focus on querying large (RDF) graphs.**
R2 **Derive actionable insights from BD systems' performance analysis.**
R3 **Consider multiple experimental dimensions'** *simultaneously* **to make sense of their** *trade-offs.***;**
R4 **Ensure the** *replicability* **of BD systems' performance while varying the experimental dimensions.**

Based on the requirements above, we can formulate a *Mezzo* question that incorporates them: *"Can we aid decision making for benchmarking big (RDF) graph processing over relational systems?"*. The goal is to guide the practitioners to select the *best-performing* configurations for an experiment out of a complex solution space of dimensions that have inherent trade-offs.

More specifically, we consider the use case of large RDF graphs querying. Several design decisions are relevant for such a use case. We opted for the most common and well-studied ones that directly impact the performance [7,9], i.e., the *schemas*, the *partitioning techniques*, and the *storage formats*[1]. Notably, these experimental dimensions are not *system-specific*, i.e., can be tested in other relational BD systems. As relational BD frameworks, we opt for *Apache Spark* which is currently the most active and widely-used large-scale data processing system in both industry and academia. In particular, *Spark-SQL* offers a prominent

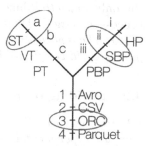

Fig. 1. Experimental solution space.

relational interface for implementing the relational schemas and answering translated SPARQL queries (i.e., into SQL). Moreover, Spark-SQL supports different partitioning techniques and multiple storage formats.

Finally, we formalize specific questions at the *Micro* level in the following sections.

4 Proposed Approach

In this section, we explain how we (plan to) answer the research questions. To begin, we explain the preliminaries of our setup.

Definition 1. *A configuration C is a combination composed of parameters/options of different experimental dimensions, i.e.,*

$$\{Schema\}.\{Partitioning\_Technique\}.\{Storage\_Format\}$$

---

[1] The relational schema impacts query joins, partitioning techniques impact data shuffling, whilst storage formats impact physical execution plans.

We define our experiments in terms of (i) a set of queries, (ii) an RDF dataset (size), and (iii) a configuration combination. Figure 1 shows the experimental design space and highlights the example of (*a.ii.3*) configuration, which is akin to Single Triples (*ST*) schema, Subject-based Partitioning (*SBP*) technique, and *ORC* file format [9]. We run the experiments using two accepted RDF benchmarks (i.e., *WatDiv* and *SP²Bench*) that include data generators and sufficiently complex workloads. In total, our experiments include 36 configuration combinations (serializing all the combinations of dimension options[2] in Fig. 1). Moreover, we consider the scalability dimension, and we conducted our experiments with different dataset sizes (i.e., $100M$, $250M$, and $500M$ triples).

4.1   Assessing the Replicability (Micro1 (μ1))

μ1: *can we replicate BD systems performance, introducing other new experimental dimensions?*

μ1 investigates the effect of changing one experimental dimension on the performance of a system. For instance, we check if the relational schema is not the only *impactful* dimension for the performance of relational BD systems for processing large RDF graphs. Thus, we investigate the following hypothesis: **HP0**: *The replicability of the BD system's performance for querying large (RDF) graphs could be affected by introducing other experimental dimensions.*

4.2   Deciding over Complex Solution Space (Micro2 (μ2))

μ2: *How can we efficiently select the best-performing configurations out of the complex experimental solution space of dimensions that emerge with querying large RDF graphs over big relational systems?*

μ2 investigates the level of abstraction required for selecting the best experimental configurations, instead of comparing a huge number of experiments' results (i.e., that sometimes are even *contradicting* due to experimental trade-offs). To guide the practitioners on this hard task, we aim to employ *ranking functions* seeking actionable prescriptions. Indeed, ranking functions show effective roles in decision support in several application domains. These ranking criteria aim to abstract from the *fine-grained* descriptive performance metrics and enable a *decision-making* model. Hence, we adopt *Single-Dimensional* (SD) as well as *Multi-Dimensional* (MD) ranking criteria for ranking the performance of dimensions' parameters. Moreover, we discuss metrics for assessing the *goodness* of the proposed ranking criteria [7].

4.3   Employing the Query Workload (Micro3 (μ3))

μ3: *Can we provide better prescriptions by incorporating the query workload?*

---

[2] We omit details about schema options (ST, VP, PT) and partitioning options (HP, SBP, PBP) due to space limits, however, still can be found in the project's GitHub page: https://datasystemsgrouput.github.io/SPARKSQLRDFBenchmarking/.

Table 1. Effect of other partitioning techniques, and storage formats on the *replicability* of the schema advancements (i.e., WPT, ExtVP) VS. baseline schemas (i.e., PT, VP).

	Technique	ExtVP VS.VP	WPT VS. PT		Format	ExtVP VS. VP	WPT VS. PT
Partitioning	V. HDFS	67.50%	58.33%	Storage	Parquet	55%	77.78%
	Horizontal	35%	47.22%		ORC	45%	74.07%
	Predicate	55%	NA		Avro	42.50%	22.22%
	Subject	30%	44.44%		CSV	42.50%	25.93%

$\mu3$ investigates how to employ the query workload while making sense of the performance analysis for providing actionable insights.

On the same note, the workload impacts the relational schema which is critical for the efficiency of the analysis [7,8]. In practice, each RDF relational schema is excellent for a particular query shape, and there is no *one-fits-all* schema for all query families [7,8,10]. Intuitively, combining multiple schemas to obtain the best of them represents a valuable research direction. Nevertheless, hybrid schema solutions require huge data engineering efforts.

To this end, we aim to provide *prescriptions* for a hybrid relational schema that dynamically adapts with the input query workload. In particular, we plan to employ the query workload to *automate* the process of eliciting a well-performing schema by tackling the issues of existing RDF relational schemas (e.g., *sparsity* and *redundancy*).

5   Results

This section discusses how we experimentally answered the questions in the research methodology according to our research plan. Notably, the results on the $\mu3$ question are still in progress.

5.1   Replicability Results

To answer the $\mu1$ question, we check the validity of the hypothesis in Sect. 4.1. Particularly, we investigate the performance improvement of *Spark-SQL* with two recent schema optimizations (i.e., *Extended Vertically-Partitioned Tables* (ExtVP) [14] and *Wide Property Tables* [13]), w.r.t. their baseline approaches (i.e., *Vertically-Partitioned (VP) Tables* and *Property Tables* (PT)). We observe if the performance of the two schemas advancements generalizes (i.e., still outperform the baseline ones) over Spark-SQL with introducing different RDF partitioning techniques and various *HDFS* storage data formats that are different from the studied *vanilla* configurations (i.e., *Vanilla HDFS* partitioning, and *Parquet* as storage format) [8,13,14]. Table 1 shows that the RDF relational schema optimizations outperform the baseline ones (table percentages) [13,14] *only* with the *vanilla* configurations (marked in green). However, this performance behavior *does not generalize* (i.e., shown in lower percentages) while introducing other new partitioning techniques (e.g., *Horizontal, Subject(Predicate)-based* partitioning), or other storage file formats (e.g., *Avro, CSV,* or *ORC*) [8].

5.2   Results of *BenchRank* Criteria

To answer the $\mu 2$ question we introduce ranking criteria for prescriptive analysis. Nonetheless, we first show the limitations of descriptive and diagnostic analyses [7,9]. Herein, Table 2 shows the *best-performing* configurations for each query and each dataset size of the SP²B datasets. The experiment results show *no decisive* configuration setting over the assessed dimensions (red configuration parameters), making the practitioner's setup selection hard, and showing the limitations of descriptive analysis. Diagnostic analysis can help understand *why* a specific dimension is outperforming. Nevertheless, due to the inherent experimental *trade-offs*, contradictions still hinder clear decisions at this level of analysis.

Table 2. Best configurations of SP²B queries.

Query	100M	250M	500M
Q1	c.i.2	c.ii.2	c.ii.2
Q2	b.ii.3	b.ii.4	b.iii.4
Q3	c.ii.3	c.ii.4	c.ii.3
Q4	a.ii.3	c.ii.1	a.ii.3
Q5	b.ii.4	b.ii.4	b.iii.3
Q6	c.ii.3	c.ii.3	c.ii.3
Q7	b.ii.1	b.ii.4	b.ii.4
Q8	c.iii.4	c.iii.4	c.iii.4
Q9	b.ii.4	b.iii.3	b.iii.4
Q10	b.iii.3	b.iii.3	b.iii.3
Q11	b.i.3	b.ii.4	b.i.3

To tackle these limitations, we seek actionable indicators via employing ranking techniques (of these dimensions) that abstract out from fine-grained performance observations and lead to actual decision making.

SD Ranking Criteria. We first generalize the ranking function proposed in [2] that ranks several RDF partitioning techniques. Thus, we attain *three* ranking criteria (one for each experimental dimension i.e., schema, partitioning, and storage). The generalized ranking function calculates *ranking scores* for the experimental dimensions' options according to the following formula:

$$R = \sum_{r=1}^{d} \frac{O_{dim}(r) * (d - r)}{|Q| * (d - 1)}, 0 < R \le 1 \tag{1}$$

R is the *rank score* of the experimental dimension (i.e., relational schema, partitioning technique, or storage backend). Such that, d represents the total number of parameters (options) under that dimension, $O_{dim}(r)$ denotes the number of occurrences of the dimension being placed at the rank r (1st, 2nd,..), and $|Q|$ represents the total number of queries. Table 3 shows an example of computing the rank scores of the schema dimension options.

Table 3. Example of *Rank Scores*.

Schema	1st	2nd	3rd	R
ST	1	3	7	0.23
VT	6	4	1	0.73
PT	4	4	3	0.55

Rank scores of the SD ranking criteria help to provide a *high-level* view of the system performance across a set of tasks (e.g., workload queries) [7]. However, our experiments show that the SD ranking prescriptions are not *coherent* across experimental dimensions [7]. Indeed, the SD ranking criteria cannot generalize as they neglect the presence of *trade-offs* as they rank (i.e., optimize) alongside a single experimental dimension [7] neglecting the other ones.

MD Ranking Criteria. The limitations of SD criteria led to extending the *BenchRank* into a *Multi-objective Optimization* (MO) problem in order to optimize all the dimensions at the same time. We adopt the standard *Pareto frontier* MO technique to consider the experimental dimensions altogether [4]. It aims at finding a set of optimal solutions if no objective can be improved without sacrificing at least one other objective. In particular, we utilized the *Non-dominated Sorting Genetic Algorithm (NSGA-II)* [4] to find the best configuration combinations in our complex experimental solution space. In our case, the algorithm operates on the ranking scores of the SD ranking criteria (i.e., we

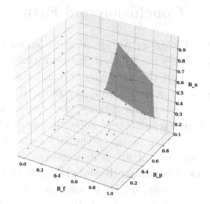

Fig. 2. Pareto-Fronts for experimental dimensions (i.e., schema R_s, partitioning R_p, and storage format R_f) of the SP^2B benchmark (500M triples). (Color figure online)

call them R_s, R_p, and R_f). The algorithm aims at maximizing the performance of the three dimensions rankings altogether. Figure 2 shows the Pareto fronts (optimal configurations, depicted by the green shaded area) of the three experimental dimensions of the BenchRank. Each point of this figure represents a *solution* of rank scores (i.e., configuration in our case).

BenchRank Criteria Goodness. In BenchRank, we consider a ranking criterion *"good"* if it does not suggest *low-performing* configurations. Herein, we discuss how we measure the ranking criteria *goodness* using *two* metrics. The first metric is the *Conformance* that measures the *adherence* of the *top-ranked* configurations w.r.t actual query rankings (i.e., positioning of those configurations[3]) [7]. The second metric is the *Coherence* which measures the level of agreement between two ranking sets that use the same ranking criterion across different experiments (e.g., different dataset scales). To measure how coherent the ranking criteria while scaling up to larger datasets, we employ *Kendall's index*[4]

Our experimental results[5] [7] show that all the ranking criteria show high *Coherence* across different scales of the datasets. Scaling the datasets does not excessively impact the rank sets' order in all the ranking criteria whereas, MD ranking criteria (i.e., Pareto fronts) show better *Conformance* results than the SD ranking criteria. Indeed, the MO Pareto ranking technique considers optimizing the performance of all dimensions simultaneously.

---

[3] Each configuration C has a rank according to its running time of the queries.
[4] Kendall's index is a common measure to compare the ordering of ranking functions.
[5] *Conformance* and *Coherence* results [7] are omitted due to space limits.

6   Conclusion and Future Work

In this paper, we discussed *how* and *why* it is important to provide prescriptive analyses for BD systems performance, focusing on ranking the complex solution space of the experimental dimensions that emerge with querying large (RDF) graphs over relational systems. Ranking criteria provide an accurate yet simple way that supports the practitioners in their evaluation task even in the existence of dimensions' *trade-offs*. In this work, we utilize SD ranking and MO techniques for making sense of the BD systems' performance analyses. In future work, we plan to (i) study additional ranking criteria and (ii) further investigate the metrics of ranking criteria *goodness*. Moreover, we plan to test the ranking *coherence* across different RDF benchmarks. Seeking fair comparisons even further, we plan to (a) uniform the challenges across benchmarks, including query complexity, or (b) go towards the level of *micro* benchmarking.

References

1. Abdelaziz, I., Harbi, R., Khayyat, Z., Kalnis, P.: A survey and experimental comparison of distributed SPARQL engines for very large RDF data. VLDB **10**(13), 2049–2060 (2017)
2. Akhter, A., Ngomo Ngonga, A.-C., Saleem, M.: An empirical evaluation of RDF graph partitioning techniques. In: Faron Zucker, C., Ghidini, C., Napoli, A., Toussaint, Y. (eds.) EKAW 2018. LNCS (LNAI), vol. 11313, pp. 3–18. Springer, Cham (2018). https://doi.org/10.1007/978-3-030-03667-6_1
3. Arrascue Ayala, V.A..: Relational schemata for distributed SPARQL query processing. In: SBD (2019)
4. Deb, K., Pratap, A., Agarwal, S.: A fast and elitist multiobjective genetic algorithm: NSGA-II. IEEE Trans. Evol. Comput. **6**(2), 182–197 (2002)
5. Ivanov, T., Pergolesi, M.: The impact of columnar file formats on SQL-on-hadoop engine performance: a study on ORC and parquet. Concurr. Comput. Pract. Exp. **32**(5), e5523 (2019)
6. Moaawad, M.R., Mokhtar, H.M.O., Al Feel, H.T.: On-the-fly academic linked data integration. In: Proceedings of the International Conference on Compute and Data Analysis, pp. 114–122 (2017)
7. Ragab, M., Awaysheh, F.M., Tommasini, R.: Bench-ranking: a first step towards prescriptive performance analyses for big data frameworks. In: IEEE Conference on Big Data (2021)
8. Ragab, M., Tommasini, R., et al.: An in-depth investigation of large-scale RDF relational schema optimizations using Spark-SQL. In: DOLAP@EDBT/ICDT (2021)
9. Ragab, M., Tommasini, R., Eyvazov, S., Sakr, S.: Towards making sense of Spark-SQL performance for processing vast distributed RDF datasets. In: SBD (2020)
10. Ragab, M., Tommasini, R., Sakr, S.: Benchmarking Spark-SQL under alliterative RDF relational storage backends. In: QuWeDa@ ISWC, pp. 67–82 (2019)
11. Ragab, M., Tommasini, R., Sakr, S.: Comparing schema advancements for distributed RDF querying using SparkSQL. In: ISWC 2020 Demos and Industry Tracks (2020)
12. Sakr, S., Bonifati, A., Voigt, H., et al.: The future is big graphs: a community view on graph processing systems. CACM **64**(9), 62–71 (2021)

13. Schätzle, A., Przyjaciel-Zablocki, M., Neu, A., Lausen, G.: Sempala: interactive SPARQL query processing on hadoop. In: Mika, P., et al. (eds.) ISWC 2014. LNCS, vol. 8796, pp. 164–179. Springer, Cham (2014). https://doi.org/10.1007/978-3-319-11964-9_11
14. Schätzle, A., Przyjaciel-Zablocki, M., Skilevic, S., Lausen, G.: S2RDF: RDF querying with SPARQL on spark. VLDB 9(10), 804–815 (2016)
15. Tommasini, R., Ragab, M., Falcetta, A., Valle, E.D., Sakr, S.: A first step towards a streaming linked data life-cycle. In: Pan, J.Z., et al. (eds.) ISWC 2020. LNCS, vol. 12507, pp. 634–650. Springer, Cham (2020). https://doi.org/10.1007/978-3-030-62466-8_39

Knowledge Extraction from Biological and Social Graphs

Mariella Bonomo$^{(\boxtimes)}$ (iD)

Department of Engineering, University of Palermo, Palermo, Italy
`mariella.bonomo@community.unipa.it`

Abstract. Many problems from the real life deal with the generation of enormous, varied, dynamic, and interconnected datasets coming from different and heterogeneous sources. This PhD Thesis focuses on the proposal of novel knowledge extraction techniques from graphs, mainly based on Big Data methodologies. Two application contexts are considered: Biological and Medical data, with the final aim of identifying biomarkers for diagnosis, treatment, prognosis, and prevention of diseases. Social data, for the optimization of advertising campaigns, the comparison of user profiles, and neighborhood analysis.

Keywords: Biological networks · Social networks · Big data

1 Introduction

Many problems of the real life can be modeled as graphs, able to take into account important relationships between interacting "actors". On the other hand, we are daily drowned in a very large amount of data, coming from different sources, that are complex in contents, heterogeneous in formats and order of Terabytes in size. These "big data" provide unprecedented opportunities to work on exciting problems, but also raise many new challenges for data mining and analysis. Indeed, most of the current analytical tools become obsolete as they fail to scale with data, especially when graphs seem to be the most suitable models to be adopted. Moreover, data are usually obtained from different information sources, and they need to be suitably integrated on the cloud. Therefore, performant technologies are required for data integration and data-intensive analysis, and algorithms need to be designed in order to be efficient and effective in this scenario.

This PhD Thesis is focused on the proposal of novel methodologies in the context of "big data" modeled as graphs in two main different application contexts:

- **Social Networks**, where users' data are often analyzed in order to learn more about their interests and connect them with content and advertising relevant to their preferences.

Supported by University of Palermo.

S. Chiusano et al. (Eds.): ADBIS 2022, CCIS 1652, pp. 648–656, 2022.
https://doi.org/10.1007/978-3-031-15743-1_60

- **Precision Medicine**, where an important source of big data is given from the biological high-throughput techniques, and the representation of interacting elements, such as cellular components, genotypic-phenotypic associations, etc., by suitable graphs is particularly relevant in order to take into account important information which would be missed by looking at each element singularly.

It is worth pointing our that, although the proposed methodologies are often general enough to be applied also in other application contexts, part of the contribution of this PhD project consists on providing satisfying solution which may be used in practice in the social and medical scenarios.

1.1 Social Networks

Automatic systems able to suggest a set of target users for advertising campaigns provide three main benefits:

1 Minimization of costs for the dissemination of the advertising campaign through social media, which is often very expensive;
2 Improvement of the user experience in OSNs, since only the possibly interested customers are contacted with advertisements which could be useful for them;
3 Avoid the spread of unuseful information through OSNs.

The proposed research consists on the proposal of novel recommendation approaches based on the comparison between the OSNs profiles associated to users (possible customers) and advertisers (brands), according to the considered campaign.

1.2 Precision Medicine

One of the most important challenges of this century is the proposal of precision therapies, that is, medical therapies adaptive with respect to specific categories of individuals, presenting well targeted features (e.g., genomic signatures, phenotypes, etc.). The recent advances in sequencing technologies have led to an exponential growth of biological data, allowing for high throughput profiling of biological systems in a cost-efficient manner. Molecules such as genes, proteins and RNA together contribute to the cellular life, and it is commonly accepted that they have to be analyzed as interacting elements when they take part in common biological processes [23]. More recently, great attention is turning towards the possible associations between cellular components and macroscopic disorders or complex diseases. In this context, we have studied two main problems: the importance of centrality measures in extracting functional knowledge from biological networks, and the prediction of long non coding RNA (lncRNA)-diseases associations (LDA).

2 State of the Art

2.1 Optimization of Advertising Campaigns

Modeling the user profiles from social media raw data is usually a challenging task. The approaches proposed in the Literature to this aim may be roughly classified in two main categories. The first category includes approaches based on the analysis of user generated contents (here referred to as *semantic approaches*). As for the approaches in the second category, individuals are characterized by "actions", e.g., visited web pages (*action-based approaches*). Our approach belongs to the first category, and we summarize below the other main related approaches. The authors of [22] use Differential Language Analysis (DLA) in order to find language features across millions of Facebook messages that distinguish demographic and psychological attributes. The framework proposed in [12] relies on a semi-supervised topic model to construct a representation of an app's version as a set of latent topics from version metadata and textual descriptions. In [11] the authors propose a dynamic user and word embedding algorithm that can jointly and dynamically model user and word representations in the same semantic space. They consider the context of streams of documents in Twitter, and propose a scalable black-box variational inference algorithm to infer the dynamic embeddings of both users and words in streams.

2.2 Node/Edge Centrality Measures in the Biological Context

Biological networks topology yields important insights into biological function, occurrence of diseases and drug design. In the last few years, different types of topological measures have been introduced and applied to infer the biological relevance of network components/interactions, according to their position within the network structure. In particular, the *Topological Overlap Measure* (*TOM*) considers only the immediate neighbors, whereas its Generalized version *GTOMm* [20,25] includes all the neighbors at distance $\leq m$. While TOM normalizes the size of common neighborhood over the smallest between i and j neighborhoods, *Edge Clustering Value* (*ECV*) by [24] is equal to 1 if and only if i and j have the same *exact* neighbors. It is worth noting that both TOM and ECV can be interpreted as a biological, neighborhood-normalized versions of Granovetter's *embeddedness* measure, historically used to characterize tie-strength in social networks [14]. Also *Dispersion* [2] extends the latter, taking into account both the size and the *connectivity* of i,j's common neighborhood. Intuitively, it quantifies how *"not well"*-connected is the i,j's common neighborhood within G_i, i.e., the subgraph induced by i and its neighbors. The authors define three enhanced versions of dispersion: *parametric* dispersion (*KB3*) and *recursive* dispersion that is divided in two versions (*KB1*) and (*KB2*) *Edge Betweenness* (*EB*) by [10] is the fraction of shortest paths in the network $\mathcal{N}$ containing the edge (i, j). *Edge Clustering Coefficient* (*ECC3*) by [19] is the number of *triangles* the edge (i, j) belongs to, divided by the number of triangles that might potentially include it. *Edge Centrality Proximity Distance* (*ECPd*) by [15] is based on computing the fraction of times a random walker traverses an edge, running through a random simple path of length at most κ.

2.3 Prediction of lncRNA-Diseases Associations

The approaches may be divided in two different categories: those that do not use already known lncRNA diseases associations, and those that do use them. Our approach belongs to the first category. In particular, the idea of not including any information on existing LDAs in the approach is based on the consideration that only a restricted number of validated LDAs is yet available, therefore a not exhaustive variability of real associations would be possible, affecting this way the correctness of the produced predictions. On the other hand, larger amounts of interactions between lncRNAs and other molecules (e.g., miRNAs, genes, proteins), as well as associations between those molecules and diseases are known. To this regard, our method is related to that presented by Chen [9], where a model of HyperGeometric distribution for LDAs inference (HGLDA) is proposed in order to predict LDAs by integrating miRNA-disease associations and lncRNA-miRNA interactions. HGLDA has been successfully applied to predict Breast Cancer, Lung Cancer and Colorectal Cancer-related lncRNAs.

3 Methods

3.1 Big Data Technologies

One of the best-known frameworks for turning raw data into useful information is known as MapReduce. MapReduce is a method for taking a large data set and performing computations on it across multiple computers, in parallel. The Map function performs the tasks of sorting and filtering, taking data and placing it inside of categories, so that it can be analyzed. The Reduce function analyze data returned by the Map in order to produce the results of the MapReduce program. Perhaps the most influential and established tool for analyzing big data is known as Apache Hadoop. Apache Hadoop is a framework for storing and processing data at a large scale. Hadoop can run on commodity hardware, making it easy to use with an existing data center, or even to conduct analysis in the cloud. Another important framework for the analysis of big data is the Resilient Distributed Datasets (RDD). It is specialized for main memory optimization, and it is useful to implement software algorithms that produce large amounts of intermediate data during their execution.

3.2 Optimization of Advertising Campaigns

The main goal of the approach we propose is to identify the most suitable k possible *buyers* to whom distributing a given advertisement campaign. The OSNs profiles are modelled as trees, which hierarchically represent the relationships among categories and sub-categories on the OSN (e.g., gender, posts, etc.). Profile matching is then applied relying on such a tree representation, and suitable similarity measures are considered for each category. The similarity between two nodes associated to textual contents may be then computed as the cosine similarity between arrays containing the TF-IDF values of the words occurring in those contents.

3.3   Node/Edge Centrality Measures in the Biological Context

The main goal is to provide an overview for the identification of significant "global" descriptors of a biological network, based on the characterization of the relevance of (nodes) edges across the network structure. We propose a methodology for the evaluation of topological ranks obtained from different measures, that relies on two different criteria: (1) *statistical significance*, via Montecarlo Hypothesis Test, and (2) *biological relevance*, quantified by comparing topological ranks against those obtained from external knowledge (e.g., gold standards). The former is a sort of *internal* criteria, which allows to discriminate the most significant ranks independently from the specific application context. The latter criteria aims to measure to what extent hidden information may be retrieved from a biological network taken as a whole, and intuitively this depends also on the specific type of information one is looking for.

3.4   Prediction of lncRNA-Diseases Associations

The main goal here is to provide a computational method able to predict novel LDA candidate for experimental validation in laboratory, given further external information on both molecular interactions and genotype-phenotype associations, but without relying on the knowledge of existing validated LDA. The main idea is to discover hidden relationships between lncRNAs and diseases through the exploration of their interactions with intermediate molecules (e.g., miRNAs) in a tripartite graph, where the three sets of vertices represent lncRNAs, miR-NAs, and diseases, respectively, and vertices are linked according to lncRNA miRNA interactions (LMI) and miRNA disease associations (MDA). Based on the assumption that similar lncRNAs interact with similar diseases [13], we aim to identify novel LDA by analyzing the behaviour of *neighbor lncRNAs*, in terms of their intermediate relationships with miRNAs. A score is assigned to each LDA (l, d) by considering both their respective interactions with common miRNAs, and the interactions with miRNAs shared by the considered disease d and other lncRNAs in the neighborhood of l. We define the *prediction-score* $S(l_i, d_j)$ for the LDA (l_i, d_j) such that $l_i \in \mathcal{L}$ and $d_j \in \mathcal{D}$ as:

$$S(l_i, d_j) = \alpha \cdot \frac{|M_{l_i} \cap M_{d_j}|}{|M_{l_i} \cup M_{d_j}|} + (1 - \alpha) \cdot \frac{|\bigcup_x (M_{l_x} \cap M_{d_j})|}{|\bigcup_x M_{l_x} \cup M_{d_j}|}$$

where M_{l_i} is the set of miRNA associated to l_i, M_{d_j} is the set of miRNA associated to d_j, α is a real value in [0,1] used to balance the two terms of the formula, M_{l_x} are all miRNA of those lncRNAs sharing at least one miRNA with l_i.

4   Results

4.1   Optimization of Advertising Campaigns

Our experimental analysis has been devoted to understand to what extent our approach is effective, in order to identify the k most convenient nodes in the input OSN to which distribute the advertisement. The main aim is to optimize two different aspects when identifying the best targets, that is, the fact that interests of considered users are related to the campaign contents, and the fact that they have "friends" on the OSN potentially interested to the distributed advertisements. The proposed approach has been implemented in Java under Apache Spark 1.6. To this respect, the use of Big Data Technologies allow to exploit the software tool also on very large OSNs. We have applied our approach to real datasets: from the repository twitter OSN (version 2010), having 90.908 vertices and 443.399 edges, which construction is part of the contribution, via web-scraping on specific topics. For each new node to be visited, a new web-page has been visited as well, following the cross-page links on the considered web-pages. We have compared our approach against a randomic one, showing that the profile matching at the basis of our approach is effective in the selection of target users for an advertising campaign [3,7]. Indeed, it performs an accuracy of 0.84% against the 0.15% of the randomic approach, computed as the number of reached target users with reference to their total number.

4.2   Node/Edge Centrality Measures in the Biological Context

Table 1 shows the best performing measures for the three considered organisms (human, worm and yeast), distinguished by those based on clustering coefficient (CC), neighborhoods (N), modularity (M) and dispersion (D) (see details in [4]). Results show that a distinct handful of best performing measures can be identified for each of the considered organisms, independently from the reference gold standard. Moreover, it seems that the proposed paradigm works better on denser networks, possibly due to the fact that the encoded information is larger than for sparse networks.

Table 1. Best performing measures for edge rank.

Organism	CC	N	M	D
H. sapiens	ECC3	GTOM2	–	–
		TOM	–	–
C. elegans	ECC3	–	EB	–
S. cerevisiae	ECC3	–	EB	KB1, KB2, KB3

4.3   Prediction of lncRNA-Diseases Associations

The approach described in Sect. 3.4 has been applied to the known experimentally verified lncRNA disease associations in the lncRNADisease database [8] according to LOOCV. We have validated the proposed approach on experimental verified data downloaded from starBase and from HMDD, resulting in 114 lncRNAs, 762 miRNAs, 392 diseases. In particular, each known disease lncRNA association is left out in turn as test sample. How well this test sample was ranked relative to the candidate samples (all the disease lncRNA pairs without the evidence to confirm their association) with respect to the considered score is evaluated. When the rank of this test sample exceeds the given threshold, this model is considered in order to provide a successful prediction. When the thresholds are varied, true positive rate (TPR, sensitivity) and false positive rate (FPR, specificity) are obtained. Here, sensitivity refers to the percentage of the test samples whose ranking is higher than the given threshold. Specificity refers to the percentage of samples that are below the threshold. Receiver-operating characteristics (ROC) curve can be drawn by plotting TPR versus FPR at different thresholds. Area under ROC curve (AUC) is further calculated to evaluate the performance of the tested methods. Our approach achieves an AUC equal to 0.82, whereas we have implemented the p-value based approach by Chen [9] and it scores AUC = 0.79, showing that the consideration of indirect relationships between lncRNAs and diseases through neighborhood analysis is more effective. Preliminary results are in [5,6].

5   Conclusion and Future Work

In this PhD project, knowledge extraction from large graphs has been investigated, with reference to two main application contexts: the social and the medical ones. Three specific problems have been solved, and for each of them we have summarized here the main preliminary results that have been achieved. As future work, we plan to explore techniques different than TF-IDF for the textual analysis referred to user profiling (e.g., [1,16,21]), as well as clustering techniques to select the only partitions of the considered graphs useful for the analysis (e.g., [17,18]. Moreover, further interesting challenges still remain open. Notably among them, the search of effective approaches able to find the best targets (e.g., social-users, patients, etc.) both in real-time and on a large scale.

References

1. Apostolico, A., Parida, L., Rombo, S.E.: Motif patterns in 2D. Theoret. Comput. Sci. **390**(1), 40–55 (2008)
2. Backstrom, L., Kleinberg, J.: Romantic partnerships and the dispersion of social ties: a network analysis of relationship status on Facebook. In: Proceedings of the 17th ACM Conference on Computer Supported Cooperative Work; Social Computing, CSCW 2014, pp. 831–841, New York, NY, USA. ACM (2014)

3. Bonomo, M., Ciaccio, G., De Salve, A., Rombo, S.E.: Customer recommendation based on profile matching and customized campaigns in on-line social networks. In: ASONAM 2019: International Conference on Advances in Social Networks Analysis and Mining, Vancouver, British Columbia, Canada, 27–30 August 2019, pp. 1155–1159. ACM (2019)
4. Bonomo, M., Giancarlo, R., Greco, D., Rombo, S.E.: Topological ranks reveal functional knowledge encoded in biological networks: a comparative analysis. Briefings Bioinform. **23**(3), bbac101 (2022)
5. Bonomo, M., La Placa, A., Rombo, S.E.: Prediction of lncRNA-disease associations from tripartite graphs. In: Gadepally, V., et al. (eds.) DMAH/Poly -2020. LNCS, vol. 12633, pp. 205–210. Springer, Cham (2021). https://doi.org/10.1007/978-3-030-71055-2_16
6. Bonomo, M., La Placa, A., Rombo, S.E.: Prediction of Disease-lncRNA associations via machine learning and big data approaches. In: Mayuri Mehta, K.P. (eds.) Knowledge Modelling and Big Data Analytics in Healthcare Advances and Applications. CRC Press (2021)
7. Bonomo, M., La Placa, A., Rombo, S.E.: Identifying the K best targets for an advertisement campaign via online social networks. In: Fred, A.L.N., Filipe, J. (eds.) Proceedings of the 12th International Joint Conference on Knowledge Discovery, Knowledge Engineering and Knowledge Management, IC3K 2020, KDIR, Budapest, Hungary, 2–4 November 2020, vol. 1, pp. 193–201. SCITEPRESS (2020)
8. Chen, G., et al.: LncRNADisease: a database for long-non-coding RNA-associated diseases. Nucleic Acids Res. **41**, D983–D986 (2013)
9. Chen, X.: Predicting lncRNA-disease associations and constructing lncRNA functional similarity network based on the information of miRNA. Sci. Rep. **5**, 13186 (2015)
10. Girvan, M., Newman, M.E.J.: Community structure in social and biological networks. Proc. Natl. Acad. Sci. **99**, 7821–7826 (2002)
11. Liang, S., Zhang, X., Ren, Z., Kanoulas, E.: Dynamic embeddings for user profiling in twitter. In: Proceedings of the 24th ACM SIGKDD International Conference on Knowledge Discovery & Data Mining, KDD 2018, pp. 1764–1773 (2018)
12. Lin, J., Sugiyama, K., Kan, M., Chua, T.: New and improved: Modeling versions to improve app recommendation. In: Proceedings of the 37th International ACM SIGIR Conference on Research; Development in Information Retrieval, SIGIR 2014, pp. 647–656. ACM (2014)
13. Lu, C., et al.: Prediction of lncRNA-disease associations based on inductive matrix completion. Bioinformatics **34**(19), 3357–3364 (2018)
14. Marsden, P.V., Campbell, K.E.: Measuring tie strength. Soc. Forces **63**, 482–501 (1984)
15. De Meo, P., Ferrara, E., Fiumara, G., Provetti, A.: Mixing local and global information for community detection in large networks. J. Comput. Syst. Sci. **80**(1), 72–87 (2014)
16. Parida, L., Pizzi, C., Rombo, S.E.: Irredundant tandem motifs. Theor. Comput. Sci. **525**, 89–102 (2014). Advances in Stringology
17. Pizzuti, C., Rombo, S.E.: *PINCoC*: a co-clustering based approach to analyze protein-protein interaction networks. In: Yin, H., Tino, P., Corchado, E., Byrne, W., Yao, X. (eds.) IDEAL 2007. LNCS, vol. 4881, pp. 821–830. Springer, Heidelberg (2007). https://doi.org/10.1007/978-3-540-77226-2_82
18. Pizzuti, C., Rombo, S.E.: An evolutionary restricted neighborhood search clustering approach for PPI networks. Neurocomputing **145**, 53–61 (2014)

19. Radicchi, F., Castellano, C., Cecconi, F., et al.: Defining and identifying communities in networks. Proc. Natl. Acad. Sci. **101**, 2658–2663 (2004)
20. Ravasz, E., Somera, A.L., Mongru, D.A., et al.: Hierarchical Organization of Modularity in Metabolic Networks. Science **297**, 1551–1555 (2002)
21. Rombo, S.E.: Extracting string motif bases for quorum higher than two. Theor. Comput. Sci. **460**, 94–103 (2012)
22. Schwartz, H.A., et al.: Personality, gender, and age in the language of social media: the open-vocabulary approach. PLoS ONE **8**(9), e73791 (2013)
23. von Mering, C., et al.: Comparative assessment of a large-scale data sets of protein-protein interactions. Nature **417**, 399–403 (2002)
24. Wang, J., Li, M., Chen, J., Pan, Y.: A fast hierarchical clustering algorithm for functional modules discovery in protein interaction networks. IEEE/ACM Trans. Comput. Biology Bioinf. **8**(3), 607–620 (2011)
25. Yip, A., Horvath, S.: Gene network interconnectedness and the generalized topological overlap measure. BMC Bioinform. **8**, 22 (2007)

Digital Twins for Urban Mobility

Chiara Bachechi[(✉)][ID]

"Enzo Ferrari" Engineering Department, University of Modena and Reggio Emilia,
Modena, Italy
chiara.bachechi@unimore.it

Abstract. Urban Digital Twins (DTs) can help tackle the challenges
of planning, monitoring, and managing modern cities. Existing mobility
systems are already inadequate, yet urbanization and population growth
will increase mobility demand still further. For this reason, urban mobility planning can benefit from DTs producing new knowledge executing automatically complex functions based on real-time data. The paper
describes two different DTs for urban mobility and their implementation.
The first one is the Traffic and Air Quality DT (TAQ) which investigates
the relationship between traffic flows and air quality conditions through
a chain of simulation models. The second DT is a multi-layered Graph-
Based Multi-Modal Mobility (GBMMM) DT to study the interaction
between different transport modes.

Keywords: Smart mobility · Digital twin · Graph database ·
Anomaly detection · Simulation models

1 Introduction

Urban DTs [12] are a virtual representation of a city's physical assets, using data,
data analytic, machine learning, and simulation models that can be updated and
changed as their physical equivalents change. Nowadays, the number of mobility
solutions is growing, enriching the variety of transport services in our cities. As
a consequence, the complexity of the urban mobility ecosystem is increasing.
In this context, DTs can be employed by planners and authorities to shape
urban mobility to manage it more efficiently. Public administration can benefit
from the use of DTs for several important goals: reducing congestion in rush
hours, identify where to add new bike-lanes, improve the public transport service,
and evaluate the impact of new mobility solutions (e.g. e-scooters). DTs allow
virtually testing different scenarios and possible solutions [17], learning more
about the interaction between the different aspects of the urban life.

As the name suggest, the DTs mirror the city itself; however, some simplification are necessary. Generally, the spatial/temporal resolution of the DT should be
determined by the purpose it serves. Developing a DT comports several challenges:
the high storage and computing resources needed, the continuous data production,
the management of the processing cycle, and the interaction with the user. Moreover, urban mobility DT have some additional technical issue: traffic related data

S. Chiusano et al. (Eds.): ADBIS 2022, CCIS 1652, pp. 657–665, 2022.
https://doi.org/10.1007/978-3-031-15743-1_61

have both a spatial and a temporal dimension, and different geometry can refer to the same road or object but can be slightly different or represented differently (e.g. a line or a polygon) and geometry matching is a challenging task. Moreover, a DT should study the interaction between mobility and other phenomena (e.g. air quality, human activities, and land use). During my research activity I have worked on the realization of a DT that, starting from traffic sensors data and integrating air quality sensors measurements, generates forecasts for urban air quality (described in Sect. 2). For smart mobility purposes, there are several different level of representation: from micro-mobility to origin-destination zones. A single DT may struggle to represent all this different resolution levels simultaneously. In big cities, several layers of transportation infrastructure work together to supply the continuously growing urban mobility demand. Since travelers usually employ more than one mode of transportation to complete their trip, the coupling of different layers of transportation infrastructure is necessary to model urban mobility. DTs provide a better insight into the spatial structure of the transportation network: for a better understanding of the interactions between neighborhoods, to identify the infrastructure needs, to better integrate different transport modes, and improve mobility services. For these reasons, I started working on a graph-based mobility DT. As described in Sect. 3, the outcome will be a composition of different graphs that simplify the road network optimizing multi-modal mobility. Finally, conclusions are provided in Sect. 4.

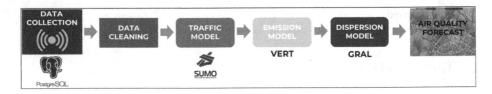

Fig. 1. TAQ DT chain of simulation models.

2   Traffic and Air Quality Digital Twin

The Trafair European project[1] aims to help decision-makers develop innovative and sustainable services and to raise awareness among citizens about the pollution caused by traffic in 6 cities [1,16]. In the scope of the Trafair project, a Traffic and Air Quality (TAQ) DT was developed. The DT provides monitoring of traffic and emissions and allows the evaluation of new traffic flow scenarios. As displayed in Fig. 1, the DT is a chain of simulation models. The TAQ DT includes: a virtual representation of the urban road network generated from Open Street Map (OSM)[2] data employing a PostgreSQL database, and the implementation of a traffic micro-simulation model based on SUMO[3]. Internet of Things

---

[1] https://trafair.eu.
[2] https://www.openstreetmap.org.
[3] https://sumo.dlr.de.

(IoT) sensors are integrated into this representation and their measurements are used to feed the model and evaluate its effectiveness. An in-depth investigation of traffic flows is conducted in order to discover trends, compare traffic conditions among working days and weekends, and identify significant deviations. The DT autonomously updates the simulated traffic, performing a daily simulation that takes into account the data collected by traffic sensors during the last 24 h. The generated traffic data can be visualized through a dashboard [7] that helps decision-makers to plan corrective actions.

2.1 Data Collection

Traffic sensors are induction loop detectors located under the surface of the street. These sensors can be faulty due to low maintenance. For this reason, the first step was to identify anomalies in traffic sensors' measurements and distinguish between sensor faults and real traffic conditions. I have implemented two different solutions for anomaly detection. The first one is based only on the temporal dimension of data and exploits the spatial dimension just for the classification of anomalies. The Seasonal-Trend Decomposition using Loess (Robust-STL [18]) decomposes the time series into three components: trend, seasonal, and remainder (or residual); then, the Inter Quartile Range (IQR), the first and the third quartiles of the residual are evaluated to define an upper and lower bound. The observations that have a residual outside of these bounds are labelled as anomalous and removed from the input of the traffic model [8]. The effect of anomaly detection on the results and performance of the traffic simulation model are also analysed showing a good improvement. However, some anomalies can reflect a real traffic condition that may occur. For this reason, I set up a classification technique that allows distinguishing between sensor faults and real traffic conditions exploiting the correlation among spatial neighboring sensors and related sensors, as described in [9]. The second methodology, instead, employs both spatial and temporal dimensions to evaluate the distance between sensor observations and detect anomalies in spatial time series. Since traffic sensors' measurements are multi-variate time series with a spatial reference, this information can be exploited to detect anomalies through the application of the Spatio-Temporal Behavioral Density-Based Clustering of Applications with Noise (ST-BDBCAN) algorithm combined with the Spatio-Temporal Behavioral Outlier Factor (ST-BOF). ST-BDBCAN is based on the distinction between spatio-temporal and behavioral attributes. The spatio-temporal attributes indicate the position of the sensors or provide temporal information about the observation; while, the behavioral attributes are all the other attributes. ST-BDBCAN creates clusters of objects with similar spatio-temporal attributes; then, ST-BOF is applied to detect anomalies by evaluating a score based on its behavioral attributes w.r.t its cluster that represents the potential outlierness of each instance. The open-source Python implementation of this solution is described in [10].

2.2 Simulation Models

Data collected from traffic sensors are employed to feed a micro-simulation model based on SUMO, as described in [3], processing both semi-real-time simulation (every 15 min) and daily simulations (once a day). Through data analysis [4], the daily simulations are employed to: evaluate statistics, extract seasonal and weekly trends, and detect congested areas. Seasonal and weekly trends are employed to estimate the traffic situation (traffic flow and speed) of tomorrow and the day after tomorrow used to feed an emission model that evaluates the deriving emissions of NO and NO_2. The emission model is a modified version of the R package VEIN (Vehicular Emissions INventories) v0.5.2 [14]. The evaluation of emissions is based on the traffic flow and the speed on each road and the fleet composition of the urban area; the employed emission factors are the ones suggested by the European Environmental Agency [15]. Since the emitted particles move in the air all around the city, and the concentrations of pollutants do not depend only on traffic, a pollutant dispersion model is needed to correctly estimate the pollutant concentration. We employed the open-source simulation software Graz Lagrangian Model (GRAL) [11] that evaluates the pollutant concentration by considering: winds, weather conditions, and the shape of the building in the urban area, and other emission sources (e.g. house heating). The output of the dispersion model is a dispersion map that shows NO_x concentration values on an urban grid of 4 cubic meters for every hour.

2.3 Data Visualization

All the models' outputs are stored in a PostgreSQL database and exposed through Geoserver[4]. Geoserver is employed as an API to let the Angular-based web applications get the information they need to generate their views. I have worked on the design and the development of two web applications considering the needs of public administration. The first application is dedicated to traffic data and it is described in [7]. The dashboard enables the visualization of possible scenarios to easily compare different vehicle fleets and understand the impact of seasonal changes. The user can select the season, the type of day (weekday or holiday), and the vehicle fleet composition; then, the NO_x concentration will be displayed for three different compositions of vehicles fleet. The users can interact with the visualization using the scroll bar and selecting the hour of the day they want to inspect. The second application, instead, shows air quality data coming from sensors and air quality forecasts generated through models. It is described in [2, 6].

3 Graph-Based Multi-modal Mobility Digital Twin

Road networks are complex structures that need to be simplified for an efficient analysis in order to gain insights into their structure. A good solution can be to

---

[4] https://geoserver.org/.

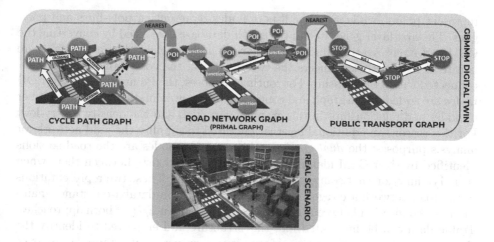

Fig. 2. GBMMM structure.

represent them with a graph and perform their analysis with a graph database; this will generate a DT of the road network that will help to find the relationship between its topology and the traffic conditions. Moreover, a comprehensive analysis of human mobility patterns requires multiple layers representing distinct transportation modes. This research activity aims to generate a framework that starting from OSM data will create a multi-layered graph that will be a DT of the urban mobility of the city. As displayed in Fig. 2, the graph database will contain several layers, one for each transport mode, and different levels of abstraction. I decide to use the ACID compliant Neo4j[5] database with the Graph Data Science tools and the Awesome Procedures on Cypher and Spatial libraries. Neo4j Spatial library allows storing different types of geometries as single properties of nodes in Well known Binary (WKB) format: the geographic geometry format employed also by PostgreSQL. Moreover, Neo4j enables spatial indexing with space-filling curves in 2D or 3D over an underlying generalized B+Tree. This allows for both equality and range queries and queries using the distance function. The road network generation has already been implemented with two different levels of abstraction [5] as described in Sect. 3.1; moreover, studies have been conducted on cycle paths and public transport to add them as additional layers, as described in Sects. 3.2 and 3.3. The purpose is to integrate vehicular, cyclist, and public transport networks in a unique graph database allowing multi-modal route planning and service integration.

3.1   The Road Network Layers

The OSM provides very high-resolution topological data regarding road networks. OSM data are represented as two main elements: nodes and ways. Nodes

---

[5] https://neo4j.com/.

represent points, while ways are lists of nodes representing poly-lines and poly-gons. The first layer generated in the graph database is realized by converting the OSM nodes representing junctions into graph nodes: a *primal graph*. When there is a road section that connects the two junctions, the corresponding nodes are connected with a relationship. For routing purposes, the Point Of Interests (POI) nodes are added to the graph and connected to the proximal streets. The number of nodes and relationships in this graph is very high and the density is very low. For this reason, an additional layer is generated to simplify the road network for analysis purposes: the *dual graph*. In this case, the nodes are the road sections identified by their OSM identifier and a relationship exists between them when there is a junction that connects the two road sections. These two representations of the road network are regenerated in the same graph database instance main-taining a connection between them to exploit the advantages of both approaches. Traffic data can be integrated into this structure and employed to identify the shortest path between two points of the city: moreover, using graph data science algorithms (e.g. community detection, centrality measures, page rank) the road network's structure is analyzed, unveiling the areas of the city prone to traffic jams.

3.2   The Cycle Paths Network Layer

A preliminary study on the distribution of cycle paths in the province of Modena was conducted integrating data from OSM and the regional geoportal[6]. QGIS was employed to perform the integration issues: matching different geometries through the geometry conversion, finding corresponding objects in the two rep-resentations through the overlapping analysis, and managing when an object is represented as a singular entity in one representation and as several entities in the other. A network of cycle paths is composed of cycleways' sections con-nected by street crossings or crosswalks. In the resulting graph, the nodes will be the sections of continuous cycle paths and a 'crossing' relationship will exist between them if there is a crossing that connects them. Moreover, the relation-ship ('proximal_to') is defined between cycle paths that are not connected but whose distance is small and can be traversed by bike or by foot. The POIs of the primal graph are connected with the cycle paths that are located nearby to define the cycle route between a POI source and a POI destination.

3.3   The Public Transport Network Layer

Public transport relies on a timetable and operates on established routes. As sug-gested in [13], the transport network can be represented with a time-expanded graph model composed of different kinds of nodes: stop nodes, time nodes, and route nodes. The relationship of a bus route that stops at a given position at

---

[6] https://geoportale.regione.emilia-romagna.it/catalogo/dati-cartografici/ cartografia-di-base/database-topografico-regionale/viabilita-e-trasporti/strade/ layer.

a certain timestamp is represented by stop nodes connected by time events to time nodes and route nodes. Therefore, an alternative solution for storing Public transport timetables is a frequency-based model that store time tables as departure times, intervals and frequencies information associated with each bus route. The resulting graph gives an indication of how tightly different parts of the city are connected through the public transportation system in different times of the day. The representation of bus routes can be simplified creating a Simplified Public Transport Graph (SPTG) with stop nodes and links between all the pairs of bus stops that are along the bus trajectory. This solution removes the time dimension and gives a better representation of the distribution of the service that the bus provides; thus, it is not for routing purposes but can be employed for analysing the coverage and the interaction between different bus routes. However, another possibility could be to generate an SPTG for each snapshot of a given time interval (e.g. 10 min) and then perform routing in the corresponding graph. Our Public Transport Graph Layer will be initially generated as the frequency-based model and then simplified in an SPTG. When modelling the PTGL it should be taken into account that the combination of bikes and public transport in a multi-modal transport node is a common solution adopted by travellers; thus, a connection between the CPGL and the PTGL layer is needed. In order to do that, the spatial library of Neo4j will be exploited generating a relationship between the bus stops and the proximal cycle path nodes.

4 Conclusions

Urban Mobility DT can help investigate the influence of traffic on other phenomena like air quality conditions or the interaction between different transport modes. This paper proposes two solutions: a chain of simulation models fed by sensors' data and a multi-layered graph model of the transport structure. While the TAQ was tested for 3 years in Modena, the GBMMM is not yet complete. The GBMMM can be employed for multi-modal routing purposes and the application of graph algorithms will help to identify the areas where services are lacking, define communities of connected regions, and detects the central associated nodes for the coupling of public transport and cycling.

References

1. Po, L., et al.: Trafair: understanding traffic flow to improve air quality. In: 2019 IEEE International Smart Cities Conference (ISC2), pp. 36–43 (2019). https://doi.org/10.1109/ISC246665.2019.9071661
2. Bachechi, C., Desimoni, F., Po, L., Casas, D.M.: Visual analytics for spatio-temporal air quality data. In: 2020 24th International Conference Information Visualisation (IV), pp. 460–466 (2020). https://doi.org/10.1109/IV51561.2020.00080

3. Bachechi, C., Po, L.: Implementing an urban dynamic traffic model. In: IEEE/WIC/ACM International Conference on Web Intelligence, pp. 312–316. WI 2019, Association for Computing Machinery, New York, NY, USA (2019). https://doi.org/10.1145/3350546.3352537

4. Bachechi, C., Po, L.: Traffic analysis in a smart city. In: IEEE/WIC/ACM International Conference on Web Intelligence - Companion Volume. p. 275–282. WI '19 Companion, Association for Computing Machinery, New York, NY, USA (2019). https://doi.org/10.1145/3358695.3361842

5. Bachechi, C., Po, L.: Road network graph representation for traffic analysis and routing. In: The 26th European Conference on Advances in Databases and Information Systems (ADBIS 2022), Torino, Italy, September 5–8, 2022 (2022, to appear)

6. Bachechi, C., Po, L., Desimoni, F.: Real-Time Visual Analytics for Air Quality, pp. 485–515. Springer International Publishing, Cham (2022). https://doi.org/10.1007/978-3-030-93119-3_19

7. Bachechi, C., Po, L., Rollo, F.: Big data analytics and visualization in traffic monitoring. Big Data Res. **27**, 100292 (2022). https://doi.org/10.1016/j.bdr.2021.100292

8. Bachechi, C., Rollo, F., Po, L.: Real-time data cleaning in traffic sensor networks. In: 2020 IEEE/ACS 17th International Conference on Computer Systems and Applications (AICCSA), pp. 1–8 (2020). https://doi.org/10.1109/AICCSA50499.2020.9316534

9. Bachechi, Chiara, Rollo, Federica, Po, Laura: Detection and classification of sensor anomalies for simulating urban traffic scenarios. Cluster Comput. **25**, 1–25 (2021). https://doi.org/10.1007/s10586-021-03445-7

10. Bachechi, C., Rollo, F., Po, L., Quattrini, F.: Anomaly detection in multivariate spatial time series: a ready-to-use implementation. In: Mayo, F.J.D., Marchiori, M., Filipe, J. (eds.) Proceedings of the 17th International Conference on Web Information Systems and Technologies, WEBIST 2021, 26–28 October 2021, pp. 509–517. SCITEPRESS (2021). https://doi.org/10.5220/0010715900003058

11. Bigi, A., Veratti, G., Fabbi, S., Po, L., Ghermandi, G.: Forecast of the impact by local emissions at an urban micro scale by the combination of Lagrangian modelling and low cost sensing technology: the Trafair project. In: 19th International Conference on Harmonisation within Atmospheric Dispersion Modelling for Regulatory Purposes, Harmo 2019 (2019)

12. Deng, T., Zhang, K., Shen, Z.J.M.: A systematic review of a digital twin city: a new pattern of urban governance toward smart cities. J. Manag. Sci. Eng. **6**(2), 125–134 (2021). https://doi.org/10.1016/j.jmse.2021.03.003

13. Huang, H., Bucher, D., Kissling, J., Weibel, R., Raubal, M.: Multimodal route planning with public transport and carpooling. IEEE Trans. Intell. Transp. Syst. **20**(9), 3513–3525 (2019). https://doi.org/10.1109/TITS.2018.2876570

14. Ibarra-Espinosa, S., Ynoue, R., O'Sullivan, S., Pebesma, E., Andrade, M.F., Osses, M.: Vein v0. 2.2: an R package for bottom-up vehicular emissions inventories. Geosci. Model. Dev. **11**(6), 2209–2229 (2018)

15. Ntziachristos, L., Samaras, Z., Kouridis, C., Hassel, D., Mccare, I., Hickman, J.: EMEP/EEA emission inventory guidebook 2009. In: European Environment Agency (EEA) (2009)

16. Po, L., Rollo, F., Bachechi, C., Corni, A.: From sensors data to urban traffic flow analysis. In: 2019 IEEE International Smart Cities Conference, ISC2 2019, Casablanca, Morocco, 14–17 October 2019, pp. 478–485. IEEE (2019). https://doi.org/10.1109/ISC246665.2019.9071639

17. Shahat, E., Hyun, C.T., Yeom, C.: City digital twin potentials: a review and research agenda. Sustainability **13**(6), 3386 (2021). https://doi.org/10.3390/su13063386

18. Wen, Q., Gao, J., Song, X., Sun, L., Xu, H., Zhu, S.: RobustSTL: a robust seasonal-trend decomposition algorithm for long time series. In: The Thirty-Third AAAI Conference on Artificial Intelligence, AAAI 2019, Honolulu, Hawaii, USA, January 27–February 1 2019, pp. 5409–5416 (2019). https://doi.org/10.1609/aaai.v33i01.33015409

Author Index